# Ecological and Environmental Physiology of **Mammals**

## Ecological and Environmental Physiology Series (EEPS)

*Series Editor: Warren Burggren, University of North Texas*

This authoritative series of concise, affordable volumes provides an integrated overview of the ecological and environmental physiology of key taxa including birds, mammals, reptiles, amphibians, insects, crustaceans, molluscs, and fish. Each volume provides a state-of-the-art review and synthesis of topics that are relevant to how that specific group of organisms have evolved and coped with the environmental characteristics of their habitats. The series is intended for students, researchers, consultants, and other professionals in the fields of physiology, physiological ecology, ecology, and evolutionary biology.

A Series Advisory Board assists in the commissioning of titles and authors, development of volumes, and promotion of the published works. This Board comprises more than 50 internationally recognized experts in ecological and environmental physiology, providing a combination of both depth and breadth to proposal evaluation and series oversight.

The reader is encouraged to visit the EEPS website for additional information and the latest volumes (http://www.eeps-oxford.com/). If you have ideas for new titles in this series or just wish to comment on EEPS, please do not hesitate to contact the Series Editor, Warren Burggren (University of North Texas; burggren@unt.edu).

**Volume 1: Ecological and Environmental Physiology of Amphibians**
*Stanley S. Hillman, Philip C. Withers, Robert C. Drewes, and Stanley D. Hillyard*

**Volume 2: Ecological and Environmental Physiology of Birds**
*J. Eduardo P.W. Bicudo, William A. Buttemer, Mark A. Chappell, James T. Pearson, and Claus Bech*

**Volume 3: Ecological and Environmental Physiology of Insects**
*Jon F. Harrison, H. Arthur Woods, and Stephen P. Roberts*

**Volume 4: Ecological and Environmental Physiology of Fishes**
*F. Brian Eddy and Richard D. Handy*

**Volume 5: Ecological and Environmental Physiology of Mammals**
*Philip C. Withers, Christine E. Cooper, Shane K. Maloney, Francisco Bozinovic, and Ariovaldo P. Cruz-Neto*

# Ecological and Environmental Physiology of **Mammals**

**Philip C. Withers**
*School of Animal Biology,*
*The University of Western Australia, Crawley 6009, Australia*

**Christine E. Cooper**
*Department of Environment and Agriculture*
*Curtin University, Bentley, Western Australia 6102, Australia*

**Shane K. Maloney**
*School of Anatomy, Physiology and Human Biology*
*The University of Western Australia, Crawley 6009, Australia*

**Francisco Bozinovic**
*Center of Applied Ecology & Sustainability and Departamento de Ecología,*
*Facultad de Ciencias Biológicas, Pontificia Universidad Católica de Chile,*
*Santiago 6513677, Chile*

**Ariovaldo P. Cruz-Neto**
*Department of Zoology, Bioscience Institute, State University of São Paulo*
*(UNESP), Campus de Rio Claro, Rio Claro 13506-900 Rio Claro, SP, Brazil*

OXFORD
UNIVERSITY PRESS

OXFORD
UNIVERSITY PRESS

Great Clarendon Street, Oxford, OX2 6DP,
United Kingdom

Oxford University Press is a department of the University of Oxford.
It furthers the University's objective of excellence in research, scholarship,
and education by publishing worldwide. Oxford is a registered trade mark of
Oxford University Press in the UK and in certain other countries



First Edition published in 2016

Impression: 1



Published in the United States of America by Oxford University Press
198 Madison Avenue, New York, NY 10016, United States of America

British Library Cataloguing in Publication Data
Data available

Library of Congress Control Number: 2016934456

ISBN 978–0–19–964271–7 (hbk.)
ISBN 978–0–19–964272–4 (pbk.)

Printed and bound by
CPI Group (UK) Ltd, Croydon, CR0 4YY

*We would like to dedicate this book to two "founding fathers" of the general field of physiological ecology and acknowledge their roles in establishing this field of research. These are, of course, George A. Bartholomew (1916–2006) and Knut Schmidt-Nielsen (1915–2007). These outstanding scientists pioneered the new research area of physiological ecology, in both the laboratory and the field, and each has spawned a considerable "phylogenetic tree" of their own via their students, their students' students, and even their students' students' students, as their academic lineages have grown over time.*

# Foreword

A physiological approach to the behaviour and ecology of organisms is fundamental to our understanding of the characteristics required of species for survival. The biochemical, morphological, and behavioural conditions of a life style may be limited or facilitated by the abundance and quality of the resources encountered in the environment, as it may be by phylogeny, i.e. the historical accumulation of character states. A physiological approach to ecology is uniquely capable of examining the mechanistic basis of the response of species to the environment.

In this treatise the response of mammals is examined with respect to temperature, water balance, gas exchange, an aquatic existence, fossorial habits, flight, sensory systems, and food acquisition and processing. A significant contribution to our knowledge of how major occupants of this planet, mammals, have responded to their limitations and opportunities. However, all groups of mammals do not use the same solutions for a shared problem, which in part relates to the geographic availability of resources and the characteristics of the clade to which they belong. An effective approach to distinguish between phyletic conservatism and functional plasticity can be explored in the consequences of evolutionary convergence.

Morphological, physiological, and behavioural convergence in mammals has occurred repeatedly with respect to food habits, habitats and climates occupied, burrowing habits, and substrates used. This raises a question of the extent to which the modification of physiology and morphology is independent of the clades in which the convergent species occur. For example, species from many clades evolved ant- and termite-eating habits, including a monotreme, a marsupial, and many eutherians. Yet, their low mass-independent energy expenditures cannot be distinguished, presumably as a result of consuming an abundant, but poor-quality food, an external determinate of a physiological process. Convergence also occurs in the energetics of marsupials and eutherians, as long as the foods used require a reduction of the expenditure. The adjustment of body size converges with respect to geography and climate, as seen in the Bergmann's and island rules. These examples give a clue to the influence of the immediacies of physiological and behavioural responses to environmental requirements.

However, many clades are morphologically, physiologically, and behaviourally uniform. This is the case in microtine rodents, all of which are grazers, live in cool to cold, wet environments, and are committed to strict endothermy, none going into torpor. In contrast, the food habits, habitats, climates, and the use of torpor show great diversity in cricetine rodents. A similar difference is found in the physiological uniformity of soricine shrews and its diversity in crocidurine shrews. What led to the differences in these clades? Are some clades 'flexible' and others 'inflexible'? If so, what is the physiological basis of such a difference and what conditions in the environment permit this difference to be expressed?

The strength of physiological analyses is their propensity to examine biological functions quantitatively. That permits a comparison of the quantitative characteristics of physiological functions with those predicted from proposed models. The effectiveness of any analysis of physiology and behaviour requires the inclusion of all potentially contributing factors, including those from the environment. The extent to which the predictions correspond to the observations is the extent to which our understanding is confirmed or rejected. Few other approaches to physiology and behaviour have this capacity.

The characteristics of physiological performance may impact the response of mammals to the warming climate, as explored in this treatise. This occurs in species that are inflexibly committed to a geographically limited set of environmental conditions. Thus, species that are found only above timberline or in a polar environment may be trapped by the geographical distribution of climatic factors. Montane species, like pika and mountain goats, cannot escape by climbing to altitudes beyond those available and polar species, including polar bears and walruses, cannot survive when conditions at the highest latitudes exceed those that are acceptable. Another geographical limit to survival is found in island endemics. Physiological ecology can determine the bases of geographic and climatic restrictions, thereby potentially identifying the species likely to be faced with extinction in the future. This ability may facilitate the transport of threatened species to sites that permit their survival.

Physiological ecology, then, is an approach to the study of organisms that deals with the basis of their performance under a variety of environmental conditions, which gives insight into the factors on which their operation, evolution, and survival depends. This treatise makes a significant approach to that goal and hopefully will spur interest to continue this effort.

**Brian K. McNab**
Emeritus Professor of Biology
University of Florida, Gainesville

# Acknowledgments

The authors wish to collectively acknowledge Oxford University Press and the various individuals who have made this book possible. Warren Burggren, as Chief Editor of the EEPS series, deserves a special acknowledgment for his foresight in planning and seeing through to production the various books in this series. We also thank the various OUP staff who assisted in the preparation of the book, including Ian Sherman, who commissioned the project, Helen Eaton, and Lucy Nash. We also thank all of the authors who generously gave permission to reproduce figures from their publications, or provided images for figures.

PCW would like to acknowledge his formative mentors, Shelley Barker, Anthony Lee, George Bartholomew and Gideon Louw, for their academic and personal contribution to his academic career. In addition, he would like to acknowledge the many fellow students, undergraduate and postgraduate, and academic colleagues who have provided so much comradery and scientific inspiration; in particular he thanks Stan Hillman, Jamie O'Shea, Michael Guppy, Don Bradshaw, Graham Thompson, Shane Maloney and Christine Cooper. Last, and by no means least, he thanks his students who have contributed more than they might individually suspect, to his academic and personal development.

CEC would like to acknowledge the outstanding mentoring provided by previous supervisors Philip Withers and Fritz Geiser, who played such important and complementary roles in the development of her scientific career and continue to be valued collaborators and advisors. She also thanks a suite of peers, colleagues, collaborators and students who have provided advice, assistance, support and importantly an avenue for scientific discussion. In particular she wishes to thank Stephen Davies, Peter Mawson, Alex Larcombe and Graham and Scott Thompson for their positive and lasting impacts on her academic career.

SKM would like to acknowledge all those who played a role in him ending up in science, after his parents Jude and Maree encouraged an inquisitive child. His PhD supervisor Terry Dawson and his Post-Doctoral supervisor Duncan Mitchell made Zoology and Physiology into a game of "How it Works", so much more interesting than learning lineages off by heart. Mark Chappell got him to Antarctica and introduced him to Cynthia, the girl who became the wife who lets him keep doing science without too many grumbles. His long time collaborators, Andrea Fuller, Phil Withers, and Dominique Blache, keep him grounded and always have some new project that is just too interesting to let go by. His students continue to inspire him and remind him of what academic life is really about. His kids, Claire and Maggie, keep him busy but also remind him what life is all about.

FB would like to acknowledge his mentor Prof. Mario Rosenmann, as well his students, post-docs, peers and colleagues for continuous inspiration and support. Phil Withers deserves the credits for this project. FB also acknowledge his parents Francisco and Rosita and his sister Pamela. Finally he would like to thank his wife Maria Jose and girls Catalina and Emma for love and support and keep him grounded.

APCN would like to acknowledge Augusto Abe, Francisco Bozinovic, Enrique Caviedes-Vidal and Phil Withers for their scientific and personal support along his academic career. Phil Withers deserves an additional credit for encouragement during his writing, and toleration of inability to meet deadlines. He also thanks his past and present students for their enthusiasm, State University of São Paulo (UNESP), Conselho Nacional de Desenvolvimento Cientifico e Tecnológico (CNPq) and Fundação de Amparo a Pesquisa do Estado de São Paulo (FAPESP) for logistic and financial support, and Alexandre Percequillo for providing a quiet space for writing. Last, and above all, he would like to deeply thank Vinicius and Julia for their unconditional love and for being by his side all the time.

# Contents

# 1
# Introduction to Mammals

Mammals (about 5,700 extant species) are found in every biome on Earth. They are readily recognizable from a few general characteristics, which define their 'bauplan' (their basic structural plan). They evolved from early reptiles, over 200 million years before the present (MYBP). Birds (Aves) also are derived from reptiles, but from a very different and advanced group (dinosaurs), more than 146 MYBP (Fountaine et al. 2005). Although there are some similarities in the bauplans of mammals and birds, reflecting both their common reptilian ancestry and the later independent evolution of traits such as endothermy and an osmo-concentrating kidney, the bauplan of mammals seems less restrictive than that of birds, which rely on their forelimbs for flight, their jaws for food acquisition, and their hindlimbs for walking. Despite the impressive radiation of birds into many niches, and their being more speciose (about 10,000 species), it is mammals that have undergone the more impressive morphological, physiological, and ecological radiations.

## 1.1 Living Mammals

The defining characteristic of extant mammals is the presence of mammae, which are used by females to suckle their young. On that basis, Linnaeus coined the term Mammalia in 1757. Other characteristics that define the mammalian bauplan are body insulation (fur), thermoregulation by internal heat production (endothermy), live birth (except for monotremes, which lay eggs), the ability to produce concentrated urine, sweat glands (in many but not all species), differentiated and specialized teeth, complex sensory systems (including chemoreception, vision, thermoreception, mechanoreception, and electroreception), three middle ear ossicles, and a complex central nervous system (including a cerebral cortex). This basic bauplan of mammals is derived from their reptilian ancestors, a group of mammal-like reptiles called the Therapsida. Therapsids diverged from the other reptile lineages relatively early in the Permian, about 280–230 MYBP (Archer 1984a) and persisted through the Triassic, initially dominated by the more successful dinosaurs and other 'ruling' reptiles. Early mammals arose from these

*Ecological and Environmental Physiology of Mammals*. Philip C. Withers, Christine E. Cooper, Shane K. Maloney, Francisco Bozinovic, & Ariovaldo P. Cruz-Neto.  Published 2016 by Oxford University Press.
DOI 10.1093/acprof:oso/9780199642717.001.0001

mammal-like reptiles in the Jurassic (230–140 MYBP), then diversified rapidly in the Cretaceous (140–65 MYBP), coinciding with the demise of the dinosaurs.

### 1.1.1 Monotremes, Marsupials, and Placentals

Extant mammals are traditionally classified into three distinctive groups: the Monotremata, Marsupialia, and Placentalia (Table 1.1). Monotremes are the only mammals that lay eggs; there is one species of platypus (*Ornithorhynchus*) and four species of echidnas in two genera (*Tachyglossus* and *Zaglossus*). Marsupials (about 300 species) give birth to live young, at a very immature stage, and have a long period of suckling, with the young typically protected in a pouch of skin, the marsupium. Placentals (about 5,400 species) also give birth to live young, but they are born at a more mature stage and have a relatively shorter suckling period. Although these three distinctive groups of living mammals are delineated by their reproductive mode, there are numerous other characteristics that also define these taxa. The terms 'marsupial' and 'placental' are in common usage to describe these two distinctive live-birth groups of mammals, but technically the term 'placental' is a misnomer because all mammalian eggs develop a 'placenta' (including monotremes). Nevertheless, the major development of and reliance on an allantoic placenta have given rise to the common usage of the term 'placental' mammal, and the taxonomic term Placentalia for extant species.

Monotremes are unmistakable mammals. The spiny short-beaked echidna, *Tachyglossus aculeatus* ('quick tongue'), was originally described as *Myrmecophaga aculeata*, as with its long naked snout it resembled the South American ant bear, and it has been likened to a porcupine (Griffiths 1989a). Its dense covering of sharp spines provides protection against predators; its compact shape and short, strong legs with digging claws are well suited for its ant/termite-eating lifestyle. The three species of long-beaked echidnas (*Zaglossus*) are larger than *Tachyglossus*, and are found in New Guinea; they primarily forage on the forest floor for insects. The semi-aquatic platypus, *Ornithorhynchus anatinus*, with a combination of duck-like beak and beaver-like tail, was originally suspected of being a fraud, concocted by combining parts of other mammals (Grant 1989). Its bill, webbed feet, tail, and dense covering of fur are well suited for its semi-aquatic lifestyle.

Many modern marsupials and placentals, in contrast, are relatively similar (apart from their very different reproductive modes) and can be quite convergent in general appearance. They provide many exceptional examples of morphological, physiological, and ecological convergence (e.g. placental and marsupial moles, 'tigers', and gliders). Nevertheless, there are a number of 'soft' and 'hard' characteristics that distinguish marsupials from placentals. Some 'soft' anatomical characteristics of marsupials are a short gestation period (usually shorter than the oestrous cycle); generally a yolk-sac placenta; often a pouch, that accommodates the young that are born at a very immature stage; birth via a temporary pseudo-vaginal canal (that persists

**Table 1.1** Orders of living mammals of the world (based on Wilson & Reeder 2005; Nowak & Dickman 2005).

| | | | | Order | |
|---|---|---|---|---|---|
| Prototheria | | | | Monotremata | Echidnas (Australia and New Guinea), and platypus (Australia) |
| Theria | Marsupialia | Australidelphia | | Dasyuromorphia | Marsupial carnivores, including planigales, antechinuses, sminthopsines, phascogales, quolls, Tasmanian devil, Tasmanian tiger (Australia and New Guinea), and numbat (Australia) |
| | | | | Diprotodontia | Kangaroos, wallabies, koala, wombats, possums, cuscuses, gliders (Australia and New Guinea) |
| | | | | Microbiotheria | Monito del Monte, *Dromiciops gliroides* (South America) |
| | | | | Notoryctemorphia | Marsupial moles (Australia) |
| | | | | Peramelemorphia | Bandicoots, bilbies, spiny bandicoots (Australia and New Guinea) |
| | | Ameridelphia | | Didelphimorphia | Mouse (pouchless) opossums, woolly opossums, bushy-tailed opossum, and pouched opossums (Americas) |
| | | | | Paucituberculata | Caenolestid 'shrew' opossums (South America) |

*(continued)*

**Table 1.1** *(Continued)*

| | | | | | Order | |
|---|---|---|---|---|---|---|
| | Placentalia | Atlantogenata | Afrotheria | | Afrosoricida | Tenrecs, golden moles (Africa) |
| | | | | | Hyracoidea | Hyraxes, dassies (Africa, Arabia) |
| | | | | | Macroscelidea | Elephant shrews (Africa) |
| | | | | | Proboscidea | Elephants (Africa, Southeast Asia) |
| | | | | | Sirenia | Dugongs, manatees, sea cows (cosmopolitan tropical) |
| | | | | | Tubulidentata | Aardvark (Africa) |
| | | | Xenarthra | | Cingulata | Armadillos (Americas) |
| | | | | | Pilosa | Sloths, anteaters (neotropical America) |
| | | Boreoeutheria | Euarchontoglires | Glires | Lagomorpha | Pikas, rabbits, hares (Eurasia, Africa, Americas) |
| | | | | | Rodentia | Squirrels, beavers, mice, rats, gopher, porcupines, gerbils, guinea pigs, chinchillas, capybaras, etc. (cosmopolitan) |
| | | | | Euarchonta | Dermoptera | Flying lemurs (Southeast Asia) |
| | | | | | Primates | Lemurs, lorises, aye-aye, bushbabies, tarsiers, monkeys, apes (Africa, Asia, Americas) |
| | | | | | Scandentia | Treeshrews (Southeast Asia) |

| | | | |
|---|---|---|---|
| Laurasiatheria | | Erinaceomorpha | Hedgehogs (Europe, Africa, Asia) |
| | | Soricomorpha | Moles, shrews, solenodons (Africa, Europe, Asia, North America) |
| | Cetartiodactyla | Artiodactyla | Even-toed ungulates: pigs, hippos, camels, giraffe, deer, antelope, cattle, sheep, goats (cosmopolitan except Australia) |
| | | Cetacea | Whales, dolphins, porpoises (Cosmopolitan) |
| | Pegasoferae | Carnivora | Canids, felids, civets, mongooses, weasels, otters, badgers, civets, skunks, red panda, bears and panda, hyaenas, raccoons, seals, walrus (cosmopolitan) |
| | | Chiroptera | Mega- and micro-bats (cosmopolitan) |
| | | Perissodactyla | Odd-toed ungulates: horses, donkeys, zebras, tapirs, rhinoceroses (cosmopolitan except Australia) |
| | | Pholidota | Pangolins (Africa, South Asia) |

in some species); males that have their scrotum located anterior to the penis (male thylacines and water opossums have their scrotum located within a pouch); a glans penis that is usually bifid; and a tail and body that merge smoothly, not abruptly.

Marsupials are also distinguished from placentals by numerous 'hard' characters, such as dental characteristics (e.g. Archer 1984b; Luckett 1993; van Nievelt & Smith 2005; Wroe & Archer 2006). Marsupials have up to four or five upper incisors whereas placentals have up to three upper incisors; marsupials primitively have three premolars and four molars whereas placentals have four premolars and three molars. Marsupials only shed one post-canine tooth (third premolar) and their premolar P3 is never molariform. Some other 'hard' characteristics, useful for distinguishing marsupials but not necessarily definitive for the group, include the following: alisphenoid bone forms the floor of the middle ear (but not in ancestral marsupials); mental foramen of the lower jaw (dentary) is often beneath molar M1; bony palate is commonly perforate; jugal bone extends posteriorly to form part of the glenoid fossa (where the dentary bone articulates with the skull); nasal bones often widen posteriorly; dentary bone generally has a markedly inflected angular process; epipubic bones are present (primitively); and tooth enamel has a microscopic tubular structure (Archer 1984b).

Extant placental mammals are classified into about 29 orders, 153 families and 1,229 genera (Wilson & Reeder 2005; Table 1.1; cf. Appendix: Classification of the families of Mammalia). There is a single order of monotremes (Monotremata), which is generally accepted to be a basal clade to the marsupial and placental mammals (e.g. Gill 1872; Simpson 1945; Novacek 1992; Janke et al. 2002; Musser 2003, 2006; Kullberg et al. 2008; Rowe et al. 2008; Figure 1.1), although the monotremes have sometimes been described as a sister clade to the marsupials, forming the Marsupionta (e.g. Gregory 1947; Kühne 1973; Penny & Hasegawa 1997). The Marsupialia consists of two basic clades, the Southern American species (Ameridelphia) and the Australian species (Australidelphia). The Ameridelphia contains two orders: the Paucituberculata (caenolestid shrew opossums) and Didelphimorphia (various opossums). The Australidelphia contains five orders: the enigmatic Microbiotheria (consisting of a single species *Dromiciops gliroides*, which is considered to be a living fossil and is found in South America but is more closely allied with the Australian than the American marsupials), and the characteristically Australian orders Dasyuromorphia (carnivorous marsupials), Notoryctemorphia (marsupial mole), Peramelemorphia (bandicoots and bilby), and Diprotodontia (koala, wombat, possums, gliders, kangaroos, and allies). The Placentalia contains the remaining 21 orders, which are generally arranged into four principal clades: Afrotheria (many African groups, such as elephants, aardvarks, tenrecs, golden moles, dassies, manatees), Xenarthra (a few mainly Southern American neotropical edentate species, such as sloths, anteaters, armadillos), Laurasiatheria (hedgehogs, moles and shrews, artiodactyls and cetaceans, bats, perissodactyls, pangolins, and carnivores), and Euarchontoglires (primates, rodents, lagomorphs).

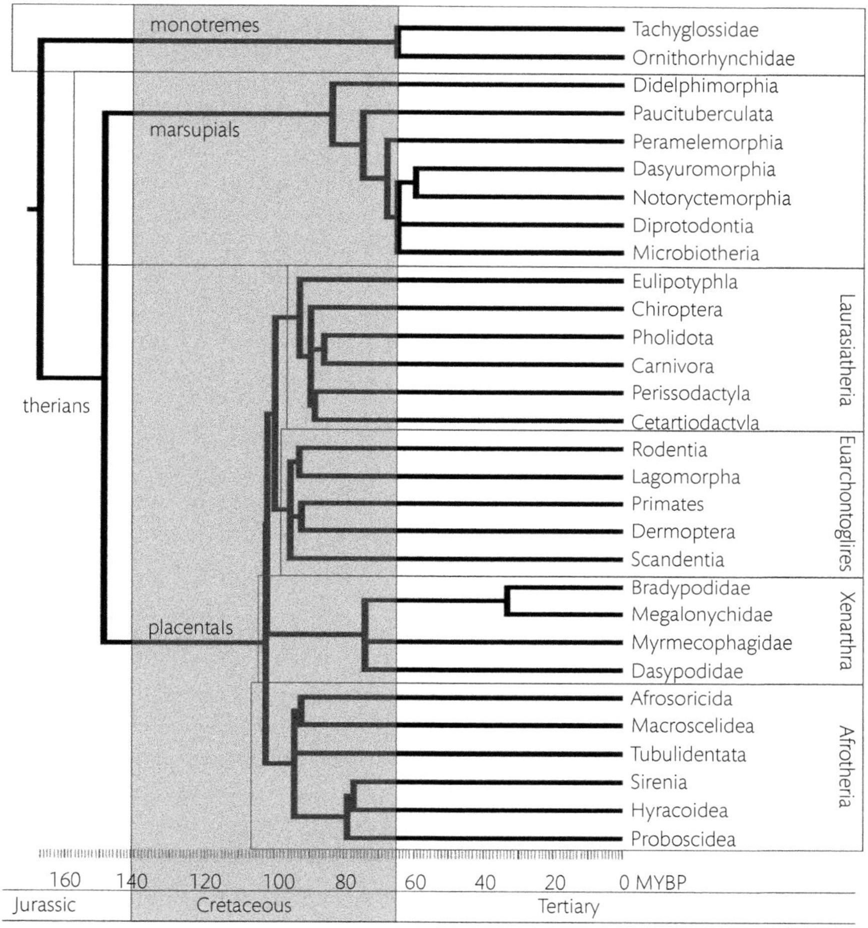


**Figure 1.1** Phylogeny of the major extant mammal clades, highlighting the Cretaceous diversification (based on Bininda-Emonds et al. 2007).

## 1.1.2 Characteristics of Living Mammals

The presence of mammae (mammary glands) that produce milk and facilitate nutrition of the young is a defining character of mammals. Female monotremes lack nipples, and their milk exudes onto the surface of a specialized area of the skin. Specialized nipples, for suckling, are found in marsupials and placentals, but the number varies widely between species (and sometimes within species) from two (e.g. primates) to more than ten (up to 29 for the tail-less tenrec, which can have a litter size of 32). Males of many mammalian species have vestigial mammary glands (and, in a few species of bats, functional mammary glands; Kunz & Hosken 2009). All living mammals are also recognizable from other anatomical characteristics. They characteristically have four limbs, although in many groups the limbs are highly specialized for locomotion or feeding, or reduced or vestigial (e.g. aquatic mammals). Mammals have hair, which generally forms a dense covering of fur providing thermal insulation, but some are essentially hairless (e.g. naked mole rats, *Heterocephalus glaber*; hairless bats, *Cheiromeles torquatus*; cetaceans). Mammalian skin characteristically has a variety of glands, including sweat glands that produce a watery secretion, sebaceous glands of hair follicles that produce an oily secretion, and various scent and musk glands with a variety of roles, including social communication (attractants, territorial marking, etc.). The mammary gland is a highly modified skin gland.

Mammals also have a common suite of physiological characteristics. General physiological principles are described in Chapter 2, basic physiological characteristics of mammals in Chapter 3, and physiological adaptations of mammals to extreme environments in Chapter 4. One important physiological characteristic of mammals is endothermy, the physiological control of a high and constant body temperature ($T_b$), using changes in metabolic heat production to defend against heat loss. The metabolic rate is also higher in mammals (and birds) than in other vertebrate groups, a characteristic named 'tachymetabolism'. Keeping $T_b$ constant (homeothermy) and high confers a number of physiological and ecological advantages, but it comes at a considerable cost, including the energetic cost of internal heat production and life history constraints (e.g. need for parental care). Many other characteristics of mammals are related to their strategy of endothermy, homeothermy, and hyper-metabolism. Fur forms an insulating body covering that is essential for maintaining homeothermy, at least for smaller mammals. A four-chambered heart completely separates the circuits of oxygenated blood (pumped to the body) and deoxygenated blood (pumped to the lungs), an alveolar lung and diaphragm provide a high capacity for gas exchange, and the loop of Henle of the kidney provides the capacity for osmoconcentration of urine. Mammals also have enucleate RBCs, lack a renal portal system, and urea is their primary nitrogenous waste product. Mammals have a limited (determinate) growth, reaching a final adult size where growth ceases, not the continuous (indeterminate) growth pattern of reptiles.

Endothermy, homeothermy, and tachy-metabolism are also characteristics of another reptilian lineage, the birds (Bicudo et al. 2010), and possibly even other reptilian lineages, such as crocodiles, a lizard, and pterodactyls (Seymour et al. 2004; Clarke & Pörtner 2010; Tattersall et al. 2016). Birds, like mammals, have limited (determinate) growth, their kidneys have loops of Henle that provide only a moderate capacity for urine osmoconcentration, and they have a four-chambered heart. Nevertheless, mammals and birds have independently evolved as two separate lineages from ancestral stem reptiles, and these and other shared physiological traits are remarkable examples of convergence. There are many differences in the details of the physiology of mammals and birds that underlie the conclusion that their adaptations are convergent rather than inheritance from a common ancestor. Birds have feathers that provide body insulation (and enable flight), not fur; the cardiac ventricular septum responsible for the four-chambered heart arises differently; the aorta is derived from the right rather than the left arch; the respiratory system is very different for mammals (elastic, alveolar lung) and birds (air sacs and rigid lung); mammals have enucleate red blood corpuscles whereas birds have nucleated red blood cells; birds have a renal portal system (lost in mammals); and birds excrete uric acid compared to urea in mammals. The reproductive pattern is also quite different. Birds lay and incubate eggs (oviparity) whereas living mammals (except for the monotremes) give birth to live young (viviparity).

### 1.1.3 Phylogeny

There are many phylogenetic reconstructions for the various clades of extant mammals, but Bininda-Emonds et al. (2007) provided a 'supertree' for 4,510 species that allows a general comparison of all extant mammalian clades. As expected, the basal split is between monotremes and marsupials-placentals (at 166.2 MYBP), with the marsupial-placental dichotomy at 147.7 MYBP (Figure 1.2). Collectively the marsupials and placentals are referred to as 'therian', which accords with the subclass Theria, whereas the monotremes are 'prototherian'. The subclass Theria is further divided into the Metatheria (marsupials) and Eutheria (placentals), although 'meta' and 'eu' imply an evolutionary sequence from 'meta' to 'eu' rather than a dichotomous split.

The divergence pattern for the major clades of marsupials is Didelphimorphia (American didelphids) at 82.5 MYBP, Paucituberculata (caenolestids) at 73.8, Peramelemorphia (bandicoots) at 66.8, Diprotodontia and Microbiotheriidae (63.6), and Dasyuromorphia (58.5). For the placentals, all four superorders diverged within about 2.5 MY, around 100 MYBP, and nearly all major placental lineages had appeared by about 85 MYBP, and all by about 75 MYBP.

The number of extant mammalian lineages appears to have increased relatively linearly over geological time, since the Theria appeared about 100 MYBP (Bininda-Emonds et al. 2007). However, there seems to have been an initial peak in the rate of mammalian diversification about 91 MYBP, then a relatively smaller second peak about the Eocene (56 to 33.96 MYBP) and Oligocene (33.9 to 23 MYBP). There doesn't appear to have been any increase in mammalian diversification rate following the Cretaceous-Tertiary mass extinction event (about 65 MYBP), when non-avian dinosaurs met their demise.

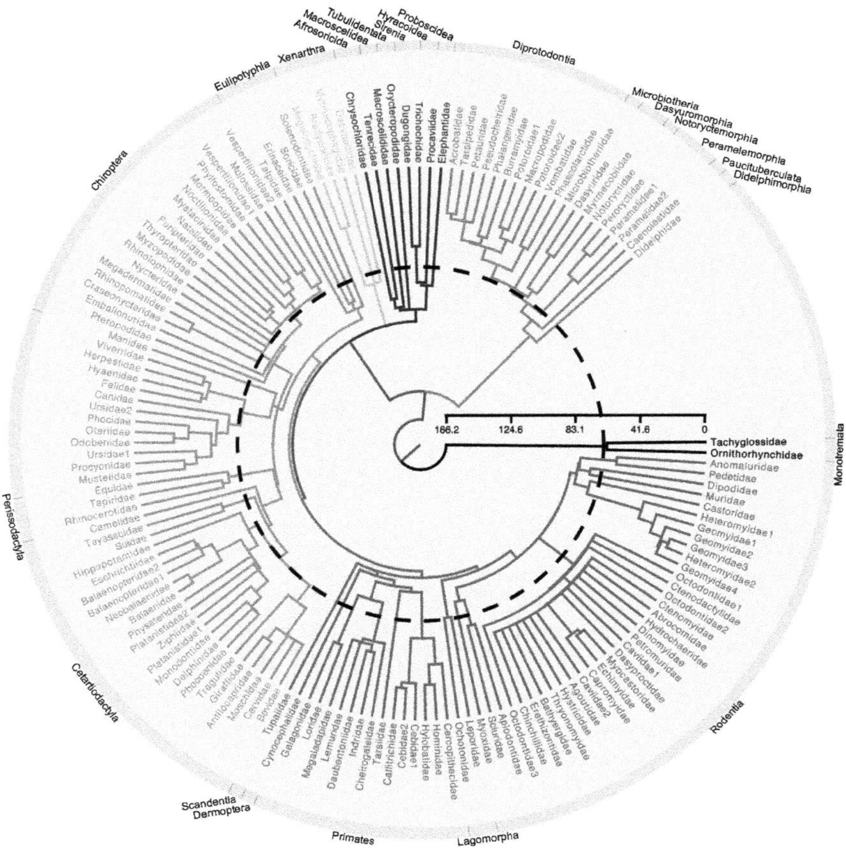


**Figure 1.2** Phylogeny of the major extant mammal clades (with Monotremata divergence at 166.2 MYBP). Reprinted by permission from Macmillan Publishers Ltd: *Nature*, Bininda-Emonds et al. The delayed rise of present-day mammals, copyright 2007.

## 1.2 The Mammalian Bauplan in an Evolutionary Context

The basic reptilian bauplan, at least for 'stem' and early mammal-like reptiles, is terrestrial; quadrupedal with long tail; anapsid skull; egg-laying with amniotic embryonic development; ectothermic with probable behavioural heliothermic thermoregulation; metanephric kidney, but with no urine-concentrating capacity; and complex sensory systems (including chemoreception, vision, thermoreception, mechanoreception, and electroreception). The basic mammalian bauplan is derived from that complex of mammal-like reptiles (Therapsida) with some key additional characteristics: fur, endothermy, renal capacity to osmoconcentrate urine, a larger and more complex central nervous system (including a cerebral cortex), and live birth (except for monotremes) with suckling of the young by mammae.

### 1.2.1 Body Mass

Body mass (M) is probably the most commonly measured trait for animals, in both descriptive and experimental research, because of its pervasive influence on virtually all aspects of an animal's biology. It is a major determinant of many life history, ecological, environmental, and physiological variables for mammals (e.g. Schmidt-Nielsen 1984; Withers 1992; Harvey et al. 1989; Promislow & Harvey 1990; Purvis & Harvey 1995; Fisher et al. 2001; Lee & Cockburn 1985; Western & Ssemakula 1982; Stearns 1992). Bergmann's bioclimatic rule is a well-known example for mammals: races of endotherms are bigger at higher latitudes (Bergmann 1847), which is more commonly interpreted as endotherms are bigger in colder climates (Rensch 1938; Mayr 1956, 1963). Body mass tends to be greatest for mammals occurring on large land masses, and smallest for mammals on small islands (Okiea & Brown 2009). Smaller mammals have a lower (in absolute mass terms) resource requirement and therefore can maintain higher population densities on a small land mass, with lower probability of extinction. However, population density and population growth rate seem to be highest at intermediate body sizes of about 100 g, rather than at the very smallest sizes, so it is not surprising that large mammals often evolve dwarf forms and small mammals evolve giant forms on islands (Foster's rule). The 'island rule' describes island dwarfism in large mammals (e.g. elephants) and gigantism in small mammals (Foster 1964; Roth 1990). Small size seems adaptive for desert mammals (e.g. foxes, Williams et al. 2002a, 2004, and little red kaluta, Withers & Cooper 2009a). Dehnel's phenomenon describes the decrease in size and mass of some small mammals in anticipation of winter (Dehnel 1949; McNab 2010). Cope's rule describes the tendency for mass to increase in evolutionary lineages over time (Cope 1896). Many of these patterns in mass can be attributed to limits to endothermy related to body mass (McNab 1983; Tracy 1977) and resource limitations (McNab 2010).

The early mammals were probably relatively small, scansorial omnivores, much like some extant didelphid marsupials. The early marsupial *Sinodelphys* and the placentals *Eomaia* and *Juramaia* were probably between 15 and 50 g in body mass (Luo et al. 2003, 2011). From these small beginnings, there has been a remarkable divergence in body size in the various taxonomic groups of mammals (see Smith et al. 2003; Figure 1.3) that has accompanied a remarkable radiation in diet (see section 1.4.3). The placental mammals show the largest divergence in body mass (and dietary niches), with extant species ranging from 1.4 g (Etruscan shrew; *Suncus etruscus*) to 190,000,000 g (190 tonnes; blue whale, *Balaenoptera musculus*), with the largest terrestrial mammal being 3,940,000 g (3.94 tonnes;

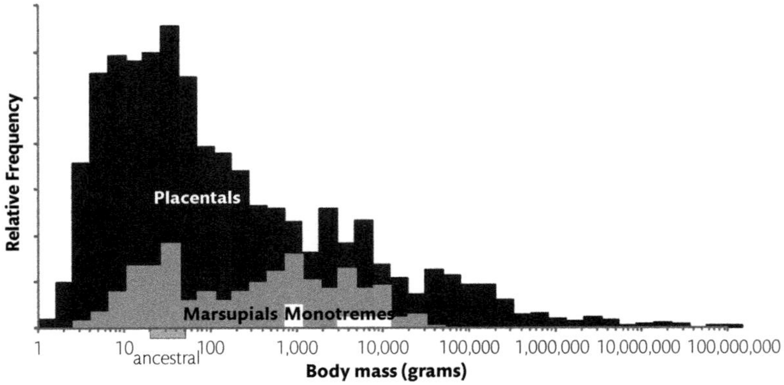


**Figure 1.3** Relative frequency distributions for body mass of extant placental (black; N = 3,877), marsupial (grey; N = 348), and monotreme (white; N = 7) mammals; likely mass range for ancestral mammals is indicated. Frequencies for monotremes, marsupials, and placentals are not to the same scale. Data from Smith et al. (2003).

African elephant, *Loxodonta africana*). It is unlikely that there were ever placental mammals much smaller than the Etruscan shrew, because this mass seems to be about the lower limit for endothermy (e.g. 1–3 g; Tracy 1977; McNab 1983). There have been larger terrestrial mammals in the past (e.g. large proboscideans, *Mammut* and *Palaeoxodon*; Larramendi 2015; the rhinoceratoid *Paraceratherium* = *Baluchitherium*, 15–20 tonnes, Fortelius & Kappelman 1993), although these are dwarfed by the largest sauropod dinosaurs (e.g. estimates of 40–90 tonnes for *Sauroposeidon* and *Argentinosaurus*; Sander et al. 2011). The extant Australian, and to a greater extent South American, marsupials show a somewhat more limited range in mass (and dietary niches), from 4.3 g for Ingram's planigale (*Planigale ingrami*) to over 50,000 g for the red kangaroo (*Macropus rufus*). Again, it is unlikely that there were ever marsupials much smaller than Ingram's planigale because of the limit in size for endothermy, but there have been larger marsupials in the past (e.g. the extinct diprotodontid *Diprotodon optatum*, 2.8 tonnes; Wroe et al. 2004). Extant monotremes vary from the 1,500-g platypus *Ornithorhynchus anatinus* to a 10,300-g echidna, *Zaglossus bruijnii*, but up to 30,000 g for a fossil echidna, *Zaglossus hacketti* (Smith et al. 2003). In contrast, the smallest extant bird is the bee hummingbird (*Mellisuga helenae*, 1.8 g), the largest is the ostrich (*Struthio camelus*, 104,000 g), and the largest fossil is an elephant bird (*Aepyornis*, 500,000 g; Blackburn & Gaston 1994; Wood 1983; Davies 2002).

Placental and marsupial mammals appear to vary along a 'slow-fast continuum', with larger body mass associated with the 'slow' end of the continuum: long gestation, large neonate size, slow growth, delayed sexual maturity, low fecundity, greater longevity, and lower mass-specific metabolic rate (Harvey et al. 1989;

Promislow & Harvey 1990; Purvis & Harvey 1995; Fisher et al. 2001; Lovegrove 2000). However, the fast-slow continuum remains even after correction for body mass and phylogeny, and is likely related to ecological and physiological effects, such as competition or predation risk, or diet and energy availability. For example, the diet of marsupials is highly correlated with key life history traits. Species with energy-rich diets such as carnivores and insectivores appear to have more energy to invest in offspring and have a higher reproductive rate than species with energy-poor diets such as browsers and grazers (Fisher et al. 2001).

### 1.2.2 Skin and Fur

The skin of mammals, like that of other vertebrates, is a large composite organ that forms the outer shape of the animal (Kardong 2009). Skin has a number of roles, including forming a physical and chemical protective barrier against invasion by pathogens, and horny scales or shields that can contribute 'exoskeletal' mechanical support and strength. It is also a barrier to the movement of water, ions, and other solutes; it is the interface for heat exchange with the environment and can act as a heat-gain surface or a heat shield; it has numerous sensory structures; and its fur provides thermal insulation.

The structure of the skin of amniote vertebrates (reptiles, birds, mammals) reflects their terrestriality, as opposed to the primarily aquatic anamniotes (e.g. fishes and amphibians). Amniote skin typically consists of a basal dermal layer derived from mesoderm, and an outer epidermis derived from ectoderm, separated by a basement membrane. The dermis is primarily fibrous (collagenous) connective tissue, with nerves, blood vessels, and chromatophores. The collagen fibres are often arranged in laminated layers of alternating bias (stratum compactum) that provides a tough but flexible layer that assists with trunk-based aquatic locomotion (e.g. cetacean mammals), but terrestrial mammals have a reduced stratum compactum that provides more flexibility for limb-based locomotion. The dermis also has bony plates in a few armoured mammals. The hypodermis is a layer of connective tissue and fat, below the dermis. Above the dermis is the epidermis, which continually produces new epidermal cells from a basal layer of cells (stratum basale) that move outwardly to the skin surface, where they degenerate and are shed. During their outward movement and degeneration, these cells accumulate various proteins (keratin) and lipids—a process called keratinization. In reptiles, birds, and mammals, a distinct (and often thick) layer of dead, keratinized cells accumulates at the skin surface as the stratum corneum. Reptilian skin is highly keratinized and has few glands; their scales are hard epidermal plates, often modified to form spines, crests, or horny structures. The mammalian epidermis forms hair, baleen, claws, nails, horns, and scales (cf. beaks and feathers in birds). In mammals, the skin contains specialized epidermal cells including Langerhans cells (possibly with an immune function), Merkel cells (mechanoreceptors), and chromatophores (pigment cells). The chromatophores are derived from neural crest cells that migrate and lodge in the epidermis, where they form the pigment melanin that is transferred to epithelial cells and incorporated into the stratum corneum or hair shafts. Mammalian skin contains three types of glands—sebaceous, eccrine, and apocrine—from which sweat glands, scent glands, and mammary glands are derived.

Hairs are slender keratinous filaments formed by a hair follicle (rooted by a small tuft of dermal tissue), with a shaft extending above the surface of the epidermis; the shaft has an inner medulla (core) covered by a scaly cuticle. The thick fur covering of most mammals consists of guard hairs, coarse long shafts that extend beyond the finer underfur. The primary function of both guard and underfur is thermal insulation; these hairs are reduced or absent in marine mammals, where insulation is provided predominantly by subdermal fat. Some hairs ('whiskers') are specialized sensory vibrissae, particularly in nocturnal, subterranean, or marine mammals that function in low light levels. The spines (quills) of echidnas and porcupines are stiff, coarse modified hairs that provide effective defence against predators. How hair evolved is necessarily speculative, with limited fossil evidence for 'proto-hair'. Hair may have evolved as surface thermal insulation structures, some of which secondarily evolved into sensory structures (vibrissae), or as small tactile sensory 'hairs', some of which secondarily evolved into insulation. The surface pits on the skulls of some fossil therapsids and early mammals have been interpreted as indicating sensory vibrissae, but in a particularly well-preserved therapsid (*Estemmenosuchus*) there was no trace of hair but indications of the presence of glands (Chudinov 1968; Blackburn et al. 1989). Early therian mammals had fur (e.g. *Eomaia*, Ji et al. 2002; *Sinodelphys*, Luo et al. 2003), and fur was present when the monotremes diverged from therians (Early Jurassic) and was presumably present much earlier in at least the Jurassic (Rougier et al. 2003; Ji et al. 2006). Although the original function of integumental hairs in early mammals is debatable, the role of a full fur coat was presumably to provide the insulation that would be required for a small endothermic mammal (see Clarke & Pörtner 2010).

## 1.2.3 Endothermy and Energetics

One of the distinguishing features of mammals is that they are endothermic. They have a high $T_b$, between about 31 and 38°C, which is precisely regulated. Thermoregulation is achieved by a combination of insulation (fur, fat), behavioural adjustments (basking, posture, huddling together), morphology (countercurrent heat exchangers), endothermy (variable internal heat production), and evaporative heat loss (licking, panting, sweating, insensible). Birds also thermoregulate, are endothermic, and have insulation (feathers), but they presumably evolved these characteristics independently of mammals (Bicudo et al. 2010). Only a few other vertebrates also thermoregulate by endothermy (e.g. female pythons shivering during incubation; some fishes with localized heat production from fat metabolism or locomotor heat production). Large vertebrates also gain a capacity for homeothermic thermoregulation as a consequence of their large body size, hence a large thermal inertia and a low surface area to volume ratio, which favours heat retention even with a reptilian level of metabolic machinery. For example, some very large ectotherms can significantly raise their $T_b$ by internal heat production (e.g. > 900-kg sea turtles) and presumably so could large dinosaurs (e.g. Paladino et al. 1990).

There are strong phylogenetic patterns in body temperature for the various mammal clades, from lower $T_b$ for the more 'primitive' monotremes (32°C) to higher values for marsupials (35.5°C) and placentals (38°C; Hulbert 1980; Withers et al. 2006; Clarke & Rothery 2008; Lovegrove 2012a,b). The phylogenetic pattern in $T_b$ seems to reflect the evolutionary divergence sequence of monotremes-marsupials-placentals, but $T_b$ is often quite variable within mammalian lineages. Having a relatively high and constant $T_b$ confers obvious advantages. Biological reaction rates increase exponentially with body temperature (see section 2.2.1) and so biochemical, physiological, and ecological processes are faster at higher temperatures. Being warm means that mammals can sustain a high activity level, which impacts positively on most aspects of their biology (feeding rates, defence, reproduction, migration, etc.). These advantages, however, come with a high energetic cost because of the nearly constant need for endothermic heat production to maintain the high $T_b$. The higher the $T_b$, the larger is the temperature difference that has to be defended between the mammal and its environment.

### 1.2.4 Respiration

The primary function of the mammalian respiratory system is to provide the high rate of gas exchange (oxygen uptake and carbon dioxide excretion) required to support a high metabolic rate. Consequently, the exchange capacity of the mammalian respiratory system is considerably higher than the ancestral reptilian respiratory system (e.g. of therapsids; see section 1.3.3). Although the basic design of the mammalian respiratory system reflects that of the ancestral reptilian system (much more so than the structure of the avian respiratory system; see Perry & Sander 2004; Bicudo et al. 2010), details of the system are highly modified in a way that increases the total surface for gas exchange and the rate at which the lungs can be ventilated. An increase in surface area resulted when the lungs became much more compartmentalized, and the diaphragm has evolved to augment lung ventilation, resulting in a close match ('symmorphosis') of respiratory gas exchange to maximum metabolic demand (Weibel et al. 1992; Jones 1998).

The ancestral amniote lung was presumably multi-chambered, with an intrapulmonary bronchus that divided into three or four longitudinal rows (Perry & Sander 2004; Figure 1.4). Synapsids had a more homogeneous multi-chambered lung. The broncho-alveolar lung structure of early cynodonts (e.g. *Thrinaxodon*) is presumed to have had more extreme lung compartmentalization, with very many tiny alveoli,

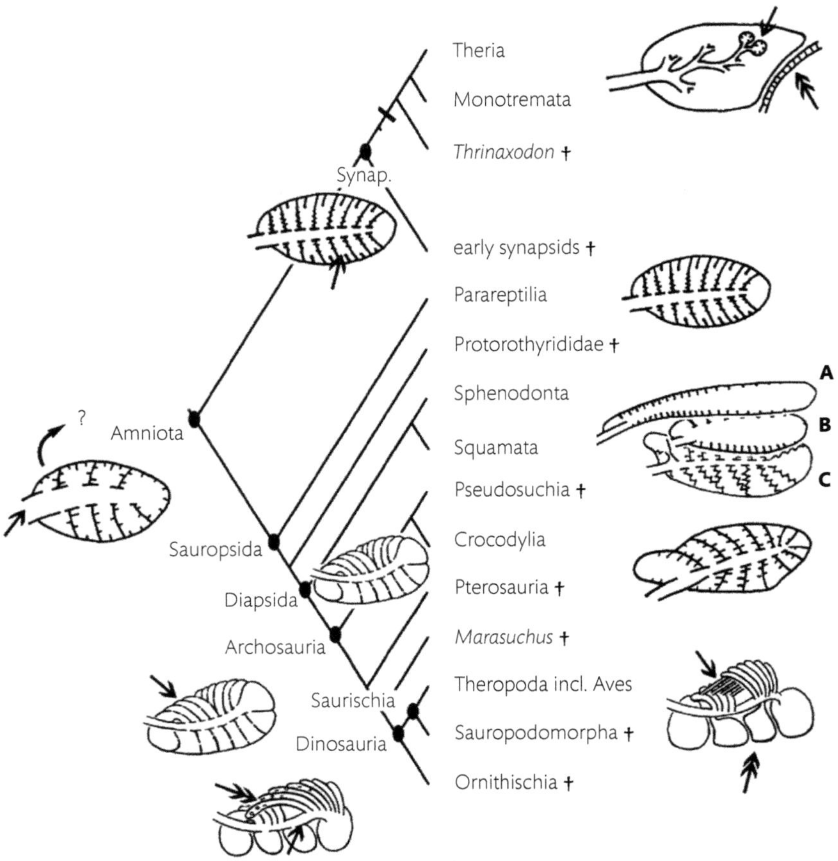


**Figure 1.4** Scenario for lung evolution in amniote vertebrates from an ancestral chambered lung to the presumably homogeneous multi-chambered lung of synapsid reptiles (arrow) and highly compartmentalized lung (arrow) and a muscular diaphragm (double arrow) of cynodont mammals (*Thrinaxodon*) and living mammals. Parareptilia (testudines) have a similar multi-chambered lung as synapsids. Sphenodontid (tuatara) and squamate reptiles often have reduced compartmentalization (A, snakes), some chamber pocketing (B, geckos and skinks), or extensive compartmentalization (C, varanoids). Basal archosaurs have four or five anterior chamber groups separated from the posterior group; crocodilians are derived from this pattern. Dinosauria have a dorsal connection (arrow) and a space between the anterior and posterior groups of chambers (double arrow). Birds have a parabronchial lung (arrow) with anterior and posterior air sacs (double arrow). Modified from Perry and Sander (2004). Reproduced with permission of Elsevier.

like living mammals. The compact, spongy structure provides for a high surface area, but means that the lungs are relatively difficult to inflate; the compliance (change in volume per change in intra-alveolar pressure, or lung stiffness) of mammalian lungs is about one-tenth that of a low-compliance lizard lung, although the body wall compliance is similar. Hence, it is likely that the highly compartmentalized mammalian lung could not have evolved except in concert with the evolution of a muscular diaphragm that contracts and creates a pressure gradient between the intrathoracic space and the environment and thus assists ventilation (Perry & Duncker 1978, 1980). Parareptilia (testudines) have a similar multi-chambered lung as synapsids. Sphenodontid (tuatara) and squamate reptiles often have reduced compartmentalization (snakes), some chamber pocketing (geckos and skinks), or extensive compartmentalization (varanoids). Basal archosaurs have four or five anterior chamber groups separated from the posterior group; crocodilians are derived from this pattern. Dinosauria have a dorsal connection and a space between the posterior and anterior groups of chambers. Theropods, including birds, have a parabronchial lung with anterior and posterior air sacs.

The dome-shaped muscular diaphragm of mammals is a unique structure amongst air-breathing vertebrates. Its primary function is to expand the thoracic cavity when it contracts, providing a negative intrapulmonary pressure, in concert with expansion of the thoracic rib cage. The diaphragm and external intercostal (and accessory) muscles provide the considerable negative intrapulmonary pressure required to overcome the low compliance of the spongy mammalian lung, resulting in inspiration (Perry & Duncker 1978, 1980; Pocock et al. 2013). The internal intercostal and abdominal muscles assist expiration. However, the original role of the mammalian diaphragm may not have been related to negative pressure generation, but to provide a physical separation between the abdominal and thoracic cavities, preventing the abdominal viscera from invading the thoracic space and compromising lung expansion (Perry & Sander 2004; Perry et al. 2010). Synapsid reptiles had gastralia, thick ventral dermal plates, that appear to have assisted lung ventilation (as also happens in archosaur reptiles) and limited the inward movement of the belly wall and visceral displacement into the thoracic region (e.g. during breathing and locomotion). Therapsids, however, typically lack gastralia, and the rib-free lumbar region of some therapsids (e.g. *Thrinaxocodon*) suggests the presence of a diaphragm that would have provided abdominal-thoracic separation and assisted ventilation. The increased flexibility of the trunk would have contributed to greater locomotor capacity. The greater ability to capture small prey (and finely chew their food) meant that a large diaphragm perforation for the oesophagus was no longer required, so more negative intrathoracic pressures could be sustained. Muscularization of the diaphragm would then provide additional inspiratory capacity, supporting a higher metabolic rate. Testudine reptiles have a post-pulmonary septum that physically separates the lungs from the viscera, as do a few other reptiles (varanids and chameleons). However, their diaphragm is not muscular and their lungs are more compliant than mammalian lungs; presumably a muscular diaphragm is not required to assist costal ventilation if their lungs are compliant. The diaphragm also allows for greater intra-abdominal pressure, providing a means for non-ventilatory expulsion of air (e.g. coughing, sneezing) and elimination of materials from the body cavity (e.g. urination, defaecation, vomiting, birth).

The high metabolic rate associated with endothermy requires a correspondingly high ventilation rate, increasing the volume of air that must be 'conditioned' during inspiration. Air reaching the alveoli that is not 100% saturated at body temperature would dry the alveolar surface and compromise of gas exchange. A consequent problem is the high respiratory water loss on exhalation. Mammals (and birds) have nasal turbinate structures in the nasal cavity that assist 'conditioning' of inspired air and cooling and recovery of water from expired air (Hillenius 1992, 1994). The turbinal bones are thin, often highly folded sheets of bone extending from the bones forming the wall of the nasal cavity (the ethmoturbinals arise from the ethmoid bone, nasoturbinals from the nasal bone, and maxilloturbinals arise from the maxilla). The naso- and ethmoturbinals are covered by olfactory epithelium and have a sensory role (olfaction) whereas the maxilloturbinals are covered with respiratory mucosa and play a role in 'conditioning' the inspired air—filtering particulates, humidifying the air, and warming it. In many mammals, particularly smaller species and those from arid regions, the maxilloturbinals also provide a countercurrent heat and water exchange mechanism, reducing respiratory water and heat loss (Jackson & Schmidt-Nielsen 1964; Collins et al. 1971; Schmidt-Nielsen et al. 1981; Schroter et al. 1989; Hillenius 1992). Well-developed turbinates are also present in several orders of birds, and have a similar olfactory and 'air conditioning' role (Schmidt-Nielsen et al. 1970a; Murrish 1973; Hillenius 1992; Geist 2000), but are not homologous structures to the maxilloturbinates of mammals (Witmer 1995; Geist 2000). Reptiles lack the specialized respiratory turbinates of mammals and birds (Geist 2000), but some lizards also conserve expired water (e.g. Murrish & Schmidt-Nielsen 1970). There is evidence for nasal turbinates in therapsid reptiles; *Morganucodon* and some multituberculates and therapsids had ridged nasal cavity walls that may have supported turbinate extensions (Hillenius 1994). The presence of a secondary palate in some advanced therapsids, in concert with the possibility of their having nasal turbinates, would have provided the structures required for nasal countercurrent water and heat exchange, concomitant with their small size and high metabolic and ventilation rates.

### 1.2.5 Circulation

The primary function of the mammalian circulatory system is to transport respiratory gases and other materials (e.g. nutrients, wastes, hormones) and heat throughout the body. The circulatory system consists of the heart, which generates the pressure required to drive blood flow, and the systemic and pulmonary circuits. The basic structure and functioning of the vascular circuits and heart are invariant amongst mammals (and remarkably convergent with birds).

The vascular circuits of tetrapod vertebrates are highly derived, relative to the six paired branchial arteries and vascular circuit of fishes, reflecting the transition from gill to lung breathing (Kardong 2009). In reptiles, the first two branchial arches were lost; the third pair functions as the external and internal carotids (connected to the systemic arches by the carotid ducts); the fourth pair of arches forms the left and right systemic arches; the fifth pair was lost; and the sixth pair forms the pulmonary trunk. In mammals, the right systemic arch feeds to the aorta (the left arch was lost) and the carotid ducts were lost. Birds have a similar arrangement to mammals, except they lost the left rather than the right aortic arch. This rearrangement of the arches in mammals (and birds) required a substantial reorganization of the heart structure and function, resulting in anatomically separated systemic (body) and pulmonary (lung) circuits—the so-called double circulation of mammals (and birds) (Figure 1.5). In contrast, modern squamate reptiles (lizards and snakes) can maintain effective

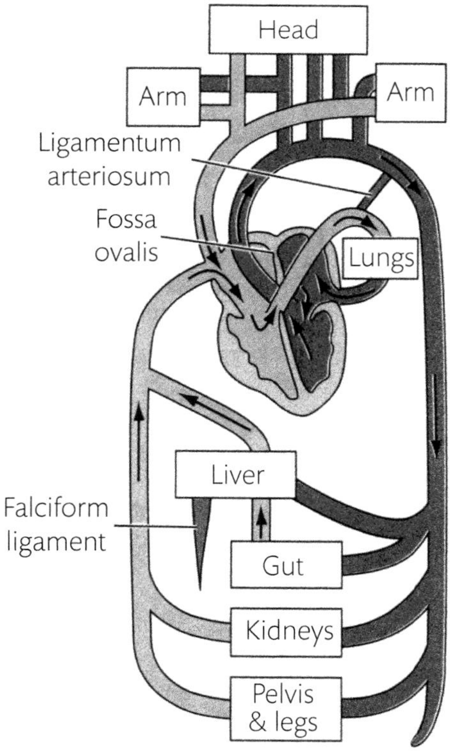


**Figure 1.5** Schematic representation of the arterial (dark grey) and venous (light grey) circuits of the adult mammalian 'double' circulatory system; the fossa ovalis, ligamentum arteriosum, and falciform ligament are remnants of the fetal circulation. Modified from Liem et al. (2001). Reproduced with permission of Elsevier.

functional separation of deoxygenated and oxygenated blood through the heart, but the separation is accomplished by a complex anatomical and temporal pattern of blood flow through essentially three incompletely divided 'ventricles' (cavum pulmonale, cavum arteriosum, cavum venosum; Kardong 2009; Withers 1992). The incompletely divided ventricle of amphibians and reptiles allows the capacity for left-to-right/right-to-left shunts (and a right-to-left shunt via the foramen of Panizza in crocodylians, despite their completely divided ventricle), during, for example, diving or other anaerobic activities (White 1968; Tazawa et al. 1979; Axelsson et al. 1996; Hicks 2002).

The double-chambered heart of mammals consists of anatomically divided right and left atria, and anatomically divided right and left ventricles. Deoxygenated blood returns from the body to the right atrium, passes through the tricuspid valve into the right ventricle, then through the pulmonary valve to the pulmonary arches, thence to the lungs (this is the pulmonary circuit). Oxygenated blood from the lungs returns to the left atrium, passes through the mitral valve into the left ventricle, then through the aortic valve to the aorta, thence to the body (this is the systemic circuit). Each circuit comprises a series of arteries and arterioles leaving the heart, capillary beds, then a series of venules and veins returning blood to the heart. During development, the lungs are non-functional with respect to gas exchange, and most blood is diverted from the pulmonary arch to the systemic arch via the ductus arteriosus (which degenerates soon after birth). There is a similar pulmonary-aortic shunt in fetal birds. Portal blood vessels are a specialized circuit that direct blood from a capillary circuit to another capillary circuit. Examples in mammals are the hepatic portal system, which conveys blood from the gut capillary drainage to the liver (facilitating the processing of nutrients), and the hypothalamic-hypophyseal portal system, which transports hormones from the hypothalamus to the anterior pituitary (see 1.3.8). The renal portal system of reptiles, which conveys blood from the hind limbs to the kidneys, is retained in birds but lost in mammals.

The mammalian 'double-circuit' heart can't be easily derived from a reptilian pattern. It likely evolved from an amphibian-like ancestral heart (e.g. Shaner 1962) or a relatively unspecialized reptilian-like heart (e.g. Holmes 1975). The 'double-circuit' heart of mammals and birds evolved independently, so their functional similarities with respect to the demands of endothermy and aerobic activity are remarkable examples of convergence. The divided ventricles of mammalian and avian hearts are likely the independent consequence of the evolution of specialized cardiac electrical conducting systems that form the interventricular septum, with thick, compact, muscular ventricle walls that provide most of the work of cardiac contraction (Jensen et al. 2012a,b). Amphibians and reptiles, in contrast, have a slow-conducting atrioventricular muscle septum, and a trabecular spongy muscular ventricle (like fishes, and embryonic mammals and birds).

The lymphatic system is an accessory circuit that returns interstitial fluid back into the circulation (Withers 1992; Saladin 2010; Kardong 2009). At the proximal end of capillaries, the high blood pressure (about 4.0 kPa) relative to the tissues (about 0.4 kPa) promotes filtration of fluid out of the capillary, despite the relatively high colloid osmotic pressure of blood (about 3.7 kPa) relative to the tissues (about 1.1 kPa); there is a net force (about 1.7 kPa) for filtration. In contrast, at the distal end of capillaries, the lower blood pressure (about 1.3 kPa) results in a net force (about 0.9 kPa) for reabsorption of fluid into the capillary. Overall, there is a net loss of fluid from capillaries, which would accumulate in the tissues and cause oedema if that fluid was not drained from the tissues by the lymphatic system. A series of blind-ended, valved, thin-walled lymphatic vessels transports this excess fluid (lymph) back into the venous circulation. The small terminal lymph vessels coalesce into larger jugular, subclavian, lumbar, and thoracic lymphatics. Lower vertebrates (including amphibians, reptiles, and embryonic birds) have lymph hearts that assist lymph return. While the mammalian lymphatic system includes valves, lymph hearts are absent; general body movements (breathing, muscle contractions, even arterial pulsations) propel lymph back to their circulatory system.

## 1.2.6 Digestion

The food that animals consume provides both chemical energy that is harnessed during metabolism and structural materials that are required for growth and reproduction (Ferguson 1985). Early mammals were generally insectivorous or carnivorous, although many lineages were specialized herbivores. Living mammals are generally insectivores, carnivores, omnivores, or herbivores (see 1.4.3 for a more detailed discussion of dietary patterns), and although the principles of digestion are similar in all of these dietary types, the herbivores are the most specialized, for digestion of insoluble carbohydrates (e.g. cellulose, hemicellulose).

Insectivores and carnivores generally have a simple digestive tract, with a relatively short small intestine, no or small caecum, and short large intestine. The small intestine is particularly long in some specialized myrmecophages (echidna) and piscivores (dolphins). In contrast, vertebrates that consume plant material tend to have an enlarged stomach (e.g. ruminants, kangaroos, sloths), caecum (e.g. iguana, grouse, rhea), or large intestine (e.g. zebra), where ingested plant matter is fermented by symbiotic microorganisms and the fermentation products are accessed by the host.

## 1.2.7 Locomotion

Mammals, like all other vertebrates, have an internal (endo-) skeleton (Ferguson 1985; Kardong 2009; Liem et al. 2001), comprising the axial skeleton (skull, vertebral column, ribs, sternum) and the appendicular skeleton (forelimbs, hind limbs, and pectoral and sacral girdles). The axial skeleton protects the central nervous system and provides points of attachment for muscles, and the appendicular skeleton provides the capacity for limb-based locomotion.

Most terrestrial mammals are quadrupedal, using both the fore- and hindlimbs for walking or running. There are two general types of posture: a 'sprawling' posture characteristic of amphibians and reptiles, and an 'erect' posture characteristic of many mammals (and dinosaurs and their avian descendents; Rewcastle 1981). The distinction between sprawling and erect is not only in posture but also in locomotor type. The sprawling posture of small (and early) tetrapods, where the limbs have essentially a lateral position and the animal moves in essentially a horizontal plane, has advantages for stability during locomotion and climbing. The body mass is supported well inside the support quadrilateral of the feet, so the posture is stable but results in a bending moment acting to collapse the limb. Not surprisingly, the body tends to lie on the ground at rest; 'limb-collapse' during locomotion, where the belly touches the ground, can reduce loading on the limbs. An erect posture is where limb movements are largely in a vertical plane, with the body mass supported directly above the limbs. The erect posture is typical for large, and especially cursorial, tetrapods (Figure 1.6).

The pattern of limb movement during locomotion (the gait) varies with size and speed (Rewcastle 1981). A slow-walking quadruped has each foot on the ground for more than 50% of the stride cycle, with the front and back of each pair of legs moving in opposite phase; the left and right front feet are 0.25 out of phase (Withers 1992). Faster locomotion becomes a symmetrical trot in salamanders, lizards, and small mammals. Increased velocity is accomplished mainly by an increase in stride length and some increase in the velocity of limb movement; the

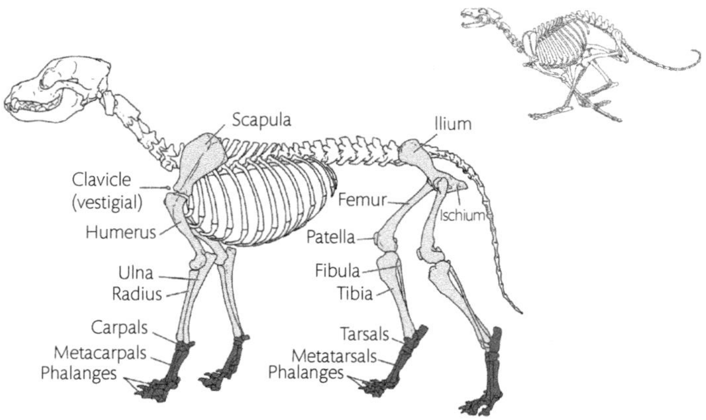


**Figure 1.6** Skeleton of a typical quadrupedal digitigrade mammal (dog), showing the axial (white) and appendicular (light grey) skeletons; foot bones are dark grey; upper inset shows the skeleton in galloping action—note the considerable flexibility of the post-cranial axial skeleton. Modified from Miller et al. (1964) and Liem et al. (2001). Reproduced with permission of Elsevier.

step frequency remains relatively constant. In mammals (and a few crocodiles), an increase in speed is accomplished by a transition to an asymmetrical four-beat gallop (each foot contacts the ground at a different time), with an increase in the duration of the suspended phase. The capacity to gallop is often related to the capacity of mammals for spinal flexion-extension in the vertical plane (unlike the horizontal spinal flexion of lizards). A galloping mammal has a highly flexed spine when the feet contact the ground (see Figure 1.6), and spinal extension increases the step length and hindlimb velocity. This vertical flexion of especially the lumbar vertebrae is facilitated by the vertical longitudinal orientation of the vertebral zygopophyseal facets, a characteristic most marked in carnivores (and to a limited extent in the sacral vertebrae of crocodiles). A vertically flexible spine is not essential for galloping—a galloping tetrapod can be viewed as two connected bipeds—but to gallop the fore and hind 'biped' limb sets must maintain a constant phase, and the fore- and hindlimbs should be of about equal length.

Early thecodont reptiles (leading to archosaurs and birds) were similar to lizards in posture. More advanced thecodonts had an erect posture, but not a vertically flexible vertebral column. Rather, their trunk column was rigid, so they apparently did not evolve a galloping gait. Why the thecodont-archosaur-bird lineage did not evolve a flexible vertebral column and galloping is unclear, but may be related to their differential evolution of lung ventilation (air sacs, thoracic rib pump) compared to mammals (diaphragm and thoracic rib pumps).

In quadrupeds, the various fore- and hindlimb bones form lever systems that permit locomotion, and in some species flexion of the axial skeleton contributes substantially to locomotion. Mammals that walk relatively slowly typically have the entire foot contact the ground when walking (plantigrade; e.g. bears, raccoons, opossums; Hildebrand 1995). Most running mammals (cursors) increase the effective length of their limbs by standing on their digits (digitigrade; e.g. carnivores and ancestral ungulates, also running birds and dinosaurs; Figure 1.6). Highly specialized cursors stand on the tips of their digits (unguligrade; e.g. perissodactyl and artiodactyl ungulates, and several extinct mammalian groups). Both digitigrade and unguligrade cursors often use vertical spinal cord flexion-extension to extend their stride and maximize running speed. Burrowing mammals typically have short and stout digging limbs that exert strong digging forces on the soil. Skeletal structures, muscle attachments, and lever principles are quite different for burrowing mammals compared to cursors (see 2.8).

Bipedal mammals use the hindlimbs for hopping or walking. Hopping bipedality has evolved independently many times amongst mammals: for example, lagomorphs (rabbits, hares), rodents (jerboas, heteromyid rodents, hopping mice), spring-hares, and macropod (kangaroos, wallabies) and dasyurid (kultarr) marsupials. Walking bipedality has also evolved in primates. The forelimbs of these mammals can assume specialized roles, such as grasping and manipulating objects, or brachiating (many primates). Hopping mammals generally have larger hind- than forequarters, a reflection of the muscular forces required for hopping. Some species (e.g. macropods) can store the energy when they land on the ground in their muscles and elastic tendons, and recover some of the energy for their next jump, somewhat like pogo sticks (Dawson & Taylor 1973; see 4.6.1).

Amongst tetrapod vertebrates, a number of lineages have evolved the capacity to glide or fly using extensions of the body surface as aerofoils (Rayner 1981). Various anuran amphibians can parachute or glide. The veined tree frog (*Phyrnohyas venulosa*) is a parachutist that can reduce its speed of descent but it cannot glide well. Other anurans (e.g. Darwin's flying frog *Rhacophorus*) use enlarged hand and foot webbing as parachutes/aerofoils; *Hyla miliaria* has an additional body webbing. Gliding lizards include a gliding gecko (*Ptychozoon*) that uses lateral body and tail flaps, and a gliding dragon (*Draco*) with rib-supported gliding membranes; even some snakes (*Chrysopelea*) use their flattened body as an aerofoil. Three lineages of living mammals have evolved a gliding membrane of skin stretched between the wrists and ankles (Byrnes & Spence 2011). These include flying lemurs (Dermoptera; a single species of *Cynocephalus*), flying squirrels (sciurids, e.g. *Petaurista*), and gliding possums (independently evolved in three genera: *Petaurus, Petauroides, Acrobates*); a number of extinct gliding mammals have also been described, and bats can glide albeit it briefly (see 1.4.3, 2.8.2, 4.7). Even a few fish are capable of gliding, possibly to escape predation or to travel faster since the drag in water is much more than in air. Some swimming mammals and penguins might also benefit from breaking the surface when they swim.

Powered flight has evolved at least three times in the vertebrate lineage. Powered flight requires considerable muscular energy input to provide sustained horizontal, rather than gliding. Powered flight also has requisite skeleto-muscular adaptations, such as expanded bone surfaces where substantial flight muscles are attached. The high metabolic requirement presumably precludes ectothermic vertebrates from powered flight. The living flying 'reptiles' are the birds (advanced theropod dinosaurs), which independently evolved flight; they appeared as such in the late Mesozoic. *Archaeopteryx* (Upper Jurassic) was the important 'missing' link between 'reptiles' and birds; early fossils, without their feathers, were not recognized as birds but were ascribed as dinosaurs. Avian wings are structurally quite different from the membraneous pterodactyl and bat wings. The extinct pterosaurs and pterodactyls were large predatory reptiles with wings largely supported by their digits, body, and hind foot (Pennycuick 1988); the early rhamphorhynchoids lived from the late Triassic to throughout the Jurassic, and the more advanced pterodactyloids from the late Jurassic to the late Cretaceous (Carroll 1988). They had musculo-skeletal adaptations very similar to birds, including a large sternum and an acromial process of the scapulocoracoid that was adapted as a pulley over which the supracoracoideus flight muscle acted. The pterodactyl *Quetzalcoatulus* is the largest known flying vertebrate, with a wingspan of 11–12 m. The only group of mammals that has evolved powered flight is the bats (Chiroptera).

## 1.2.8 Excretion

The regulation of body water, ions, other solutes, and waste products is a fundamental role of the excretory system, and the primary excretory organ of vertebrates is the kidney (Withers 1992). Some reptiles and birds also have salt glands that excrete specific solutes (principally $Na^+$, $K^+$, and $Cl^-$). Sweat and other glands have only a minor excretory role in mammals.

From an evolutionary viewpoint, the excretory and reproductive systems are inextricably linked because both arose from similar structures with similar roles—the 'excretion' of wastes and eggs/sperm from the body cavity. In all vertebrates, embryonic development of the kidney begins with a few pronephric renal tubules that differentiate from the nephric ridge, at the anterior end of the archinephric duct, forming what is known as the pronephric kidney (Liem et al. 2001; Figure 1.7). In amniotes, only one to three pronephric tubules develop, and these are lost in the adult. As the pronephros regresses, more distal mesonephric tubules develop to form the mesonephric kidney, which is the functional embryonic kidney of amniotes.

In mammalian embryos, the role of the mesonephric kidney is reduced in proportion to the functional role of the placenta in embryonic excretion. The mesonephric kidney tubules develop an association with the developing gonad. When the mesonephric kidney degenerates, the archinephric duct retains a role in sperm transfer from the gonad (testis). In amniotes, a ureteric bud forms later in development from the archinephric duct (caudal to the mesonephric kidney) and forms the collecting ducts and ureter (which separates from the archinephric duct), and induces the development of metanephric nephrons in the nephric ridge. This metanephric kidney drains via collecting ducts and ureter to the dorsal-lateral wall of the cloaca in monotremes (and reptiles and birds), so urine must move across

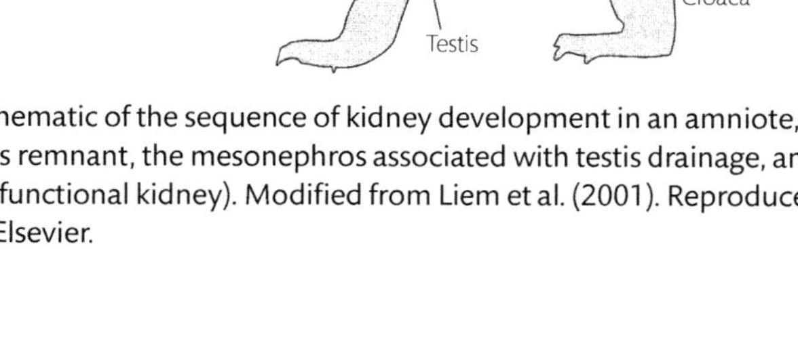

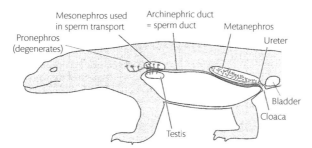


**Figure 1.7** Schematic of the sequence of kidney development in an amniote, showing the pronephros remnant, the mesonephros associated with testis drainage, and the metanephros (functional kidney). Modified from Liem et al. (2001). Reproduced with permission of Elsevier.

the cloaca to enter their urinary bladder. The ureters connect directly to the urinary bladder in therian mammals (how this direct connection occurred embryologically has major significance to reproduction in marsupials compared to placentals; see 3.8.2). In females, the oviduct (Müllerian duct) forms separately as a groove along the coelomic surface of the mesonephric kidney (Liem et al. 2001).

Water and solute balance varies greatly in mammals, reflecting their habits, geographic distribution, and phylogenetic history. Mammals are primarily terrestrial (although some are secondarily aquatic) but occur in a wide range of geographic localities that differ markedly in water availability. The ability of mammals to volume-, iono-, and osmoregulate in such diverse environments depends on a combination of physiological and behavioural adaptations, some of which are unique to mammals (Bentley 2002). Two evolutionary events profoundly changed the volume-, iono-, and osmoregulatory strategies of mammals: first, becoming terrestrial (a trait shared with reptiles and birds); second, becoming endothermic (a trait shared with birds but independently evolved).

The paired (metanephric) kidneys of adult mammals are the principal excretory organ, and play a major role in the regulation of water, ions, and other solutes, including nitrogenous wastes (ammonia, urea, uric acid). The renal tubules (nephrons) of early freshwater vertebrates (e.g. lampreys and freshwater teleosts) and amphibians consist of a vascular glomerulus where urine is formed by filtration from the blood; then a long tubule—divided into proximal tubule, intermediate segment, distal tubule, and collecting duct—drains into the ureter. The enigmatic marine hagfish has a simple nephron tubule structure, reflecting its minimal need to excrete urine (as it is essentially iso-osmotic and iso-ionic with seawater). Marine teleost fishes also have a reduced requirement for urine formation compared to freshwater and terrestrial vertebrates, and have a reduced renal tubule (proximal tubule and collecting duct); some even lack glomeruli. Terrestrial reptiles retain essentially the same tubular structure as amphibians (and cannot produce concentrated urine), whereas both mammals and birds have independently modified the intermediate segment of the tubule into a loop of Henle, enabling the osmoconcentration of urine.

The mammalian kidney is divided into an outer cortex and an inner medulla. The cortex contains the cortical nephrons, which are similar in structure and function to those of other amniotes and anamniotes. These nephrons perform the basic functions, filtration, reabsorption, and secretion, of a vertebrate nephron that evolved in freshwater or estuarine environments (Griffith 1987; Ditrich 2007). Like the nephrons of other vertebrates, these cortical nephrons cannot produce urine more concentrated than their blood. In contrast, juxtamedullary nephrons, with glomeruli located near the junction of the cortex and the medulla, form urine more concentrated than blood. Their intermediate segment is modified to form a long loop of Henle, which descends into and ascends out of the renal medulla. The long loops of Henle establish an osmotic gradient along the renal medulla via the countercurrent multiplier system, whereby solutes are actively transported out of the ascending limb of Henle's loop, and diffuse into the interstitial fluid, descending limbs and vasa recta. Tubular fluid leaving the ascending limb is carried to the distal convoluted tubule, then the collecting duct, where it can be osmotically concentrated. Because only the long-looped nephrons contribute to the medullary osmotic gradient, the presence of short-looped nephrons may limit the capability to which the kidneys can produce hyperosmotic urine, but determine the overall rate of urine formation. This novel osmoconcentrating role of the mammalian juxtamedullary nephron is shared only with birds, but seems to have arisen independently from a reptilian kidney incapable of osmoconcentration (Dantzler & Braun 1980).

The mammalian kidney is capable of considerably greater urine-concentrating capacity than the bird kidney, presumably reflecting the excretion of urea as the primary nitrogenous waste product by mammals, compared to uric acid excretion by birds. In mammals, urea plays a central role in establishment of the medullary osmotic gradient, hence urine-concentrating capacity, and it is an important osmotic component of mammalian urine. In contrast, NaCl contributes the major portion of the osmotic concentration in the medullary cones of birds. Uric acid is relatively insoluble and can precipitate, so avian urine must be sufficiently liquid and dilute to flow down the ureters. The avian kidney therefore has only a modest osmoconcentrating capacity; its concentrating ability is poorer than the mammalian kidney. In birds, ureteral urine is subsequently modified by water reabsorption in the hindgut, after urine entering the cloaca is reflexed into the distal segment of the hindgut (Skadhauge 1981; Withers 1992). In mammals, ureteral urine is not subsequently modified in the bladder or hindgut (except bears; see 4.1.2).

Mammals also differ from birds (and reptiles) in not having a renal portal circulation (Kardong 2009). The renal arteries are the only afferent blood supply to the kidneys, and after providing blood flow to the glomeruli for filtration, the efferent glomerular blood forms a peritubular capillary network (vasa recta) surrounding the renal tubules. In contrast, reptilian and avian kidneys have an additional afferent supply, blood returning from the hind limbs and tail (renal portal system).

## 1.2.9 Neurobiology

Evolutionary changes in the nervous system of mammals are largely associated with the size, structure, and function of the central nervous system (brain and spinal cord) rather than the peripheral nervous system (peripheral nerves and ganglia). Brain size has increased considerably, relative to body mass, in mammals compared to other amniote vertebrates; there has been a considerable reorganization of the relative proportions (hence function) of the different parts of the brain; and there has been a spectacular development of 'higher' brain functions in mammals. Both sensory and motor functions have been considerably developed in mammals, although mammals have not developed any new sensory systems or motor capacities compared to their reptilian ancestors.

Mammals have a considerably larger brain mass than reptiles and dinosaurs, but there is substantial overlap with birds (Butler & Hodos 2005, after Jerison 2001). For mammals, brain size is predicted to equal 0.055 $M^{0.74}$ (Eisenberg 1981), but there are considerable differences between clades, as indicated by considerable variation in the 'encephalization quotient' (EQ), which is the ratio of the actual brain mass to the predicted value for a species from the regression line for all mammals (i.e. EQ = 1 indicates that the species falls on the mammalian regression line). There are broad patterns in EQ amongst mammals (Eisenberg 1981; Streidter 2005). Long-lived species tend to have a high EQ; mammals with a passive antipredator strategy (spines, armour) have a low EQ compared to mammals with active antipredator strategies (directed flight or counter-attack); mammals with complex locomotor patterns (e.g. arboreal or aquatic species) have a high EQ. Mammals that rely on three or more sensory modalities typically have a higher EQ than mammals relying on one or two sensory modalities. High EQ is generally associated with late sexual maturation and high parental care whereas low EQ is associated with minimal parental care and reliance on a ubiquitous food source (e.g. ants or termites; Eisenberg & Wilson 1978) and low learning capacity (Jerison 1973).

The limited information available for the brain size of early mammals suggests that relative brain size increased in three major pulses (Rowe et al. 2012; Figure 1.8). Cynodonts had an EQ of about 0.2, with small olfactory bulbs, consistent with poor olfactory discrimination, poor vision and hearing, low tactile discrimination, and simple motor coordination. The first EQ pulse was most likely related to increased olfaction and tactile sensitivity (from body hair) and neuromuscular coordination; *Morganucodon* had an EQ of about 0.3, with an expanded olfactory bulb and cortex, and expanded mechanoreceptor (tactile) sensation. The second EQ pulse was related to further olfactory enlargement, whereby the brain size reached mammalian levels; *Hadrocodium* represents this second encephalization pulse, with an EQ about 0.5 (which is within the current mammalian distribution), reflecting expanded olfactory bulbs and cortex. The third EQ pulse was a further increase in olfactory sensation associated with an expanded olfactory epithelium associated with turbinate bones, seen in the 'crown' mammals; that is, the 'true' Mammalia: the monotremes (but apparently not other prototherians), marsupials, and placentals. More recently, acute visual and auditory systems evolved in various mammals, olfaction was further enhanced in some, or supplanted by yet other sensory modalities such as echolocation or electroreception.

Not only have profound differences in brain mass occurred during mammalian evolution, but there have also been changes in the relative proportions of different areas of the brain, particularly those related to specialized functions, which are often enlarged, whereas those brain structures associated with reduced (or lost)

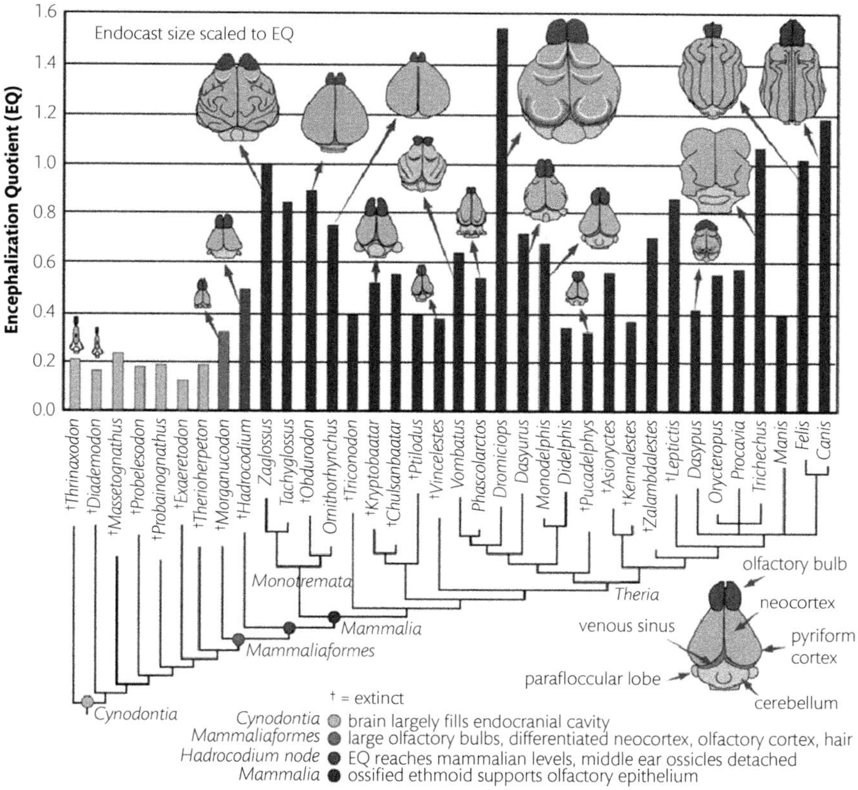


**Figure 1.8** Evolution of brain size in mammals, showing the progression in encephalization quotient (EQ) from cynodonts (EQ = 0.2), through pulses I (EQ = 0.3; morganucodonts) and II (EQ = 0.5; *Hadrocodium*) in olfactory enlargement to pulse III (extant mammals; EQ = 1; Mammalia). From Rowe et al. (2012). Reproduced with permission from Science (AAAS).

functions are reduced (Eisenberg 1981; Streidter 2005). Arboreal and aquatic mammals often have a reduced cerebellum (whose role is largely motor coordination). Bats specialized for diets other than insectivory have larger brains. Early mammals had relatively large olfactory lobes and projection areas for the 1st cranial nerve (olfactory) and 7th cranial nerve (auditory), reflecting the major roles of chemosensation and hearing, as would be expected for small, nocturnal mammals (Eisenberg 1981). An increasing reliance on vision that accompanied the evolution of diurnal activity was accompanied by increasing importance of the 2nd cranial nerve (optic) projections associated with vision. Mammals that have very agile paws to assist with food manipulation, have their paw sensory and motor cortex areas correspondingly highly represented (e.g. racoons; Welker & Campos 1963).

## 1.2.10 Reproduction and Development

The developmental biology of tetrapod vertebrates reflects an evolutionary progression that is generally related to increasing terrestriality, reflected by the appearance of the amniotic cleidoic ('closed') egg in early reptiles (compared to the anamniote egg of amphibians) and the consequent ability to lay the eggs on land. A subsequent reproductive modification in mammals and some reptiles (but not birds) is the retention of eggs during development and maternal nutrient provision to the developing embryo, with live birth.

During maturation, the egg becomes encased by eggshell membranes that are secreted by the maternal oviduct. These membranes form a flexible gelatinous coat in amphibians and a leathery or calcareous shell in reptiles, birds, and monotreme mammals. The embryo also has an extra-embryonic membrane, the yolk sac (formed by the maternal Müllerian duct), which surrounds the yolk and connects to the digestive tract of the developing embryo. For early reptiles, the development of a cleidoic (amniote) egg with a calcareous or leathery egg shell was one of the major adaptions that facilitated terrestriality. The amniote egg is characterized by the further development of three extra-embryonic membranes (amnion, chorion, allantois; Figure 1.9). During early development, a projecting lip of the yolk sac begins to surround the developing embryo, and then fuses to form an outer chorionic membrane and an inner amniotic membrane, within which the embryo develops, protected inside the amniotic fluid environment. A bud of yolk sac membrane develops into the allantoic sac, surrounded by the allantoic membrane.

Early reptiles, and most extant reptiles, are oviparous; the egg is laid and deposited in the environment, and further development of the embryo relies on nutrients stored in a large yolk. Ovoviviparous reptiles retain the egg in the reproductive tract until just before it will hatch, or just after it hatches and the young is born 'alive'; however, in ovoviviparity, the developing young still receives nutrients only from the egg yolk (like oviparity and unlike viviparity), although the embryo relies on gas

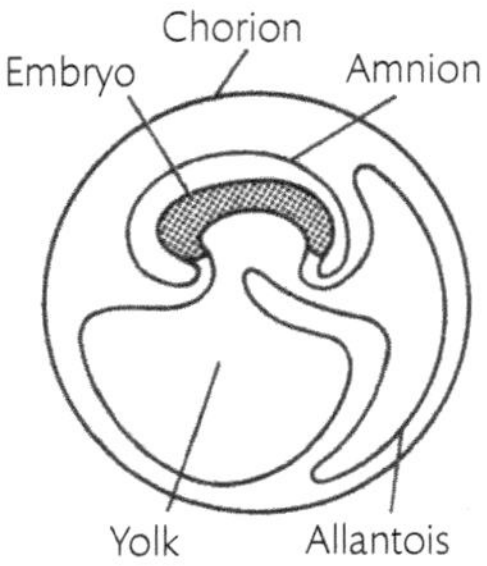


**Figure 1.9** Development of the extra-embryonic membranes (amnion, chorion, allantois) in a developing amniote egg. From Goin et al. (1978). Reproduced with permission of G. Zug.

exchange with maternal tissues. Viviparity occurs, in some reptiles, when the egg is retained in the female's reproductive tract, the developing young receives significant nutrients from the maternal tissues (e.g. a placenta), and the young is born 'alive'.

Monotremes are the only living mammals that lay eggs. The egg is formed in the uterus over about 28 days, then is laid and held in a rudimentary pouch (echidna) or deposited in a nest (platypus). A considerable portion of embryonic development occurs in the egg while it is in the uterus (about 28 days). The eggs hatch at an early stage of development (after about 3 weeks for echidnas, 10 days for platypus), and the young are then nourished by milk from the mammary glands, at the 'milk patch' (Griffiths 1989b; Grant 1989). Milk is simply exuded to the surface at the areolar area (which is covered by fur—no nipple is present), and the young vigorously sucks milk from the skin surface as it is exuded. Marsupials and placentals (the therians) are clearly a more derived group than monotremes (prototherians), giving birth to live young (Dawson 1983, 1989). Marsupials give birth to very immature young after a short gestation; the young then attach to a teat to suckle, often in a well-developed pouch (marsupium, hence the term marsupial). Lactation is prolonged (compared to placentals), lasting up to about 500 days in some large macropods. Placental mammals have a relatively long gestation period, with gas and nutrient exchange and excretion being facilitated by the development of an allantoic placenta in close contact with the uterine wall; lactation is relatively short.

While the placenta is the etymological root of the common name, 'placentals', given to the eutherians, the developing embryos of both marsupial and placental mammals are nourished *in utero* from the maternal blood supply via a placenta. Placental structure provides valuable insights into the evolutionary history of therians, particularly the marsupial-placental dichotomy (Wildman et al. 2006). Although marsupial embryos develop a functional yolk sac placenta, and sometimes a rudimentary allantoic placenta (see 3.8.2), they are still commonly called 'marsupials' as opposed to 'placental' mammals, where the embryos develop and rely primarily on their allantoic placenta.

Lactation is a characteristic of all living mammals, and probably reflects an early evolutionary development in the mammalian lineage (Blackburn et al. 1989; Blackburn 1991; Hayssen 1993). Mammalian milk is a complex mix of water, proteins, carbohydrates, lipids, salts, and vitamins. The production of milk from mammary glands presumably evolved in early synapsid mammals, and the transformation of primitive lactation fluids—which may have provided moisture, and had antimicrobial and immune functions—into nutritious milk accompanied the evolution of the suite of other 'mammalian' bauplan characteristics (Oftedal 2002; Kawasaki et al. 2011). Benefits of the additional parental care from lactation include the ability of females (and potentially males) to provide a transitional and highly nutritious diet for their young between hatching and independence, leading to increased growth rate and survival. Some birds have a functionally similar 'milk' system for nourishment of their young. Male and female pigeons and doves

(Columbidae), and male emperor penguins (*Aptenodytes forsteri*), produce a crop 'milk' for nourishment of their young, and some passerine birds secrete stomach oils; presumably some of their dinosaur ancestors had also evolved 'milk production' for nourishment of their young (Oftedal 2002; Else 2013). Pigeon milk is highly nutritious, but lacks the mammalian calcium-sequestering casein proteins and is low in carbohydrates. Pigeon 'lactation' and mammalian lactation are surprisingly similar in function but differ in their secretory mechanism and structure, and have clearly evolved independently (Gillespie et al. 2012).

## 1.3 Early Mammals

For living species, the presence of mammae has been a sufficient defining definition for a mammal (with endothermy, fur, hyper-metabolism, etc. being other useful characteristics). However, defining a mammal becomes much more complex for fossil species, especially those close to the reptile/mammal transition. It is not necessarily a given that mammae were present in the reptilian ancestors of mammals prior to the development of other mammalian characteristics that might have provided a selective advantage for those reptiles. None of the distinctive anatomical and physiological characteristics that define living mammals are likely to be preserved in the fossil record (except, occasionally, fur), and therefore cannot be definitive characteristics for the entire group Mammalia, including fossil forms. Evolutionary lineages are a continuum, and it can be very difficult to define a major transitional point, such as the 'first' mammals.

Mammalia is a monophyletic group that includes the common ancestor of monotremes, marsupials, and placentals (Rowe 1988, 1993). Huxley (1880) divided mammals into Prototheria ('first beasts'; monotremes) and Theria, with Theria divided into Metatheria ('later beasts'; marsupials) and Eutheria ('true beasts'; placentals). Various other fossil groups (e.g. multituberculates, triconodonts, and docodonts) were often placed with monotremes in the Prototheria, but it is now apparent that this conglomerate 'Prototheria' is not a monophyletic group (Johanson 2006).

So, how do we know that the early mammals that appeared in the Jurassic were indeed mammals? How do we decide at which step along the lineage from therapsids to modern mammals was the 'first' mammal? We must look to the bony characteristics of mammals for more definitive characters of other than living mammals. And it depends on our definition of 'Mammalia'.

### 1.3.1 Characteristics of Early Mammals

The fragmentary and incomplete fossil material for mammals typically consists of only the most durable parts of the skeleton, particularly the teeth. Cranial and post-cranial skeletal material is often incomplete or completely lacking. Unfortunately, many of these skeletal characters reflect a graded evolutionary sequence from mammal-like reptiles to mammals, and this sequence is not exclusive to mammals but also occurred to some extent in other mammal-like reptile lineages, so it is quite difficult to list definitive mammal characters. Nevertheless, there are a number of traditional 'hard' characteristics that are useful for defining mammals (Table 1.2).

Monotremes have a mix of pleiseomorphic (primitive, reptilian) and apomorphic (derived, mammalian) characteristics, but the living monotremes are highly specialized for a semi-aquatic (platypus) or ant/termite-eating (echidna) lifestyle, so it is sometimes difficult to discern plesiomorphic characters. The skulls of platypus and echidnas share many common pleiseomorphic characters but differ in beak/bill morphology. The snout has a large bone (the septomaxilla) that is found in reptiles, therapsids, some early mammals, and some living xenarthrans (anteaters).

**Table 1.2** Traditional 'hard' characteristics useful for defining mammals (e.g. Archer 1984a).

1. Single bone (dentary) in the lower jaw that provides a more powerful bite (the reptilian jaw, from which the mammalian jaw is derived, consists of four bones: the dentary, angular, surangular, and articular).
2. Dentary-squamosal articulation of the lower jaw (dentary) with the skull (squamosal bone) allowed a more powerful bite and improved hearing (the reptilian jaw has an articular-quadrate articulation).
3. Three middle ear bones—incus, malleus, and stapes—provide for improved hearing. Incus is derived from the quadrate of reptiles, malleus from the articular, and stapes from the columella (original middle-ear bone of amphibians and reptiles); the middle-ear incus/malleus articulation of mammals is equivalent to the reptilian jaw articulation.
4. Differentiation of teeth into incisors, canines, premolars, and molars, which provide occlusion for cutting and grinding (reptilian teeth are generally uniformly simple and conical).
5. Tribosphenic molars with a triangular pattern of a cusp set between an anterior and posterior cusps, good for cutting/grinding occlusion.
6. Diphyodont dentition (incisors, canines, and premolars are replaced only once, and molars are not replaced at all) to optimize occlusion.
7. Enlarged braincase surrounding larger and more complex brain.
8. One lateral skull foramen (synapsid rather than diapsid condition).
9. Loss of prefrontal bones, resulting in a stronger skull.
10. Secondary palate that allows for simultaneous breathing and feeding.
11. No lumbar ribs, permitting greater expansibility of the gut, and not required for breathing (mammals have a diaphragm).
12. Seven cervical vertebrae.
13. Bones of the pelvis fused providing greater strength.
14. Ilium of the pelvis enlarged allowing more muscle attachment; the ileum lacks the 'reptilian' posterior flange.
15. Skull articulates with the first cervical vertebra (atlas) via a double occipital condyle, allowing better control of head movement.
16. Phalangeal formula reduced to 2-3-3-3-3 (from 2-3-4-5-3) reflecting the forward-facing foot.

The ear opening of the skull lacks an auditory bulla; the incus-malleus-stapes organization is typical of mammals, but the cochlea is only semi-coiled (270°) compared to marsupials and placentals (360°). The ventral surface of the skull has ectopterygoid bones, which may be remnants of the reptilian pterygoids. A post-temporal canal extends from the temporal fossa to above the ear (also present in mammal-like reptiles and some early mammals). The postcranial characteristics of monotremes also are generally plesiomorphic, although echidnas have specializations for digging and the platypus for swimming and diving. The sprawling posture of monotremes is more likely derived from an ancestral adaptation for fossoriality rather than reflecting the sprawling reptilian posture (Jenkins 1973; Musser 2006). The cervical vertebrae have ribs (like reptiles and unlike marsupials/placentals), as do more of the lumbar vertebrae than in marsupials/placentals. The shoulder girdle is similar to that of therapsid reptiles but has some features similar to other mammals. The pelvis also resembles that of therapsid reptiles and early mammals, with some specializations. Monotremes (like many other early mammals, most marsupials, and early but not living placentals) have epipubic bones, extending anteriorly from the pubis, and they have ankle spurs associated with a poison gland (like many other early mammals). Living monotremes have highly specialized (platypus) or absent (echidnas) teeth, and the teeth of fossil monotremes are reduced and specialized, making it difficult to relate their tooth structure to that of other mammals, living or fossil (Johanson 2006).

Therian mammals, which include marsupials and placentals but also other lineages, characteristically have tribosphenic molars, a specialization providing a cutting and grinding action of the opposing upper and lower molars. Tribosphenic molars have three major cusps arranged in a reversed triangular pattern. The upper molar has paracone, protocone, and metacone cusps; the lower molar has paraconid, protoconid, and metaconid cusps, as well as three additional cusps (hypoconid, hypoconulid, and entoconid) that are located on a posterior shelf of the molar that forms the talonid basin (Figure 1.10). The upper molar occludes with the lower molar, forming a shearing edge and a grinding action with the talonid basin.

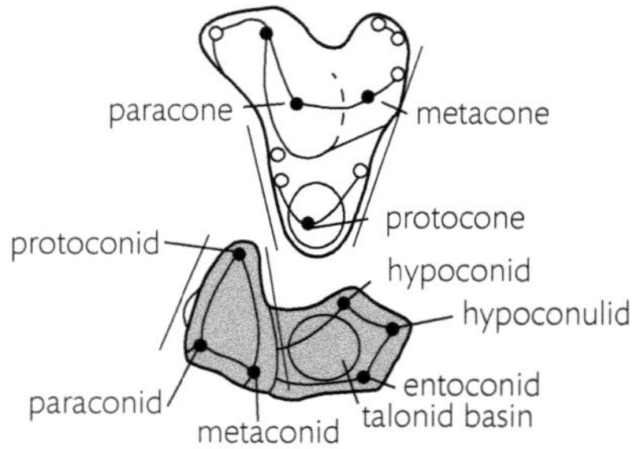


**Figure 1.10** Upper (top) and lower (bottom, shaded) tribosphenic molars of an early mammal (*Pappotherium*); anterior is to the left, lingual to the top. Thin lines show shearing surfaces; circles show grinding surfaces. Modified from Hopson (1994) and Johanson (2006).

The 'hard' characters, especially those related to teeth, are very important for distinguishing fossil marsupials from placentals because these are the best-preserved characters in the fossil record. However, some of these traditional 'hard' tooth characteristics are not necessarily definitive characters for all marsupials including the 'ancestral' forms (e.g. Archer 1984b; Wroe & Archer 2006). Useful 'hard' molar characters include the following: metacone is subequal or taller than the paracone (but subequal is a relative term); hypoconulid and entoconulid are closer together than are the hypoconulid and hypoconid of the lower molars; and stylar cusp 'C' of the upper molars is large (but probably was not present in the 'ancestral' marsupials). Another useful 'hard' character is the pattern of tooth differentiation and replacement. Tooth differentiation is consistent through the evolution of mammals—incisors, canines, premolars, and molars. The primitive eutherian and metatherian dental formulae were long considered to be $I^3_3\ C^1_1\ P^4_4\ M^3_3$ and $I^5_4\ C^1_1\ P^3_3\ M^4_4$, respectively (superscripts, number of upper jaw teeth; subscripts, number of lower jaw teeth; Rose 2006), but early eutherians were $I^5_4\ C^1_1\ P^5_5\ M^3_3$ (*Eomaia, Juramaia*; Ji et al. 2002; Luo et al. 2011), and early metatherians were $I^4_4\ C^1_1\ P^3_3\ M^4_4$ (*Kokopellia; Sinodelphys*; Cifelli & de Muizon 1997; Luo et al. 2003).

Diphyodonty, having only two sets of teeth, is characteristic of mammals (or at least therians), but it is not as simple as a complete replacement of the first set of 'milk' teeth with a second set of 'adult' teeth (Luckett 1993; van Nievelt & Smith 2005). In mammals, teeth develop from the dental lamina, and a replacement tooth develops from the stalk of the original 'milk' tooth. In eutherian mammals, the dental lamina forms 'milk' incisors, canines, and premolars, but not molars. The milk teeth are lost, and replaced by the adult set of teeth. In metatherian mammals, the 'milk' incisors, canines, and premolars generally do not develop. This presumably reflects the retention of the nipple in the mouth during prolonged lactation, and selection for suppression of anterior tooth development. Only the third premolar forms a milk tooth, which is shed and replaced by an adult tooth. It has been suggested that the dental plate of marsupials forms precursor 'milk' incisors, canines, and premolars (not molars), but these do not develop and are never shed as milk teeth (except for the third premolar); the adult 'replacement' teeth include the incisors and canines, and one premolar (Luckett 1993). Alternatively, it has been suggested that some marsupials retain only a single developmental generation of teeth (van Nievelt & Smith 2005).

## 1.3.2 Transition from Mammal-Like Reptiles to Mammals

Early Mesozoic mammals evolved in successive episodes of diversification by lineage-splitting, with many separate evolutionary experiments and ecological specializations (Luo 2007). Recently discovered fossils show that the evolution of some key characters, such as the middle ear and tribosphenic molars, was possibly quite labile amongst these Mesozoic mammals. The main episodes of diversification were the following: Late Triassic and Early Jurassic (200 MYBP) diversification of pre-mammalian mammal-like reptiles; Middle Jurassic (170 MYBP) diversification of docodonts and other pre-mammalian groups; Late Jurassic (150 MYBP) diversification of monotremes and other mammalian groups that are closer to marsupials and placentals; and the Early Cretaceous (125 MYBP) divergence of the marsupial and placental lineages. These episodes were related to many morphological advances, including molar teeth.

The reptilian group that was ancestral to mammals, the Synapsida, diverged from the basal amniote terrestrial vertebrates more than 300 MYBP. The synapsid reptiles (subclass Synapsida) are differentiated from the other reptilian groups by the development of a single opening in the lateral brain case below the squamosal-postorbital bar, in contrast to basal anapsids (historically but not currently represented by turtles; Zardoya & Meyer 1998), which lack a foramen; euryapsids (extinct plesiosaurs and ichthyosaurs), which have a foramen above the squamosal-postorbital bar; and diaspsids (dinosaurs; pterodactyls; living reptiles, probably including turtles; and birds), which have both lower and upper foramina (Carroll 1988; Rose 2006). Their lower jaw consisted of four bones (dentary, angular, surangular, articular), and the articular bone formed the articulation with the quadrate bone of the upper jaw. The order Pelycosauria was an early radiation of synapsids (Early Pennsylvanian to the Late Permian, 300 MYBP) that included more basal and generalized ophiacodonts and later sphenacodonts, that appeared in the middle of the Pennsylvanian and were the primary carnivores by the Early Permian. These pelycosaurs gave rise to the therapsids. All sphenacodonts had tall neural spines (e.g. *Haptodus*), but the 'sail-back' forms (e.g. *Dimetrodon, Edaphosaurus*) had specialized elongate neural spines that supported a 'sail' that might have had a thermoregulatory function (Bramwell & Fellgett 1973; Bennett 1996). Pelycosaurs were most likely very reptilian in their overall skeletal anatomy and general biology, with a small lateral temporal fenestra being the only portent of their ancestry of mammals.

Therapsid reptiles arose in the Middle Permian (260 MYBP) from sphenacodontid pelycosaurs (probably similar to *Haptodus*) and diverged into the carnivorous theriodonts and the herbivorous anomodonts. The most successful anomodonts were the dicynodonts (e.g. *Dicynodon*) that had a wide distribution by the Lower Triassic, and persisted to the end of the Triassic (208 MYBP). Early eotitanosuchians (e.g. *Biarmosuchus*) typify the early carnivorous therapsids. Dinocephalians (e.g. *Titanophoneus*) were initially carnivorous, but more advanced lineages were herbivorous. Gorgonopsids (e.g. *Lycaenops*) were dominant carnivores. They had massive skulls with expanded adductor muscles, a reduced orbit and large canines but reduced cheek teeth, and they had a cursorial skeletal morphology. They are not closely related, however, to the more advanced carnivorous therapsids, and disappeared by the end of the Permian (280 MYBP). By then, two more advanced carnivorous therapsid groups had appeared, the therocephalians and the cynodonts. Therocephalians included small, possibly insectivorous, forms; larger carnivores; and herbivores (e.g. *Bauria*). They had a much-expanded jaw musculature, resulting in a sagittal crest, and a secondary palate. They were approaching a mammalian-grade in many respects, but the *Bauria*-like therocephalians are no longer considered possible ancestors for modern mammals. It was the cynodonts that developed the specialized dentition, braincase, lower jaw, and other characteristics of modern mammals.

By the Late Permian-Triassic (245 MYBP), the Cynodontia were a diverse mammal-like radiation of therapsids. Through the Permian and Triassic, the cynodonts (e.g. *Procynosuchus, Cynognathus, Thrinaxodon, Probainognathus*) progressively acquired mammal-like characteristics, including heterodont dentition, post-canine teeth with three cusps aligned longitudinally, enlargement of the dentary and reduction of the other lower jaw bones, paired occipital condyles, secondary palate and turbinate bones, differentiation of vertebrae into cervical/thoracic/lumbar/sacral, restriction of ribs to the thoracic region, modified limb girdles with well-developed joints, and a less-sprawling posture. Of considerable importance was the evolution of the dentary-squamosal jaw articulation (the 'mammalian' jaw joint) and loss of the 'reptilian' jaw joint between the articular-quadrate, freeing up these bones to become middle ear ossicles in mammals. Associated with the new jaw articulation, the post-canine teeth became more complex, and a reorientation of the jaws and their musculature allowed more precise occlusion of these teeth, leading to improved shearing and sometimes grinding. The reorganization in these reptiles of respiratory ventilation, from buccal pumping to aspiration, allowed the head and neck region to become specialized for feeding. In mammal-like reptiles, the evolution of a secondary palate and nasal turbinate bones, loss of lumbar ribs and likely evolution of a diaphragm, and a more upright and active posture, all suggested a greater respiratory capacity. This suite of changes suggests a greater upright posture, agility, biting strength, and food-processing capacity, and increased ventilatory capacity—hence capacity for a sustained and elevated metabolic rate and endothermy.

These mammal-like features appear to have arisen repeatedly in cynodontids, making the precise origins of mammals difficult to determine. The cynodont group most closely related to mammals is often considered to be the Tritheledontidae (ictidosaurs), although Tritylodontidae are also considered by some to be closest to mammals (or at least to have independently evolved some mammal-like traits), or the Dromatheriidae, or recently *Brasilodon* and *Brasilitherium* (which seem closer to *Morganucodon*) than either tritheledonts or tritylodonts. After mammal-like reptiles emerged from cynodonts, they coexisted with dinosaurs during the Mesozoic (from 208 to 65 MYBP) for about 150 million years. The various Mesozoic radiations of mammal-like reptiles largely became extinct, but the morganucodonts were generally considered to be near the base of the dichotomy between non-mammals and mammals.

### 1.3.3 Mammalian Evolutionary History

The Cynodontia is the group of 'mammal-like' reptiles thought to be most closely related to the Mammaliamorpha, the group that made the transition from mammal-like reptiles to mammals and was the ancestral group including monotremes, marsupials, and placentals (Rowe 1993; Johanson 2006; Figure 1.11). Fossils such as *Adelobasileus, Sinoconodon*, morganucodontids (e.g. *Morganucodon*), and docodonts represent various grades of these mammaliamorphs.

The pleiseomorphic (primitive) and apomorphic (derived) mix of characteristics of monotremes, combined with the possibility that some features may have been acquired convergently with other mammals and the extreme specializations of living monotremes, makes the definition and phylogenetic placement of these 'primitive' mammals somewhat problematic (e.g. Musser 2006). The specialized/reduced/absent molar teeth of monotremes make comparison with other mammal groups difficult, although homologies with the Holotheria have been suggested (Johanson 2006). The skulls of platypus and echidnas share many common primitive characters: lack of an auditory bulla, semi-coiled cochlea, septomaxilla in the snout, ectopterygoid bones, and post-temporal canal. The postcranial characteristics of monotremes are also generally plesiomorphic (reptile-like cervical and some lumbar vertebral ribs, shoulder girdle, pelvis, epipubic bones), but show derived specializations related to digging (echidnas) and swimming and diving (platypus). Their sprawling posture is more likely derived for semi-fossoriality and swimming rather than reflecting the sprawling reptilian posture (Jenkins 1973; Musser 2006). Ankle spurs are generally considered to be a monotreme specialization, but the symmetrodont *Zhangheotherium* (a holotherian mammal) had similar spurs (Hu et al. 1997).

The Theriiformes excludes the monotremes but includes the multituberculates and other lineages such as some triconodonts (which is not a monophyletic group) and symmetrodonts leading to therians (Johanson 2006). Triconodonts have the three main cusps of the teeth aligned in an antero-posterior line; the status of the monotremes in this regard is problematic because there are difficulties in resolving cusp patterns for their highly derived teeth. In the Holotheria, these molar cusps have become arranged as reverse triangles (e.g. Symmetrodonta). The next evolutionary progression is seen in the Prototribosphenida, which consists of the common ancestors of *Vincelestes* and the therians, wherein the lower molar with its reversed triangle dentition has developed a small heel, which is the precursor to the true talonid basin of the Tribosphenida. The Tribosphenida, which includes the fossil *Aegialodon* (about 140 MYBP) and the therians, have a true tribosphenic dentition where the protocone cusp of the upper molar occludes with the talonid basin of the lower molar (see Figure 1.10). One lineage of therians, the Metatheria, diverged into the living marsupials, and the other therian lineage, the Eutheria, into living placentals.

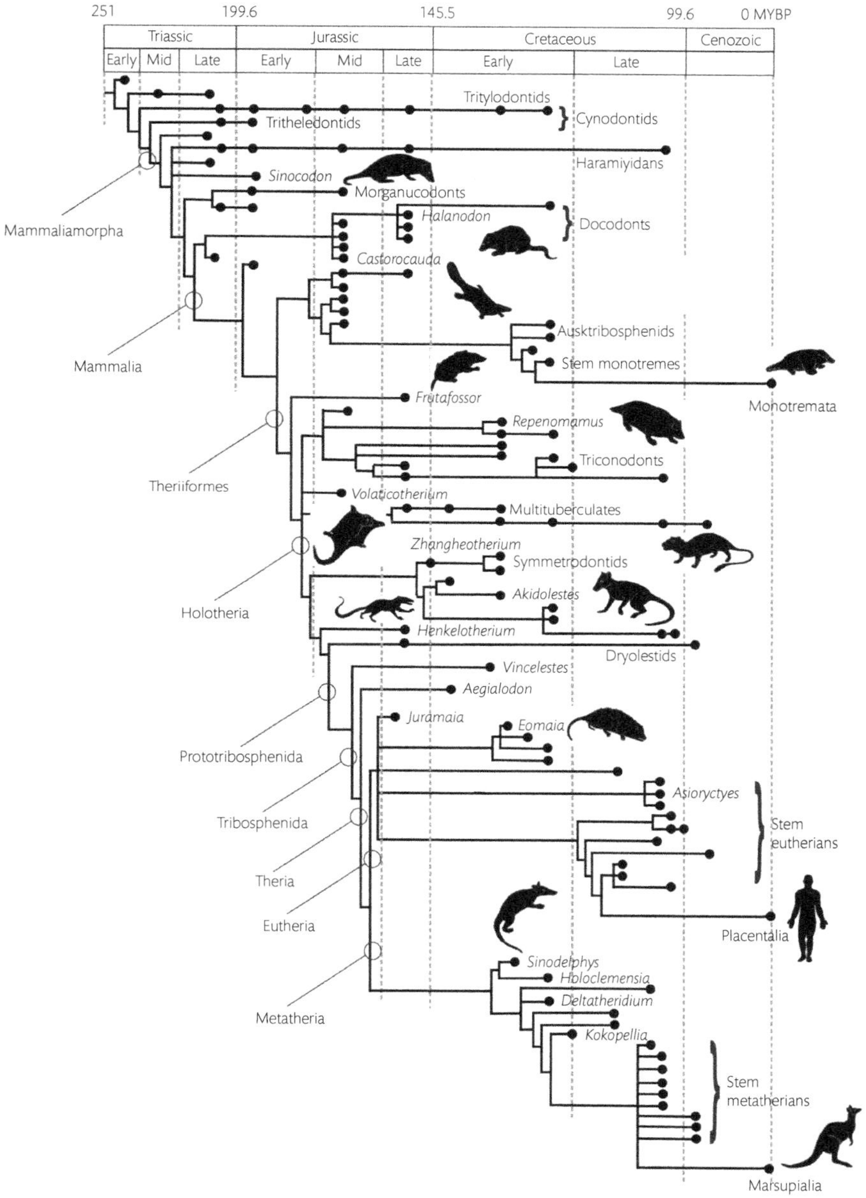


**Figure 1.11** Phylogeny of the major extant mammal clades, highlighting the Cretaceous diversification (based on Bininda-Emonds et al. 2007).

The divergence of metatherian and eutherian mammals occurred before about 160 MYBP, a date that is currently set by the age of the oldest known eutherian mammal, *Juramaia* (Luo et al. 2011). *Juramaia* was small (about 16 g), insectivorous, and scansorial, as were the slightly larger placental *Eomaia* (125 MYBP; 25–40 g; Ji et al. 1999) and the earliest metatherian *Sinodelphys* (125 MYBP; about 25–40 g; Luo et al. 2003), suggesting that the earliest therians were small scansorial insectivores. Contemporary Mesozoic mammals were mostly terrestrial, so the early evolutionary diversification of therians was apparently accompanied by a major ecomorphological shift to a scansorial lifestyle, which made possible the exploitation of, and diversification into, arboreal niches.

Currently, the eutherian lineage begins with *Juramaia* (160 MYBP; Luo et al. 2011), but the origin of the clade that includes living placental mammal, the Placentalia, is less clear. Perhaps none of the 40 or so extinct eutherian genera from the entire Cretaceous period can be assigned to the Placentalia (Luo 2007), whereas most of the more than 4,000 extinct genera of eutherians from the Cenozoic period (the last 65 million years) can be ascribed to the modern Placentalia. Molecular, placental structure, and other data suggest that there are four basic clades of placental mammals (Figure 1.1): the Afrotheria (elephants, hyraxes, sirenians, aardvarks, elephant shrews, golden moles, tenrecs), Xenarthra (armadillos, anteaters, sloths), Euarchontoglires (primates, tree shrews, flying lemur, lagomorphs, rodents), and Laurasiatheria (moles, shrews, hedgehogs, bats, pangolins, canids, felids, horses, tapirs, rhinoceroses, llamas, pigs, ruminant bovids, hippopotamuses, dolphins, whales).

Of these four clades, the Euarchontoglires and Laurasiatheria are generally considered to be the most closely related (as the Boreoeutheria). The resolution of Afrotheria, Xenarthra, and Boreoeutheria is controversial. For example, placental structure suggests that Afrotheria are the basal clade with Xenarthra diverging first, then Boreoeutheria (Wildman et al. 2006), but molecular data suggest that Afrotheria is basal, with Xenarthra diverging, then Boreoeutheria (Springer et al. 2003), or that Afrotheria and Xenarthra are sister clades (Murphy et al. 2007), as are Euarchontoglires and Laurasiadelphia (as Boreoeutheria). Wible et al. (2007) suggest that Laurasiatheria are the basal clade, with Euarchontoglires diverging next, with Afrotheria and Xenarthra as the final sister clade; this scenario is more consistent with a Laurasian origin of eutherians than the other patterns. However, the divergence of these basal placental lineages probably occurred in a relatively narrow window of time, around the end of the Cretaceous, so the order of their divergence is difficult to resolve, hence contentious (Springer et al. 2003; Wildman et al. 2006; Murphy et al. 2007; Wible et al. 2007). Further divergence extended into the Cenozoic, consistent with the 'short-fuse' model of placental evolution (major divergences from the mid-Cretaceous into the Cenozoic), rather than the 'long-fuse' (major divergences from the Early Cretaceous into the Cenozoic) or 'explosive' model (major divergences in a short period mostly after the start of the Cenozoic; Archibald & Deutschman 2001; Springer et al. 2003; Wible et al. 2007).

The oldest fossil in the marsupial lineage was for many years the North American *Holoclemensia* (from about 110 MYBP), but other Cretaceous marsupials as old or older have now been described (Wroe & Archer 2006). *Sinodelphys* is a small (about 25–40 g), scansorial marsupial fossil from the Lower Cretaceous (125 MYBP) of China (Luo et al. 2003). South American marsupials (Ameridelphia) appear about 80 MYBP. The Sparassodonta includes many extinct groups, including borhyaenids (large predators) and thylacosmilids (sable-tooth marsupial 'cats'). The Didelphimorphians include the extant South American didelphids (Didelphidae) and fossil groups. The Paucituberculata include the extant Caenolestoidea (shrew opossums, with procumbent lower incisors) and the fossil Argyrolagoidea (rodent-like, seed-eating marsupials). The Microbiotheriidoidea includes a single extant species, the diminutive monito del monte (*Dromiciops gliroides*); this South American group is allied with the Australian marsupials (Australidelphia) rather than the South American marsupials (Ameridelphia). Antarctic marsupials (dating from the Middle Eocene) include polydolopids, didelphids, and possibly microbiotherians.

In Australia, fossils from Murgon (south-eastern Queensland) include a small and generalized marsupial, *Djarthia*; a marsupial, *Thylacotinga*, closely allied with South American marsupials; and possibly a peramelid (bandicoot) and a microbiotheriid (Godthelp et al. 1992; Wroe & Archer 2006). Beck (2008) and Beck et al. (2008) suggest that the Australidelphia diverged about 65–70 MYBP, perhaps in Australia from a *Djarthia*-like ancestor, or in eastern Gondwana (meaning that Australidelphia did not originate in South America). With either scenario, microbiotheriids back-dispersed from eastern Gondwana. The Murgon deposit includes *Tingamarra*, which might possibly be a placental condylarth; if so, this would argue against the zoogeographic hypothesis that marsupials reached Australia, whereas placentals didn't, before Australia separated from Antarctica (see later).

## 1.3.4 Historical Zoogeography

The complex evolutionary history of mammals has been accompanied by an equivalently complex pattern of zoogeographic dispersal of the various groups of mammal-like reptiles, early mammals, monotremes, and therian mammals. Before the general acceptance of continental drift by the 1970s, concepts of mammalian zoogeography were dominated by theories relating to dispersal along existing connections between continents, or island-hopping routes, and the formation of land bridges (some feasible and well accepted, others more fanciful). Explaining the current distribution of the different mammalian groups generally relied on an assumption of the location of the earliest mammal groups, and the nature of permanent and temporary faunal links between continents. Faunal corridors are substantial connections allowing most types of organisms to disperse through them; filter routes are more limited in habitat diversity, and allow only a limited number of organisms to disperse; sweepstakes routes provide very difficult conditions for dispersal, such that very few organisms are able to traverse them, more or less by chance. The general acceptance in the 1960s and 1970s of plate tectonics and continental drift (based on Wegener's original concept; Wegener 1924) offered a new framework for understanding the evolution and dispersal of organisms (Jardine & McKenzie 1972). The continuing refinement of our understanding of plate tectonics and patterns and timing of continental drift, climate and sea levels, and discoveries of more fossil material means that theories of mammalian dispersal are constantly being revised and refined to more closely coincide with the confirmed presence of monotremes, marsupials, and placentals on the moving continents.

Marsupials have been suggested to have evolved in, and dispersed from, Australia (Fooden 1972) or Africa (Cox 1970), but they now are generally considered to have evolved in western North America or Asia, since that is where the oldest marsupial fossils have been found (see earlier). How then did marsupials reach Australia? An early hypothesis recognized the significance of a Gondwanan floral and faunal pattern, and suggested a southern entry route via land bridges between South America, Antarctica, and Australia (e.g. Hedley 1899; Haswell 1914; Harrison 1921), but the likelihood of there having been intercontinental land bridges joining South America, Antarctica, and Australia was problematic then. And, why didn't placentals also get to Australia (or did they?). Perhaps placentals hadn't reached Gondwana before it split from Laurasia and fragmented, or they had reached Gondwana but went extinct there.

A competing hypothesis for the dispersal of marsupials was the northern route, whereby marsupials dispersed from Asia (China) throughout south-east Asia, then island-hopped to northern Australia (e.g. Matthews 1915; Longman 1921; Darlington 1957, 1965; Simpson 1961; Clemens 1968); that scenario would require the subsequent disappearance of marsupials from south-east Asia, but the unlikely existence of southern continental land bridges was avoided. Again, why didn't placentals also get to Australia? Perhaps there was a low probability of any mammals dispersing to Australia by island-hopping from the north, so it was a low-chance event that marsupials and not placentals arrived in Australia via a sweepstakes route from the south or north (Simpson 1961).

The acceptance of continental drift—with plate tectonics as the mechanism for continental movement—changed interpretations of zoogeographic origins and dispersals. A speculative continental drift theory explained the presence of marsupials in Australia and the Americas by suggesting that marsupials evolved on a mid-Pacific continent (over the 'Darwin Rise') that split and diverged, carrying marsupials to Australia and North America (Martin 1970). However, current concepts of continental drift, in the geological time periods relevant to the evolution and dispersal of early mammals, provide less fanciful explanations.

Some other early explanations based on continental drift suggested a southern origin for marsupials. Cox (1970) suggested an African origin for marsupials, which entered Australia via Antarctica. Fooden (1972) suggested that marsupials evolved in Australia and moved to Antarctica, then South America, where placentals diverged and spread via Africa to North America and Asia. Other theories have a North American/Asian origin and divergence of marsupials and placentals: marsupials radiated from North America into South America, then Antarctica and Australia, whereas placentals radiated from Asia to Europe and North America, then South America, but had not reached Australia before it separated from Antarctica (e.g. Clemens 1971; Keast 1971; Hofstetter 1972; Cracraft 1974).

The currently accepted scenario for mammalian origins and dispersal is that placentals and marsupials diverged in Asiamerica (Asia and western North America) about the time that Pangaea separated into the northern Laurasia and southern Gondwana, 130–100 MYBP (Cox 2000; Beck 2008; Beck et al. 2008; Figure 1.12). Marsupials diverged into three lineages of didelphid-like mammals and dispersed about 80 MYBP from western North America to South America (along with two to four placental lineages), probably via an island-hopping (filter) route, where they diversified into six families. One family (Didelphidae) back-radiated to North America about 2–4 MYBP. Three of the marsupial families spread to Antarctica (along with three placental families) about 45–65 MYBP. One lineage, the Australidelphia, back-dispersed to South America as the Microbiotheria and also dispersed to Australia, from which six families have subsequently diverged. No placentals appear to have spread from Antarctica to Australia, suggesting that the faunal link between Antarctica and Australia was a sweepstakes island-hopping route (Australia first separated from Antarctica about 90 MYBP, but remained close until about 45 MYBP).

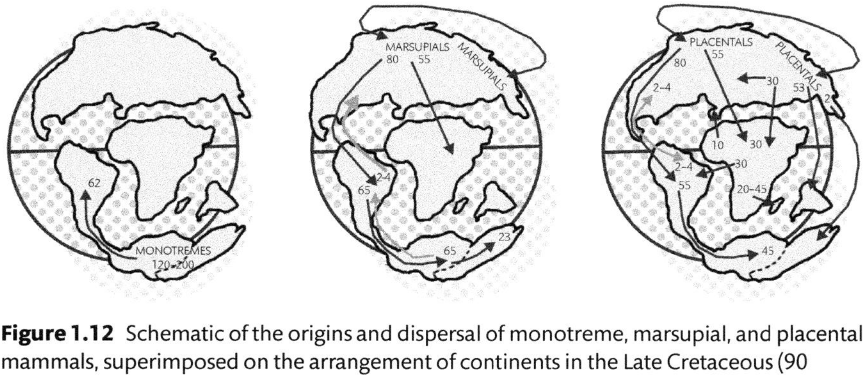


**Figure 1.12** Schematic of the origins and dispersal of monotreme, marsupial, and placental mammals, superimposed on the arrangement of continents in the Late Cretaceous (90 MYBP). Numbers indicate approximate times of movement. Based on Cox (2000), Krause (2001), Beck (2008), Beck et al. (2008), Luo (2007), and Cox and Moore (2000).

The most parsimonious explanation for the absence of placentals in Australia at this time is that they did not reach Australia, rather than being present and becoming extinct from competition with marsupials (something that has not occurred anywhere else that the two groups have been sympatric). However, the possibility that *Tingamarra* is an Australia placental condylarth (Wroe & Archer 2006) argues against this explanation. The early separation of India from Antarctica (about 130 MYBP; Krause 2001) explains its lack of marsupials, but the presence of a marsupial in Madagascar in the Late Cretaceous suggests that there was a biotic connection between Madagascar and Antarctica (about 70 MYBP), after Madagascar had separated from India (about 90 MYBP); this was presumably a sweepstakes connection, since Antarctic placentals apparently did not also reach Madagascar (or India). The didelphid marsupials of North America crossed into Europe via Greenland (about 60 MYBP) and North Africa, but later became extinct in Europe and Africa.

Placental mammals radiated in Asiamerica (about six families by 100 MYBP) and spread to South America/Antarctica, and to Euramerica (Europe and eastern North America), thence to Africa, Asia, and India (Cox & Moore 2000; Cox 2000). The mammal fauna of western and eastern North America rapidly merged once the mid-continental seaway dried up (about 60 MYBP) and they also crossed into Europe via Greenland (as did didelphid marsupials). Asian mammals also moved into Europe and Africa about 30 MYBP. The climatic cooling of the ice ages led to extreme impoverishment of the mammalian fauna of the northern hemisphere. About 80 MYBP, two to four placental lineages of Asiamerican origins radiated from western North America to South America and diverged into about 10 families by 55 MYBP, and eventually into nearly 30 families (including ungulates, armadillos, sloths, and anteaters). New World monkeys and caviomorph rodents appeared about 30 MYBP, presumably arriving via an oceanic route from Africa.

The Panama isthmus joined North and South America by about 3.5 MYBP, allowing the Great American Biotic Interchange (GABI), but there appear to have been significant and complex earlier interchanges of some fauna at about 20 and 6 MYBP (Bacon et al. 2015). Fifteen placental families (and the didelphid marsupial family) entered North America from South America, but only the armadillo and porcupine persisted. Sixteen placental families also entered South America from North America, and nearly all of these persisted; one of the most successful groups was the cricetine rodents, which diversified into 45 genera; the elephants and horses did not persist in South America (and also subsequently became extinct in North America). Similar faunal interchanges are evident for tropical forest birds, although the exchange was primarily south to north (Weir et al. 2009). Three placental families reached Antarctica from South America, but none apparently reached Australia or India/Madagascar. The Australian placental mammal fauna consists of bats and rodents, which have recently moved into Australia from northern islands, and the dingo, which has very recently arrived in Australia, about 4,000 years ago (Newsome & Coman 1989).

When India separated from Antarctica about 130 MYBP, it didn't carry any placental (or marsupial) lineages that persisted. Its extant placental fauna appeared when it was close enough to Asia (about 53 MYBP) for island-hopping, then corridor, faunal links, and perhaps some mammals arrived even earlier from Africa. The placental fauna of Africa also reflects its early separation from South America (about 100 MYBP) and its already close proximity to Europe. Early insectivores, primates, and carnivores were present by about 55 MYBP, with elephants/hyraxes, elephant-shrews, aardvarks, and golden moles apparent by 30 MYBP, and hoofed mammals, carnivores, and rodents soon after. Further interchanges of mammals occurred when Africa collided with Asia about 19 MYBP (pigs, bovids, cricetid rodents, and African carnivores crossed into Africa, and elephants, African carnivores, and primates crossed into Asia), and 19 MYBP, when the subsequent intervening seaway receded (horses spread into Africa, and rhinoceros, hyaenas, and sabre-tooth cats spread into Asia).

The zoogeographic origins and dispersal of monotremes are less clear than that of marsupials and placentals. Simpson (1961) suggested that monotremes evolved in Australia and never radiated out of the region. Subsequent records of fossil monotremes in South America suggest that they might have arisen from early mammals in Australia or Antarctica (since the oldest fossils are in Australia), and then radiated into South America (Musser 2006). For example, Fooden (1972) suggested that monotremes evolved in the Australian region of Gondwana and dispersed to South America but not farther, and later became extinct there; Patterson & Pascual (1968) suggested the opposite—that monotremes may have originated in and dispersed from South America to Australia. The current consensus is similar: monotremes are a relictual taxon from an ancient mammal diversification within Gondwana that at present appear to have been restricted to Australia and South America (Luo 2007), as well as Antarctica (although no monotreme fossils have yet been found there).

## 1.4 Ecological and Environmental Diversity of Mammals

Mammals are an exceedingly diverse group of animals with respect to their ecological and environmental characteristics. They are found in essentially all climates and all biomes on Earth, in all of the major environments (water, land, and air), and encompass essentially all of the conceivable trophic (diet) niches for animals bigger than about 1 g.

### 1.4.1 Climate and Biomes

The Earth's climate zones are characterized using a combination of general environmental conditions, particularly precipitation and temperature. The Köppen-Geiger climate system (Kottek et al. 2006) has main climate zones (A, equatorial;

**Table 1.3** The Earth's terrestrial and aquatic biomes (Olson et al. 2001).

| Environment | Biome |
|---|---|
| Terrestrial | 1. Tropical and subtropical moist broadleaf forest |
| | 2. Tropical and subtropical dry broadleaf forest |
| | 3. Tropical and subtropical coniferous forest |
| | 4. Temperate broadleaf and mixed forest |
| | 5. Temperate coniferous forest |
| | 6. Boreal forest/taiga |
| | 7. Tropical and subtropical grasslands, savannahs, shrublands |
| | 8. Temperate grasslands, savannahs, shrublands |
| | 9. Flooded grasslands and savannahs |
| | 10. Montane grasslands and shrublands |
| | 11. Tundra |
| | 12. Mediterranean forests, woodlands, scrub/sclerophyll forests |
| | 13. Deserts and xeric shrublands |
| | 14. Mangrove |
| Aquatic | 15. Freshwater |
| | 16. Marine |

B, arid; C, warm temperate; D, snow; E, polar), as well as precipitation (e.g. W, desert; S, steppe) and temperature patterns (e.g. h, hot arid; f, polar frost). Climate zones are defined by combinations of these; e.g. Af is equatorial, fully humid; BWk is arid, steppe, cold arid; and Dsb is snow, steppe, warm summer.

Biomes (ecosystems) are geographic areas defined by having similar climate and biological communities. The most fundamental division for biomes is terrestrial and aquatic (including marine and freshwater). One classification system defines 16 biomes, or major habitat types (Olson et al. 2001), 14 terrestrial biomes, and 2 aquatic biomes (Table 1.3).

Within these 14 terrestrial biomes are 867 ecoregions (Olson et al. 2001). The freshwater biome consists of 12 major habitat types (FEOW 2013). These include large lakes; large river deltas; polar freshwaters; montane freshwaters; temperature coastal rivers; temperate floodplain rivers and wetlands; temperate upland rivers; tropical and subtropical coastal rivers; tropical and subtropical floodplain rivers and wetlands; tropical and subtropical upland rivers; xeric freshwaters and endorheic basins; and oceanic islands. These major habitat types are divided into 426 ecoregions (Abell et al. 2008). The marine biome consists of 12 geographic provinces (Spalding et al. 2007): Arctic; temperate North Atlantic; temperate Northern Pacific; tropical Atlantic; western Indo-Pacific; central Indo-Pacific; eastern Indo-Pacific; tropical Eastern Pacific; temperate South America; temperate Southern Africa; temperate Australasia; southern Ocean. These provinces are further divided into 232 ecoregions.

## 1.4.2 Zoogeography

The current distribution of extant mammals reflects the location of their origins, subsequent patterns of continental drift, sea-level changes, climate changes, wholesale faunal exchange by corridor routes, selective faunal exchange by filter routes, chance dispersal by sweepstakes routes, and the vagaries of competition and extinction. The current zoogeographic zones of the world reflect these various processes.

The zoogeographic zones of the world, based on faunal regions originally proposed by Wallace (1876), include the Palearctic, Nearctic, Neotropical, Ethiopian, Oriental, and Australian Regions (Vaughan 1986; Figure 1.13). The Palearctic Region, which includes the northern Old World, is the largest zoogeographic zone. Its climate is largely temperate but ranges from desert to polar. There are about 35 families of mammals in the Palearctic, including 1 endemic family. The Nearctic Region comprises most of the New World and also includes a wide range of climates. It has about 34 families, 1 being endemic. The Neotropical Region consists of the New World, from tropical Mexico south. It is mainly tropical savannah and grasslands, with desert, montane, and tropical rain forest areas. It has 46 families, with 20 endemics. The Ethiopian Region, including Africa and Madagascar, is the most diverse, with about 50 families, of which 14 are endemic. The Oriental Region, which essentially connects the southern Palearctic and Australian Regions, has about 43 families, with 4 endemic. The Australian Region, which includes the Australian mainland, Tasmania, and New Guinea and various other islands,

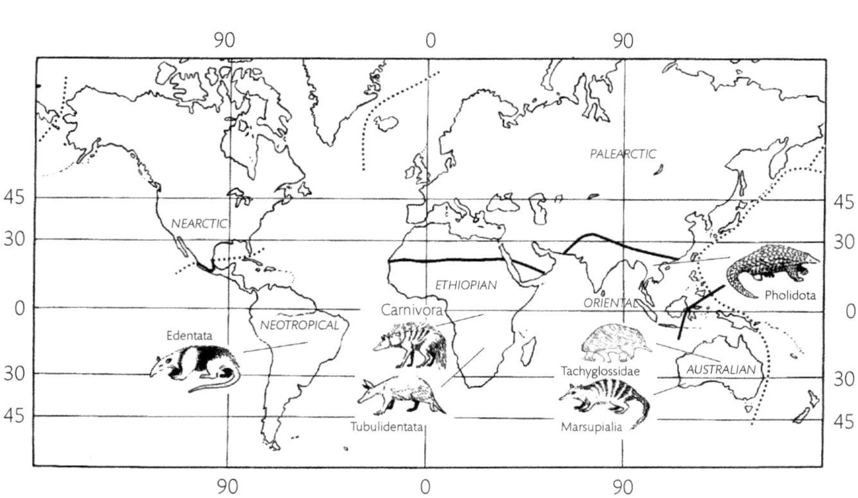


**Figure 1.13** Zoogeographic zones of the world, and examples of convergence of ant-eating mammals from the southern continents. Modified from Vaughan (1986). Reproduced with permission of Elsevier. Echidna image from http://www.supercoloring.com/pages/walking-echidna.

has 20 families, with 12 being endemic. The southern zoogeographic regions (Neotropical, Ethiopian, Oriental, and Australian) generally have a higher number of mammalian families, and especially more endemic families, than the northern regions (Palearctic and Nearctic), reflecting their greater isolation since the break-up of Gondwana.

Zoogeographic zones are coarse biogeographic regions, reflecting the product of current geographic barriers that restrict the movement of animals and plants, and recent and earlier patterns of climate change (Cox 2000). Mammals are found in all climate zones, but there are definite patterns in species richness (numbers of species), especially with latitude (Buckley et al. 2010). For all mammals, there is a strong decrease in species richness with latitude—i.e. more species near the equator (hotter climates) than the poles (colder climates)—and this pattern is often interpreted as being linked to energy availability or geographically determined rates of evolution. However, the evolutionary history of mammalian clades needs to be accounted for when attempting to determine whether species richness is impacted or caused by environmental variation (Buckley et al. 2010). For example, many evolutionarily young clades have a negative species richness-temperature relationship (i.e. fewer species in the warmer latitudes), since the age of these clades coincides with the Late Eocene expansion of temperate climate zones. Some carnivoran clades (e.g. felids), in contrast, have a positive species richness-temperature relationship that seems to be related to their tropical origins and restricted distributions. The very speciose bat clade has a very strong positive species-richness temperature relationship, presumably reflecting the conservative link between temperature and energetic costs; few bat lineages have overcome the energetic constraints of being outside the tropics, although vespertilionids have been quite successful in temperate climates because of reproductive and physiological adaptations (such as torpor). Overall, the general latitudinal species richness pattern of mammals is highly biased by the very strong pattern of the highly speciose bat clade.

A striking feature of mammalian evolution is the remarkable convergence of independent lineages in the various zoogeographic regions in terms of morphological, physiological, and ecological adaptations. For example, 14 ant- and termite-eating mammal species have independently converged with respect to their morphology, including body size; powerful digging forelimbs and claws; and physiology, including low body temperature and metabolic rate (McNab 1984; Figure 1.13). These include one monotreme (echidna), one marsupial (numbat), an Afrotherian mammal (aardvark), an African carnivore (aardwolf), a South American edentate (anteater), and an Oriental pholidotan (pangolin). Many other examples of striking convergence include cursorial herbivorous mammals (e.g. some marsupials, rodents, lagomorphs, artiodactyls, and perissodactyls), and small insectivores (Insectivora, Edentata, Rodentia, Marsupialia).

### 1.4.3 Habitats and Diet

Habitat has a strong influence on essentially all aspects of the life history, physiology, and ecology of mammals. Mammals are found in aquatic, terrestrial, and aerial habitats. Most mammals are terrestrial, but there are many strictly aquatic species, and bats are aerial during their activity phase but terrestrial during their resting phase. These general habitat types can be categorized more finely into a total of eight habitat types (Eisenberg 1981). These are aquatic, semi-aquatic, fossorial, semi-fossorial, terrestrial, scansorial, arboreal, and volant. Monotremes are found only in two of these (semi-aquatic, terrestrial), marsupials in six (no aquatic or volant species), and placentals in all eight habitats.

The terrestrial environment of mammals can be subdivided into fossorial (underground), semi-fossorial, terrestrial, and above-ground habitats. Many mammals are highly specialized for living almost completely underground (e.g. reduced eyes, ears, specialized forelimbs or teeth that are adapted for digging, short neck and tail, reduction in underfur, and relaxed thermoregulatory strategy); these highly specialized fossorial species include the marsupial moles (*Notoryctes*), many golden moles (e.g. *Eremitalpa*), and some rodents (e.g. *Heterocephalus, Talpa, Geomys*). Semi-fossorial mammals are less specialized in their adaptations to living underground, frequently using burrows as a refuge and being primarily active above ground; many small marsupials (e.g. dunnarts; numbats, *Myrmecobius fasciatus*) and large wombats (*Vombatus*) are quite semi-fossorial, as are many placentals (e.g. rodents, badgers, rabbits).

Terrestrial species forage above ground, with limited or no digging or climbing ability (e.g. macropod marsupials, artiodactyls, many felids and canids). Scansorial species spend considerable time climbing in trees or on rocks, hence have some adaptations for climbing (e.g. marsupial phascogales, *Phascogale*; placental rodents, tree-shrews, and marmosets). The most adept arboreal species spend most or nearly all of their lives above ground, in trees; they often have more specialized morphology, physiology, and diets than the semi-arboreal mammals (e.g. the marsupial koala, *Phascolarctos cinereus*, and the placental sloths). The marsupial gliding possums and various gliding placentals (some rodents; flying lemurs, *Cynocephalus*) are proto-volant, but bats are the only volant mammals, capable of powered flight. If we consider the mega- and micro-bats to be a mono-phyletic clade, then powered flight has evolved only once amongst the mammals. Aquatic environments can be the exclusive habitat for many mammals, which do not need ever to come onto land (e.g. cetaceans, sirenians), or the primary habitat for species that do not need to come onto land every day (e.g. pinnipeds). Many mammals are semi-aquatic, spending part of every day in water but coming onto land for the rest of the day (e.g. platypus, *Ornithorhynchus anatinus*, water shrews, otters).

Mammals have specialized to such an extent that there is almost no food source that they cannot exploit in some way. The diet of many mammal species varies considerably, and so it is challenging to categorize them into a single feeding category, but many mammal species have a very restricted diet and are more easily classified into a dietary niche. In all, we can define 16 diet categories for mammals (Table 1.4; Eisenberg 1981). Living monotremes fit into 2 of these, marsupials into 9, and placentals into all 16 (Eisenberg 1981; Lee & Cockburn 1985).

These 8 habitats and 16 dietary categories define 128 niche combinations (Table 1.5), of which 64 are possible and 64 are excluded as incompatible combinations (e.g. planktonivores can only be aquatic). Of these 64 possible combinations, placental mammals occupy 53 (Eisenberg 1981), marsupials occupy 17 (Lee & Cockburn 1985), and monotremes 2 (terrestrial myrmecophage; semiaquatic insectivore/omnivore). The only combination occupied by a marsupial but not a placental is semi-aquatic insectivore/omnivore (the water opossum *Chironectes*), although semi-aquatic water shrews (*Neomys*) would seem to also fit this category. The vampire bat is the only example of a volant sanguivore.

Although many early mammals were small (10–100 g), generalized, and terrestrial or scansorial (e.g. *Morganucodon, Yanoconodon, Zhangheotherium, Sinodelphys, Eomaia*), it is not surprising, given the diversity of habitats and diets of extant mammals, that many equivalent habitat and dietary specializations are apparent for various fossil mammals (Luo 2007; Figure 1.14). Various Mesozoic mammals reveal a diversity of ecological bauplans involving locomotory and skeletal adaptations, derived from the early small, generalized forms. Scratch-digging and fossoriality were widespread for some pre-mammalian cynodonts, multituberculates, and docodonts. *Fruitafossor* (150 MYBP), for example, is highly convergent with modern scratch-diggers (echidna, *Tachyglossus aculeatus*; aardvark, *Orycteropus afer*; pangolin, *Manis*; armadillo, *Dasypus*). The large limbs of burrowing docodonts represent an adaptation for swimming; for example, the docodont, *Haldanodon* (cf. semi-aquatic moles), and *Castoricauda* (with a beaver-like tail; Ji et al. 2006). Another line of ecological specialization is carnivory/scavenging associated with large body size, evident for multiple lineages of Jurassic and Cretaceous predatory species (e.g. 500 g *Sinoconodon*; 5–12 kg gobiconodontids). Climbing adaptations were common amongst early mammals (e.g. *Henkelotherium, Sinodelphys*, Eomaia; cf. tree shrews; Ji et al. 2002). These scansors had elongate digits and other limb specializations. Scansoriality would seem preadaptive for the evolution of gliding, and the general scansorial habit of many early small mammals led to gliding in various extant mammals and at least one extinct mammal (*Volaticotherium*; Meng et al. 2006). The latter has a gliding membrane (patagium) and elongate limbs, similar to those of extant marsupial gliders and the placental flying squirrels and lemurs. There were also Jurassic and Early Cretaceous equivalents to extant terrestrial insectivores (*Morganucodon, Yanoconodon*; cf. short-tailed opossums) and terrestrial ambulatory carnivores (*Sinoconodon, Repenomamus*; cf. racoons).

**Table 1.4** Mammalian diet categories (see Eisenberg 1981; Lee & Cockburn 1985).

| Diet category | Diet and examples |
|---|---|
| 1. Piscivores and squid-eaters | Consume large (about 3 kg or more) aquatic vertebrates (e.g. fishes and birds) and invertebrates (e.g. squid); examples: large otters and dolphins/porpoises. |
| 2. Carnivores | Consume terrestrial vertebrates; examples: large dasyurid marsupials, and placental felids, canids and mustelids |
| 3. Nectarivores | Forage on flowers for nectar and pollen; examples: small terrestrial (e.g. some rodents), arboreal (e.g. the marsupial honey possum), and aerial (e.g. blossom bats) mammals. |
| 4. Gumivores | Forage on plant exudates, such as gum, sap, and resin, but may also take invertebrates, fruit, or small vertebrates; examples: arboreal mammals, such as some marsupial gliders and lemurs. |
| 5. Crustacivores and clam-eaters | Consume marine crustaceans, echinoderms, clams, and oysters; example: walrus. (This diet category differs from no. 1 in that these prey are less mobile than those of piscivores and squid-eaters.) |
| 6. Myrmecophages | Feed primarily on ants and/or termites; examples: echidna, numbat, pangolins, and neotropical anteaters. |
| 7. Aerial insectivores | Feed on insects in flight; example: insect-eating bats. |
| 8. Foliage-gleaning insectivores | Feed by searching plant surfaces or under bark for insect prey, in flight; example: insect-eating bats. |
| 9. Insectivore/omnivores | Consume insects, other arthropods, earthworms, molluscs (e.g. snails), and sometimes small vertebrates and fruit; examples: medium-size dasyurid marsupials and badgers. |
| 10. Frugivore/omnivores | Specialize on feeding from the reproductive parts of plants, particularly the fleshy fruit, but may also consume the seeds and, if available, small invertebrates and vertebrates; examples: woolly opossums and marmosets. |
| 11. Frugivore/granivores | Similar to frugivores/omnivores, but also specialize on nuts and seeds; examples: the extinct argyrolagid marsupials, chipmunks. (Granivory is a dietary niche that is absent in at least extant marsupials.) |
| 12. Herbivores/frugivores/ fungivores | Feed on leaves, storage roots, and the reproductive parts of plants and fungi; examples: bettongs, musky rat kangaroo, and agoutis. |
| 13. Herbivores/browsers | Mainly feed on plant stems, twigs, buds, and leaves; typically have particular ruminant/pseudoruminant/post-gastric adaptations that facilitate the digestion of plant material by symbiotic microorganisms; examples: koala and tree sloths. |
| 14. Herbivores/grazers | Feed particularly on grasses, but this dietary niche grades into browsing depending in large part of the size of the mammal; typically have particular ruminant/pseudoruminant/post-gastric adaptations; examples: kangaroos and antelope. |
| 15. Planktonivores | Feed primarily on zooplankton by filter-feeding; example: highly specialized baleen whales. |
| 16. Sanguivores | Feed on the blood of warm-blooded vertebrates; this is the most highly specialized dietary niche, consisting of only three species of phyllostomatid bats (*Desmodus*, *Diaemus*, and *Diphylla*). |

**Table 1.5** Matrix of the habitats and diet of mammals, showing the 2 realized niches for monotremes (P, prototherian), 17 for marsupials (M, metatherian), and 53 for placentals (E, eutherian). Modified from Eisenberg (1981) and Lee and Cockburn (1985). Numbers indicate Eisenberg's numeric classification for habitats and diets; e.g. 609 = terrestrial (6) + insectivore/omnivore (09). Empty cells are Eisenberg's incompatible niches; * indicates potentially compatible but unrealized niches.

| | Aquatic | Semi-aquatic | Fossorial | Semi-fossorial | Terrestrial | Scansorial | Arboreal | Volant |
|---|---|---|---|---|---|---|---|---|
| | (3) | (4) | (1) | (2) | (6) | (7) | (8) | (5) |
| Planktonivore (15) | E | | | | | | | |
| Crustacivore/ clam-eater (05) | E | E | | | | | | |
| Piscivore/ squid-eater (01) | E | E | | | | | | E |
| Herbivore/ browser (13) | | E | E | E | ME | ME | ME | |
| Herbivore/ grazer (14) | E | E | * | ME | ME | * | | |
| Herbivore/ frugivore (12) | | E | E | E | E | E | E | |
| Myrmecophage (06) | | | * | E | PME | E | E | |
| Gumivore (04) | | | | | | * | ME | |
| Nectarivore (03) | | | | | | * | ME | E |
| Sanguivore (16) | | | | | | | | E |
| Frugivore/ granivore (11) | | E | * | E | ME | E | E | |
| Frugivore/ omnivore/fungivore (10) | | E | * | * | ME | E | ME | E |
| Insectivore/ omnivore (09) | | PM | ME | E | ME | ME | ME | |
| Foliage-gleaning insectivore (08) | | | | | | | | E |
| Aerial insectivore (07) | | | | | | | | E |
| Carnivore (02) | E | * | * | E | ME | E | E | E |

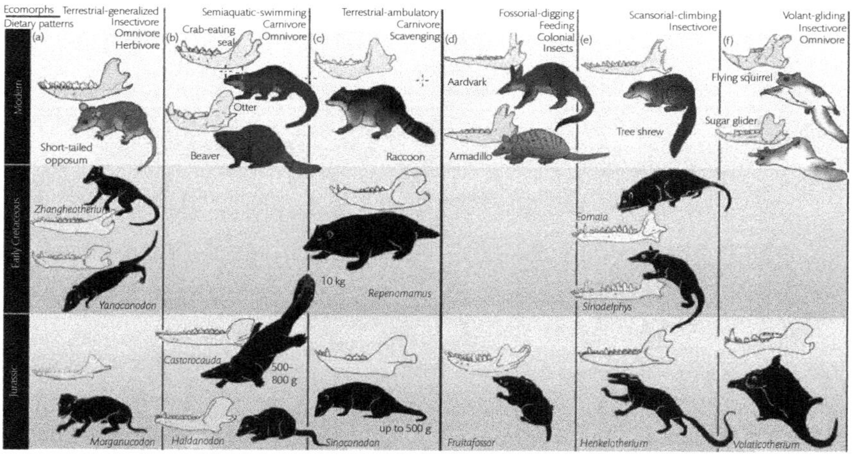


**Figure 1.14** Diversity of Mesozoic mammals, showing convergences in various ecomorphs with differing dietary patterns. Reprinted by permission from Macmillan Publishers Ltd: *Nature*, Luo (2007), copyright 2007.

## 1.5 Importance of Mammals

Mammals are, from our anthropocentric point of view, the most important vertebrate taxa because they include us and most of our major protein sources. To best understand our own ecology, physiology, behaviour, and general position in the tapestry of life, the obvious comparisons are first with other primates, then other placental, other therian, and other mammals. Although birds have in many ways achieved an equivalent evolutionary pinnacle, having independently evolved many of the traits that we as mammals consider highly evolved (endothermy, high aerobic metabolic capacity, double circulation, urine-concentrating kidney, extensive parental investment, etc.) they have not achieved the same level of behavioural or neural complexity as mammals.

### 1.5.1 Pinnacle Taxon

It is not surprising that early evolutionary biologists had an anthropocentric point of view. Haeckel's gnarled oaktree-like hypothetical phylogeny shows a clear progression to humans (Menschen) with our more distant ancestors depicted lower down the tree as minor branches (Figure 1.15). Even Haeckel himself accepted that his trees were provisional and often controversial.

We humans interpret ourselves to be the pinnacle taxon of animal evolution, and we increasingly exploit and endanger other mammals more than we conserve them. Most mammal-based research is directed at improving our own general health, welfare, and comfort.

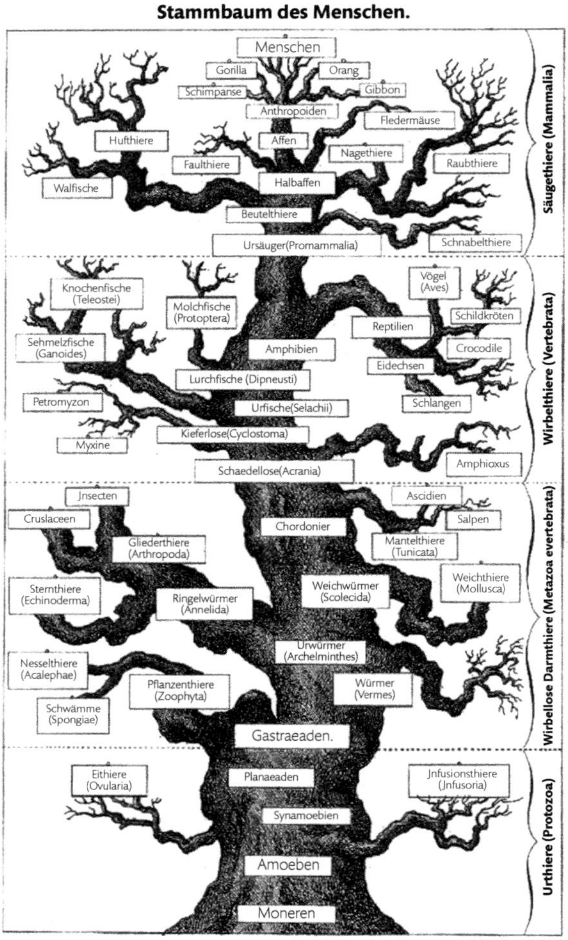


**Figure 1.15** Haeckel's anthropocentric view of animal evolution, culminating in mammals, thence to humans (Menschen), as the pinnacle taxon. Lithograph by J. G. Bach of Leipzig, from Haeckel (1874). Image courtesy of T. Buklijas and N. Hopwood, 'Evolution', *Making Visible Embryos* [http://www.hps.cam.ac.uk/visibleembryos/s5_3.html].

## 1.5.2 Conservation

Despite our being mammals ourselves, the lack of success by human conservation efforts to preserve the numbers and diversity of mammalian species is not much different from our poor success in conserving other taxa. In fact, humans have long been more interested in relentlessly exploiting other species, and the long-term impact of our hunting, gathering, agriculture, and mechanization has been devastating on the Earth's biotic and abiotic environments. Even in prehistoric times, humans are thought by many scientists to have been responsible for extensive mammal and bird extinctions, particularly the post-Pleistocene extinction events (Baillie et al. 2004). The strong correlation between the arrival of humans in Australia, the Americas, Madagascar, and many islands with the extinction of large mammals and birds suggests a causal link, but climate change has also been implicated as a possible causal factor (Wroe et al. 2013).

Sadly, non-human mammals are not faring much better in the present. In historical times (i.e. since about 4000 BC), many mammal species have been despatched remarkably swiftly. Marine mammals in particular have been impacted by human hunting (e.g. whaling and sealing). Steller's sea cow (*Hydrodamalis gigas*) was pushed to extinction by European hunters in the Bering Sea only 27 years after it was first discovered (Vaughan 1986). Sea otters (*Enhydra lutris*) were hunted mercilessly along the Pacific coast of North America; more than 200,000 were killed between 1786 and 1868 for their fur, and they became rare over most of their distribution by 1900. Fortunately, sea otters were protected by legislation in the early 1900s and their population is now healthy. Mammals elsewhere have not fared much better. In Southern Africa, quaggas (*Equus quagga quagga*; a subspecies of plains [Burchell's] zebra) were originally common across Africa but were extinct in the wild by about the 1870s (Lowenstein & Ryder 1985). The Tasmanian tiger, or thylacine, was hunted as a 'sheep killer' with a bounty paid by the Tasmanian government until 1909, and the last known specimen died of neglect in the Hobart Zoo in 1936. The Tasmanian government did not declare it endangered until 1936 and prevaricated on announcing its official extinction until 1986 (Dixon 1989; Owen 2003). Fortunately, some mammals deemed extinct have been rediscovered, such as Miss Waldron's Red Colobus monkey (*Piliocolobus badius waldronae*; Roberts & Kitchener 2006) and Gilbert's potoroo (*Potorous gilbertii*; Sinclair et al. 1996), but these examples are unfortunately the exceptions rather than the rule.

Of the 5,487 mammals species listed by the International Union for Conservation of Nature (IUCN), nearly 25%—a staggering 1,219 species—are threatened or extinct (IUCN Red List of Threatened Species; Baillie et al. 2004); 76 species are extinct; a further 2 are extinct in the wild; more than 188 are considered critically endangered; 448 are endangered; and 505 are considered vulnerable. Further, 323 species are near threatened (5.9%) and 3,109 species are of least concern (56.7%); 836 species are data deficient. Primates (414 species) are one of the mammalian orders most at risk; 49% are considered threatened or extinct. Many less speciose taxa have a higher percentage of threatened or extinct species; Sirenia (100% of 5 species), Perissodactyla (81% of 16 species), Monotremata (60% of 5 species), and Proboscidea (50% of 2 species).

Humans are widely recognized as the cause of the Earth's sixth mass extinction event, which started about 50,000 years ago and has intensified in the last 500 years (Brook et al. 2008). There are various reasons for this parlous state of mammalian conservation (and conservation of other animals). The main threat for mammals is habitat loss or change. For example, in tropical rainforest, the threatening processes include habitat loss, overharvesting, fragmentation, fire, and climate change; these interact in various ways upon the population size. A major secondary threat is overexploitation (Baillie et al. 2004; Greyner et al. 2006); many mammal species are overexploited for food, medicines, and products such as leather and ivory. Climate change is also a major threat to mammals (Helmuth et al. 2010). The combination of different threatening processes can have simple additive effects (the overall threat is the sum of the individual threats) but there can also be synergistic interactions, such that the total threat is more than the sum of the individual threats (Brook et al. 2008).

Effective conservation strategies are essential for about 25% of mammalian species that are threatened. Although we might like all species to have equal conservation priority, the reality is that we must prioritize those species most requiring conservation management. Di Marco et al. (2012) have proposed a new metric, the Extinction risk Reduction Opportunity (ERO), for likelihood of recovery opportunity. Using this metric, they found that 65–87% of all threatened but potentially recoverable mammalian species are overlooked by current prioritizations, suggesting that current conservation efforts might have to be refocused to address this.

### 1.5.3 Human Perspective

As mammals ourselves, we have long been interested in, and exploited, our mammalian cousins. The extinct hominid, *Australopithecus*, more than a million years ago exploited (ate) other mammals such as baboons and antelope, and modern humans have continued to use mammals for food, as beasts of burden, as research subjects, and as pets, and in some instances we have revered them as gods.

Domestication of mammals has contributed greatly to human development (Diamond 2005). Domesticated mammals have provided meat, milk, fertilizer, wool and fur clothing, transport (carts, ploughs, etc.), and military assistance (e.g. Hannibal's elephants, horses). Perhaps surprisingly, few large mammals (i.e. > 50 kg or so) have been successfully domesticated. The five large herbivorous mammals of greatest significance to human societies through domestication are sheep (from West and Central Asian mouflon sheep), goats (from bezoar goats of West Asia), cattle (from Eurasian and North African aurochs, now extinct), pigs (from Eurasian and North African wild boar), and horses (from Russian wild horses). Another nine large herbivorous mammals have also been domesticated: Arabian (one-humped) camel, Bactrian (two-humped) camel, llama/alpaca, donkey, reindeer, water buffalo, yak, Bali cattle, and mithan (Indian and Burmese cattle). Asian elephants have also been domesticated as work animals. Wolves were domesticated in Eurasia and North America, first as camp followers, then as hunting dogs and as companions. Numerous other small mammals have been domesticated as food sources (rabbits, guinea pigs, giant rats), for hunting (ferrets, cats), for their pelts (mink, foxes, chinchillas), or simply as pets (hamsters), but none of these provide the important societal roles of the larger domesticated mammals (pulling sleds, ploughs, or wagons; carrying riders) or as food. The global number of domestic mammals is staggering; there were more than an estimated 33,000 million domesticated mammals in 2013, including more than 6,900 million cattle, 5,900 million sheep, and more than 14,000 million buffalo, goats, and pigs (Figure 1.16). All of these species provided substantial meat production, and considerable milk production from cattle, sheep, goats, buffalo, and camels.

Large domesticated mammals have had, and continue to have, an immensely important role by providing food and useful by-products (leather, fibre for clothing, bone and sinew tools, etc.) for humans; some of the smaller mammals also contribute. Per capita meat consumption has plateaued in developed countries, but is continuing to rise in developing countries (Speedy 2003). In 2013, the total global production from mammals was about 198 million tonnes of meat (more than 110 million tonnes of pig and 63 million tonnes of cattle) and 768 million tonnes of milk (more than 630 million tonnes from cattle and 100 million tonnes from buffalo) (Figure 1.16). Wild mammals were and are still exploited for these human needs and as incidental by-catch from other food-gathering activities (e.g. fishing), contributing little compared to the production from domesticated species but impacting greatly on conservation. Marine mammals, for example, continue to be hunted throughout the world (more than 87 species in more than 114 countries) and provide a significant food source and economic benefit in more than 50 of these countries (Robards & Reeves 2011).

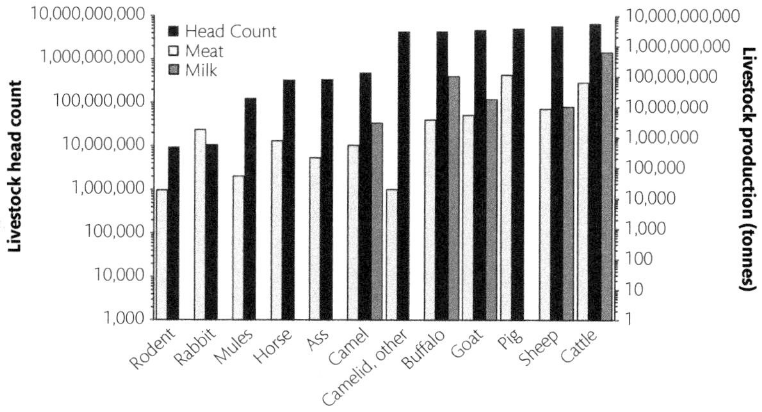


**Figure 1.16** Estimate of total global head count for livestock mammal, and tonnes of meat and milk production, for 2013. Data from faostat3.fao.org.

**Table 1.6** Summary of animal usage for scientific research in 2014.

| Mammal category | USA[1] | Spain[2] | Switzerland[3] | Finland[4] | UK[5] | NZ[6] |
|---|---|---|---|---|---|---|
| Mice | * | 457,267 | 390,144 | 60,988 | 3,058,821 | 45,018 |
| Rats | * | 61,388 | 83,000 | 17,655 | 278,386 | 10,806 |
| Rodents, other | 291,458 | 7,688 | 2,689 | 4,358 | 19,380 | 2,209 |
| Cats | 21,083 | 52 | 788 | 452 | 247 | 676 |
| Dogs | 59,358 | 765 | 3,286 | 4,260 | 4,843 | 1,437 |
| Non-human primates | 57,735 | 489 | 251 | 0 | 3,020 | 0 |
| Rabbits | 150,344 | 23,881 | 951 | 177 | 13,866 | 1,386 |
| Sheep, goats | 10,315 | 199 | 1,496 | 1,379 | 44,529 | 46,799 |
| Pigs | 45,392 | 8,043 | 4,627 | 431 | 3,379 | 236 |
| Other farm animals | 27,393 | 3,553 | 0 | 335 | 14,041 | 54,509 |
| All other | 171,375 | 410 | 1,331 | 11 | 533 | 3,825 |
| Total | 834,453 | 563,735 | 488,563 | 90,046 | 3,441,045 | 166,901 |

*numbers not available

[1]www.aphis.usda.gov/animal_welfare/downloads/7023/Animals%20Used%20In%20Research%202014.pdf

[2]http://www.magrama.gob.es/es/ganaderia/temas/produccion-y-mercados-ganaderos/informesobreusosdeanimalesen2014_tcm7-402651.pdf

[3]tv-statistik.ch/de/statistik/index.php

[4]speakingofresearch.files.wordpress.com/2015/11/finish-animal-research.pdf

[5]for 2013: Home Office (2013). Annual Report. http://www.understandinganimalresearch.org.uk/files/3314/4552/1574/2014_Home_office_animals_stats.pdf

[6]National Animal Ethics Advisory Committee (2014). Annual Report January to 31 December 2013. Wellington, NZ: Ministry for Primary Industries. https://speakingofresearch.files.wordpress.com/2016/02/new-zealand-animal-research-statistics-2014.pdf

From a scientific point of view, mammalian research (especially medical) attracts more funds, involves more researchers, and results in more publications than does research into other taxa. Although it is difficult to estimate the total number of mammals used in research, due to different accounting procedures in different countries (and none in some countries), many countries do provide detailed summaries (see Table 1.6). The major research countries are the United States of America (about 20 million procedures in 2010), the European Union (about 12 million), Japan (about 5 million), and Canada (about 2 million); Switzerland and Australia are less than 1 million per year (Understanding Animal Research 2012). If the rest of the world contributed 10 million procedures, then the total is probably less than 60 million.

In the United Kingdom, more than 4 million procedures on mammals were reported for 2012 (Home Office 2013); 97.6% involved rats, mice, and other rodents; 0.43% were other small mammals (rabbits, ferrets, etc.); 1.8% were sheep, cattle, pigs, and other large mammals; 0.15% were laboratory-bred cats and dogs; and 0.09% were primates (e.g. marmosets and macaques). Although the total number of procedures per year is continuing to rise, the increase is less than the increase in UK biomedical research funding, so fewer procedures are undertaken per British pound. This increase is primarily attributed to the increase in procedures with genetically modified (GM) animals, being about one half of all procedures. It is informative to compare these UK numbers for animal research with more the than 7 million animals killed per year as vermin and 1 billion eaten (more than 2.5 billion if fish are included).

Mammals are also of religious importance in many human cultures. Religious rituals involving animals (zoolatry) were especially common in premodern societies, involving the worship of animal deities or animal sacrifice. There are various hypotheses for the origins of zoolatry, but the appreciation by early humans of animals with admirable or desirable traits could lead to adoration and eventually animal worship (Weissenborn 1906). Alternatively, animal worship could have arisen from families naming themselves after animals with desirable traits, and this respect developed into worship of the family totemic animal (Lubbock 1868). Egyptian religion was particularly zoolatric, with many animals sacred to particular gods (e.g. cats to Bastet; baboons to Thoth; mongoose, shrews, and birds to Horus; dogs and jackals to Anubis; and bulls to Apis). These animals were often mummified, reflecting their religious significance. Dietary laws prohibiting the consumption of some species are common in some religions, perhaps being developed from the belief that particular animals are sacred, or customs that particular animals are unclean and not to be consumed.

# 2
# General Physiological Principles

This chapter focuses on the basic physiological processes that are common to mammals, but also more widely to other vertebrate and invertebrate animals for comparison. We start with the general principles of the scaling of physiological variables with body mass (M) because of the overarching significance of body size to the structure and function of animals (Schmidt-Nielsen 1975, 1984), and the functioning of control systems, because of their fundamental role in homeostasis. Thermal balance is important because it determines body temperature ($T_b$), which in turn affects all physiological functions. Energetics describes the supply of energy (ATP) by aerobic and anaerobic cellular processes. We then describe the main physiological systems of animals. Digestion involves the process of acquisition of food, then its breakdown into substrates for energy provision. For aerobic animals—which all vertebrate animals are, for most of the time—the respiratory and circulatory systems deliver oxygen ($O_2$) to mitochondria and remove carbon dioxide ($CO_2$) from the cells. The principles of the regulation of body water and solutes are then presented, to complete the main physiological systems of animals. Biophysical principles of terrestrial locomotion by lever systems during walking, running, and hopping are described, then the fluid-dynamic principles of locomotion by flight and swimming are presented. Chapter 2 concludes with a description of reproduction by egg-laying (oviparity) and live birth (viviparity).

## 2.1 Scaling

Body size has an all-pervasive influence on physiological, and indeed most other biological functions, and as such is of fundamental significance to the structure and function of all organisms (Schmidt-Nielsen 1975, 1984; Peters 1983; McMahon & Bonner 1983; Calder 1984; Bonner 2006). The importance of body size to life functions is reflected in the considerable literature concerning mass and its relationship with other biological traits. The study of scaling relationships, or the 'structural and functional consequences of changes in size or scale among otherwise similar organisms' (Schmidt-Nielsen 1984, p. 7) is widely applied to

*Ecological and Environmental Physiology of Mammals*. Philip C. Withers, Christine E. Cooper, Shane K. Maloney, Francisco Bozinovic, & Ariovaldo P. Cruz-Neto.  Published 2016 by Oxford University Press.
DOI 10.1093/acprof:oso/9780199642717.001.0001

various sub-disciplines within physiology. Despite the fundamental importance of body size to biological functioning, the nature of the scaling relationship—and in particular the mechanisms causing those scaling relationships—is poorly understood, and the subject elicits some of the most debated questions in biology.

### 2.1.1 Isometry and Allometry

In general, the relationship between body mass (M) and a physiological (or other) function (Y) can be expressed as a power curve $Y = aM^b$ where $a$ is the scaling coefficient and $b$ is the scaling exponent. If $b = 1$, then there is a linear relationship between mass and the variable, and the relationship is known as isometric. If $b \neq 1$, then the relationship is curvilinear, often referred to as an allometric relationship. For convenience (and for mathematical and statistical reasons), the relationship can be linearized by logarithmic transformation to $\log(Y) = \log(a) + b\log(M)$ where $\log(a)$ becomes the intercept constant for a line and $b$ is the slope (scaling exponent). Any base can be used for the logarithmic transformation, but base 10 (i.e. $\log_{10}$) or the natural logarithm (ln) are the most common transformations in biology. Note that the scaling exponent is independent of the base used, but $a$ and $\log(a)$ vary with the base; we will use $\log_{10}$ throughout this book, unless otherwise specified.

The metabolic rate (MR) of all organisms scales allometrically with body mass, with an allometric exponent $b$ of about 0.75. There are three major 'grades' of metabolic intensity reflected in differences in the allometric coefficient $a$. This variation in $a$ between animal taxa reflects differences in body complexity (unicellular and multicellular organisms) and temperature regulation (ectotherms and endotherms). The coefficient is lowest for unicells, intermediate for ectothermic multicellular animals, and highest for endothermic multicellular animals (Hemmingsen 1950; Phillipson 1981). The enormous range of mammalian body mass, from 1.3 g for the Etruscan shrew (*Suncus etruscus*) to 200 tonnes for the blue whale (*Balaenoptera musculus*) means that allometric studies are of particular significance for understanding mammalian function and the limits to performance.

For physiological and other variables, there is considerable variation in the scaling exponent, although the scaling exponents are generally similar for different taxa for any particular variable. Some physiological traits change in proportion to mass and are therefore isometric ($b = 1$). For example, the body water content of most animals is about 75% of body mass, so total body water content (TBW) increases linearly with mass ($TBW = 0.75\,M^1$). The exponent is also about 1 for respiratory tidal volume ($V_T$; Stahl 1967) and heart stroke volume (Calder 1984).

Isometric scaling, however, seems to be the exception rather than the rule, and many physiological parameters scale allometrically ($b \neq 1$). Allometric scaling exponents generally fit into three ranges: $b > 1$, $0.5 < b < 0.8$, and $b < 0$. Only a few physiological variables scale with $b > 1$, such as skeletal mass that scales with $b = 1.13$ (Schmidt-Nielsen 1975, 1984). Physiological functions that are dependent on surface area (SA) should scale with $M^{0.67}$ (see later). Basal metabolic rate (BMR) scales close to $mass^{0.67}$, at about $M^{0.75}$ (section 2.1.2). Some physiological rate functions, such as breathing and heart rate, scale inversely with body mass, reflecting an interactive scaling of related parameters. For example, respiratory minute volume ($V_I$) is the product of breathing rate ($f_R$) and tidal volume ($V_T$); $V_I = f_R . V_T$; since $V_I$ scales at about $\mathbf{b} = 0.75$ (reflecting the scaling of BMR), and $V_T$ scales with $\mathbf{b} = 1$, then $f_R$ must scale with $\mathbf{b} = -0.25 (0.75 - 1)$; similarly, heart rate is expected to scale with $\mathbf{b} = -0.25$.

Many transport variables are thought to scale with SA, because area directly influences the exchange rate (e.g. diffusion, heat exchange). The SA of isometrically scaled objects increases when mass (or volume) increases, but not proportionally; the SA for a cube with $1 \times 1$ cm sides is 6 cm$^2$ and the volume is $1\,cm^3$ $(SA/V = 6)$. Doubling the size of the cube to $2 \times 2$ cm sides results in a SA of 24 cm$^2$ and a volume of $8\,cm^3$ $(SA/V = 3)$. Since mass is proportional to volume (the density of water and thus animal bodies is about 1), then $M \propto x^3$ and $x \propto M^{1/3}$; because $SA \propto x^2$ it follows that $SA \propto M^{2/3} [(M^{1/3})^2]$. Consequently, the scaling exponent for the surface area of both a cube and sphere is ⅔ (or 0.67), with $SA = 6.00\,M^{0.67}$ for a cube and $4.84\,M^{0.67}$ for a sphere.

For animals with more complex and variable shapes, the scaling exponent $b$ for surface area is still about 0.67, but the coefficient $a$ is greater than that for a sphere or a cube, depending on the shape. The Meeh relationship is commonly used to express the surface area of animals as a function of mass, based on the theoretical exponent of 0.67; i.e. $SA = K\,M^{0.67}$, where K is a species-specific constant (Meeh 1879; Dawson & Hulbert 1970; Walsberg & King 1978; Reynolds 1997). K is typically about 10 for 'compact-shaped' mammals and birds but $> 10$ for species that are elongate (e.g. weasels, *Mustela* spp.) or have extensive membranous areas (e.g. sugar glider, *Petaurus breviceps*); $SA = 10.95\,M^{0.650}$ for a pooled mammal-bird data set (Figure 2.1).

The surface area to volume ratio is the mass-specific surface area (i.e. surface area per unit mass) and is proportional to $M^2/M^3$ (or $M^{2/3}$) and clearly decreases with increasing mass. The scaling exponent $b$ for $M^{2/3}$ as a function of $M^1$ is $-0.33 (= 0.67 - 1.0)$. Consequently, physiological variables that are proportional to mass-specific surface area should scale with $M^{-0.33}$. For example, Rubner (1883) found that the MR of dogs was independent of mass when corrected for surface area; this SA-dependent scaling relationship is often called 'Rubner's law' (but see 2.1.2).

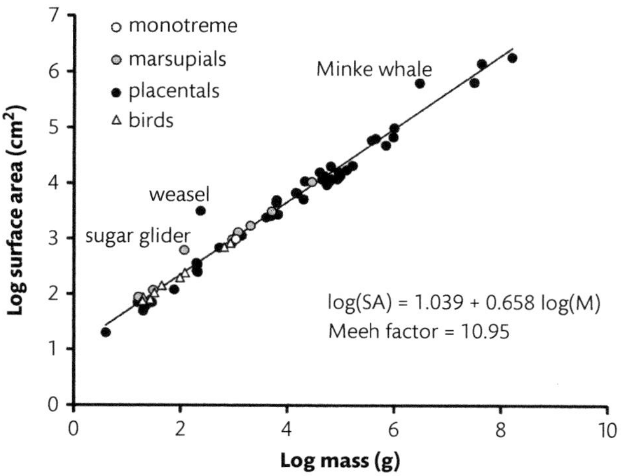


**Figure 2.1** Scaling of surface area (SA) to body mass (M) for monotreme, marsupial, and placental mammals, and birds. Meeh factor is the anti-log value of the regression intercept. Three species with particularly high SA are identified. Data from Dawson and Hulbert (1970), Walsberg and King (1978), and Reynolds (1997).

Physiological effects that are scaled for body mass ('mass-specific' units) are only independent of mass if they scale isometrically. For example, BMR per body mass, or mass-specific BMR, has a scaling $b$ of about $-0.25(=0.75-1)$. When using mass-specific units, it is essential to keep in mind the following: (1) if $0 < b < 1$, then the scaling relationship is reversed (mass-specific exponent $= b - 1$), and (2) mass-specific units do not remove the influence of M unless $b = 1$. Mass-specific BMR is inversely related to mass ($\alpha$ $M^{-0.25}$), whereas absolute BMR increases with mass ($\alpha$ $M^{0.75}$); mass-specific BMR is not mass-independent (Packard & Boadman 1988). Mass-specific BMR (ml $O_2$ $g^{-1}$ $h^{-1}$) is lower for an elephant than a mouse, whereas absolute BMR (ml $O_2$ $h^{-1}$) is higher for an elephant than a mouse.

## 2.1.2 Physiological Variables

Most physiological parameters scale allometrically, including $T_b$ (although weakly; Stahl 1967; Calder 1984; Withers et al. 2000; Clark et al. 2010; Lovegrove 2012b), thermal conductance and other thermal parameters (Bradley & Deavers 1980; Aschoff 1981; Withers et al. 2000, 2006; Riek & Geiser 2013), respiratory ventilation (Stahl 1967; Calder 1984; Frappell & Baudinette 1995), evaporative water loss (Hinds & MacMillen 1985, 1986; Withers et al. 2006; Williams 1996; Van Sant et al. 2012), and urine-concentrating capacity (Greenwald & Stetson 1988; Beuchat 1990, 1996). Physiological variables that are related to BMR—such as maximal metabolic rate (MMR), field metabolic rate (FMR), respiratory variables, cardiac variables, and digestive functions—have strong allometric relationships with M, reflecting metabolic scaling.

A strong allometric effect for resting metabolic rate (RMR) is recognized for all animals, but has been most thoroughly examined for mammals, particularly for the precisely defined state of BMR, measured when the mammal is at rest, not digesting, not growing, and is in a thermally neutral environment during its inactive phase (Benedict 1915; Kleiber 1932, 1975; see 3.1.1). The foundations for the study of animal energetics were established by the early work of Sarrus and Ramaeux (1839), Bergmann (1847), Bergmann and Leukart (1852), Rubner (1883), Meeh (1879), and Richet (1889), who experimentally examined the 'surface law' and related metabolism to size and thermoregulation (cited by McNab 1992). These early scientists suggested that the scaling of BMR with body size reflected the geometric and physical processes of heat dissipation. Because surface area scales with $M^{0.67}$, and metabolic heat is predominantly lost across the body surface, metabolic heat production (thus MR) was expected to scale with $M^{0.67}$. That scaling became known as Rubner's surface law, and was widely accepted for nearly a century. However, Brody and Proctor (1932), Kleiber (1932, 1947, 1975), Benedict (1938), and Brody (1945) were fundamental in developing a more widely accepted paradigm that the BMR of mammals scales with an exponent closer to 0.75. Brody's (1945; Figure 2.2) and Kleiber's (1975) 'mouse-to-elephant' curves famously captured the majority of terrestrial mammal body mass

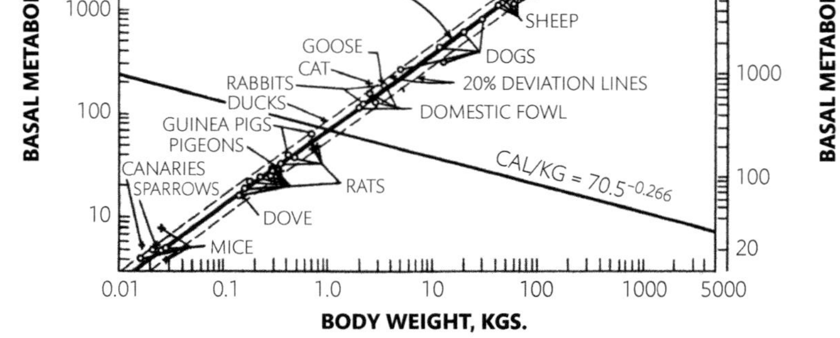

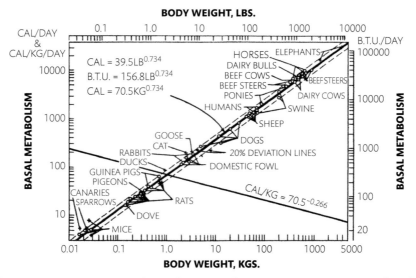


**Figure 2.2** 'Mouse-to-elephant' curve for scaling of basal metabolism. From Brody (1945) Bioenergetics and Growth with Special Reference to the Efficiency Complex in Domestic Animals. New York: Reinhold Publishing Corporation.

variation, supporting the 0.75 scaling theory, and 0.75 has been the most widely accepted scaling exponent for BMR since.

Basal ventilatory variables such as $f_R$, $V_T$, and $V_I$ all scale with mass. Predictably, the scaling of $V_I$ reflects that of BMR, as $V_I$ must accommodate the metabolic demands for gas exchange. For placental mammals, $V_I$ scales as 379 $M^{0.8}$ (M in kg; Stahl 1967), whereas marsupials have a lower intercept but similar slope ($V_I = 218M^{0.8}$; Larcombe 2002); the 40% lower $V_I$ reflects the 30% lower BMR of marsupials. Methodological issues such as non-basal measurements and small sample sizes may have overestimated the slope for these relationships (i.e. 0.8 rather than the expected 0.67–0.75). More recent scaling relationships for $V_I$ suggest a scaling exponent of 0.65–0.75 (Savage et al. 2004a; Cooper et al. 2010), similar to BMR. The scaling of $V_T$ seems to reflect the scaling of lung volume. Lung volume is directly related to body size and $V_T$ scales with an exponent very close to 1 (Stahl 1967; Dawson & Needham 1981; Schmidt-Nielsen 1997). There is a negative scaling relationship of approximately −0.23 for $f_R$ (small mammals breath faster than larger mammals; Stahl 1967; Dawson & Needham 1981; Cooper & Withers 2004a), reflecting the overall scaling relationship of BMR and the contribution of $f_R$ to $V_I$ (given $V_T$ scales at 1).

Cardiac output is another physiological variable that should scale allometrically, like BMR, because cardiac output accommodates the body's energy demands. Cardiac output is the product of stroke volume and heart rate, and since stroke volume is proportional to heart size (which scales isometrically with mass), it is heart rate that scales allometrically (Schmidt-Nielsen 1984; Weibel et al. 1991; Weibel & Hoppeler 2004). Small mammals have higher heart rates than large mammals, and resting heart rate scales reciprocally with mass (slope approximately −0.25; Stahl 1967; Kinnear & Brown 1967; Weibel et al. 2004), and maximal heart rate with MMR (slope approximately −0.19; Baudinette 1978; Weibel & Hoppeler 2004).

Allometric scaling is also apparent for digestive functions. The duration of the mammalian chewing cycle scales with an exponent of 0.25 to 0.33 (Gerstner & Gerstein 2008), and the length of various gut components (small and large intestine, caecum, colon) of rodents also scales with mass (exponents of 0.26 to 0.40; Lovegrove 2010). The nominal and actual gut surface area of mammals, and other animals, scales with a mass exponent of approximately 0.75 (Chivers & Hladik 1980; Chivers 1989; Ricklefs 1996). Accounting for the surface area contributed by villi and microvilli increases the elevation of this scaling relationship compared to the nominal surface area (by a 6.7 and 364 increase in surface area, respectively), but does not alter the slope (Karasov & del Río 2007). Gut volume scales isometrically (i.e. slope = 1) with mass (Calder 1984; Karasov & del Río 2007), but digesta flow rate scales proportionally to $M^{0.75}$, and, conversely, gut retention time scales close to $M^{0.25}$ (Calder 1984; Karasov & del Río 2007). These scaling exponents for energy digestion are consistent with estimates of the scaling of basal and field energy requirements.

### 2.1.3 Life History Variables

Considering the wide array of physiological functions that are influenced by body mass, it is not surprising that many life history parameters, such as age at maturity, gestation period, litter size, age at weaning, parental investment, and maximum lifespan potential, that all have an underlying physiological basis, also scale allometrically (Millar 1977; Bleuweiss et al. 1978; Western 1979; Lindstedt & Calder 1981; Western & Ssemakula 1982; Peters 1983; Gittleman 1986; Promislow 1993; Robbins 1993; Duncan et al. 2007; Lovegrove 2003). The gestation period of mammals has a low allometric exponent ($b = 0.16$), neonatal body mass is close to isometric (0.94), litter mass scales similarly to MR (0.79), age at first breeding has a low allometric exponent (0.17), and lifespan scales with 0.21. Corresponding values for birds are quite similar: incubation period (0.16), egg mass (0.77), clutch mass (0.72), age at first breeding (0.23), and lifespan (0.19).

The maximum lifespan potential (MLP) has attracted particular interest, since explanations for patterns of longevity may be relevant to the human aging process. Since Sacher's (1959) early work on the allometry of lifespan, numerous workers have examined the scaling of MLP and M (e.g. Calder 1984; Jürgens & Prothero 1991; Charnov 1991, 1993), finding that MLP scales at approximately $M^{0.25}$ (Duncan et al. 2007). Explanations have ranged from Lindstedt and Swain's (1988) physiological clock that regulates physiological variables in a size-dependent manner ('periodengeber'), to Calder's (1984) 'live fast, die young' theory whereby all animals accomplish the same number of physiological events (e.g. total number of heartbeats) in their lifetime, to Charnov's (1991, 1993) trade-off between growth (that scales with $M^{0.75}$) and mortality (for which he derived a $-0.25$ scaling exponent). However, the co-correlation between MLP, mass, and almost all other physiological and biological traits means that a search for specific mechanistic physiological explanations for MLP (e.g. rates of DNA oxidative damage, levels of DHEA and fatty acid desaturation in heart phospholipids, rates of urinary excretion of DNA repair products; Pamplona et al. 2006; Barja 2004; Foksinski et al. 2004) is always confounded by this co-dependence on mass (Promislow 1993; Speakman 2005). Clearly, shared patterns of allometric scaling must be accounted for before we can gain insight into potential physiological correlates of MLP (Speakman 2005).

Ultimately, the effects of allometry on physiological factors and their flow-on to life history parameters lead to allometric implications for broadscale ecological functions. McNab (1963) noted that there was no statistical difference between the allometric scaling exponents of home range and BMR, and suggested that home range was determined by an animal's energy requirements (although other authors subsequently found significantly different scaling exponents for home range and energy use, e.g. Harestad & Bunnell 1979; Gompper & Gittleman 1991). Lindstedt et al. (1986) identified physiological time, Haskell et al. (2002)

fractal resource distributions, and Jetz et al. (2004) resource defence as factors mediating the relationship between home range and the metabolic requirements of animals. Millar and Hickling (1990) examined fasting endurance as a factor driving the evolution of mammalian body size. Importantly, the maximum population growth rate ($r_m$) scales allometrically with an exponent of −0.25 (e.g. Bleuweiss et al. 1978; Peters 1983; Savage et al. 2004b), providing a link between processes acting at the level of the individual to those at higher population or community levels (Duncan et al. 2007; Figure 2.3).

Broader scaling relationships have subsequently been developed in an attempt to provide some uniform, mechanistic explanation of ecological processes. Ecologists have widely used allometry to describe patterns of abundance and the diversity of individuals (e.g. Brown & West 2000). The scaling of species density varies for mammals of differing body mass, from 0.25 (not significantly different from 0) for mass $< 0.1$kg, to −0.70 (significantly less than 0) for $0.1 <$ mass $< 100$kg, to −0.11 (not different from 0) for mass $> 100$kg (Silva & Downing 1995).

Brown et al. (2004) proposed a metabolic theory of ecology (MTE), a framework that uses the effects of body mass and temperature to explain the metabolism of individual animals (i.e. $MR = b_0 M^b e^{-E/kT}$; where $b_0$ is an empirically fitted normalization constant (independent of mass and temperature), b is the allometric scaling exponent, k is Boltzmann's constant, E is mean activation energy estimated empirically for metabolism from enzyme kinetics, and T is temperature), which in

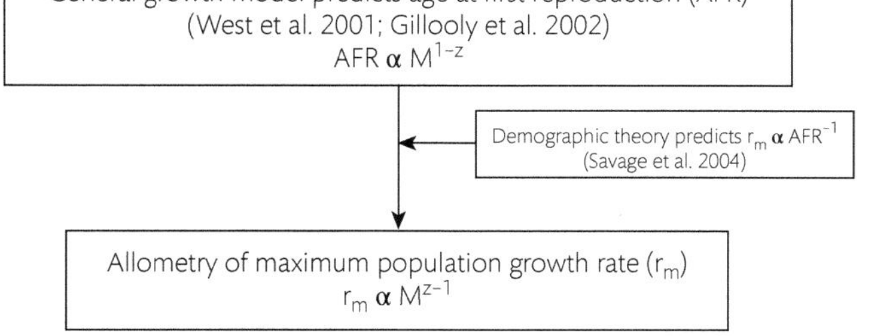

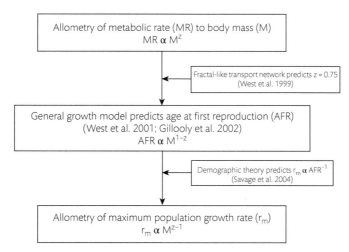


**Figure 2.3** Schematic for application of metabolic theory of ecology (applied to mammals, which have a relatively constant body temperature) to predicting the maximum population growth rate based on body mass. Modified from Duncan et al. (2007).

turn characterizes the energy pools and energy flow in populations, communities, and ecosystems. According to the MTE, constraints on transportation networks predict that whole-animal MR will scale with an exponent of 0.75. In turn, MR determines an animal's rate of resource uptake, and thus also life history parameters such as growth, reproduction, and survival, which ultimately determine processes at the population, community, and ecosystem level (Brown et al. 2004; Figure 2.3). Thus the MTE provides a mechanistic basis to explain an array of allometric relationships from the animal to the ecosystem. This theory has subsequently been tested and applied to an array of ecological questions, such as BMR determining growth rate (Lovegrove 2009a) and predicting global patterns of abundance (Duncan et al. 2007; Hawkins et al. 2007; McCoy & Gillooly 2008; Allan & Gillooly 2009; Munch & Salinas 2009). However, controversy surrounding both the exponent and underlying mechanistic basis for BMR scaling, non-uniformity of temperature effects on metabolism, debate concerning the scaling exponents of various life history parameters, modelling inadequacies, questions concerning functionality of the MTE at a variety of scales, and an inability to empirically test the theory have all raised doubts about the universal applicability of this theory (see the review by O'Connor et al. 2007 and references therein).

Another theory relating the allometry of energy use to broadscale ecological functions is the energetic equivalence rule (EER). The EER is a macro-ecological hypothesis that integrates the allometry of MR and species density to infer that the total population energy flux of species (i.e. kJ per unit area) should be independent of the mass of the individuals, because the total population energy use equals MR (e.g. MR $\alpha$ $M^{0.75}$) multiplied by species density ($\alpha$ $M^{-0.75}$), hence $M^{0.00}$ (Damuth 1981, 1987, 2007; Munn et al. 2013a). Similar to the MTE, the EER has been controversial, with varying degrees of support from a range of studies and databases (e.g. Cotgreave 1993; Marquet et al. 1995; Silva and Downing 1995; Ernest 2005; Loeuille & Loreau 2006; Damuth 2007). For Australian marsupials, neither BMR nor FMR coupled with species density supported the EER (Munn et al. 2013a).

## 2.2 Control Systems

The cell is the fundamental unit for living organisms. For unicellular organisms (e.g. protozoans), many aspects of the intracellular environment are regulated by various membrane control processes; e.g. the intracellular fluid is generally low in $Na^+$ and high in $K^+$ compared to the external environment, and this ion imbalance is maintained by the active transport of $Na^+$ out of the cell and $K^+$ into the cell, by a membrane $Na^+/K^+$ pump. Many other solutes are also regulated by control systems (e.g. biochemicals, $H^+$ concentration), but many physical aspects of the environment are not (e.g. temperature, light, respiratory gases). Multicellular organisms (like mammals) are more complex, consisting of many (often millions, billions, or more) different cells of many different kinds (e.g. nerve, muscle, gland, epidermal, connective). In addition, all of these cells are located in a 'new' environment, the extracellular environment of the organism, in which all of the cells exist and which can differ (often markedly) from the external environment. This extracellular environment was first recognized and named the 'milieu intérieur' (interior medium) by Claude Bernard in 1878 (Bradshaw 2003).

The maintenance of constancy of particular aspects of the extracellular environment is called 'homeostasis' (to stay relatively constant), a term coined by Walter Cannon in 1929 (Bradshaw 2003). Homeostasis is a fundamentally important physiological process because it provides a suitable environment for the cells of the body (Cannon 1932, 1935). Internal constancy can sometimes be apparent in environments that are constant, or where a rate process is relatively constant, without the intervention of an actual control system (Withers 1992). Passive homeostasis occurs when the external environment is stable and the animal's extracellular environment is in equilibrium with it, called equilibrium homeostasis. An example of equilibrium homeostasis is the $T_b$ of a mammal during passive torpor, when it comes into thermal equilibrium with the $T_a$ (ambient temperature) of its environment (so long as $T_a$ is not below its torpor $T_b$ setpoint, in which case the mammal resumes thermoregulation; see 3.2.4.2). Steady-state homeostasis occurs when a particular reaction has a constant rate, which results in an apparent constancy of a physiological variable; for example, an animal that synthesizes a waste product (e.g. ammonia) at a constant rate and loses it at a constant rate to a medium with a low concentration, would appear at equilibrium to be maintaining a constant internal ammonia concentration. However, if the external environment changes (water temperature or ammonia concentration increases) then the physiological variable changes, indicating the absence of a regulatory control system. For complex animals in changing environments, the homeostasis of their milieu intérieur is actively maintained by a variety of regulatory control processes, each involving a negative feedback regulatory system.

## 2.2.1 Regulation of Homeostasis

Active regulatory control systems maintain homeostasis of physiological variables in the face of external disturbances that would otherwise result in a change in that physiological variable. Overall, there are hundreds to thousands of negative and positive regulatory control systems in mammals, which serve to maintain homeostasis of most aspects of the internal environment and thereby provide an extremely constant milieu intérieur in which the cells live. As Cannon famously stated, 'So long as this personal, individual sack of salty water, in which each of us lives and moves and has his being, is protected from change, we are freed from serious peril' (Cannon 1935, p. 2).

An active regulatory control system for a physiological variable, at its simplest (Withers 1992; Guyton & Hall 1996), consists of the following: (1) a setpoint at which the variable is regulated, (2) a sensor that monitors a particular physiological variable, (3) a comparator that compares the actual value (affected by a disturbance) with the setpoint, (4) a resulting 'error' in the physiological variable, and (5) an effector system that produces a response that compensates for the disturbance (Figure 2.4). The actual value of the physiological variable is equal to the setpoint + disturbance + compensation. Some effector systems have a simple on-off response (like turning a light switch on or off), but proportional effector control systems have a more sophisticated response, with the magnitude of the response changing with the magnitude of the error (like a light with a dimmer switch). The response is the error term multiplied by a gain (or amplification) term. The effector in most control systems has the opposite effect to the disturbance, hence these are termed negative feedback control systems; the gain of these control systems is negative (e.g. gain is about −32 for human thermoregulation; see later). There are, however, some important positive feedback control systems.

Most physiological control systems are considerably more complex and sophisticated than the simple schematic in Figure 2.4. There are often multiple effector systems operating in parallel, with their own different setpoints,

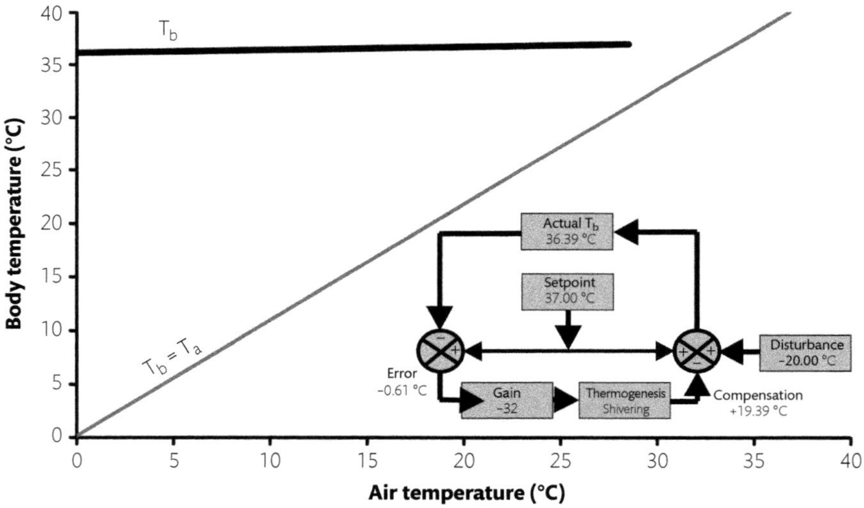


**Figure 2.4** Schematic relationship between body temperature ($T_b$) and disturbance (air temperature below thermoneutrality, $T_a$), for a human; $T_b$ is 37°C at the lower critical temperature of the thermoneutral zone (28.5°C) and slope is 0.030, which is equivalent to a gain of about −32 (data from Hardy & Soderstrom 1938). Inset shows schematic of thermoregulatory control by negative feedback for a human; setpoint = 37°C, disturbance = −20°C, gain = −32; at equilibrium, $T_b$ = 36.39°C and error = −0.61°C.

inputs to the comparator, and effector mechanisms. For example, mammalian thermoregulation in response to cold has numerous effector systems (e.g. shivering and non-shivering thermogenesis, cutaneous vasoconstriction, pilo-erection, and postural control) that increase heat production and decrease heat loss. There is also a separate sophisticated control system for cooling the body, again involving multiple effector systems with their own setpoints and mechanisms (e.g. evaporative heat loss, peripheral vasodilation, pilodepression, and postural changes that maximize the exposed surface area). Many control systems incorporate a 'feed-forward' control capacity, where the response is modified by previous 'experience' regarding whether the response was adequate to achieve the required degree of regulation. If not, then the next time the response is required, the compensation is adjusted based on the prior response to achieve a better outcome. Thus, feed-forward control is an adaptive control process that 'learns' from previous actions.

In positive feedback control systems, the effector response is in the same direction as the disturbance. Positive feedback control systems might seem inappropriate since they lead to instability of physiological variables rather than stability, and sometimes they are inappropriate. For example, blood loss can lead to sudden circulatory collapse (a highly undesirable response); excessive blood loss leads to a critical drop in venous return to the heart, hence cardiac output and blood pressure fall and coronary blood flow declines, further decreasing cardiac output. A positive-feedback cycle suddenly ensues resulting ultimately in circulatory collapse (Guyton & Hall 1996).

Nevertheless, there are useful, in fact essential, examples of positive feedback control in mammalian physiology (Withers 1992; Guyton & Hall 1996). Blood clotting is an example: clot formation in response to blood vessel damage releases clotting factors (enzymes) that positively induce the activation of further clotting factors, leading to a positive feedback cycle resulting in the maximum response to blood vessel damage. The contraction of body spaces (e.g. in the processes of urination, defaecation, and parturition) rely on a positive feedback of the initial contraction resulting in the full contraction of the body space. Another example of an essential positive feedback loop is the generation of an action potential. An initial depolarization of the membrane of an excitable cell, often via the entry of $Na^+$ through $Na^+$ channels, opens further voltage-gated $Na^+$ channels and increases sodium conductance across the membrane. This causes further depolarization and opening of more $Na^+$ channels, which, if the initial depolarization was sufficiently great, develops into a positive feedback loop that essentially opens all $Na^+$ channels and completely depolarizes the membrane to the maximum extent possible. Hence, an action potential is an 'all-or-none' phenomenon wherein the membrane depolarization is either maximal ('all'), or insufficient, and the membrane potential returns to rest ('none').

## 2.2.2 Neural Control

The nervous system is the most rapid and direct control system. It consists of the central nervous system (CNS) and the peripheral nervous system (PNS). The CNS and PNS receive sensory information from somatic tissues (muscle, skin, and their derivatives) and visceral tissues (involuntary muscles and glands), and transmit responses to various effectors, such as muscles and glands. The integration of input signals to effector responses can be simple (e.g. spinal reflexes) or complex (e.g. integration in higher levels of the brain).

Spinal reflexes are a relatively rapid means of control, particularly of peripheral muscles in response to specific sensory inputs (Pocock et al. 2013). The spinal reflex consists of a neural circuit of sensory neurons from peripheral sensory receptors, leading into the spinal cord via the dorsal root, synapsing with one or more association neurons located within the spinal cord, then exiting via the ventral root from where neurons innervate peripheral motor organs (muscle, glands). The somatic and visceral reflex arcs differ in the detail of their neural circuitry.

The somatic reflex arc consists of a sensory neuron, either no or one association neuron, and a motor neuron that innervates the target skeletal (voluntary) muscle via a cholinergic (acetylcholine neurotransmitter) synapse (Figure 2.5). The simplest two-neuron reflex arc controls postural reflexes; for example, an inadvertent change in posture stimulates muscle stretch receptors, which reflexly contract an antagonistic muscle, restoring the posture (Figure 2.5A). Inclusion of an association interneuron provides more complexity in the reflex response, but also slows the reflex (Figure 2.5B). The withdrawal reflex is elicited if, for example, a foot touches a sharp object. Sensory receptors stimulate an interneuron in the spinal cord, which stimulates a motor response to muscles in that limb that retract the foot, as well as to muscles of the opposite limb that will support the body when the original limb is retracted. The visceral reflex arc is more complex in circuitry and pharmacology; the effector organ is cardiac muscle, smooth muscle, or glands (i.e. involuntary effectors). The sensory axon passes through a peripheral sympathetic chain ganglion (see later), enters the dorsal root via the ramus communicans, and synapses with an interneuron in the spinal cord. The motor output consists of two neurons: the first preganglionic neuron synapses with a postganglionic neuron (in a sympathetic chain ganglion, collateral ganglion, or at the motor organ). If the synapse is located in a sympathetic ganglion, then the postganglionic axon normally passes through a spinal nerve to the effector organ. Both somatic and visceral reflex arcs can be modulated by other levels of the spinal cord and by higher brain centres.

The autonomic nervous system involves visceral sensory and motor effects that are not under voluntary control (Guyton & Hall 1996; Pocock et al. 2013). Sensory modalities monitor the internal environment (e.g. blood pressure, $O_2$ and $CO_2$ levels, pH, central and surface $T_b$, and visceral activity). Motor effects control

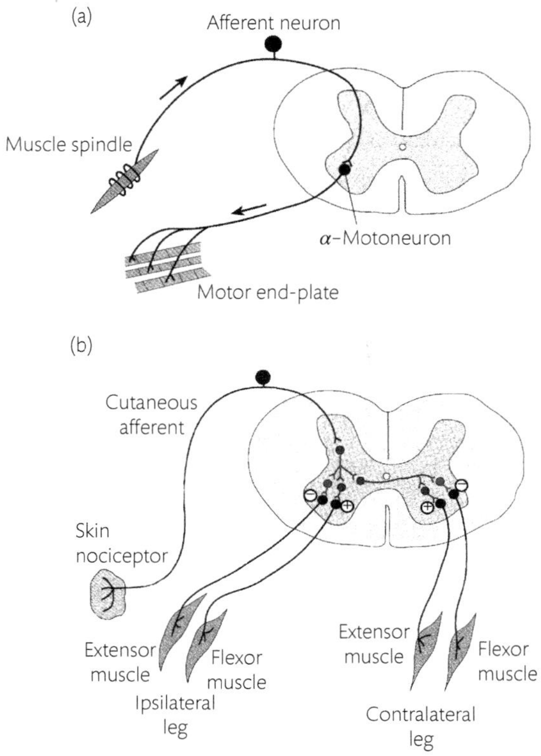


**Figure 2.5** Schematic of somatic (voluntary) reflex arcs, (a) without and (b) with association interneurons. From Pocock et al. (2013) Human Physiology. Reproduced with permission from Oxford University Press.

involuntary peripheral effectors (e.g. cardiac muscle, smooth muscle, glands) involving the heart, digestive tract, reproductive tract, blood vessels, respiratory tract, bladder, etc. The autonomic nervous system is divided into the sympathetic and parasympathetic systems, which generally have antagonistic actions and differ in the location of their site of exit from the CNS.

Sympathetic nerves leave the CNS in the thoracolumbar region of the spinal cord, whereas parasympathetic neurons exit the craniosacral regions of the CNS. Sympathetic preganglionic axons are generally short and synapse at a sympathetic chain or peripheral ganglion, with long postganglionic axons (Figure 2.6). The neurotransmitter is acetylcholine at the preganglionic synapse but adrenaline or noradenaline at the postganglionic synapses in the effector tissue. The actions of the sympathetic nervous system generally elicit responses that increase the activity of the target organ. The activation of the sympathetic motor system is often summarized as the 'fight or flight' response that increases heart rate and blood pressure, dilates coronary blood vessels, contracts the spleen, mobilizes glucose from glycogen stores, and slows digestive processes (Table 2.1). While sympathetic activation is often described as the fight or flight response, in everyday situations

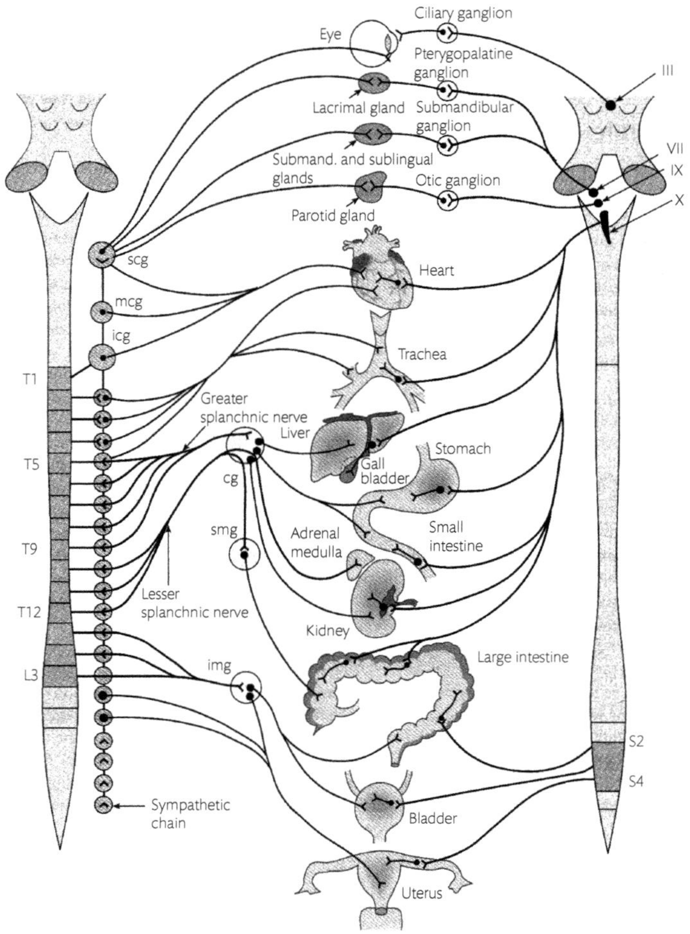


**Figure 2.6** Schematic of the sympathetic (left) and parasympathetic (right) nervous systems of a mammal. From Brodal (1992) and Pocock et al. (2013) Human Physiology. Reproduced with permission from Oxford University Press.

the activation of the sympathetic system is more nuanced, and the activation of one sympathetic effector system does not mean that all systems are activated. For example, during heat exposure the sympathetic activation of vascular smooth muscle in the skin is inhibited, while sympathetic activation of vascular smooth muscle in the gut is enhanced.

The parasympathetic nervous system maintains the general resting state, but stimulates the digestive system. Its effects are generally antagonistic to those of the sympathetic nervous system (Table 2.1). Preganglionic axons of the

**Table 2.1** Functional effects of the sympathetic and parasympathetic divisions of the autonomic nervous system. Modified from Kardong (2009) and Pocock et al. (2013).

| Organ and action | Sympathetic effect | Parasympathetic effect |
|---|---|---|
| Eye | | |
| Ciliary muscle | Relaxation | Contraction |
| Pupillary muscle | Dilatation | Constriction |
| Heart | | |
| Muscle | Increased rate and force | Slowed rate |
| Coronary arteries | Dilatation | Constriction |
| Lung bronchi | Dilatation | Constriction |
| Spleen | Contraction | Relaxation |
| Blood vessels | Constriction | None |
| Adrenal gland | Secretion | None |
| Glands | | |
| Salivary | Vasoconstriction, minor secretion | Vasodilation, copious secretion |
| Gastric | Inhibition of secretion | Stimulation of secretion |
| Pancreas | Inhibition of secretion | Stimulation of secretion |
| Lacrimal | None | Stimulation of secretion |
| Sweat | Stimulate sweating | None |
| Digestive tract | | |
| Sphincters | Increase tone | Decrease tone |
| Walls | Decrease motility | Increase motility |
| Liver | Glucose mobilization | None |
| Bladder | | |
| Wall | Relaxation | Contraction |
| Sphincter | Contraction | Relaxation |
| Sexual organs | | |
| Penis | Ejaculation | Erection |
| Clitoris | ? | Erection |
| Metabolism | Increased | None |

parasympathetic nervous system are generally long, and synapse at the peripheral target organ from where a short postganglionic axon stimulates the effector tissue. The neurotransmitter is acetylcholine at both the pre- and postganglionic synapses. The adrenal gland is an exception because it is innervated by the preganglionic axons of the sympathetic nervous system; it is essentially modified postganglionic nerve cells.

### 2.2.3 Chemical Control

Chemicals that regulate biological processes are collectively termed bioregulators (Norris 2007); examples include secretions of the nervous system (neurocrines), of specialized endocrine glands (hormones), and of the immune system (cytocrines). Some cells release 'local hormones' that affect the cell itself (such bioregulators are called autocrines) or nearby cells (paracrines). Bioregulators released from an animal into the external environment can affect conspecific animals (pheromones) or other species; interspecies bioregulators that benefit the signal producer are termed allelomones, and those that benefit the signal receiver are kairomones (Sbarbati & Osculati 2006; Rajchard 2013).

Neurons secrete a variety of neurocrines. Those neurocrines that act through synapses on other nerve cells, or other target cells (e.g. muscle, glands) are called neurotransmitters or neuromodulators. Some neurons also secrete neurohormones, which are released by neurosecretory cells into the blood and distributed to target cells. The more traditional concept of an endocrine organ is a non-neural organ/gland that releases a hormone into the blood. The hormone is then distributed throughout the body and has its effect on specific target organs. Water-soluble hormones bind to specific cell surface receptors, whereas lipid-soluble steroid hormones bind to intracellular receptors; a hormone binding to its receptor elicits a cellular response. The half-life of neurohormones and hormones is much longer than that of neurotransmitters.

In an evolutionary sense, the first bioregulators were neurohormones, produced by the neuroendocrine system, since the nervous system predates both true endocrine glands (that require a circulatory system to disperse their hormone secretions) and the immune system. In vertebrates, the main bioregulators are from the neuroendocrine system (brain and pituitary gland) and regulate the outputs of the classical endocrine glands (pituitary, thyroid, adrenal, gonads, and liver). Other endocrine glands are not under direct neural control (parathyroid glands, thymus, pancreas, gastrointestinal tract, pineal gland, kidney). The numerous neuroendocrine and endocrine glands of mammals release a wide range of neurohormones/hormones that regulate either other endocrine glands or stress (e.g. hypothalamo-hypophyseal axis), or effects of various target cells (Selye 1956; Bentley 1998, 2002; Norris 2007). These glands and hormones show an evolutionary progression in both the structure of the glands, the structure of the hormones, and the role of the hormones (Bentley 1998).

Pheromones are used by almost all animals for intraspecific communication. Amongst mammals, their function has been observed in many taxa including marsupials, insectivores, carnivores, ungulates, and primates, but they are best studied in rodents. They contribute to communication in a variety of contexts such as social, alarm, and reproductive signalling (Bigiani et al. 2005). Pheromones that have an acute influence on behaviour are termed releaser pheromones, while those that impact developmental or physiological processes are known as primer pheromones (Petrulis 2011).

The sensory pathways involved in pheromone detection and processing include the main olfactory system and the accessory olfactory or vomeronasal system. Vomeronasal organs do not occur in fish or archosaurs (birds and crocodilians), but are present in amphibians, squamates, and most mammals, although they are absent or vestigial in dolphins and some primates, including Old World monkeys, apes, and humans (Bigiani et al. 2005). In mammals, the vomeronasal organ is located on the anterior nasal septum, and is a cylindrically shaped, bilaterally symmetrical structure encased in a bony capsule that opens anteriorly into the nasal cavity or mouth via the vomeronasal duct. The lumen of the organ is filled with mucus that may function as a pheromone carrier, and the medial epithelium is sensory (Dulac & Torello 2003; Bigiani et al. 2005). The main olfactory and vomeronasal systems have discrete neural pathways to the main and accessory olfactory bulbs, respectively, with the pheromone stimulus ultimately acting on the hypothalamus, influencing aggressive and reproductive behaviour and also the animal's endocrinology (Bigiani et al. 2005; Dulac & Torello 2003).

Proteins and an array of small molecules function as pheromones. These are often secreted by glands or in urine, under endocrinological control (Dulac & Torello 2003). Reproductive functions of pheromones have been most widely studied, particularly in rodents. For females, exposure to pheromones from mature males generally increases secretion of lutinizing hormone and reduces prolactin by modification of the hypothalamic-pituitary axis, with consequences including advancing the onset of puberty, promoting ovulation, and terminating pregnancy. Females investigate, scent-mark, and are attracted to the odours of reproductively active male mice, but these behaviours can be moderated by cues relating to the male's status, immunological, and dietary condition. Exposure to the pheromones of other females may delay puberty, suppress oestrous cycles, and potentially synchronize ovulation, presumably due to increased prolactin secretion. Pheromones may also influence female/young interactions and aggression in lactating females (Dulac & Torello 2003; Petrulis 2011). For males, the onset of puberty; sperm density and allocation; endocrine function; attractive, aggressive, and investigative behaviour; vocalizations; and copulatory behaviour are all influenced by pheromones (Petrulis 2011).

Allomones influence interspecific interactions for mammals in a variety of ways. They may involve the defensive responses of mammalian hosts to ectoparasites (e.g. waterbuck, *Kobus ellipsiprymnus*, produce allomones that have a repellent effect on tsetse flies; Gikonyo et al. 2003) or to disease. Mammalian allomones can disrupt the quorum-sensing mechanism used by bacteria to monitor their population density, and therefore function as an important innate defence against infection (Sbarbati & Osculati 2006). However, other organisms may use allomones against mammals. Ticks produce immunosuppressive allomones in their saliva (Barriga 1999), and many plants produce volatile organic compounds in their foliage that deter herbivores. Koalas (*Phascolarctos cinereus*) are folivorous marsupials that feed predominantly on *Eucalyptus* leaves. *Eucalyptus* trees produce formylated phloroglucinol compounds; at high concentrations, these plant secondary metabolites impose time, energetic, and nutritional costs on folivores consuming their leaves and so are an important defensive strategy for these plants (see 4.8.2). Under controlled conditions, koalas reduce their leaf intake in proportion to the concentration of plant secondary metabolites, and wild koalas are less likely to visit trees characterized by high concentrations of plant secondary metabolites (Moore & Foley 2005).

Kairomones play an important role in many aspects of mammalian biology, including disease, parasitism, and predator-prey relationships. Kairomones have been most widely studied for their role in mediating parasite and host relationships. Parasites may interpret kairomonal cues emitted by hosts as attractants. For example, kairomones emitted by the elephant shrew, *Elephantulus myurus*, appear to stimulate attachment and feeding of the ticks *Rhipicephalus warburtoni* and *Ixodes rubicundus* (Harrison et al. 2012), and female yellow fever mosquitoes (*Aedes aegypti*) are attracted to kairomones of mice (McCall et al. 1996). These parasite/host interactions may be further mediated by kairomones that carry information concerning the host's immunological condition (Rajchard 2013). Many prey species use kairomones that presumably have important roles in intraspecific communication amongst predatory species as a cue to assess potential predation risk (Sbarbati & Osculati 2006). There is a considerable body of literature examining the behavioural and physiological responses of prey to predator odours, more correctly termed kairomones. Many carnivorous mammals produce a biogenic amine (2-phenylethylamine) that appears to be widely recognized by many prey species as representing potential predation risk (Rajchard 2013). However, prey response to carnivore kairomones can be mediated by a variety of factors, including experience, evolutionary history, motivation, and personality (Apfelbach et al. 2005; Mella et al. 2016).

### 2.2.4 Glands

Glands are cells or organs that secrete chemicals that have effects elsewhere in the body. Glands are generally derived from epithelial tissues, by invagination of a surface epithelium. Exocrine glands generally maintain their contact with the epithelial surface via a duct, whereby they transfer their secretions. Endocrine glands lose their contact with the surface, have no ducts, and generally rely on the circulatory system to transport their secretions. This distinction is sometimes not clear-cut. The liver, for example, is an exocrine gland that secretes bile into the small intestine via the bile duct, but also secretes hormones (and other substances, like albumin) into the bloodstream. Several other glands also have both exocrine and endocrine functions (e.g. pancreas, kidney, testis, ovary).

### 2.2.4.1 Exocrine Glands

Exocrine glands have a duct that conveys their secretions to either the skin surface (e.g. sweat, tear, odoriferous, and mammary glands) or the cavity of another organ (e.g. salivary glands, pancreas, liver). Simple glands have an unbranched duct whereas compound glands have a branched duct (Saladin 2010). A tubular gland has a similar diameter in both the secretory and duct portions, whereas an acinar (or alveolar) gland has an expanded terminal secretory portion. The mechanism for secretion also differs. Serous glands produce a thin, watery secretion (e.g. sweat, milk, tears), whereas mucous glands secrete a thicker glycoproteinaceous mucin (e.g. mucus from goblet cells). Some glands are both serous and mucous (e.g. some salivary glands). The testes and ovaries are cytogenic glands, releasing whole cells (sperm, eggs). Exocrine glands also differ in the cellular mechanism involved in the formation of the secretion. The secretory cells of merocrine (or eccrine) glands release the secretion via exocytosis of vesicles (e.g. tear and gastric glands, pancreas); apocrine glands (e.g. axillary sweat glands) are primarily merocrine but differ in histological appearance (the secretion seems to form by the outer region of the secretory cell disintegrating) and function. Apocrine glands are found in the skin, eyelids, ears, and breasts. Holocrine glands synthesize their product, then the entire cell disintegrates, releasing a mixture of the secretion and other cellular debris (e.g. oil glands of the scalp and eyelid).

Mammalian skin contains an array of exocrine glands (Table 2.2), some of which are unique to mammals (Blackburn 1991; Liem et al. 2001). Sebaceous glands are specialized epithelial cells, usually associated with a hair follicle, that produce an oily sebum by holocrine secretion, that helps to keep the skin soft, flexible, and waterproof; areolar (nipple) sebaceous glands (and some others) open directly onto the skin surface. Sebaceous glands are especially important in marine mammals to

**Table 2.2** Summary of the mammalian exocrine glands, including their secretions and functions.

| Organ | Exocrine gland | Secretion | Function |
|---|---|---|---|
| Skin | Sebaceous | Sebum | Soften, flexible, waterproof skin |
| | Apocrine sweat glands | Sweat | Odorous, ear wax, evaporative cooling |
| | Eccrine sweat glands | Sweat | Lubrication, friction, evaporative cooling |
| Eye/eyelids | Meibomian, Moll's, Zeis, Henle's, Krause's | Oily and water secretions | Protection, lubrication |
| | Lachrymal | Tears | Moisten, lubrication, cleansing |

*(continued)*

**Table 2.2** (*Continued*)

| Organ | Exocrine gland | Secretion | Function |
|---|---|---|---|
| Gastrointestinal tract | | | |
| – membranes | Goblet cells | Mucus | Lubrication, protection against autodigestion |
| – oral cavity | Parotid, zygomatic, mandibular, Suzanne's | Saliva | Lubrication, binding food into bolus, amylase |
| – tongue | Ebner's, sublingual, Weber's, Bauhin's | Saliva (serous, mucous) | Lubrication, binding food into bolus |
| – stomach | Parietal cells, chief cells | HCl, pepsinogen | Acid digestion, protein digestion |
| – liver | Hepatic cells | Bile | Lipid emulsification, pH buffering |
| – pancreas | Pancreatic cells | Pancreatic juice | Digestive enzymes, pH buffering |
| – small intestine | Lieberkuhn's, Brunner's, Paneth cell, Peyer's patch | | Enzymes, pH buffering, antimicrobial, defence |
| Respiratory membranes | Goblet cells | Mucus | Lubrication |
| Nasal cavity | Bowman's (olfactory) | Mucous secretion | Dissolve odorant molecules |
| Reproductive organs | | | |
| – male | Prostate, bulbourethral, Cowper's | Pre-ejaculate | Nutrition, sperm activation |
| – female | Vagina, vulva, cervix | Mucus | Lubrication |
| | Mammary gland | Milk | Suckling young |
| | Montgomery's glands (mammary areola) | | Lubrication of the nipple for suckling |

keep the fur waterproof. The tarsal (Meibomian) glands of the eyelids are modified sebaceous glands, whose oily secretion protects the corneal surface and prevents the tears from overflowing the rim of the eyelid.

The two forms of sweat glands, eccrine and apocrine (sudoriferous), are common in the footpads of mammals, but are more widely distributed in some species (e.g. humans, *Homo sapiens*). Eccrine sweat glands secrete a watery fluid directly onto the skin surface; the fluid contains various ions (e.g. $Na^+$, $K^+$, $Cl^-$) as well as other minor solutes (e.g. urea). Sweat helps provide frictional adhesion with the substrate and can enhance tactile sensation. In some species, it also contributes to evaporative heat loss and hence thermal balance, but species with limited eccrine sweat glands, or an absence of these glands (e.g. small rodents, bats, aquatic mammals) rely on other mechanisms for heat loss (e.g. panting, immersion in water or mud). Apocrine sweat glands are usually associated with hair follicles and produce an odorous secretion, usually into the hair follicle. Specialized scent-producing glands are common; examples include sternal, axillary, urogenital, ear, wrist, oral, and frontal glands, as well as nasal glands, sebaceous and serous ocular glands, and lachrymal glands (Russell 1985). The anal glands of mustelids are specialized to squirt a noxious irritating secretion that can be projected several metres. Ceruminous (wax) glands are modified apocrine sweat glands that keep the tympanic (middle ear) membrane flexible. Mammary glands are also thought to be specialized apocrine sweat glands (or perhaps sebaceous glands) that secrete milk; the ducts of the mammary glands converge at the nipple where the milk is released (monotremes lack a nipple and the milk is secreted onto the skin surface).

The other major group of exocrine glands is associated with the digestive tract (mouth, tongue, oesophagus, stomach, small and large intestine, pancreas; Table 2.2). Most mammals have well-developed parotid glands that empty, along with the zygomatic gland (in the floor of the orbit) of some species, into the roof of the buccal cavity, and mandibular and sublingual glands that empty into the floor of the buccal cavity (Liem et al. 2001). Some have buccal and molar glands associated with the mucous membrane of the lips. Saliva from these glands is a mix of serous and mucus secretions that lubricate and bind the food as it is masticated and then swallowed. The saliva of some mammals contains amylase that assists carbohydrate digestion. Some insectivorous mammals have a toxin in their saliva, and vampire bat saliva contains an anticoagulant, called draculin. Salivary glands of specialist myrmecophage mammals (e.g. echidnas, *Tachyglossus aculeatus*; numbats, *Myrmecobius fasciatus*; anteaters Pilosa) are especially enlarged to produce copious saliva that coats the tongue and aids in harvesting ants and termites. The stratified squameous epithelium of the oesophagus has some mucus and serous secretory cells, but generally lacks larger glands. The goblet cells in the mucosa of the gastrointestinal tract produce a mucus that lubricates and protects against autodigestion. The stomach mucosa has goblet cells (mucin) and simple, tubular gastric glands that secrete HCl (from parietal cells) and pepsinogen (from chief cells). Some mammals, such as insect-consuming carnivores and insectivorous bats, and krill-consuming whales, produce chitinase that allows them to digest chitin which is found in insect and crustacean cuticle (Cornelius et al. 1975; Souza et al. 2011; Strobel et al. 2013; see 4.8.1). Young mammals often secrete an enzyme, rennin, that curdles milk and increases its retention time in the stomach, facilitating better digestion. In the small intestine, there are goblet cells that secrete mucus, and many simple tubular glands (crypts of Lieberkuhn) at the bases of the villi. Exocrine secretions of the liver (bile) and pancreas (pancreatic juices) empty into the small intestine where they play important roles in pH buffering and in digestion, especially of lipids. The lining of the large intestine has mainly mucous glands that secrete lubricants; enzymes involved in digestion here are derived from earlier in the digestive tract.

The male reproductive tract has several accessory sex glands, prostate, bulbourethral, and vesicular glands, which secrete seminal fluid (Liem et al. 2001). The seminal fluid transports spermatozoa along the sperm passages, and provides nutrients and activating factors for sperm. The female reproductive tract also has several accessory sex glands. Vaginal and vulval glands provide lubrication and Montgomery's glands provide lubrication of the nipple area during suckling.

#### 2.2.4.2 Endocrine Glands

In contrast to exocrine glands, endocrine glands are ductless and secrete hormones into the bloodstream. The output of these glands is transported throughout the body by the circulatory system and influence the actions of other cells within target organs that have the appropriate receptors for these hormones. The endocrine system plays an important role in a variety of body functions, including osmoregulation, ionoregulation, reproduction, growth, thermoregulation, cardiovascular control, and metabolism.

In mammals, as in other vertebrates, the pituitary arises as an association between a downgrowth of neural tissue (infundibulum) from the brain and an upgrowth of tissue (hypophysis) from the roof of the mouth, which encloses a segment of mesodermal tissue (Bentley 1998). The pituitary lies near the hypothalamus, and in mammals is usually enclosed in a bony chamber (the sella turcica). The infundibulum forms the neurohypophysis (posterior pituitary), which consists of the median eminence (part of the wall of the hypothalamus, with a system of portal blood vessels), the infundibular stalk that remains connected to the brain, and the neural lobe (or pars nervosa). The neurohypophysis secretes two neurohormones: arginine vasopressin (AVP; or analogues), also known as antidiuretic hormone (ADH), and oxytocin (OXY; Table 2.3). Magnocellular neurons in the supraoptic and paraventricular nuclei (outside the hypothalamus) synthesize AVP and OXY, which travel along axons to the neural lobe, from where they are released into the bloodstream. The hypophysis differentiates partly into the adenohypophysis (anterior pituitary), which consists of the pars tuberalis (associated with the portal blood vessels from the median eminence), the large pars distalis, which secretes a number of 'trope' hormones (e.g. thyrotropes, gonadotropes, somatotropes), and the pars intermedia, which secretes melanocyte-stimulating hormone (MSH) and endorphins. The secretions of the adenohypophysis are regulated by hormones (such as corticotropin-releasing hormone, thyrotropin-releasing hormone, gonadotropin-releasing hormone, and somatostatin) that are formed by parvicellular cells in the hypothalamic nuclei, and released from their nerve terminals at the median eminence into the hypophysial-portal blood vessels, thence transported to the adenohypophysis. This portal system is absent in 'lower' vertebrates (cyclostomes, teleost fishes).

**Table 2.3** Summary of the mammalian endocrine systems, including neuroendocrine and endocrine organs, their secretions, major target organs, and actions. Modified from Bentley (1998, 2002) and Norris (2007). Neurohormones in bold; immune cytocrines in italic.

| Endocrine gland | Secretion | Target | Action |
|---|---|---|---|
| **Hypothalamus** | **Thyrotropin-releasing hormone TRH** | **Anterior pituitary** | **TSH release** |
| | **Gonadotropin-releasing Hormone GnRH** | **Anterior pituitary** | **LH/FSH release** |
| | **Corticotropin-releasing Hormone CRH** | **Anterior pituitary** | **ACTH release** |
| | **Somatostatin GH-RIH or SST** | **Anterior pituitary** | **GH inhibition** |
| | **Somatocrinin GHRH** | **Anterior pituitary** | **GH release** |
| | **Prolactin release-inhibiting Hormone PRIH** | **Anterior pituitary** | **PRL inhibition** |
| | **Prolactin-releasing hormone PRH** | **Anterior pituitary** | **PRL release** |
| | **Melanotropin release-inhibiting Hormone MRIH** | **Anterior pituitary** | **MSH inhibition** |
| | **Melanotropin-releasing Hormone MRH** | **Anterior pituitary** | **MSH release** |
| | **Endorphins/enkephalins** | **Pain neurons** | **Desensitization to pain** |
| **Posterior pituitary** | **Arginine vasopressin AVP** | **Kidney, brain** | **Water reabsorption, drinking** |
| | **Oxytocin OXY** | **Uterus, vas deferens** | **Smooth muscle contraction** |
| **Pineal gland** | **Melatonin** | **Brain** | **Puberty, thyroid, adrenal, reproductive rhythms** |
| Anterior pituitary | Thyroid-stimulating hormone TSH | Thyroid gland | Thyroid hormone release ($T_3$, $T_4$) |
| | Luteneizing hormone LH | Gonads | Androgen and progesterone synthesis |
| | Follicle-stimulating hormone FSH | Gonads | Gamete formation, oestrogen synthesis |
| | Growth hormone GH | Liver, connective tissue, muscle | IGF and protein synthesis |
| | Prolactin PRL | Mammary gland, epididymis | Protein synthesis |
| | Adrenocortico-stimulating hormone ACTH | Adrenal cortex | Corticosteroid synthesis |
| | Melanotropin MSH | Melanin-producing Cells | Melanin synthesis |
| Thyroid gland | Thyroid hormones $T_3$, $T_4$ | Most tissues | Metabolism, cell differentiation, development |
| | Calcitonin CT | Bone | Bone resorption prevented (antagonizes PTH) |

*(continued)*

**Table 2.3** (*Continued*)

| Endocrine gland | Secretion | Target | Action |
|---|---|---|---|
| Adrenal gland | | | |
| – cortex | Aldosterone | Kidney | Sodium reabsorption, potassium secretion |
| | Corticosterone | Liver, muscle | Protein-to-carbohydrate conversion |
| | Cortisol | Liver, muscle | Glycogen-to-glucose breakdown |
| – medulla | Adrenaline/nonadrenaline | Smooth muscle (various organs) | Vaso-constriction, vasodilation, heart rate, stimulates metabolism |
| Parathyroid gland | Parathyroid hormone PTH | Bone, kidney | Bone resorption, calcium reabsorption, phosphate excretion |
| Pancreas endocrine | Insulin | Liver, muscle | Glycogen storage, glucose uptake, fat mobilization inhibition |
| | Glucagon | Liver, adipose tissue | Glycogen breakdown, glucose release, fat mobilization |
| | Pancreatic polypeptide PP | Brain | Appetite suppression |
| | Somatostatin | Endocrine pancreas | Pancreatic hormone inhibition |
| Kidney | Erythropoetin | Bone marrow | Red blood corpuscle formation |
| | Renin | Blood renin substrate | Angiotensin formation |
| | 1,25-dihydroxy cholecalciferol 1,25-DHC | Small intestine | Calcium absorption stimulation |
| Adipose tissue | Leptin | Brain | Feeding inhibition |
| GI system | | | |
| – stomach | Gastrin | Stomach gastrin glands | Acid secretion ($H^+$) stimulation |
| – small intestine | Ghrelin | Brain | Feeding stimulation |
| | Secretin | Exocrine pancreas | Pancreatic juice release |
| | Cholecystokinin CCK | Exocrine pancreas | Enzyme release |
| | Gastrin-releasing peptide GRP | Stomach gastrin cells | Gastrin release |
| | Gastric-inhibiting peptide GIP | Endocrine pancreas | Insulin release |
| | Motilin | Stomach | Pepsinogen secretion stimulation, gastric Motility |
| | Somatostatin | Small intestine | Inhibition of release of other regulators |
| | Vasoactive intestinal peptide VIP | Blood vessels | Increase of blood flow to intestines |

| | | | |
|---|---|---|---|
| Liver | Insulin-like growth factors IGF-I IGF-II | Many tissues | Cell mitogenesis effects |
| | Synlactin (?) | Mammary gland | Affects action of PRL on mammary gland |
| Gonads | | | |
| – ovary | Oestrogens (e.g. oestrogen) | Sexual structures, brain | Stimulates development, reproductive behaviour |
| | Progesterone | Uterus | Uterine gland secretion stimulation |
| | Inhibin | Anterior pituitary | Blocks FSH release |
| – testis | Androgens (e.g. testosterone) | Sexual structures, brain | Stimulates development, reproductive behaviour |
| | Inhibin | Anterior pituitary | Blocks FSH release |
| *Immune system* | | | |
| *– thymus* | *Thymosins* | *Lymphocyte-producing tissue* | *Lymphocyte production, stimulation* |
| *– macrophages, lymphocytes* | *Interleukin 1, Interleukin 2* | *Helper T cells, cytotoxic T cell* | *Activation* |

The thyroid gland has the longest 'phylogenetic history' of vertebrate endocrine glands (Bentley 1998). Protochordates (*Amphioxus* and ascidians) have homologous tissues to vertebrate thyroid tissue. In vertebrates, thyroid tissue consists of follicles of secretory epithelial cells surrounding a central cavity. In lower vertebrates (cyclostomes, some teleosts), thyroid follicles are scattered along blood vessels near the pharynx, but in other groups of vertebrates the follicles are either aggregated into a single glandular mass (chondrichthyeans, some teleosts, lungfishes, coelacanth), two separate glandular masses (amphibians, some reptiles, birds) or two lobes joined together (lizards, mammals). The thyroid gland secretes tri-iodothyronine ($T_3$) and the less potent tetra-iodothyronine ($T_4$, thyroxin), which have a range of metabolic effects, including a general stimulation of metabolism (thermogenic effect) and more specific effects on carbohydrate, lipid, and protein metabolism (Norris 2007). In placental mammals, thermogenesis is stimulated by $T_3$ (but not $T_4$) in brown adipose tissue (BAT) via uncoupling protein 1 (UCP1; see 3.2.3.3). Mammalian thyroid tissue is morphologically associated with the parathyroid glands and C cells.

Parathyroid glands first appear in vertebrates with amphibians, and in mammals there are one or two pairs of glands associated with the thyroid (Bentley 1998, 2002). The parathyroids secrete parathormone (PTH), which is hypercalcaemic (increases plasma $Ca^{2+}$ levels) through its action on bone (promoting resorption), the kidney (increasing $Ca^{2+}$ resorption and $PO_4^{3-}$ excretion), and gut (increasing $Ca^{2+}$ and $PO_4^{3-}$ absorption). Ultimobranchial bodies (formed from the most posterior branchial pouches) are present in all vertebrates (except cyclostomes), usually between the oesophagus and heart, but in mammals they are incorporated into the thyroid tissue as C cells. The C cells synthesize and release calcitonin, a hormone that is hypocalcaemic (reduces plasma $Ca^{2+}$) through its action on bone (promoting deposition).

There is a phylogenetic trend in vertebrates towards a close association of adrenal glands from two separate glandular tissues, adrenocortical tissue (of mesodermal origin) and chromaffin (neural) tissue that forms the adrenal medulla in mammals. In lower vertebrates, the adrenal cortical and medullary tissues are quite separate structures (Bentley 1998). In amphibians they are typically closely associated, they are intermingled in reptiles and even more so in birds, and in mammals they are intimately associated as an outer cortex of adrenocortical tissue cortex and an inner medulla of chromaffin tissue (except in echidnas, whose adrenals resemble those of reptiles; Wright et al. 1957).

The adrenal cortex secretes a variety of steroid hormones (Table 2.3), whereas the adrenal medulla secretes a combination of adrenaline (epinephrine) and noradrenaline (norepinephrine), reflecting its neural (sympathetic ganglion) origin. The adrenal cortex generally has three zones, which secrete aldosterone (zona glomerulosa), cortisol and corticosterone (zona fasciculata), and androgenic steroids (zona reticularis), although the difference between the zones seems related more to their control (e.g. by corticotropin) than their inherent biochemical capabilities (Bentley 1998; Norris 2007). Aldosterone is associated with sodium regulation, stimulating $Na^+$ reabsorption from nephrons in the kidney into the blood and promoting $K^+$ excretion, in concert with the renin-angiotensin system (see later). Cortisol and corticosterone affect energy metabolism and protein-to-carbohydrate conversion (inhibiting the peripheral use of glucose, while stimulating amino acid entry into liver cells for gluconeogenesis, and mobilizing fat stores) related to the fight or flight response. The androgenic steroids promote the onset of sexual development.

The fight or flight responses evolved early in chordate evolution (Chang & Hsu 2004). The corticotropin-releasing hormone (CRH) peptide family divided (predating teleost fishes and tetrapods) into CRH/urocortin and stresscopin/urocortin branches, then by a second gene duplication into CRH, urocortin/urotensin, stresscopin (SCP)/urocortin III, and stresscopin-related (SRP)/urocortin II genes. Each of these four genes is highly conserved from teleosts through to mammals. Thus, mammalian fight or flight responses have evolved over more than 550 million years of vertebrate evolution. In mammals, urocortins I, II, and III have physiological effects, including gastric secretion, hypotension, and reduced inflammation (Fekete & Zorilla 2007; Norris 2007).

The vertebrate gut is a long organ with diffuse and complex endocrine functions (Bentley 1998; Norris 2007). Its various peptide hormones (including gastrin, secretin, somatostatin, glucagon-like peptide I, cholecystokinin-pancreozymin, gastric inhibitory peptide, and vasoactive intestinal peptide) generally control digestion (Table 2.3); some such as vasoactive intestinal peptide also have roles in the central nervous system, including in the control of appetite. The primordial endocrine cells seem to have originated in the gut lining (as in some molluscs, protochordates, and cyclostomes), with specialized adjacent glands (e.g. pancreas) derived from them. The gut and pancreatic cells of some fishes secrete these various peptide hormones, as well as insulin, glucagon, and pancreatic polypeptide. Many of these are also secreted by neural tissues in the brain. In mammals and most other vertebrates, the alpha and beta cells of the pancreas secrete insulin and glucagon, respectively (Table 2.3); the cells that secrete them are usually associated as the islet of Langerhans in amongst the exocrine cells of the pancreas that secrete various digestive enzymes.

The gonads (ovaries, testes) have the related functions of egg/sperm production and reproductive hormone secretion (Bentley 1998). In all vertebrates, the testis (formed from the medulla of the primordial gonad) contains Leydig cells, in the interstitium between the seminiferous tubules, which secrete androgens (male sex steroid hormones) that act on the testis and the brain and stimulate gonadal development and reproductive behaviour. Inhibin is also secreted by Leydig cells and blocks follicle-stimulating hormone (FSH) release from the adenohyphysis. The ovary (formed from the cortex of the primordial gonad) secretes hormones that control gonadal development and secondary sexual characteristics, including reproductive behaviour (e.g. oestrogens and mammary gland development), stimulates uterine gland secretion (progesterone), and blocks FSH release (inhibin). During follicle development, the theca cells (surrounding the developing egg) secrete steroids and inhibin. After ovulation, the theca collapses to form the corpus albicans, or, in mammals, proliferates as the corpus luteum that secretes especially progesterone (PRO) but also oestradiol and other progestogens and oestrogens. The latter also occurs in ovoviviparous and viviparous (live-bearing) non-mammalian vertebrates.

In addition to its osmoregulatory and waste management roles, the kidney has an endocrine role (Bentley 1998, 2002). In mammals, the juxtaglomerular (JG) apparatus consists of cells that secrete renin (JG cells) and the macula densa, a thickening of the distal convoluted tubule where it contacts the afferent/afferent arteriole of the glomerulus. Non-mammalian vertebrates lack a macula densa (some chondrichthyeans and birds may have a homologous structure). The JG cells release renin in response to changes in the $Na^+$ content of the fluid in the macula densa. Renin is an enzyme that is released into the blood where it converts angiotensinogen (from the liver) into angiotensin I, which in turn is transported in the blood to the lungs, where angiotensin-converting enzyme (ACE) is present on the endothelium and converts it to the much more bioactive angiotensin II (Norris 2007). Angiotensin II constricts blood vessels, causes ADH release from the neurohypophysis, and aldosterone release from the adrenal cortex; it thus regulates blood pressure, $Na^+$ / $K^+$balance, and blood volume. The lower vertebrate hormone stanniocalcin has been identified in the renal tubules of mammals, but it is unclear if it has a functional role or is an embryological vestige of the corpuscles of Stannius—small endocrine groups of cells in some fishes that secrete stanniocalcin, which lowers plasma $Ca^{2+}$ in marine fishes. Cells of the macula densa also have a role in regulating glomerular flow rate (GFR) in mammals. An excess of $Na^+$ in the distal convoluted tubule (DCT) of the nephron causes constriction of the afferent arteriole of the nearby glomerulus, whereas a decrease causes dilation of the afferent arteriole.

The pineal gland arises as an outgrowth of the dorsal midbrain (diencephalon); its primitive role is as an endocrine organ and photoreceptive eye, a role retained in many living non-mammalian vertebrates (many fishes, amphibians, reptiles, and perhaps birds). The photo-neuroendocrine pinealocytes secrete melatonin, which signals the daily light cycle (high circulating levels at night and lower levels during the day). In mammals, the photoreceptive pinealocytes are modified as endocrine-secreting cells but retain vestigial signs of their photoreceptive role.

A variety of 'unconventional' endocrine tissues include the heart, blood vessels, adipose tissue, mammary glands, and the placenta (Bentley 1998). The atrial muscle of the heart synthesizes and releases atrial naturetic peptide (ANP) in response to stretch caused by an increase in central venous pressure; ANP then acts on the DCT of the kidney and increases $Na^+$ excretion. Adipose tissue synthesizes and releases leptin in proportion to body adiposity; leptin inhibits appetite and increases fat metabolism. The stomach releases ghrelin when it is empty, and ghrelin stimulates appetite in the CNS. Other reproductive-related hormones include a variety (e.g. oestrogens, GH, PTH, PRO) from the mammary glands, and those related to maintaining pregnancy (e.g. oestrogens, progestogens, gonadotropins, lactogens) from the placenta.

In vertebrates, the thymus develops from pharyngeal pouches, and in mammals the thymus is derived from the third, and to a lesser extent the fourth, pouches (Liem et al. 2001). The epithelial thymus bud aggregates stem cells from the spleen and bone marrow, which differentiate into T lymphocytes (T cells) that are released and circulate in the blood and populate lymph nodes and other lymphoid tissues. These T cells are regulated by the thymus hormone thymopoetin. T cells are involved with immune responses, both cell-mediated, where T cells directly attack infected cells (by viruses, fungi, and other foreign cells), and responses involving lymphocytes.

## 2.3 Energy Balance

An animal's energy budget is an integration of its energy gain, storage, and loss. Like all animals, mammals are heterotrophs: they are unable to directly convert the sun's energy into potential chemical energy, but utilize the chemical energy stored in food, obtained by eating and drinking. Energy can be stored by tissue growth or production, and energy is lost in faeces, urine, external work (e.g. moving substrate, overcoming drag, etc.), and as metabolic heat (Withers 1992). As well as energy, heterotrophs consume essential compounds such as proteins and amino acids, minerals and trace elements, and vitamins (Schmidt-Nielsen 1997).

Energy-consuming processes in the body do not directly use the energy stored in the food, but generally use the energy stored in the terminal phosphate bond of adenosine triphosphate (ATP). The chemical energy acquired in food is converted to ATP by cellular metabolism, a complex series of chemical reactions. The hydrolysis of the terminal phosphate linkage on ATP to form adenosine diphosphate (ADP) and inorganic phosphate ($P_i$) releases about 36.8 kJ mole$^{-1}$, which can be used to power cellular activities (e.g. muscle contraction, ion pumping, glandular secretion, biosynthesis of other macromolecules). The hydrolysis of ADP to adenosine monophosphate (AMP) and $P_i$ releases a further 36.0 kJ mole$^{-1}$, but the hydrolysis of AMP to adenosine and $P_i$ yields only about 12.6 kJ mole$^{-1}$ (McGilvery & Goldstein 1983). Anaerobic metabolism is cellular metabolism and ATP synthesis in the absence of oxygen, whereas cellular metabolism involving oxygen is termed aerobic metabolism. The latter is the most important metabolic process in mammals. Aerobic metabolism requires $O_2$ as well as the biochemical substrates for metabolism, but it results in the greatest conversion of chemical energy to ATP. Carbohydrates, lipids, and proteins can all be metabolized by aerobic metabolism, although carbohydrates and lipids are the most common metabolic substrates.

## 2.3.1 Anaerobic Metabolism

Glycolysis is the first step in the process of both aerobic and anaerobic metabolism (McGilvery & Goldstein 1983); it is an ancient anaerobic metabolic pathway, occurring in almost every living cell. The series of reactions that form the glycolysis pathway occur in the cell cytoplasm, and involve the conversion of glucose to pyruvate, with a net yield of ATP (Figure 2.7). Glucose is a six-carbon ($C_6$) monosaccharide, which can be derived from most dietary and some stored carbohydrates, and can be synthesized from amino acids by gluconeogenesis in the liver. Nine enzyme-mediated reactions, the Embden-Meyerhof pathway, convert one glucose molecule into two pyruvate molecules. One ATP is required to convert glucose to glucose 6-phosphate (which is then converted to fructose 6-phosphate) and another ATP is used to convert fructose 6-phosphate to fructose 1,6-bisphosphate; these energy-requiring reactions have a large positive free-energy change ($\Delta G^{\circ\prime}$), and ATP supplies that energy. Fructose 1,6-bisphosphate is then split into two interconvertible $C_3$ compounds. The conversion of each of these $C_3$ molecules (as glyceraldehyde 3-phosphate) to 1,3-bisphospho-glycerate frees two hydrogen atoms (H), one of which is transferred to the electron receptor nicotinamide adenine

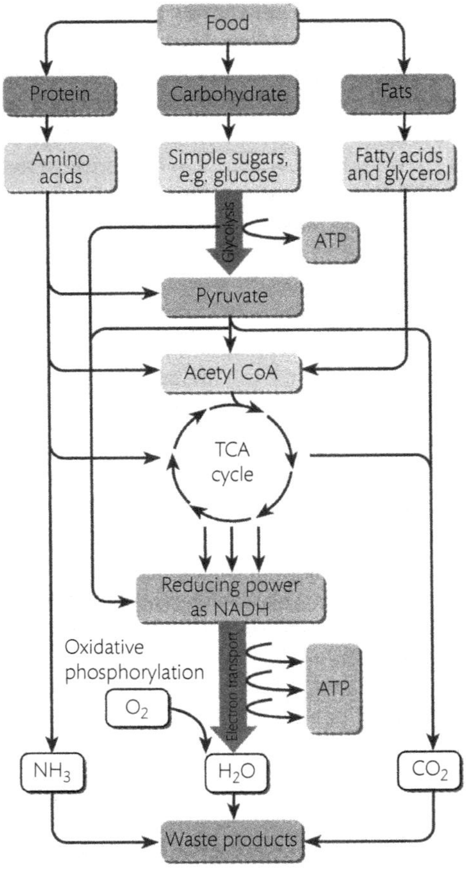


**Figure 2.7** Schematic of the breakdown of ingested food into protein, carbohydrate, and lipids, and their subsequent aerobic metabolism via glycolysis, the TCA cycle, and the electron transport system to produce ATP and waste products ($CO_2$, $H_2O$, and $NH_3$). From Pocock et al. (2013) Human Physiology. Reproduced with permission from Oxford University Press.

dinucleotide ($NAD^+$) with two electrons ($e^-$); the other enters the cellular proton pool as $H^+$; i.e. two $NAD^+$ and four H form two NADH and two $H^+$ per glucose. The subsequent conversions of 1,3-bisphosphoglycerate to 3-phosphoglycerate, then phospho-enol-pyruvate to pyruvate, have negative $\Delta G^{o\prime}$ and each reaction produces an ATP; i.e. a total of two ATP per $C_3$ or four ATP per glucose.

Glycolysis therefore has a net production of two ATP, two NADH, and two $H^+$ per glucose, and forms two pyruvate molecules. The ATP produced at this stage stores about 61 kJ $mole^{-1}$ of chemical energy, which is only about 6% of that potentially available by the complete oxidation of glucose (Withers 1992; Sherwood et al. 2005; Guppy 2007). The $NADH/H^+$ has an important role in the mitochondrial electron transport chain (see 2.3.2), where NADH is oxidized to $NAD^+$, which diffuses back to the cytoplasm, allowing glycolysis to continue. If $O_2$ is available, then the two pyruvates can enter the mitochondria, where further ATP synthesis via the tricarboxylic acid (TCA) cycle and the mitochondrial electron transport chain occurs (see 2.3.2). In the absence of oxygen—such as during periods of environmental hypoxia or intense exercise, when the circulatory and respiratory systems can't match $O_2$ supply with demand—the reactions of glycolysis are limited by the supply of $NAD^+$, but NADH can be oxidized to $NAD^+$ when pyruvate is converted to lactate (lactic acid), catalyzed by the enzyme lactate dehydrogenase, allowing glycolysis to continue.

$$\textbf{pyruvate} + \textbf{NADH / H}^+ \rightarrow \textbf{lactate} + \textbf{NAD}^+$$

Thus, in vertebrate animals, if oxygen is unavailable, anaerobic metabolism is based on glycolysis, even though it is relatively inefficient (two ATP per glucose). During anaerobic metabolism, lactate gradually accumulates in the tissues, which results in end-product inhibition of glycolysis and reduces the pH, leading to muscle fatigue.

Sometimes other substrates can be produced from pyruvate in anaerobic metabolism. For many invertebrates and some fish, a variety of other anaerobic end-products are produced (e.g. succinate, propionate, ethanol), and anaerobic metabolism can sustain life for considerable periods of environmental anoxia. However, endothermic mammals and birds can only use anaerobic metabolism to sustain their high MRs for short periods; i.e. during intense exercise or transient oxygen restriction, such as diving (section 4.5.1).

Once oxygen becomes available after transient hypoxia, the lactate is removed from tissues via the blood supply, and is primarily converted back to pyruvate. The pyruvate can then enter the TCA cycle and undergo the final two steps of aerobic metabolism, releasing the remainder of the 36 ATP held in its chemical bonds (see 2.3.2). Alternatively, lactate can be converted to glucose (e.g. in the liver or skeletal muscle) via gluconeogenesis (the formation of 'new' glucose from non-carbohydrate precursors, including lactate; McGilvery & Goldstein 1983). Glycolysis is essentially an irreversible series of reactions, so the reconversion of lactate to glucose requires some additional reactions and the input of energy, particularly to 'reverse' the phosphorylation of glucose to glucose 6-phosphate. In theory, aerobic metabolism of about 17% of the lactate provides sufficient energy to convert the remaining 83% to glucose (Withers et al. 1988). In practice, ectotherms with low MR, such as frogs, aerobically metabolize much less of the lactate than this (< 10%). In contrast, endothermic mammals aerobically metabolize higher fractions of accumulated lactate (20–90%; Gleeson 1996).

## 2.3.2 Aerobic Metabolism

If sufficient $O_2$ is available in cells, then aerobic metabolism is used to further metabolize the pyruvate produced by glycolysis (Withers 1992; Sherwood et al. 2005). The further biochemical reactions required for aerobic metabolism occur primarily in the mitochondria, via the tri-carboxylic acid (TCA) cycle (or Krebs or citric acid cycle) and the electron transport chain (Figure 2.7). Oxygen is required as the final electron acceptor in the electron transport chain that regenerates $NAD^+$ from $NADH/H^+$ (and FAD from $FADH_2$) and allows glycolysis and the Krebs cycle to continue.

In the mitochondria, two carbons of pyruvate are condensed with coenzyme A (CoA) to form acetyl CoA, and the other is freed as a carbon dioxide molecule, with one $NAD^+$ being converted to $NADH / H^+$. Acetyl CoA then enters a seven-step cycle, which has three roles: (1) oxidizing acetyl CoA and the four-carbon compound oxaloacetate to form citrate and storing the chemical energy as reduced electron carriers ($NADH / H^+$and $FADH_2$), (2) converting the acetyl component of acetyl CoA to carbon dioxide (an end-product that does not accumulate and is not rate-limiting), and (3) reforming oxaloacetate that then begins the next cycle (Withers 1992; Guppy 2007). At three stages of the cycle, an $NADH / H^+$ is formed from $NAD^+$, accepting two protons, and at one stage another proton accepter, flavin adenine dinucleotide (FAD), accepts protons and forms $FADH_2$. When oxalosuccinate is converted to $\alpha$-ketoglutarate, $CO_2$ is produced and, with the $CO_2$ formed during the condensation of pyruvate with CoA, diffuses out of the mitochondria, to be eliminated via the circulatory and gas exchange systems. A single guanosine triphosphate (GTP) is formed by phosphorylation from guanosine diphosphate (GDP) during the conversion of succinyl-CoA to succinate and CoA; GTP's high-energy phosphate bond is then used to form an ATP. Since two molecules of pyruvate are produced from each glucose molecule, the cycle occurs twice for every glucose metabolized, resulting in four $CO_2$, eight $NADH/H^+$, two $FADH_2$, and two ATP.

The final stage of aerobic metabolism is the electron transport chain. Here, $NAD^+$ and FAD are regenerated from $NADH/H^+$ and $FADH_2$, providing ATP and allowing glycolysis and the TCA cycle to continue. This regeneration process accounts for almost 90% of the ATP produced during aerobic metabolism. The electron transport chain consists of a series of electron carrier molecules situated on the inner mitochondrial membrane that accept and donate electrons. These carriers are organized such that each successive carrier has a higher redox potential than the previous carrier. Electrons from $NADH/H^+$ and $FADH_2$ therefore move through this series of electron carriers, with energy released at each transfer. The final electron acceptor, with the highest electron affinity, is oxygen derived from the environment. Oxygen acquires electrons, becomes negatively charged, and then binds to positively charged hydrogen ions (which originally donated electrons to the transport chain), forming water. The energy that results from electrons passed to carriers with successively greater redox potentials is used to transport $H^+$ ions across the mitochondrial membrane, resulting in the accumulation of more $H^+$ on one side of the inner membrane than the other. Since the membrane is relatively impermeable to $H^+$, the accumulation generates a membrane potential. That $H^+$ electrochemical potential is used to generate ATP. A protein channel in the inner mitochondrial membrane, called ATP synthetase, allows $H^+$ to diffuse down its electrochemical gradient, and in the process phosphorylates ADP to ATP (McGilvery & Goldstein 1983; Guppy 2007). The electron transport system generates a total of about 34 ATP for each glucose metabolized.

In total, the biochemical reactions of aerobic metabolism are used to harness 11,641 kJ mole$^{-1}$ of energy to usable chemical energy (about 38 ATP) per glucose, which is 41% of the total of 2854 kJ mole$^{-1}$ of free energy that is available from the complete oxidation of glucose to $CO_2$ and $H_2O$. The remaining 59% of the chemical energy is lost as heat. The overall reaction can be summarized as follows:

$$C_6H_{12}O_6 + 6O_2 \rightarrow 6CO_2 + 6H_2O + 38\,ATP$$

Substrates other than glucose, such as lipids and proteins, may be metabolized instead of glucose. Lipid metabolism has a conversion efficiency of around 60%, and protein metabolism about 30%.

The stoichiometry of $CO_2$ released by animal metabolism to $O_2$ consumed (moles $CO_2$/moles $O_2$) is termed the respiratory quotient, RQ (see 2.3.3). The ratio of the rate of $CO_2$ excreted by an animal to rate of $O_2$ uptake ($VCO_2/VO_2$) is termed the respiratory exchange ratio (RER). The RER reflects the RQ, but can transiently differ from it by internal 'buffering' of the $O_2$ consumed (e.g. use or replenishment of $O_2$ stores) or especially via the 'buffering' of the $CO_2$ produced (e.g. $pCO_2$ or temperature changes, $HCO_3^-/pH$ changes).

### 2.3.3 Joule Equivalents of Food

Most biological materials have an energy content of between 17 and 30 kJ g$^{-1}$ of ash-free dry mass (Table 2.4). Energy contents are given as ash-free dry mass to account for the large and variable water content and variable inorganic content of different materials. The energy content of a particular material will depend on its relative composition of carbohydrate, lipid, and protein; carbohydrates contain about 17 kJ g$^{-1}$, lipids about 38 kJ g$^{-1}$, and proteins about 23 kJ g$^{-1}$. The determination of the total energy content of ingested food and the energy content of faeces and urine that are produced enables the indirect calculation of an individual's MR using an energy balance model: MR = energy intake − energy in urine and faeces (Withers 1992; Schmidt-Nielsen 1997).

The energy content of various foods can be determined by combustion (bomb calorimetry), and varies considerably between foodstuffs, reflecting their differing biochemical composition (i.e. different proportions of carbohydrate, lipid, and protein), and water and ash contents (Table 2.4). Water content has a major effect on energy content. For example, jellyfish have a very low energy content per wet mass ($0.1 - 0.2\,kJ\,g_{wet}^{-1}$), reflecting their very high water content (96.1–96.5%). Their energy content per dry mass is much higher ($2.4 - 4.2\,kJ\,g_{dry}^{-1}$). Jellyfish also have a very high ash content (77.8% of dry mass) so their ash-free energy content is much higher, at about $18.2\,kJ\,g_{ash\text{-}free,\,dry}^{-1}$ (Doyle et al. 2007).

**Table 2.4** Average energy content (kJ $g^{-1}$ ash-free, dry) for various organisms, and biochemical constituents of organisms. Values from Withers and Dickman (1995), Kemp (1999), Cooper et al. (2002), Cooper and Withers (2004), and Doyle et al. (2007).

| Organism | Energy content kJ $g^{-1}$ | Biochemical | Energy content kJ $g^{-1}$ |
|---|---|---|---|
| Cod | 17.1 | Glycine | 13.0 |
| Soft corals (30–50% ash) | 17.3 | Glucose | 15.5 |
| Gastropod tissues | 17.7 | Saccharose | 16.5 |
| Lichens | 18.1 | Starch | 17.5 |
| Jellyfish | 18.2 | Cellulose | 17.6 |
| Mosses | 18.5 | Protein | 23.0 |
| Herbs | 19.3 | Leucine | 27.2 |
| Squid | 20.7 | Ethanol | 29.7 |
| Spiders | 20.9 | Tripalmitin | 38.9 |
| *Hakea* seeds | 21.1 | Fat | 38.9 |
| Herring | 21.3 | Palmitic acid | 39.3 |
| Sloughed snake skin | 21.5 | | |
| Arthropods (average) | 21.7 | | |
| *Eucalyptus* seeds | 21.7 | | |
| *Banksia* seeds | 22 | | |
| Yeasts | 22.3 | | |
| Chrysomelid beetle | 22.6 | | |
| Bacteria | 22.7 | | |
| Marine benthic community | 22.7 | | |
| Termites | 22.9 | | |
| Polychaete annelids | 23 | | |
| Fungi | 23.5 | | |
| Pollen–insect-dispersed, dicots | 24.2 | | |
| Holothurian echinoderms | 25 | | |
| *Pinus* seeds | 25.4 | | |
| Freshwater copepods | 27.9 | | |
| Gecko–regenerated tail | 28.0 | | |
| *Crematogaster* ants | 29.2 | | |
| Wax moth–imago | 37.5 | | |

The energy content of the biochemical constituents of food (primarily carbohydrates, lipids, and proteins) varies considerably, as does the exact chemical stoichiometry of combustion (Table 2.5). Glucose, for example, yields 15.9 kJ $g^{-1}$ when completely combusted to $CO_2$ and $H_2O$. Protein has a slightly higher energy content and lipids have a much higher energy content.

When glucose is metabolized, RQ is 1 and 21.4 kJ is released for every litre of oxygen consumed and every litre of carbon dioxide produced. For most lipids the RQ is closer to 0.7, and the joule equivalents are $19.5\,kJ\,l\,O_2^{-1}$ and $27.9\,kJ\,l\,CO_2^{-1}$. Protein has a considerable nitrogen (N) and sulphur (S) content, making the stoichiometry of protein metabolism more complicated than for glucose or lipids. The final form of the nitrogenous waste product complicates the issue, because energy is required to convert ammonia to urea or uric acid (see 3.6.4). For mammals, which are predominantly ureotelic, protein metabolism is typically characterized by an RQ of 0.84, and a joule equivalent of $17.6\,kJ\,l\,O_2^{-1}$ (Withers 1992; Schmidt-Nielsen 1997). The metabolism of glucose produces more metabolic water per kJ of energy than the other substrates, but the physiological consequences for water balance are complex (see 2.7.1).

Mammals rarely metabolize pure carbohydrate, lipid, or protein, so the RQ—and hence joule equivalents for metabolism and metabolic water production from metabolism—depend on the ratio of carbohydrate:lipid:protein that is being metabolized. Carbohydrate and lipids are more routinely metabolized than protein, so the typical range for joule equivalents is $19.5 - 21.4\,J\,ml\,O_2^{-1}$ and $0.565 - 0.663\,mg\,H_2O\,ml\,O_2^{-1}$; interpolated values for J equivalents and metabolic water production are given in section 5.3.1.

**Table 2.5** Energy content and chemical stoichiometry for major body biochemicals, by complete combustion. Adapted from Withers (1992).

| | Glucose | Glucosyl unit | Lipid | Protein |
|---|---|---|---|---|
| Energy content kJ $g^{-1}$ | 15.9 | 17.7 | 39.2 | 20.1 |
| kJ $mole^{-1}$ | 2,870 | 2,870 | 10,042 | 64,400 |
| kJ l $O_2^{-1}$ | 21.4 | 21.4 | 19.5 | 18.8 |
| l $O_2$ $g^{-1}$ | 0.75 | 0.83 | 2.01 | 1.07 |
| l $CO_2$ $g^{-1}$ | 0.75 | 0.83 | 1.40 | 0.91 |
| RQ l $CO_2$ l $O_2^{-1}$ | 1.0 | 1.0 | 0.70 | 0.84 |
| Metabolic water g $H_2O$ $kJ^{-1}$ | 0.038 | 0.031 | 0.029 | 0.021 |
| Mole urea $mole^{-1}$ | 0 | 0 | 0 | 19.5 |
| kJ g $N^{-1}$ | 0 | 0 | 0 | 126 |

## 2.4 Thermal Balance

Temperature is arguably the most pervasive environmental variable that influences an animal, due to its effects on macromolecular structure and biochemical reactions, which have striking effects on the rates of physiological processes. The $T_b$ of an animal is determined by a complex interaction of the various avenues for heat exchange with its environment (conduction, convection, radiation, evaporation) as well as its own internal heat production, which is generally negligible for ectothermic animals but is substantial for endotherms such as mammals and birds.

### 2.4.1 Temperature

Temperature is a measure of the thermal kinetic energy of molecules. For an ideal gas, the relationship between the mean kinetic energy (E; J mole$^{-1}$) and absolute temperature (T, °K) is $E = 0.5\,mv^2 = 1.5\,kT$, where m = molecular weight, v = mean molecular velocity (m s$^{-1}$) and k = the Boltzmann constant (1.381 $10^{23}$ J molecule$^{-1}$ °K$^{-1}$). The rate of physical, chemical, and biochemical reactions increases with temperature, because at higher temperatures the mean kinetic energy of molecules increases. Reactions occur when reactant molecules with energy higher than the activation energy collide with one another. When the kinetic energy increases, so does the probability that such reactant molecules will collide and therefore react (Withers 1992). The effect of temperature on the rate of a physiological process or reaction can be described by the $Q_{10}$, which quantifies the effect of a 10°C change in temperature on a reaction rate (biochemists use an Arrhenius plot and activation energy to quantify the effect of temperature on reaction rates). The $Q_{10}$ for physical processes such as diffusion is around 1, while the $Q_{10}$ is typically 2 to 3 for biochemical reactions and most physiological processes (e.g. metabolism; Withers 1992; Schmidt-Nielsen 1997; Guppy & Withers 1999). The $Q_{10}$ can be calculated for any two temperatures (not necessarily 10°C apart), as $Q_{10} = (R_2 / R_1)^{(10/(T_2-T_1))}$ where $R_1$ and $R_2$ are the reaction rates at temperatures $T_1$ and $T_2$ respectively. The $Q_{10}$ function can be rearranged to calculate the increased reaction rate, as $R_2 = R_1\, Q_{10}{}^{(((T_2-T_1)/10)}$.

### 2.4.2 Thermal Exchange

The thermal environment of an animal is complex, especially for terrestrial species. Animals are unable to avoid thermal exchange with their environment, but they can manipulate the avenues of heat loss and gain to achieve effective heat balance, and thus thermoregulation. The four avenues of heat exchange are conduction, convection, radiation, and the evaporation/condensation of water.

Conduction is the exchange of heat energy between two solid objects in direct physical contact, by kinetic energy transfer between adjacent molecules. Heat exchange between a solid object and a stationary fluid is also conduction, but if the fluid is moving, heat exchange is by convection. The rate of conductive heat flow ($Q_{cond}$) is described by Fourier's law of heat transfer, $Q_{cond} = k\,A\,(T_2 - T_1)\,/\,L$, where k is the thermal conductivity, A is the surface area of contact ($cm^2$), $T_2$–$T_1$ is the temperature difference (°C), and L is the distance (cm) between $T_2$ and $T_1$ (McNab 2002). Thermal conductivity is a material property: it is 0 $J\ s^{-1}\ m^{-1}\ °C^{-1}$ for a vacuum (since there are no molecules to transfer kinetic energy), and varies from 0.024 for still air to several hundred for metals (Withers 1992). The terms (k A/L) are sometimes combined into a conductive heat transfer coefficient ($h_{cond}$).

Conduction is often considered the least important mechanism of thermal exchange, at least for active mammals, because only a small area (the soles of the feet) is usually in direct contact with the ground. For example, the hoof surface area of muskoxen (*Ovibos moschatus*) is only 1.1–1.2% of their total body surface area, and therefore even when standing on ice, the conductive heat loss accounts for only 6–10% of total heat loss (Munn et al. 2009a). However, resting or burrowing mammals can have a much greater proportion of their surface in direct contact with their substrate and are subjected to greater conductive heat exchange (McNab 2002). For example, for sheep sitting on cold, poorly insulated ground, the loss of heat by conduction can approach 30% of their heat production (Gatenby 1977).

Convection is the transfer of heat due to the movement of a fluid (gas or liquid), especially at the interface with a solid object (Incropera & de Witt 1981). Convection can be free or forced. Free convection occurs when the fluid surrounding an animal gains heat, resulting in a change in the density of the fluid; the warmer fluid rises. Forced convection results from an external force, such as wind, water currents, or animal movement (Bicudo et al. 2010). A boundary layer develops at the surface of the object, such that there is a gradient in fluid velocity and temperature extending from the surface of the object (where fluid temperature is the same as the object, and velocity is zero) to the free stream (where temperature and velocity equal that of the free stream). Conceptually, convection is conductive heat transfer across the average boundary layer thickness. Convective heat transfer ($Q_{conv}$) is dependent on the thermal conductivity (k), the area exposed to fluid movement (A), the temperature differential ($T_b$–$T_a$), and the average boundary layer thickness ($\delta$); $Q_{conv} = k\,A\,(T_b - T_a)\,/\,\delta$. The convective heat transfer coefficient ($h_c = k\,A\,/\,\delta$; $J\,cm^{-2}h^{-1}°C^{-1}$) is a complex variable that depends on factors such as the characteristics of the fluid, including density, viscosity, and velocity, and the dimensions and shape of the solid object. While it can be predicted on the basis of theory, it is best determined empirically (Withers 1992).

Convection has important consequences for heat exchange of mammals. For example, when the air temperature is lower than the skin temperature, an increase in the wind speed results in an increase in convective heat loss and requires a thermoregulatory response (increased heat production) (Walsberg & Wolf 1995a). The fur of mammals (like the feathers of birds) reduces heat loss by trapping a layer of still air close to the skin (and still air is the best insulative material available to a mammal). Denser, thicker pelts trap a thicker layer of still air within the pelt and therefore provide better insulation to heat loss (or gain; Scholander et al. 1950b). Free and forced convection within the fur's air layer increase heat exchange, and high wind speeds further disturb the air trapped within the fur and increase heat exchange. For example, increasing the wind speed from 0.5 to 3 m $s^{-1}$ progressively decreased the pelt insulation (boundary layer, pelt, and total resistance) of numbats (*Myrmecobius fasciatus*) (Cooper et al. 2003a; Figure 2.8). For aquatic mammals, the high thermal conductivity of water (5.9 $10^{-1}$ J $s^{-1}$ $m^{-1}$ °$C^{-1}$) compared to air (2.4 $10^{-2}$ J $s^{-1}$ $m^{-1}$ °$C^{-1}$) compromises their capacity to maintain a considerable ($T_b - T_{water}$) temperature difference, but large size and subcutaneous blubber (which is effective insulation underwater) facilitates maintaining a high $T_b - T_{water}$ in water.

Radiative heat exchange occurs between objects via electromagnetic waves (photons; McNab 2002). All objects with a surface temperature ($T_{surf}$) higher than absolute zero (0°K) emit radiation, with the predominant wavelength ($\lambda_{max}$; nm) being inversely related to $T_{surf}$. Wien's law quantifies this relationship as $\lambda_{max} = 2.898\ 10^6 / T_{surf}$. At a biologically relevant temperature of 300°K (27°C), $\lambda_{max} = 9{,}700\,\text{nm}$, which is in the infrared region of the spectrum; for the surface of the sun (5,800°K), $\lambda_{max} = 500\,\text{nm}$, which is in the visible region of the spectrum. The radiant energy emitted by an object is $Q_{rad} = \sigma\ \varepsilon\ A\ T_{surf}^4$, where $\sigma$ is the Stefan-Boltzmann constant ($2.04 \times 10^{-8}\,\text{J}\,\text{cm}^{-2}\text{h}^{-1}\,{}^{\circ}\text{K}^{-4}$) and $\varepsilon$ is the emissivity, which ranges from 0 (no energy is radiated) to 1 (the maximum possible radiation, for a 'black body'). At a given wavelength, the emissivity is equal to absorbance (1 - reflectance). Most biological materials have an $\varepsilon$ of 0.90–0.95 in the infrared region of the spectrum. Mammals and other biological materials and objects at biologically relevant temperatures emit primarily long wavelength

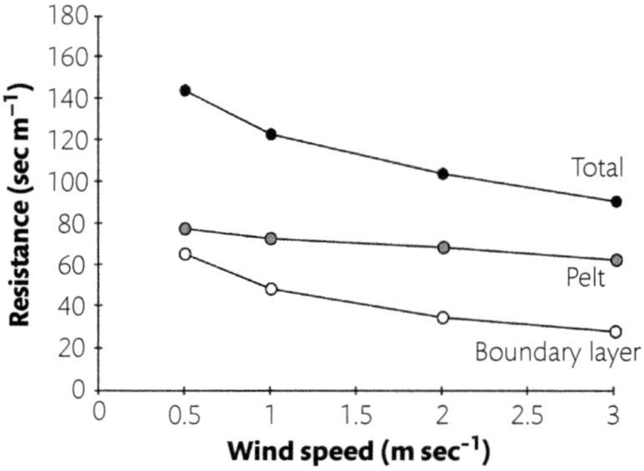


**Figure 2.8** Thermal resistance of the pelt of numbats (*Myrmecobius fasciatus*) as a function of wind speed, showing the environmental boundary layer resistance with the fur erect (modified from Cooper et al. 2003).

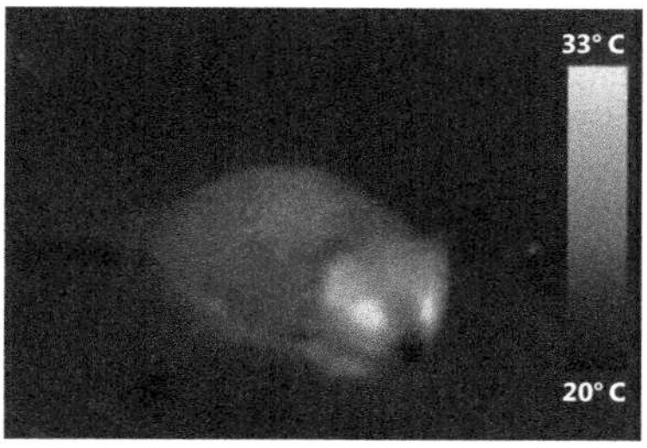


**Figure 2.9** Infrared image showing surface temperatures for a mulgara (*Dasyuroides byrnei*); inset shows temperature scale (°C). Photograph courtesy of P. Withers & C. Cooper.

infrared radiation (Figure 2.9) and relatively little energy (300–500 J $s^{-1}$ $m^{-2}$), in comparison to the sun, which has a very high $T_{surf}$ and emits primarily shorter wavelength radiation and much more energy ( 6.1 $10^7$ J $s^{-1}m^{-2}$ ; Withers 1992).

Animals not only emit radiation but they also absorb radiation that is emitted by the objects in their environment. Consequently, they have a net radiative heat exchange ($Q_{rad,\ net}$) reflecting the balance between radiative heat gain and loss: $Q_{rad,net} = \varepsilon_2\ \sigma\ A_2\ T_{s2}^{\ 4} - \varepsilon_1\ \sigma\ A_1\ T_{s1}^{\ 4}$, where $\varepsilon_1$ and $\varepsilon_2$ are the mean emissivities over a range of wavelengths of the mammal's average surface ($T_{s2}$) and surroundings ($T_{s1}$). Solar radiation is an important source for heat gain, and hence thermoregulation, for terrestrial ectotherms. Mammals can behaviourally exploit radiative heat sources to reduce the energy cost of maintaining an elevated body temperature (Walsberg et al. 1997). For example, normothermic fat-tailed dunnarts (*Sminthopsis crassicaudata*) reduced their MR by 74% at a $T_a$ of 15°C by basking under a heat lamp, and torpid dunnarts were able to passively rewarm from $T_b$ = 20°C to $T_b$ = 35°C without the dramatic increase in MR characteristic of endogenous arousal in the absence of basking (Warnecke & Geiser 2010).

Condensation of water vapour can occur when skin temperature is lower than the dew point temperature (such as walking from an air-conditioned room into a hot and humid tropical environment), but is generally not an avenue whereby non-human animals exchange (gain) heat, so condensation is not considered further. The evaporation of water dissipates a significant amount of heat. The latent heat of vaporization is 2.5 kJ $g^{-1}$ $H_2O$ for water at $T_a$ = 0°C (2.4 kJ $g^{-1}$ $H_2O$ at $T_a$ = 40°C), which is about five times as much energy as is required to heat water from $T_a$ = 0°C to boiling (Schmidt-Nielsen 1997). The evaporation of body water is therefore potentially a major avenue of heat loss for terrestrial mammals (but not aquatic species). Heat loss by evaporation ($Q_{evap}$) depends on the difference in water vapour density between the animal ($\rho_b$) and the air ($\rho_a$), the effective surface area over which evaporation occurs (A), and the resistance to water loss from the animal's surface (r), as described by the equation $Q_{evap} = A(\rho_b - \rho_a) / r$

(Monteith & Campbell 1980; Withers 1992). Generally it is assumed that the air at the animal's skin surface is saturated at $T_{skin}$, in which case $\rho_b$ may be calculated as $\rho_b = (9.16\ 10^8)\,e^{-(5,218/Tskin)}$, where the constant 5,218 is the molar heat of vaporization divided from the universal gas constant (McNab 2002). The rate of evaporation will depend on air movement. In still air, the boundary layer is thicker than in moving air, and the rate of evaporative water loss (and heat loss) is lower.

Evaporative heat loss (EHL) is particularly important when $T_a$ approaches $T_b$, and especially when $T_a$ exceeds $T_b$, since evaporation is the only mechanism by which mammals can dissipate metabolic heat production (MHP) once their convective/conductive/radiative heat exchange is no longer negative. Evaporation can occur from a mammal's body surface as cutaneous evaporative water loss ($EWL_{cut}$) and from the respiratory passages as respiratory evaporative water loss ($EWL_{resp}$). Some mammals spread saliva, or sweat, as mechanisms to increase $EWL_{cut}$ and augment EHL at high $T_a$; some pant to increase $EWL_{resp}$ and augment EHL; some use various combinations of these strategies. For example, long-nosed potoroos (*Potorous tridactylus*) can dissipate more than 200% of their MHP via panting and sweating, and thus maintain $T_b$ well below $T_a$, at $T_a$ up to 40°C (Hudson & Dawson 1975). Eastern grey kangaroos (*Macropus giganteus*) and red kangaroos (*Macropus rufus*) can dissipate, respectively, 160 and 173% of their MHP via EHL at $T_a$ >43°C; about 60–70% of this EHL was achieved by panting, and the remainder by skin licking (Dawson et al. 2001).

## 2.4.3 Body Temperature Regulation

Endotherms (mammals, birds, and a few reptiles and insects) use endogenous metabolic heat production to maintain $T_b$ relatively independent of ambient temperature ($T_a$). For mammals, the resting $T_b$ varies considerably for different taxa, from about 31 to 39°C (Lovegrove 2012a,b; see also Chapters 1 and 3), and this contributes along with mass to much of the variation in BMR (Clarke et al. 2010). Many mammalian species (e.g. apes, large macropods) are strict homeotherms: their $T_b$ is relatively constant over a wide range of $T_a$. However, the maintenance of homeothermy creates a considerable energetic demand, so many mammals (e.g. some rodents, micro-bats, dasyurid marsupials) use heterothermy (i.e. torpor or hibernation) to reduce their energetic requirements (see 3.2.4.2, 4.1.2).

The principle of thermoregulation involves physiological feedback mechanisms that facilitate the maintenance of a relatively constant core body temperature despite changes in environmental temperature. Temperature and heat are related via the heat capacity of body tissues (4.2 J $g^{-1}$ °$C^{-1}$ for water, about 3.5 J $g^{-1}$ °$C^{-1}$ for animal tissue; Jessen 2001a), and an imbalance between heat gain and heat loss results in a quantifiable change in $T_b$. Homeothermic animals have the capacity to control (to some extent) both heat gain and heat loss.

Ultimately, energy that is processed by an animal ends up either stored as body tissue, as external (physical) work, or as heat. That heat can contribute to the maintenance of $T_b$. When an animal is exposed to an air temperature that is lower than its $T_b$, and in the absence of an external heat source (such as solar radiation or hot substrate), the animal will lose heat constantly by conduction, convection, radiation, and evaporation (Jessen 2001a). A constant $T_b$ is achieved when an endotherm's rate of metabolic heat production is altered to balance the rate of heat loss, by literally generating heat (thermogenesis). In addition to the heat production associated with the basal or RMR, an endotherm increases heat production by shivering or non-shivering thermogenesis when it is required (section 3.2.3).

The range of environmental temperatures over which a homeotherm can maintain $T_b$ can be divided into three 'zones' (Figure 2.10). The thermoneutral zone (TNZ) is that range of ambient temperatures where heat balance can be achieved at basal (or resting) metabolism and evaporative heat loss is minimal (IUPS Thermal Commission 2003). For a small mammal, the TNZ can be reduced to a thermoneutral point. Below the TNZ, heat loss exceeds basal heat production, so to maintain $T_b$ a mammal must increase heat production by either shivering or non-shivering thermogenesis (cold challenge; see 3.2.3). The lower end of the TNZ is

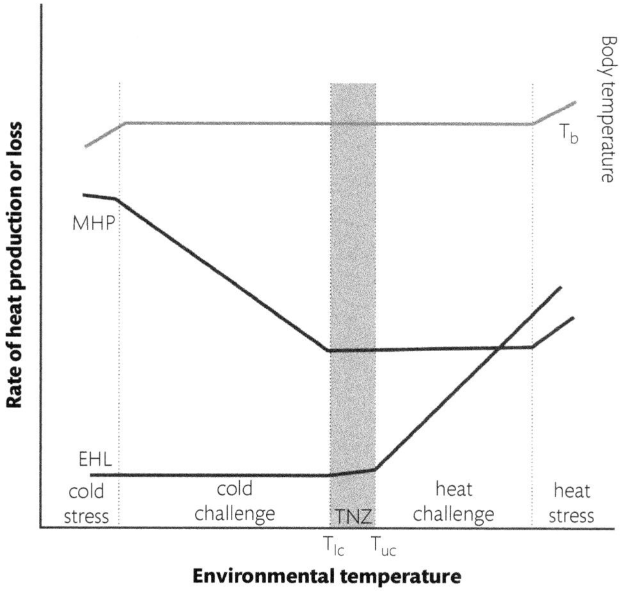


**Figure 2.10** Heat balance diagram showing the schematic relationship between metabolic heat production (MHP), evaporative heat loss (EHL), and body temperature ($T_b$) for an endothermic mammal. TNZ: thermoneutral zone; $T_{lc}$: lower critical temperature; $T_{uc}$: upper critical temperature.

known as the lower critical temperature ($T_{lc}$, °C). Above the TNZ ($T_a > T_{uc}$), heat loss by conduction, convection, and radiation is insufficient to balance metabolic heat production, and a mammal must increase evaporative heat loss to maintain $T_b$ (heat challenge). Within the TNZ, heat balance is achieved by non-metabolic changes in insulation that can occur by changes in skin blood flow (altering heat delivery to the skin), changes in the fur such as piloerection, or changes in posture that open or close heat windows and alter the exposed surface area.

The process of afferent temperature reception and efferent thermal effector signalling is controlled by the hypothalamus. The 'regulated variable' $T_b$ is controlled by a combination of central and peripheral inputs from thermally sensitive neurons throughout the body, although the highest density of thermosensitive neurons is in the hypothalamus. 'Cold sensitive' sensory neurons in the skin relay information to the preoptic area of the hypothalamus via the external lateral subnucleus of the lateral parabrachial nucleus. 'Warm sensitive' sensory neurons in the skin project to third-order sensory neurons and relay via the dorsal subnucleus of the lateral parabrachial nucleus (Figure 2.11; Morrison & Nakamura 2011). Each thermal effector has its own control system (Romanovsky 2007). Three primary control pathways that drive thermogenesis in the cold originate in the medial preoptic subnucleus. Two of these relay via the dorsomedial hypothalamus and via the rostral raphe pallidus to $\alpha$-motor and $\gamma$–intrafusal muscle fibres, and (at least in placental mammals) brown adipose tissue (BAT); the other relays via the rostral raphe nucleus that, when activated, causes vasoconstriction of cutaneous blood vessels.

In the laboratory rat, there are four separate control pathways for responses to the cold—constriction of tail blood vessels, constriction of back blood vessels, activation of BAT, and activation of fusimotor fibres in skeletal muscles that promote shivering—that vary in their responsiveness to core temperature compared to cutaneous temperature (McAllen et al. 2010). The pathway controlling sweat glands in a warm environment presumably involves cutaneous warm thermal receptors and spinal sympathetic preganglionic neurons. Salivation is complex, involving both increased secretion of saliva via salivatory premotor neurons in the paraventricular hypothalamic nucleus, and spreading of saliva on the body surface. Control of panting also has a physiological (autonomic) component that results in an increase in fluid secretion in the respiratory tract, and a somatic component that increases respiratory air flow (see 4.2.1.3).

## 2.5 Gas Exchange

Most animals, including mammals, rely on aerobic metabolism to meet their everyday energy requirements, with anaerobic metabolism used only during short periods of oxygen limitation (section 2.3). Therefore, systems that supply sufficient

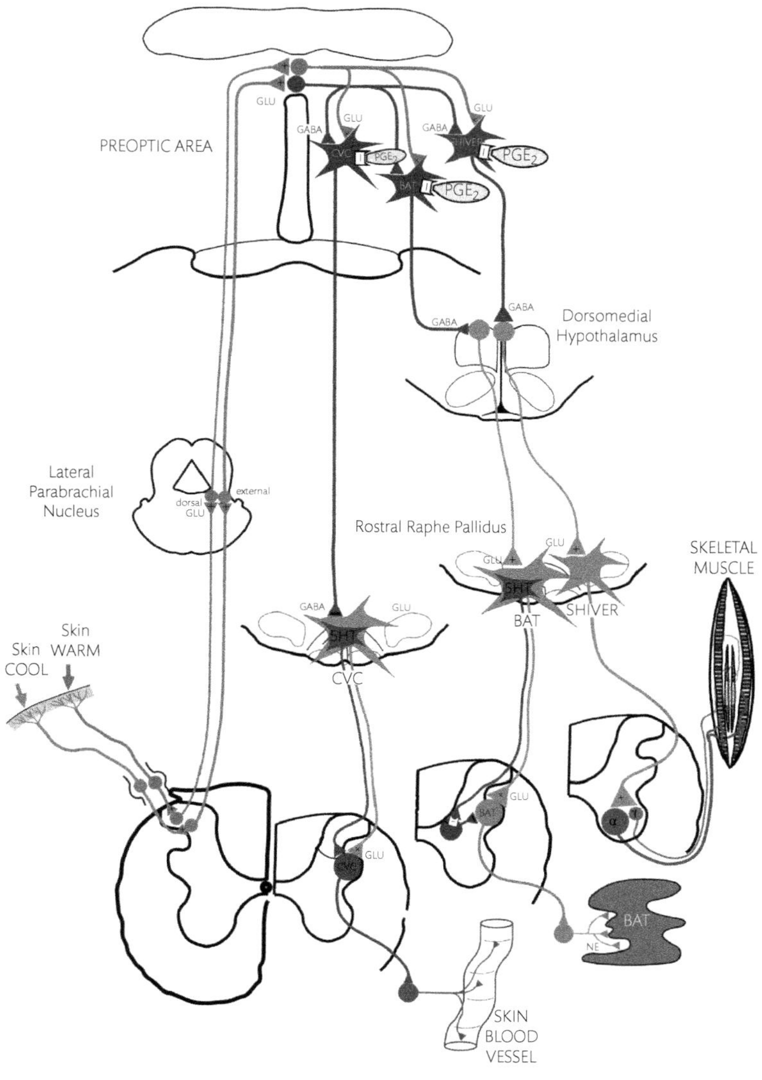


**Figure 2.11** Schematic of pathways for thermogenesis in the cold, for shivering ($\alpha$ motoneurons and $\gamma$ intrafusal fibres), brown adipose tissue (BAT) and cutaneous vasoconstriction, controlled by cool and warm cutaneous thermoreceptors. Neurotransmitters are GLU, glutamate; GABA, gamma-amino-butyric acid; NE, norepinephrine; 5HT, serotonin; $PGE_2$, prostaglandin 2. Modified from Morrison and Nakamura (2011). Reproduced with permission of Frontiers in Bioscience 2011 16:74 Morrison & Nakamura.

$O_2$ to maintain aerobic metabolism, and remove the resulting $CO_2$, are required. Two different mechanisms, molecular diffusion and convection, are used to transport $O_2$ along the 'oxygen cascade' from ambient air to mitochondria, and $CO_2$ in the reverse direction.

The relatively impermeable integument of mammals reduces evaporative water loss in a terrestrial environment, but also precludes cutaneous gas exchange, so almost all gas exchange occurs across specialized internal lungs. The significant cutaneous gas exchange of marsupial neonates is an exception, reflecting their extreme precocial condition at birth (see 3.3.4; Frappell & MacFarlane 2006). Tidal ventilation of the lungs is achieved when negative pulmonary pressure—generated by a combination of external intercostal muscles and the muscular diaphragm that forms a septum between the abdominal and thoracic cavities—causes inspiration, and positive pulmonary pressure—generated by elastic recoil of lung structures and internal intercostal muscles—causes expiration. The relatively high $O_2$ content and low density of air means that tidal ventilation of a pool of respiratory gas is possible. The loss of heat and water via expiration is reduced by nasal countercurrent heat and water exchange. All reptiles, birds, and mammals, including secondarily aquatic taxa (e.g. pinnipeds and cetaceans), ventilate their lungs with air rather than water; water has a lower $O_2$ concentration and is much more dense and viscous than air. The high metabolic rates of mammals and relatively inefficient tidal flow, which results in the mixing of inspired and expired air in the bronchi and trachea, means that lung ventilation is usually continuous (although hibernating mammals may have considerable periods of apnoea).

## 2.5.1 $O_2$ and $CO_2$ Cascades

Prior to the evolution of respiratory and circulatory systems, animals (and plants) were restricted to a small size because the fundamental requirements for cell function cannot be satisfied by simple diffusion over distances larger than a few millimetres. In large, multicellular animals, like mammals, there is a complex, multistep pathway for the delivery of $O_2$ from the ambient air to the mitochondria, where it is used at the last step in the electron transport chain to form $H_2O$. This pathway for $O_2$ delivery at ever-decreasing $pO_2$s through the transport steps from ambient air to mitochondria is called the $O_2$ cascade (Figure 2.12), analogous to the movement of water cascading down a waterfall. There are four steps in the $O_2$ cascade: respiratory ventilation (convection), the alveolar membrane (diffusion), blood circulation (convection), and transport to the mitochondria inside cells (diffusion). A $CO_2$ cascade occurs in the reverse direction.

The first step in the $O_2$ cascade is convective, involving the tidal respiratory system (Guyton & Hall 1996). The ambient $pO_2$ (about 21.2 kPa) is reduced within the alveoli to about 13.9 kPa, reflecting the mixing of inspired tidal air into the residual lung volume (sometimes called the dead space) and saturation with water vapour

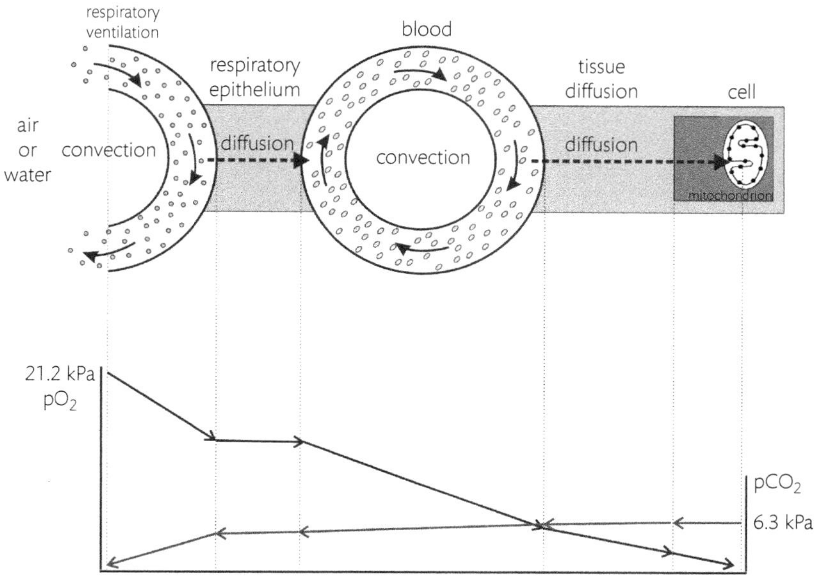


**Figure 2.12** The $O_2$ cascade from ambient air or water to mitochondrion, and reverse $CO_2$ cascade. Top: sequence of convection (respiratory ventilation), diffusion (respiratory epithelium), convection (blood circulation), and diffusion (tissues) steps. Bottom: $O_2$ and $CO_2$ partial pressures along the cascades. Modified from Weibel (1964); see Table 2.6 for example $pO_2$ and $pCO_2$ values.

(see Table 2.6). The second step is oxygen diffusion across the alveolar membrane. This is so effective that the $pO_2$ in blood leaving the alveoli is nearly the same as the alveolar $pO_2$ (about 13.9 kPa), but this oxygenated blood mixes with blood that supplies $O_2$ to the pulmonary tissues, so the $pO_2$ of blood leaving the lungs is only about 12.7 kPa. The third step is convective movement of arterial blood to the capillaries, where the $pO_2$ declines (on average) to about 5.3 kPa at the venous end. The final step is diffusion of $O_2$ from the capillaries to inside cells. The intracellular $pO_2$ is lower than capillary $pO_2$, providing the diffusive gradient for $O_2$ to move from the capillaries into the cells; intracellular $pO_2$ varies in different cells and under different physiological conditions, and is between 1.5 and 2.7 kPa for mammalian cells (Nunn 1993). The mitochondrial $pO_2$ must be maintained above a critical level (about 0.13 kPa) for oxidative phosphorylation to occur.

There is an equivalent multistep pathway for the excretion of $CO_2$ from the cells where it is produced to the ambient air, which we could equivalently term the $CO_2$ cascade. The respiratory convection, alveolar diffusion, and tissue diffusion steps are conceptually the reverse of these steps for the $O_2$ cascade. The blood convective step is slightly different for the $CO_2$ cascade, because $CO_2$ is transported mostly

**Table 2.6** Comparison of partial pressures of $O_2$, $CO_2$, $N_2$, and $H_2O$ between ambient air and mitochondria, via the $O_2$ cascade. Values are for humans, based on Guyton and Hall (1996).

| Partial pressure (kPa)[a] | $pO_2$ | $pCO_2$ | $pN_2$ | $pH_2O$ |
|---|---|---|---|---|
| Ambient air[b] | 21.2 | 0.04 | 79.6 | 0.49 |
| Expired air | 16.0 | 3.60 | 75.8 | 6.26 |
| Alveolar air | 13.9 | 5.33 | 75.4 | 6.26 |
| Arterial blood | 13.9 | 5.33 | 75.4 | 6.26 |
| Venous blood | 5.3 | 6.00 | 75.4 | 6.26 |
| Intracellular | 2 | 6.26[c] | 75.4 | 6.26 |
| Intra-mitochondrial | 0.2 | 6.26 | 75.4 | 6.26 |

[a]101.3 kPa = 760 mm Hg = 1 atm;
[b]average cool, clear day;
[c]calculated as $^1/_{20}$ of the corresponding $pO_2$ difference.

(about 57%) as $HCO_3^-$ (formed from $CO_2$ and $H_2O$ by carbonic anhydrase; see later), bound to proteins (including haemoglobin) as carbamino compounds (about 27%) and also as dissolved $CO_2$ (about 10%). On the other hand, $O_2$ is primarily transported as $O_2$ bound to the $Fe^{2+}$/haem complex of haemoglobin with almost none as dissolved $O_2$. The source for $CO_2$ is the cell cytoplasm, whereas the $O_2$ sink is the mitochondrion.

## 2.5.2 Diffusion

Diffusion is one of the processes by which $O_2$ and $CO_2$ are exchanged between the animal's mitochondria and its external environment. The diffusion of respiratory gases (or other molecules, or heat) is caused by the continuous movements of molecules as a result of random (Brownian) thermal motion. This constant random motion of molecules means that any initial unequal distribution of molecules (e.g. $O_2$ concentrations across a membrane) will eventually become homogeneous. While there is a concentration difference, there will be a constant movement of $O_2$ across a membrane by random thermal motion, with the magnitude of exchange determined by the difference in concentration. The rate of exchange by diffusion (Q) is described by Fick's law:

$$Q = \Delta P A D / \Delta X$$

where $\Delta P$ is the partial pressure difference, A is the surface area, D is the diffusion coefficient, and $\Delta X$ is the diffusion distance; $\Delta P / \Delta X$ is the diffusion gradient (Withers 1992).

Diffusion is sufficient to meet the respiratory gas exchange requirements for very small animals, because they have a high surface area to volume ratio (and thus a relatively large A and small $\Delta X$). However, diffusion is inadequate to meet the gas exchange requirements of larger animals, such as mammals, that have a relatively low surface area to volume ratio, a longer diffusion path length, a complex and relatively impermeable integument, and a high MR. Another potential problem that limits diffusion is that exchange across a respiratory membrane forms a boundary layer of depleted $O_2$ and elevated $CO_2$, which locally reduces the partial pressure difference for gases, and effectively increases the diffusion path length ($\Delta X$), thereby reducing the rate of exchange.

So, although diffusion is the fundamental process by which gas is exchanged across respiratory membranes, large animals require a more complex approach to gas exchange than diffusion alone. Convective transport physically moves respiratory gases dissolved in the fluid over long distances by bulk flow of fluid, and minimizes the boundary layer thickness. Respiratory ventilation and internal circulation provide convective transport to achieve the required rates of gas exchange for aerobic metabolism in large animals.

### 2.5.3 Convection

Convection is the transfer of gases (or biochemicals or heat) by the physical, or bulk, movement of fluid (a liquid or a gas). The amount of the material that is exchanged depends on the rate at which the material is moved (Q; e.g. ml per minute) and the concentration of the material (C; e.g. mg per ml). The net movement of the material when the fluid is moved between two locations, as occurs with gas exchange in the lungs (air is inspired then expired) or the circulatory system (blood is moved from the lungs to the tissues then back to the lungs) is described by

$$Q = V(C_2 - C_1)$$

where V is the volume flow (e.g. ml $min^{-1}$) and $C_2$ and $C_1$ are the respective concentrations. Partial pressures can be used instead of concentrations by incorporating the appropriate solubility coefficient into the equation. Heat movement is described if $C_2$ and $C_1$ reflect the respective temperatures, and the appropriate heat capacity coefficient is included in the equation (see 2.4.2).

### 2.5.4 Gas Laws

Gases are compressible; their volume depends on pressure and temperature, unlike liquids such as water, which are essentially incompressible and have a relatively constant volume. Consequently, the important physical relationships between gas volume with pressure and temperature are of physiological relevance to animals, particularly those that live at low ambient pressures (i.e. high altitudes) or high ambient pressures (i.e. underwater). These important relationships are described by various gas laws.

Boyle's law describes the relationship between the pressure (P) and volume (V) of a gas at constant temperature (T; absolute temperature in °K); PV = constant (Guyton & Hall

1996; Pocock et al. 2013). The law of Charles and Gay-Lussac describes the relationship between the volume and temperature of a gas, at constant pressure; $V/T = \text{constant}$. The volume of a gas at constant pressure and temperature depends on the number of moles (n) of the gas. For a gas at standard temperature and pressure and dry (STPD; 1 atm pressure and 273 °K), 1 mole of gas ($6.022 \times 10^{23}$ molecules; Avogadro's number) occupies 22.4 l. A mole of gas has a mass equal to its molecular weight; for example, 1 mole of $O_2$ has a mass of 32 g ($32 = 2 \times 16$). These relationships are summarized by the ideal gas law, $P V = n R T$, where R is the universal gas constant (8.31 J $mole^{-1}$ $°K^{-1}$) and n is the number of moles. Although no gas is 'ideal', most conform closely to the ideal relationships (especially at low pressures).

Dalton's law describes how the total pressure of a gas is the sum of the partial pressures of its constituent gases. The partial pressure of a gas is the fraction of the total pressure that is due to that gas, and depends on the proportion of molecules that are a particular gas. For example, the partial pressure of $O_2$ ($pO_2$) in normal dry air (which is 20.9% $O_2$) at 1 atmosphere pressure (= 101.3 kPa) is $(20.9 / 100) \times 1 = 0.209$ atm (= 21.2 kPa). Normal air at a pressure of 0.5 atm has half of this $pO_2$, i.e. 0.105 atm, which is the same as the $pO_2$ of air with half the normal $O_2$ concentration at 1 atm. The partial pressure of gases, rather than concentration, is the physiologically relevant determinant of diffusive exchange, since air and water have very different gas concentrations at the same partial pressure, and pressure is not necessarily constant in biological systems. For example, there would be no net movement of $O_2$ between air and water equilibrated with air, since they would have the same $pO_2$ despite having different $O_2$ concentrations. When a gas contains water vapour, the water molecules act like any other gas in having a partial pressure, the water vapour pressure ($pH_2O$). Air that is in equilibrium with liquid water has a saturation water vapour pressure ($pH_2O_{sat}$) that increases dramatically and curvilinearly with temperature (Withers 1992) in the biological temperature range, from about 0.609 kPa at 0°C to 2.33 kPa at 20°C and 7.36 kPa at 40°C.

The law of Laplace describes the physics of the stability of an air space (e.g. a balloon; Withers 1992). The internal pressure (P) depends on the tension in the wall ($\gamma$; e.g. the elasticity of the rubber wall of a balloon) and the internal radius (r); $P = 4\gamma / r$. The significance of this relationship to biological air spaces, such as lung alveoli, is that the internal pressure will vary between alveoli of differing radii. The law of Laplace predicts that a smaller alveolus will have a higher pressure than a larger alveolus, hence the smaller alveolus will empty into the larger alveolus (with a lower internal pressure), rather than the more intuitive (but incorrect) expectation that the volumes of the alveoli will adjust to become equal. Because the collapse of alveoli (known as atelectasis) creates problems for gas exchange, this consequence of the law of Laplace means that it is necessary for smaller alveoli to counter their tendency to empty and coalesce into a large alveolus. A component of the wall tension is determined by the surface tension of the water film lining the alveolus, and the evolutionary solution to this problem was selection for special cells lining the alveoli that secrete a surfactant (a lipoprotein) that reduces surface tension. The outcome of the secretion of surfactant is that surface tension increases as the alveolus is expanded. This means that not only does the increase in surface tension with alveolar expansion counteract the tendency for small alveoli to collapse into larger ones, but also that it is easier to expand an initially collapsed alveolus (which is essential for newborn mammals when they first inflate the lung; see 3.3.4).

Henry's law describes the relationship between the concentration of a gas ([A] for gas A) that is dissolved in a fluid (e.g. water) with its partial pressure (pA) and solubility ($\alpha_A$), at a particular temperature; $[A] = pA\ \alpha_A$ (Withers 1992). For example, the molar concentration of $O_2$ dissolved in water that is in equilibrium with normal atmospheric air at 20°C is 290mM ($pO_2 = 21.2$ kPa; $\alpha_{O2} = 13.7\,mM\,l^{-1}kPa^{-1}$). The solubility coefficient varies for different gases, at different temperatures (e.g. $\alpha_{O2}$ is about 21.7 $\mu M\,l^{-1}\,kPa^{-1}$ at 0°C and 10.2 at 40°C), and with differing osmotic concentrations (e.g. $\alpha_{O2}$ is about 11 for seawater at 20°C). The solubility also varies for different media; the high solubility of $N_2$ in fat versus body fluids can have serious physiological consequences for air-breathing mammals when they dive underwater at high pressures (section 4.5.3).

The concentration of a gas in air directly reflects its fractional concentration. Gases in air do not have a 'solubility coefficient', but they do have a capacitance ($\beta$), which is the concentration of a gas per partial pressure; $[A] = pA\beta_A$. All gases have the same capacitance coefficient (9.88 ml $l^{-1}$ $kPa^{-1}$ at 20°C; Withers 1992). $O_2$ has a much lower concentration in water (6.98 ml $l^{-1}$ at 20°C) than air (209.5 ml $l^{-1}$) because $\alpha_{O2}$ in water (0.331 ml $l^{-1}$ $kPa^{-1}$) is much lower than $\alpha_{O2}$ in air (9.88 ml $l^{-1}$ $kPa^{-1}$). In contrast, $C_{O2}$ has about the same concentration in water and air because $\alpha_{CO2}$ (9.30 ml $l^{-1}$ $kPa^{-1}$) and $\alpha_{CO2}$ (9.88) are similar.

## 2.5.5 Flow through Vessels

Mammals, as do all vertebrates, have a closed circulatory system with the blood contained in a series of vessels. The biophysics of blood flow is determined by the physical principles for fluid flow through tubes (Guyton & Hall 1996; Levick 1996). Darcy's law describes the fundamental relationship between flow (Q) through a tube, the pressure difference two points ($\Delta P$, or $P_1 - P_2$), and the resistance to flow (R) between those points: $Q = \Delta P / R$. The resistance to flow depends on the tube radius (r), length (l), and a physical property of the fluid, its viscosity ($\mu$); $R = 8\eta l / \pi r^4$. Poiseuille's law (or the Poiseuille–Hagen flow formula) replaces the general resistance term in Darcy's law with the preceding expression for R (Guyton & Hall 1996): $Q = \Delta P \pi r^4 / 8 \eta l$. Importantly, flow is not proportional to the radius, or even to the cross-sectional area of the tube ($\pi r^2$), but is proportional to $r^4$; this means that even small changes in vessel radius have a large effect on resistance, and hence on flow. While the Poiseuille–Hagen flow formula is crucial to understanding how the cardiovascular system works, it can cause confusion when it is applied to some situations (e.g. constrictions in vessels, gravitational effects). These problems are overcome once it is appreciated that fluid flows not down a pressure gradient, but down an energy gradient, and that pressure is just one of the three forms of energy for a flowing fluid (along with kinetic and hydrostatic energy).

Bernoulli's theorem states that the steady-state flow rate between two points along a tube is proportional to the difference in the energy of the fluid at the two end points. Energy at any point along the tube is the sum of pressure energy, potential energy, and kinetic energy. Pressure energy is pressure times the volume (P V). Potential energy is the capacity of mass to do work in a gravitational field by virtue of the height of the fluid, and equals mass (density times volume = $\rho V$) times height (h) times the force of gravity (g); that is, $\rho V h g$. Kinetic energy is the energy that a moving fluid possesses by virtue of its momentum, and equals mass ($\rho V$) times the square of velocity ($v^2$), divided by two; that is, $\frac{1}{2}\rho V v^2$. Hence, total energy per unit volume $= E / V = P + \rho hg + \frac{1}{2}\rho v^2$.

One apparent 'violation' of Darcy's law occurs when a fluid enters a constriction, as occurs in a stenosed artery (Levick 1996). The velocity of the fluid increases and its pressure decreases in the constriction (Figure 2.13). After the constriction, when the radius increases (and thus the resistance decreases), the reverse happens; velocity decreases and pressure increases. Fluid flows out of the constriction, even though pressure is higher there than at the constriction. The explanation is that pressure energy has been converted to kinetic energy in the constriction, and the total mechanical energy of the blood remains higher within the constriction than in the distal vessel.

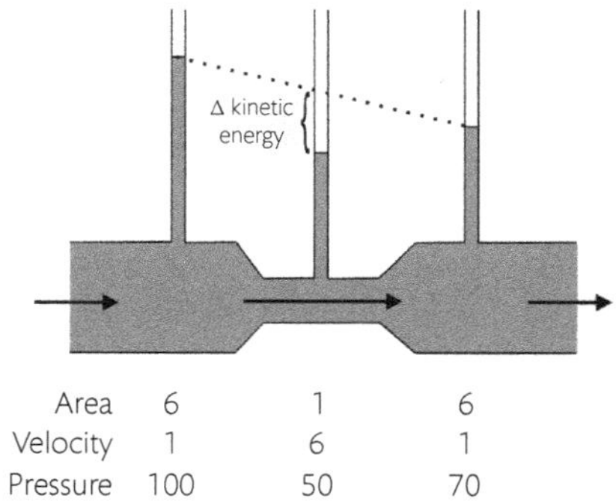


**Figure 2.13** Example illustrating Bernoulli's theorem for conservation of energy illustrating change in fluid kinetic energy at a constriction in a horizontal tube, relative to the pressure energy, cross-section area, and fluid velocity. Modified from Levick (1996).

The other puzzling aspect of blood flow that can be confusing when framed by Darcy's law, but satisfied by Bernoulli's theorem, is the flow of blood from the heart to the lower extremities in large animals, across which there is a relatively large range of potential energy. For example, in a human standing upright, the mean aortic pressure (at the level of the heart) is about 12.6 kPa above atmospheric pressure, and in the foot the pressure is about 24 kPa above atmospheric pressure. Yet blood still flows from the aorta to the feet. The explanation is that the gravitational potential energy of blood in the aorta is much higher than that of the blood in the foot, about 12 kPa more, so the total energy of blood in the aorta is equivalent to about 24.6 kPa compared to the foot. Thus there is a net energy gradient of 0.6 kPa driving blood from the aorta to the foot.

Darcy originally developed his flow formula to describe the movement of water in the gravel beds of fountains in Dijon, but the biophysics of fluid flow through fountains also applies to flow through a branching network of blood vessels. In the systemic circulation, flow is equal to the cardiac output (CO), and the pressure difference driving CO is, strictly speaking, the difference between mean arterial pressure ($P_a$) and central venous pressure (CVP; Guyton & Hall 1996). However, since CVP is close to zero (atmospheric pressure) in most situations, the pressure difference is taken as $P_a$.

The resistance to systemic flow is the total systemic peripheral resistance ($TPR_{sys}$), a 'catch-all' term. Consequently, $CO = P_a / TPR_{sys}$ (cf. $Q = \Delta P / R$). This formula describes the relationship between the three main variables of the cardiovascular system, but is a little misleading in that $P_a$ does not physically 'cause' cardiac output (although baroreceptor control of the circulation can modify CO when $P_a$ changes). Rather, it is the contractile work done by the heart (CO) that determines $P_a$; so, more correctly, $P_a = CO\,TPR_{sys}$. For a resting human, $TPR_{sys}$ is 1 peripheral resistance unit (PRU); systemic blood flow is about 100 ml $s^{-1}$ and $P_a$ is 100 mmHg, so $TPR_{sys} = 100/100$ ($TPR_{sys}$ is not such a convenient value if we use SI units for $P_a$ of 13.3 kPa; $TRP_{sys} = 13.3/100 = 0.133$). Strong peripheral vasoconstriction can increase $TPR_{sys}$ from 1 to about 4 PRU, and vasodilation can decrease $TPR_{sys}$ to about 0.2 PRU (Guyton & Hall 1996).

Fluid flow is described similarly for the pulmonary circuit; the pressure gradient driving pulmonary blood flow is the right arterial pressure ($P_{ra}$; 2.13 kPa) minus the left atrial pressure ($P_{la}$; 0.27 kPa), so $(P_{ra} - P_{la}) = CO\,TPR_{pul}$, where $TPR_{pul}$ is the pulmonary resistance. CO is the same for the systemic and pulmonary circuits, reflecting the complete left and right separation (except in the developing fetus; see 3.4.5). Consequently, $TPR_{pul}$ is about 0.14 PRU. During exercise, CO can increase to 400 ml $s^{-1}$ and $P_{ra}$ to 4.0 kPa, meaning that $TRP_{pul}$ decreases to 0.075 PRU, reflecting an increase in the number of open pulmonary capillaries and increased distension.

As blood flows through the circulatory system, its pressure changes dramatically. For example, in humans the left ventricular pressure cycles from 0 kPa during rest (diastole) to about 16 kPa during contraction (systole). In the aorta, the elasticity of its walls dampens the cycle to about 10.6–16 kPa. The blood pressure cycle decreases in magnitude and amplitude through the large arteries, small arteries, and arterioles, to a relatively constant pressure of about 4.0 kPa at the arterial end of the capillaries, and 1.3 kPa at the venous end (Saladin 2010). Pressure continues to decline through the venules and veins, to the vena cavae and sinus venosus.

Vascular pressures are particularly important determinants of fluid balance across capillaries. Fluid movement into or out of a capillary is determined by the balance of hydrostatic (pressure) forces and colloid osmotic forces, commonly referred to as the Starling forces (Guyton & Hall 1996; Pocock et al. 2013). On average, the capillary hydrostatic pressure (in humans) is about 2.3 kPa, and the interstitial hydrostatic pressure is about –0.4 kPa, so the net hydrostatic pressure pushing fluid out of the capillary is about 2.71 kPa (Guyton & Hall 1996). Plasma has a high concentration of protein and, since the capillary is relatively impermeable to protein, this draws water into the capillary from the interstitium. The plasma colloid pressure, drawing water into the capillary, is about 3.73 kPa. The interstitial colloid osmotic pressure is lower (1.07 kPa), reflecting its lower protein concentration. Therefore the net colloid pressure is 2.67 kPa, drawing water into the capillary. Overall, the net balance of Starling forces is a filtration pressure of about 0.04 kPa, which results in lymph formation. This balance of Starling forces is, however, the average along a capillary. Hydrostatic pressure is higher at the arterial end (about 4.0 kPa) than the venous end (1.33 kPa), so the balance at the arterial end is a net filtration force of $1.73\,\text{kPa}\,(4.40 - 2.67)$, and at the venous end is a net reabsorption of $-0.93\,\text{kPa}\,(1.73 - 2.67)$. The difference in the balance of forces at the arterial and venous ends of the capillary has the effect of bulk movement of fluid (including permeable solutes) out of, then back into capillaries, facilitating exchange between capillary blood and tissue interstitial fluid, with a slight loss forming the lymph fluid that circulates in the lymphatic system and eventually returns to the venous system in the thorax.

### 2.5.6 Acid–Base Balance

Water molecules spontaneously and reversibly dissociate to form hydrogen ions ($H^+$) and hydroxyl ions ($OH^-$):

$$H_2O \longleftrightarrow H^+ + OH^-$$

The balance of this chemical reaction (like chemical reactions in general) is reflected by its apparent ion product ($K_{eq}$), which is calculated as the multiplication of the concentrations of the 'products' on the right-hand side of the equation, divided by the product of the concentration of the 'reactants' on the left-hand side. For dissociation of water, $K_{eq}$ is $[H^+][OH^-]/[H_2O]$, which is simplified to $[H^+]\ [OH^-]$, since $[H_2O]$ is considered to be 1 for $K_{eq}$ calculation. The $K_{eq}$ is about $10^{-14}$, indicating that the concentrations of H + ($[H^+]$) and $OH^-$ ($[OH^-]$) are small; for neutral water at 25°C, $[H^+]$ and $[OH^-]$ are about $10^{-7}$ M (compared to $[H_2O]$ which is about 55 M). The $[H^+]$ is the measure of acidity, but pH is a more convenient unit; pH is $-\log_{10}[H^+]$, so pH is about +7 for neutral water at 25°C (so is pOH). For acid solutions, $[H^+] > 10^{-7}$ M and pH < 7; for basic (alkaline) solutions, $[H^+] < 10^{-7}$ M and pH > 7. Intracellular fluids are about neutral pH, being more acidic than extracellular fluids, which are generally slightly alkaline (about pH 7.4).

Temperature significantly alters the physiochemical dissociation of water: higher temperature increases the dissociation, and hence increases $[H^+]$ (and $[OH^-]$) and decreases pH; the pH of neutral water (= pOH) is about 7.47 at 0°C and 6.68 at 50°C (Harned & Robinson 1940). This change in pH units (U) for neutral water is about $-0.017$ U °C $^{-1}$ (Rahn 1974; Rahn 1975); this is called the α-stat pattern for the effect of temperature on pH. In contrast, the pH-stat pattern is maintaining a relatively constant pH despite changes in temperature. The body fluids of many animals have a similar $\Delta pH/\Delta T$ as water, of about $-0.017$ to $-0.020$ U $C^{-1}$, with the physiological consequence that their ratio of $[OH^-]:[H^+]$ is fairly constant even when their temperature changes. This is important, because it maintains the net charge of proteins, and their molecular structure and functionality. Ectothermic animals can maintain relative alkalinity over a range of $T_b$, using precise regulation of pH by controlling respiratory, metabolic, and excretory functions. The $T_b$ of endothermic mammals decreases during torpor, requiring appropriate changes in their body acid–base status, generally to maintain a relatively constant pH rather than relative alkalinity ($\Delta pH / \Delta T = 0 \text{ to } -0.009\,U\,°C^{-1}$; Withers 1977a). The higher pH needed to maintain relative alkalinity at lower $T_b$ requires a decreased blood $pCO_2$ and/or increased $HCO_3^-$, caused by a physiological change in the respiratory ventilation/metabolism ratio. For the little pocket mouse (*Perognathus longimembris*), pH declines from 7.28, with $pCO_2$ = 4.9 kPa and $[HCO_3^-] = 17.3$ mM when normothermic ($T_b$ = 37°C), to pH = 7.51, with $pCO_2$ = 1.9 kPa and $[HCO_3^-] = 18.8$ mM when torpid ($T_b$ = 10°C; Withers 1977a). Interestingly, ectothermic and endothermic pouch young of tammar wallabies (*Macropus eugenii*) maintain pH constant (i.e. pH-stat model) rather than have pH change by about $-0.017$ U °C $^{-1}$ (α-stat model; Andrewartha et al. 2014).

The $CO_2$ cascade plays a major role in acid–base (pH) balance through the reversible chemical reaction between $CO_2$ and water:

$$\mathbf{CO_2 + H_2O \longleftrightarrow H_2CO_3 \longleftrightarrow H^+ + HCO_3^-}$$

where $H_2CO_3$ is carbonic acid and $HCO_3^-$ is bicarbonate (Withers 1992). The first reaction (to form $H_2CO_3$) is relatively slow, whereas the second reaction (to form $H^+$ and $HCO_3^-$) is very rapid. The rate of the first reaction is increased by carbonic anhydrase, an enzyme present in erythrocytes (and many other tissues) where rapid conversion between $CO_2$ and $HCO_3^-$ is advantageous. For example, small mammals generally have a high concentration of erythrocyte carbonic anhydrase, facilitating $CO_2$ transport in the blood. At high pH (> 8.5), the above reaction can proceed further to $CO_3^{2-}$ (carbonate).

The general equation given earlier for $K_{eq}$ can be applied to the $CO_2$- $HCO_3^-$ acid–base system (known also as the bicarbonate buffer system), leading to the Henderson–Hasselbalch equation:

$$pH = pK_{eq} + \log[HCO_3^-]/[CO_2]$$

with $pK_{eq}$ = 6.1. This equation describes the relationship between acid–base balance and $CO_2$ concentration; an increase in dissolved $CO_2$ will decrease pH, whereas an increase in $[HCO_3^-]$ will increase pH. The buffering capacity of a solution (β) is measured as the amount of acid or base that is required to change the pH by 1 unit ($\beta = \Delta mmol/\Delta pH$). Although the $pK_{eq}$ for the bicarbonate buffer system (6.1) is not very close to the normal extracellular pH (about 7.4), and the concentrations of $HCO_3^-$ and $CO_2$ are not high, it is nevertheless an important buffer system because the concentration of $CO_2$ is readily controlled by the respiratory system, and the concentration of $HCO_3^-$ can be controlled by the kidneys. The phosphate buffer system

$$H_3PO_4 \longleftrightarrow H^+ + H_2PO_3^- \longleftrightarrow 2H^+ + HPO_4^{2-} \longleftrightarrow 3H^+ + PO_4^{3-}$$

has a $pK_{eq}$ (6.8) closer to normal pH, but has a much lower concentration than the bicarbonate buffer system, and hence a much smaller buffering capacity in extracellular fluids. However, phosphate buffering is more important in nephron tubules and intracellular fluid, where the phosphate concentration is higher and the pH is closer to its $pK_{eq}$. The greatest chemical buffer system of the body is Proteins (amino acids), which often have high concentrations and a $pK_{eq}$ closer to pH, are the body's greatest chemical buffering system. A few buffer systems have a similar thermal dependence of pH as does water (e.g. imidazole), but the biologically important bicarbonate ($-0.005$ U °C$^{-1}$) and phosphate buffer systems are relatively independent of temperature.

Rapid regulation of body fluid pH depends on both the chemical buffering systems of extracellular fluids (e.g. bicarbonate, phosphate, proteins) and intracellular fluids (e.g. proteins, including haemoglobin, organic phosphates, bone minerals; Pitts 1963). Longer-term pH regulation is provided by the respiratory system through excretion or retention of $CO_2$ via the lungs. Still longer-term regulation is provided by the kidneys through renal excretion of acidic or alkaline urine.

Ammonia ($NH_3$) is a primary waste product of nitrogen metabolism (see 3.6.4.1) and is a very strong base:

$$NH_3 + H_2O \longleftrightarrow NH_4^+ + OH^-$$

with a $pK_{eq}$ between 9 and 10, so that at physiological pH (about 7.4), about 99% of any $NH_3$ is actually present as $NH_4^+$, and only 1% as $NH_3$. This means that the concentration, hence partial pressure, of $NH_3$ is very low and the $[NH_4^+]$ is high. Both $NH_3$ and especially $NH_4^+$ are very toxic, but it is generally impractical for terrestrial animals to excrete gaseous $NH_3$ down a $pNH_3$ gradient, and exposure to even low ambient $NH_3$ concentrations is lethal because of high associated $[NH_4^+]$. Nevertheless, some bats are able to tolerate remarkably high ambient $NH_3$ concentrations arising from bacterial breakdown of urea in their confined roost environments (section 3.6.4.1).

## 2.6 Digestion

Mammals, as do all animals, consume organic material to provide the fuel for their metabolic processes. Ingested food is a source of energy, and also provides the structural materials necessary for body maintenance, growth, and reproduction. For a period after birth, all mammals feed on a liquid diet of milk, but as adults they switch to an array of specialized feeding strategies (see 3.5.1), obtaining their energy and other nutritive requirements from a wide range of sources. Mammals can broadly be classified as herbivores (feeding on plant material), carnivores (feeding on other animals), and omnivores (feeding on both plant and animal material). Some mammals have considerable trophic specialization, feeding on few or even only one food source, while others are generalists, feeding on a wide range of materials (see 1.4.3; McNab 2002).

Digestion is the process of breaking down large, often insoluble, organic food molecules into smaller, simpler, soluble molecules that can be absorbed across the gut into the body and made available for metabolism, growth, and maintenance. Both physical and chemical processes play important roles in digestion.

### 2.6.1 Digestive Tract

Vertebrate animals have a complex digestive tract, including the mouth and teeth, pharynx, gut (oesophagus, stomach, small and large intestines, rectum), and anus, along with accessory organs (e.g. salivary glands, liver, pancreas), all of which can be highly specialized for the digestive requirements of a particular diet (Kardong 2009). A complex, compartmentalized digestive tract ensures that the digestive process is sequential, subjecting food to a coordinated array of step-like processes as it passes along the tract. The autonomic nervous system controls the overall function and coordination of digestive processes via hormones released into the bloodstream and into the tract itself (Ferguson 1985).

Teeth assist with the acquisition and mechanical breakdown of food into pieces that are small enough to ingest. Mastication grinds food particles and increases their surface area for chemical digestion. A key characteristic of mammals is their heterodont dentition, with incisors, canines, premolars, and molars (Figure 2.14; see 1.3.1, 3.5.1). Generally, incisors cut or snip, canines puncture and hold, and

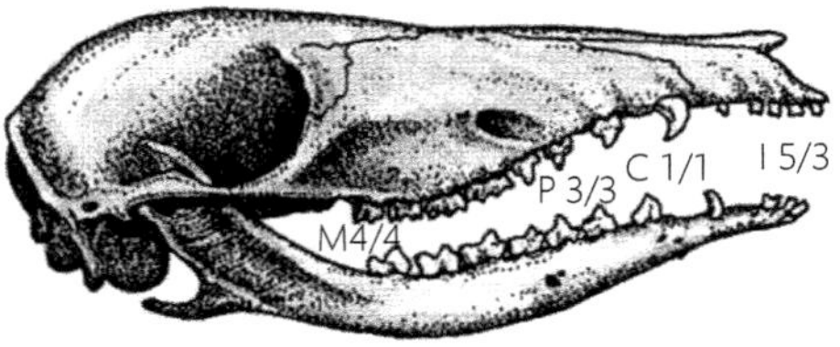


**Figure 2.14** Mammalian heterodont dentition illustrated by a marsupial bandicoot (*Perameles*); incisors (I), canines (C), premolars (P), and molars (M). Dental formula $I^5/_3\,C^1/_1\,P^3/_3\,M^4/_4$. Modified from Macdonald (2010). Reproduced with permission from Oxford University Press.

premolars and molars crush or grind. The dental formula of mammals indicates the number of upper and lower incisors (e.g. $I^5/_3$), canines (e.g. $C^1/_1$), premolars (e.g. $P^3/_3$) and molars (e.g. $M^4/_4$). Mammals have an array of dental specializations, including reduction and sometimes increase in the dental formula, related to phylogeny and specific diets. Mammals are generally diphyodont: they develop a deciduous dental set that is replaced just once during growth to adulthood (i.e. they have only two sets of teeth during their life). The deciduous dentition sometimes has little resemblance to the adult dentition. Diphyodonty increases the importance of the functional maintenance of the teeth because they are not replaced. In contrast, a few polyphyodont mammals (e.g. manatees, a wallaby, and a mole rat) generate new teeth as required throughout life.

Chemical digestion commences in the oral cavity during mastication. The parotid and sublingual salivary glands produce lubricating saliva that is rich in glycoproteins. Saliva also contains amylase enzymes that hydrolyze the internal α-1,4-glucoside bonds in starch and glycogen to form dextrins and maltose. Ebner's gland on the tongue produces a lipase that acts on triglycerides (McGilvery & Goldstein 1983; Ferguson 1985).

Once food is chewed and swallowed, the bolus of masticated food leaves the mouth and travels along the oesophagus to the stomach. The movement of food along the oesophagus and other portions of the gut occurs via propulsive peristalsis. The gut wall typically consists of an inner layer of circular smooth muscle and an outer layer of longitudinal smooth muscle, and coordinated waves of contraction and relaxation of these smooth muscle layers squeeze the food in an anterior–posterior direction. Segmentation peristalsis (rhythmic contractions of the circular muscle) enhance the mixing of the gut contents (Withers 1992). The mammalian stomach generally consists of several regions. The cardia is a transitional region between the oesophagus and stomach, characterized by cardiac glands that secrete mucus. The fundus is usually the largest region of the stomach; it contains parietal cells and chief cells. Parietal cells secrete hydrochloric acid, reducing the pH within the stomach to 2–3, and chief cells secrete pepsin, an endopeptidase enzyme that catalyzes the initial breakdown of protein. The pyloric region of the stomach occurs before the junction with the small intestine, and is characterized by pyloric glands that secrete a mucous that neutralizes the acid chyme before it enters the small intestine (Ferguson 1985; Withers 1992; Kardong 2009). In some mammals, a non-glandular region develops at the base of the oesophagus (e.g. some herbivores); in other taxa (e.g. rodents), the stomach mucosa is non-glandular and the stomach has storage and mixing roles (Kardong 2009).

The intestine can generally be divided into the small and the large sections, based morphologically on the diameter of the lumen (Kardong 2009). The small intestine is the major site of digestion and absorption in non-ruminant mammals (Withers 1992; Pocock et al. 2013). It typically has a narrower diameter than the large intestine, and consists of three regions: the duodenum, jejunum, and ileum. Chyme from the stomach enters the duodenum, along with digestive secretions from the pancreas and liver. The digesta leaves the stomach as a semi-liquid and is neutralized to a pH of about 7 by bicarbonate from the pancreatic juices and bile. The stimulus for the release of bicarbonate-rich pancreatic secretions is the presence of acid chyme in the small intestine. The discovery that the process occurs independently of the nervous system (to the consternation of Pavlov, who won the 1904 Nobel Prize for nervous control of the digestion) led to the discovery of the first 'hormone' (secretin) and to the development of the field of endocrinology (Modlin & Kidd 2001).

Bile produced in the liver is released from the gall bladder into the duodenum (Kardong 2009). Bile acids (mainly cholic and chenodeoxycholic acids) form powerful detergents when connected via an amide linkage with glycine or taurine to form bile salts. Bile emulsifies lipids, together with a small protein (colipase) from the pancreas, and enables lipases secreted by the pancreas to hydrolyze triglycerides into fatty acids and glycerol. The formation of micelles, aggregations of fatty acids, cholesterol, fat-soluble vitamins, lysolecithins, and bile salts facilitates the transport and absorption of digested lipids by forming a hydrophobic droplet with a hydrophilic shell that enables suspension in an aqueous environment (Caspary 1992; Withers 1992; Sherwood et al. 2005).

The products of digestive processes, along with vitamins, ions, and water, are absorbed through the intestinal wall into the blood, which transports them to the tissues for metabolism, biosynthesis, storage, or excretion (Withers 1992; Pocock et al. 2013). Most absorption occurs in the small intestine, particularly in the jejunum and ileum, but the large intestine is also an important site for absorption, particularly for water, water-soluble vitamins, and volatile fatty acids. The inner surface of the intestine is highly folded (Kerckring's folds) and covered with a series of villi (motile, finger-like projections) that increase the surface area by approximately tenfold and promote the mixing of the intestinal contents (Caspary 1992). The inner ends of these endothelial cells are in close proximity to the circulatory system, facilitating the transfer of absorbed nutrients to the blood and lymph (Crane 1975). Microvilli projecting from the outer surface of the endothelial cells, forming the brush border, further enhance the intestinal surface area by another twenty-fold. This brush border provides a barrier between the gut lumen and the body, separating the digestive functions of enzymes in the gut from the metabolic functions of the cells, as well as contributing an absorption role, facilitating the uptake of nutrient molecules into the circulatory system (Crane 1975; Caspary 1992).

## 2.6.2 Digestive Function

Pepsin hydrolyzes non-terminal peptide bonds next to aromatic or dicarboxylic amino acids in the stomach, forming small peptides (Figure 2.15). Proteins are hydrolyzed in the small intestine into their constituent amino acids. The pancreas secretes endopeptidases such as trypsin, chymotrypsin, and elastase into the lumen of the duodenum, along with exopeptidases such as carboxypeptidases, aminopeptidases, and dipeptidases. The endopeptidases cleave peptide bonds between specific amino acids. For example, trypsin hydrolyzes non-terminal peptide bonds between carboxyl groups, targeting arginine or lysine, and chymotrypsin hydrolyzes peptide bonds next to aromatic amino acids. The exopeptidase carboxypeptidase cleaves the peptide bond adjacent to a free carboxyl group, producing a free amino acid. Amino-peptidases break the bonds of terminal, free amino acids, and dipeptidases cleave dipeptides.

Proteases must be synthesized and secreted in an inactive form, to avoid the catastrophic effects on secretory cells and tissues that would occur if the enzymes were active in the intracellular environment. These inactive proenzymes, or zymogens, are activated when they reach the site of digestion. Zymogens are activated by exposure to specific ions; for example, $H^+$ converts inactive pepsinogen to active pepsin in the stomach. Specific enzymes are also released to activate other zymogens. Enterokinase is secreted by the intestinal mucosa and activates trypsinogen by hydrolyzing a lysyl peptide bond, allowing the remaining trypsin protein to attain a functional configuration. Active peptide enzymes then undergo autocatalysis

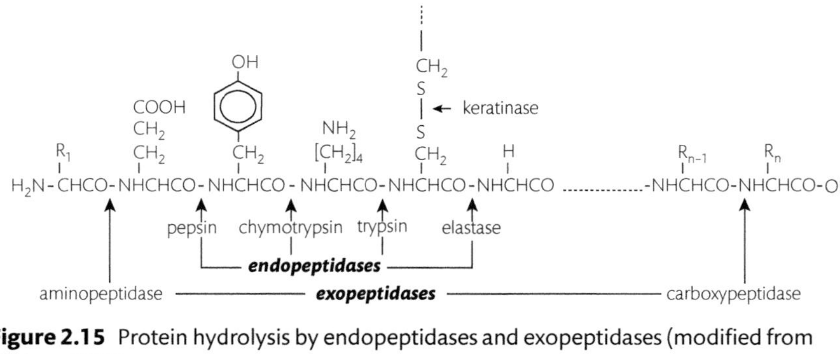


**Figure 2.15** Protein hydrolysis by endopeptidases and exopeptidases (modified from Withers 1992).

(e.g. trypsin activates trypsinogen, and pepsin activates pepsinogen) and activate other inactive peptidases (e.g. trypsin activates chymotrypsinogen).

The amino acids formed from protein digestion are then absorbed into the bloodstream and contribute to the body's amino acid pool (along with hydrolyzed tissue proteins). These free amino acids of the body pool are either catabolized to synthesize new proteins or metabolized by various metabolic pathways, providing energy for cellular metabolism and forming nitrogenous wastes. The liver is a key organ in amino acid/protein metabolism and catabolism. It synthesizes most of the plasma proteins and gluconeogenesis from amino acids, which involves deamination that produces ammonia. Along with ammonia from intestinal microorganismal deamination, ammonia is detoxified in the liver by conversion to urea (via the urea cycle; see 3.6.4.2). Amino acids are generally a minor source of metabolized energy, but can sometimes be substantial (e.g. during starvation). The body has limited capacity to store protein. Non-essential amino acids can be synthesized in the liver but essential amino acids must be derived from the diet.

There are nine essential dietary amino acids (histidine, isoleucine, leucine, lysine, methionine, phenylalanine, threonine, tryptophan, valine) for animals, because they cannot be synthesized (Karasov & del Río 2007). There are five universally non-essential amino acids (alanine, asparagine, aspartic acid, glutamic acid, serine) that all animals can synthesize. There are seven remaining 'conditionally essential' amino acids (arginine, cytosine, glutamine, glycine, proline, taurine, tyrosine) whose synthesis is limited under some conditions or impossible in some species. Mammals vary considerably in which amino acids they can and cannot synthesize. For example, arginine and proline are synthesized in the gut, primarily from dietary precursors, so species that reduce their intestine size while migrating or fasting might experience arginine or proline deficiency. Arginine appears to be essential for strict carnivores (e.g. ferrets and mink; Karasov & del Río 2007).

The enzymatic digestion of carbohydrates also occurs mainly in the small intestine, via polysaccharidases and oligosaccharidases located in the brush border of the intestine. Most dietary carbohydrates consist of polysaccharides (e.g. glycogen, starch) and disaccharides (e.g. sucrose and lactose). Amylase enzymes hydrolyze the non-terminal α-1,4 and α-1,6-glucosidic bonds in polysaccharide starches and

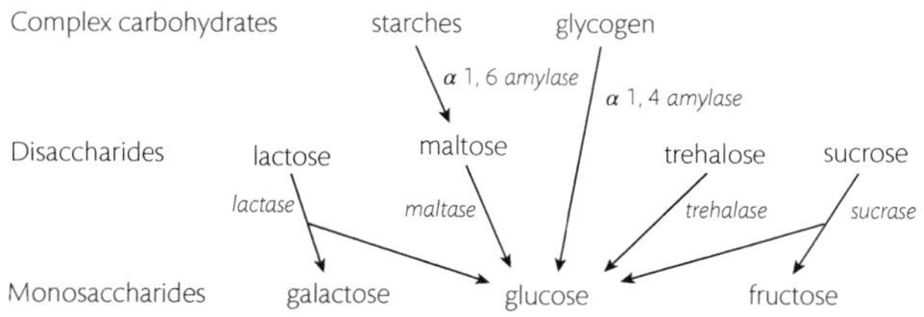


**Figure 2.16** Schematic for digestion of complex carbohydrates (glycogen, starches) to disaccharides, then various important monosaccharides (modified from Withers 1992).

glycogen, forming dextrins and maltotriose, and ultimately the disaccharide maltose and the monosaccharide glucose (Figure 2.16). Oligosaccharidases hydrolyze trisaccharides and disaccharides such as maltose, sucrose, and lactose. For example, α-glucosidase hydrolyzes glucose molecules linked by an α-bond to a second monosaccharide (i.e. maltose is hydrolyzed to form two glucose molecules). If glucose is linked to another monosaccharide by a β-bond, then β-glycosidases hydrolyze the bond. For example, lactose is hydrolyzed by lactase (β-galactosidase), forming glucose and galactose; sucrose is hydrolyzed by sucrase, forming fructose (McGilvery & Goldstein 1983; Ferguson 1985; Withers 1992).

Triglycerides are an important energy source, as a dietary component, and as a body storage lipid (e.g. 'white' fat stores). They are first hydrolyzed to fatty acids and glycerol (Figure 2.17); the fatty acids are transported to tissues where they are resynthesized into triglyceride and stored (e.g. in adipose tissue) or metabolized to $CO_2$ and $H_2O$ by β-oxidation, with phosphorylation of ADP to form ATP (McGilvery & Goldstein 1983; Withers 1992).

Following a meal, the blood glucose level is kept relatively constant by conversion of monosaccharides to glucogen for storage, or to fatty acids and triglycerides for metabolism or storage (Withers 1992). These processes are stimulated by the hormone insulin (released from pancreatic beta cells) that result in the lowering of glucose in the plasma. Glycogen breakdown or glucose synthesis from amino acids (gluconeogenesis) are stimulated by the hormone glucagon (released from pancreatic alpha cells) and help to maintain the plasma glucose between meals. Gluconeogenesis is particularly important for ruminants and pseudoruminants since they are normally 'starved' of glucose by their symbiotic gastric microorganisms (see 4.8.2).

Nutrients are transported from the gut lumen into the cytoplasm of gut epithelial cells by passive and facilitated diffusion, active transport, and pinocytosis. The absorption of lipids occurs mostly in the distal duodenum and proximal jejunum. Bile salts are secreted from the liver via the biliary tract and associate with lipids in the intestine to form micelles, which transport lipids from the lumen of the intestine through the unstirred boundary layer to the surface of epithelial

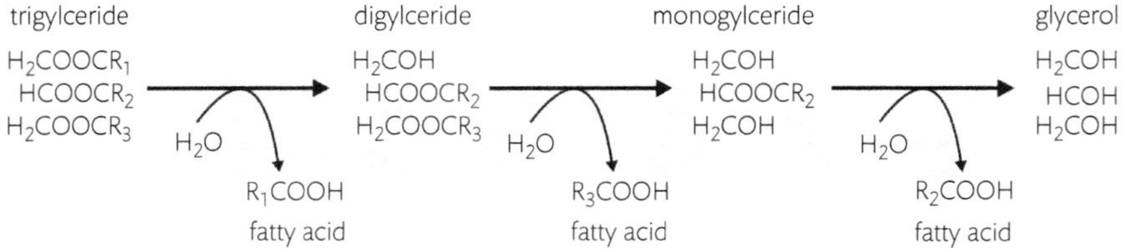


**Figure 2.17** Schematic for digestion of triglycerides by lipases into diglycerides, then monoglycerides, and finally glycerol, releasing fatty acids that are metabolized further by β-oxidation.

cell membranes. The bile salts become dissociated from the micelles at the brush border membrane as a result of its acidic nature, and are recycled to form new micelles in the gut lumen, eventually being absorbed in the terminal portion of the ileum and returned to the liver. Hydrophobic monoglycerides and fatty acids move through the lipid bilayer of epithelial cell membranes by passive diffusion into the intracellular fluid (Withers 1992). Within the epithelial cells, long-chain fatty acids and monoglycerides are bound by a cytosolic protein that transports them to the smooth endoplasmic reticulum, where triglycerides are resynthesized. These triglycerides are then combined with cholesterol, cholesterol esters, fat-soluble vitamins, and apolipoproteins, forming vesicles containing chylomicrons and low-density lipoproteins via the Golgi apparatus. The vesicles then migrate to the plasma membrane and the chylomicrons are released to the lymphatic system by exocytosis (Caspary 1992). Short-chain fatty acids and glycerol are sufficiently polar to simply diffuse across the basal membrane into the bloodstream (Moyes & Schulte 2008).

Most carbohydrates are absorbed as simple sugars, such as glucose and galactose, via carrier-mediated active transport, wherein the potential energy stored in the sodium gradient across the cell membrane is harnessed to transport other molecules. The ubiquitous sodium-potassium pump ($Na^+$-$K^+$ ATPase) on the basolateral cell surface uses ATP to actively transport $Na^+$ (out) and $K^+$ (in), against their concentration gradients, thus maintaining a strong electrochemical gradient across the membrane. $Na^+$ then diffuses down its concentration gradient into the cell via a carrier molecule (SGLT-1) that has two binding sites, one for $Na^+$ and one for the sugar molecule. Increased levels of glucose, detected by SGLT-1, trigger the synthesis and transport to the microvilli of a second form of carrier molecule (GLUT-2) that increases the rate of glucose uptake. Facilitated diffusion via the carrier molecule GLUT-2 in the cell membrane transports sugars such as fructose that have relatively low blood concentrations, from the high concentration in the gut lumen to the low concentration inside the cell (Crane 1975; Caspary 1992; Moyes & Schulte 2008). Glucose can also cross the epithelial barrier and enter the circulatory system via a paracellular pathway, through junctions between the epithelial cells, because glucose uptake continues to increase with increasing lumen glucose concentration even after transcellular pathways are saturated. Paracellular transport appears to be facilitated by a change in the zonula occludens (tight junctions between epithelical cells), triggered by activation of carrier molecules by glucose, which make it leaky to small molecules. This change is maximized at nutrient concentrations that saturate transmembrane pathways (Madara 1989).

The absorption of proteins involves similar processes to those for carbohydrates, and occurs in the jejunum and proximal ileum. Oligopeptides in the intestinal lumen are cleaved into single amino acids by brush border enzymes. Active transport of these amino acids then occurs via carrier-mediated active transport linked to sodium transport, with specific carrier molecules for the transport of neutral, basic, and acidic amino acids, and for glycine, proline, and hydroxyproline. Small di- and tripeptides may also be transported by carriers, and are subsequently cleaved within the cell cytoplasm to form amino acids. Amino acids then diffuse into the capillary network within the villus. Amino acids may also be absorbed via a paracellular pathway through a leaky zonula occludens, as for glucose (Madara 1989; Caspary 1992).

The breakdown of proteins to amino acids before entry into the circulatory system is important to prevent an immune response to foreign proteins. Some proteins, however, are absorbed intact into the bloodstream via two pathways: proteins may bind to specific receptors that transport them across the intestinal epithelium, or they may be transported by pinocytosis, where they are engulfed in an invagination of the cell membrane on the luminal side of the epithelial cell and transported through the cytoplasm in a vesicle, before being exocytosed from the cell into the bloodstream. This transport occurs in enterocytes and specialized cells in the proximal small intestine. The absorption of macromolecules into the circulatory system is particularly important in facilitating the absorption of growth factors and immunoglobulin in postnatal mammals from maternal colostrum and milk, the latter being crucial for some mammals, such as ungulates, which are nearly agammaglobulinemic at birth (Pácha 2000).

### 2.6.3 Digestibility

Not all foods are composed of material that can be readily digested. Materials such as cellulose and hemicellulose (carbohydrates), chitin (a polysaccharide), the proteins keratin and collagen, and wax (a lipid) are difficult for most vertebrates to digest. Specialized enzymes or gut morphology are required to exploit these materials (see 4.8). For example, no vertebrates have the enzymes necessary to break the β bonds in cellulose, but some mammals (and other animals) have evolved mechanisms to digest cellulose, via specialized digestive systems and symbiotic microorganisms (see 4.8). Ant- and termite-consuming animals have difficulty digesting the chitin in their prey's exoskeleton. The numbat, a marsupial termitivore, digests about 81% of the energy content of their diet, but chitinous exoskeleton fragments are a conspicuous component of their faeces (Cooper & Withers 2004b). Baleen whales can digest the chitin and waxes in their crustacean (krill) diet, using symbiont-derived or endogenously produced enzymes (Nordøy 1995; Souza et al. 2011). So, a combination of the chemical composition of food and the digestive specializations of the mammal determine how much energy can be assimilated by the mammal from different diets.

Generally, carnivores and granivores have highly digestible diets (digestibility of 80–95%; Grodzinski & Wunder 1975; Hume 1982; Green & Eberhard 1983), insectivores have a less digestible diet (around 60–80% digestibility; Nagy et al. 1978; Bell 1990; Cooper & Withers 2004b), while herbivores have a poorly digestible diet (typically 50–70%; Grodzinski & Wunder 1975; Hume 1982). Specialized frugivores and sanguivores have very digestible diets (approximately 90%; Breidenstein 1982; Korine et al. 1996; Wenninger & Shipley 2000). Digestibility varies for fungivores from 60 to 93%, depending on the extent of their digestive adaptations (Cork & Kenagy 1989; Claridge & Cork 1994).

### 2.6.4 Specific Dynamic Action

Animals have an increase in MR following the ingestion of food, that is referred to by numerous terms, including darmarbeit, generic dynamic action, thermic energy, thermic effect of food, diet-induced thermogenesis, heat increment, post-prandial calorigenesis, and calorigenic effect. The term specific dynamic action (SDA), adapted from the German *specifisch-dynamische wirkung* used by Rubner (1902), is the most commonly used term (McCue 2006; Secor 2009). For mammals, SDA results in an increase in MR of between one and two times BMR or RMR, lasts less than 24 hours, and has a coefficient (100 × energy of SDA/energy of the meal) between 4 and 50. However, for other animals—particularly those that feed infrequently, such as pythons—SDA may lead to an increase in MR by as much as 17 times RMR and last for up to 18 days (McCue 2006; Secor 2009). The characteristics of the meal influence the magnitude of the SDA; SDA is generally higher for high- compared to low-protein meals, large compared to small meals, and hard-to-digest compared to readily digestible meals.

The mechanisms for SDA can be divided into three categories: (1) pre-absorptive processes, such as chewing, enzyme and acid secretion, peristalsis, protein catabolism, intestinal re-modelling and food warming; (2) absorptive processes, such as intestinal absorption and active transport of nutrients across the intestinal epithelium and the metabolic effect of digestive hormones; and (3) post-absorptive processes, such as the cost of protein, ketone, and glycogen synthesis, urea production and renal excretion, amino acid oxidation, and growth. Clearly, the mechanisms involved in SDA are complex and inter-related, and despite more than 100 years of research, we still have a poor understanding of this phenomenon (McCue 2006; Secor 2009).

## 2.7 Water and Solute Balance

Water is the universal solvent, and is the primary constituent of animals, comprising 60–80% of a mammal's body (e.g. Richmond et al. 1962; Withers 1992; Schulte-Hostedde et al. 2001; Schmid et al. 2003). Most body water is found in the largest organs (skin and muscle), which also have a high percentage water content. Adipose tissue and the skeleton have a lower percentage water content and thus a smaller proportion of total body water (Pitts 1963). Body water is distributed between two body water spaces, the intracellular and extracellular fluid spaces. The intracellular fluid content is 30–40% of the body (about two-thirds of the total body water). Extracellular fluid consists of interstitial fluid (16%), blood plasma (4.5%), and lymph (2%), and is important for maintaining a constant, isoosmotic environment that buffers the intracellular environment from a mammal's external environment. Transcellular fluid is a specialized extracellular fluid space, separated from other extracellular fluids by a continuous layer of epithelial cells. It includes pleural, intra-occular, peritoneal, synovial, and cerebrospinal fluids. Transcellular fluids are a small proportion of the body (Pitts 1963).

There are three general patterns of solute regulation amongst animals: (1) iono- and osmo-conform, (2) iono-regulate and osmo-conform, and (3) iono- and osmo-regulate. All mammals regulate both their ionic composition and their osmolarity independent of that of their external environment. Generally, the osmotic concentration of mammalian body fluids is 300–400 mOsm (Waymouth 1970; Oritz 2001). Since the cell membrane is quite elastic and permeable to water, intracellular and extracellular fluid has the same osmotic concentration. While animal cells are relatively permeable to water, they are relatively impermeable to solutes, so that when an osmotic difference does exist across a cell membrane, the cell volume will change by water movement rather than remain constant by solute exchange, at least initially. Animal cells in a hypertonic environment will lose water and shrink, whereas those in a hypotonic environment will gain water and expand, and—without the rigid cell wall that is present in plant cells—may burst. Over the long term, intracellular mechanisms (such as amino acid pool dynamics) can maintain cell volume constant, or at least limit volume change (Hoffman et al. 2009). Animals that rapidly drink large amounts of dilute water can place their tissues (especially blood cells) at hypo-osmotic risk. Camels (Perk 1963, 1966) and elephants (Silva & Kuruwita 1994) have red blood corpuscles with a low sensitivity to hypotonic exposure (low osmotic fragility). Goats do not, but their rumen retains the ingested water and releases it gradually (Chosniak & Shkolnik 1977).

The ionic compositions of extra- and intracellular fluids differ markedly, despite having the same osmotic concentration. The intracellular fluid varies with cell type, but generally has a relatively high organic content, low ionic content, and a neutral pH. Potassium and magnesium are the main cations and protein and organic phosphate the major anions (Pitts 1963; Withers 1992). In contrast, in the extracellular fluid, sodium is the primary cation, and chloride, bicarbonate, and protein the main anions. The protein concentration is generally lower in interstitial fluid than plasma and, as a consequence, interstitial fluid has a slightly lower concentration of sodium and a higher concentration of anions such as chloride, in accordance with the Gibbs-Donnan rule (Pitts 1963).

## 2.7.1 Water and Solute Intake

For mammals, as for most terrestrial animals, water is gained via three pathways: from drinking, from preformed water in food, and from metabolic water production. The sum of these avenues for intake must balance the sum of water losses (see 3.6.1) for homeostasis of hydration state, but short-term inbalance is buffered by water storage or loss.

Most mammals drink if free water is available, but in general they do not rely on drinking to the same extent as birds. The ability of birds to fly large distances to water probably means that small mammals (except bats) have experienced more selection pressure to limit drinking requirements. Large terrestrial mammals can regularly travel long distances (as much as 50 km) to drink water (Schmidt-Nielsen 1964), or seasonally migrate to maintain access to drinking water (and food; Western 1975; section 4.1.2). Large mammals can ingest large quantities of water when it is available; for example, camels (*Camelus dromedarius*) and elephants (*Loxodonta africana*) can consume more than 100 l in minutes (Schmidt-Nielsen 1964; Silva & Kuruwita 1994). Mammals in very mesic and aquatic environments generally have a ready supply of drinking water, although marine mammals must be able to cope with the salt load if they ingest seawater or consume food that is isosmotic with seawater. In contrast, many mesic and most small, arid-habitat mammals cannot reliably obtain drinking water, and must balance their water budget with preformed and metabolic water.

Preformed water is the water that is present in food, but preformed water content varies markedly for different diets. Insectivorous and carnivorous diets have a relatively high preformed water content, reflecting the high (60–90%) water content of animal tissue. Herbivorous diets vary considerably in preformed water content; dry vegetation may contain only a few percent water, whereas succulent plants and fruits have high water contents. Nectar, sap, and blood have very high water contents. The preformed water content and digestibility of food have a major effect on the water requirements of an animal, relative to its energy requirements, at least in the absence of drinking. The water economy index (WEI)—the water turnover relative to energy turnover—and the relative water economy (RWE)—the ratio of metabolic water production to evaporative water loss (EWL)—are influenced by the water content, digestibility, and chemical stoichiometry of the diet (MacMillen & Hinds 1983; Cooper & Withers 2004b). The WEI is highest for invertebrativores (0.11–0.36), intermediate for carnivores (0.07–0.17) and herbivores (0.15–0.27), and lowest for granivores (0.04; Nagy & Peterson 1988; Cooper & Withers 2004b). This range in WEI necessitates adjustment of the avenues of water loss (e.g. drinking, kidney function) for the maintenance of water homeostasis.

Metabolic water production (MWP) is water synthesized by aerobic metabolism, and the amount of water produced depends on the substrate being metabolized

(Withers 1992; McNab 2002). The metabolism of fat produces the most water per gram of substrate (1.14 g $H_2O$ $g^{-1}$), carbohydrates are intermediate (0.55 g $H_2O$ $g^{-1}$), and proteins are lowest (0.42 g $H_2O$ $g^{-1}$; Table 2.7). Because MWP requires molecular oxygen, which must be supplied via the respiratory system, MWP is coupled to the respiratory component of evaporative water loss ($EWL_{resp}$). Consequently, the ratio of metabolic water formed relative to the oxygen consumed (MWP / $VO_2$) may be a better measure of the effective MWP for various foods than MWP per gram of food. Carbohydrates produce slightly more water per gram of oxygen metabolized ($0.66\,mg\,H_2O\,ml\,O_2^{-1}$) than fats ($0.57\,mg\,H_2O\,ml\,O_2^{-1}$) and proteins ($0.39\,mg\,H_2O\,ml\,O_2^{-1}$). Mammals with a higher MR will have a higher MWP and potentially a better water balance; birds are potentially 'better' than mammals because they have a higher respiratory extraction of $O_2$.

The preformed water content of the energy substrate adds an extra level of complexity to the interpretation of the role of substrate metabolism to water balance. Fat stores typically have a low water content, whereas the water content of protein is higher and the water content of carbohydrate is higher than both fat and protein. Consequently, the total water made available by the metabolism of a substrate is much higher than the metabolic water production per se for carbohydrates, and considerably higher for protein, while for lipids, metabolism remains the primary source of water. There are trade-offs from the associated increases in MHP and EWL that must be balanced if MWP is to make a significant contribution to the overall water budget. Nonetheless, MWP is the primary source of water for many mammals; it and preformed water are the only water sources required by various granivorous rodents that can survive without drinking (e.g. kangaroo rats, *Dipodomys*, Schmidt-Nielsen 1964; gerbils and jerboas, Kirmiz 1962; El-Husseini & Haggag 1974; Australian hopping mice, *Notomys*, MacMillen & Lee 1967; some Southern African rodents, Withers et al. 1980; but not South American rodents, Gallardo et al. 2005).

**Table 2.7** Metabolic water production (MWP) from substrate metabolism, the 'preformed' water content of substrates, and the total water made available by substrate metabolism. See Withers (1992).

| | MWP g $H_2O$ per g substrate | MWP/$VO_2$ g $H_2O$ per l $O_2$ | Preformed water g $H_2O$ per g substrate | Total water g $H_2O$ per g substrate |
|---|---|---|---|---|
| Fat | 1.14 | 0.57 | 0.03[a] | 1.17 |
| Carbohydrate | 0.55 | 0.66 | 2.70[b] | 3.25 |
| Protein | 0.42 | 0.39 | 0.40[c] | 0.82 |

[a]triglycerides have about 30 mg $H_2O$ $g^{-1}$; calculated from Thorsteinsson et al. (1976) and Herring (2002)
[b]glycogen contains about 2,700 mg $H_2O$ $g^{-1}$ (see Sherman et al. 1982)
[c]minimum hydration for functioning is about 400 mg $H_2O$ $g^{-1}$ (Rupley & Careri 1991)

Ion balance is conceptually similar to, but simpler than, water balance because animals obtain ions only from drinking and food; there is no 'metabolic' gain of ions. Drinking dew or rainwater results in a trivial ion intake, but marine mammals and those that drink from saline inland water bodies incur a substantial ion load from drinking (or from inadvertently imbibing water with food). Ions in seawater are predominantly sodium and chloride, but inland water bodies may have quite different ionic compositions due to the leaching of salts from the earth, and the intake of high levels of ions such as magnesium, calcium, sulphate, and phosphate can lead to physiological problems (e.g. diarrhoea and kidney dysfunction). Ion intake via food varies with the diet. Carnivorous and invertebrativorous mammals ingest food with an ionic composition and concentration similar to their own body fluids. Mammals feeding on marine invertebrates have a higher dietary ion load than terrestrial carnivores or insectivores, as most marine invertebrates iono- and osmo-conform to seawater. Herbivores have a relatively high potassium and low sodium intake.

### 2.7.2 Water and Solute Loss

For mammals, as for most terrestrial animals, water is primarily lost via three routes: in the urine, in the faeces, and by evaporation from the skin and the respiratory tract. There can be other minor avenues for water loss (e.g. saliva, sweat), but they normally contribute little to total loss. Mammals do not have extra-renal avenues for salt regulation, such as the cephalic salt glands of birds and reptiles, the cutaneous ion transporters of amphibians, or gill ion transporters of fish, so the kidney is essential for osmoregulatory and ionoregulatory homeostasis, making it the major organ responsible for the control of water and ion balance through its production of urine. The kidney is also an important avenue for the excretion of toxic materials, such as nitrogenous wastes generated from protein metabolism, and foreign compounds.

Urine is formed by the kidney. The functional unit of the kidney is the nephron, and the arrangement of nephrons within the kidney defines its gross structure, an outer renal cortex and central renal medulla, the latter consisting of an outer medulla and an inner medulla that often forms a large cone-shaped renal papilla (Figure 2.18). Kidney function is closely related to the anatomical organization of these various segments of the nephron and their arrangement in the kidney (see 3.6.3; Withers 1992). There are four basic functions of the nephron: (1) filtration of blood (plasma), forming what will eventually be urine; (2) reabsorption of useful solutes and water; (3) secretion of specific wastes and toxins; and (4) in mammals and some birds, osmo-concentration of urine to higher than plasma osmolality. These four functions are associated with different segments of the nephron. The proximal end of the nephron is

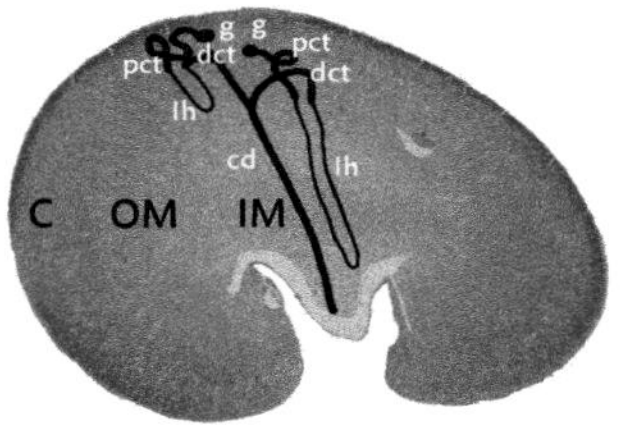


**Figure 2.18** Kidney cross-section of a numbat (*Myrmecobius fasciatus*) showing the overall anatomical arrangement of renal cortex (C), outer medulla (OM), and inner medulla (IM), and the general segments of a nephron (g, glomerulus; pct, proximal convoluted tubule; lh, loop of Henle; dct, distal convoluted tubule; cd, collecting duct) for a cortical nephron (short lh) and juxtamedullary nephron (long lh). Image courtesy of C. E. Cooper and P. C. Withers.

the glomerulus, consisting of a Bowman's capsule that surrounds the glomerular capillaries. There is close association of the inner capsule layer (podocytes) with the capillaries, forming filtration slits, where glomerular filtration forms glomerular filtrate. The next segment, the proximal convoluted tubule, is an important site of both solute and water reabsorption, by active and passive transport mechanisms. The next segment is the loop of Henle, which creates an osmotic gradient in the renal medulla that is used to osmotically concentrate urine. The distal convoluted tubule then forms part of the juxta-glomerular apparatus that is important in feedback control of the glomerular filtration rate and provides reabsorption of many solutes, but is relatively impermeable to water (hence it is often called the diluting segment of the nephron). The final nephron segment is the collecting duct, which passes through the renal medulla and drains urine into the renal pelvis.

The mammalian kidney (and to a lesser extent the avian kidney) is highly specialized to produce urine that is more concentrated than blood plasma (Gottschalk 1987; section 3.6.3). The collecting duct traverses the medulla; the permeability of its walls is under hormonal control. The presence of anti-diuretic hormone (ADH) increases the permeability of the collecting duct walls and allows water to be osmotically drawn out of the duct into the surrounding medullary tissues, forming an osmotically concentrated urine. Birds are the only other vertebrate group that can produce concentrated urine, but they do not achieve the high urine concentrations of many mammals. Birds generally have a urine osmotic concentration approximately 2 times the plasma concentration, whereas some mammals can concentrate urine to 30 times their plasma concentration (Schmidt-Nielsen 1997; MacMillen & Lee 1967; McNab 2002).

Water is also lost via the faeces, with the amount depending on the percentage water content of faeces and the total amount of faeces eliminated. The faecal water content can vary from $98 - 280\,g\,H_2O\,g_{dry}^{-1}$ for a range of mammals when fully hydrated to $68 - 212\,g\,H_2O\,g_{dry}^{-1}$ when dehydrated (Maloiy 1973). The volume of faeces eliminated depends on the amount of food ingested and its digestibility. For a small desert rodent (*Perognathus parvus*) with ad lib. water, the water lost in faeces varied markedly with digestibility, from 5 mg $H_2O\ g^{-1}\ day^{-1}$ for a highly digestible diet (millet seed, 94.5% digestibility) to 53 mg $g^{-1}\ H_2O\ day^{-1}$ for a poorly digested diet (bran, 66.1%) at a faecal water content of 51–62% (Withers 1982). The mice consuming bran had a slightly lower food intake, hence energy consumption, but a higher faecal production reflecting the lower digestibility, so their faecal water loss per energy consumed of 34 mg $H_2O\ g^{-1}\ kJ^{-1}$ for bran was higher than 4.0 mg $H_2O\ g^{-1}\ kJ^{-1}$ for millet. However, these pocket mice had ad lib. water, so the mice consuming bran drank more, to sustain their much higher faecal water loss. Water-deprived mice reduce their food intake, so it is difficult to make these same calculations. Instead, a simple model of faecal water loss per metabolic rate (as mg $kJ^{-1}$) at varying faecal water contents for diets of differing digestibility indicates a marked effect of digestibility on faecal water loss; for example, 18.4 (bran), 6.1 (soy), 2.7 (millet), and 0.3 (nectar) mg $kJ^{-1}$ for 40% faecal water content (Table 2.8). Faecal water loss increases dramatically with faecal water content, by about two times for 60% and six times for 80% compared to 40% water content.

Total evaporative water loss ($EWL_{tot}$) is an important component of the water budget, accounting for up to 75% of total water loss in mammals (Schmidt-Nielsen & Schmidt-Nielsen 1952; Hinds & MacMillen 1985; Tracy & Walsberg

**Table 2.8** Energy content and digestibility of bran, soy, or millet seed diets for a granivorous rodent (*Perognathus parvus*; data from Withers 1982), and estimates for a highly digestible nectar diet, and the calculated faecal water loss per metabolic rate calculated for these diets.

| | Bran | Soy | Millet | Nectar |
|---|---|---|---|---|
| Energy content (kJ $g^{-1}$ dry weight) | 18.6 | 21.8 | 17.1 | 25 |
| Digestibility (% dry mass) | 66.1 | 83.4 | 94.5 | 99 |
| **Metabolic-specific Faecal Water Loss (mg $H_2O\ kJ^{-1}$)** | | | | |
| 40% digestibility | 18.4 | 6.1 | 2.7 | 0.3 |
| 60% digestibility | 41.4 | 13.7 | 5.1 | 0.6 |
| 80% digestibility | 110.3 | 36.5 | 13.6 | 1.6 |

**Table 2.9** Cutaneous resistance of various tetrapod vertebrate animals relative to free water, including typical, cocooned, and 'waterproof' amphibians, reptiles, birds, and mammals (in bold type). Adapted from Withers (1992, 1998b) and Lillywhite (2006); mammal values estimated from the literature.

| Species | Resistance ($R_{H_2O}$) ($s\ cm^{-1}$) |
|---|---|
| Free water surface | ≈ 1 |
| Trilling frog (*Neobatrachus centralis*) | 2 |
| Water-holding frog (*Cyclorana australis*) | 2 |
| Leopard frog (*Rana pipiens*) | 2 |
| Cane toad (*Bufo marinus*) | 2 |
| Water-holding frog (*Cyclorana australis*) | 2 |
| Spiny soft-shell turtle (*Trionyx spiniferus*) | 6 |
| Blue-sided tree frog (*Agalychnis annae*) | 10 |
| Striped treefrog (*Rhacophorus leucomystax*) | 14 |
| Bocage's burrowing frog (*Leptopelis bocagei*) cocooned | 40 |
| Trilling frog (*Neobatrachus centralis*) cocooned | 57 |
| Mallard (*Anas platyrhyncos*) | 61 |
| Alligator (*Alligator mississippensis*) | 65 |
| **Brown big-eared bat (*Plecotus auratus*)** | **81** |
| **Daubenton's bat (*Myotis daubentonia*)** | **87** |
| Zebra finch (*Poephila castanotis*) | 112 |
| Dainty green tree frog (*Litoria gracilenta*) 'waterproof' | 118 |
| Poorwill (*Phalaenoptilus nuttali*) | 125 |
| Roadrunner (*Geococcyx californicus*) | 139 |
| Llanos frog (*Lepidobatrachus llanensis*) cocooned | 163 |
| **Merriam's kangaroo rat (*Dipodomys merriami*)** | **166** |
| **Pygmy gerbil (*Gerbillurus paeba*)** | **175** |
| **Monito del Monte (*Dromiciops australis*)** | **187** |
| Anole (*Anolis carolinensis*) | 196 |
| **California ground squirrel (*Spermophilus beechyi*)** | **199** |
| White-winged dove (*Zenaida asiatica*) | 200 |
| **House mouse (*Mus musculus*)** | **201** |
| **Striped mouse (*Rhabdomys pumillio*)** | **208** |
| Water-holding frog (*Cyclorana australis*) cocooned | 214 |
| **Bailey's pocket mouse (*Chaetodipus baileyi*)** | **223** |

| | |
|---|---|
| Sharp-nosed tree frog (*Hyperolius nasutus*) 'waterproof' | 257 |
| **Cape short-eared gerbil (*Desmodillus auricularis*)** | **283** |
| Chukar partridge (*Alectoris chukar*) | 309 |
| **Brush-tail possum (*Trichosurus vulpecula*)** | **333** |
| Monkey frog (*Phyllomedusa hypochondrialis*) 'waterproof' | 364 |
| **Humans (*Homo sapiens*)** | **377** |
| Foam-nest tree frog (*Chiromantis rufescens*) 'waterproof' | 404 |
| Mexican burrowing frog (*Pternohyla fodiens*) cocooned | 457 |
| Diamond-backed rattlesnake (*Crotalus atrox*) | 1,011 |
| Desert night lizard (*Xantusia vigilis*) | 3,310 |

2000; Withers & Cooper 2012). The $EWL_{tot}$ can be partitioned into two components, evaporation across the skin ($EWL_{cut}$) and from the respiratory tract ($EWL_{resp}$). The skin resistance to evaporation is relatively high for mammals, about 50–300 s $cm^{-1}$ (Withers 1992; see 5.2.3; Table 2.9), because the stratum corneum (the outer layer of keratinized cells of the epidermis) reduces the skin's permeability to water. The stratum corneum, together with a layer of fur, which acts as a barrier to water loss, and the boundary layer of moist air above the fur surface increases the resistance and reduces the $EWL_{cut}$. Birds also have a low $EWL_{cut}$ because of high cutaneous resistance and their feathers (Webster et al. 1985).

The terminal alveoli of the lungs have a high surface area and are kept moist, enhancing their gas exchange properties; air in the alveoli is 100% saturated at $T_b$. Water evaporates from the nasal cavity and upper respiratory tract as a free water surface (resistance of 1 s $cm^{-1}$); the high surface area (endothermic mammals and birds have turbinates that increase the nasal surface area; section 1.2.4) of the upper respiratory tract ensures that inspired air is saturated, at $T_b$, before it enters the alveoli (Schroter & Watkins 1989). $EWL_{resp}$ can be reduced by countercurrent heat and water exchange in the nasal passages. Nevertheless, $EWL_{resp}$ is a considerable fraction of $EWL_{tot}$ in mammals; estimates range from less than 25% to more than 80% for various mammals (Cooper 2003; Withers & Cooper 2012). However, it is methodologically difficult to accurately partition total EWL into its respiratory and cutaneous components, and estimates can vary even for the same species (e.g. Merriam's kangaroo rat, *Dipodomys merriami*, from less than 44% to more than 70%; Chew & Dammann 1961; Schmidt-Nielsen 1964; Tracy & Walsberg 2000).

## 2.8 Locomotion

Vertebrates have an internal bony skeleton that provides support and allows for the generation of locomotory forces, with an axial skeleton (skull, vertebral column, ribs, sternum) and an appendicular skeleton (forelimbs, hindlimbs, and pectoral and sacral girdles). The appendicular skeleton provides the primary capacity for limb-based locomotion on land, based on a lever system operating with voluntary skeletal muscles that act on the limb bones. The axial skeleton can also contribute substantially to locomotion (e.g. running and swimming). Mammals are unique amongst vertebrates in having species that inhabit all available environments: terrestrial, fossorial, aerial, and aquatic habitats (see 1.4.3). Birds have a similar variety of adaptation, but there are no truly fossorial or completely aquatic birds. As such, mammals have evolved an impressive array of adaptations for moving in the particular habitat(s) they live in.

In the limbs, rigid bones articulate about a fixed fulcrum point with other appendicular or axial bones, and lever movement is powered by the contraction of skeletal muscle. There are three classes of lever systems, defined by the relative locations of the muscular force and the load force, relative to the fulcrum. A simple, first-class lever system (e.g. a see-saw) consists of a rigid lever (the see-saw) and a fulcrum (Figure 2.19). A force at one end (i.e. the 'muscle' force) acts against a load at the other end. When in balance (equilibrium), the 'muscle' and load torques are equal; that is, $ML_M = LL_L$ where M is the muscle force, $L_M$ is the distance at which the muscle force acts on the fulcrum ($M\,L_M$ is the muscle torque), L is the load force,

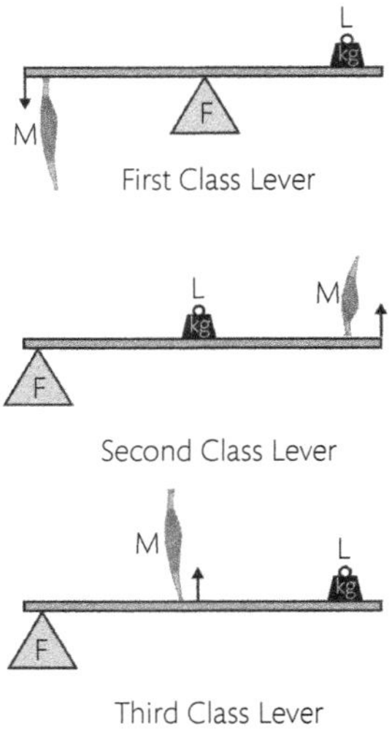


**Figure 2.19** Lever systems vary in relative positioning of the load (L), muscle action (M), and fulcrum (F).

and $L_L$ is the distance at which the load force acts on the fulcrum ($L\ L_L$ is the load torque). An example is the neck muscles pulling on the back of the skull to lift the head. A second-class lever system (e.g. a wheelbarrow) has both the muscle and load forces on the same side of the fulcrum, with the muscle force applied farther from the fulcrum than is the load. An example is a plantigrade mammal (human) standing on tip-toe; the gastrocnemius and soleus muscles lift the body, with the toes being the fulcrum. A third-class lever system (e.g. a crane and jib) has both the muscle force and load acting on the same side of the fulcrum, but with the muscle force applied between the fulcrum and the load. The forelimb is a good example: the fulcrum is the elbow joint, about which the muscle force acts on the radius-ulna near the elbow joint to lift a weight held in the hand (farther from the elbow joint).

In any lever system, the farther the muscle effort is located from the fulcrum, compared to the load from the fulcrum, the greater is the force that can be exerted on the load by the muscle (Withers 1992). But there is a trade-off between the force that can be applied and the distance that the load can be moved. A muscle that inserts farther from the fulcrum than is the load will have a considerable mechanical advantage (MA) or a high ratio of load (L) to muscle effort (ME); that is, $MA = L / ME = \Delta ME / \Delta L$, where $\Delta ME$ is the distance from the muscle insertion to the fulcrum and $\Delta L$ is the distance from the load to the fulcrum. Muscles that insert much closer to the fulcrum than $\Delta L$ have a low MA but a high range of motion (ROM). The range of motion is the distance that a muscle moves a load when the muscle contracts a distance of $\Delta x$. The distance of the load from the fulcrum is proportional to the ROM, and ROM is inversely proportional to MA. The velocity of movement of the load is also proportional to ROM, and inversely proportional to MA.

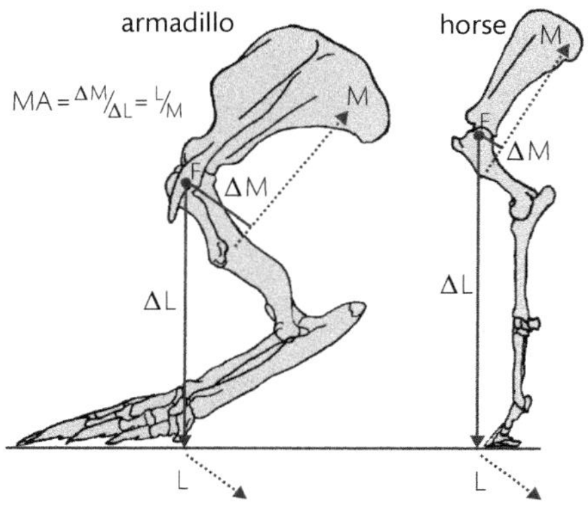


**Figure 2.20** Forelimbs of a burrowing armadillo (*Dasypus*) and cursorial horse (*Equus*), showing their different mechanical advantages ($L/M = \Delta M/\Delta L$) for the teres major muscle; F is fulcrum, L is load, and M is muscle. From Withers (1992) after Maynard Smith and Savage (1956). Reproduced with permission from Elsevier.

The digging forelimb of an armadillo (Figure 2.20) is an example of a third-class lever system geared to provide maximum digging force. The fulcrum is the shoulder joint, about which the in-force (retractor muscles that insert on the humerus) and out-force (contact of the terminal digit with the ground) act. The in-lever arm is the distance from the point that the triceps muscle acts on the ulna to the fulcrum ($\Delta M$); the out-lever arm is the distance from the fulcrum to where the foot contacts the ground ($\Delta L$). For a digging limb, such as this armadillo, the in-lever arm is relatively long (cf. other mammals), which confers a high mechanical advantage—the triceps force exerts a relatively big force on the ground, compared to the situation if the in-lever arm were shorter. In contrast, the forelimb of a cursorial mammal, such as a horse, is adapted in a way that provides an increased range of motion (speed), at the expense of mechanical advantage. Essentially the entire forelimb (humerus, radius/ulna, foot bones) acts as a rigid lever.

Swimming and flying are fundamentally different forms of locomotion from walking, because they depend on fluid dynamic forces, drag, and lift, rather than limbs and lever mechanics (although limbs and lever mechanics are still important for moving the limbs; Nachtigall 1983). The relative roles of drag and lift are important for swimming and flying. Drag is primarily a viscous force ($F_{viscous}$) that depends on body size (e.g. length, l), velocity (v), and the viscosity of the fluid ($\eta$); $F_{viscous} = l\,v / \eta$. Lift is primarily an inertial force ($F_{inertial}$) that depends on the square of body length, the square of velocity, and the density of the fluid ($\rho$); $F_{inertial} = l^2 v^2 / \rho$. Whether an animal swims or flies depends on the relative roles of viscous forces and inertial forces, indicated by Reynold's number ($R_e$; Figure 2.21),

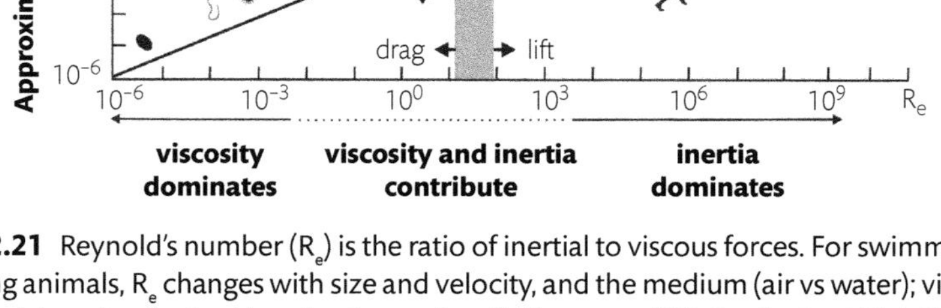

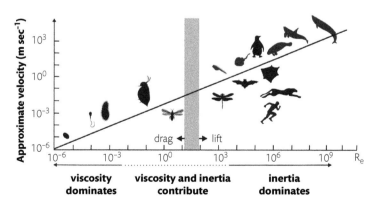


**Figure 2.21** Reynold's number ($R_e$) is the ratio of inertial to viscous forces. For swimming and flying animals, $R_e$ changes with size and velocity, and the medium (air vs water); viscous forces and drag dominate at low $R_e$ whereas inertial forces and lift dominate at high $R_e$. Silhouettes indicate the approximate $R_e$ range for some swimmers, fliers, and runners. Modified from Nachtigall (1983) and Gazzola et al. (2014).

the ratio of inertial to viscous forces; $R_e = (l^2v^2 / \rho) / (l\, v / \eta) = l\, v\eta / \rho = l\, v / \nu$, where the ratio $\eta/\rho$ is a physical constant called the kinematic viscosity ($\nu$; $10^{-6}$ for water and $16\ 10^{-6}\ m^2\ s^{-1}$ for air; Nachtigall 1983). $R_e$ varies dramatically, from $< 10^{-6}$ for unicells to $> 10^8$ for large animals; its significance for animal locomotion is that $R_e$ reflects whether the dominant force is drag (at $R_e < 10$) or lift (at $R_e > 100$). For small animals in water, the length and velocity are low while the kinematic viscosity is high, meaning that drag predominates; for very large animals in water (e.g. marine mammals and birds), lift predominates. Thus, large swimming animals (like dugongs and cetaceans) and medium-size aerial animals (like flying lemurs and bats) fly, relying on lift for propulsion.

### 2.8.1 Walking and Running

Regardless of the details of the lever systems used for terrestrial locomotion, the energy required for locomotion is provided by muscle contraction, moving the centre of mass forwards (and generally in the process of walking or running, of moving the centre of mass up and down). Most terrestrial mammals are quadrupedal, using both the fore- and hindlimbs for walking or running. The various fore- and hindlimb bones form lever systems and for some species, flexion of the axial skeleton contributes substantially to locomotion (e.g. vertebral flexion of running cheetah; Hildebrand 1961). Bipedal mammals use the hindlimbs for hopping or walking, and the forelimbs of these mammals can have specialized roles, such as grasping and manipulating objects (e.g. kangaroos) or brachiating (many primates).

Locomotor gait, or footfall pattern, is defined by the relative phase (RP; the timing through a stride when each foot contacts the ground) and duty factor (DF; the fraction of the stride that a foot is on the ground). By convention, the front left foot is the reference for the relative phase of each foot (McGhee 1968; McNeill Alexander 1981, 1986). For a quadruped, the relative phases of the four feet are defined by $(0, \gamma, \varepsilon, \delta)$, where RP is 0 for the left front foot, while $\gamma$ is the front right foot, $\varepsilon$ is the back right, and $\delta$ is the back left. Examples of various gaits are an amble (0, 0.5, 0.25, 0.75), trot (0, 0.5, 0, 0.5), canter (0, 0.3, 0, 0.7), and transverse gallop (0, 0.1, 0.6, 0.5). For a biped, a symmetrical walk is (0, 0.5) and a hop is (0, 0).

When walking slowly, a quadruped can maintain equilibrium (and not tip over) if it has at least three feet on the ground at all times (i.e. the duty factor is at least 0.75); the relative phases must always support the centre of mass within a triangle formed by the feet that are on the ground. If the duty factor is 0.75, then the optimal relative phases are (0, 0.5, 0.25, 0.75). Faster-walking animals do not need to maintain equilibrium, and the duty factor and gait vary, presumably being optimized for various reasons such as stability and minimum cost of transport. For running quadrupeds, there is also a change in the gait. Most mammals trot, camels pace, and elephants amble when running slowly. Ambling is a gait used by some large mammals (e.g. elephants) and primates, that is a moderate speed gait where there is no phase with all four feet aerial (Schmitt et al. 2006). An amble is the fastest gait for elephants, which maintain the same footfall pattern at higher speeds, without an aerial phase (Hutchinson et al. 2006). For most mammals, canter is the common gait for intermediate speeds, and gallop or bound achieves high speed.

## 2.8.2 Gliding and Flying

Gliding and powered flight have evolved a number of times amongst animals (Dudley et al. 2007). A variety of mammals have fore- and hindlimbs that are specialized for gliding (flying lemurs, flying squirrels, marsupial gliders, *Volaticotherium*) or powered flight (bats). Gliding mammals, such as the sugar glider (*Petaurus breviceps*, Figure 2.22), have gliding membranes or wings, generally extending from their wrists to their ankles, which form a cambered airfoil structure. Bats' wings are a naked membrane, supported primarily by the elongate digits of the hands, and in many bats also the hind limbs. Regardless of whether they have gliding or powered flight, the wings act as an aerofoil that generates the aerodynamic forces, lift and drag. The difference between gliding and powered

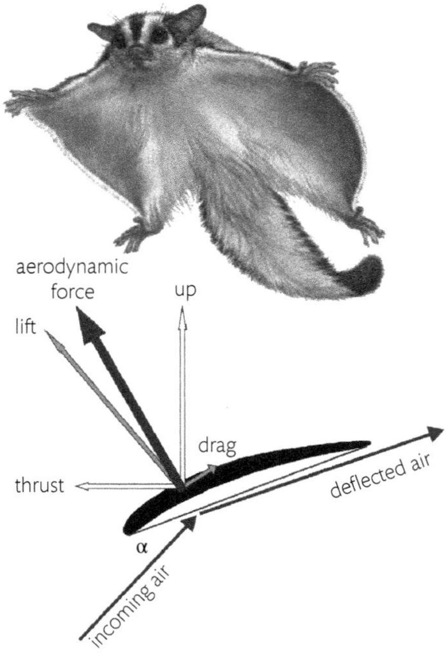


**Figure 2.22** Top: Gliding marsupial, the sugar glider, *Petaurus breviceps*. Modified from Macdonald (2010). Reproduced with permission from Oxford University Press. Bottom: Aerodynamic forces for an aerofoil. The aerodynamic force is resolved into lift (perpendicular to incoming airstream), drag (parallel to incoming airstream), thrust, and upwards forces. $\alpha$ is the angle of attack.

flight is whether muscular contraction is used to overcome drag and provide indefinite horizontal flight (as well as ascent, descent, and sometimes incredibly acrobatic flight manoeuvres). 'Ballistic' leapers, in contrast to gliders, are animals that simply jump, then accelerate downwards under the influence of gravity; they do not generate any lift to counteract gravity. Aerodynamic modelling indicates that gliding confers on average 2.5 times greater flight distance than leaping for the flying lemur (Malayan colugo, *Galeopterus variegatus*), without a constraint on vertical landing velocity, which can be excessive. Gliding distance is 20 times greater than leaping distance with the additional constraint of a typical vertical landing velocity (Byrnes et al. 2015). Thus, gliding not only increases the horizontal distance travelled, but also helps to prevent excessive landing forces.

The effectiveness of gliding is indicated by the glide angle (the minimum angle of gliding from the horizontal) in still air, or the glide ratio (the ratio of the forward distance travelled relative to the vertical distance dropped; glide ratio = cotangent of glide angle). The glide ratio is also equal to the lift/drag ratio; L/D = glide ratio = cotangent (glide angle). Parachuting can be defined by the glide angle being greater than 45°. Gliding, however, is defined as a glide angle less than 45° (Oliver 1951), or if there is active control of the aerodynamic forces (e.g. to turn, change velocity/glide angle, or stall for landing; Dudley et al. 2007; Byrnes & Spence 2011). Glide angle decreases from gliding frogs to lizards and snakes, gliding mammals, and is lowest (best) for specialized soaring birds (Table 2.10).

A wing (aerofoil) deflects the incoming airstream, and a cambered aerofoil is more effective and less likely to stall than a flat aerofoil. The momentum change

**Table 2.10** Glide angle and glide ratio (= lift/drag ratio) for various gliding vertebrates. Values from Lekagul and McNeely (1977), Rayner (1981), Lindhe Norberg et al. (2000), Alexander (2002), and Park and Choi (2010). Glide ratio = cotangent of glide angle.

| | Glide angle (°from horizontal) | Glide ratio (= lift/drag ratio) |
|---|---|---|
| Gliding gecko (*Ptychozoon*) | 45 | 1 |
| Flying frog (*Hyla*) | 18 | 3 |
| Flying fish (*Cypselurus*) | 15 | 4 |
| Gliding dragon (*Draco*) | 15 | 4 |
| Gliding snake (*Chrysopelea*) | 15 | 4 |
| Flying squirrel (*Petaurista*) | 11 | 5 |
| Gliding possum (*Petaurus*) | 11 | 5 |
| Flying lemur (*Cynocephalus*) | 5 | 11 |
| Gliding fruit bat (*Pteropus*) | 4 | 13 |
| Thermal-soaring vulture (*Cathartes*) | 4 | 16 |
| Dynamic-soaring albatross (*Diomedea*) | 3 | 19 |

that occurs when the airstream is deflected imposes an equal and opposite force on the aerofoil, and this aerodynamic force is perpendicular to the deflected airstream. This aerodynamic force can be resolved into lift (perpendicular to the incoming airstream) and drag (parallel to the incoming airstream; Figure 2.22). For a gliding mammal, like a sugar glider, the incoming airstream is directed at an angle upwards, because of the downward glide path. The glider does not change velocity, but continues on its glide path until it reaches the ground. The glider could maintain elevation if the aerodynamic force was resolved into an upward force that equalled body mass, but thrust would be insufficient to sustain its horizontal speed. Soaring does not appear to be very important for mammalian gliders, or bats, whereas many birds are highly proficient at gliding using updrafts off objects (incline soaring), waves over the ocean (dynamic soaring), or thermal updrafts (thermal soaring). Although bats use flapping (powered) flight and can glide for short times and distances, they generally have a poor gliding performance compared to soaring birds (Lindhe Norberg et al. 2000), and any use of thermal or dynamic soaring is restricted because they are mainly nocturnal and none are marine.

The dynamics of lift and drag forces are changed if the airstream is not horizontal, but is directed up and back by the aerofoil being pulled down relative to the body by muscular contraction: the downstroke of powered flight (or when gliding downwards or into an upward-directed airstream; e.g. deflected off a hill; Nachtigall 1983; Withers 1992). Since aerofoils—at least for fliers that are large enough and fast enough at sufficiently high Reynold's numbers—produce much more lift than drag, the vertical force can counterbalance the weight, and horizontal flight can be sustained. Now the incoming airstream is directed upwards by the downwards motion of the wing (not the body, which moves horizontally), so the vertical force can balance the weight and there is a net forward thrust. Although a flier using a powered downstroke can maintain its forward velocity and its vertical height, it actually has to gain height or forward velocity during the powered downstroke to make up for the loss in height and velocity during the upstroke.

The evolutionary pathway for gliding, and especially the spectacular capacity for the powered flight of bats (and the other extant vertebrate fliers, birds), is largely conjectural, but there are two main theories. The 'arboreal' theory (Marsh 1880) suggests that gliding, then powered flight, evolved in arboreal animals from ballistic leaping; this seems to be a reasonable scenario for the evolution of mammalian gliders, all of which are essentially arboreal leapers that glide (e.g. Dudley et al. 2007; Byrnes & Spence 2011). The 'cursorial' theory of flight (Williston 1879) suggests that birds evolved flight in a sequence beginning in running bipeds with a series of jumps with the forelimbs used as wings for lift, thrust, and stability (Ostrom 1974). It is now widely accepted that birds evolved from bipedal theropod dinosaurs (Heers & Dial 2012). Variations on the cursorial theory include the 'wing-assisted incline running' model (Dial 2003a; Bundle & Dial 2003) and the 'pouncing proavis' model, where a pouncing raptorial predator initially used the forelimbs for drag-based then lift-based leaping and the control of body position (Garner et al. 1999).

Powered flight evolved independently in bats and birds, and in both groups has the same suite of morphological, metabolic, and aerodynamic constraints and adaptations, so it is not surprising that there are many similarities between flying bats and birds (Hedenström et al. 2009). Size range overlaps considerably, about 2 g to 1.5 kg (bats) or 15 kg (birds), and the power requirements, and hence the MRs, needed for flying are similar. Birds and bats both have reduced cell DNA content, and can accumulate fat as a hibernation and migration fuel. Similar habitat-related adaptations in wing shape (aspect ratio) and wing-loading (mass per wing area) are apparent. However, the wing beat kinematics differ for bats and birds: their different wing structures (stretched skin membrane in bats vs feathered bird wing) and differing patterns of wing flexing indicate that bats have a more complex wing stroke and more complex wake than birds, and seem better adapted for slow manoeuvrable flight (section 4.7).

## 2.8.3 Swimming and Diving

Semi-aquatic mammals often have their limbs and/or tail modified in a way that enhances swimming ability (e.g. large feet, webbing between digits, and a flattened tail), but they retain some capacity for terrestrial locomotion. More-aquatic mammals, such as seals and sea lions, have specialized flippers that are of more limited use in land-based locomotion. Completely aquatic mammals, such as dolphins and whales, have forelimbs (flippers) and tails (flukes) that are highly specialized for swimming (actually 'flying'; see later). Their hindlimbs have been functionally lost as their body has become elongated (in some species, vestigial hindlimb bones are retained, but are no longer connected to the axial skeleton); they are incapable of terrestrial locomotion (Bejder & Hall 2002).

Aquatic animals do not have to support their weight against gravity; animals are nearly neutrally buoyant when submerged because the density of animal tissue is generally about the same as water. Nevertheless, swimming animals still have some issues related to their buoyancy. At different times, an animal might need to float at the surface (positive buoyancy), sink underwater (negative buoyancy), or remain horizontal in the water column (neutral buoyancy). Remaining at the

**Table 2.11** Comparison of fluid-dynamic physical properties of water and air (at 20°C). Values from Nobel (1999).

| Physical Property | Water | Air (dry) |
|---|---|---|
| Viscosity ($\eta$; kg $m^{-1}$ $s^{-1}$) | 1.002 $10^{-6}$ | 0.813 $10^{-5}$ |
| Density ($\rho$; kg $m^{-3}$) | 998.2 | 1.205 |
| Kinematic viscosity ($\nu = \eta / \rho$; $m^2s^{-1}$) | 1.004 $10^{-6}$ | 1.505 $10^{-5}$ |

water surface is generally accomplished by the positive buoyancy provided by air pockets in the fur of mammals or feathers of birds, or subcutaneous fat (blubber). Air pockets and subcutaneous fat also provide thermal insulation. To dive underwater, animals generally rely on muscle-powered swimming movements of the appendages (appendicular—limbs or fins; axial—flukes or tails; Nowacek et al. 2001). Many air-breathing animals that dive deeply also experience a decrease in buoyancy because the elevated pressure compresses air spaces in the body, and hence decreases buoyancy (Williams et al. 2000; Williams 2001; section 4.5).

The hydrodynamic principle of swimming underwater is often based on the generation of drag forces rather than lift (cf. gliding and powered flight), reflecting the very different physical properties of water compared to air (Table 2.11). The drag force acting on an animal as it moves through water is substantial (compared to air): it is the sum of friction drag, from friction between the water and the animal's surface, and pressure drag, caused by a low-pressure turbulent wake behind the object (and induced drag if there is any generation of lift). The propulsive force required to overcome this drag is often provided by webbed hands and/or feet and tail, which act as 'oars'. The drag of these structures provides forward thrust. However, large swimmers can also use lift to power their locomotion (see later).

Swimming at the water surface, which is common for semi-aquatic and shallow-diving animals, particularly mammals (and birds and reptiles) that must come to the surface to breathe, entails an additional drag due to wave (wake) formation (Biewener 2003). Because this wave drag can be much higher (up to five times) than the hydrodynamic drag, it can considerably increase the metabolic cost of swimming (Williams 1989; see 3.1.2.1). This often considerable wave drag results from the metabolic cost of raising the mass of water in the wave against gravity (the energy is then dissipated as heat or sound). The ratio of the relative importance of an object's inertia and wave drag is the Froude number ($F_r$); relatively small waves are produced at low $F_r$ (large size and low velocity) and wave drag is relatively small, but wave size and drag reach a maximum at $F_r = 0.45$. At higher $F_r$, swimming becomes 'planing' if the shape of the object is suitable, and wave drag is reduced; otherwise, an $F_r$ of 0.45 is a limit on velocity. Few animals probably can benefit from a sufficiently high $F_r$ to plane (possibly aquatic birds landing on the water). At low swimming speeds, the bow wave moves faster than the animal, but at high speeds the animal overtakes its bow wave and is essentially swimming 'uphill' at an elevated metabolic cost. Some animals swim close to the water surface, alternating between submerged swimming and having the body partly out of the water ('porpoising') or even completely out of the water (e.g. some 'flying' fish). The advantage here is that friction is lower in air than water, so that above a certain 'crossover speed', the overall metabolic cost of porpoising or gliding in air is lower than swimming underwater (Au & Weihs 1980; Blake 1983).

For large animals, the hydrodynamic principles of swimming underwater are quite similar to the aerodynamic principles of flight, being based on the same fluid dynamics principles. Reynold's number of large animals is sufficiently high that lift predominates over drag (see Figure 2.21); that is, large swimming mammals (e.g. cetaceans) and birds (penguins) 'fly' underwater.

While they are swimming or diving, some animals can obtain $O_2$ from the water (e.g. via gills). Mammals cannot obtain $O_2$ from water, and so rely on $O_2$ stored either in the lung or in body tissues (e.g. blood, muscle), or use anaerobic metabolism (section 4.5.2). Semi-aquatic and shallow-diving tetrapods can remain aerobic by relying on $O_2$ stores (lungs, blood, muscles) and avoid physiological problems such as 'the bends', but deep-diving mammals, such as many cetaceans and some seals, require particular respiratory, cardiovascular, and metabolic adaptations to survive extended periods of diving and avoid the physiological consequences of exposure to high pressure (e.g. the 'bends', nitrogen narcosis and $O_2$ toxicity; section 4.5.3).

## 2.9 Reproduction and Development

Most vertebrates, including mammals, reproduce sexually, where fertilization of an egg by a sperm forms a diploid embryo, which develops through various developmental stages into an adult. The adult gonads (ovaries and testes) form haploid eggs and sperm. The fertilization of mammalian eggs is internal (a trait shared with some amphibians as well as reptiles and birds). Mammalian mating systems differ from avian systems in that more than 90% of mammalian species are typically polygynous, while more than 90% of bird species are usually monogamous (Clutton-Brock 1989). Mammalian mating patterns include male monogamy, polygyny, and promiscuity, and female long-term and serial monogamy, polyandry, and promiscuity. Female home range, group size and stability, and density and distribution, together with the importance of male assistance in young rearing, are typically associated with these mating patterns for mammals (Clutton-Brock 1989).

After insemination, sperm are typically held in the female's oviduct, either in mycosal crypts (marsupials and insectivores) or bound to the oviductal epithelium (most other mammals). When ovulation occurs, sperm undergo capacitation, are released gradually and are guided to the oocyte by various secretions (Suarez 2008a,b). Some mammals may store sperm for prolonged periods; up to 200 days for some bats (Hosken 1997). This prolonged storage of sperm allows for advantageous timing of reproduction (e.g. by delaying pregnancy and production of young until environmental conditions are more favourable) by separating copulation from fertilization, and may also promote sperm competition (Birkhead 1993). Sperm competition occurs when the sperm of more than one male are present in the reproductive tract of a female during a single oestrus period (Ginsberg & Huck 1989). Large social group size and long periods of sperm storage are factors that contribute to the likelihood of sperm competition. Sperm competition may be an important mechanism by which females can improve the quality of males fertilizing their ova (Birkhead 1993).

Sex is determined genetically in mammals, as it is in most vertebrates (although it is temperature-dependent in some groups, including crocodilians, most turtles, the tuatara, and some squamates). The gonads develop into ovaries (female, genotype XX) unless a Y chromosome is present (i.e. genotype XY) and a gene product from the Y chromosome differentiates the gonads into testes; this is male heterogamety. Some amphibians and reptiles also have XY male heterogamety, but there are also other patterns (ZW, ZZW, ZWW, XXY, XO; Pough et al. 2004; Sarre et al. 2004). Pathenogenesis, where diploid female eggs develop without fertilization by male sperm, occurs in many invertebrates, as well as 70 vertebrate taxa, but has never been observed to occur naturally in mammals. After ovulation, mammalian oocytes typically require sperm penetration to progress past the second metaphase stage of meiosis. If sperm penetration does not occur, they usually degenerate, and if spontaneous activation does occur, development generally does not continue beyond one or two cleavage divisions (Whittingham 1980; Simon et al. 2003).

The developmental biology of tetrapod vertebrates shows an evolutionary progression that has aided adaptation to terrestriality. Lodé (2012) defined five stages in a reproductive development sequence: (1) Ovuliparity is characterized by external fertilization, where the female deposits the eggs externally into the environment. (2) Oviparity occurs when there is internal fertilization of lecithotrophic eggs (eggs with abundant yolk) that then develop outside the female, without any direct nutritive interaction between the female and embryo. (3) Ovoviviparity occurs when there is a period of prolonged internal incubation, often facilitated by the retention of fertilized eggs within the female's oviduct. There is no exchange of nutrients between the female and embryo. (4) Histotrophic viviparity is characterized by matrotrophy, a supply of nutrients to the embryo from the female, in addition to nutrients provided by the yolk. (5) Haemotrophic viviparity occurs when the embryo develops from a yolkless egg, and relies entirely on nutrients supplied by the mother's bloodstream. A placenta (or equivalent) facilitates the exchange of nutrients and wastes between the embryo and the female, and there is direct contact between embryonic and maternal tissues. Mammals are characterized by two of these forms of reproduction: oviparity, as seen in the monotremes (echidna and platypus), and haemotrophic viviparity, as seen in marsupial and placental mammals.

### 2.9.1 Egg-Laying

Egg-laying (oviparity) is typical for vertebrates. The amphibian egg has a flexible gelatinous coat; it is an anamniote egg. In contrast, reptile eggs are amniote, with a leathery or calcareous shell. The evolution of the amniotic (cleidoic) egg by early reptiles freed them from having to reproduce by laying eggs in water. For amniotes, extra-embryonic membranes—the amnion, allantois, and chorion—maintain the embryo in its own pool of water, store metabolic wastes, and provide mechanical protection, respectively (see Figure 1.9). These developments allowed for eggs to be deposited in dry terrestrial environments; the extra-embryonic membranes also form the embryonic components of the placenta of haemotrophic viviparous animals.

Crocodilians, turtles, tuatara, and many squamates are generally oviparous, with internal fertilization. Chelonid, sphenodontid, and crocodilian reptiles lay eggs very early in development; turtles and the tuatara when the embryo has reached the gastrula stage, and crocodiles at the neurula stage. This early oviposition (presumably related to gas exchange limitations within the female) appears to have limited the evolution of reproductive strategies within these groups, as no chelonid, sphenodontid, or crocodilian reptile is viviparous. There is much variation in the stage of development at which oviposition occurs amongst oviparous squamate reptiles, ranging from gastrulation to last-stage embryos within days of hatching. It is within this group that viviparity has evolved numerous times (Andrews & Mathies 2000).

Birds are also oviparous (Causey Whittow 2000). Birds, as may be expected from their phylogenetic affinities with the crocodylians as archosaurs, also lay eggs at an early stage of embryological development, the gastrula stage. They may be similarly prevented from evolving viviparity by limited embryonic gas exchange, due to a thick, hard eggshell (Andrews 2004). Monotremes are the only oviparous mammals. After laying, their eggs have a gestation period of 15–23 days, in the female's pouch (echidnas) or a burrow (platypus). After the egg hatches, a young echidna (puggle) further develops in a pouch for 40–50 days; then it is ejected, but parental care continues for about 200 days. After hatching, a baby platypus remains in the burrow and is suckled for 3–4 months (Temple-Smith & Grant 2001).

### 2.9.2 Live Birth

Viviparity, live birth, is a life history strategy that eliminates the free-swimming larval stage (amphibians) or the need to rely on external ambient conditions for the incubation of aquatically or terrestrially deposited eggs (amphibians, reptiles, birds, and monotremes). Viviparity has evolved independently in a number of amphibians (Pough et al. 2004; Vitt & Caldwell 2009), more than a hundred times in reptiles (Stewart & Thompson 2000; Thompson & Speake 2006; Murphy & Thompson 2011), but presumably only once in mammals (Renfree et al. 2013; Motani et al. 2014). Viviparity has often evolved in cold climates, suggesting an advantage to the female thermally incubating the developing embryos, but other advantages might be related to a short activity season, reduction of egg predation, an arid environment, or a marine lifestyle (Guillette 1993; Lodé 2012). While there are obvious advantages, viviparity has costs: the retention of young limits reproductive output, often to one litter per year; it can reduce litter size; and it can increase predation risk to gravid females (Pough et al. 2004).

Viviparity generally requires the development of specialized embryonic and maternal structures that facilitate the exchange of nutrients as well as respiratory gases. For viviparous amphibians, the jelly capsule surrounding the egg is reduced in thickness, and the contact area for exchange between maternal and embryonic tissues is promoted by blood vessel elaboration and highly vascularized gills and tails. In reptiles, simple or complex (and multiple) placentae provide for nutrient exchange. Most viviparous reptiles essentially retain a large, shell-less egg in the reproductive tract, and the embryos are sustained largely by yolk nutrients (lecithotrophy) with little nutritional gain across a placenta. There is, however, substantial gas exchange across a region of close contact between the chorion and the allantois, the chorio-allantois (e.g. Thompson & Speake 2006). The evolutionary step from a simple to complex placenta has occurred about four times (possibly five) in scincid lizards, and once in mammals. Here, eggs are small, and gain substantial nutrients from maternal secretions across a comparatively well-developed placenta (matrotrophy).

The retention of eggs in viviparity requires adjustment to the endocrinological cycle associated with egg retention and maturation. The reproductive endocrine cycle is well understood in mammals. Although there is some diversity—for example, in patterns of ovulation during the annual cycle—the general endocrinological cycle is similar amongst species. The cycle is integrated and controlled by the hypothalamus, which secretes gonadotropin-releasing hormone (GnRH) that travels via the hypophyseal portal system to the anterior pituitary, where it promotes the release of two gonadotropins, FSH and LH (Liem et al. 2001). These, in turn, control the secretion of sex hormones from the gonads.

In females, an increase in both LH and FSH is required for follicle growth and the synthesis of oestrogen; LH stimulates thecal cells to secrete testosterone, which is the precursor for oestrogen synthesis by granulosa cells in response to FSH. Oestrogen promotes the growth of follicles and, in some species, the thickening of the uterine lining. Oestrogen also stimulates LH receptor expression in granulosa cells, which initiates a positive feedback cycle influence on the hypothalamus, causing a surge in LH secretion and ovulation. LH then transforms the ovulated follicles into corpora lutea, which secrete primarily progesterone. Progesterone increases vascularity and secretion of the uterine lining in preparation for implantation of a fertilized embryo (and inhibits GnRH secretion by the hypothalamus in eutherian mammals). The lutea persist if pregnancy occurs, but otherwise regress; progesterone levels decline; the uterine lining is shed; and the inhibition of FSH and LH declines, allowing the pituitary to restart secretion in poly-oestrous species, or a period of anoestrous commences.

In mammals, the placenta has a major endocrinological role during development. Although this role has traditionally been considered a hallmark of mammalian placentae, similar patterns are also apparent in other viviparous species (including many lizards, but not crocodilians or birds; Guillette 1993; Cruze et al. 2012). In viviparous lizards, the extra-embryonic membranes and maternal uterus (i.e. placentae) develop endocrine roles in the synthesis and secretion of various steroid reproductive hormones. For example, during development in a sceloporine viviparous lizard (*Sceloporus jarrovi*), there is a decrease in the diameter of the corpora lutea but a marked increase in the plasma progesterone level, at about the time that there is a rapid onset of embryonic growth and formation of the chorio-allantoic membrane (Guillette et al. 1983). It appears that amniote vertebrates share the unifying characteristic of the ability of their extra-embryonic membranes to synthesize and respond to a variety of steroid hormones, including the oviparous crocodilians and birds (Cruze et al. 2012).

# 3

# Physiological Characteristics of Mammals

The basic bauplan of mammals (see 1.2) is defined by a set of morphological and physiological characteristics that have been subsequently shaped by the evolutionary history of this taxon. In this chapter, we examine the functioning of physiological systems specifically for mammals in an ecological context, building upon the general physiological principles presented in Chapter 2. In the next chapter, we will elaborate on specific physiological adaptations of mammals that facilitate survival in extreme environments.

## 3.1 Energetics

The metabolic rate (MR) of a mammal is the overall sum of its myriad of biochemical reactions (see 2.3). Because all but short-term metabolism depends on oxygen as an electron acceptor in the mitochondrion, the typical units for MR are oxygen consumption rate ($VO_2$), either absolute (e.g. ml $O_2$ $h^{-1}$) or mass-specific (e.g. ml $O_2$ $g^{-1}$ $h^{-1}$), but various other measures (e.g. the carbon dioxide production rate, $VCO_2$, ml $CO_2$ $h^{-1}$; or heat production, J $h^{-1}$) are often used. The specific units are all interconvertible, although some assumptions are required to relate, for example, $VCO_2$ to $VO_2$, which will depend on the substrate being metabolized and the resulting respiratory quotient (RQ; see 2.3.2, 5.3.1) or to convert to heat production units (generally a value of 20.08 kJ l $O_2^{-1}$ is used as an average for metabolized substrates if the RQ is not known). Metabolic-specific metabolic rate is the MR expressed in allometrically corrected units (e.g. ml $O_2$ $g^{-0.75}$ $h^{-1}$).

Metabolic rate can vary markedly over the lifetime of an individual, changing with developmental or reproductive stage and increasing with the level of activity from basal metabolic rate (BMR; 3.1.1) to maximal (or summit) metabolic rate (MMR), and for many species, down to torpid or hibernating metabolic rate. Between species, MR varies with phylogenetic position, generally being lowest for monotremes, intermediate for marsupials, and highest for placental mammals (see 2.3, 3.1.1). For most mammals, MR will generally be at some increment above BMR, with the magnitude of the increment dependent on many factors, including

*Ecological and Environmental Physiology of Mammals*. Philip C. Withers, Christine E. Cooper, Shane K. Maloney, Francisco Bozinovic, & Ariovaldo P. Cruz-Neto.  Published 2016 by Oxford University Press.
DOI 10.1093/acprof:oso/9780199642717.001.0001

locomotory activity, digestion, developmental and reproductive state, variation in ambient temperature ($T_a$; below and above thermoneutrality), and body temperature ($T_b$) via the $Q_{10}$ effect (see 2.4.1). For some mammals that use torpor, MR is lowered to some fraction below BMR, depending primarily on the magnitude of the reduction in $T_b$ (3.2.4.2).

### 3.1.1 Basal Metabolic Rate

BMR is the lowest metabolic rate measured for an endothermic mammal (or bird) under tightly defined conditions, and is used for comparative studies. It is often compared between mammal species of differing body mass (metabolic-mass-specific MR is especially useful for these comparisons; for example, marsupials, Withers et al. 2006, all mammals, McNab 2008), or the same species from different locations or at different times of the year (kangaroo rats, *Dipodomys merriami*, Tracy & Walsberg 2000; numbats, *Myrmecobius fasciatus*, Cooper & Withers 2012), or even the same individual at different times of the year or under different environmental conditions (bank voles, *Clethrionomys glareolus*, Labocha et al. 2004; deer mice, *Peromyscus maniculatus*, Russell & Chappell 2007).

Basal metabolic rate is defined as the rate of energy expenditure measured for a normothermic, post-absorptive, non-reproductive mammal (or bird), during its inactive circadian phase and within its thermoneutral zone (Benedict 1915; Kleiber 1932; Scholander et al. 1950a; McNab 1997; Cooper & Withers 2009, 2010). While it is unlikely that any mammal in its natural setting would ever be within the stringent set of conditions necessary to measure BMR, it remains the most common measurement of metabolic rate for mammals (McNab 2015). For some species, some of these conditions cannot be achieved or are mutually exclusive even under controlled laboratory conditions (Cruz-Neto & Bozinovic 2004). For example, a fasting condition might be impossible to achieve for ruminants (Artiodactyla), pseudoruminants (e.g. macropod marsupials), and post-gastric fermenters (e.g. lagomorphs; White & Seymour 2003, 2004). Similarly, the criteria to be post-absorptive and inactive, while defending a normothermic $T_b$, is problematic for some small mammals, such as shrews and insectivorous bats that enter torpor when they are fasted (Speakman et al. 1993; White & Seymour 2003). It is also doubtful if it is possible to achieve these standard conditions for large marine mammals such as cetaceans, seals, and sea lions (Gallivan 1992).

As a consequence, it has been suggested that resting metabolic rate (RMR) is a more practical measure, as it meets the criteria that the animal be resting within its thermoneutral zone but allows for conditions that violate other criteria for BMR (Speakman et al. 2003, 2004). Perhaps because of this caveat, some authors use RMR to assess the increment in energy expenditure associated with different levels of activity (thermoregulation, reproduction, digestion, locomotion) to characterize and contrast different metabolic states (torpor vs hibernation; Geiser & Ruf 1995), to explore the underlying causes that led to the evolution of endothermy (Koteja 1987), and as a template for dissecting the cellular and molecular basis of the costs of maintenance in mammals (Rolfe & Brown 1997; Hulbert & Else 2000, 2004). However, when the more stringent set of conditions is observed, BMR provides a unified measurement that can be compared between different species of mammals.

The most conspicuous feature of BMR measurements in mammals is its enormous variability (McNab 2008), ranging from 6.0 ml $O_2$ $h^{-1}$ for the sooty moustached bat (*Pteronotus quadridens*) to 389,400 ml $O_2$ $h^{-1}$ for the killer whale (*Orcinus orca*). A fruitful avenue in mammalian comparative energetics has been the exploration of the underlying causes and the behavioural, ecological, and physiological consequences associated with this enormous variability in BMR. Such analyses have been pivotal in increasing our knowledge of the allometric, thermal, and phylogenetic effects that underly rates of energy expenditure of mammals (Dawson 1973a; Dawson & Grant 1980; Savage et al. 2004a,b; White & Seymour 2005; Clarke & Rothery 2008; Sieg et al. 2009; Capellini et al. 2010; Clarke et al. 2010; White 2011), as well as unravelling the proximate and ultimate factors responsible for the correlations of BMR with climatic variables, patterns of species richness and distribution, and with several life history traits (e.g. Rezende et al. 2004; Cruz-Neto & Jones 2006; Lovegrove 2000, 2003; Mueller & Diamond 2001; White & Seymour 2004; Withers et al. 2006; Raichlen et al. 2010; Naya et al. 2013; Agosta et al. 2013; McNab 2015). BMR is highly repeatable (Nespolo & Franco 2007; but see White et al. 2013), and heritable (Konarzewski et al. 2005; Sadowska et al. 2009, 2015; but see Bacigalupe et al. 2004; Nespolo et al. 2005). BMR is highly correlated with (Cooper et al. 2003b), and is up to 50% of, the field metabolic rate (FMR), which is the total energy expenditure of free-ranging mammals (Speakman 1997). These factors give BMR an overt ecological and evolutionary significance that normally would not be expected from a laboratory-based measurement (McNab 2012). The literature on mammalian BMR can be roughly divided into studies that attempt to describe and explain allometric trends and those that are more focused on the ecological and evolutionary analysis of the residual variability (i.e. the variability in BMR after the main effects of body mass, temperature, and phylogeny have been accounted for).

Body mass (M) accounts for about 95.9% of the variance in BMR between mammalian species (McNab 2008, 2015). The curvilinear relationship between M and BMR($BMR = aM^b$) can be linearized by logarithmic transformation to $\log(BMR) = \log(a) + b\log(M)$, where $\log(a)$ is the intercept and $b$ is the scaling exponent (see 2.1). A continuing and pervasive discussion in the literature revolves around two related questions: what is the precise value for the exponent $b$, and what are the mechanistic causes of the allometric scaling of BMR? The former debate is centred on whether $b$ is $\frac{2}{3}$(0.67), which was first suggested, based on the idea that surface area dictates heat loss in mammals (Rubner's law), or proportional to ¾(0.75), which has come to be known as Kleiber's law (see McNab 2012 and White & Kearney 2014 for a historical overview). The scaling exponent *b* for mammals also varies with the criteria used to select data for analysis, and whether the effects of $T_b$ and phylogenetic relatedness are taken into account (Savage et al. 2004; White & Seymour 2003, 2004, 2005; Glazier 2005; O'Connor et al. 2007; Clarke & Rothery 2008; Sieg et al. 2009; White et al. 2009; Capellini et al. 2010; Clarke et al. 2010; White 2011; McNab 2008, 2015). The ongoing debate has led a number of authors to suggest that there is no universal scaling exponent for animals in general, or even mammals specifically, and that allometric effects on BMR differ with taxonomic group, geographic zone, and other factors that influence BMR (Economos 1982; McNab 1988, 2002; Withers 1992; West et al. 2002a,b, 2003; West & Brown 2004; Glazier 2005, 2008; White et al. 2007, 2009; Sieg et al. 2009).

Kleiber's classic scaling analysis (Figure 2.2; Kleiber 1932), based on a few and mostly domesticated mammals and birds, has been supplanted by numerous studies with much larger data sets. For example, $BMR(Jh^{-1}) = 69.7M^{0.720\pm0.006}$ for 637 species of mammal (McNab 2008; Figure 3.1). This *b* is significantly different from both Rubner's ($P < 0.001$) and Kleiber's ($P < 0.001$) scaling exponents.

The inclusion of some data points that might not represent true measurements of BMR in the analysis can affect the value of the scaling exponent. A more 'conservative' subset of metabolic data can be chosen that excludes lineages in which BMR might not be accurately measured (White & Seymour 2003, 2004; White

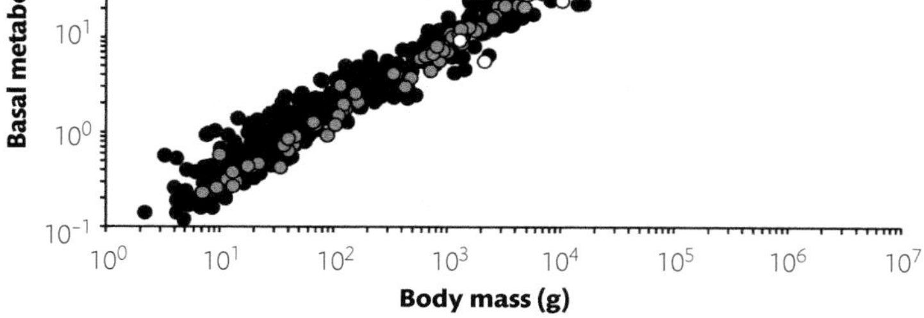

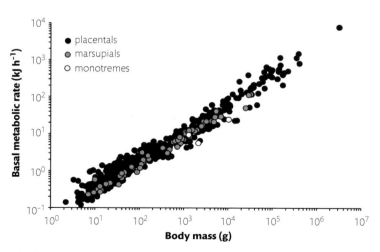


**Figure 3.1** Allometry of mammalian basal metabolic rate for monotremes, marsupials, and placentals. Data from McNab (2008).

et al. 2009). The exclusion of artiodactyls can be justified because the fasting durations (typically 2–3 h) were probably insufficient to produce a post-absorptive state; this might require 2–7 days, or even be unachievable. Similar arguments can be made for macropod marsupials and lagomorphs. Shrews might also be excluded because they might never be post-absorptive when inactive. Although excluding certain lineages such as these can be criticized, if their data are basal measurements, then the scaling relationship would be unaltered by their exclusion, and if their exclusion alters the value of the exponent substantially, then this suggests either that their data are not basal measurements, or that there is no universal scaling exponent for mammalian BMR and different lineages differ in their metabolic scaling relationships (see later). White et al. (2009) found essentially the same allometric relationship (including phylogenetic informed analysis; see later) for a large data set ($N = 585; BMR \alpha M^{0.68}$) and the 'conservative' data set ($N = 537; BMR\ \alpha\ M^{0.69}$).

Body temperature has a substantial effect on the metabolic rate of mammals, and animals in general (see 2.4.1). The $Q_{10}$ for metabolic rate varies, but is generally about 2.5 (Guppy & Withers 1999) (i.e. metabolic rate increases by 2.5 times with a 10°C increase in $T_b$). After correcting BMR to a typical placental $T_b$ of 37°C to account for the variation in $T_b$ of different species and lineages (e.g. montremes vs marsupials vs placentals), the BMR scaling for McNab's (2008) 'all-mammals' data set is $BMR(J\,h^{-1}) = 81.1M^{0.697\pm0.006}$ (Table 3.1). The intercept value is slightly higher after correction because 37°C is higher than the average $T_b$ of the original data. Various lineages of mammals differ slightly in their scaling of $T_b$ BMR, with *b* ranging from 0.68 to 0.827 (Table 3.1).

Body mass and various morphological, physiological, and behavioural traits of mammals (and other organisms) are generally not independent of phylogeny, but tend to be more similar in closely related species (e.g. Alroy 1998; Blomberg et al. 2003; Withers et al. 2006). The phylogenetic signal is highly significant for body mass and BMR of mammals in general; for example, Blomberg's K* (see 5.1.2) is 0.793 ($P < 0.05$) for log M and 0.586 ($P < 0.05$) for log BMR (using the data of McNab 2008 and phylogeny of Bininda-Emonds et al. 2007). Similar phylogenetic patterns are also observed within several mammalian clades, including marsupials (Withers et al. 2006), bats (Cruz-Neto & Jones 2006), rodents (Lovegrove 2003; Rezende et al. 2004), primates (Nunn & Barton 2000), and carnivores (Muñoz-Garcia & Williams 2005).

While variation in M explains about 95.9% of the variability in BMR (McNab 2008), the remaining 4% or so can be substantial, with nearly a tenfold difference in BMR for some species with similar M (McNab 2015), even for closely related species. For example, two cricetid rodents, the western heather vole (*Phenacomys intermedius*) and the cactus mouse (*Peromyscus eremicus*), have a similar M (21.5 g), but their BMR differs by 100% (1.35 vs 0.65 kJ $h^{-1}$; McNab 2008). Various analyses of the residuals of BMR on M have suggested relationships with diet

**Table 3.1** Selected allometric relationships for BMR ($Jh^{-1} = aM^b$) of mammals, $Q_{10}$ corrected to a common body temperature of 37°C. Data for BMR are from McNab (2008); data for body temperature are from Kolokotrones et al. (2010). Lineages with ten or more species are presented separately for marsupials and placentals (N = 3 for monotremes). $Q_{10}$ is assumed to be 2.5 (Guppy & Withers 1999).

| | | *a* | $Log_{10}$ *a*±se | *b* ± se | $r^2$ | N |
|---|---|---|---|---|---|---|
| Mammals | All | 81.1 | 1. 909 ± 0.016 | 0.697 ± 0.006 | 0.964 | 447 |
| | Monotremes* | 68.8* | 1.838* | 0.698* | - | 3 |
| | Marsupials | 65.3 | 1.815 ± 0.029 | 0.718 ± 0.011 | 0.986 | 60 |
| | Placentals | 82.9 | 1.919 ± 0.018 | 0.695 ± 0.007 | 0.961 | 384 |
| Marsupials | Dasyuromorphs | 63.9 | 1.806 ± 0.043 | 0.708 ± 0.020 | 0.984 | 21 |
| | Didelphids | 68.6 | 1.837 ± 0.050 | 0.733 ± 0.019 | 0.994 | 10 |
| | Diprotodonts | 73.3 | 1.865 ± 0.063 | 0.702 ± 0.020 | 0.986 | 19 |
| Placentals | Carnivores | 27.0 | 1.431 ± 0.155 | 0.827 ± 0.041 | 0.923 | 35 |
| | Primates | 39.1 | 1.593 ± 0.198 | 0.779 ± 0.062 | 0.940 | 12 |
| | Artiodactyls | 52.5 | 1.720 ± 0.167 | 0.771 ± 0.036 | 0.982 | 10 |
| | Bats | 65.5 | 1.816 ± 0.033 | 0.743 ± 0.020 | 0.951 | 71 |
| | Rodents | 89.6 | 1.592 ± 0.031 | 0.682 ± 0.015 | 0.921 | 190 |
| | Shrews | 173.8 | 2.241 ± 0.105 | 0.568 ± 0.092 | 0.693 | 19 |

* For monotremes, ***a*** was calculated using the pooled slope for all three mammal lineages (***b*** = 0.698) forced through the mean monotreme values for log mass and log BMR.

and other factors such as climate, habitat, biogeography, substrate, use of torpor, exclusively island or high-altitude distribution, demography, and life history traits. In general, mammals that consume vertebrate prey, nectar, or fungi have a higher BMR than species that consume invertebrates, fruits, or plants (e.g. McNab 1986; Muñoz-Garcia & Williams 2005; Withers et al. 2006). Mammals from hot, dry environments such as deserts have a lower BMR than those from more mesic or cold environments (e.g. Lovegrove 2003; Withers et al. 2006; Careau et al. 2007; Raichlen et al. 2010), and mammals that burrow, are armoured, or are heterothermic are also characterized by comparatively low BMRs (e.g. McNab 1979; Lovegrove 1986, 2001; Cooper & Geiser 2008). McNab's (2008) equation $BMR\,(kJ\,day^{-1}) = 0.064MISTCHEFM^{0.694}$ uses various coefficients to reflect the role of food type (F), infraclass taxonomy (E), habitat (H), climate (C), use of torpor (T), substrate (S), and exclusive island (I) or high-altitude (M) distribution on mammalian BMR (see Table 3 of McNab 2008 for values of these coefficients).

Despite some debate concerning the actual scaling exponent for mammalian BMR and factors that influence the variation around this scaling relationship, a more challenging and acrimonious debate is centred on the functional explanation for the allometric relationship between BMR and M. Despite the enormous body of literature devoted to the allometric scaling relationship for BMR of mammals (and other animals, and indeed other organisms), the mechanisms responsible for the observed scaling relationships remain poorly understood. A myriad of potential mechanistic explanations have been proposed, with little consensus. Contenders include the following: (1) the geometry of nutrient supply networks (West et al. 1997; West 1999; Banavar et al. 1999, 2002a,b); (2) four-dimensional scaling (with time or the ratio of quantities required for mechanical stability as the potential fourth dimension; Blum 1977; West et al. 1999); (3) an allometric cascade that links cellular and whole-animal metabolism (Darveau et al. 2002; Hochachka et al. 2003; Suarez & Darveau 2005); (4) metabolic constraints resulting from increased resistance to blood flow and size-dependent oxygen unloading at the tissues (Spatz 1991); (5) increased accumulation of metabolically inert body components in larger animals (e.g. connective tissue, skeleton, body fluids, and fat; Ultsch 1974; Spaargaren 1992); (6) scaling factors of other body systems, such as gas-exchange surfaces or mitochondrial number (Ultsch 1973, 1974; Else & Hulbert 1985); (7) additive scaling reflecting an intermediate between isometric ($b = 1$) and geometric ($b = 0.67$) scaling (Swan 1972; Withers 1992); (8) metabolic costs of regulation and tissue composition (Glazier 2015); (9) structural support (McMahon 1973); (10) fractal scaling (Sernetz et al. 1985); and (11) universal temperature dependence (Gillooly et al. 2001).

As if the lack of consensus on the 'cause' of metabolic scaling were not enough, controversy also continues as to whether this log–log relationship is best described by ordinary least-squares linear regression (OLS) or by alternative structural regression models, such as reduced major axis (RMA) or least squared variance–oriented residuals (LSVOR; O'Connor et al. 2007; Packard & Boardman 2008; Sieg et al. 2009; Glazier 2010; White 2011). Even the utility of linear models to describe the relationship between BMR and M has been questioned. For example, some authors argue that nonlinear equations give a better fit at high M for mammals (White 2011; White & Kearney 2014), or that the log–log relationship between BMR and M is curvilinear (with a convex curvature) and better described by a quadratic polynomial function than a power function (Kolokotrones et al. 2010; Bueno & Lopez-Urrutia 2014). Even the usefulness of log–log transformed analyses has been questioned (Packard & Birchard 2008; Packard 2012).

### 3.1.2 Incremental Metabolic Rate

The metabolic rate of free-living mammals is generally higher than BMR, but can be lower during torpor (see 3.2.4.2), due to many different factors: wakefulness, locomotor and other forms of activity, digestion, lowered ambient temperature ($T_a$) requiring thermoregulatory thermogenesis, or elevated $T_a$ with elevated body temperature and hence MR via the $Q_{10}$ effect. There is a continuum of MR

**Table 3.2** Metabolic rate (watts) during various activities in humans, from basal to summit (modified from Passmore & Durnin 1955).

| | |
|---|---|
| Basal/lying at ease | 98 |
| Sitting | 112 |
| Standing | 126 |
| Watching football | 139 |
| Cleaning gun | 258 |
| Horse ploughing | 411 |
| Horse riding gallop | 565 |
| Cleaving birch billets | 621 |
| Cross-country running | 739 |
| Tree-felling with saw | 746 |
| Furnace slag removal | 809 |
| 51 perpendicular axe blows $min^{-1}$ | 1,681 |
| Climbing 90° ladder, carrying 50 kg at 12 m $min^{-1}$ vertical | 1,771 |

from basal to resting to summit (maximal). For humans (*Homo sapiens*), MR varies from basal at about 98 watts (J $s^{-1}$) to a sustained summit of about 1800 W (Table 3.2). The factorial metabolic scope (i.e. increase relative to BMR) is about 18 times basal.

### 3.1.2.1 Locomotion

Despite the considerable diversity in structure and function of terrestrial locomotory systems, the metabolic cost of terrestrial locomotion is remarkably uniform for mammals and other terrestrial vertebrates. For a particular species, there is a linear relationship between the net metabolic rate (metabolic rate when running – resting metabolic rate) and speed (Figure 3.2). The net cost of transport (NCOT; kJ $kg^{-1}$ $km^{-1}$) is the slope of the line relating the net metabolic rate when running (kJ $kg^{-1}$ $h^{-1}$) with the running velocity (km $h^{-1}$). Many large mammals that change their gait with velocity—for example, horses that progressively move from walk to trot, canter, then gallop to achieve faster velocities—show a more complex and variable relationship between NCOT and speed (Hoyt & Taylor 1981).

The linear relationship between speed and metabolic rate, although apparent for walking/running and sand-swimming mammals, is not observed in some hopping mammals (e.g. macropod marsupials; Dawson & Taylor 1973; Baudinette et al. 1992). Kangaroos (*Macropus*) have a linear increase in metabolic rate while they locomote slowly (e.g. when pentapedal, using their tail as well as fore- and hindlimbs to walk slowly, or during slow hopping) in a gait that

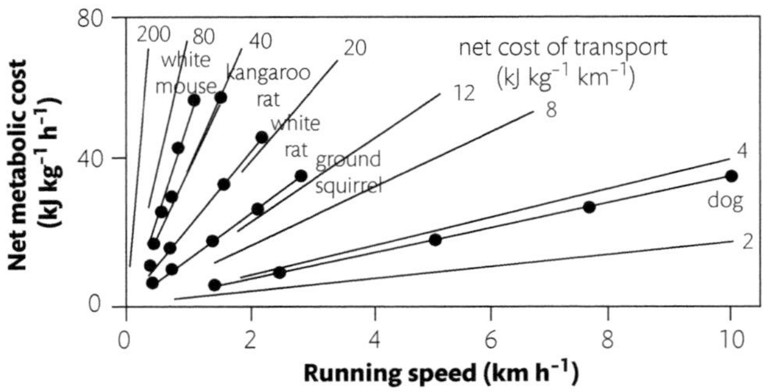


**Figure 3.2** Linear relationships between velocity and net metabolic cost of running (i.e. running metabolic rate - resting metabolic rate) for mammals of differing body mass. Modified from Taylor et al. (1970) and Withers (1992).

has a much higher NCOT than typical walking/running mammals (O'Connor et al. 2014; see 4.6.1). However, MR does not continue to increase with speed when the kangaroo transitions from pentapedal locomotion and begins to hop bipedally, but remains essentially constant. This means that their NCOT actually decreases as speed increases. The mechanisms that underlie this remarkable capacity to disassociate metabolic rate from hopping speed reflects a capacity to recover energy elastically stored in the hindlimbs, primarily tendons and activated muscle cells (Morgan et al. 1978; Biewener & Baudinette 1995; Alexander 2002; Higham & Irschick 2012). The hopping gait and changes in posture of these macropods enables them to increase stride length but keep stride frequency constant while speed increases (Webster & Dawson 2003), and they have a 'big motor' of aerobic locomotory muscle and large heart that provides the power for hopping at high speeds (Webster & Dawson 2012). Smaller bipeds can't store as much elastic energy (Biewener et al. 1981), but can reduce the metabolic cost of hopping by reducing the cost of redirecting the centre of mass while hopping (Gütmann et al. 2013).

There is a remarkably consistent relationship between the metabolic cost of locomotion and body mass in typical terrestrial mammals (and other animals) when walking or running. While NCOT increases with mass in absolute terms (e.g. kJ m$^{-1}$), mass-specific NCOT (kJ g$^{-1}$ m$^{-1}$) is lower in heavier mammals. There are significant exceptions to these general rules (e.g. for mammals that change gait with velocity). Mammals that burrow in soil have a much higher NCOT than walkers/runners (see later). The high NCOT reflects both the high costs of digging through the soil (with forelimbs or teeth) as well as the cost of removing the soil from the tunnel. A few specialized mammals 'sand-swim' through loose aeolian sand (Namib Desert golden mole, Seymour et al. 1998; marsupial mole, Withers et al. 2000). Their MR increases linearly with sand-swimming speed, but NCOT is independent of speed. They have a much higher MR and NCOT than when walking, but a considerably lower MR and NCOT than equivalent-mass burrowers.

For flying mammals (bats), MR is typically higher than for walking/running mammals (of equivalent M), and the high MR requires physiological specializations to sustain their high rate of gas exchange (see 4.7). However, the cost of transport is lower for flying than walking/running (Tucker 1970; Withers 1992), reflecting the much higher speeds of flying. For swimming mammals, the metabolic cost of locomotion is lower than for both fliers and walkers/runners. This pattern for mammals, that NCOT is lower for fliers than runners, and lowest for swimmers, is true for animals in general (Tucker 1975; Withers 1992).

### 3.1.2.2 Digestion

The ingestion, digestion, absorption, and assimilation of a meal can substantially increase the rate of energy expenditure above pre-feeding levels for mammals. This increase—referred to as specific dynamic action (SDA) or postprandial metabolic response—can be divided into a mechanical and a biochemical component, each comprising a series of processes that occurs prior to, during, and after the absorption of the food (Secor 2009; see 2.6.3). The few studies that have partitioned SDA into mechanical and biochemical components show that the latter always accounts for most of the energy expended (Campbell et al. 2000; Hindle et al. 2003). No studies of SDA in mammals, however, have accounted for the mechanical costs associated with the act of feeding (handling, chewing, and swallowing food). These energetic costs related to feeding are expected to be high for mammals ingesting food that requires extensive manipulation and mastication, such as herbivores (Robbins 1993). For example, the act of feeding elevates the metabolic rate above pre-feeding levels by 22% in the elk (*Cervus elaphus nelsoni*; Wickstrom et al. 1984), 32% in the bighorn sheep (*Ovis canadensis*; Chappel & Hudson 1978), and up to 75% in the Amazonian manatee (*Trichechus inunguis*; Gallivan & Best 1986). Although these feeding costs appear to be high, they represent between 0.44 and 3.0% of the metabolizable energy of the ingested food for domestic ruminants (cattle, *Bos taurus*, and sheep, *Ovis aries*; Blaxter 1989), and only about 6% of the gross energy intake for the Amazonian manatee (Gallivan & Best 1986). Another potentially important pre-absorptive component of the SDA is the cost associated with warming the food to core body temperature prior to digestion. MacArthur and Campbell (1994) estimated that for muskrats (*Ondatra zibethicus*) the cost of warming 60 g of aquatic vegetation from ambient temperature (14°C) to core $T_b$ is 5.9 kJ, about 35% of the total SDA.

The postprandial metabolic response has usually been quantified as the absolute and relative costs (factorial scope = peak postprandial metabolic rate/pre-feeding metabolic rate); the time necessary for the postprandial metabolic rate to return to pre-feeding levels; or the SDA coefficient, which is 100 times the energy cost of the postprandial response/meal energy content (McCue 2006; Secor 2009). Except for manatees, where the post-feeding metabolic rate did not differ significantly from pre-feeding values (Gallivan & Best 1986), all mammals studied thus far show a measurable and significant increase in metabolic rate after feeding.

The duration of SDA can vary from 1 h for domestic dogs (*Canis lupus familiaris*) fed a formulated diet, to up to 70 h for livestock consuming straw (Secor 2009). For wild species, SDA lasts 1.8 h for the muskox (*Ovibus moschattus*; Lawler & White 2003) to 8–10 h for marine mammals such as seals (*Phoca vitulina* and *Phoca groenlandica*; Gallivan & Ronald 1981; Markussen et al. 1994) and Steller sea lions (*Eumetopias jubatus*; Rosen & Trites 1997). The energetic cost of SDA can vary markedly, from 0.51 kJ for the rodent *Phyllotis darwini* feeding on a formulated diet to more than 18 MJ for livestock feeding on corn (Eisemann & Nienaber 1990; Nespolo et al. 2003). The absolute energy cost varies allometrically, with body mass explaining up to 87% of the variance in SDA (Secor 2009).

Meal size and energy content also explain some of this variability (McCue 2006; Secor 2009). McCue (2006), however, cautioned about the interpretations of the effects of meal size and energy content on the SDA response. First, expressing diets in terms of percentage of body mass, without taking into account the allometric variation, is valid only when the SDA response is compared between conspecifics of the same size, eating the same meal. Second, diets that represent the same percentage of an animal's body mass, even if the diets are iso-energetic, can still elicit markedly different SDA responses if the diets differ in composition. For mammals, as for other vertebrates, the major component of the SDA response is associated with the post-absorptive, anabolic processes of protein turnover and synthesis (Blaxter 1989; Barboza et al. 2009). Protein-based diets elicit a larger response than diets composed of lipids or carbohydrates.

Secor (2009) reported that the factorial scope of postprandial metabolism averaged 1.37 for mammals. For wild mammals, it varied from 1.22 for the rodent *Phyllotis darwini* (Nespolo et al. 2003) to 4.3 for the fish-eating bat (*Myotis vivesi*) (Welsh et al. 2015). This scope is substantially smaller than those reported for snakes, but is well within the range reported for other vertebrates (McCue 2006; Secor 2009). Although allometric trends have not been analysed for this variable, with the caveat of expressing meal size as percentage of body mass, it seems that diet composition affects the SDA scope in mammals. For example, the highest scope reported for mammals (4.3) was measured for a 28-g bat feeding on a protein-based meal that represented 10% of its body mass (Welsh et al. 2015). In contrast, the scope for a muskrat eating a carbohydrate-based meal, which also represented nearly 10% of its body mass (about 800 g), was only 1.42 (MacArthur & Campbell 1994).

The SDA coefficient for mammals ranges from 4 to 50% (Secor 2009). As with other variables that characterize SDA, much of the variability is thought to be accounted for by differences in the characteristics of the meal. However, for mammals of the same body mass consuming diets with the same composition, an increase in meal size does not necessarily lead to an increase in the SDA coefficient by the same proportion. For example, for fish-eating bats, doubling the meal size (from 5 to 10% of the bat body mass) proportionally increases the SDA costs but the SDA coefficient remains unchanged at 20% of the meal energy content (Welsh et al. 2015). Campbell et al. (2000) reported that SDA costs to the star-nosed mole (*Condylura cristata*) increased from 1.35 kJ after ingesting a meal representing 6.5% of its body mass to 1.77 kJ after ingesting a meal representing 15% of its body mass, but the SDA coefficient decreased from 10.2 to 6.3% of the meal energy content. It should be noted that these values for the SDA coefficient are based on gross energy intake, not on metabolizable energy intake. Digestive efficiency, which is affected by diet composition (chemical composition, deterrents, presence of indigestible components) and ingestion rate (Robbins 1993; Barboza et al. 2009) are other factors that should be, but rarely are, taken into account by SDA studies (McCue 2006).

In terms of energy balance, SDA energy is lost and cannot be used to support other physiological processes, but heat loss associated with the postprandial response in mammals is not wasted if it can be used for thermoregulation. Small insectivorous bats (e.g. *Myotis lucifugus*) maintain a normothermic body temperature at ambient temperatures below the thermoneutral zone when fed, whereas unfed individuals readily enter torpor when exposed to the same conditions (Matheson et al. 2010). For shrews (*Blarina brevicauda*; Hindle et al. 2003), the slope and intercept of the relationship between MR and $T_a$ below the thermoneutral zone (TNZ) did not differ between fed and unfed individuals, suggesting that this small insectivore can use the heat generated during SDA to offset thermoregulatory costs. However, the intercept for this relationship was higher for fed star-nosed moles (Campbell et al. 2000) than unfed individuals, indicating that this species does not substitute SDA heat for thermogenesis at low $T_a$.

Similarly varied results have been reported for large, aquatic mammals. Costa and Kooyman (1984) found that post-absorptive sea otters (*Enhydra lutris*) increase activity to generate extra heat to offset thermoregulatory heat loss, but once fed they reduced these bouts of activity. The authors suggested the heat generated by SDA could substitute for the heat generated during exercise to offset thermoregulatory costs. For the muskrat (*Ondatra zibethicus*), the heat increment associated with feeding did not retard body cooling, but MR of fed and unfed individuals did not differ when these animals entered water (MacArthur & Campbell 1994). Rosen and Trites (2003), however, observed that the metabolic rate of fasted and fed Steller sea lions increased as temperature decreased, but the difference between the two levels of metabolic rate remained unchanged. This lack of change led the authors to conclude that the heat released by SDA did not substitute for thermoregulatory heat production in this species.

### 3.1.2.3 Summit Metabolism

Summit (or maximum) metabolic rate (MMR) can be induced by exercise (e.g. Weibel et al. 2004) or cold exposure for endotherms such as mammals (e.g. Rosenmann & Morrison 1974a; Rosenmann et al. 1975; Hinds et al. 1993), although exercise-induced values are generally higher. The factorial metabolic scope (i.e. MMR/BMR) elicited by cold exposure (often in concert with exposure to a helium-oxygen atmosphere that augments heat loss) is generally lower than that elicited by exercise (Hinds et al. 1993). For monotremes, the aerobic factorial scope is about 5.4 for cold exposure vs 9.7 for exercise; for marsupials, 8.3 vs 16.8; and for placentals, 5.1 vs 13.6. These differences presumably reflect the different muscles and other tissues that are involved in locomotion compared to those involved in cold-induced thermogenesis. Lactation is also a high-energy-demanding physiological activity; the sustained MR at peak lactation is 7.2 × BMR, which is close to the ceiling for cold exposure (Hammond & Diamond 1992).

The allometric exponent for MMR is significantly higher than the exponent for BMR, being near 1 (Koteja 1987; Hinds et al. 1993; Bishop 1999; Savage et al. 2004a; Weibel et al. 2004; Weibel & Hoppeler 2005; but see Bozinovic 1992). Unlike BMR, the mechanism for scaling of MMR is apparent: it is directly related to the aerobic capacity of skeletal muscle, which accounts for the majority of oxygen consumption under conditions of exercise-induced MMR (Mitchell & Blomqvist 1971). The scaling exponents for total mitochondrial volume and the volume of capillary blood, which determine muscle aerobic capacity, are consistent with that of MMR (Weibel et al. 2004; Figure 3.3). The MMR/BMR ranges from < 6 (for some rodents) to > 50 for horse (*Equus caballus*) and pronghorn antelope (*Antilocapra americana*); small mammals have a comparatively low scope compared to large species (Koteja 1987; Bishop 1999; Weibel et al. 2004), and marsupials have higher scopes than placentals (Dawson & Dawson 1982).

## 3.1.3 Field Metabolic Rate

Field metabolic rate (FMR) is the overall average metabolic rate of a free-living animal. Depending on the period of time that it is measured, FMR can reflect a daily maintenance energy requirement (e.g. daily FMR) or a seasonal FMR that might include the costs of reproduction. Traditionally, it was difficult to measure FMR, but the advent of modern isotopic methodologies has made the measurement FMR almost routine (see 5.3.2).

There is a strong allometric scaling of FMR in mammals: $\mathrm{FMR}\,(\mathrm{kJ\,day^{-1}}) = 4.93\,\mathrm{M}^{0.721\pm0.020}$; log mass explains 94.1% of the variation in log FMR (Figure 3.4). Riek (2008) reported an allometric relationship of $\mathrm{FMR} = 6.68\,\mathrm{M}^{0.67\pm0.03}$ after accounting for variability between different FMR studies. Thus FMR scales with an exponent between 0.67 and 0.75, depending on the data set and statistical model

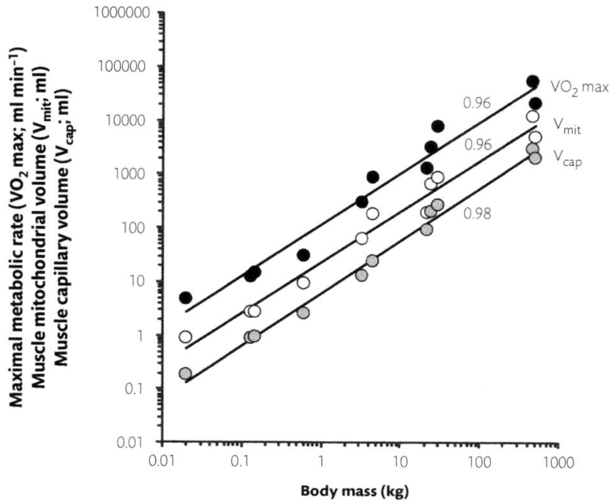


**Figure 3.3** Allometry is similar for total muscle mitochondrial volume (white symbols), total muscle capillary volume (grey symbols), and maximal metabolic rate (black symbols) for mammals; values are the scaling exponents. Data from Weibel et al. (2004).

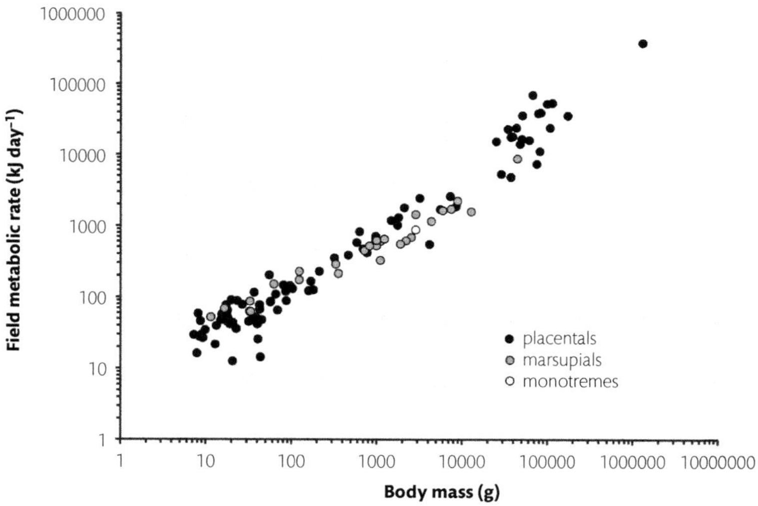


**Figure 3.4** Allometry of field metabolic rate for mammals. Data from Nagy et al. (1999), Nagy and Bradshaw (2000), Cooper et al. (2003), and Munn et al. (2012).

(Nagy et al. 1999; Nagy 2005; Savage et al. 2004a; Riek 2008; Munn et al. 2012; Riek & Bruggeman 2013), as might be expected, since BMR is a significant component of FMR. Indeed, there is a strong correlation between FMR and BMR in mammals, although the scaling exponent b is not 1 and the ratio FMR:BMR ratio is not constant with M (Koteja 1991; Degen & Kam 1995; Cooper et al. 2003b). As smaller species have higher costs of thermoregulation and higher activity levels than larger species, BMR presumably comprises a larger proportion of the FMR of larger than smaller species, resulting in a smaller FMR/BMR as body mass increases (Degen & Kam 1995; Cooper et al. 2003b). Marsupials appear to have a lower FMR scaling exponent than placentals (i.e. small marsupials have higher, and larger marsupials lower, FMRs than equivalently sized placental mammals; Nagy 2005; Riek 2008).

Some of the residual (5.9%) variation in mammalian FMR is related to taxonomy (e.g. monotremes-marsupials-placentals), with the intercept for marsupials being higher than for eutherians, although the difference in intercept is confounded by the slope for marsupials (0.59) being lower than that for placentals (0.77; Nagy et al. 1999; Nagy 2005). The FMR of a monotreme, the short-beaked echidna (*Tachyglossus aculeatus*, 4–5 kg), is within the 95% prediction limits of both the marsupial and placental regressions (C. E. Cooper, personal observations). Accounting for phylogenetic patterns reduces the phylogenetic differences between marsupials and placentals, but otherwise does not change the interpretation of patterns of FMR (Nagy et al. 1999).

Further variability in FMR can be explained by habitat and diet, but there does not appear to be a significant effect of season (summer-winter), suggesting that FMR is not higher when it is colder. However, there is no significance difference in the allometric scaling or level of FMR for desert and non-desert terrestrial mammals, but if phylogenetic history is accounted for, then there is a difference in the scaling exponent (Munn et al. 2012). While the allometric scaling of FMR differs for carnivores, granivores, herbivores, insectivores, and omnivores, the analysis also appears to be confounded by phylogeny, because there are no dietary differences when marsupials and placentals are analysed separately (Nagy et al. 1999).

Riek and Bruggerman (2013) used the strong phylogenetic pattern for FMR in marsupials to predict (from phylogeny) the FMR of marsupial species, finding a close correspondence between actual and predicted FMR (predicted = 101% of actual, ranging from 84 to 125% of predicted). They reported an overall allometry of $5.27\ M^{0.68}$, but substantial differences between carnivores and herbivores. This is an interesting, if somewhat circular approach, to predict the FMR of species that haven't been measured—it takes into account mass and phylogeny, and potentially other factors such as diet, but not other possible unaccounted-for species adaptations.

## 3.2 Thermoregulation

Endothermy is one of the most important physiological changes that occurred in the mammalian bauplan, and its evolution poses a paradox if evolution selects for 'efficiency', because it can be extremely energetically inefficient (see 3.2.2). Nevertheless, the capacity of mammals to increase metabolic heat production in the cold, even when at rest, can balance increased heat loss, and allow homeothermy. Various other physiological and behavioural mechanisms are also used by mammals to maintain a high and relatively constant body temperature, in both the cold and the heat.

### 3.2.1 Body Temperature

Amongst mammals, $T_b$ can vary between species by more than 10°C (Table 3.3; Dawson 1973a; Clarke & Rothery 2008; Lovegrove 2012a). Monotremes have

**Table 3.3** Normothermic body temperature (± standard error; se) for monotremes and various orders of marsupial (italic) and placental (bold) mammals, ranked by mean $T_b$. Modified from Lovegrove (2012a).

| Order | Body temperature (°C ± se) |
|---|---|
| Monotremata | 31.2 ± 0.6 |
| **Afrosoricida** | 32.9 ± 0. 5 |
| **Pholidota** | 32.9 ± 0.2 |
| **Philosa** | 33.1 ± 0.3 |
| **Cingulata** | 34.3 ± 0.3 |
| *Didelphimorphia* | 34.7 ± 0.3 |
| *Peramelemorphia* | 35.0 ± 0.3 |
| **Chiroptera** | 35.3 ± 0.2 |
| **Erinaceomorpha** | 35.4 ± 0.4 |
| *Dasyuromorphia* | 35.5 ± 0.2 |
| *Diprotodontia* | 35.8 ± 0.2 |
| **Soricomorpha** | 36.7 ± 0.3 |
| **Rodentia** | 36.8 ± 0.1 |
| **Primates** | 36.9 ± 0.3 |
| **Macroscelidea** | 37.2 ± 0.2 |
| **Carnivora** | 37.9 ± 0.2 |
| **Artiodactyla** | 38.5 ± 0.2 |
| **Lagomorpha** | 39.0 ± 0.3 |

the lowest mean $T_b$ (31.2°C), marsupials a higher mean $T_b$ (35.5°C), and placentals the highest mean $T_b$ (36.6°C). However, $T_b$ diversity varies within marsupials, from about 31°C (marsupial moles, *Notoryctes*) to 36.8°C (macropods), and within placentals from about 32–33°C (naked mole rats, *Heterocephalus glaber*; anteaters, pangolins, and tenrecs) to 39.0°C (lagomorphs). The low $T_b$s of monotremes might reflect pleisiomorphy ('primitiveness'), but the overlap of $T_b$ for marsupial and placental groups suggests that marsupials have not retained that plesiomorphic characteristic. The slightly lower $T_b$ for bats than most other placentals might reflect their considerable thermolability, hence difficulty in measuring normothermic $T_b$ (e.g. Hosken & Withers 1999; Willis & Cooper 2009). In fact, the main difference between marsupial and placental $T_b$ seems to be a wider range for placentals compared with marsupials (Figure 3.5), presumably related to the more constrained BMR of marsupials, with an apparent metabolic ceiling that few marsupials exceed (the honey possum, *Tarsipes rostratus*, is an exception; Withers et al. 1990; Cooper & Cruz-Neto 2009). This in turn presumably reflects the greater ecological diversity and competitive success of placentals and the thermal adaptations of individual species.

A weak overall allometric relationship has been reported for mammalian $T_b$ (Clarke & Rothery 2008; Figure 3.5). For marsupials overall, $T_b = 34.0\ M^{0.580\pm0.129}$ ($r^2 = 0.21$), but there is no significant slope for dasyuromorphs (0.10), diprotodontids (0.11), or didelphids (0.21) if analysed separately. For placentals overall, $T_b = 35.9\ M^{0.318\pm0.067}$ ($r^2 = 0.043$), but the slope is negative for erinaceomorphs (–0.81), artiodactyls (–0.25), and carnivores (–0.17), and positive for bats (0.56).

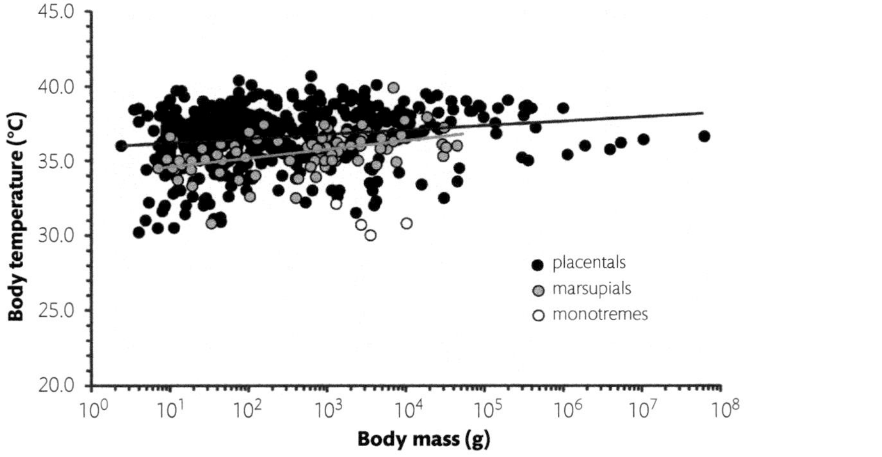


**Figure 3.5** Allometric relationships for body temperature and body mass of mammals. Data provided by A. Clarke (Clarke & Rothery 2008).

In addition to these slight phylogenetic and allometric patterns, there is considerable diversity in $T_b$ based on species adaptations to particular environments. For example, the Namib Desert golden mole (*Eremitalpa granti namibensis*) and the marsupial mole (*Notoryctes caurinus*) have a low and thermolabile $T_b$, presumably reflecting the relatively constant temperature of their sandy habitat and the necessity to limit heat production and conserve energy in a fossorial habitat (Seymour et al. 1998; Withers et al. 2000; see 4.3.2). Various burrowing rodents have similarly low and thermolabile $T_b$ (e.g. naked mole rats, *Heterocephalus glaber* (McNab 1966; Withers & Jarvis 1980; Buffenstein & Yahav 1991; see 4.3.2). These mammals are poor thermoregulators and border on being 'poikilothermic'.

Various other mammals are quite thermolabile, with $T_b$ varying substantially with $T_a$. For example, the $T_b$ of a small marsupial, the monito del monte (*Dromiciops gliroides*), varies from about 30 to 35°C over an ambient temperature range of 14–35°C (Withers et al. 2012); the $\Delta T_b/\Delta T_a$ is about –0.32°C °C$^{-1}$. This thermolability is not indicative so much of a poor thermoregulatory ability (the standard error of $T_b$ is about 0.4–1.4°C), but rather a thermal strategy that conserves energy in the cold: at $T_a$ = 14°C, lowering $T_b$ from the thermoneutral value (34.6°C) to 30.1°C results in a 22% energy saving compared to maintaining $T_b$ constant. This mild heterothermia is an important physiological strategy that can be used to balance the energy budget by reducing energy expenditure and is of considerable advantage when thermoregulatory costs are high, even in a relatively warm and productive environment. Another small marsupial, the little red kaluta (*Dasykaluta rosamondae*) is even more thermolabile (0.32°C °C$^{-1}$), perhaps as an adaptation for water conservation in its arid environment (Withers & Cooper 2009a). The hyrax (*Procavia capensis*) is also very thermolabile, perhaps reflecting its intra-abdominally located testes (Louw et al. 1972).

A phylogenetic analysis of the body temperature of mammals (Lovegrove 2012a; Figure 3.6) shows a complex pattern of increase from a $T_b$ of about 31–32°C for monotremes, to about 34–37°C for marsupials and 34–39°C for placental mammals. Reflecting this general phylogenetic trend, Lovegrove (2012b) proposed that endothermy evolved from the ancestral (plesiomorphic) pattern of baso-endothermy ($T_b$<35°C) and heterothermy (the seasonal, facultative, or obligate reduction in $T_b$) of small Cretaceous mammals to meso-endothermy (35<$T_b$<37.9°C) and then supra-endothermy ($T_b$>37.9°C). His plesiomorphic-apomorphic endothermy (PAE) model suggests that there have been various reversions (apomorphies) towards baso-endothermy and returns to meso- and supra-endothermy. Many extant mammalian orders include baso-endothermic and/or heterothermic species, but orders that comprise large body mass species (e.g. artiodactyls, perissodactyls) or species with a high locomotor metabolic capacity (e.g. lagomorphs) do not.

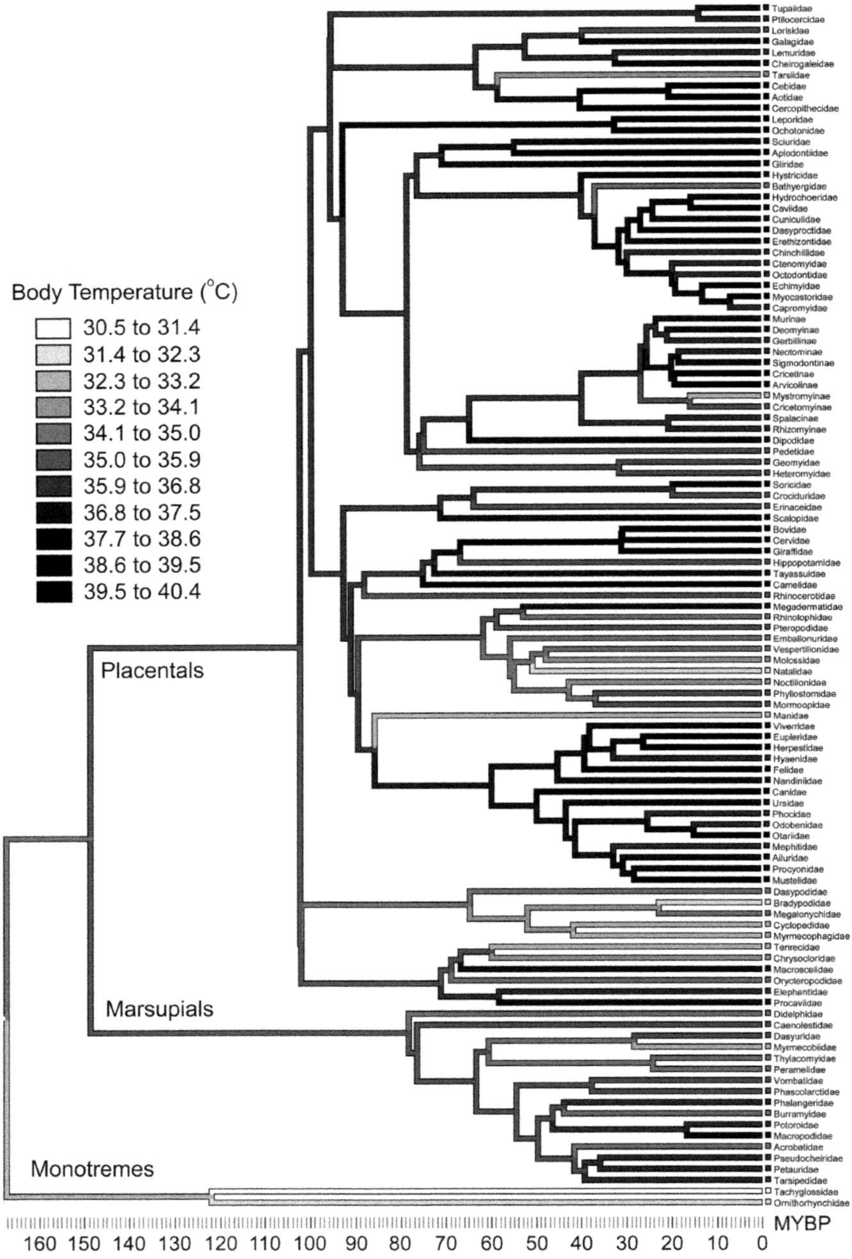


**Figure 3.6** Phylogenetic reconstruction for evolution of $T_b$ by mammals. Modified from Lovegrove (2012a).

## 3.2.2 Evolution of Endothermy

The evolutionary sequence for the evolution of endothermy and thermoregulation by mammals is largely speculative because the basic physiological mechanisms for thermoregulation (hyper-metabolism, muscle shivering, brown adipose tissue, etc.) are not preserved in the fossil record, although evidence of insulation can be. While the original function of various integumental structures of small early mammals and birds, and well as some dinosaurs, is speculative, there is some evidence that it provided insulation (see Clarke & Pörtner 2010). The presence of an insulating cover in diverse pre-mammalian/pre-avian reptilian lineages suggests that those animals had developed the control of heat exchange, but does not mean that they were endothermic. Early therian mammals had fur (e.g. *Eomaia*, Ji et al. 2002; *Sinodelphys*, Luo et al. 2003), and fur was presumably present in the Early Jurassic when the monotremes diverged from the therians (Rougier et al. 2003; Ji et al. 2006).

A number of scenarios have been proposed for the evolution of endothermy (Clarke & Pörtner 2010; Lovegrove 2012a; Figure 3.7). Some suggest that an elevated metabolic capacity was the primary advantage of endothermy. The aerobic capacity model (Bennett & Ruben 1979) is the most commonly espoused hypothesis. Here, the first step in the evolution of endothermy was selection for a high aerobic capacity that would support sustained locomotion and other activities. BMR was 'dragged' up along with this increased aerobic capacity. The elevated BMR that then facilitated homeothermy was not itself selected for, and any elevation of $T_b$ in the early stages of the evolution of endothermy was a useful by-product, rather than a selected trait. The assimilation capacity model (Koteja 2000, 2004) proposes that endothermy evolved to increase the availability of energy for the developing young, by increasing the capacity of the parent for food consumption and energy assimilation.

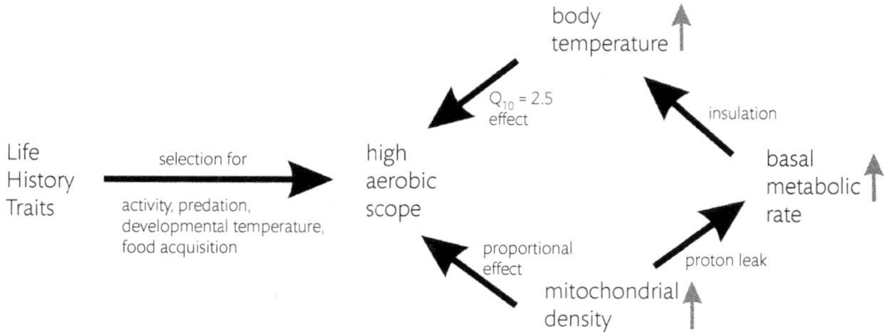


**Figure 3.7** Schematic of the synergistic effects of metabolic rate, insulation, and mitochondrial density on elevated body temperature and aerobic scope, and the evolution of endothermy by mammals.

While these and similar scenarios suggest that elevated metabolic capacity was the primary advantage of endothermy, other scenarios suggest that an elevated $T_b$ was the primary advantage. The parental care model (Farmer 2000) hypothesizes that endothermy evolved to provide a high and constant $T_b$ for gestation, meaning that a constant high $T_b$ was the selected trait and that the increased aerobic metabolic capacity was a useful consequence (Clavijo-Baquet & Bozinovic 2012). The niche expansion hypothesis (Crompton et al. 1978) regards the evolution of endothermy as a two-step process, with first the ability of nocturnal species to maintain a fairly high and relatively constant $T_b$ (facilitated by body insulation) during activity at night, and then an increased metabolic capacity and higher $T_b$ in lineages that became diurnal. The internal thermal environment hypothesis (Heinrich 1977) proposes that the relatively stable $T_b$ of proto-endotherms provided a stable and optimal environment for enzyme function, which was beneficial to biochemical, physiological, and ecological efficiency, and a high $T_b$ was then selected for, enhancing the magnitude of the advantages. McNab (1978) suggested that ancestral mammals were sufficiently large (30–100 kg) to have been inertial homeotherms ('gigantotherms'), and that there was a decrease in size over evolutionary time with the retention of a high $T_b$ and the required metabolic machinery.

However, several lines of evidence now suggest that endothermy first evolved in small, active, carnivorous, predatory proto-mammals (Seebacher 2003; Clarke & Pörtner 2010). Recently, more incremental and integrated models have been proposed for the evolution of endothermy. Kemp (2006) argued that small increments in a loose linkage of all of the required structures and functions resulted in the progressive evolution of endothermy. Clarke and Pörtner (2010) suggested that the acquisition of insulation that resulted in an increased $T_b$ was the primary mechanism that increased aerobic scope, and that an increased muscle and visceral mitochondrial density was a key requirement for the evolution of endothermy through intermediate stages from ectothermy. The synergistic effects of increased mitochondrial density and $T_b$ on aerobic scope provided the selective advantage for the progressive evolution from ectothermy to endothermy (Figure 3.7).

### 3.2.3 Thermogenesis

Endothermic mammals (and birds) have a substantial obligatory thermogenesis due to their relatively high BMR. The basal heat production is sufficient to match heat loss within the thermoneutral zone (TNZ), but below the lower critical temperature for thermoneutrality ($T_{lc}$) they need to elevate metabolic heat production to match the increased heat loss (Figure 3.8). Metabolic rate is constant through the TNZ, then increases above the upper critical temperature ($T_{uc}$), due to the use of energy-requiring mechanisms for cooling (e.g. salivation, licking, sweating, panting) and also possibly the $Q_{10}$ effect of increased $T_b$ on metabolism.

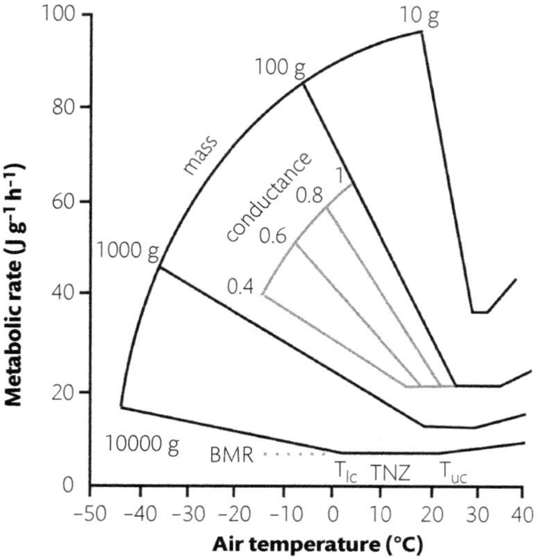


**Figure 3.8** Schematic of the Scholander-Irving model of endothermy, showing the thermoneutral zone (TNZ) from the lower critical temperature ($T_{lc}$) to the upper critical temperature ($T_{uc}$), effects of body mass and the basal metabolic rate (BMR), the effect of body mass on the position of the relationship, and the effect of changing thermal conductance for a 100-g mammal. Modified from Withers (1992).

There is a weak allometry for the $T_{lc}$ ($=T_b - 3.14M^{0.19}$), TNZ($=2.57M^{0.17}$), and $T_{uc}$ ($=36.8M^{-0.027}$) of mammals (Riek & Geiser 2013). Smaller mammals have a high mass-specific basal heat production but they also have a high mass-specific thermal conductance and heat loss, so they have a narrow TNZ and a steep increase in metabolic heat production and heat loss as $T_a$ declines. This can be ameliorated by a relatively low thermal conductance (e.g. good insulation; see 3.2.2.4).

The digestion of food and locomotory activity may also contribute heat production that aids thermoregulation. Skeletal muscles are inherently inefficient, and about 80% of the energy liberated for the purposes of locomotion and other work is converted into heat during exercise (Hohtola 2004). SDA can increase mammalian metabolic rate by 1–2 times BMR, and in some species can replace thermogenic increases in metabolic rate at low temperatures, although for other species an increase in peripheral blood circulation associated with SDA dissipates this additional heat, so it is not of thermoregulatory value (Jansky 1973).

When an increase in heat production is required and in the absence of locomotion or SDA, endotherms can also facultatively increase their metabolic heat production via shivering or non-shivering thermogenesis (NST). Some authors argue that mammalian NST occurs solely via uncoupling protein 1 (UCP1) in brown adipose tissue (BAT; Nedergaard et al. 2001; Cannon & Nedergaard 2004; Golozoubova et al. 2006), but many argue that other organs, such as the liver and skeletal muscle, are also important contributors. Indeed, not all of the energy generated by NST can be attributed to heat production by BAT (Block 1994); it generally accounts for about 60% of the increased whole-animal oxygen consumption

(Foster & Frydman 1978; Puchalski et al. 1987; Schaefer & Staples 2006). The very existence and/or thermoregulatory role of BAT and UCP1 is debated for some mammals, such as marsupials, that do not possess BAT but seem to also achieve thermoregulatory heat production via NST.

### 3.2.3.1 Shivering Thermogenesis

Shivering is the rapid, repeated, uncoordinated, non-locomotory contraction of skeletal muscle, with metabolic heat as a by-product. As muscle contractions are uncoordinated, these contractions do not achieve any external work, and so all of the energy liberated by ATP de-phosphorylation to achieve muscle contraction generates heat within the muscle, which is then convectively transported throughout the body by the circulatory system (Hohtola 2004). Shivering is the only universal facultative thermogenic mechanism in vertebrates and is the primary thermogenic mechanism in both mammals and birds. The relatively large proportion of body mass made up of skeletal muscle, its large metabolic scope, and its inherent biochemical inefficiency means that skeletal muscle is potentially a substantial source of heat. In addition, instantaneous neural control and the capability for long-term activation mean that skeletal muscle can be considered to be preadaptive for facultative thermogenesis (Hohtola 2004).

Shivering thermogenesis by mammals consists of two stages. First, pre-shivering muscle tone is increased, characterized by a continuous, low-amplitude pattern of electromyogram (EMG). However, higher shivering intensity usually results in micro-vibrations and ultimately shivering tremor (Meigal et al. 2000; Meigal 2002). Interestingly, increased muscle tone, micro-vibrations, and tremor do not contribute to heat production; they are simply a consequence of static muscle activation. Indeed, tremor may increase heat loss by convection—the more intense the tremor, the greater the convective heat loss (Kleinebeckel & Klussmann 1990). The anatomy and function of the motor units involved, particularly their synchronization of contraction, determines the amplitude of micro-vibration and tremor during shivering. In mammals, small muscle units commence heat production, resulting in the initial pre-shivering tone. True shivering results when motor units discharge as groups at higher contraction intensities. In contrast, birds smoothly recruit larger and larger units as shivering intensity increases. As a consequence of these differences (and a more compact body form), birds rarely produce visible shivering tremors, whereas shivering tremors are clearly visible for most mammals. As a consequence, small birds are more cold-tolerant than small mammals (Hohtola 2004).

As shivering can occur for considerable periods, only the most aerobic muscles and motor units are involved in shivering; muscle types that are less fatigue-resistant are not suitable. This restricts the potential maximal metabolic rate for shivering to oxidative (rather than glycolytic) fibres, and the maximum thermogenic capacity to approximately 4–5 times resting MR, compared to activity that may involve a greater proportion of muscle types and increase metabolic rate substantially more (Hohtola 2004). Shivering also occurs primarily in the large central muscles rather than the smaller peripheral muscles, ensuring that the produced heat contributes to core $T_b$ and is not lost from the extremities where the surface area to volume ratio is high. In humans, 71% of shivering heat is generated in the central trunk muscles, 21% in thigh muscles, and only 8% in small peripheral muscles (Bell et al. 1992). Shivering may be continuous or may occur in bursts; it is fuelled by carbohydrates, lipids, and proteins supplied by intramuscular reserves or by other tissues and delivered to the muscles by the circulatory system. The ratio of different fuels used during shivering can be modified by the use of varying metabolic pathways within the muscle fibres, by employing different fibres within the same muscle, or by recruiting different muscles characterized by different fibre types (Haman 2006).

Shivering is initiated to prevent hypothermia during cold exposure, and as such is under precise neural control, with peripheral and deep $T_b$ receptors providing the afferent signal that stimulates the efferent motor neural drive to the muscles (Schönbaum & Lomax 1990; Meigal et al. 2000; Meigal 2002). In this way there is proportional control of shivering heat production in relation to both the rate of heat loss and rates of other, non-thermoregulatory rates of heat production. Although central nervous system control of shivering thermogenesis is poorly understood, it appears that cutaneous signals are sent via the lateral parabrachial nucleus to the integrating preoptic area of the hypothalamus. Efferent signals from the posterior hypothalamus to the muscle occur by activation of fusimotor neurons that depend on the activity of neurons in the rostral ventromedial medulla. Shivering is ultimately achieved via rhythmic activity bursts of the $\alpha$-motoneurons, which directly innervate the skeletal muscle fibres (Bicego et al. 2007; Morrison et al. 2008). Shivering is suppressed when other forms of heat production, such as digestion or activity, are increased. Shivering can continue during voluntary activity despite a sharing of the neural pathways for these processes, and shivering individuals are still capable of some fine motor control. Shivering may be temporarily suppressed voluntarily to enhance skilled motor performance, but limiting of shivering to the large central muscles is probably an important factor in maintaining motor function during shivering (Meigal et al. 2000; Meigal 2002).

The relationship between shivering thermogenesis and muscle EMG activity and varying $T_a$ is modified by thermal acclimation. For example, in the golden hamster (*Mesocricetus auratus*), after acclimation to 28°C there is a fairly linear relationship between muscle EMG and thermogenic metabolic rate, although MR begins to decrease at $T_a < -20$°C (Pohl 1965; Figure 3.9). In contrast, when acclimated to 6°C, hamsters have a lower EMG at $T_a < 10$°C, yet MR increases in the same manner at lower $T_a$ as it does after acclimation to 28°C, and does not level, then drop off, at $T_a < -10$°C, suggesting an increased role of non-shivering thermogenesis.

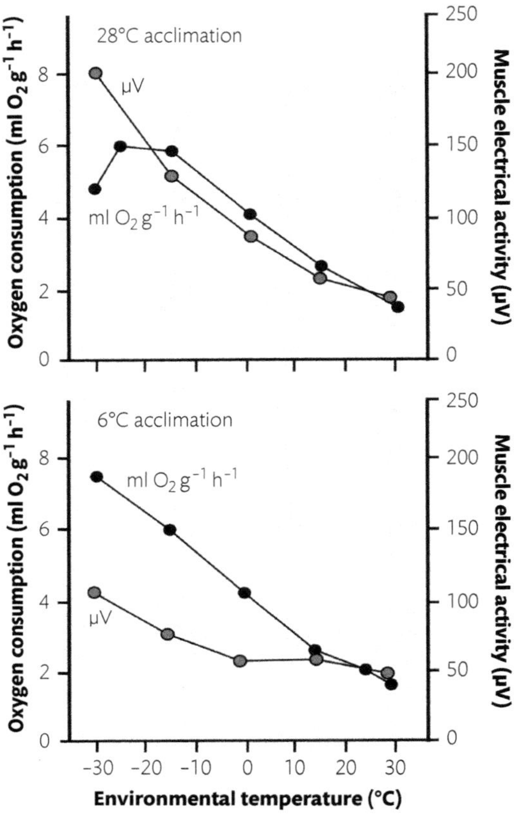


**Figure 3.9** Effect of thermal acclimation at 6 and 28°C on thermogenesis (metabolic rate; black symbols) and shivering (muscle EMG activity; grey symbols). Modified from Pohl (1965).

### 3.2.3.2 Non-shivering Thermogenesis

Non-shivering thermogenesis (NST) is metabolic heat production in the cold that does not involve shivering. In most mammals, the major organ for NST is BAT. Birds, which almost certainly lack BAT, use skeletal muscle for NST, achieved by either release and reuptake of $Ca^{2+}$ from the sarcoplasmic reticulum, in a futile cycle that uses ATP without producing muscle contraction, or via uncoupling of electron transport and ATP production in the skeletal muscle via the avian uncoupling protein (avUCP; Bicudo et al. 2010). Cytochrome C oxidase activity also increases in muscle, heart, kidneys, and liver of cold-acclimated birds, suggesting that these organs also contribute to non-shivering thermogenesis (Zhou et al. 2015). However, the function of skeletal muscle and other organs in NST is controversial for mammals.

Skeletal muscle represents a large percentage of body mass. After acclimation to cold in mammals, there is an increase in the oxygen demand of skeletal muscle, an elevation of serum free fatty acids, and enhanced effects of thyroid hormones and catecholamines on metabolism in skeletal muscle, all suggesting that skeletal muscle may well contribute substantially to NST in mammals (Block 1994). Mammals express uncoupling protein 1 (UCP1) in BAT (see later), and two additional uncoupling proteins are expressed: UCP2 in white adipose tissue, smooth muscle, skeletal muscle, liver, spleen and pancreatic islets; and UCP3 in skeletal muscle (Kabat et al. 2003a; Mozo et al. 2005). The role of mammalian UCP2 and UCP3 in thermogenesis is unclear (Brand & Esteves 2005; Mozo et al. 2005). Avian skeletal muscle avUCP is more similar to mammalian UCP2 and UCP3 than mammalian UCP1 (Raimbault et al. 2001). Futile $Ca^{+}$ cycling in skeletal muscle, initiated by catecholamines and regulated by free fatty acids and thyroid hormones, is one proposed mechanism for NST. Sarcolipin is a regulator of muscle-based thermogenesis in mammals, via sarco/endoplasmic reticulum $Ca^{2+}$-ATPase activity (Bal et al. 2012). A number of studies suggest that considerable heat can be generated by skeletal muscle in response to noradrenaline. Macropods show a 30–50% increase in metabolic rate in response to $\alpha_1$-adrenergic stimulation by noradrenaline, naphrolazine, or phenylephrine, and this has been interpreted as NST in the skeletal muscle in the absence of BAT (Jansky 1973; Ye et al. 1996; Nicol et al. 1997; Rose et al. 1999). However, other studies, often using UCP1 knockout mice, suggest that UCP1 in brown fat is essential for adrenergic cold-acclimated NST (Golozoubova et al. 2006).

### 3.2.3.3 Brown Adipose Tissue

Brown adipose tissue (BAT), or brown fat, is a specialized thermoregulatory tissue that is present only in placental mammals (see later). The functional thermogenic units of BAT are small (10–50 μm) adipose cells that contain several droplets of triacylglycerol and are packed with many large mitochondria with well-developed cristae (Himms-Hagen 1985; Rothwell & Stock 1985; Figure 3.10). BAT also contains endothelial cells of the many capillaries that perfuse the tissue, interstitial cells, mast cells, and preadipocytes that can differentiate and divide, increasing the number of brown adipocytes (Cannon & Nedergaard 2004). BAT is innervated by sympathetic nerves, and noradrenaline activates heat production in the tissue. Brown adipocytes are connected by gap junctions that may allow for electrical

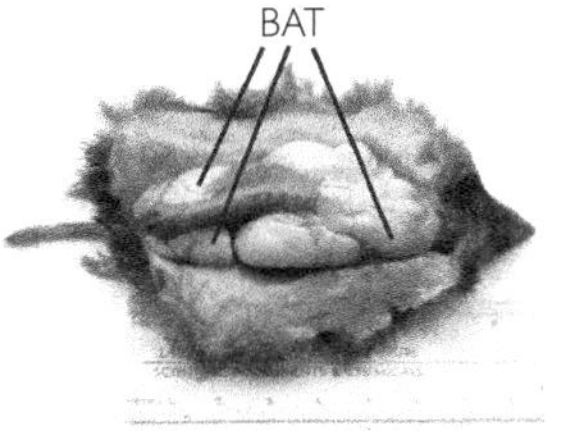


**Figure 3.10** Brown adipose tissue (BAT) in a shrew (*Blarina brevicauda*). Photograph courtesy of T. J. Dawson.

coupling (Himms-Hagen 1985). BAT deposits are located throughout the body, including the axilla, subscapular, and interscapular regions, between the cervical muscles, and there may also be periaortic and perirenal deposits. BAT can also be found in clusters within white adipose deposits (Girardier 1983; Rothwell & Stock 1985; Cannon & Nedergaard 2004).

Brown adipose tissue can generate considerable heat by oxidizing endogenous or exogenous substrates at a high rate. Heat is produced without the intermediary formation of ATP. Lypolysis of endogenous triacglycerides into glycerol and free fatty acids within BAT provides the primary fuel for thermogenesis. However, exogenous triacylglycerol and glucose can also be utilized in thermogenesis, and replenish endogenous lipid stores. Fatty acids are bound by fatty acid-binding proteins and are channelled to the mitochondria where they drive β-oxidization, the citric acid cycle, and finally the electron transport chain, the catabolic pathways typical of cells.

However, BAT produces more heat by uncoupling the movement of protons through the inner mitochondrial membrane from the formation of ATP. Rather than harnessing the electrical potential that is created across the inner mitochondrial membrane by the electron transport chain to phosphorylate ADP to ATP via ATP synthase, the inner mitochondrial membrane uncoupling protein one (UCP1, or thermogenin, a 32,000-Da GDP binding protein) allows protons to 'leak' through the membrane, dissipating the potential energy as heat rather than being harnessed in ATP. This heat is then rapidly transported to the rest of the body by the circulatory system, owing to the high vascularity of BAT (Himms-Hagen 1985; Cannon & Nedergaard 2004). Cold acclimatization induces an increase in BAT tissue in rodents. In rats (*Rattus norvegicus*) with high levels of BAT, UCP1 can account for as much as 5% of the total mitochondrial protein in brown adipocytes, indicating the important role of UCP1 in functional BAT, and the requirement for high concentrations to achieve effective non-shivering thermogenesis (Klingenspor et al. 2008). In the presence of noradrenaline, the proton leak across the inner mitochondrial membrane is stimulated, and that leak is enhanced in the presence of fatty acids derived from lipolysis and inhibited by the presence of purines, most importantly GDP (Rial et al. 1983).

Despite being a relatively small component of total body mass, the high thermogenic capacity of BAT means it can be a substantial source of heat production. Heat production is approximately 400 W $kg^{-1}$ of tissue, and oxygen consumption may be as high as 80 times BMR (Girardier 1983; Rothwell & Stock 1985). Blood flow and the arteriovenous differences in blood oxygen content in cold-acclimated rodents suggest that BAT can account for up to 60% of NST (Foster & Frydman 1978; Puchalski et al. 1987). Venous drainage of BAT to the heart and spinal cord allow for distribution of heat to vital organs (Girardier 1983). Bat is therefore an important aspect of thermoregulation for many mammals, and is of particular importance for those mammals that require considerable and rapid heat production, such as neonates with a high surface area to volume ratio and poor insulation, and hibernators arousing from torpor.

There is considerable controversy concerning the presence of BAT in mammals other than placentals, and in other vertebrate groups. This controversy arises due to the difficulty in unequivocally identifying BAT. Histological characteristics of BAT and white adipose tissue can be similar under some circumstances. For example, BAT with high lipid stores is white in appearance (Himms-Hagen 1985; Rothwell & Stock 1985; Hope et al. 1997), and developing white adipose tissue, or that from animals subjected to nutritional or cold stress, may have similar multilocularity and increased vascularity and mitochondrial density to BAT (Olson et al. 1988; Hayward and Lisson 1991). GDP binding is not sufficiently specific to clearly identify the presence of BAT (Trayhurn 1993; Hope et al. 1997; Nicol et al. 1997). More recently, the presence of UCP1 is considered to be evidence of the presence of BAT (but see later).

The general consensus is that BAT does not occur in fish, amphibians, reptiles, or birds (Hayward and Lisson 1991) despite the presence of UCP1 in these groups (see later). Some authors suggest that BAT is a feature of all mammals, or at least marsupial and placental mammals, but many consider that BAT is restricted to placental mammals. BAT has been unequivocally described in various placental mammals (e.g. Hayward & Lisson 1991; Bicego et al. 2007). It had been well accepted that functional BAT is present in the human neonate, but that it was lost during development and did not exist in adults. However, functional BAT has more recently been described for adult humans (Nedergaard et al. 2007). There is no evidence of BAT in monotremes (Hayward & Lisson 1991; Grigg et al. 2004), and its presence in marsupials is equivocal. Using morphological and histological techniques, Hayward and Lisson (1991) failed to find unequivocal evidence of BAT in 38 species of marsupial, representing all extant families, or monotremes. Kabat et al. (2003a,b) found no evidence of UCP1 in two marsupials, the Tasmanian bettong (*Bettongia gaimardi*) or the Tasmanian devil (*Sarcophilus harrisii*), and it was also absent in the marsupial *Antechinus flavipes* (Jastroch et al. 2008). However, Loudon et al. (1985) reported evidence of brown fat in the young of Bennett's wallaby (*Macropus rufogriseus*). Hope et al. (1997) identified BAT by electron microscopy, the presence of GDP binding, and immunological detection of UCP in the fat-tailed dunnart (*Sminthopsis crassicaudata*) after nearly two weeks of cold acclimation. Jastroch et al. (2008) demonstrated UCP1 gene expression in juvenile monodelphis (*Monodelphis domestica*) and adult fat-tailed dunnarts. In some respects, this equivocal situation in marsupials reflects the earlier controversy surrounding BAT in adult humans.

Even if BAT is present in some marsupials, its role in thermoregulation seems minimal. Hope et al. (1997) suggested for dunnarts that BAT did not appear to assist rewarming from torpor or in maintaining $T_b$, based on low GDP binding, and Nicol et al. (1997) reported that Australian and South American marsupials did not have $\beta_3$-mediated thermogenesis, as would be expected if BAT were an important thermoregulatory tissue. Polymeropoulos et al. (2012) further supported this idea, suggesting that a lack of increased thermogenic capacity in response to noradrenaline in cold- vs warm-acclimated dunnarts indicates that UCP1 in marsupials may have an alternative physiological role than the non-shivering thermogenesis of placental mammals. Identification of UCP1 orthologues in fish (Jastroch et al. 2005), amphibians, 'old' Afrotherian placental mammals (rock elephant shrew, *Elephantulus myurus*), and marsupials (fat-tailed dunnart) led Klingenspor et al. (2008) to suggest that UCP1 may have more recently evolved a thermogenic role in placental mammals, and retained its original, unknown ancient function in other groups, including marsupials. Indeed, the branch-length between marsupial and placental UCP1 is almost twice the length of that between marsupials and amphibians, indicating accelerated evolution of placental UCP1.

Beige fat is a second type of thermogenic fat that accumulates in white fat deposits (e.g. in rodents and adult humans; Kazak et al. 2015). It expresses UCP1, like brown fat, but its thermogenic mechanism is based on creatinine-based substrate cycling.

### 3.2.3.4 Insulation

Fur was presumably an essential prerequisite for the evolution of mammalian endothermy (see 3.2.2), possibly as were feathers for birds (see Bicudo et al. 2010). Fur appears early in the mammalian lineage, presumably having a role as thermal insulation that enabled small nocturnal mammals to retain more of their metabolic heat production and elevate $T_b$ significantly above $T_a$. The insulative role of fur is determined primarily by its thermal conductive properties; conduction is heat transfer that occurs between solid materials or non-moving fluids (e.g. still air; see 2.4.2). Thermal exchange across fur is also substantially modified by convection and radiation. Heat exchange is only one of many functions of fur; crypsis, social and warning signalling, physical protection against injury, sensory reception, and waterproofing are other functions (Dawson et al. 2014).

The rate of heat transfer across a material (Q/Δt; joules per second) depends directly on the surface area (A, $m^2$) for exchange and the temperature difference (Δt) and inversely with its thickness (x, m), as well as a material property termed the thermal conductivity (k; $J\,s^{-1}\,m^{-1}\,°C^{-1}$); $\Delta Q / \Delta t = -k A \Delta °C / x$. Thermal conductivity varies markedly for different materials (Table 3.4), and the insulation value of materials is inversely proportional to thermal conductivity. The insulation value of materials (clo units), or thermal resistance units ($s^{-1}\,m^{-2}$), includes their material thickness, and needs to be expressed per thickness for comparison with conductivity. The least conductive (best insulative) material is a vacuum; the absence of air molecules prevents heat flux by conduction or convection. Still air is a very effective insulator, fat less so, and water is a relatively poor insulator, with glass and various metals even poorer. The thermal conductance value of fur (or its insulative value;

see later), is more a property of the air held within the fur layer than the fur elements themselves. Helium is a much poorer insulator than air (air is predominantly nitrogen), and this has been exploited as a method to thermally challenge endotherms by placing them in a helox atmosphere (21% oxygen, 79% helium) rather than exposing them to cold ambient temperatures that can cause tissue damage.

Mammalian fur consists of an underlayer of hair, which provides its insulative value, with scattered longer guard hairs that have a sensory role and function to maintain pelt structure. Heat is exchanged between the skin and the fur surface by conductive transfer through air held within the underfur layer, by conduction along the length of the hair shafts, by convective heat exchange if there is any free or forced air movement over or within the fur, and by radiative exchange (Cena & Monteith 1975a,b; Walsberg 1988a,b; Wolf & Walsberg 2000). Measurements of conductivity for mammalian fur range from about 0.035 to 0.150 W $m^{-1}$ $°C^{-1}$ (cf. still air 0.024; Table 3.4), so it is clear that there is considerably more heat exchange

**Table 3.4** Thermal conductivity (k) of various biological and non-biological materials; k is a material property that reflects the rate of heat transfer (J) per unit time (s) and surface area ($m^2$) per thickness of the material layer (m). Thermal resistance (r[a]) per thickness and insulation (clo[b]) per thickness are reciprocal functions of conductivity. Modified from Cena and Monteith (1975a) and Withers (1992).

| | Conductivity (K) (W $m^{-1}$ $°C^{-1}$) | Resistance/m $m^{-1}(s^{-1}\ m^{-2})$ | Insulation/m $m^{-1}(°C\ m\ W^{-1})$ |
|---|---|---|---|
| Vacuum | 0 | - | - |
| Still air | 0.024 | 50,000 | 269 |
| Red fox fur | 0.036 | 33,000 | 179 |
| Husky dog fur | 0.041 | 29,000 | 157 |
| Goose down | 0.053 | 23,000 | 122 |
| Sheep wool | 0.063 | 19,000 | 102 |
| Hair shaft material | 0.125 | 9,600 | 51.6 |
| Galloway cattle | 0.13 | 9,200 | 50 |
| Helium | 0.14 | 8,600 | 46 |
| Fat | 0.17 | 7,100 | 38 |
| Water | 0.59 | 2,000 | 11 |
| Glass | 1.0 | 1,200 | 6.5 |
| Ice | 2.2 | 550 | 2.9 |
| Steel | 46 | 26 | 0.14 |
| Aluminium | 240 | 5 | 0.027 |

[a] r $m^{-1}$ = 1,200/k;

[b] Clo $m^{-1}$ = 6.45/k

through fur than just by still air conduction, reflecting contributions from conduction along hair shafts, free convective exchange, and/or radiative exchange (Cena & Monteith 1975a). Conduction along hair shafts is unlikely to contribute more than about 0.0025 W $m^{-1}$ $°C^{-1}$ of heat exchange, since their cross-sectional area is only about 0.01 of the skin area and the conductivity of hair shaft material is unlikely to be more than about five times that of air. Radiation from the animal can potentially contribute about 0.005 – 0.02 W $m^{-1}$ $°C^{-1}$ to heat exchange. So, free convection would appear to be significant for fur coats with conductance greater than about 0.050 W $m^{-1}$ $°C^{-1}$. For avian plumage, which is less open than mammalian fur, thermal conductance is about 0.0075 W $m^{-1}$ $°C^{-1}$ (Walsberg 1988a,b; Wolf & Walsberg 2000). About half of this is conductive and free convective, and the increment of this conductance component above that calculated for an unstirred air layer of equivalent thickness suggests that about 30–50% of heat exchange is free convective (Wolf & Walsberg 2000). Most of the remaining heat exchange is conduction along the feathers, with only about 0.0025 (5%) of the total plumage conductance being radiative.

The insulation provided by a mammal's fur pelt depends directly on the pelt's depth (thickness from the skin to pelt surface) and hair density, which in turn vary between species (and seasons for mammals that have a distinctive summer and winter coat; Scholander et al. 1950a,b; Hart 1956; see 4.1.1). Insulation generally increases in proportion with fur depth (Figure 3.11), but the conductance of deeper pelts is not necessarily the same as that of thinner pelts; for example, deep polar bear (*Ursus maritimus*) fur is a poorer insulator relative to its length than thinner fur of other mammals (and in comparison to dense cotton). There is no clear overall correlation between body mass and pelt depth, and therefore insulation. Small mammals from all habitats are restricted to relatively short fur, but the fur of large mammals can vary dramatically in length—some Arctic species like polar bears have thick fur, but many species from warmer environments have short or no fur; for example, elephants (*Loxodonta africana* and *Elephas maximus*), pigs (*Sus scrofa*), hippopotamus (*Hippopotamus amphibius*), and four of the five species of rhinoceros (*Ceratotherium simum, Diceros bicornis, Rhinoceros unicornis*, and *R. sondaicus*), while the Sumatran rhinocerus (*Dicerorhinus sumatrensis*) retains some hair.

For mammals in their natural environment, wind speed is an important determinant of heat loss through the pelt, as it determines convective exchange. Wind reduces the still air boundary layer above the pelt surface, enhancing heat loss, and can disrupt the pelt structure and reduce the layer of insulating still air within the pelt. The structure of the pelt determines the impact of wind speed; for example, heat flux through the thin, sparse pelt of the numbat (*Myrmecobius fasciatus*) is more influenced by wind speed than the thicker, denser pelts of other marsupials such as macropods and the koala (*Phascolarctos cinereus*; Cooper et al. 2003a); wind speed has less effect on the conductance of African ungulates with deeper

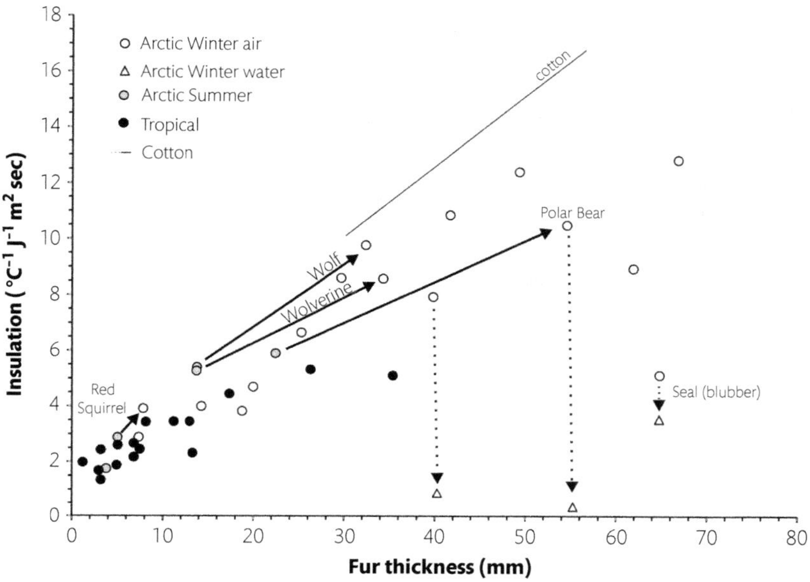


**Figure 3.11** Relationship between fur thickness and insulation for tropical and Arctic mammals, and aquatic mammals that rely on subcutaneous fat for insulation. Modified from Scholander et al. (1950a,b), Hart (1956), and Withers (1992).

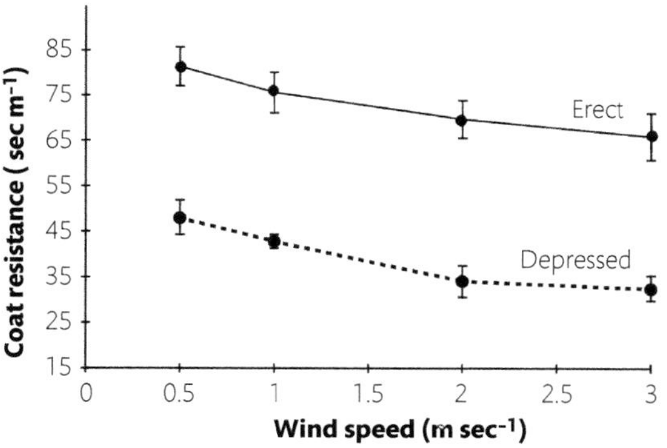

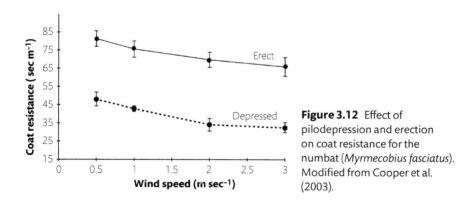

**Figure 3.12** Effect of pilodepression and erection on coat resistance for the numbat (*Myrmecobius fasciatus*). Modified from Cooper et al. (2003).

pelts than those with thinner pelts (Hofmeyr & Louw 1987). Besides seasonal changes in pelt structure (see 4.1.1.), mammals can acutely modify their pelt by piloerection. Erect fur is deeper, and consequently offers increased insulation. Erect numbat pelts have twice the thermal resistance of depressed pelts, and are similarly impacted by wind speed (Cooper et al. 2003; Figure 3.12). Grooming is also important in maintaining the structural characteristics and insulative properties of mammal fur. Sea otters rely on extensive grooming to maintain pelt insulation and therefore $T_b$ in their cold-water habitat (see 4.1.1), but grooming is also important for terrestrial mammals. Social grooming by primates maintains pelt cleanliness and maintains loft (pelt depth). Back-combed vervet monkey (*Chiorocebus pygerythrus*) pelts, with increased loft, have significantly better insulation than flat pelts, and also provide improved protection from environmental radiative heat loads, suggesting that social grooming can have an important thermoregulatory function (McFarland et al. 2016).

Pelt properties not only influence heat loss from the animal to the environment, they also impact on heat gain to the animal from the environment when environmental temperature exceeds $T_b$. Solar radiation is a significant source of heat for diurnal mammals, and pelt structural and optical properties as well as colour determine how much of this radiation becomes a heat load on the skin. For some mammals, such as sheep, the pelt can function as a thermal shield, minimizing heat gain (Priestley 1957), while for other mammals, such as numbats and ground squirrels (Walsberg 1988a,b; Cooper et al. 2003), the pelt facilitates solar heat gain (SHG), which can be an important supplement to metabolic heat for thermoregulation. For example, simulated solar radiation of 780 W $m^{-2}$ reduced metabolic heat production of the ground squirrels *Spermophilus lateralis* and *S. saturatus* by 23–42% and 19–29% at wind speeds of 0.5–4 m $s^{-1}$ (Walsberg & Wolf 1995a).

Solar heat gain through the pelt is generally a trade-off with resistance to heat loss; shallower, sparser pelts tend to have higher rates of SHG. For example, there is a significant negative relationship—$\text{SHG}(\%\,\text{irradiance}) = 64.3 - 0.36 \times \text{resistance}(\text{s m}^{-1}); r^2 = 0.97$—between the SHG and pelt thermal resistance of various species of ground squirrel (Cooper et al. 2003). However, this general relationship is complicated by colour, which determines reflectance, and, more importantly, by pelt optical characteristics and structure that determine the level of penetrance of solar radiation into the pelt. For mammals with very deep, well-insulating pelts, colour has little impact on SHG despite its obvious effect on reflectance. For example, white polar bear pelts have a reflectance of 66%, compared to 10% for dark koala pelts, but at low wind speeds both have a low SHG of about 19% of incident radiation (Dawson et al. 2014). For mammals with less well-insulating pelts, there can be complex interactions between pelt colour, structure, and optics, and environmental conditions such as wind speed.

Colour determines overall pelt reflectivity, with light pelts reflecting more incident solar radiation than dark pelts. Therefore, for mammals with identical pelt characteristics, but differing only in colour, darker pelts will result in higher incident solar heat loads at the level of the skin (e.g. springbok, *Antidorcas marsupialis*; see later). However, varying pelt characteristics impact on the fate of absorbed radiation. Pelt structure—such as hair angle, density, diameter, and length—determines the probability that a photon of incoming radiation is intercepted by a hair shaft. Hair optical properties of absorbance, reflection (not to be confused with overall pelt reflectance that determines colour), and transmission determine if intercepted photons are absorbed by the hair shaft, transmitted deeper into the pelt, or reflected away from the skin. These determine the penetrance of radiation and therefore SHG at the level of the skin (Walsberg 1983). Therefore, a mammal with a dark pelt may absorb a considerable proportion of incoming radiation, but it may be absorbed near the pelt surface, and therefore the heat will be readily lost to the environment, rather than acting as a heat load on the skin. A mammal with a white pelt may only absorb a relatively small fraction of incoming radiation, but that which is not reflected may penetrate deep within the pelt before it is absorbed, resulting in a considerable heat load on the skin (Walsberg 1983).

Various pelt and environmental factors can impact on SHG. For example, wind speed can interact with pelt colour, structure, and optical properties. A light-coloured mammal may have a lower SHG than a dark mammal at low wind speeds, but at higher wind speeds the disruption to insulative properties of the pelt means that the absorbed radiation in the outer part of the pelt of the dark mammal is more readily lost to the environment than that absorbed deeper in the light pelt, so the darker mammal may have lower SHG. Mammals may acutely manipulate their SHG, just as they do their pelt insulation, by altering the optical and structural properties of the fur (e.g. by piloerection and by postural changes). For example, rock squirrels (*Spermophilus variegatus*) were able to increase their SHG by more than two times as wind speed increased from 0.25 to 4 m $s^{-1}$ (Walsberg & Wolf 1995b). Seasonal alternation in pelt structure and optics can also maintain optimal thermal characteristics with changing environmental demands, without compromising the other requirements of pelt colour, such as crypsis. Rock squirrels maintain the same coat appearance in summer and winter, but increase SHG at low wind speed by 20% in winter by increasing forward scattering, decreasing backward scattering and absorptivity, and decreasing thermal resistance of the inner pelt (Walsberg & Schmidt 1989). Conversely, two species of golden-mantled ground squirrel differing in pelt colour (*S. lateralis* and *S. saturatus*) and hence reflectivity were able to achieve similar, presumably optimal, SHG by differences in their pelt structure and optics. The darker species reduced SHG by having greater interception of radiation in the outer pelt resulting from greater hair density, while the lighter species enhanced SHG with optical properties in the middle and outer pelt, enhancing forward scattering and decreased back scattering (Walsberg 1990).

Impacts of pelt characteristics on the radiative heat exchange of mammals can have serious implications for their thermoregulation, activity, and potentially survival. Springbok, which have particularly shallow pelts (Hofmeyr & Louw 1987), have three colour morphs, normal beige, white, and black. Despite having the same depth and density, pelts of black springbok have lower reflectance and higher SHG than those of normal and white springbok (Hetem et al. 2009). During winter, white springbok have lower mean $T_b$ (37.4°C) and reduced activity compared with the other two colour morphs (mean $T_b$ = 38.1, 38°C for black and beige, respectively), but in warmer weather, black springbok have higher $T_b$ and greater daily temperature cycles. This suggests that the radiative heat gain facilitated by black pelts is advantageous for springbok in winter, but white springbok benefit from a lower solar heat load in hot weather. These observations are consistent with anecdotal reports of differential survival of colour morphs in the wild (Hetem et al. 2009).

## 3.2.4 Heterothermy

A high and regulated $T_b$ is one of the basic physiological characteristics of mammals, but some specialized small mammals that live in thermally stable environments are poor endotherms or essentially ectotherms (e.g. naked mole rats, McNab 1966; Namib Desert golden moles, Seymour et al. 1998; and marsupial moles, Withers et al. 2000). Even for normally endothermic mammals, a constant $T_b$ is not necessarily regulated constant all the time or for all body parts. For example, many mammals have a reduced $T_b$ in peripheral body parts in cold environments, and many have a slightly reduced (e.g. down to 25–30°C) to dramatically reduced (down to 5°C or even less) core $T_b$ during torpor, hibernation, or aestivation. The former pattern of reduced peripheral temperature is termed 'regional heterothermy' as it occurs spatially; the latter is termed 'temporal heterothermy' as it occurs over time.

### 3.2.4.1 Regional Heterothermy

The temperature of any region of the body—whether it is an internal organ (e.g. the brain) or an external organ or tissue (e.g. skin)—depends on the balance between the relative rates of heat loss from, and heat gain to (including local metabolic heat production), that region (Jessen 2001b). Vascular 'retes' are specialized anatomical associations of numerous small arterioles and venules, whose high surface area of contact facilitates heat exchange between the arterial and venous bloodstreams (Scholander 1957). Because the direction of flow in the veins is opposite to the arteries, the rete is a countercurrent heat exchanger.

Many terrestrial and aquatic mammals have vascular retes at the origin of their limbs or tails that allow the extremities to be perfused without excessive loss of heat to the environment. In the rete there is almost complete countercurrent transfer of heat from the arteries to veins, resulting in arterial heat being transferred to venous blood returning to the body and not lost from the periphery. When heat dissipation is required, blood can bypass the rete and facilitate heat dissipation. These countercurrent thermal vascular retes are particularly important for aquatic animals because water has a much higher heat capacity and thermal conductivity than air, promoting substantial heat loss in water. Similarly, the carotid rete of some mammals lies within the venous cavernous sinus, facilitating heat transfer from arterial blood on its way to the brain to cooler blood returning from the nasal mucosa; this results in 'selective brain cooling' (Baker 1979). The adaptive significance of selective brain cooling has been the subject of some controversy (see 4.2.1; Strauss et al. 2015).

The heat balance of an organ depends on two main processes: conduction from cell to cell and tissue to tissue (or tissue to the external medium) and convection by the blood. Heat loss will occur by conduction to tissues that are in direct contact with, and cooler than, the region of interest, and by convection when the blood entering is cooler than the region of interest. Conversely, heat gain will occur by conduction from tissues that are in direct contact with, and warmer than, the region of interest, and by convection when the blood entering is warmer than the region of interest. For example, tissues with high relative heat production, such as the brain, liver, kidneys, and heart, are cooled by the blood perfusing them, and some heat flows by conduction from these organs to nearby cooler parts of the body (Jessen 2001a). The metabolic heat of these organs is lost primarily by convection in the blood. On the other hand, regions of the body that are exposed to a cooler environment, such as the skin and the nasal mucosa, lose heat to the external medium and are warmed by the blood perfusing them.

Large blood vessels have a relatively small surface area relative to their volume, and the magnitude of heat exchange between them and surrounding tissues is minimal, especially given that the temperature differences are usually small (Lemons et al. 1987). When vessels are smaller than 50–100 μm, their surface area to volume ratio is large enough that thermal equilibrium is reached between vessels and surrounding cells. The result is that blood flowing through small vessels is a great homogenizer of temperature. The capillaries in most tissues have a diameter of 5–10 μm, so the venous blood exiting tissues is in equilibrium with the temperature of the surrounding cells (Chato 1980).

When homeothermic mammals are exposed to low environmental temperature, there is advantage to be gained from reducing heat loss. A practical solution would be to reduce the flow of warm arterial blood from the core to peripheral tissues where heat loss to the environment is high, but that would also reduce the supply of oxygen and nutrients to those tissues. A specific vascular arrangement, the previously mentioned rete (or rete mirabile, 'wonderful net'; Scholander 1957), in the extremities of some mammals—especially marine mammals, because they commonly live in cold water—helps in this regard. In many marine mammals, a central artery that supplies blood to peripheral areas (like fins and flukes) is surrounded by veins, such that warm arterial and cool venous blood flow in opposite directions and close proximity. By this 'countercurrent' flow of blood, heat is transferred from the artery to the veins, and retained in the venous blood rather than lost to the environment when the warm arterial blood reaches the extremity. The peripheral tissues are still perfused but less heat is lost (Scholander & Schevill 1955). When the animal becomes active, vasodilation of the artery is thought to compress the deep veins in proximity to the artery, forcing venous blood to return via superficial veins and bypassing the countercurrent arrangement. In this way, heat loss is facilitated in the active animal.

There are three main variations in the structure of countercurrent heat exchangers, with variation in vessel size and presumably heat exchange capacity. In an artery-vein (model I) rete, both the artery and veins anastomose into finer vessels that are in close contact; for example, human (*Homo sapiens*), manatees (*Trichechus*), and sloth (*Choloepus didactylus*) limbs, beaver (*Castor canadensis*) tail. In a central (model II) rete, a central artery is surrounded by 10–20 small veins (e.g. fins and flukes of whales and dolphins; Mitchell & Myers 1968; Withers 1992). In vena comitantes (model III), a central artery is surrounded by a small number of anastomosing veins (e.g. beaver hindlimb). The magnitude of the decrease in blood temperature in the rete artery and vein depends on the relative thermal conductance between blood vessels, and between the blood vessels and the ambient medium (Mitchell & Myers 1968; Figure 3.13). High thermal conductance between the artery/veins and between the vessels and the ambient medium results in arterial blood approaching ambient temperature along the rete, and venous blood rewarming to nearly body temperature, for both model I and II retes. All these countercurrent heat exchangers essentially subserve the maintenance of core $T_b$; the 'shell' is allowed to cool and heat is retained in a central 'core'. However, some examples of countercurrent heat exchange subserve the

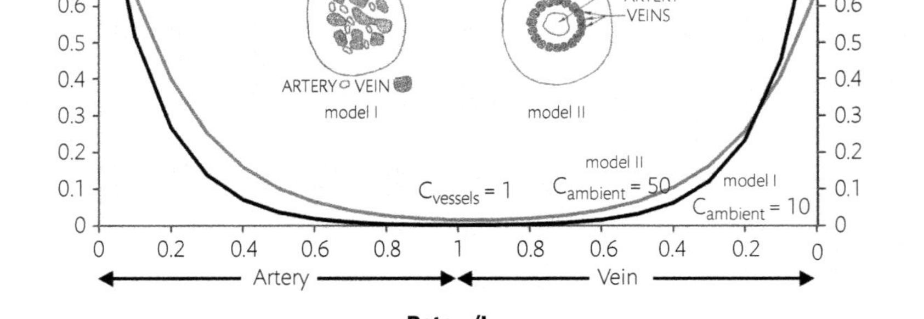

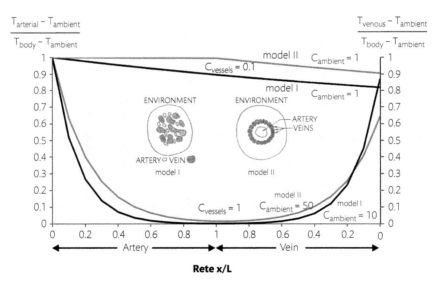


**Figure 3.13** Model of arterial and venous temperatures (as a fraction of $T_b - T_{ambient}$) along a model I and model II rete, as a function of relative thermal conductance between the artery/veins ($C_{vessels}$) and between the vessels and the ambient medium ($C_{ambient}$). Blood cools considerably more in the rete if $C_{vessels}$ is higher, and $C_{ambient} < C_{vessels}$. Modified from Mitchell and Myers (1968).

regulation of specific tissue or organ temperatures independently of the regulation of core temperature.

During development, the testes of many mammals migrate into a pendulous scrotum that is maintained at a temperature generally 2–3°C lower than core $T_b$ (Waites 1976). Testicular temperature seems to be maintained in the face of environmental or organismal perturbation, and as such is regulated (Maloney & Mitchell 1996). For example, at a $T_a$ of 21°C, the tissue and vascular temperatures decrease from core $T_b$ in rams (*Ovis aries*) of 39.6°C to 39.0°C at the internal spermatic artery and 38.6°C at the internal spermatic vein, 34.8°C at the terminal coil of the spermatic artery to 34.1°C in the testis tissue, and 33.0°C in the subcutaneous scrotum (Waites 1970). Raising the temperature of the testes and scrotum via external heat exchange or insertion of the testes into the abdomen results in infertility (Waites 1976). The regulation of scrotal or testicular temperature depends on the operation of a countercurrent heat exchanger in the neck of the scrotum. The arrangement varies from a central rete where the single testicular artery is surrounded by a venous arrangement (the pampiniform plexus) in many artiodactyls (Waites & Moule 1961), to more complex artery-vein retes in rabbits (*Oryctolagus cuniculus*) and kangaroos (*Macropus fuliginosus* and *M. rufus*; Setchell 1977). In elephants, which are testicond, testicular temperature is the same as core body temperature (about 36°C), but other testicond mammals tend to have a low $T_b$, hence low testis temperature (Werdelin & Nilsonne (1999). The bottlenose dolphin (*Tursiops truncatus*) has internal testes, presumably selected to facilitate a streamlined shape for swimming (Rommel et al. 1994), and they have an unusual countercurrent heat exchanger that cools these internal testes. The spermatic arteries run countercurrent to veins returning cool blood from the dorsal fin and flukes, thereby cooling the testes and surrounding deep abdominal cavity by about 1.3°C.

A reverse artery-vein rete is present in the blood supply to the brain of artiodactyls, felids, and canids. Here, the carotid artery supplying blood to the brain divides into a carotid rete, a fine (200–300 μm diameter) plexus of arteries (Baker 1979), before coalescing again into a single artery that then flows into the circle of Willis at the base of the brain (Jessen 2001b). The carotid rete lies within a venous sinus—the intracranial cavernous sinus in artiodactyls and the extracranial pterygoid sinus in felids—that contain venous blood returning from the evaporating surfaces of the nasal cavity. The large surface area of contact between arterial blood in the rete and the venous blood within the sinus, and the thin walls of the rete vessels, facilitates heat transfer from the arterial blood to the venous blood, resulting in cooler arterial blood reaching the brain than would be case if no heat exchange occurred.

Because the major determinant of brain temperature is the temperature of the blood reaching it, the activation of heat exchange in the carotid rete results in cooler blood reaching the brain, and thus a reduction in brain temperature—a process known as selective brain cooling. The process uncouples brain temperature from arterial blood temperature, and in many ungulates a brain temperature more than 1°C lower than arterial blood temperature has been recorded (Fuller et al. 2011a). In species that do not have a carotid rete, notably perissodactyls (odd-toed ungulates), rodents, and primates, the carotid artery does not ramify as a rete, although it does traverse a venous sinus, and brain temperature is coupled directly to arterial blood temperature (Fuller et al. 2011b).

### 3.2.4.2 Temporal Heterothermy

The maintenance of a high, stable $T_b$ by mammals is energetically costly, particularly when there is a large thermal gradient between body and environmental temperature. Therefore many mammals, potentially as many as 40% of extant species (Geiser & Turbill 2009), allow $T_b$ to decline to a new, lower setpoint when the energetic costs of maintaining endothermy becomes prohibitive; they enter a period of adaptive hypothermia or torpor (Figure 3.14). Torpor can be distinguished from pathological hypothermia by the mammal's ability to spontaneously arouse to a normothermic $T_b$ independent of external heat sources, using endogenous metabolic heat production. Torpor is therefore not an inability to thermoregulate;

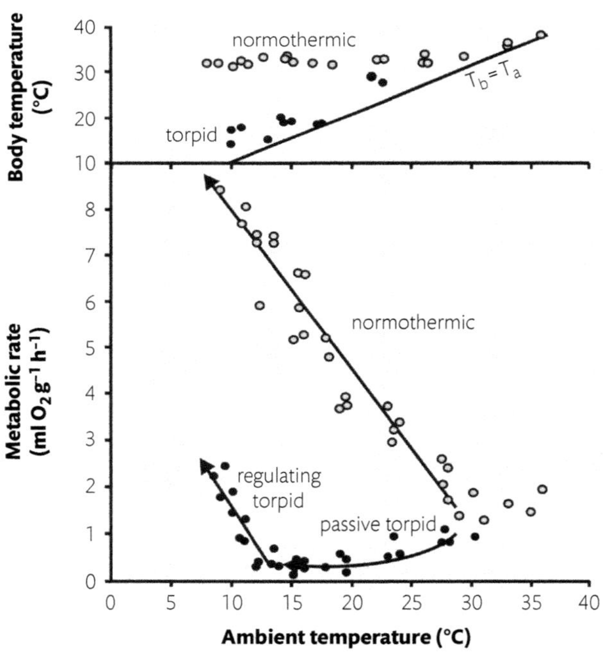


**Figure 3.14** Body temperature and metabolic rate of a small dasyurid marsupial (*Planigale gilesi*) when normothermic, passive torpor ($T_a$ >15°C), and regulated torpor ($T_a$ <15°C). Modified from Geiser and Baudinette (1988).

rather it represents an adaptive response that reduces energy expenditure when environmental conditions (temperature, food, water, predator activity, etc.) are not conducive to remaining active and homeothermic.

Mammals enter torpor during periods of short wave sleep (SWS), and maintain a greater proportion of SWS compared to rapid eye movement (REM) sleep with increasing torpor depth; a mammal in deep torpor does not enter REM sleep (Heller et al. 1978). During torpor entry, $T_b$ and MR decline until a new torpor $T_b$ setpoint is reached; if $T_a$ is above this torpor setpoint, then the mammal simply thermoconforms $T_b \approx T_a$ for the duration of the torpor bout; but if ambient conditions are lower than the torpor setpoint, then proportional thermoregulation occurs, just as during normothermia with $T_b > T_a$. A similar slope of the relationship between MR and $T_a$ for torpor and normothermia indicates a similar thermal conductance during torpor as for normothermia.

Two general forms of torpor are recognized: shallow daily torpor and deep multi-day torpor (hibernation). Shallow daily torpor lasts for < 24 h, during which time $T_b$ typically falls to 11–28°C (mean 18.1°C), with an associated decrease in metabolic rate to 10–60% (mean 30%) of BMR. In contrast, deep hibernation lasts for multiple days, weeks, or months; $T_b$ is typically < 10°C, often as low as 0–5°C, and metabolism can fall to 2–6% (mean 4.4%) of BMR (Ruf & Geiser 2014; Figure 3.15). Exceptions to these generalizations, such as the honey possum, which appears to be restricted to torpor bouts of < 24 h duration but has torpor $T_b$ as low as 5°C (Withers et al. 1990), are evidence supporting the idea that torpor is better viewed as a continuum of thermoregulatory states from strict homeothermy to deep, long-term hibernation, rather than discrete physiological states (Cooper & Withers 2010; but see Ruf & Geiser 2014).

Arousal from torpor is typically achieved by increased shivering or non-shivering thermogenesis, or metabolism of BAT, or basking (Opazo et al. 1999; Geiser & Drury 2003; Cannon & Nedergaard 2004; Ruf & Geiser 2015). However, some mammals use passive rewarming, such as basking in solar radiation or utilizing hibernacula with daily temperature fluctuations, because passive warming reduces the metabolic cost of arousal (Mzilikazi et al. 2002; Geiser & Drury 2003; Dausmann et al. 2009; Warnecke & Geiser 2009, 2010). At one time, all deep hibernators were observed to undergo periodic arousals to normothermic body temperature during the hibernation period, usually for periods < 24 h. These short arousals actually account for the majority of energy use during the hibernation period. For example, the mountain pygmy possum (*Burramys parvus*) uses up to 1.85 g of fat during a single day of normothermy, more than half that used during 155 days of torpor (Geiser & Broome 1991). The reason for these costly arousals is not clear, but they appear essential to restore some aspect of physiological homeostasis that is perturbed during hibernation, such as a need for restorative sleep to re-establish neural pathways and memory, or a need to drink due to dehydration (Humphries et al. 2003). Some of the energetic costs of interbout arousals can be

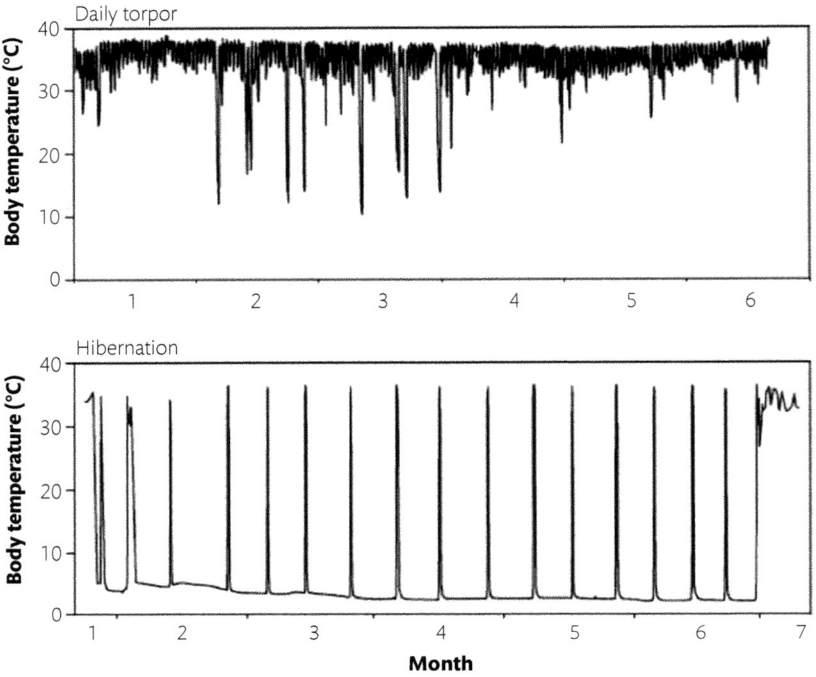


**Figure 3.15** Body temperature during daily torpor by a sugar glider (*Petaurus breviceps*; top panel) and hibernation by a mountain pygmy possum (*Burramys parvus*; bottom panel). From Geiser (2013). Reproduced with permission of Elsevier.

mitigated for hibernating mammals in warm climates, if their hibernacula have sufficient temperature fluctuations to allow periodic passive warming to $T_b$ approximating normothermia (e.g. dwarf lemurs, *Cheirogaleus medius*; Dausmann et al. 2009). However, later studies of wild, free-living tenrecs (*Tenrec ecaudatus*; Lovegrove et al. 2015) found no evidence of periodic interbout arousals for individuals hibernating for a period of 8–9 months at moderately warm ambient temperatures of > 22°C. The $T_b$ of these tenrecs remained between 22 and 27.5°C during the hibernation period; this is presumably sufficiently high to prevent the typical disruption of physiological homeostasis associated with long-term hibernation, negating the need for periodic arousals.

A number of environmental cues initiate entry into torpor, including food and water availability, environmental temperature, and photoperiod (Geiser 1988a,b, 1994; Lovegrove et al. 1999; Mzilikazi & Lovegrove 2004). Patterns of daily torpor are generally less dependent on seasonal cues than for long-term hibernation, and can occur at any time throughout the year. Most species use daily torpor predominantly during their rest phase, and lean individuals are more likely to use daily torpor. Some species readily enter daily torpor when food is withheld or when it is restricted to levels below the minimum necessary for maintenance; this is facultative torpor (McNab 1983). For facultative torpidators, the frequency, depth, and duration of torpor increases when food availability is lower compared to individuals for which food availability exceeds daily energy requirements (e.g. Mzilikazi & Lovegrove 2004; Bosinovic et al. 2007). Facultative torpor is a flexible and opportunistic response to unpredictable environmental conditions. For example, fat-tailed dunnarts increase the frequency and length of torpor bouts in response to the unpredictability of food availability rather than to food restriction per se (Munn et al. 2010a). On the other hand, some species use torpor even when food is provided *ad libitum* (McNab 1983). Many species hibernate seasonally, in response to seasonal cues such as photoperiod, and fatten before hibernation, relying on stored fat as an energy source throughout a prolonged period of inactivity, sometimes supplemented by stored food. Many of these species will not hibernate if they do not have appropriate fat stores (Geiser 2013).

Energetic savings during torpor occur primarily due to the reduction or abolishment of the thermoregulatory increment above BMR, and the further reduction in MR because of the $Q_{10}$ effect (about 2.5) of reduced $T_b$ on MR (Figure 3.16). $Q_{10}$ values of $> 3$ have been reported for MR in some mammals during torpor (Geiser 1988a), suggesting additional metabolic reduction during torpor, or intrinsic

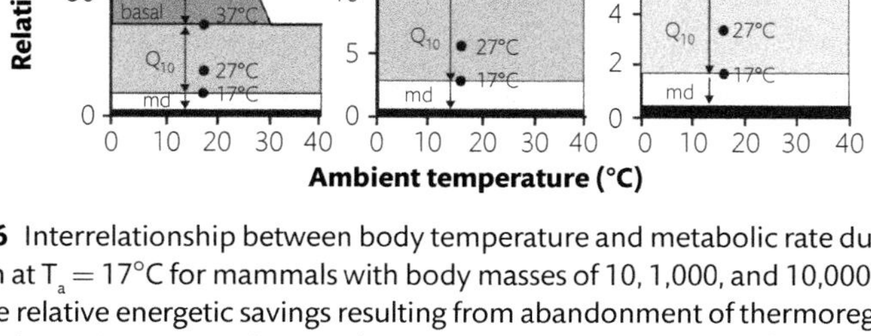

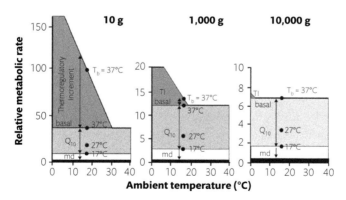


**Figure 3.16** Interrelationship between body temperature and metabolic rate during hibernation at $T_a = 17$°C for mammals with body masses of 10, 1,000, and 10,000 g, showing the relative energetic savings resulting from abandonment of thermoregulation (thermoregulatory increment; dark grey), $T_b$ reduction and the $Q_{10}$ effect (light grey), and from possible intrinsic metabolic depression (white), collectively contributing to the metabolic rate reduction during hibernation. Modified from Withers (1992).

metabolic depression, reflecting perhaps changes in acid–base status or molecular changes (Guppy & Withers 1999; Malan 2014; Martin & Yoder 2014). However, the potential energetic consequences of intrinsic metabolic depression during torpor would be small compared to thermoregulatory and $Q_{10}$ effects (Withers 1992; Guppy & Withers 1999; Withers & Cooper 2010).

Torpor has been described for species from many mammalian lineages, including monotremes, marsupials, bats, rodents, insectivores, and primates (Geiser 2013). Torpor is currently considered a plesiomorphic trait amongst mammals (Grigg and Beard 1993; Lovegrove 2012a; Ruf & Geiser 2014), with homeothermy a derived characteristic of those mammals for which periods of very low $T_b$ and inactivity are incompatible with their ecological niche (e.g. artiodactyls, which rely on microbial fermentation and large body size to feed on poor-quality plant matter). For rodents, a low BMR is associated with increased incidence of torpor use, but this is not the case for marsupials or bats. For these groups, there are instead significant phylogenetic patterns in the occurrence of heterothermy and the use of short-term torpor versus multi-day hibernation (Cooper & Geiser 2008). Torpor, and especially hibernation, appear much more widespread amongst mammals than for the convergently endothermic birds; there is experimental evidence for only one species of bird, the poorwill (*Phalaenoptilus nuttali*) using multi-day torpor (Jaeger 1948). This difference may reflect the increased mobility (e.g. flight) of birds; they are better able to avoid unfavourable environmental conditions, exploit available resources, and withstand environmental extremes than mammals. However, one of the most heterothermic groups of mammals are bats, which share with birds the capacity for powered flight.

McNab (1983) suggested that patterns of heterothermy for endotherms could be predicted as a function of body mass and metabolic rate, and proposed a 'minimal boundary curve' for homeothermy based on the allometry of BMR; species with a BMR that falls below this boundary were considered obligate heterotherms. Cooper and Geiser (2008) reviewed this concept and concluded that McNab's boundary curve and similar analyses of the relationship between mass and basal metabolic rate were not reliable predictors of the occurrence and type of heterothermy for mammals. Nevertheless, the propensity for heterothermia is associated with several general characteristics, one of which is body mass. The advantages of heterothermy are considerably greater for small compared to large species. Smaller mammals have a higher surface area to volume ratio and so have proportionally greater heat loss than larger species, resulting in more rapid cooling rates, higher mass-specific energy requirements, greater relative reductions in energy expenditure during torpor, and smaller costs of rewarming—and are therefore more likely to use torpor (Cooper & Geiser 2008).

Heterothermia, particularly long-term hibernation, was traditionally viewed as a characteristic of mammals from temperate and Arctic environments, but it is now recognized that torpor is an important aspect of the physiology of mammals from a wide range of habitats, including those in tropical and hot desert environments (Geiser 2004; Dausmann et al. 2009). Indeed, torpor has many benefits beyond the energetic advantages of a low $T_b$ reducing thermoregulatory costs during periods of low $T_a$ (Geiser & Brigham 2012). Reduced energy requirements are advantageous in a variety of situations where food may be limited, including deserts with overall low primary productivity and unpredictable resource availability (Lovegrove 2000; Munn et al. 2010a). Torpor also reduces the rate of evaporative water loss (EWL), which is particularly advantageous in arid habitats. For example, EWL of the stripe-faced dunnart (*Sminthopsis macroura*) during torpor was reduced to 23.5–42.3% of the normothermic rate, saving 50–55 mg of water per hour (Cooper et al. 2005).

Torpor may be beneficial for reducing energy consumption during life history stages that are energetically challenging, such as development, pregnancy, and lactation, or storing energy for upcoming energetically demanding events (Geiser & Brigham 2012). Pregnant female mulgara (*Dasycercus* spp.) use torpor more extensively than males, and gain mass during pregnancy, presumably in preparation for the energetically demanding lactation phase, where they remain normothermic (Geiser & Masters 1994). For some heterothermic species, torpor plays an important role in reproductive processes, such as facilitating sperm storage and optimizing reproductive timing. Torpor during reproduction appears particularly important for species inhabiting unpredictable environments and/or those that rely on unpredictable or highly seasonal food resources such as nectar or invertebrates (McAllen & Geiser 2014). Torpor may facilitate small mammals withstanding natural disasters such as droughts, storms, and fires (e.g. Stawski et al. 2015). Sugar gliders (*Petaurus breviceps*) increase their use of torpor in response to inclement weather, with minimum daily $T_b$ falling well below the $T_b$ threshold for torpor in response to increased rainfall from a storm (Nowack et al. 2015; Figure 3.17).

The ability of small heterothermic mammals to remain in safe refugia, presumably underground, for a considerable period is a hypothesis explaining how they may have survived, by entering torpor, the consequences of catastrophic meteorite impacts that occurred approximately 65.5 MYBP (Kikuchi & Vanneste 2010; Lovegrove et al. 2015). Torpor may also aid in withstanding human-induced habitat modification (Liow et al. 2009). Lovegrove et al. (1999) suggested that torpor facilitates avoidance of risky foraging behaviour in small mammals, and observations of torpor patterns in dormice (*Glis glis*) and long-eared bats (*Nyctophilus bifax*) support the hypothesis that predator avoidance is an important role of torpor (Bieber & Ruf 2009; Stawski & Geiser 2010). Torpor has also been proposed to prolong lifespan for small heterothermic species (Turbill et al. 2012; Ruf et al. 2012). These numerous positive consequences of torpor confer overall advantages to the long-term persistence of heterothermic species. Liow et al. (2009) and Geiser & Turbill (2009) provide evidence that heterothermic mammals are less likely to become extinct than homeothermic species.

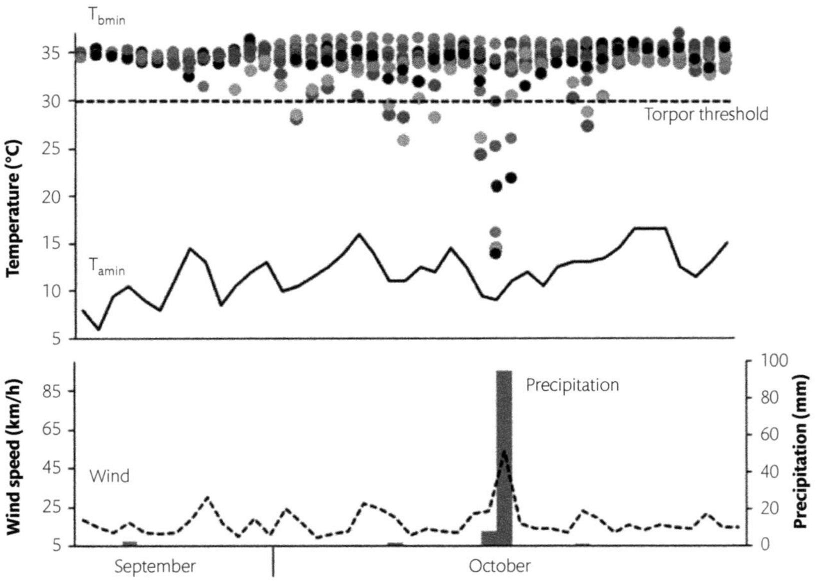


**Figure 3.17** Sugar gliders (*Petaurus breviceps*) increase torpor use ($T_{bmin}$; different symbols are for different individuals) in response to a high-rainfall storm (increased precipitation and wind speed), despite consistent minimum ambient temperatures ($T_{amin}$). From Nowack et al. (2015). Reproduced with permission of J. Nowack.

There are, however, disadvantages associated with heterothermia, which may limit, or in some cases, prohibit torpor use. Energetically costly periodic arousals to normothermia by most hibernators are evidence of these costs. For large mammals, the energetic costs of arousal from torpor are prohibitive, and explain why torpor is restricted to relatively small mammals (< 5,000 g). Even for small mammals, there is a risk of having insufficient energy supplies or too low a $T_b$ for successful arousal, or of thermal death (Withers & Cooper 2010). Cryan and Wolf (2003) reported that female hoary bats (*Lasiurus cinereus*) defend normothermia during gestation, presumably to expedite embryonic growth, while males use torpor. For some mammals, such as sciurid and cricetid rodents and some bats, reproduction and heterothermia are mutually exclusive, and torpor is confined to the non-reproductive period. For other mammals, however, torpor may occur during pregnancy or lactation, but this slows the rate of development of offspring (McAllen & Geiser 2014). Torpor may lead to osmotic imbalance: although absolute rates of EWL are reduced during torpor, the relative water economy (RWE; metabolic water production/evaporative water loss) is unfavourable; that is, more water is lost by evaporation than is produced metabolically.

This is because metabolic rate and thus metabolic water production decline more than EWL during torpor. For example, the RWE of normothermic monito del monte is > 1 at $T_a = 15°C$, but decreases to only 0.28 for torpid individuals at the same $T_a$ (Figure 3.18). Torpid monitos therefore lose water equivalent to approximately 5% of their body mass per week, and this may necessitate periodic arousals (Withers et al. 2012). The low $T_b$ associated with torpor can also result in impaired digestive, immune, and muscle function, and the lack of restorative REM sleep leads to memory loss from accumulated declines in neuronal connectivity (Humphries et al. 2003). Ground squirrels (*Spermophilus citellus*) that hibernated

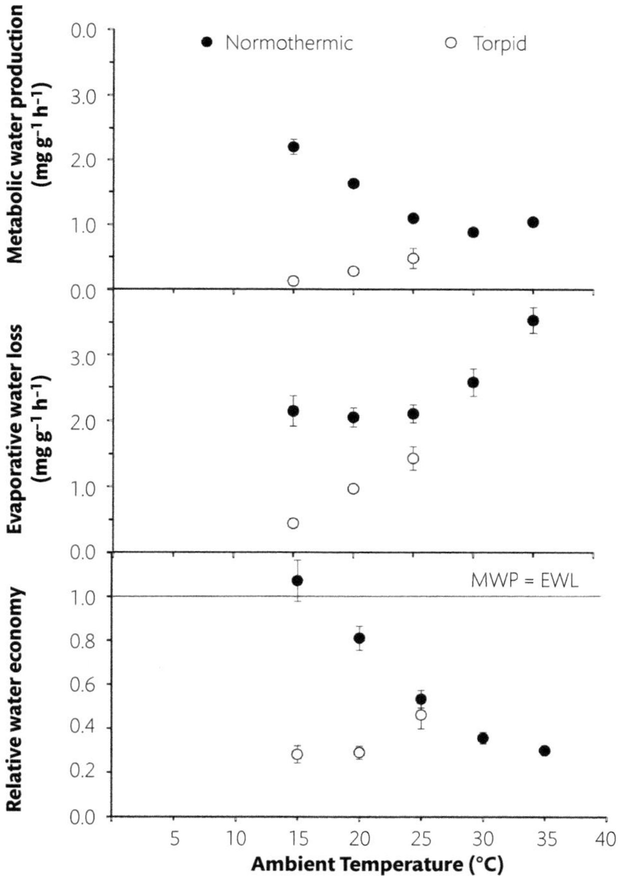


**Figure 3.18** Metabolic water production, evaporative water loss, and relative water economy of the monito del monte (*Dromiciops gliroides*) during normothermia (black symbols) and torpor (white symbols). RWE is significantly less favourable during torpor despite absolute reductions in EWL. Redrawn from Withers et al. (2012).

had significantly lower performance for tasks relating to spatial memory and operant conditioning than those that didn't (Millesi et al. 2001). Torpor may also increase the risk of predation by impacting on sensory and motor function and therefore limit predator detection and escape mechanisms, and can adversely impact on competition for resources (Humphries et al. 2003). Alternatives to heterothermia include the following: development of a large body mass to reduce the surface area to volume ratio, and therefore relative rates of heat loss, and also facilitate increased fur length and thickness and therefore insulation; use of nests and burrows to reduce heat loss; and exploitation of abundant, reliable, energy-rich food, or food hoarding (Ruf & Geiser 2014).

### 3.2.5 Heat Loss

Mammals lose heat by direct physical routes (e.g. conduction, convection, radiation) and also by evaporation. Non-evaporative heat loss relies on a favourable thermal differential (i.e. $T_b > T_a$) for conductive, convective, or evaporative heat loss, but is ineffective if $T_b < T_a$ because this promotes passive heat gain from the environment. Evaporative heat loss (EHL) is especially effective at high $T_a$ because it does not rely on a thermal differential favouring heat loss (i.e. $T_b < T_a$).

#### 3.2.5.1 Non-evaporative Heat Loss

The exchange of heat by non-evaporative ('dry') routes of conduction, convection, and radiation depends on skin ($T_{skin}$) and surface ($T_{surface}$) temperature, the resistance to heat flow from the skin to the surrounding medium, and the environmental conditions. The $T_{skin}$, in turn, depends on the balance between heat loss from the skin to the environment and heat delivery to the skin, predominantly via heat convection in the blood. Within the animal the bulk of heat production at rest takes place in the viscera and brain, that is, the body core (Stainer et al. 1984). Blood flowing throughout the body convects heat away from regions of high heat production out to the skin or the nasal mucosa (see later), where heat loss to the environment occurs. By controlling blood flow to the periphery, an animal alters heat exchange by dry routes since convective and radiant heat exchange are proportional to the difference between $T_{skin}$ and $T_a$, and an increase in blood flow to the skin increases $T_{skin}$. Organs producing large amounts of heat are hotter than arterial blood, and so heat flows from the organ (cooling it) to the blood. On the other hand, the skin and nasal mucosa continually lose heat to the environment, and blood flowing to these sites loses heat. Core temperature then depends on the mix of these warmer and cooler venous returns to the right side of the heart. If heat loss is compromised, then the blood returning to the right heart is warmer than arterial blood, and core temperature will rise.

Skin blood flow is controlled by the hypothalamus, and changes predictably from very low levels when an animal is below the TNZ to maximal levels when the animal is above the TNZ. To some extent, the surface area available for heat exchange can respond to environmental conditions: as adults, rats raised under cold conditions have smaller ears and tails and have a lower tail-to-body-length ratio than those raised at moderate ambient temperature, whereas those raised under warm conditions are smaller and have a higher tail-to-body-length ratio than those raised at moderate ambient temperature (Villarreal et al. 2007). These results show that young rats respond to chronic exposure to environmental air temperatures with developmental plasticity in the expression of morphological traits that may promote thermal homeostasis.

Given the importance of skin blood flow in the control of heat exchange, it would seem logical that the capacity for blood supply to the skin might also exhibit some variation with exposure. There has been only one study on the development of skin blood vessels in response to environmental conditions, showing that pigs raised at 35°C had a higher density of cutaneous blood vessels than pigs raised at 5°C (Ingram & Weaver 1969). Unfortunately, this study lacked a control intermediate $T_a$ group, so it is not clear whether the differing capillary densities were due to inhibition by the cold, stimulation by the heat, or both.

Changes in body posture also contribute to variation in dry heat loss. Adopting a spherical posture minimizes the surface area to volume ratio and heat loss, whereas adopting a sprawling posture increases the surface for non-evaporative (and evaporative) heat loss. For example, the non-evaporative thermal conductance ($C_{dry}$) of sandhill dunnarts (*Sminthopsis psammophila*) remains constant at low $T_a$, but increases at high $T_a$, reflecting the facilitation of heat dissipation (e.g. posture and peripheral blood flow changes; Withers & Cooper 2009b; Figure 3.19). In contrast, the total thermal conductance, which includes evaporative heat loss ($C_{wet}$), is

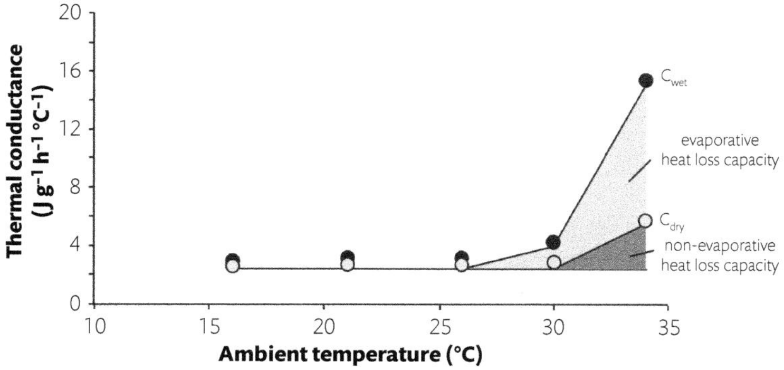


**Figure 3.19** Thermal conductance of sand-hill dunnarts (*Sminthopsis psammophila*) for non-evaporative heat exchange capacity ($C_{dry}$) at varying $T_a$ compared with total heat exchange capacity, including evaporation ($C_{wet}$). Redrawn from Withers and Cooper (2009b).

higher than $C_{dry}$ and increases more, and at lower $T_a$, reflecting the increased EHL at $T_a = 30$ and 35°C.

Non-evaporative heat exchange can be augmented by conductive heat exchange with cool or warm substrates. In the cold, heat can be gained by contact with warm substrates, including water. For example, the Japanese 'snow monkey' macaque (*Macaca fuscata*), the northern-most (non-human) primate, is renowned for resting in hot springs, presumably to facilitate thermoregulation, although they are physiologically capable thermoregulators in the cold, with thick, well-insulating fur (Hori et al. 1977; Zhang et al. 2007; see 4.1.1). In the heat, contact with cold substrates can facilitate body cooling. For example, koalas (*Phascolarctos cinereus*) can hug cool tree trunks; the substantial increase in passive heat loss reduces the requirement for evaporative cooling (Briscoe et al. 2014; Briscoe 2015).

### 3.2.5.2 Evaporative Heat Loss

At high $T_a$, EWL (see 3.6.2) is the most effective means of dissipating heat, particularly as it is the only mechanism for dissipating heat if $T_a > T_b$, since other avenues of heat exchange (conduction, convection, radiation) would all result in heat gain from the environment. The high latent heat of evaporation for water (about 2,400 J $g^{-1}$; McNab 2002) provides for effective heat loss. EHL has traditionally been divided into an insensible component, reflecting 'baseline' cutaneous and respiratory EHL, and a sensible component, reflecting enhanced EHL by regulatory mechanisms, such as salivation, sweating, and panting, that occur under more extreme environmental conditions.

At low $T_a$ ($<T_{lc}$) and within the TNZ, EHL of mammals is generally low and relatively constant, but often begins to increase towards the upper TNZ and then increases substantially at $T_a > T_{uc}$. Consequently, the wet thermal conductance is typically higher than the dry thermal conductance, and becomes progressively greater at higher $T_a$ (e.g. Figure 3.19) as EWL, and hence EHL, increases. There is little effect of body mass on the dissipation of metabolic heat by evaporation for mammals in their TNZ, as indicated by the similar allometry of BMR ($69.7\,M^{0.720}$ J $h^{-1}$; McNab 2008) and EWL ($18.5\,M^{0.696}$ *J* $h^{-1}$; Van Sant et al. 2012). That is, the allometry for percentage basal metabolic heat dissipation by evaporation is essentially independent of M ($26.5\,M^{-0.024}$)—there is little difference between the 23% of metabolic heat dissipated by evaporation for a 300-g mammal and the 18.5% for a 3-tonne mammal. However, large mammals have a low surface to volume ratio and a low mass-specific metabolic rate, hence they gain heat much more slowly than small mammals in hot environments. A 300-g mammal storing its metabolic heat production would experience a 3°C increase in $T_b$ after 46 min, whereas a 3-tonne mammal would experience a 3°C increase in $T_b$ only after 11 h (McNab 2002). Small mammals are therefore much more reliant on EHL in the heat than are large mammals (see 4.2).

### 3.2.6 Fever

Fever is recognized as an increase in core $T_b$ above the normal range, and is the oldest and most universally known hallmark of disease or infection (Kluger et al.1975; Kluger 1978; Dinarello 2004; Saunders et al. 2015). During a fever, the thermoregulatory system operates as normal, but defends a higher $T_b$. Bacterial products (such as lipopolysaccharide, LPS, from the cell walls of gram-negative bacteria) that can stimulate fever in a host are known as exogenous pyrogens. It was recognized in the mid-20th century that exogenous pyrogens do not directly stimulate fever, but that the process is transduced by endogenous pyrogen (EP; Atkins 1960). EP turned out to be several cytokines (most importantly, interleukin-1β, interleukin-6, and tumour necrosis actor) that are released by immune cells when toll-like-receptors on the surface of those cells recognize components of exogenous pyrogens (sometimes referred to as pathogen-associated molecular patterns). Thus, a large range of invading pathogens are recognized by the innate immune system, and EP mediates a common response.

Fever occurs in a wide range of ectothermic and endothermic animals. In ectotherms, $T_b$ increases from 1.5 to over 5°C, and in endotherms about 1.8°C in pigeons (*Columba*) and for placental mammals about 1.2°C (dog) to 3.9°C (humans). This increased $T_b$ is accomplished by changes in behavioural thermoregulation in ectotherms, and increased metabolic heat production in endotherms (mammals and birds). In marsupial and placental mammals, the fever response is similar, with many aspects of immune system function having evolved before their divergence; these include the main cytokine peptide families (interleukins, interferons, and chemokines (Morris et al. 2010). The echidna also becomes febrile in response to pyrogen, but apparently has a unique (compared to marsupials and placentals) response to prostaglandins, where $T_b$ and MR are decreased (Baird 1974).

Why animals develop fever has long been the subject of debate. On one side is the argument that the higher $T_b$ augments the body's defence against infection, and on the other is the argument that fever is stimulated by invading pathogens and is of benefit to those pathogens. The acute phase response—which includes fever and a decrease in plasma iron—reduces the proliferation of many microorganisms (Kushner & Rzewnicki 1997); the inhibition of fever in reptiles (that thermoregulate behaviourally to reach a febrile body temperature) leads to increased mortality relative to reptiles that are given access to environmental heat (Kluger et al. 1975), suggesting that fever is of benefit to the host. On the other hand, randomized clinical trials are equivocal on the survival value of fever in critically ill patients (Harden et al. 2015). The introduction of antibiotics has to some extent reduced the 'need' for fever during bacterial infections, but the widespread use of antipyretics during viral fever does not seem to have led to a large increase in morbidity or mortality during viral infections, suggesting that fever is not a requirement for the defence against viral pathogens (Harden et al. 2015).

The production of a fever is metabolically costly, with each 1°C increase in body temperature requiring about a 10% increase in metabolic rate (Kluger 1991). The extent of the increase in core $T_b$ during fever depends on the cause of the fever. For example, malaria is notorious for the high fevers that it causes in humans. In general, fevers higher than 41°C are rare and, at least in humans, fever's upper limit seems to correspond to the temperature limits for the thermal damage of tissues of about 41 – 42°C (Mackowiak & Boulant 1997). In this regard, there are signalling molecules, termed cryogens (to distinguish them from the pyrogens that cause fever) that are released during fever and play a role in limiting the increase in $T_b$ (Kluger 1991).

To alter the thermoregulatory setpoint, the increase in circulating EP might be expected to affect neural circuits in the hypothalamus, requiring the peripherally produced cytokines to cross the blood–brain barrier (which normally restricts transfer of large molecules and cells between the blood and brain tissues). However, cytokines that are injected peripherally only reach the brain several hours to days after the febrile episode (Dinarello 2004), whereas fever is stimulated in a much shorter time frame, so it is now thought that cytokines act on their specific receptors on endothelial cells (i.e. blood side of the blood–brain barrier) in the organum vasculosum laminae terminalis (OVLT), near the hypothalamus. This stimulates the release of arachidonic acid, which is rapidly converted to prostaglandins on the brain side of endothelial cells. Ultimately the cause of fever is an increase in prostaglandin (specifically PG-$E_2$) on the hypothalamus, acting on the receptor EPR-3 that inhibits warm-sensitive neurons and excites cold-sensitive neurons (Boulant 1997), resulting in a regulated rise in core $T_b$.

While fever is well studied in the laboratory in response to pathogens or specific molecules (e.g. LPS), fever has also been described for free-ranging mammals. For example, in a herd of springbok (*Antidorcas marsupialis*) in South Africa, Fuller et al. (2005) noted increases in the core $T_b$ of some individuals to 41°C (compared to the normal $T_b$ of about 39°C) that lasted for several days, indicative of a fever response to pathogen infection. On one occasion, fever developed consecutively in several members of the herd, suggesting movement of a pathogen through the herd. Later, Hetem et al. (2008) reported fever and sickness behaviour in response to a naturally acquired bacterial pneumonia for free-ranging greater kudu (*Tragelaphus strepsiceros*). Their response included increases in core $T_b$ of about 1.5°C accompanied by the selection of warmer microclimates (as assessed by mini-globe temperatures recorded on collars), and an increase in the core-to-skin temperature difference indicative of peripheral vasoconstriction and heat conservation. Based on these data, it seems that the physiological mechanisms of fever that have been measured in the laboratory occur also in free-ranging mammals when they are infected naturally.

### 3.2.7 Development

The $T_b$ of placental mammals *in utero* is coupled to the temperature of the mother, with a difference of 0.5°C between the fetus and the mother by the time of parturition. This thermal differential is established by the balance between metabolic heat production of the fetus(es) and loss of heat to the mother (Abrams et al. 1969; Laburn et al. 1992). About 85% of the heat produced by the fetus is lost via the placental circulation and the other 15% flows by conduction through the placental wall (Gilbert et al. 1985).

The stability of the uterine environment (except in some marsupials that use torpor while pregnant; Geiser & Masters 1994) renders it unnecessary for the developing fetus(es) to thermoregulate, and *in utero* the fetus is incapable of defending its $T_b$ against a decrease in temperature. However, this situation changes dramatically at birth, with the fetus now exposed to the external environment, wet from the birth process, and being small with a large surface area to volume ratio. A question then arises how the fetus adapts to the sudden change in thermal requirements, given that many species are born with apparently fully developed thermoregulatory responses (see later).

Bruck and Hinckell (1996) used three criteria to define the maturity of the thermoregulatory system at birth: (1) the capacity of effector systems relative to the adult, (2) the qualitative and quantitative (threshold/gain) of the system to thermal stimuli, and (3) qualitative or quantitative differences in central processing of thermal information. Based on those criteria, they distinguished three categories of mammal at birth. The first category is able to maintain $T_b$ within a certain range of ambient conditions, having essentially all the thermal responses of the adult. This group includes humans, guinea pigs (*Cavia porcellus*), pigs and miniature pigs (*Sus scrofa*), sheep, and larger mammals such as cattle. Pigs are a special case because they have no thermogenic BAT and lack the gene for UCP-1 (Berg et al. 2006). Newborn piglets therefore rely on shivering thermogenesis and thermoregulatory behaviour for the defence of $T_b$, while for the other species, thermogenesis by BAT is an important component of cold defence. In the second category, thermoregulatory responses can be evoked at birth, but either the capacity of the effector systems is much less or the thresholds for activation of those systems are lower (requiring $T_b$ to decrease more before they are stimulated) than for the adult. The result is a decrease in $T_b$ at $T_a$ just below thermoneutrality. Included in this category are the low-birth-weight humans, cats (*Felis catus*), dogs (*Canis familiaris*), rabbits (*Oryctolagus cuniculus*), and the rat (*Rattus rattus*). In the third category, no thermoregulatory responses can be evoked, and $T_b$ decreases on exposure to even mild cold. Rather than metabolic rate increasing in defence against a decrease in $T_b$, it decreases with $T_b$ in accord with the $Q_{10}$ effect. Species in this category include the ground squirrel and the golden hamster (*Mesocricetus auratus*).

Marsupials are born developmentally much sooner than placentals, and they fall in the third category. Young marsupials are therefore dependent on the pouch environment for the maintenance of $T_b$. During the early stages of pouch life, the metabolism of the young marsupial is quite low, with a mass-specific metabolic rate similar to reptiles and well below that of adult marsupials or placentals (Hulbert 1988). During development, metabolism switches to the normal marsupial level. As the young develops inside the pouch, it increases in mass (with a consequent decrease in surface area to volume ratio) and grows fur, contributing to its thermoregulatory ability. After accounting for these physical changes in the kowari (*Dasyuroides byrnei*), Geiser et al. (1986) concluded that vasoconstriction and the metabolic response were functional by day 70–80 in development, with weaning at day 110. Similarly, the metabolic response to cold is present in the quokka (*Setonix brachyurus*) at about day 100, although the response is transient, and persistent defence develops only by day 144 (Shield 1966). By the time of permanent pouch emergence, the young marsupial is a competent endotherm (Tyndale-Biscoe 2005).

How is it that some mammals are incapable of generating a heat defence response *in utero*, but are born with a fully developed thermoregulatory ability? Gunn et al. (1993) showed that while lambs *in utero* developed no metabolic response to cooling applied via a heat exchanger wrapped around the fetus, either oxygen supply to the fetal lung or the occlusion of the umbilical cord caused an increase in oxygen consumption and an increase in plasma free fatty acids and glycerol in the cooled fetus, suggesting that a signal of placental origin inhibited thermoregulatory responses in the fetus. They went on to show that the blockade of prostaglandins permitted thermoregulatory responses in the fetus, and the infusion of prostaglandin $E_2$ after umbilical occlusion prevented thermoregulatory responses in the fetus. Therefore, in lambs at least, thermoregulatory responses are inhibited by the placental circulation, and by prostaglandins of placental origin. At birth, that inhibition is removed, and the newborn has a full response to cold exposure (Schröder & Power 1997). There is no need for the fetus to develop a thermogenic capacity prior to birth, as this role is subsumed by the mother, and any metabolic thermoregulatory response would be an additional demand on placental gas exchange anyway.

### 3.2.8 Bergmann's and Other 'Rules'

Body size has a fundamental impact on the physiology and ecology of mammals in different environments (e.g. at varying $T_a$). Mass constrains almost every aspect of morphology and physiology, and influences the nature of most ecological relationships. Patterns of within-species phenotypic variation along environmental gradients, or the lack thereof, are useful starting points to formulate general hypotheses and undertake mechanistic studies aimed at understanding the ecology and evolution of species (Endler 1986; Conover et al. 2009). This information may be useful to understand the ecological and evolutionary responses of populations or species to current and future climatic conditions. Bergmann's rule is one of the 'classic' bioclimatic rules that describe effects of latitude (a proxy for environmental temperature) on physiology and morphology. Other 'rules' (or 'laws') include Allen's (appendage size), Wilson's (insulation thickness), Gloger's (colour), Hesse's (heart size), Rapoport's (latitudinal breadth), Cope's (increase in body mass), island (dwarfism/gigantism) rules, and Dehnel's (winter mass reduction) phenomenon (Withers 1992). There is also a pattern of increased body mass along an aridity gradient in small mammals, with larger individuals found in the driest regions (Storz et al. 2001).

The general interpretation of Bergmann's rule is that species of mammal at higher latitudes are larger than closely related species at lower latitudes (Bergmann 1847), or that populations of a mammal species are bigger at higher latitudes (Mayr 1956). This rule is commonly interpreted as 'mammals are bigger in colder climates' (Rensch 1938; McNab 2002; Meiri 2011; Blackburn et al. 1999). The mechanistic rationale for Bergmann's rule, based on modern physiological principles, is that a larger mammal would have a lower surface area to volume ratio than would a smaller individual, providing a reduced rate of heat loss in colder environments (McNab 1971; Atkinson & Sibly 1997). Heat loss (across the body surface, SA) would be lower relative to its mass even though there is an absolutely higher SA; i.e. $SA/M \alpha r^2/r^3 \alpha r^{-1} \alpha M^{-0.33}$. So long as mass-specific surface area ($\alpha M^{-0.33}$) decreases faster than mass-specific metabolic rate ($\alpha M^{-0.25}$), heat is conserved relative to metabolic heat production. Whether mass-specific SA and MR (as opposed to absolute SA and MR) can be under selection pressure is arguable (McNab 1999).

Bergmann's rule is applicable to most, but not all, mammals (and birds), with narrow-to-long latitudinal ranges (Sullivan & Best 1997), over even relatively small temperature gradients (Smith et al. 1998), and is maintained for species that cross the equator (Martínez et al. 2013). We expect that climate change will increase global temperatures (see 6.2), and Bergmann's rule predicts that species will evolve a smaller body size in a warmer climate. There is already weak evidence for this (Teplitsky & Millien 2013). But not all mammals conform to Bergmann's rule. In North America, most mammal species don't follow Bergmann's rule, although carnivores and granivores do, perhaps due to resource availability or character displacement (McNab 1971). In Australia, house mice (*Mus musculus*) conform to Bergmann's rule, but sandy inland mice (*Pseudomys hermannsburgensis*) do not (Tomlinson & Withers 2008). Malagasy strepsirrhine primates do not conform to Bergmann's rule, and there is no resource seasonality/scarcity/quality or primary productivity pattern, but there is a strong phylogenetic pattern for mass (Kamilar et al. 2012). Lovegrove and Mowoe (2013) found no evidence for Bergmann's rule in North American Caenozoic mammals. Many ectotherms also conform to

Bergmann's rule (Ashton & Feldman 2003), which is surprising if the explanation for the rule depends on thermal balance and surface area to volume ratio (see later), so alternative explanations have been proposed for a wider generality of Bergmann's rule, such as 'food scarcity' (Langvatn & Albon 1986; Ochocińska & Taylor 2003) and productivity (Rosenzweig 1968) models.

Allen's rule (Allen 1877; see Serrat 2013) states that appendage size is inversely correlated with environmental temperature and latitude: smaller appendages and limbs have a lower surface area to volume ratio, hence lower rates of heat loss. For example, ear size is greatest for kit foxes (*Vulpes macrotis*), intermediate for red foxes (*Vulpes vulpes*), and smallest for Arctic foxes (*Vulpes lagopus*). Allen's and Bergmann's rules formulate the general observation that mammals in cold climates have a larger and stockier body, with shorter limbs and appendages.

Wilson's rule states that insulation (a combination of fur length and adipose tissue thickness) is greater in cooler climates (McLean 1981). For example, Arctic mammals have thicker pelage than tropical species, and thickness changes seasonally (Scholander et al. 1950; Withers 1992; see 4.1.1). Gloger's rule relates plumage colour of birds to habitat humidity: birds are darker in more humid environments, and it has also been applied to mammals (e.g. primates, including humans; Gloger 1883, see Withers 1992). However, examples such as polar bears being white in the drier Arctic than grizzly bears in the more humid tundra can also be ascribed to crypsis of their colouration (Feldhamer et al. 2015).

Hesse's (heart size) rule suggests that heart mass increases with increasing latitude (i.e. in colder climates; Hesse 1924; Müller at al. 2014). For example, heart mass of two small rodent species (*Apodemus flavicollis, Myodes glareolus*) increases with altitude (hence cold), as does lung mass (but not other internal organs; Müller et al. 2014). Rapoport's rule suggests that mammal species from higher latitudes (colder climates) are better adapted to survive a wider range of environmental conditions hence latitudes than are species from lower-latitude, warmer climates (Rapoport 1982; Stevens 1989, 1992). There is support for Rapoport's rule from some North American mammals and neotropical bats, but it might reflect historical climate patterns (e.g. long-term glaciation patterns; Gaston et al. 1998; Feldhamer et al. 2015).

Cope's rule describes the tendency for mass to increase in evolutionary lineages over time (Cope 1896). This evolutionary pattern is widely known for fossil dinosaurs, but applies to mammals as well. The rapid increase in the range of body mass (2 g to over 10 tonnes) for mammals during the Caenozoic, after the demise of the large dinosaurs, is consistent with both Cope's rule and Bergmann's rule (Saarinen et al. 2015). The 'island rule' describes the occurrence on islands of dwarfism for large mammals (e.g. elephants) and gigantism for small mammals (Foster 1964; Roth 1990). Dehnel's phenomenon indicates that body mass declines in winter, not just because of reduced fat stores but also reduced skeleton, braincase, brain mass, etc. (Dehnel 1949). It has been observed for some meadow voles (*Microtus pennsylvanicus*) and shrews (McNab 2002). The reduced body mass during periods of low food availability is consistent with Geist's (1987) food resources model.

## 3.3 Ventilation

The ventilatory system provides the first step in the $O_2$ cascade (and the last step in the $CO_2$ cascade) necessary for aerobic metabolism. Although the basic structures and processes in mammals are the same as for 'lower' vertebrates, the ventilatory system of mammals operates at a higher capacity that matches their high metabolic demands (as does the very different ventilatory system of birds). Consequently, the structure and function of the ventilatory system is adapted to provide very high rates of gas exchange.

### 3.3.1 Airways and Lungs

The ventilatory system of mammals has a number of roles; gas exchange is one, but other roles include vocalization, olfaction (see 3.7.2.1), acid–base regulation (via $CO_2$ balance), synthesis of angiotensin II (a potent vasoconstrictor), lymph return (via thorax-abdomen pressure differences), and thermoregulation (via respiratory evaporative water loss). Breath-holding is associated with urination, defaecation, and childbirth, via the Valsalva manoeuvre. The primary structures are the nose, pharynx, larynx, trachea, bronchi, and lungs.

In the mammalian thoracic cavity, the trachea branches into two bronchi, which further divide into secondary, and finally tertiary bronchi, or bronchioles, which deliver the air to the tiny sac-like alveoli, the primary site of gas exchange. Alveolar lungs are a characteristic of mammals (see 1.2.4). Alveoli are numerous tiny compartments at the tips of the bronchial tree, which provide a much greater surface area than the faveolar lungs of other vertebrates (except birds), which consist of larger compartments formed by numerous septae (Kardong 2009). Alveolar walls consist of a single cell layer of Type I alveolar cells. The alveoli are enveloped by a dense network of pulmonary capillaries (also with walls a single cell thick), that cover 80–90% of the alveolar surface, with very little interstitial space between the capillary and alveolar wall. Alveolar structure provides for a high diffusion capacity, by minimizing the distance between air and blood (small $\Delta X$), and maximizing the diffusion area (A). However, a potential problem for these very small alveoli is that cohesive forces between water molecules lining their interior surface generate a high surface tension that tends to collapse them (see 2.5.4 and the law of Laplace). The tendency of small alveoli to collapse is reduced by the presence of pulmonary surfactants (phospholipids combined with protein) that reduce the surface tension, and therefore the pressure required to keep the alveoli inflated.

### 3.3.2 Ventilatory Mechanics

Mammalian lungs are suspended in the pleural (chest) cavity. Inspiration occurs when the rib cage expands outwards and anteriorly, and the diaphragm contracts towards the abdomen, expanding the thoracic cavity. The subsequent decrease in intrapleural pressure causes an increased trans-pulmonary pressure, or an increased pressure gradient across the wall of the alveoli. The lungs expand in response to this pressure gradient, decreasing alveolar pressure and drawing air into the lungs via the trachea. The trachea is held open against the negative pressure by cartilaginous rings. Expiration is generally passive when an animal is not active and occurs when the diaphragm and external intercostal muscles relax, resulting in the elastic recoil of tissues of the thorax, increased pulmonary pressure, and an increase in alveolar pressure, resulting in gas leaving the lungs via the trachea. Active expiration is also possible during periods of high $O_2$ demand, such as during exercise. During forced expiration, the internal intercostal muscles of the rib cage, and other thoracic muscles, contract, causing an acceleration in the volume depletion of the pleural cavity, enhancing the increase in pleural pressure and causing an increase in the convective flow. For cursorial and saltatory mammals, such as dogs, horses, rabbits, kangaroos, and even humans, ventilation may be coupled to locomotory gait, as the visceral organs move in rhythm with the pattern of limb oscillation, working like a piston on the diaphragm that compresses and expands the thoracic space (Alexander 1989; Bramble 1989; Baudinette et al. 1987; Kardong 2009). The visceral organs of cetaceans move in a similar manner, working like a piston on the diaphragm that compresses and expands the thoracic space (Piscitelli et al. 2013).

The lung capacity is the total volume of air that can be in the lungs at any time, and is partitioned into four 'pulmonary volumes'. When breathing normally, the lungs are neither filled to capacity on inspiration nor emptied on expiration. The tidal volume ($V_T$) is the volume of air that is normally inspired per breath; it ranges from about 0.106 ml for the 5-g honey possum (Cooper & Cruz-Neto 2009), to 278 ml for the red kangaroo (*Macropus rufus*; Dawson et al. 2000), to about 500 ml for humans, to 36,000 ml for Asian elephants. The inspiratory reserve volume is the additional volume of air that can be forcibly inspired after a normal inspiration; it is about 3,000 ml for humans. The expiratory reserve volume is that volume of air that can be forcibly expelled after a normal expiration (about 1,100 ml for humans). The remaining air is the residual volume (about 1,200 ml), a volume of air that cannot be exhaled and that remains in the lungs thereby preventing the alveoli from collapse (deep-diving marine mammals rely on complete lung collapse to avoid the bends and nitrogen narcosis; see 4.5.3).

There are functional limitations on the gas exchange efficiency for the mammalian tidal ventilatory system. Because air enters the alveoli via the bronchial tree and is a 'dead-end' system, there is a considerable anatomical dead space—not all the inspired air actually reaches the alveolar gas exchange surfaces (i.e. air remains in the trachea and bronchi). Furthermore, there is physiological dead space (known as alveolar dead space in mammals) where alveolar gas exchange could potentially occur, but some factor, such as under-perfusion with blood, prevents maximal exchange in this specific area (Pocock et al. 2013). As the alveolar system involves cyclic ventilation of terminal air spaces, convective flow does not completely renew all the air in the lungs with each breath. Consequently, mammalian oxygen extraction efficiency ($EO_2$) rarely exceeds 30% of inspired $O_2$ (Stahl 1967; Chappell 1985, 1992; Withers 1992; Cooper & Withers 2004). However, this is not really an 'inefficiency', because if fresh air did reach the alveoli with each breath, there would be a considerable variation in pulmonary $O_2$ and $CO_2$ partial pressures (alveolar $pO_2$ and $pCO_2$) during the respiratory cycle. The high functional residual lung capacity of mammals (about 460% of tidal volume for humans; from 200 to 600% for 10 g to 1,000 kg mammals, calculated from Stahl 1967) means that most of the air in the lungs remains from breath to breath, with only a small proportion of the air exchanged with each breath. That low turnover stabilizes alveolar $pO_2$ and $pCO_2$ throughout the respiratory cycle, and provides for a more constant pressure gradient between air and blood throughout the respiratory cycle.

In the alveoli, $O_2$ passively diffuses down its partial pressure gradient across the alveolar membrane and into the blood capillaries, at a rate described by Fick's law (see 2.5.2). There is no active transport of either $O_2$ or $CO_2$, a fact that was the subject of intense research in the early 20th century and led to infamous debates between Christian Bohr and August Krogh (Gjedde 2010). The structure of the alveoli is adapted to maximize diffusion (Guyton & Hall 1996; Pocock et al. 2013). The alveolar surface area is remarkably high in mammals; in humans, the ≈300 million alveoli each have a diameter of about 0.25 mm, providing a total lung surface area of about 70 $m^2$. The alveolar membrane is very thin (0.2 – 0.5μm) despite its consisting of a layer of fluid lining the lung, the alveolar epithelium cell, epithelial basement membrane, interstitial space, capillary basement membrane, capillary endothelium cell, and fluid lining the capillary. Because the average diameter of the pulmonary capillary is less than the diameter of an erythrocyte, the erythrocytes are squeezed through the capillaries and this makes for a very close contact of the erythrocyte surface with the capillary cell, minimizing the distance from the alveolar source and haemoglobin sink for $O_2$. The pulmonary diffusion capacity reflects the various morphological components that contribute resistance to the diffusive transfer of $O_2$ (and $CO_2$) between the alveolar air and the capillary blood. It is about 2.8 ml $O_2$ $min^{-1}$ $kPa^{-1}$ for humans, or 230 ml $O_2$ $min^{-1}$ for a $pO_2$ difference of 1.5 kPa at rest. The $CO_2$ pulmonary diffusion capacity is considerably higher, because of its high diffusing capacity, and the $pCO_2$ difference is about 0.13 kPa.

From the alveolar capillary, convective transport by the blood circulates $O_2$ to tissue capillaries (see 2.5), where the $O_2$ partial pressure gradient drives passive diffusion, out of the capillaries into the cells of the perfused tissue (Guyton & Hall 1996; Pocock et al. 2013). Within the cells, cytoplasmic streaming may be considered another convective transport mechanism that aids in the transport of $O_2$ from the cell membrane to the mitochondrial membrane more rapidly than simple diffusion. The transport process is reversed for the movement of $CO_2$ from the cells to the external air (with the difference that $O_2$ is transported bound to haemoglobin in red cells, whereas $CO_2$ is transported mostly as $HCO_3^-$ in both the red cell and plasma). The partial pressure difference in gases between ambient air and the cell/mitochondrion (see 2.54) reflects the various convective and diffusive steps in the $O_2$ and $CO_2$ cascade. These partial pressures and the capacity for $O_2$ and $CO_2$ transport can be modified in underground environments, and are greatly modified by pressure changes in extreme environments (altitude, depth) and often require extensive adaptation of the respiratory (and circulatory) system to maintain gas exchange (see 4.3–4.5).

There is a strong allometry for many respiratory, and related physiological, variables in mammals (Table 3.5; Stahl 1967). Mass and volume variables (e.g. lung mass, tidal volume, stroke volume, blood volume) generally scale with $M^1$, rate variables (e.g. respiratory rate, heart rate) generally scale with $M^{-0.25}$, and metabolic-related variables (e.g. metabolic rate, minute volume, cardiac output) generally scale with $M^{0.75}$. Body surface area scales with $M^{0.65}$. The respiratory $O_2$ extraction efficiency ($EO_2$) is nearly independent of mass, at about 14.6% (i.e. 14.6% of the inspired $O_2$ is removed from the alveoli to the blood).

**Table 3.5** Allometry of mammalian respiratory variables; variable = $aM^b$ (from Stahl 1967).

| Respiratory variable | Allometric constant *a* | Allometric coefficient *b* |
|---|---|---|
| Lung mass (g) | 11.3 | 0.99 |
| Total lung volume (ml) | 53.5 | 1.06 |
| Tidal volume (ml) | 7.69 | 1.04 |
| Respiratory rate ($min^{-1}$) | 53.5 | -0.26 |
| Minute volume (ml $min^{-1}$) | 379 | 0.80 |
| $O_2$ extraction (ml $O_2$ ml $O_2$ in $air^{-1}$)[a] | 14.6 | -0.04 |
| Blood volume (ml) | 65.6 | 1.02 |
| Heart mass (g) | 5.8 | 0.98 |
| Heart rate ($min^{-1}$) | 241 | -0.25 |
| Stroke volume (ml)[b] | 0.78 | 1.06 |
| Cardiac output (ml $min^{-1}$) | 187 | 0.81 |
| Body surface area ($m^2$) | 0.11 | 0.65 |
| Metabolic rate (ml $O_2$ $min^{-1}$) | 11.6 | 0.76 |

[a] calculated as (100/0.2095) metabolic rate/minute volume; 0.2095 is the fractional $O_2$ content of air;
[b] calculated as cardiac output/heart rate

### 3.3.3 Ventilatory Control

An increase in the metabolic demand for gas exchange is generally accommodated by an increase in the convective flow (or minute volume, $V_I$), due to the limited scope to increase $EO_2$ from alveolar air. Minute ventilation can be increased by an increase in respiratory frequency ($f_R$), by an increase in tidal volume ($V_T$), or by some combination of the two. The importance of $f_R$ and $V_T$ to increasing $V_I$ are related to body mass and differences in the mechanics and energetics of ventilation for large and small species (see Table 3.5; Stahl 1967). Generally, small species increase $f_R$ in response to increased $O_2$ demand, while large species increase $V_T$ (Larcombe 2002; Cooper & Withers 2004).

The basic medullary respiratory rhythm is generated by rhythmic neuronal activity in the respiratory centre, located in the medulla of the brainstem (Guyton & Hall 1996; Pocock et al. 2013). Some neurons in this region have pacemaker activity, rhythmically generating action potentials that create and drive the respiratory rhythm. Inspiratory neurons innervate the diaphragm and external intercostal muscles, stimulating inspiration when activated, and passive expiration when inactive. Expiratory neurons are not usually active when the animal is resting, but during periods of activity these neurons stimulate the internal intercostals and thus active expiration. Expiratory neurons also play a role in increasing ventilation when $O_2$ demand is high. The basic respiratory rhythm can be modified by other inputs to the central nervous system (CNS). The brainstem's pons region contains the pneumotaxic and apneustic centres, which are essential to the maintenance of a normal, smooth respiratory rhythm. The apneustic centre prolongs inspiration by stimulating the inspiratory neurons, while the pneumotaxic centre stimulates the expiratory neurons and is dominant over the apneustic centre, halting inspiration and allowing expiration to occur (Guyton & Hall 1996; Sherwood et al. 2005). Numerous other CNS centres modify the function of the respiratory system in response to variables such as temperature, pain, blood pressure, and mechanical stimulation, including stretch of the lung, but the major driver is the partial pressure of respiratory gases in the extracellular fluid.

Matching the metabolic demand for $O_2$ and $CO_2$ with ventilation requires a negative feedback system with sensors and effectors (Guyton & Hall 1996; Sherwood et al. 2005; Pocock et al. 2013). Receptors consist of chemoreceptors of two types: peripheral chemoreceptors in the aorta and carotid arteries that respond to blood $pO_2$, $pCO_2$, and $H^+$ concentrations, and central chemoreceptors located near the respiratory centre of the medulla that respond to the $H^+$ concentration of the cerebrospinal fluid (CSF). The $H^+$ concentration of the CSF is not determined by the blood $H^+$ concentration, because the blood–brain barrier is impermeable to $H^+$. However, the cerebral capillaries that form the blood–brain barrier are permeable to $CO_2$, meaning that increased blood $pCO_2$ results in an increase in $CO_2$ in the CSF, which dissolves to form carbonic acid in the CSF. The formation of carbonic acid, and its subsequent dissociation into bicarbonate ions and $H^+$, naturally occurs very slowly but is catalyzed by carbonic anhydrase, an enzyme present in the cytosol of most cells, notably red blood cells, and in the medulla. Thus an increase in blood $CO_2$ results in an increase in the $H^+$ concentration of the CSF. The primary respiratory stimulant for mammals is $pCO_2$, as its effect is immediate and allows for the precise regulation of ventilation to achieve homeostasis of $pO_2$ and $pCO_2$. The $pO_2$ is not as important a respiratory stimulant, because $O_2$ is transported bound to haemoglobin in the blood, and the sigmoid shape of the haemoglobin dissociation curve results in haemoglobin maintaining high saturation until $pO_2$ falls below 50 mmHg. As a result, short-term modification of ventilation has little effect on $O_2$ transport to the tissues until haemoglobin saturation decreases to around 40–60%.

An increase in $H^+$ at the central chemoreceptors causes an increase in ventilation and a subsequent increase in $CO_2$ excretion, reducing extracellular $CO_2$ and $H^+$, and helping in the maintenance of acid/base homeostasis. Other stimuli also increase ventilation. In mammals that pant, an increase in hypothalamic temperature causes an increase in ventilation. The increase in ventilation results in an increase in respiratory evaporative heat loss, which tends to reduce hypothalamic temperature and thus aids thermal homeostasis, but the elevation in ventilation also increases $CO_2$ excretion that can lead to the development of respiratory alkalosis (Robertshaw 2006). There has been selection for changes in the ventilatory pattern during thermal panting that ensure that the development of alkalosis is minimized. Generally at low to moderate elevations in hypothalamic temperature, respiratory frequency is increased but tidal volume is decreased, resulting in hyperventilation of the respiratory dead space without an increase in alveolar ventilation, and extracellular $CO_2$ and $H^+$ are not disturbed (Hales 1976). This type of panting is called Type I panting. If hypothalamic temperature increases further, the pattern changes to Type II panting, in which tidal volume increases and the alveoli are ventilated more. Type II panting results in $CO_2$ washout and respiratory alkalosis (see 4.2.1).

### 3.3.4 Fetal and Newborn Ventilation

The fetal lungs develop from a bifurcation at the terminus of the developing trachea, and the alveoli begin to develop in a pseudo-glandular fashion (at about day 9.5 in mice, Have-Opbroek & Antonia 1981; by 6–16 weeks in humans, Lissauer et al. 2006). In humans, the distal airways develop in the canalicular phase at 17–24 weeks and then the alveolar ducts and terminal sacs develop in the saccular phase after 24 weeks, while in mice these two stages occur concurrently after day 17. At birth, about 20% of the adult number of alveoli have developed. Surfactant is produced late in the second and during the third trimester by the alveolar Type II cells. The fetal lungs, of course, are not ventilated.

Marsupials present an interesting case, being born at a relatively underdeveloped stage and then maturing in the mother's pouch. At birth, the lung of many marsupials is at the canalicular phase and the diffusion capacity of the lung cannot support their gas exchange requirements. In the quokka, the transition from the canalicular to the saccular phase occurs only four days after birth (Makanya et al. 2001). To varying degrees, newborn marsupials are dependent on gas exchange across the skin, a process that compensates for the relative underdevelopment of the lungs (Frappell & MacFarlane 2006). In the Julia Creek dunnart (*Sminthopsis douglasi*) that is born after a gestation of 13 days and at a birthweight of 17 mg, nearly 95% of the gas exchange in the newborn is across the skin (Frappell & MacFarlane 2006). The monotremes are similar to marsupials in that their lungs are quite immature at birth (Ferner et al. 2009), meaning, presumably, that the skin acts as an accessory gas exchange organ in the newborn, but there are no data on skin gas exchange.

When the newborn lung is first expanded, considerable exertion is required, because the alveoli are initially collapsed; considerable intrapleural pressure is required to break the surface tension forces that tend to keep the alveolar surfaces in close contact (see law of Laplace; Withers 1992; 2.5.4). In the absence of surfactant, an intrapleural pressure of −2.8 to −4.2 kPa is required (in humans) to separate the alveolar surfaces (Guyton & Hall 1996). Newborn mammals, especially premature ones, lack sufficient surfactant to reduce the required intrapleural pressure to physiologically possible values, and cannot properly expand their lungs without great effort. This condition is called infant respiratory distress syndrome (RDS), or hyaline membrane disease (Sabogal & Talmaciu 2005; Pillow & Jobe 2008). Without treatment (e.g. administration of artificial surfactants), infants that are surfactant deficient die soon after birth from inadequate gas exchange.

## 3.4 Circulation

The circulatory system of mammals provides the second, convective step in the $O_2$ cascade (and the second-to-last step in the $CO_2$ cascade). Although the basic structures and processes are the same as in 'lower' vertebrates, a major difference is that the four-chambered heart of mammals maintains complete separation of deoxygenated and oxygenated blood through both the right and left sides (except in the fetus; Guyton & Hall 1996; Pocock et al. 2013). The systemic circulation distributes oxygenated blood to the tissues and the gas transport function of the mammalian circulatory system matches the high metabolic demands (as does the convergent cardiovascular system of birds). Consequently, the mammalian cardiovascular system's structure and function have been selected to provide a very high capacity for rates of gas transport.

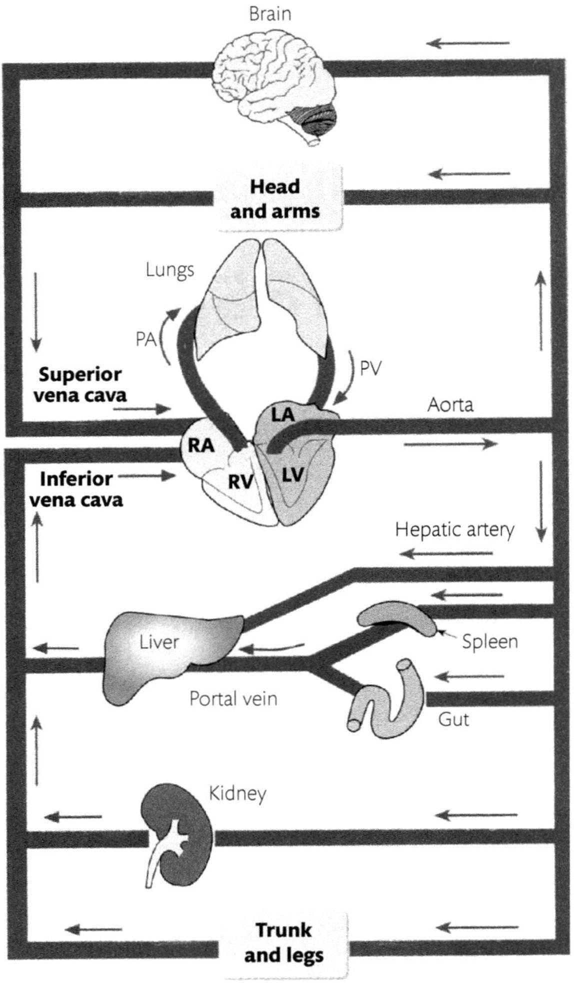


**Figure 3.20** Schematic of the mammalian circulatory system. From Pocock et al. (2013). Reproduced with permission of Oxford University Press.

The 'plumbing system' of the systemic and pulmonary circuits (Figure 3.20) consists of a complex series of different-sized tubes. In the systemic circuit, the aorta divides into large then small arteries, which divide into arterioles that ultimately branch into capillaries. Large arteries, also known as windkessel vessels, carry blood away from the heart, either oxygenated blood to the body or deoxygenated blood to the lungs. Arteries are under high pressure, and have thick, elastic walls consisting of an inner endothelial layer, a layer of thick connective tissue, a thick layer of smooth muscle, and a thin outer fibrous layer. Between these layers are elastic membranes, and the elasticity of the structure is important for maintaining a more constant blood pressure by damping the fluctuations resulting from contractions of the heart.

Arterioles are small arteries, surrounded by smooth muscle, that link arteries to capillaries. Arteries, and particularly the arterioles, are the primary 'resistance vessels' of the circulation, regulating and distributing flow to various tissues, and determining the blood pressure at the level of the capillaries. Capillaries are the smallest blood vessels where exchange occurs between the blood and tissues. They consist of only a single layer of endothelial cells and have a very small diameter (e.g. $5-10\,\mu m$ in humans). Since erythrocytes are about 8 μm in diameter, they must elongate and bend to squeeze through the smallest capillaries. This maximizes their surface area relative to internal blood volume, but capillaries are also quite short, so each capillary has a small exchange capacity, in absolute terms. But there are many of them that sum to provide adequate exchange for all of the body tissues. For example, in humans there are about a billion capillaries in the body, with a total surface area of about 6,300 $m^2$ (cf. the body surface area is about 2 $m^2$, and the lung surface area is about 70 $m^2$). Most cells in the body are within $60-80\mu m$ of a capillary, so the diffusion path is relatively short. Capillaries coalesce into venules (which consist a single layer of endothelial cells surrounded by collagen), then small veins, large veins, then the venae cavae, which drains into the right atrium. Veins contain blood with a much lower pressure than in arteries and they are much thinner-walled than arteries. Actions of surrounding muscles and internal valves that prevent backflow are responsible for achieving venous return. Veins contain the majority (approximately 75%) of the blood volume.

## 3.4.1 Blood

Blood consists of water, dissolved solutes, and cells. The non-cellular fraction of blood (the plasma) is a complex solution of water and various solutes, including proteins, various nutrients, electrolytes, nitrogenous wastes, hormones, and gases. Cells in the blood include leucocytes (white blood cells) and erythrocytes (red blood corpuscles). The percentage of the blood volume that is occupied by erythrocytes is known as the haematocrit, and is between 35 and 50% in mammals. Leucocytes are less numerous than erythrocytes, and make up about 1% of the blood volume.

Leucocytes include neutrophils, eosinophils, and basophils (collectively called granulocytes because of their granular-staining cytoplasm), as well as monocytes and lymphocytes (collectively called agranulocytes because they lack any granular cytoplasmic staining; Saladin 2010). The primary role of neutrophils is defence against bacteria. Eosinophils are involved with allergic reactions and responses to parasitic infections and various diseases. Basophils are the least common leucocyte. They secrete histamine (a vasodilator) that increases blood flow to regions with tissue injury and increases capillary permeability (that then allows blood components to penetrate into tissues) and heparin (an anticoagulant) that helps to maintain blood flow in regions of tissue injury. Lymphocytes are involved with immune function, and monocytes are active anti-viral cells that leave the circulation, where they become tissue macrophages, highly phagocytic cells that remove dead or dying cells, microorganisms, and pathogenic materials. Finally, platelets are an important blood constituent, being the second-most abundant 'cellular' component of blood. Platelets are actually circulating fragments of cytoplasm from megakaryocytes (bone marrow cells), and are responsible for various aspects of blood clotting, including vasoconstriction, the formation of platelet plugs, the secretion of procoagulants (clotting factors), and eventually clot dissolution.

Mammalian erythrocytes are enucleate corpuscles (unlike the nucleated erythrocytes of reptiles and birds) that contain large amounts of haemoglobin, the respiratory pigment that transports oxygen. In mammals, erythrocytes are generally small, biconcave disks, although their size and shape vary (e.g. camelids have elliptical erythrocytes, and mouse deer have spherical erythrocytes; Bartels et al. 1963; Snyder & Weathers 1977). There is considerable variation in the haematocrit (Hct, percentage of blood consisting of red blood corpuscles and white blood cells) between species and in the oxygen-binding properties of haemoglobin (Promislow 1991), which determine the $O_2$ transport capacity of the blood.

While a higher Hct increases the blood's capacity to transport $O_2$, there are functional limitations on the Hct. Viscosity increases with Hct and is an important haemodynamic property of the blood because it influences the blood pressure (and power) required to circulate the blood. At a given Hct, viscosity is inversely related to erythrocyte size, but there are not any obvious effects of erythrocyte shape (Stone et al. 1968). The trade-off between increased Hct, increased oxygen transport capacity, and increased blood viscosity means that there is an optimal Hct for blood oxygen transport capacity that maximizes the $O_2$ transport by the cardiovascular system (Crowell & Smith 1967). The transport of $O_2$ (OT) is calculated as m Hct F, where m is a constant of proportionality (the amount of $O_2$ per Hct) and F is the blood flow rate. In tubes, $F = e^{-k\,Hct} F_p$, where k is a constant of proportionality and $F_p$ is the plasma flow rate, so $OT = m\,Hct\,e^{-k\,Hct} F_p$ and differentiation shows that the optimal haematocrit ($H_{opt}$) is $1/k$. In more complex biological systems, F seems to depend on Hct as $(1 - a\,Hct)\,F_p$, (where a is a constant), so $OT = m\,Hct\,(1 - a\,Hct)\,F_p$ and differentiation indicates that the optimal haematocrit ($H_{opt}$) is ½ a. If for example, a = 0.025 (Richardson & Guyton 1959), then $H_{opt}$ is 40%. In practice, both models yield similar results, a $H_{opt}$ of about 40% (k = 0.025). Optimal Hct seems quite consistent for very different mammals (e.g. about 35–36% for rabbits and elephant seals; Hedrick et al. 1986), suggesting little evolutionary adaptive capacity to modify $H_{opt}$. Marine mammals need to trade-off the roles of blood $O_2$ content for $O_2$ storage during diving with blood viscosity. Many have a Hct of 43–45%, consistent with their 40–45% $H_{opt}$, but killer and beluga whales have a higher Hct (to more than 60%), suggesting enhanced storage at the expense of blood viscosity (Hedrick & Duffield 1991).

The functional unit of the red cell is the haemoglobin molecule, which consists of four subunits, each made up of a ferrous ($Fe^{2+}$) ion, embedded within a porphyrin ring, and surrounded by a globin protein (Withers 1992; Guyton & Hall 1996). Mammalian haemoglobin typically consists of two $\propto$ subunits and 2 $\beta$ subunits, which form the tetrameric $\propto_2\beta_2$. Because each $Fe^{2+}$ ion can bind one $O_2$ molecule, each haemoglobin molecule can bind four $O_2$. There is relatively little variation in the maximum $O_2$-binding capacity of mammalian haemoglobins and myoglobins (about 1.32–1.34 ml $O_2$ $g^{-1}$), reflecting minor differences in the molecular weight of mammalian haemoglobins. However, the amount of $O_2$ bound to haemoglobin is markedly dependent on the $pO_2$. The subunit structure of haemoglobin creates a functional relationship between the $O_2$ binding sites, and once haemoglobin binds a single $O_2$, the probability of another $O_2$ binding at another site increases, resulting in a cooperative $O_2$-binding interaction between subunits. The result is a sigmoidal (S-shaped) relationship between the partial pressure of $O_2$ and the percentage saturation of haemoglobin (Figure 3.21). The $pO_2$ at which haemoglobin is 50% saturated is termed the $P_{50}$, which is about

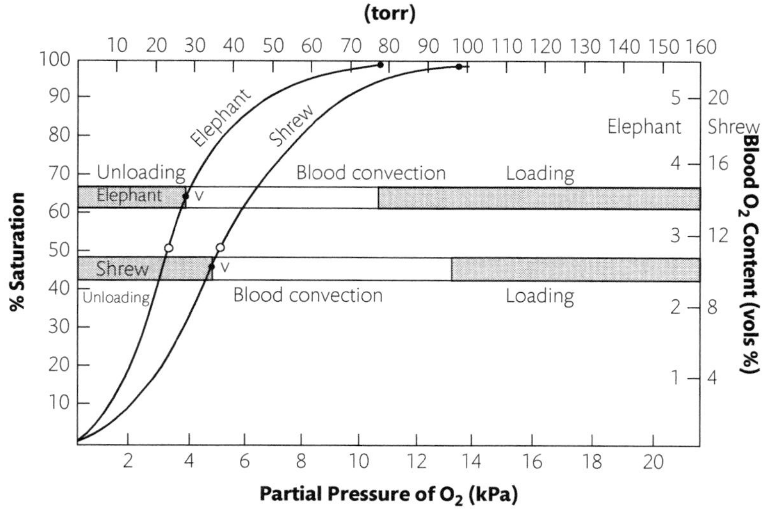


**Figure 3.21** Haemoglobin oxygen loading curves for a small mammal (shrew) and a large mammal (elephant), showing a similar sigmoidal relationship for both species between the % saturation of blood (or blood $O_2$ content, right axis) and partial pressure of oxygen ($pO_2$ in kPa and torr) but a considerable difference in $P_{50}$ ($pO_2$ for 50% saturation; open circles), $O_2$ content of blood (right axis legend), and $O_2$ differentials for $O_2$ loading in the lungs, blood convection, and tissue unloading; v indicates venous blood. Modified from Withers (1992). Reproduced with permission of Elsevier.

3.6 kPa for humans; a low $P_{50}$ corresponds to a strong $O_2$ binding capacity and favours loading of $O_2$ from the lungs, whereas a high $P_{50}$ favours $O_2$ delivery to the tissues. There are considerable adaptive differences in the $P_{50}$ of mammalian haemoglobins, reflecting both body mass (e.g. shrews have a higher $P_{50}$, 4.8 kPa, than elephants, 2.7 kPa, that facilitate unloading to the tissues in the shrew) and environment (e.g. altitude: vicuna, *Vicugna vicugna*, haemoglobin has a $P_{50}$ of about 2.8 kPa; see 4.4.1; diving: grey whale, *Eschrichtius robustus*, $P_{50}$ is about 4.8 kPa; see 4.5.2). In contrast, myoglobin, which is an important intracellular $O_2$ store (see 4.5.2), consists of a single haemoglobin unit, so its saturation curve is hyperbolic, not sigmoid, and it has a much lower $P_{50}$ (about 0.3–0.4 kPa) than haemoglobin.

Fetal mammals express a different form of haemoglobin, with $\Upsilon$ rather than $\beta$ subunits (i.e. $\propto_2\Upsilon_2$). The fetal haemoglobin is 'left-shifted' relative to adult haemoglobin, with a higher $O_2$ affinity and a lower $P_{50}$ than adult haemoglobin (Darling et al. 1941). The adaptive significance is that the left-shifted haemoglobin of the fetus can obtain $O_2$ even from saturated maternal blood (see later).

There is a considerable variation between mammals in many haematological variables. Variation in Hct reflects variation in erythrocyte numbers (red cells per litre of blood), and there is also considerable variation in erythrocyte size and haemoglobin content. Taken together, this variation results in quite a range in the $O_2$-carrying capacity of blood (Table 3.6). Much of the haematological variation is adaptive and related to small size and high mass-specific metabolic rate (Sealander 1964; Bartels et al. 1969), high levels of activity (e.g. flying; Jürgens et al. 1981; see 4.7.1), high altitude (Storz 2007; see 4.4), diving (Snyder 1983; see 4.5), or fossorial habit (Jelkmann et al. 1981; see 4.3). Neither haemoglobin concentration nor Hct are significantly related to body mass (Withers et al. 1979a). Erythrocyte count and size are inversely related, and vary considerably (e.g. mean cell volume varies from $< 10$ fl for goats, *Capra aegagrus hircus*, and mouse deer to $> 100$ fl for whales and elephants), with a significant effect of body mass (Withers et al. 1979a). Erythrocyte size and shape seem to be related to osmotic fragility (the capacity of erythroctyes to swell but resist bursting when exposed to low osmotic concentration). Camels (*Camelus dromedarius*) and other camelids have ovoid-shaped erythrocytes that are particularly resistant to osmotic lysis (Perk 1963, 1966), as are the large erythrocytes of elephants (Silva & Kuruwita 1994). This is presumably an adaptation that prevents lysis when these animals drink large amounts of water when rehydrating (see later). Mean corpuscular haemoglobin content (MCH) is strongly correlated with MCV and body mass. In contrast, mean corpuscular haemoglobin concentration (MCHC) is quite uniform amongst mammals (311–494 g $l^{-1}$) and independent of body mass (Withers et al. 1979a), reflecting the similar close-packing of haemoglobin within erythrocytes.

**Table 3.6** Haematological parameters for some mammals; haemoglobin concentration (Hb), haematocrit (Hct), erythrocyte count (RBC), mean corpuscular volume (MCV), mean corpuscular haemoglobin content (MCH) and mean corpuscular haemoglobin concentration (MCHC). Values from Bartels et al. (1979), Withers (1992), Clark (2004), and Carneiro et al. (2010).

| Species | Hb g $l^{-1}$ | Hct % | RBC$10^{12}$ $l^{-1}$ | MCV[a] fl | MCH[b] pg | MCHC[c] g $l^{-1}$ | Habitat |
|---|---|---|---|---|---|---|---|
| Goat (*Capra*) | 104 | 29 | 16.1 | 18 | 6.5 | 356 | Terrestrial |
| Opossum (*Didelphis*) | 110 | 34 | 4.32 | 81 | 25.5 | 324 | Terrestrial |
| Mouse deer (*Tragulus*) | 117 | 31 | 55.9 | 6 | 2.1 | 380 | Terrestrial |
| Hairy-nosed wombat (*Lasiorhinus*) | 128 | 40 | 4.7 | 85 | 27.4 | 324 | Fossorial |
| Koala (*Phascolarctos*) | 130 | 42 | 3.9 | 110 | 34.4 | 313 | Arboreal |
| Bottlenose dolphin (*Tursiops*) | 143 | 41 | 3.6 | 114 | 39.7 | 350 | Marine |
| Human (*Homo*) | 146 | 44 | 5.0 | 90 | 29.2 | 335 | Terrestrial |
| African elephant (*Loxodonta*) | 153 | 43 | 3.0 | 141 | 51.0 | 356 | Terrestrial |
| Tasmanian devil (*Sarcophilus*) | 153 | 42 | 6.5 | 64.8 | 23.8 | 369 | Terrestrial |
| Bryde's whale (*Balaenoptera*) | 154 | 48 | 3.6 | 133 | 42.8 | 321 | Marine |
| Shrew (*Crocidura*) | 156 | 44 | 11 | 41 | 14.2 | 354 | Terrestrial |
| Camel (*Camelus*) | 158 | 35 | 11.0 | 319 | 14.4 | 494 | Desert |
| House mouse (*Mus*) | 159 | 50 | 8.3 | 60 | 19.2 | 320 | Terrestrial |
| Red kangaroo (*Macropus*) | 160 | 47 | 4.7 | 101 | 34.3 | 340 | Desert |
| Brown antechinus (*Antechinus*) | 161 | 45 | 10.8 | 42 | 14.9 | 354 | Terrestrial |
| Brush-tail possum (*Trichosurus*) | 161 | 50 | 7.4 | 67 | 29.1 | 326 | Arboreal |
| Killer whale (*Orcinus*) | 163 | 45 | 4.1 | 111 | 39.8 | 361 | Marine |
| Bilby (*Macrotis*) | 171 | 55 | 7.5 | 73 | 22.9 | 311 | Fossorial |
| Shrew (*Suncus*) | 174 | 50 | 18 | 26 | 9.7 | 348 | Terrestrial |
| Echidna (*Tachyglossus*) | 176 | 49 | 8.7 | 62 | 22.2 | 359 | Terrestrial |
| Little red flying fox (*Pteropus*) | 184 | 53 | 10.4 | 51 | 17.7 | 349 | Flight |
| Swamp rat (*Rattus*) | 189 | 48 | 7.1 | 70 | 27.2 | 393 | Terrestrial |
| Platypus (*Ornithorhynchus*) | 190 | 49 | 10.0 | 50 | 19.5 | 395 | Semi-aquatic |

*(continued)*

**Table 3.6** (*Continued*)

| Species | Hb g $l^{-1}$ | Hct% | RBC$10^{12}$ $l^{-1}$ | MCV[a] fl | MCH[b] pg | MCHC[c] g $l^{-1}$ | Habitat |
|---|---|---|---|---|---|---|---|
| Beluga (*Delphinapterus*) | 193 | 46 | 3.3 | 134 | 58.5 | 427 | Marine |
| Pocket mouse (*Perognathus*) | 193 | 54 | 11.8 | 45 | 16.4 | 367 | Terrestrial |
| Common bent-wing bat (*Miniopterus*) | 194 | 48 | 10.9 | 44 | 17.8 | 406 | Flight |
| Southern elephant seal (*Mirounga*) | 224 | 54 | 12.5 | 213 | 88.2 | 413 | Marine |

[a] MCV = 10Hct / RBC; [b] MCH = Hb / RBC; [c] MCHC = 100Hb / Hct

### 3.4.2 The Heart

The fundamental function of the heart is to transfer blood from the low pressure venous side of the circulation to the high pressure arterial side. In turn, the pressure energy in the arteries moves blood through vessels, against the vascular resistance, to the capillaries where tissues are supplied with oxygen and nutrients, and back into the venous circulation to the heart. The simplest hearts are tubular and use peristaltic contractions of their muscular wall and one-way valves to produce a unidirectional flow of blood. The hearts of vertebrates become progressively more complex as the tubular heart develops specializations along its length (e.g. atrium and ventricle in fishes), then partially or divided atria that accommodate the pulmonary return to the heart (e.g. lungfish, amphibians), then completely divided atria and ventricles in the mammalian and avian circulatory systems (and crocodilians; Kardong 2009).

In mammals (and birds), the heart has four 'chambers'—two atria and two ventricles—and requires a more complex pattern of ventricular contraction (Guyton & Hall 1996; Pocock et al. 2013). The double circulation means that an imbalance is created if the right and left ventricles do not eject the same amount of blood, a function that is facilitated by mechanisms that alter the pumping performance of each ventricle with the volume of blood in the ventricle prior to contraction (the Frank–Starling law of the heart). The complete functional separation of right and left sides of the heart is correlated with the high metabolic rates and much higher left than right ventricular blood pressure in mammals and birds; an incomplete morphological separation of especially the ventricles would presumably compromise functional separation of deoxygenated and oxygenated blood flows and be detrimental to sustaining a high metabolism.

The heart muscle also requires gas exchange to support its metabolism. In many fishes and 'lower' tetrapods, direct gas exchange with the blood flowing through the chambers of the heart is sufficient; this exchange can be facilitated by a spongy trabecular structure of the inner wall of the heart. Diffusion from the chambers becomes insufficient when wall size increases; coronary arteries supplying the myocardium are particularly well developed in elasmobranch fishes, crocodiles, birds, and mammals. The coronary arteries supply the outer myocardium with oxygenated blood (and drain into the sinus venosus).

The function of the heart (the ventricles in particular) is aided by a rapid spread of depolarization (which stimulates muscle activation) that allows for the almost simultaneous contraction of the entire myocardium and the generation of high pressure. A slower spread of excitation would limit the pressure. Vertebrate hearts have regional specializations related to their electrical activity, which drives the cardiac contraction cycle. Many cardiac muscle cells have an intrinsic electrical cycle of depolarization and contraction, but the entire heart beats in synchrony because there is electrical propagation between cardiac cells. The cardiac contraction cycle is typically initiated by a part of the sinus venosus, called the pacemaker or sinoatrial (SA) node. The SA node has a higher frequency of depolarization than other pacemaker cells in the heart, and it thereby 'paces' and coordinates the contraction of all the cardiac muscle cells; its depolarization cycle spreads from the sinosus venosus through a system of conducting fibres (that are modified muscle cells, not neurons) to the other contracting regions of the heart (atrium, then ventricle). Electrical conduction is further modified in mammals and birds. A second conducting node, the atrioventricular (AV) node and Purkinje fibres (cardiac muscle cells modified for rapid electrical conduction, not contraction) rapidly spread excitation from the SA node through the interventricular septum to the base of the ventricle, then upward through the ventricle walls. This pattern of electrical propagation ensures synchronized contraction, and propels blood from the base of the ventricles towards the outflow vessels (aorta, pulmonary arteries).

The variations seen in cardiac morphology across mammalian lineages reflect their functional capacity. Heart mass varies adaptively, reflecting in part different metabolic demands. For example, mammals from low $T_a$ or high altitude tend to have a large heart, to accommodate their high metabolic demand for thermogenesis (Hesse's 'law of the heart'; see 3.2.8). Heart mass is generally higher for marsupials ($g=7.5\,kg^{0.94}$) than placentals ($g=6.0\,kg^{0.97}$), suggesting a higher aerobic capacity (Dawson et al. 2003).

### 3.4.3 Lymphatic System

Because the net capillary hydrostatic pressure exceeds the net oncotic pressure in capillaries (2.5.5), there is a net loss of fluid along the capillary. Sometimes that accumulation of filtered fluid in tissue can be observed; for example, in the expansion of exercising muscle, or in swollen feet after long periods standing. But under normal conditions, that fluid does not accumulate in the tissues and cause oedema, because the lymphatic system—a series of blind-ended, valved, thin-walled tubes (like veins)—transports the excess fluid back into the venous circulation as lymph. The small terminal lymph vessels coalesce into larger jugular, subclavian, lumbar, and thoracic lymphatics. Lower vertebrates (including reptiles and embryonic birds) have lymph hearts that assist lymph return, but these are absent in mammals, so mammals rely on general body movements (breathing, muscle contractions, even arterial pulsations) to propel lymph back to the circulatory system. The lymphatic system also plays an important role in immunity, allowing the immune system at lymph nodes to 'sample' the extracellular fluid. The activation of the immune system when infections are detected results in the classical sign of infection in swelling of the lymph nodes.

### 3.4.4 Gas and Heat Transport

The $O_2$ delivery capacity of the cardiovascular system is summarized by the Fick principle, which is often used to calculate the rate of oxygen consumption of an animal (or an organ or tissues). For a mammal, $O_2$ consumption ($VO_2$; ml $O_2$ min$^{-1}$) is determined by the cardiac output (CO; ml min$^{-1}$) and the difference in $O_2$ content between the arterial blood ($CA,O_2$; ml $O_2$ ml$^{-1}$) and the venous blood ($CV,O_2$; ml $O_2$ ml$^{-1}$), the latter known as the arterio-venous (A–V) difference in blood $O_2$ content, and symbolized as $C(A–V)O_2$: $VO_2 = CO(CA,O_2 - CV,O_2) = CO\,C(A - V)O_2$. For example, Ohmura et al. (2002) measured cardiac output of trained thoroughbred yearlings (*Equus caballus*) before and after a year of racing, using the Fick method. Before training, $VO_2$ was 59.7 l min$^{-1}$, $CA,O_2$ was 254 ml $O_2$ l$^{-1}$, and $CV,O_2$ was 37 ml $O_2$ l$^{-1}$, so CO was 278 l min$^{-1}$; corresponding values for horses after one year of racing were 73.3 l min$^{-1}$, $CA,O_2$ was 256 ml $O_2$ l$^{-1}$, $CV,O_2$ was 26 ml $O_2$ l$^{-1}$, so CO was 322 l min$^{-1}$. Murdaugh et al. (1966) measured the cardiac output of harbour seals (*Phoca vitulina*) with the Fick method (2,500 – 5,100 ml min$^{-1}$) to validate the dye-dilution method (2,940 – 3,810 ml min$^{-1}$), which was used to examine cardiac output pre-dive (5,180 ml min$^{-1}$), dive (620 ml min$^{-1}$), and post-dive (8,270 ml min$^{-1}$).

The $O_2$ content of arterial and venous blood depends on the $pO_2$ and the $O_2$ carrying capacity of blood. The $O_2$ carrying capacity of blood is determined by the haemoglobin content of the blood (g Hb ml$^{-1}$) and the $O_2$ binding capacity of the haemoglobin (about 1.3 ml $O_2$ g Hb$^{-1}$). The haemoglobin content of the blood is reflected in the simple measure of Hct. The $O_2$ transport capacity of the circulation can be increased by increased cardiac output (increased heart rate, stroke volume, or both) and/or increased A–V difference (increased $pO_2$, which is usually not possible, or decreased venous $pO_2$, or increased haemoglobin content of the blood). The hormone erythropoietin stimulates red blood cell production and increases the haemoglobin content of the blood; for example, in mammals exposed to high altitude. Some human athletes cheat by using the hormone to increase their oxygen transport capacity and improve performance.

While the type of haemoglobin does not change once a mammal passes infancy, the position of the haemoglobin $O_2$ loading curve can change dynamically in response to many conditions ($pCO_2$, pH, temperature, cofactors, body mass). One particularly important cofactor is 2,3 diphosphoglycerate (DPG), which moves the $O_2$ loading curve to the right. During glycolysis, DPG is produced from 1,3 bisphosphoglycerate (see 2.3). In humans, exercise and exposure to high altitude increase DPG levels, resulting in a right-shift of the $O_2$ loading curve, which presumably facilitates $O_2$ delivery under these conditions (see 4.4). Mammals can be divided into two groups based on their Hb properties: most have Hb with a high $O_2$ affinity (low $P_{50}$) and modulated by DPG, whereas ruminant mammals, felids, and a lemur have a low $O_2$ affinity haemoglobin that is relatively insensitive to DPG (Giardina et al. 2004).

Typically, an increase in $pCO_2$ or decrease in pH moves the $O_2$ loading curve to the right, increasing the $P_{50}$. The adaptive value of the shift is that it decreases the binding of $O_2$ to haemoglobin in venous blood (compared to arterial blood), where the $pCO_2$ is increased by $CO_2$ uptake from the tissues and pH is reduced. This right-shift in the haemoglobin saturation curve is called the Bohr shift (Giardina et al. 2004). The Bohr shift decreases the $O_2$ content of venous blood, thus increasing the A–V difference and $O_2$ delivery to the tissues. The magnitude of the Bohr shift is expressed as the Bohr effect factor, the $\Delta \log(P_{50}) / \Delta$ pH. Increased temperature also moves the $O_2$ loading curve to the right (increasing the $P_{50}$). The effect of temperature can facilitate the unloading of oxygen when $T_b$ increases during activity. The high pH dependence of $P_{50}$ can counterbalance the reduced $O_2$ unloading to tissues for mammals in the cold and at high altitude (Giardina et al. 2004). Mammals that use torpor or hibernation (see 4.1.2) experience large variation in $T_b$; hence their $P_{50}$ and $O_2$ delivery characteristics change with $T_b$.

The $pO_2$ in the tissues determines how much $O_2$ is unloaded from the haemoglobin at the tissues, and thus the difference in $O_2$ content between arterial and venous blood (A–V difference). The relative $O_2$ differentials required for $O_2$ loading onto haemoglobin in the lungs, $O_2$ transport by the blood, and $O_2$ delivery to the tissues are indicated by the differences in ambient and arterial $pO_2$, arterial and venous $pO_2$, and venous and tissue $pO_2$, respectively. The oxygen binding characteristics of haemoglobin vary markedly between mammals, particularly with body mass and altitude (Withers 1992; Storz 2007).

Body mass influences almost all aspects of $O_2$ delivery in mammals, including the role of the circulatory system. Small mammals have a right-shifted loading curve (higher $P_{50}$), a higher blood $O_2$ content, a smaller respiratory $O_2$ differential, and higher blood convective and tissue unloading $O_2$ differential (Figure 3.21). A right-shifted $O_2$ loading curve means that to provide the same amount of $O_2$ to the tissues, the $pO_2$ needs to be higher, but this provides a large venous blood-cell $pO_2$

difference that facilitates the diffusion of $O_2$ to the tissues. That scenario matches the much higher mass-specific metabolic requirement of tissues in smaller mammals. In contrast, large mammals like elephants have a left-shifted $O_2$ loading curve. To unload the same amount of $O_2$ from the blood requires a lower $pO_2$, which translates into a smaller venous blood-cell $pO_2$ difference driving the diffusion of $O_2$ into the tissues. Larger mammals require a larger ambient–arterial $pO_2$ difference than small mammals for $O_2$ loading in the lungs, but at sea level the normal $pO_2$ in the alveoli is high enough to saturate even quite left-shifted haemoglobin.

The resting A–V $pO_2$ difference does not vary much between small and large mammals, with the difference in $O_2$ requirement being made up by differences in cardiac output. During activity, the increased aerobic metabolic rate, up to MMR, must be supported by concomitant increases in $O_2$ delivery and $CO_2$ excretion by each stage in the $O_2/CO_2$ cascade. The nature of the ultimate limitation to the maximal rate of $O_2$ delivery has been extensively investigated and debated for mammals, as well as other vertebrates (e.g. Bishop 1997; Di Prampero 2003). From these analyses the concept of 'symmorphosis' emerged, stating that in any biological cascade system (such as the $O_2$ supply cascade), each step in the cascade would be matched, each step having a capacity that matched the limit of the functional performance of the cascade (Weibel et al. 1991). The concept was linked to the efficiency of 'design': each component of the system had just the right resources to do the overall job. The conclusion was that the hypothesis held true for all but the lungs, which appear to be 'over-designed', but limits to rates of $CO_2$ excretion are relatively poorly considered (e.g. Withers & Hillman 1988; Hillman et al. 2013). A comparative meta-analysis of maximal aerobic metabolic rate for vertebrates supports some previous studies that have suggested that the cardiovascular system limits maximal $O_2$ transport, but that the respiratory system limits maximal $CO_2$ transport. This explains the conundrum that the respiratory system seems to be 'over-designed' for $O_2$ transport.

Diving mammals have profound cardiac and microvascular adaptations that conserve essential nutrients (especially oxygen), and isolate lactate that is produced in active muscle while diving (see 4.5). The heart rate is decreased (diving bradycardia) and there is a major redistribution of blood flow away from many peripheral tissues (gut, kidneys, muscle), resulting in the conservation of oxygen for hypoxia-sensitive tissues such as the heart and brain. In natural dives, the bradycardia and redistribution of blood flow is less pronounced, and whole-body metabolism remains largely aerobic, so the primary role of the diving response seems to be to regulate the level of hypoxia in the skeletal muscle for efficient use of blood and tissue oxygen stores.

One of the essential roles of the circulatory system in mammals is the circulation of heat, and it is largely the micro-circulation (especially arterioles) that is responsible for the control of heat exchange. Although the basic structure and functioning of the vascular circuits and heart are invariant amongst mammals (and remarkably convergent with birds), there are circulatory specializations related to heat transfer (retention in the cold, dumping in the heat) and nutrient sparing/waste isolation (during diving); for instance, vascular retes (see 3.2.4.1) and 'thermal windows'.

Many mammals have thermal windows that facilitate heat loss when the animal is heat-challenged. These areas are usually barely insulated, or not at all insulted, and increased blood supply to the skin enhances heat loss. Thermal windows are especially important for those mammals that lack sweat glands. For example, many rodents have extensive subdermal vascular beds that can function as thermal windows (e.g. ears, dorsal skin, tail; Rand et al. 1965). Elephants can use their large ears as effective thermal windows (Williams 1990; Weissenböck et al. 2010). In bats, the unfurred wing membrane and some hairless regions located beneath the forearms are the main thermal windows for heat exchange with the environment (Lancaster et al. 1997; Reichard et al. 2010). Dolphins and other cetaceans typically use their fin and flipper retes, and thick blubber, for heat conservation, but blood can bypass the fin and flipper retes, and the skin can vasodilate to become a thermal window thereby facilitating heat dissipation (Meagher et al. 2002). Hauled-out seals can vasodilate their trunk to lose heat in conjunction with increased evaporation from wet skin (Mauck et al. 2003), and can use their flippers as heat exchangers to dissipate heat in air (or conserve heat in water; Erdsack et al. 2012).

## 3.4.5 Fetal and Newborn Circulation

Mammalian fetuses do not have functional lungs before birth; oxygen is supplied to fetal tissues from the maternal blood via the placenta. The transfer of $O_2$ from maternal to fetal blood is facilitated by the left-shifted haemoglobin of the fetus (3.4.1). Further, the complete separation of deoxygenated and oxygenated blood through the right and left sides of the heart is not necessary because the oxygen supply derives from the placenta, entering the fetus through the umbilical vein that supplies oxygenated blood to the fetal vena cava via the ductus venosus. Not surprisingly, there are fetal modifications in haemoglobin function and cardiovascular modifications that provide right-to-left shunts such that blood bypasses the lungs.

To bypass the non-functional (for gas exchange) pulmonary circuit, the foramen ovale connects the right and left atria and allows blood returning to the right atria of the heart to flow directly into the left atrium, bypassing the pulmonary circuit (Figure 3.22). The right ventricle still pumps some blood into the pulmonary artery and the ductus arteriosus connects the pulmonary artery to the aorta, bypassing the pulmonary circuit. Thus, oxygenated blood returning from the placenta in the umbilical vein enters the ductus venosus and the vena cava, and to the right atrium. From the right atrium blood enters the systemic circuit (via the foramen ovale and ductus arteriosus bypasses) and the pulmonary circuit (via the right ventricle and pulmonary artery).

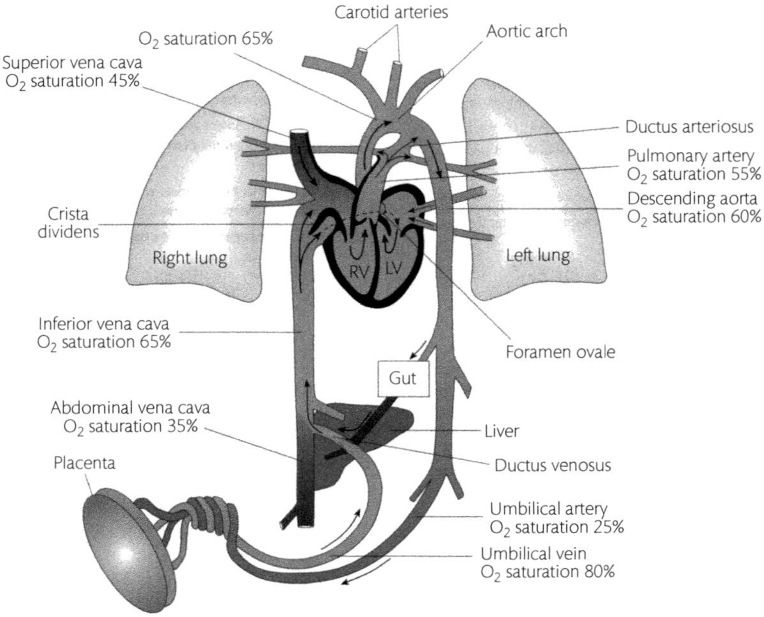


**Figure 3.22** Schematic of the fetal circulatory systems in a placental mammal, showing the right-to-left bypasses (foramen ovale and ductus arteriosus) and the umbilical artery and vein with ductus venosus bypass to return placental blood to the venous circulation. From Pocock et al. (2013). Reproduced with permission of Oxford University Press.

In eutherian mammals, oxygenated blood returning from the placenta via the umbilical vein mostly passes through the ductus venosus (in the liver) to the venous return. This blood enters the right atrium and preferentially passes through the foramen ovale (which is valved) into the left atrium (a 'right-to-left' shunt), from where it is pumped via the left ventricle to the body. Deoxygenated blood returning to the right atrium is preferentially passed into the right ventricle and towards the lungs; however, most of this blood bypasses the lungs and passes through the ductus arteriosus to the aorta. Why have both a ductus arteriosus and a foramen ovale? The ductus arteriosus is considered to be an 'exercise' bypass for the right ventricle, since it allows blood to pass through the right ventricle, which facilitates the anatomical and functional development of the right ventricle, while still allowing blood to bypass the lung. Similarly, the foramen ovale is the 'exercise channel' for the left ventricle, facilitating the anatomical and functional development of the left ventricle (Liem et al. 2001).

Because the Hb-$O_2$ dissociation curve of the fetus is left-shifted relative to the maternal haemoglobin, the transfer of $O_2$ to the placental (umbilical) vein from the maternal (uterine) artery is facilitated. As a result, fetal umbilical blood can have as high (or higher) percentage saturation as maternal arterial blood despite needing to have a lower $pO_2$ (Barron & Meschia 1954).

In marsupials, the fetal haemoglobins (there are several) have a higher $P_{50}$ than the haemoglobin of the pouch young and adults. The right-shifted haemoglobin makes it difficult to efficiently transfer $O_2$ from the maternal to fetal blood (Calvert et al. 1994; Holland et al. 1994; Tibben et al. 1991; Tyndale-Biscoe 2005). However, a high maternal placental blood flow presumably provides sufficient $O_2$ delivery to the fetal blood to sustain the metabolic demand of the small and immature fetus.

After birth, the placental newborn transitions from its fetal circulatory pattern to the adult double-circulation pattern. The lungs inflate, the pulmonary resistance decreases to less than the systemic resistance, and pressure in the circulatory system changes. The ductus arteriosus begins to constrict and blood is preferentially passed to the lungs; the ductus is sealed off within a few hours to days after birth, and eventually is replaced by connective tissue (the adult ligamentum arteriosum). The initial patency of the ductus arteriosus after birth allows some left ventricular blood to recirculate through the lungs, ensuring that the pulmonary blood is fully saturated while the transition is made between fetal haemoglobin and adult haemoglobin (the lower $P_{50}$ of fetal haemoglobin that helps to promote loading of $O_2$ from the maternal placental blood retards $O_2$ loading in the newly functional lungs). The increased pulmonary return after birth increases the pressure in the left atrium, which closes the valve of the foramen ovale, which soon seals onto the inter-atrial septum (but remains throughout life as the fossa ovalis).

In marsupials, the circulatory adjustments are slightly different, reflecting their shorter gestation and different placentation (Runciman et al. 1995). The fetus has a ductus arteriosus that bypasses the pulmonary circuit, but the septum that partially closes the inter-atrial foramen ovale is fenestrated (whereas it is solid in placentals), and does not seal for days after birth. In some marsupials, the inter-atrial septum itself is actually incomplete at birth.

## 3.5 Feeding and Digestion

To a large extent, what an animal eats defines its biology. Dietary habits are associated with specific structural, physiological, behavioural, and even life history traits (Bozinovic & del Río 1996). Consequently, and not surprisingly, the study of animal diets is central to evolutionary physiological ecology. Several dietary groups can be recognized within extant mammalian species (see 1.4.3). Pineda-Munoz and Alroy (2014) have identified seven general dietary categories for mammals: insectivory, carnivory, granivory, gummivory, frugivory, fungivory, and herbivory (Table 3.7). Not surprisingly, the species within these dietary niches

**Table 3.7** Percentage composition of various dietary categories, relative to wet mass. Adapted from Jordano (2000) and Knox et al. (2014).

| Diet type | Water | Protein | Lipids | Carbohydrate (non-structural) | Carbohydrate (structural) | Minerals |
|---|---|---|---|---|---|---|
| Insects | 64 | 24 | 5 | 5 | 0 | 2 |
| Mammals | 70 | 18 | 6 | 1 | 0 | 5 |
| Seeds | 11 | 11 | 2 | 69 | 3 | 2 |
| Fruits | 71 | 3 | 7 | 25 | 7 | 0.1 |
| Leaves | | | | | | |
| young | 72 | 7 | 1 | 6 | 13 | 2 |
| mature | 59 | 5 | 1 | 2 | 31 | 2 |

have corresponding specializations in dentition and the structure and function of the digestive tract.

### 3.5.1 Foods and Consumers

Mammals have an array of dental specializations related to phylogeny and specific diets, and dentition is an important character for identifying living and fossil mammals (see 1.3.1; Kardong 2009). Dentition is also strongly indicative of diet (Table 3.7). Carnivorous mammals typically have enlarged canines, retain incisors, and have numerous premolars and molars (Table 3.8). The carnivorous pattern is the primitive dental condition, reflecting the insectivorous diet of early mammals. Omnivores, and especially specialized herbivores, have reduced incisors and canines, and large premolars and molars that facilitate the cutting and grinding of their plant diet. Piscivorous or carnivorous marine mammals have numerous conical teeth (lacking clear homology to standard mammalian incisors, canines, premolars, or molars). The long-snouted spinner dolphin (*Stenella longirostris*) has 184–260 such teeth; this high number of teeth is strongly correlated with the elongate shape of the jaws (Werth 2006). The numbat has the most numerous teeth of all terrestrial mammals, with always more than seven post-canine teeth. It is unclear if these extra teeth are retained deciduous premolars or additional molars that can be accommodated in the elongated palate. Variation in the number and structure of the teeth, and their simplified form and limited wear in aged animals suggest that the teeth are under little selection pressure for this exclusively termitivorous species (Cooper 2011). Supernumerary teeth (especially additional molars), resulting from developmental, post-trauma, or other effects, have been reported for individuals representing most orders of mammals (e.g. Libardi & Percequillo 2014).

**Table 3.8** Examples of dental formulae for ancestral, monotreme, marsupial, and placental mammals, of varying diets. Data from Green (1937); Perrin (1963); Olds and Shoshani (1982); Poole (1982); Ziegler (1982); Heyning and Dahlheim (1988); Wells (1989); Best and Hill Henry (1994); Best and Skupski (1994); Greenhall and Schutt (1996); Gompper and Decker (1998); Adam (1999); Jones and Rose (2001); Seebeck 2001; Pappas (2002); Luo et al. 2003, 2011; Gillihan and Foresman (2004); Kielan-Jaworowska et al. (2004); Cooper (2011); Hayssen (2011); Averianov and Lopatin (2014); Renovoisé and Michon (2014); Feldhamer et al. (2015).

| Taxon | Species | Dental formula | Diet |
|---|---|---|---|
| Early mammals | *Morganucodon, Sinoconodon* | $I\frac{5}{5}C\frac{1}{1}PM\frac{8}{8}$ (no differentiation of P and M) | Insectivorous, small vertebrates |
| Monotremes | | | |
| Ancestral | | $I\frac{5?}{5?}C\frac{1}{1}P\frac{5}{5}M\frac{3}{3}$ | |
| Platypus | *Ornithorhynchus anatinus* | $I\frac{0}{5}C\frac{1}{1}P\frac{2}{2}M\frac{3}{3}$ (embryo)<br>$I\frac{0}{0}C\frac{0}{0}P\frac{1}{0}M\frac{2}{3}$ (nestling)<br>$I\frac{0}{0}C\frac{0}{0}P\frac{0}{0}M\frac{0}{0}$ (adult) | Insects, small crustaceans |
| Short-beaked Echidna | *Tachyglossus aculeatus* | (adults lack teeth) | Ants, termites, insect larvae |
| Marsupials | | | |
| Ancestral | *Sinodelphys* | $I\frac{4}{4}C\frac{1}{1}P\frac{3}{3}M\frac{4}{4}$ | Insectivorous |
| Eastern quoll | *Dasyurus viverinus* | $I\frac{4}{3}C\frac{1}{1}P\frac{2}{2}M\frac{4}{4}$ | Carnivore |
| Numbat | *Myrmecobius fasciatus* | $I\frac{4}{3}C\frac{1}{1}P\frac{4}{4}M\frac{4}{4}$<br>or $I\frac{4}{3}C\frac{1}{1}P\frac{3}{3}M\frac{5}{5}$ | Termitivore |
| Bandicoot | *Perameles gunni* | $I\frac{5}{3}C\frac{1}{1}P\frac{3}{3}M\frac{4}{4}$ | Omnivore |
| Honey possum | *Tarsipes rostratus* | $I\frac{2}{1}C\frac{1}{0}P\frac{1}{0}M\frac{3}{3}$ | Nectarivore |
| Eastern grey kangaroo | *Macropus giganteus* | $I\frac{3}{1}C\frac{0}{0}P\frac{2}{2}M\frac{4}{4}$ | Herbivore (pre-gastric) |
| Hairy-nosed wombat | *Lasiorhinus latifrons* | $I\frac{1}{1}C\frac{0}{0}P\frac{3}{3}M\frac{4}{4}$ | Herbivore (post-gastric) |
| Placentals | | | |
| Ancestral | *Juramaia* | $I\frac{5}{5}C\frac{1}{1}P\frac{5}{5}M\frac{3}{3}$ | Insectivorous |
| Anteater | *Tamandua tetradactyla* | $I\frac{0}{0}C\frac{0}{0}P\frac{0}{0}M\frac{0}{0}$ | Myrmecophagous |
| Shrew | *Sorex vagrans* | $I\frac{3}{1}C\frac{1}{1}P\frac{3}{1}M\frac{3}{3}$ | Insectivore |
| Rodent | *Perognathus merriami* | $I\frac{1}{1}C\frac{0}{0}P\frac{1}{1}M\frac{3}{3}$ | Granivore |

*(continued)*

**Table 3.8** (*Continued*)

| Taxon | Species | Dental formula | Diet |
|---|---|---|---|
| Brown-nosed coati | *Nasua nasua* | $I\frac{3}{3}C\frac{1}{1}P\frac{4}{4}M\frac{2}{2}$ | Omnivore |
| Vampire bat | *Diaemus youngi* | $I\frac{1}{2}C\frac{1}{1}P\frac{1}{2}M\frac{2}{1}$ | Blood |
| Blossom bat | *Syconycteris naias* | $I\frac{2}{2}C\frac{1}{1}P\frac{3}{3}M\frac{2}{3}$ | Nectarivore |
| Eland | *Taurotragus oryx* | $I\frac{0}{3}C\frac{0}{1}P\frac{3}{3}M\frac{3}{3}$ | Herbivore (ruminant) |
| Two-toed sloth | *Choloepus didactylys* | $\frac{5}{4}$ * | Herbivore (pseudoruminant) |
| Rock hyrax | *Procavia capensis* | $I\frac{1}{2}C\frac{0}{0}P\frac{4}{4}M\frac{3}{3}$ | Herbivore (post-gastric) |
| Arctic hare | *Lepus arcticus* | $I\frac{2}{1}C\frac{0}{0}P\frac{3}{2}M\frac{3}{3}$ | Herbivore (coprophagous) |
| Killer whale | *Orcinus orca* | $\frac{10-14}{10-14}$ * | Carnivore |
| Bottlenose dolphin | *Tursiops truncatus* | $\frac{20-21}{20-21}$ * | Piscivorous (fish) |
| Spinner dolphin | *Stenella longirostris* | $\frac{44-64}{42-62}$ * | Piscivorous (fish) |

* a dental formula cannot be determined because of their unknown homology with incisors, canines, premolars, and molars.

The height of the tooth crown also shows phylogenetic and functional patterns. The reduction in tooth number and less replacement (diphyodonty rather than polyphyodonty, and no replacement of molars) increases the importance of the functional maintenance of the reduced number of teeth (O'Leary et al. 2013; Renovoisé & Michon 2014). Only a few mammals, including three manatee species (*Trichechus*), one wallaby (*Petrogale concinnus*), and one mole-rat (*Heliophobius argenteocinereus*), have polyphyodonty of molars by continuously replacing anterior molars that are shed by new molars that develop at the posterior of the tooth row (Gomes Rodrigues et al. 2011; Tucker & Fraser 2014). A shift from brachydont (low crown height) to hypsodonty (high crown height) or hypselodonty (continuous tooth growth) has occurred a number of times (for all tooth types: incisors, canines, premolars, molars) in mammals, often in response to abrasive diets.

There is a corresponding specialization of digestive tract structure (Figure 3.23), particularly the stomach and hindgut, for mammals, reflecting their diet (Hume 1982; Stevens & Hume 1995). Even closely related species can differ markedly in

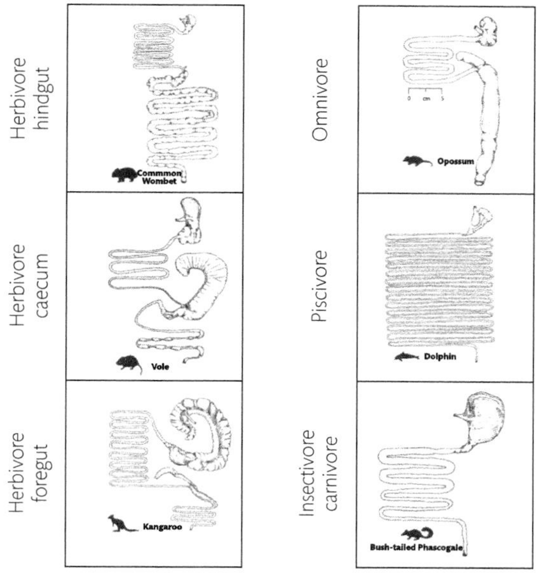


**Figure 3.23** Representative digestive tracts of mammals; simple guts for insectivores, carnivores, piscivores, generalized omnivores, and specialized herbivores (foregut, caecum, and hindgut fermenters). Modified from Stevens and Hume (1995). Reproduced with permission of Cambridge University Press.

gastrointestinal structure, and distantly related species can be convergently similar, depending on their diet. Insectivores and carnivores typically have a simple gastrointestinal structure, with a simple stomach and short small and large intestine. Piscivores (e.g. dolphins) have a compartmentalized stomach: a fore-stomach lined by stratified squameous epithelium, a main stomach with the typical gastric mucosa, and a pyloric stomach with mucous-secreting mucosa. Whales have one to three stomach compartments, and some have a proximal duodenal compartment containing stones, that probably functions as a grinding gizzard. The intestine is typically long for dolphins (up to 14 times the body length) but shorter for toothed whales (6–11 times) and baleen whales (4–6 times). Omnivorous mammals have a relatively simple gastrointestinal structure, with a simple stomach, short small intestine, and somewhat enlarged large intestine. Herbivores generally have considerably more complex gastrointestinal structures, being specialized for fermentation of relatively indigestible plant cell wall components, such as cellulose, hemicellulose, and lignin (see 4.8.1). Foregut fermenters (e.g. ruminant bovids and cervids, various pseudoruminant placentals and macropods) have a large and compartmentalized fore-stomach, whereas hindgut fermenters (e.g. perissodactyls, lagomorphs, koalas and wombats, possums and gliders) have an enlarged hindgut or caecum, or both.

Food selection, intake, and digestion by mammals presents trade-offs between competing factors (Chilcott & Hume 1985). For instance, nutrient turnover and extraction are directly related to energy metabolism and the amount of food transported through the digestive tract (Bozinovic 1993). Besides food quality, digestibility and resource availability affect the rate of energy metabolism (Batzli 1985). An increase in dietary fibre diminishes digestibility (Hume et al. 1993), thereby affecting rates of energy metabolism and allocation (Van Soest 1982), which, in turn, influence energy budget and fitness. Subtle dietary chemical differences are also important. Indeed, slight differences in the chemical structure of nutrients can have profound consequences for the physiology of animals (e.g. the impact of $\propto$ or $\beta$ glycosidic bonds on enzymatic breakdown of starches versus cellulose). At a more complex level, Geiser and Kenagy (1987) and Frank (1994) have shown that during prehibernation fattening, chipmunks tend to choose food containing unsaturated lipids over those containing saturated lipids. The increased ingestion of unsaturated fatty acids seem to be correlated with lower $T_b$, lower rates of metabolism, significantly longer torpor bouts, and consequently higher energy savings from torpor (Geiser et al. 1992). The molecular mechanisms for these physiological responses seem to be the incorporation of dietary lipids into cell and organelle membranes, which changes their permeability and viscosity at low temperatures (Geiser & Kenagy 1987).

Complexity in dental variation, gastrointestinal structure, and dietary selection is matched by complexity in behaviourally matching dietary intake with nutritional requirements. This matching of dietary intake with requirements can be described by a geometry in two (or more) dimensions of nutrient intake that compares the proportions of nutrients for various possible combinations of diets consumed with the 'ideal' combination of dietary intakes that meets the animals nutrient requirements (Cheng et al. 2008). The linear relationships for nutrient content of combinations of different diets are termed 'rails', and animals seem adept at balancing different dietary intakes to fit on the 'rail' that best meets their target nutrient intake. For example, mink (*Mustela vison*) consuming diets with different combinations of lipid and protein content effectively vary the proportion of different diets to meet their target lipid and protein intake (Mayntz et al. 2009; 35% protein rail in Figure 3.24). When provided only a single diet of 20, 25, 30, 35, or 40% protein, mink maintain 200–300 g of lipid consumption, but only the 35% protein diet provides close to the protein consumption target. When given various choices of two diets, mink approach the target consumption rail except for the 20 and 30% protein choice, since neither reaches the target 35% protein rail. Carbohydrate is also present in their diet, but presumably in excess, since dietary selection clearly is directed at attaining the target lipid/protein nutritional intake.

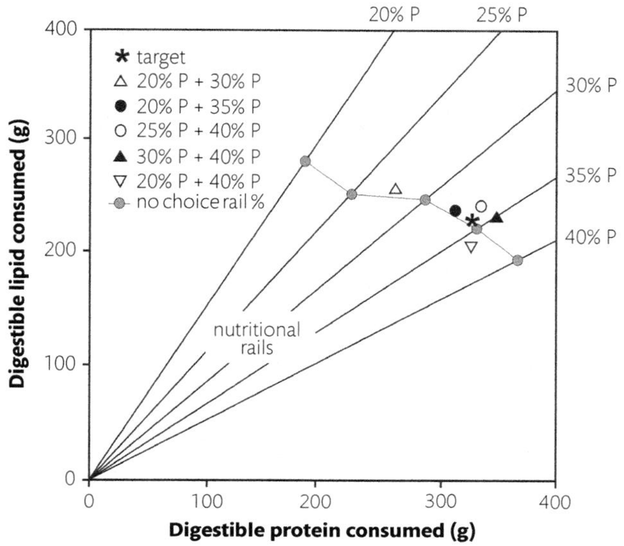


**Figure 3.24** Five nutritional rails of mink (*Mustela vison*) for lipid and protein dietary intake based on diet protein content (% P); the 35% protein rail is closest to the target consumption (*) for protein and lipid. Grey symbols show the actual lipid and protein intake for each rail. Other symbols show the intakes for mink offered the choice of two different diets. Modified from Mayntz et al. (2009).

### 3.5.1.1 Insectivores and Carnivores

Insectivores and carnivores usually have a relatively short and simple digestive tract, reflecting the generally similar body composition of them and their prey, and the relatively straightforward requirements for their digestion (Ferguson 1985; Figure 3.23). Specializations are generally related to feeding mechanisms, with only minor variations in digestive processes. The main digestive modification is the ability of many carnivores to quickly consume large amounts of food. For example, spotted hyaena (*Crocuta crocuta*) consume up to 62 g $kg^{-1}$ $day^{-1}$ at a large carcass, which is similar to that predicted for large carnivores ($g\,kg^{-1}\,day^{-1}$ consumption $= 806\,M^{-0.61}$). Wolves (*Canis lupus*) and lions (*Panthera leo*) have even higher rates of consumption than predicted from body mass, and wild dogs (*Lycaon picus*) have the highest mass-specific consumption (about 150 g $kg^{-1}$ $day^{-1}$; Henschel & Tilson 1988). Carnivores secrete large amounts of bile that emulsifies fat, and proteases that digest protein, and have a considerable capacity for amino acid deamination in the liver. Perhaps not surprisingly, they excrete considerable concentrations of urea as their nitrogenous waste.

Mammals that feed on insects and other small arthropods are defined as insectivorous. The insectivorous niche represents the primitive mammalian condition (Feldhamer et al. 2015). Insectivory is apparent in several subclasses and orders of mammals, including Monotremata (e.g. echidnas), Marsupialia (e.g. small dasyurids), and placentals such as Insectivora (e.g. shrews), Chiroptera (bats), Xenarthra (armadillos), Tubulidentata (aardvarks), and even Carnivora (e.g. aardwolves). In general, the dentition of insectivores is typified by numerous sharp lower incisors that are slightly procumbent. Their digestive tract is short and simple, and most insectivores lack a caecum (Feldhamer et al. 2015). Mysticete cetaceans have a reduced dentition and use keratinous sheets of baleen to filter feed on small crustaceans (e.g. krill; see 4.8.1). The crabeater seal (*Lobodon carcinophagus*) is also a major krill predator in the Antarctic, but unlike baleen whales they are selective, not filter, feeders.

Carnivores feed primarily on vertebrate prey. Extant members of this feeding guild represent many members of the order Carnivora (including canids, mustelids, felids, pinnipeds), the larger dasyurid marsupials (quolls, *Dasyurus* spp., and the Tasmanian devil), and odontocete cetaceans. Carnivores have a diverse array of feeding types and dental morphologies (Van Valkenburgh 1989), ranging from obligate carnivores with reduced dentition and specialized carnassial teeth, such as felids, to scavengers with powerful crushing dentition, such as hyaenas and Tasmanian devils. Because animal material is generally highly digestible compared to plant material (e.g. grasses, fruits), the digestive tract of carnivores (like insectivores) is short and has a small or absent caecum. However, some seals have a surprisingly long small intestine, nearly 40 times the body length. The reason for their long gut is unknown, but the area of the small intestine provides a rapid absorption capacity for nutrients while the seals are at the surface and the gut is well perfused with blood (Krockenberger & Bryden 1994; Mårtensson et al. 1998). As with other carnivorous mammals, the caecum and colon of seals are comparatively short.

Carnivores feed in various habitats, including land, air, and water. Canids are, in general, opportunistic hunters. Usually small species capture their prey singly or in pairs, whereas larger species (e.g. *Canis lupus*) hunt in packs (Feldhamer et al. 2015). Felids, in contrast, are typically solitary hunters, except lions (*Panthera leo*), which are remarkably convergent with wolves in the size of the group during hunting. Interestingly, *Panthera* apparently evolved at about the same time as group-foraging canids in response to the same selective pressures (Nudds 1978). Most mustelids inhabit and feed on land, with the exception of the otters (Lutrinae) and some mink (*Mustela vison*), which are semi-aquatic carnivores.

Some bats are aerial carnivores, feeding on small ectothermic and endothermic vertebrates, even other bats, and a few feed on fish (e.g. *Pizonyx*). The diet of the false vampire bat (*Vampyrum spectrum*), which was thought to be a true vampire bat, is not blood but birds, bats, rodents, insects, and fruit (Gardner 1977). The truly (and only) sanguinivorous mammal, the vampire bat (*Desmodus rotundus*), has a reduced rostrum and small cheek teeth but large upper incisors and canines, and a tongue that functions like a drinking straw that facilitates puncturing and lapping blood. The long tubular stomach but small intestine allows for rapid digestion of large volumes of easily digested blood (Greenhall 1972). The saliva of the vampire bat contains an anticoagulant protein known as draculin, which immediately and irreversibly inhibits the activated form of coagulation factors IX and X, without impacting on thrombin, trypsin, or chymotrypsin (Apitz-Castro et al. 1995), allowing the bat to consume more than 20 g of blood at a time, approximately 60% of their body mass (Wimsatt 1969).

Aquatic carnivores that eat fish and large invertebrates (such as squid) include seals, sea lions, and the odontocete cetaceans (toothed whales, porpoises, and dolphins). They have abundant, small, simple, homodont teeth. The largest predator of other mammals, the killer whale (*Orcinus orca*), consumes pinnipeds, baleen whales, and smaller cetaceans, as well as fish, squid, invertebrates, penguins, and other aquatic birds. It uses social cooperation during foraging, with pods varying from only a few to more than 20 individuals; adults and juveniles cooperatively herd their prey (Heyning & Dahlheim 1988).

### 3.5.1.2 Omnivores

Many mammals, particularly small species, are omnivorous—they opportunistically consume both animal and plant material and so have a greater variety of food sources than carnivores. They include primates (chimpanzees, *Pan troglodytes*), carnivorans (e.g. procyonids), suids, rodents (rats and mice), and marsupials (e.g. peramelids, didelphids, small possums, and gliders). Digestion is relatively straightforward, and their digestive tract is relatively simple (Figure 3.23). The stomach is comparatively simple, the small intestine is relatively long, and the colon is large with many folds. The caecum of most omnivores is poorly developed, but is sometimes large enough to process some plant material. Some have limited bacterial fermentation in the large intestine (cf. hindgut fermenters), but this is of minor digestive significance.

The primates are an interesting group because of their dietary diversity; most are omnivorous, but some are quite specialized herbivores and even carnivores (Stevens & Hume 1995). The 'lower' Strepshirhini, including the lemurs (e.g. the indri, *Indri indri*), aye-aye (*Daubentonia madagascariensis*), potto (*Perodicticus potto*), lorises (*Loris* spp.), and bush babies (Galagidae), are variously omnivorous, insectivorous/carnivorous, frugivorous, folivorous, or consume plant exudates, and their gastrointestinal structures vary somewhat. Reflecting their mixed diet, ruffed lemurs (*Varecia* spp) and lorises have a long caecum, and indri have a long intestine and caecum. The aye-aye has such chisel-like incisor teeth that it was once classified as a rodent. The 'higher' primates (Tarsiidae, Cercopithecidae, Cebidae, Callitrichidae, Hylobatidae, and Hominidae) vary considerably in omnivory. Tarsiers eat insects and small vertebrates, and have a consequently simple digestive tract. Cercopithecines (Old World monkeys) are generally herbivorous or folivorous, although baboons can also be predatory carnivores; most have cheek pouches for food storage, a simple stomach and small intestine, and a pocketed large intestine. Langurs (Colobinae) have a more complex stomach (likely for food storage, as they lack cheek pouches), and a longer small intestine and caecum; colobus monkeys (*Colobus* spp., *Piliocolobus* spp.) are similar. The cebid New World monkeys tend to be omnivorous but some feed on fruit and leaves and can have an enlarged pocketed colon; night monkeys (*Aotus* spp.) can be insectivorous and carnivorous (eating bats; Vaughan 1986). Within the Hominidae (great apes and humans), gorilla (*Gorilla* spp.) are strict herbivores; gibbons (Hylobatidae) and orangutans (*Pongo* spp.) are arboreal folivores; chimpanzees are herbivorous, insectivorous (termites), and carnivorous (small monkeys); and humans are omnivores. Overall, carnivorous primates tend to have a simple stomach, a long small intestine, and a simple hindgut; folivorous primates have a large stomach and/or caecum and colon; and frugivorous primates are intermediate (Chivers & Hladik 1980). The coefficient of gut differentiation (CGD; size of stomach+caecum+colon/small intestine) differentiates these dietary categories; based on surface area, CGG $< 0.1$ for strict carnivores and $> 3.0$ for strict folivores.

### 3.5.1.3 Herbivores

Mammalian herbivores feed on a wide diversity of plant foods, including grasses, leaves, fruit, seeds, nectar, pollen, and gums. Plants are more abundant than animals, but in general have a comparatively low energy content and digestibility, and a different ionic composition to animal tissues (e.g. comparatively high $K^+$ and low $Na^+$). Plant products, such as seeds, fruit, nectar, gums, and fungi, may in some cases provide a more nutritious plant-based diet for herbivores than leaves and stems, but often vary in availability seasonally.

Gum is produced by plants in response to mechanical damage. It seems to be a low-quality, difficult-to-digest food that is low in fat, carbohydrate, and vitamins. Gums can also increase requirements for water and protein by binding them, resulting in increased faecal loss. However, they form an important food source for mammals such as some primates (e.g. the needle-clawed bush-babies *Euoticus* spp.) and the marsupial petaurid possums (Charles-Dominique 1977; Hume 1982; Stevens & Hume 1995; Burrows & Nash 2010; Power 2010). Sap has a high energy yield due to its soluble sugar content, but can be difficult to access, requiring mechanical severing of phloem ducts. Some mammals, such as red squirrels

(*Sciurus vulgaris*), porcupines, and bats, feed at 'sap-wells' (Eastman 2000), and the yellow-bellied glider (*Petaurus australis*) and sugar glider use specialized teeth to incise tree trunks and then lap up the exuded sap (Hume 1982). Manna is encrusting dried sap resulting from mechanical wounds to plants from insects. This, together with honeydew excreted by sap-sucking insects, can also form an important component of the diet of arboreal mammals, particularly on a seasonal basis (Hume 1982).

Herbivorous mammals play important roles in pollination and seed dispersal. A variety of mammals are significant dispersers of pollen for plants, particularly tropical and arid species (Fleming & Sosa 1994). Mammals (and birds) probably began pollinating flowering plants around 50–60 MYBP, and while few plants rely on a single pollinator, there are examples of interdependence, with floral characteristics more suited to mammalian than avian or insect pollinators. Many mammals include both nectar and pollen as components of a broad diet, but very few specialize on them (e.g. some chiropterans and the marsupial honey possum, *Tarsipes rostratus*). Honey possums feed on the flowers of banksias such as *Banksia nutans, B. quercifolia, B. meisneri*, and *B. candolleana*, which are generally dull-coloured, located within a bush or low to the ground, have a strong odour, and have a crepuscular or nocturnal pattern of nectar production (Wooller & Wooller 2013). Nectar is easily digestible and provides a high energy return, but is lacking in other essential nutrients, particularly protein, for which pollen or invertebrates may be an important supplement. Nectar is 75–80% water, so its consumption also results in high water intake; this must be excreted via the kidneys. Honey possums have a kidney with a relatively large cortex that facilitates the production of copious dilute urine. They can also differentiate between dilute and concentrated nectar solutions, and avoid very dilute solutions (Wooller & Wooller 2013).

Many different angiosperm plant lineages have evolved fleshy fruits that encourage consumption, hence seed dispersal, by birds and mammals, and many mammals consume fruit opportunistically, while some are frugivore specialists. Mammalian frugivores include elephants; carnivores (palm civets, *Paradoxurus hermaphrodites*, and raccoons, *Procyon lotor*); chevrotains (Tragulidae); various primates, including lemurs, capuchins, night monkeys, sakis (*Pithecia* spp.), howlers and spider monkeys; chiropterans (leaf-nosed and fruit bats); rodents (spiny rats and agoutis); and tapirs (Fleming & Kress 2011). There is some overlap in the fruits consumed by primates with those consumed by birds, leading to the suggestion that frugivory by primates might have been facilitated by the evolution of fruits adapted for foraging by birds. On the other hand, there is little overlap in fruit choice between pteropodid and phyllostomid bats with other frugivores. Frugivores must often contend with a diet low in protein. Most fruits are 'nutrient poor'—they are high in carbohydrates and provide abundant energy, but low in protein, so frugivores consuming these fruits often supplement their diet with insects, seeds, or meat (Thomas 1984). Phyllostomid bats overcome the limitation by including insects in their diet, while pteropodid bats increase their overall fruit intake to about double the energy gain of phyllostomids. How pteropodids expend this excess energy, however, is unclear (Thomas 1984; McNab 2002). In contrast, some 'nutrient-rich' fruits are high in lipids and protein, and some obligate frugivores specialize on these fruits, primarily birds (e.g. oilbirds and bellbirds). Although frugivorous mammals are relatively inefficient dispersers of seed, they are nevertheless the most important seed dispersers for many tropical and arid plants (Fleming & Sosa 1994).

Granivores, whose diet exceeds 50% seed, are mainly rodents, but a variety of other rodents and other mammals consume small amounts of seed; for example, foxes, badgers (*Taxidea taxus*), raccoons, brown bear (*Ursus arctos*), shrews, moles, elephant shrews, and a few marsupials, including the mountain pygmy possum, bilby (*Macrotis lagotis*), and monito del monte (Pineda-Munoz & Alroy (2014). Seeds are a well-balanced diet, containing some water (11%), considerable protein (11%), and soluble carbohydrate (69% dry weight; Table 3.7). Seed is highly digestible; it produces considerable metabolic water and minimizes faecal water loss (see 2.7.2). Consequently, it is not surprising that a range of small granivorous desert rodents can exist on a seed diet without requiring drinking water (e.g. Australian hopping mice, *Notomys alexis*; MacMillen & Lee 1967; North American kangaroo rats, *Dipodomys merriami, Perognathus fallax*, and the nearly independent *Peromyscus crinitis*; MacMillen 1983; South American desert mice, *Phyllotis gerbillus*; Koford 1968; North African gerbils, *Dipus aegypticus, Jaculus orientalis*; Kirmiz 1962; and Southern African rodents, *Aethomys namaquensis, Petromyscus collinus, Mus minutoides, Desmodillus auricularis*; Withers et al. 1980).

There are marked intercontinental differences in the roles of mammals, birds, and ants on seed consumption in semi-arid and arid ecosystems (Hulme & Benkman 2002). In Israel, mammals are the primary seed consumers, ants are significant consumers, and birds are not, whereas in South Africa, ants are the primary consumers, mammals are about half as important, and birds are not. The greater abundance of small granivorous rodents in North American compared with Australian deserts generally reflects small-mammal diversity, rather than size, evolutionary age, or physiography of the deserts (Morton 1979). Interestingly, there are no specialist marsupial granivores in Australia, although omnivorous species such as the bilby and mountain pygmy possum include seeds in their diet (Hume 1982). Seeds are less readily consumed by mammals (and also birds and ants) in South American deserts (Chile, Argentina), perhaps reflecting the evolutionary consequences of the extinction of granivorous argyrolagid marsupials in South America (Mares & Rosenzweig 1978).

Another important group of 'herbivores' feed on fungi (which are not plants). Fungi contain complex carbohydrates associated with cell walls, and consequently are difficult to digest. In spite of this, fungi are important in the diet of many rodents and marsupials, such as the woylie (*Bettongia penicillata*) and potoroos (*Potorous* spp.), which disperse fungal spores and thus contribute to maintaining forest ecological function and biodiversity.

Accessing the protein and other nutrients within plant cells is difficult due to the fibrous cell wall of plants, as well as their often complex physical and chemical defences. It is the herbivorous mammals that have the greatest specializations for digestion, because their plant food is often high in carbohydrates such as cellulose that are difficult or impossible for normal vertebrate digestive enzymes to hydrolyze (see 4.8.2). Herbivores that feed on leaves, stems, and grasses can be divided into two main groups: (1) browsers and grazers, such as Perissodactyla (e.g. horses) and Artiodactyla (e.g. deer), and (2) gnawers, such as Rodentia and Lagomorpha (Feldhamer et al. 2015). They share many tooth and digestive tract characteristics that are adaptations to feeding on fibre-rich plant material. In general, their teeth are specialized to break up the cell walls of plants. This considerably wears the teeth, especially for grazers that feed on grasses containing siliceous crystals. As a consequence, herbivorous mammals have dental specializations that strengthen the teeth, including the addition of hydroxyapatite, thickened enamel (e.g. some primates), large tooth size (e.g. elephants, camels, hippopotamus, pigs), increased tooth height (e.g. ungulates, some rodents), or continuous tooth growth (e.g. wombats, *Vombatus ursinus*, edentates, some rodents; McNab 2002).

The digestion of fibre (cellulose) is problematic for mammals, because only certain specialized enzymes can digest cellulose, and mammals do not produce these cellulolytic enzymes. Rather, mammals usually rely on symbiotic microorganisms that break down and hydrolyze the cellulose of plants by anaerobic fermentation, and also produce fatty acids that can be absorbed by the mammal (see 4.8.2). The exception to this is the 85-kg giant panda (*Ailuropoda melanoleuca*) that, although it feeds almost exclusively on the leaves and stems of bamboo (*Sinarundinaria* and *Fargesia* spp.), has a typical carnivore-type gut unspecialized for the fermentation of plant materials. It survives on its highly indigestible diet by consuming vast quantities (typically 18–20 kg day$^{-1}$, but as much as 38 kg day$^{-1}$) of bamboo, passing it through the gut rapidly and digesting only about 19% of the dry matter (Dierenfeld et al. 1982). Most browsers and grazers, however, have evolved specialized structures that facilitate the maintenance of a population of microorganisms that ferment and break down cellulose, either in the foregut (rumination and pseudorumination) or hindgut (Robbins 1993).

## 3.5.2 Digestive Function and Flexibility

Food must be broken down in the digestive tract into small units that can be absorbed across the gut lining. Carbohydrates (e.g. starch and cellulose) are hydrolyzed to monosaccharides, lipids (e.g. triglycerides) to fatty acids and glycerol, and proteins to amino acids. The process of breakdown is either by chemical (enzymatic) digestion or by microbial fermentation; digestion is preferable because the animal gets a larger energy return. Animal digestion is fundamentally similar in process to physical chemistry (Alexander 1999). Penry and Jumars (1987) and Martínez del Río et al. (1994) used principles from chemical engineering to model the design and function of the digestive system. According to these models, the gut can be viewed as large chambers (like the mammalian stomach) and narrow tubes (like the small intestine); some physical (mixing) and chemical (reaction) processes are carried out in the large chambers and others in the small tubes.

Alexander (1999) nicely exemplified the use of reactor theory to model mammalian digestive function. He considered that in each kind of vessel there is a reagent of concentration $C_i$ as it enters the reactor; it remains in the reactor where a chemical reaction occurs for time $t$; and then leaves with concentration $C_o$. The large chamber (whose contents are kept stirred) is represented as a continuous-flow stirred tank reactor (CSTR) where a reagent entering the reactor is immediately diluted to the concentration at which it will leave (Penry & Jumars 1987; Jumars 2000; Figure 3.25). The concentration of the reagent is $C_o$ throughout the mixed contents of the reactor, and the reaction proceeds at a rate $rC_o$. The change in concentration during the time $t$ in the reactor will be $C_i - pl\, C_o = rC_o t$. Alternatively, a long narrow tube through which the reagents flow (with little mixing) is equivalent to a plug flow reactor (PFR). On entry, the concentration is $C_i$ and the reaction proceeds at a rate $rC_i$, but as the reagent flows along the tube, its concentration falls to $C_o$ and the reaction rate falls to $rC_o$; the final concentration as the reagent leaves the reactor is $C_o = C_i e^{-rt}$. Thus, reactions are faster in PFRs than in CSTRs because of the higher average concentration. For example, when $rt = 1, C_o / C_i$ is 0.5 for a CSTR but 0.37 for a PFR; 50% of the reagent is broken down in a CSTR and 63% in a PFR (Alexander 1999).

This line of reasoning suggests that PFRs should be preferable, so why do CSTRs exist? There are other constraints involved. For instance, if a PFR had a fermentation process due to microorganisms, then new microorganisms would need to be added continuously at the inlet, as the original ones were washed out. Contrarily, mixing in a CSTR ensures that despite some microorganisms being washed out of the reactor, some remain in it. If the remaining microorganisms reproduce fast enough, the population will be maintained within the reactor. Nevertheless, the weakness of CSTRs—that they work slower than PFRs—can partly be overcome by arranging several CSTRs in series. In fact, as suggested by Martínez del Río et al. (1994) and Alexander (1999), a sequence of more than two CSTRs would do better, though still not quite as well as a PFR.

In summary: (1) food can be broken down by digestion or fermentation, but digestion is preferable because the animal gets a larger energy return; (2) a PFR is better than a CSTR for digestion; (3) for fermentation, only a CSTR will work; and (4) a series of CSTRs is better than a single one of the same total volume. These conclusions suggest that mammals with no fermentable component in the diet,

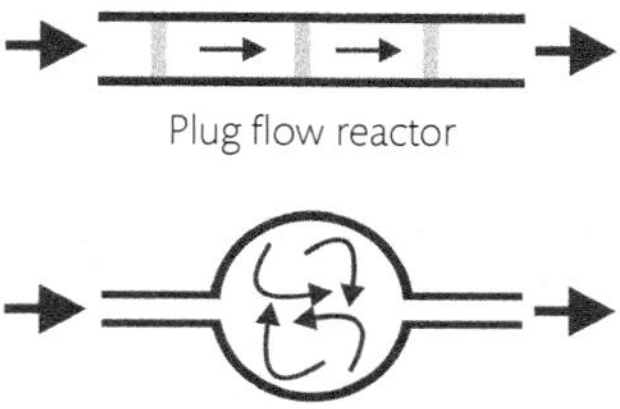


**Figure 3.25** Schematic of gut reactor models, comparing a plug flow reactor (PFR) with a continuous-flow stirred tank reactor (CSTR). Adapted from Jumars (2000).

such as carnivores, should have solely a PFR model of digestion, but herbivores should have a gut that includes both a PFR in which readily digestible foods are broken down, and one or more CSTRs in which fermentable foods are broken down by microorganisms. Digestion and fermentation cannot generally occur in the same part of the gut, because a fermentation chamber must not contain enzymes that would destroy the symbiotic microorganisms.

According to optimal digestion theory, the digestive strategy that maximizes fitness will be favoured (Sibly & Calow 1986). The relationship between food ingestion ($i$) and digesta retention time ($t$) for the digesta mass ($m$) transported by a continuous flow digestive system is $m = it$. Then, $i$ yields $k$ units of energy by digestion after $t$, so the net rate of energy gain ($e$) is $e = ik(t)$, where $k(t)$ indicates that the quantity $k$ is a function of $t$ (Sibly and Calow (1986). Substituting the equation $m = it$, and into $e = ik(t)$, then $e = mk(t) / t$. Thus, for a given $m$, the optimal strategy to maximize $e$ is to maximize $m$, $k(t)/t$, or both. The prediction from this optimal digestive strategy would be that an increase in the size of digestive organs to increase the amount of food ingested ($i$) and optimize absorption and fermentation processes ($k(t)$) for digesta mass ($m$) would result in higher energy yield ($e$). Theoretically, mammals consuming a low-quality diet will develop a longer digestive tract, thereby increasing $m$.

Resource partitioning by browsing and grazing African antelope is an interesting example of possible interrelationships between diet, gastrointestinal structure, and body mass. Demment and Van Soest (1985) found a relationship between the 'quality' (inverse of fibre content) of grasses and body mass; since gut volume scales with $M^{1.03}$ and energy requirements with $M^{0.75}$, then body mass potentially influences diet selection, with larger mammals able to select 'lower-quality' food. However, these relationships have been challenged; although there may be a relationship between body mass and diet quality, there may not be a relationship for digestive efficiency. De Long et al. (2011) found considerable overlap in diet composition for African savannah herbivores, indicating patterns for mass, diet quality, and diet composition in the dry season but not the wet season.

Physiological flexibility refers to reversible changes in the traits of organisms due to changes in internal or external environmental conditions (Piersma & Drent 2003). Several lines of evidence suggest that phenotypic flexibility in digestive function is adaptive and that it evolves through natural selection. Indeed, several studies have reported that mammals, mainly omnivores—and despite some allometric constraints on digestion and metabolism—compensate for low-quality diets via digestive flexibility, use of mechanisms such as more rapid turnover time of fibrous food, changes in gut capacity, increased nutrient uptake by the small intestine, and use of fermentation chambers (Foley & Cork 1992; Veloso & Bozinovic 1993). During nutritional bottlenecks—for example, during a summer drought, or in habitats with low productivity—only low-quality, high-fibre food is available to herbivorous mammals (Bozinovic 1995). In the field during the dry season, some species include fibrous plant tissues as a major dietary item, but avoid it when better-quality forage is available, most likely because they are able to obtain sufficient energy from fibre to satisfy their maintenance energy requirements.

Mammalian nutrition depends not only on food type but also on digestive strategy. Comparatively low-quality, energy-diluted food, such as fruits, have low energy density (Cipollini 2000). Therefore, the expected foraging behaviour and digestive strategy for fruit-eating mammals to maintain energy balance—when consuming fruits or other low-quality food items—is an increase in the rate of food intake and a decrease in the digesta transit time. However, a critical point is eventually reached ('critical digestibility', in the sense of Cork 1994) as a consequence of constraints on the utilization of low-quality foods. Theoretically, mammals should compensate for such constraints by behaviourally complementing their diet with higher-quality, though less available, food items. As pointed out earlier, fruit as well as other food items such as fungi, are not high quality, but nonetheless are consumed by mammals, including carnivores, herbivores, and insectivores (Elmhagen et al. 2000). Westoby (1978) noted that individuals should consume a mixed diet to respond to changes in diet quality, to reduce searching cost, to sample items, and to balance nutrient intake.

Pennings et al. (1993) proposed that, if consumers attempt to maximize the intake of different nutrients simultaneously, and if different dietary items are sources of variable levels of limiting nutrients, then consumers might be able to survive consuming just one item, but would obtain better short-term (energetic/physiological or survival) and long-term (reproductive) success with a mixture of different items. Indeed, many carnivorous and insectivorous mammals commonly ingest fleshy fruits (Gittleman 1989).

Nevertheless, the role of fruit in mammalian nutritional ecology and energetics has remained virtually unexplored (but see Silva et al. 2005). Many authors speculate that an increase in the consumption of fruits by terrestrial carnivores, such as foxes, is a consequence of low availability of vertebrate prey, but with negative impact on their nutritional ecology. Indeed, a mixed diet may yield higher assimilation efficiencies and hence higher nutrient intakes than those predicted from the ingestion and assimilation of pure diets (Bjorndal 1991). Silva et al. (2005) found that culpeo (*Pseudalopex culpaeus*) did nutritionally well on a mixed diet, and hypothesized that a mixed diet should yield a positive energy/mass balance for the fox during nutritional bottlenecks or low availability of mammalian prey. These authors calculated that the minimum values of dry-matter, energy, and nitrogen intake of mixed diets are around 210 g $day^{-1}$, 1,500 kJ $day^{-1}$, and 25 g $day^{-1}$, respectively, theoretically allowing foxes to survive for at least 10 days. On a meat (e.g. rats) diet, a fox should be able to satisfy its basic requirements with a dry-mass intake of 70 g $day^{-1}$ or a wet-mass intake of about 123 g $day^{-1}$. Thus, temporal and spatial variation in nutrient, energy, and water contents of available prey in a given habitat can have an important effect on carnivores' nutrition, energy use, and mass balance.

It is clear that large- and medium-sized mammalian carnivores cannot survive on fruits for more than a week. Why then do large carnivores feed on such low-quality items as fruits? It may be argued that these mammals are doing the best of a poor job, and eating whatever is available when preferred prey abundance is low, even though it may not satisfy their most basic nutritional requirements. Foxes, for instance, may be eating fruits just to satisfy their hunger. Alternatively, fruits may represent sources of water or as yet unidentified vitamins or minerals (Bozinovic & del Río 1996). In addition, fruit consumption may contribute to the associative effect that results from feeding on a mixed diet, because the nutritional value of dietary items may vary with the compounds with which it is consumed.

One way mammals can respond to changes in external conditions is by modifying their digestive attributes. In this sense, digestive flexibility has been suggested to be one of the most important and widely used physiological adjustments to changes in both internal and external conditions. In terms of digestive tract adjustments, 'hypertrophy is caused by the increased food intake and is an adaptive response of the alimentary canal which results in the maintenance of a constant coefficient of digestibility at increased levels of food consumption' (Campbell & Fell 1964, p. 96).

### 3.5.3 Lactation

Lactation, the production of milk from the mammary glands that is used for nourishment of the young, is a characteristic of all living mammals, and probably reflects an early evolutionary development in the lineage (Blackburn et al. 1989; Blackburn 1991). Lactation probably evolved in early synapsid mammals. Genomic analysis suggests that the complex lactation system evolved gradually through the Triassic in the therapsid lineage, and was probably evident for Mammaliaformes and well established by the time of the monotreme-therian divergence (Lefévre et al. 2010). Recent molecular characterization of the highly conserved casein gene clusters in monotremes, marsupials, and placentals suggests a common ancestral state at least 166 MYBP and perhaps as early as 240 MYBP.

Mammary glands resemble other skin glands. They may have evolved from a combination of different skin glands (Blackburn 1993) or ancestral apocrine-like glands associated with hair follicles (Oftedal 2002). There is a clear association of mammary gland ducts with hairs in monotremes, which lack nipples (the presumably ancestral condition), and a transient association in marsupials, but in placentals the development of hair follicles is actively suppressed near the developing mammary gland (Lefévre et al. 2010). This close association of mammary glands with hair suggests their coevolution in therapsid mammals.

Primitive lactation fluids may have provided moisture for egg hydration or thermoregulation, or antimicrobial, immune, or pheromonal functions. The transformation into a highly nutritious milk accompanied the evolution of a suite of other 'mammalian' characteristics, such as high metabolic rate and aerobic capacity, endothermy, rapid nutritive processing, and high growth rates (Oftedal 2002; Lefévre et al. 2010; Kawasaki et al. 2011). Benefits of the additional parental care from lactation include the capacity to provide a transitional and highly nutritional diet for their young between hatching and independence, as well as the capacity to feed their young and increase their growth rate and survival, without having to transport food from external sources (e.g. birds).

Extant mammals have a diverse array of lactation strategies (Lefèvre et al. 2010). Monotreme young are extremely altricial and rely for nutrition on milk whose composition changes over the long lactation period (compared to the short gestation period). Marsupial young are also highly altricial and totally dependent on milk, whose composition can also change over the extended lactation period (compared to a short gestation period). Placental mammals, in contrast, have an extended intrauterine development and shortened (but variable between species) lactation period (Hayssen 1993), with relatively constant milk composition (except for the initial colostrum). Lactation length can vary from as little as 4–5 days (rodents and some phocid seals) to more than 900 days (some primates such as chimpanzees and orangutans). Although body mass is a major determinant of lactation period in placentals, phocid seals have very short lactation periods (reflecting their breeding on unpredictable and transient ice floes), as do whales (0.5–1 year, a disproportionately short lactation relative to their large mass). The exceptionally long lactation periods of some placentals likely reflects additional important roles of lactation, such as protection, mother-young social bonding, and foraging development.

Mammalian milk is a complex mix of water, proteins, carbohydrates, lipids, salts, vitamins, and various bioactive molecules. For the first few days of lactation, eutherian mammals produce colostrum, a dilute milk with high concentrations of antibodies that prevent oral and intestinal microbial infections in the neonate, and various other bioactive proteins such as lactalbumen, lactoglobulin, lactoferrins, lactoperoxidases, and growth factors (Korhonen 2013). Milk composition varies markedly between species, and sometimes within species over the lactation period (Lefévre et al. 2010). Fat content varies from almost 0% in rhinoceros to more than 40% in pinnipeds. Milk composition changes considerably over the long lactation period of marsupials. For example, the changes in the composition of the milk of the quokka (*Setonix brachyurus*), a medium-sized macropodid marsupial, are synchronous with critical developmental milestones in the pouch young. The most dramatic changes in milk composition occur when young first exit the pouch, associated with an increase in protein concentration; at permanent pouch exit, associated with increased lipid concentration; and when young start feeding on plant matter, associated with decreased carbohydrate concentration (Miller et al. 2009). Some marsupials (e.g. kangaroos) show concurrent asynchronous lactation, where different mammary glands produce different milk compositions for pouch young born from separate pregnancies and births, generally with one newly born young and the other approaching the end of its lactation period, and the different milk compositions reflecting the different growth stages of the young.

Caseins are a major milk protein that form a casein micelle complex with calcium phosphate, which is important for the mineralization of bones and teeth. The genes for calcium-binding phosphoprotein (SCPP) have only been identified in mammals, but arose from the SPARC-like gene family that is present in early vertebrates (Kawasaki et al. 2011). Milk provides nutrients for growth and development of the young, but it also has many other functions, including antimicrobial activity that protects the young and the mammary gland from infection, as well as direct effects on the mammary gland and the gut (Kuruppath et al. 2012). The milk that monotremes secrete just prior to hatching is thought to protect the imminent hatchling from microbial infection, and the cathelicidin and whey acidic proteins of marsupial milk have antimicrobial and immunological roles in the young and the mammary gland itself (Oftedal 2002; Kuruppath et al. 2012). There are also particular bioactive milk proteins that promote growth and organ (e.g. stomach) development in the young. There are similarly diverse milk proteins in placental mammals, with equivalent nutritional, antimicrobial, immunological, hormonal, and regulatory roles (Gillespie et al. 2012).

There are some examples of 'milk' production in other than the mammalian lineage. Male and female pigeons and doves, and male emperor penguins, produce a crop 'milk' that nourishes their young, and some passerine birds secrete stomach oils. Presumably some of the dinosaur ancestors of birds had also evolved 'milk' production (Oftedal 2002; Else 2013). Pigeon milk is highly nutritious, consisting of keratinocytes shed from the epithelium of the crop sac. While this curd-like 'milk' is high in lipid and protein, minerals, and bioactive proteins, including growth factors, prebiotics, and immunoglobulins, it lacks the mammalian calcium-sequestering casein proteins and is low in carbohydrates. Pigeon 'milk' production and related behaviours, like mammalian milk production, are regulated by the lactogenic hormone, prolactin (Horseman & Buntin 1995). Pigeon 'lactation' and mammalian lactation are surprisingly similar in function but differ in their secretory mechanism and structure, and have clearly evolved independently (Gillespie et al. 2012).

## 3.6 Water and Solutes

Mammals, like other terrestrial vertebrates, normally have a constant body water content of about 55–75% of body mass, and also maintain constant extra- and intracellular ion concentrations. For mammals, the allometry of body water content is $0.605\ M^{1.00}$ (Calder & Braun 1983), so the average water content is 60.5%. However, body water content can vary considerably. Humans, for example, have about 57% water content, varying from about 45% in an obese adult to 75% in a newborn. Tissue water content ranges from about 10% for adipose tissue to more than 80% for kidneys and blood (Table 3.9). The extracellular solute concentrations differ slightly for plasma and interstitial fluid, reflecting the ultrafiltration of plasma across the capillaries to form interstitial fluid, which is similar in most constituents to plasma, but lacks blood cells and has a very low protein concentration. The solute concentrations differ markedly between interstitial and intracellular fluids, reflecting the higher intracellular organic concentrations and different ion concentrations responsible for the cell membrane potential. The solute composition of extracellular and intracellular fluids is remarkably constant amongst different mammalian taxa, reflecting their overall homeostatic constancy of water, solutes, and most physiological variables and functions. Plasma [$Na^+$] is about 145–165 mM, [$K^+$] is 4–6 mM, [$Ca^{2+}$] is 2–10 mM, [$Mg^{2+}$] is 1–2 mM, [$Cl^-$] is 100–120 mM, and [protein] is 65–75 g $l^{-1}$.

### 3.6.1 Water and Solute Balance

Water and solute homeostasis is determined by the balance between intake and loss (Withers 1992; Bentley 2002). A budget approach considers the balance of various avenues of water and solute intake, various avenues of water and solute loss, and, for mammals, the limited capacity for storage of water and solutes in body fluids. Generally, intake and loss are balanced, and homeostasis of body water volume and solute concentrations is maintained. The turnover of water and solutes are the rates at which they enter and leave the body. In balance, water intake = water loss = water turnover rate, and similar balance equations apply to simple solutes such as $Na^+$, $K^+$, $Cl^-$.

Avenues for water intake include drinking, preformed water in food, and metabolic water production. Drinking supplies potentially fresh water and can make the water balance very positive, or make up for considerable water losses to

**Table 3.9** Mass and water content of various tissues for a 70-kg human, and extracellular solute concentrations (plasma, interstitial fluid) and intracellular fluid. Data from Skelton (1927), Pitts (1963), and Guyton and Hall (1996).

| | Water composition of body tissues | | | Solute composition of body fluids (mM and mOsm) | | | |
|---|---|---|---|---|---|---|---|
| **Tissue** | **% Water** | **% Body Mass** | **l Water per 70 kg** | **Solute** | **Plasma** | **Interstitial** | **Intracellular** |
| Skin | 72 | 18 | 9.07 | $Na^+$ | 146 | 142 | 14 |
| Muscle | 75.6 | 41.7 | 22.1 | $K^+$ | 4.2 | 4.0 | 140 |
| Skeleton | 22 | 15.9 | 2.45 | $Ca^{2+}$ | 2.5 | 2.4 | 0 |
| Brain | 74.8 | 2.0 | 1.05 | $Mg^{2+}$ | 105 | 108 | 31 |
| Liver | 68.3 | 2.3 | 1.03 | $Cl^-$ | 105 | 108 | 4 |
| Heart | 79.2 | 0.5 | 0.28 | $HCO_3^-$ | 27 | 28.3 | 10 |
| Lungs | 79.0 | 0.7 | 0.39 | $H_2PO_4^-$, $HPO_4^{2-}$ | 2 | 2 | 11 |
| Kidneys | 82.7 | 0.4 | 0.25 | $SO_4^{2-}$ | 0.5 | 0.5 | 1 |
| Spleen | 75.8 | 0.2 | 0.10 | Amino acids | 2 | 2 | 8 |
| Blood | 83.0 | 8.0 | 4.65 | Protein | 1.2 | 0.2 | 4 |
| Intestine | 74.5 | 1.8 | 0.94 | Urea | 4 | 4 | 4 |
| Adipose tissue | 10 | ~10 | 0.70 | Glucose | 5.6 | 5.6 | - |
| Total | 57 (45–75) | | 39.9 (31.5–52.5) | Osmotic concentration | 282.6 | 281.3 | 281.3 |

maintain water balance. Small desert mammals (e.g. rodents) do not generally have access to free water, and rely on physiological and behavioural means to balance their water budget (e.g. nocturnality, fossoriality, low metabolic rate, high urine-concentrating capacity; MacMillen 1972; McNab 2002). Larger desert mammals (e.g. antelope, deer, rhinoceros, elephants, camels) can travel often large distances (50 km or more) to drink, and can consume a large amount of water every few days (e.g. camels can consume about one-third of their body mass in a few minutes; Schmidt-Nielsen 1964). Various East African ungulates differ markedly in their reliance on drinking despite having similar metabolic water production and pre-formed water intake (Taylor 1968). Some mammals can drink brackish water or even seawater and still maintain water balance (e.g. tammar wallabies, *Macropus eugenii*, sea otters, kangaroo rats, seals, dolphins). The diet of insectivores and carnivores provides considerable water, as their diet is generally about 70% water content, like their own bodies. The consumption of animal material, however, brings with it a high protein nitrogen load. The excretion of that nitrogen as urea takes with it an amount of water that depends on how concentrated the mammal can make its urine. The water content of plant material can vary from quite low (5% in dried vegetation) to high (95% in succulent vegetation). Dry food can often absorb substantial amounts of water by hygroscopicity from humid air. Metabolism of food produces metabolic water (fat, 1.1 ml $g^{-1}$; carbohydrate, 0.6 ml $g^{-1}$; protein, 0.4 ml $g^{-1}$). The water gain from metabolism needs to be balanced against the associated loss of water through gas exchange, because if metabolism increases, so does ventilation of the lungs and the associated evaporative water loss.

Avenues of water loss include respiratory and cutaneous evaporation, urine, and faeces, as well as minor routes such as saliva, regurgitation, and vomiting. Insensible evaporation from the skin is an obligate source of water loss under even water-conserving conditions; the rate of evaporation depends on the capacity of the stratum corneum to reduce water flux, as well as the ambient temperature and humidity. Sensible water loss is the enhanced evaporation during heat challenge as a consequence of sweating, panting, or salivating. Ventilatory gas exchange involves an obligate water and heat loss, despite the capacity of the nasal counter-current exchanger to reduced expired air temperature below body temperature (Schmidt-Nielsen et al. 1970a; Schmid 1976). There is considerable variation amongst mammals (and birds) in their capacity to reduce expired air temperature, hence to reduce respiratory water loss, related to the role and development of nasal turbinate structures (see 1.2.4). Urine is also an obligate water loss, related to the necessity to excrete various solutes and waste products (primarily urea). There is considerable variation amongst mammals in the relative size of the renal cortex and medulla, urine-concentrating capacity, and magnitude of urinary water loss (Sperber 1944; Brownfield & Wunder 1976; Beuchat 1990, 1996). Mammals with a very moist diet (e.g. nectarivores, sanguivores) typically have kidneys with a larger cortex than medulla, and produce urine with a very low concentration, whereas many desert mammals have a large medulla and high urine osmotic concentrating capacity (see 2.7.2, 3.6.3). Faecal water loss depends on the amount of faeces produced, and the water content of the faeces (Withers 1982; Degen 1997; Bentley 2002). Faecal water content is generally about 50–55% but can be as little as about 40% of total wet faecal mass in arid-adapted mammals (e.g. the dik-dik, *Madoqua* spp.; Maloiy 1973), and can be much higher in mammals with moist food, access to drinking water, or a liquid diet (e.g. 68% for white laboratory rats; about 75% for waterbuck, *Kobus ellipsiprymnus*, and Hereford steers; 51–74% for vampire bats; 85% for fruit bats).

The maintenance of constant body water content is important for maintaining the solute concentration of body fluids as well as the hydrostatic pressures in various fluid compartments. An imbalance in the water budget (e.g. net loss of water by evaporation) decreases the extracellular fluid volume, and potentially both the blood volume and the interstitial volume. As water freely crosses cell membranes and flows passively to areas of high solute concentration, any change in extracellular solute concentration will impact on intracellular fluid volume and solute concentration. Blood pressure depends in large part on central venous pressure and venous return, because the heart cannot pump more blood than is returning to it from the venous circulation. An important determinant of cardiac output is the filling pressure of the heart, which determines the stroke volume (i.e. the Frank–Starling mechanism, see 3.4.2). In turn, the filling pressure of the heart is determined by the central venous pressure, which depends on the 'mean circulatory pressure', that is, the pressure in the vascular system when there is no flow. The major determinant of the mean circulatory pressure is the blood volume, which decreases when water balance is negative.

The maintenance of solute concentration in body fluid depends also on maintaining the balance between the intake and loss of solutes. For example, $Na^+$ influx via drinking and food must (in balance) equal loss via faeces and urine. The turnover rate of water, or solutes, reflects the continual influx via various avenues for intake and continual efflux via various avenues for loss. The measurement of turnover rates, particularly for mammals in the field, has been facilitated by the availability of radioisotopes for water (e.g. doubly labelled water) and various solutes (e.g. $Na^{22}$; see 5.7).

The allometric, physiological, and ecological determinants of water turnover rate (WTR) have been examined for a wide range of mammals in the field (Nagy & Peterson 1988; Figure 3.26). There is a significant phylogenetic pattern to field WTR of mammals (Munn et al. 2012), but, in general, the water turnover rate (ml $day^{-1}$) increases allometrically with body mass in marsupials (2.76 $M^{0.593}$) and placentals (0.23 $M^{0.885}$), reflecting a close association with metabolic rate. Various ecological factors may influence WTR, including diet and habitat (Nagy & Peterson 1988). For monotremes, WTR is in the lower range of that for other mammals for echidnas (170 ml $day^{-1}$; Schmidt et al. 2003) and the upper range

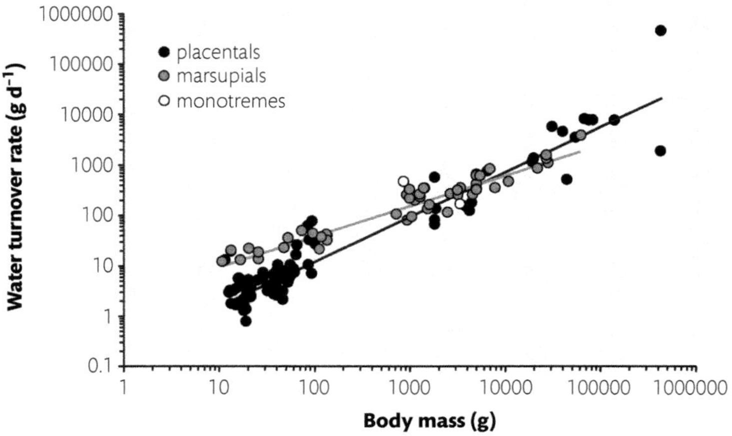


**Figure 3.26** Allometry of water turnover rate for monotreme, marsupial, and placental mammals. Data from Nagy and Peterson (1988).

for the semi-aquatic platypus (*Ornithorhynchus anatinus*, 478 ml $day^{-1}$; Hulbert & Grant 1983). For placental mammals, there is no significant effect of diet, but there is a significant effect of habitat. The residuals of WTR on body mass are lowest for species reliant on seawater (0.62 ml $day^{-1}$), intermediate for xeric (0.80 ml $day^{-1}$) and mesic species (1.31 ml $day^{-1}$), and highest for mesic herbivores (8.77 ml $day^{-1}$). For marsupials, there is a significant effect of diet, with WTR being lowest for omnivores (0.755 ml $day^{-1}$), intermediate for herbivores (0.89 ml $day^{-1}$), and highest for carnivores (1.29 ml $day^{-1}$), but there is no effect of habitat. For a combined data set of non-volant terrestrial mammals, Munn et al. (2012) found no significant difference in the scaling exponent or elevations for field WTRs of arid and mesic species, although after accounting for phylogenetic history, the allometric scaling relationships were significantly steeper for arid than mesic species.

### 3.6.2 Evaporative Water Loss

Evaporative water loss (EWL) is an obligate avenue of water loss for terrestrial mammals (except under the unlikely condition that the ambient air has the same water vapour pressure as the mammal's evaporating surface). For aquatic mammals, there is osmotic exchange of water across the skin: a loss for marine mammals and a gain for freshwater mammals.

The rate of evaporative water loss from mammals varies with body mass and $T_a$. Standard EWL (when mammals are measured at the lower critical temperature of their thermoneutral zone, where there is no enhanced EWL for thermoregulation) provides a useful comparative measure of EWL (as does BMR for metabolism).

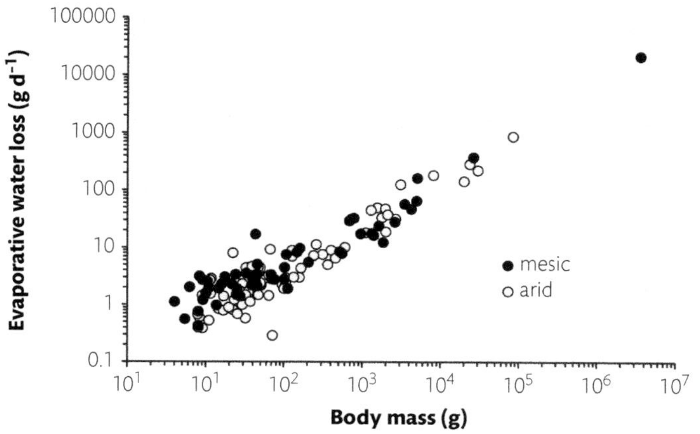


**Figure 3.27** Allometric relationship between evaporative water loss of mammals from mesic and arid environments. Data from Van Sant et al. (2012).

For marsupials, allometry of EWL ($\text{ml day}^{-1} = 0.218\,\text{M}^{0.68}$) has a similar slope as BMR (0.74; Withers et al. 2006), and M accounted for 95% of the variability in EWL, some of which was significantly related to habitat rainfall variability both before and after accounting for phylogenetic history, but not diet or habitat. An all-mammal conventional regression analysis for 136 mammalian species indicated an overall allometric relationship of $\text{EWL}(g\ day^{-1}) = 0.185\,\text{M}^{0.696}$, becoming $\text{EWL} = 0.190\,\text{M}^{0.693}$ after accounting for phylogeny (Van Sant et al. 2012). There was no EWL difference between marsupials and placentals, but species from arid zones had lower EWL than species from mesic environments: $\text{EWL} = 0.22610^{-0.143\text{H}}\,\text{M}^{0.695}$, where H is 0 for mesic species and 1 for arid species (Figure 3.27). There were also significant negative relationships with various continuous environmental variables, including a measure of environmental aridity, as well as several temperature and precipitation variables.

The rate of evaporative water loss theoretically depends on the water vapour pressure difference ($\Delta$wvp) between the evaporating surface (e.g. skin or surface of the ventilatory system) and the ambient air. The water vapour pressure at the animal surface ($\text{wvp}_{\text{surface}}$) is assumed to be the saturation wvp at the surface temperature (i.e. air at the evaporating surface is assumed to have 100% relative humidity); in reality, the relative humidity (RH) in equilibrium with physiological tissues is very slightly less than 100% (e.g. 99.5% RH at 277 mOsm osmotic concentration; Withers 1992), but this slight discrepancy is generally ignored. The ambient wvp is determined by the saturation water vapour pressure at $T_a$ and RH.

The EWL of mammals can be partitioned into two avenues, from the general skin surface (cutaneous EWL; CEWL) and from the airways (respiratory EWL; REWL). Cutaneous EWL theoretically depends on skin temperature, ambient temperature, relative humidity, and the cutaneous resistance ($R_{wvp}$; s $cm^{-1}$) to evaporation; $R_{wvp} = (\chi_{skin} - \chi_{air}) / CEWL$, where $\chi_{skin}$ is the water vapour density (g $cm^{-3}$) at the skin temperature, assuming 100% relative humidity; $\chi_{air}$ is the water vapour density (g $cm^{-3}$) at the ambient air temperature and relative humidity; and CEWL is the cutaneous component of water loss (g $cm^{-2}$ $s^{-1}$; Webster et al. 1985; Withers 1992). The resistance to evaporation is determined by a combination of the roles of the strateum corneum layer of the epidermis and the diffusive barrier role of the fur layer (Cena & Monteith 1975b). The majority of the resistance to EWL is due to the intercellular matrix of various lipids in the stratum corneum (Elias et al. 1981; Menon et al. 1986; Elias 2004; Williams et al. 2012), but the extent of keratinization of the stratum corneum varies between mammals and amongst regions of the body (e.g. it is thicker in the footpads to protect against abrasion; Feldhamer et al. 2015). A thicker epidermis presumably has a higher $R_{wvp}$. Similarly, bird feathers have little influence on $R_{wvp}$ (Webster et al. 1985; Williams et al. 2012). Whole-animal $R_{wvp}$ for mammals ranges from about 100 to 300 s $cm^{-1}$ (Withers et al. 2012). In comparison, $R_{wvp}$ is about 1 s $cm^{-1}$ for a free water surface.

Some mammals have a nictitating membrane, or 'third eyelid' (e.g. platypus, rabbits, rodents, aardvark, camels, polar bears, cats, ungulates); many do not (the plica semilunaris of humans is a vestigial nictitating membrane; Stibbe 1927–1928). The mammalian nictitating membrane, unlike the avian and reptilian membrane, lacks closure muscles and seems to function for cleaning of the cornea. For those species with a nictitating membrane, the tear film (secreted by the orbital Harderian gland) provides lubrication for its movement (Tarpley & Ridgway 1991). The tear film also reduces the rate of ocular water loss. Although it is generally thought that the eye evaporates as a free water surface (e.g. Kimball et al. 2010) and the aqueous tear film over the mammalian corneal surface has an $R_{wvp}$ of only about 1 s $cm^{-1}$, a superficial lipid layer retards EWL ($R_{wvp}$ = 12.9 s $cm^{-1}$) and the corneal epithelium itself has a relatively high resistance (82.5 s $cm^{-1}$) compared to the underlying stroma cells (1 s $cm^{-1}$; Iwata et al. 1969). The eyelashes of mammals are remarkably uniform in length, about one-third of the eye width; they not only provide physical protection against airborne materials but also reduce evaporation from the tear film by about two times (Amador et al. 2015).

Respiratory EWL is determined by the rate of air movement in and out of the respiratory system (minute ventilation, $V_I$ and $V_E$ for the volume of air inspired and expired, respectively, per minute) and the difference between the water content of the inspired and expired air (Figure 3.28). The inspired water content is determined by the ambient air temperature and humidity (as for cutaneous EWL), but expired air content is determined by the efficacy of the nasal countercurrent heat exchanger, which determines the expired air temperature ($T_{exp}$; °C). Inspired air is humidified to 100% RH, at $T_b$, primarily during passage over the high-surface-area respiratory turbinates. These nasal bony projections markedly increase

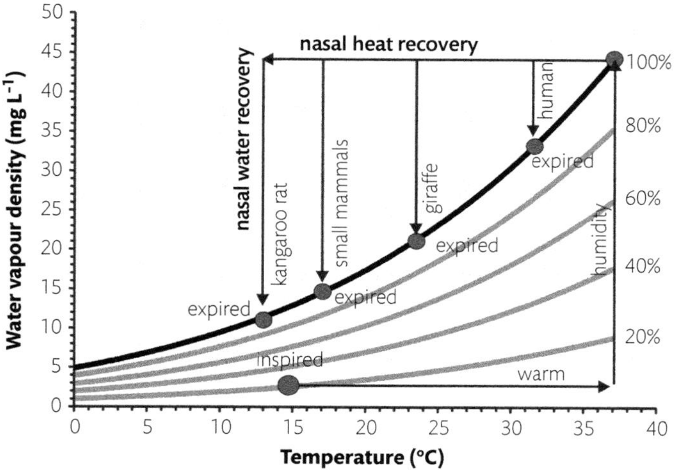


**Figure 3.28** Schematic showing water vapour density of air as a function of temperature and relative humidity (20, 40, 60, 80, and 100% saturation), calculated after Parish and Putnam (1977). Examples are given for changes in temperature and humidity of inspired air at 15°C and 20% RH, 'air-conditioned' to alveolar conditions (37°C, 100% RH), then expired at varying air temperatures, are illustrated for a kangaroo rat, typical small mammals, a giraffe, and a human. Adapted from Walker et al. 1961; Schmidt-Nielsen et al. 1970; Schmid 1976; Langman et al. 1979; Withers 1992.

the surface area of the respiratory epithelium, ensuring that the inspired air comes into close contact with a moist epithelial surface and is fully saturated and at body temperature by the time it reaches the alveoli (Walker et al. 1961; Schmidt-Nielsen et al. 1970; Langman et al. 1979). Otherwise, the alveolar epithelium would be subjected to desiccation and tissue damage. If the turbinates are not highly perfused with arterial blood, then the surface remains cool during the ventilatory cycle. On expiration, dry heat flows from the air to the turbinate surface and water condenses onto the turbinate surface, with the result that the expired air is cooled and the water vapour pressure of the expired air can be much lower than that of air in the lungs.

In general, expired air is assumed to be 100% saturated at $T_{exp}$, but some mammals expire air at RH < 100% e.g. camels (Schmidt-Nielsen et al. 1980); sheep (Johnson et al. 1988), and humans (Turner et al. 1992), as well as a bird, the ostrich, *Struthio camelus* (Withers et al. 1981). The mechanism by which they do this is unknown. The effect of lowering $T_{exp}$ below $T_b$ is to reduce REWL. The extent of water conservation depends directly on the reduction in $T_{exp}$. For example, kangaroo rats can reduce $T_{exp}$ to below $T_a$, but they still have some REWL because expired air has a higher water content (it is 100% saturated at $T_{exp}$) compared to ambient air (Schmidt-Nielsen et al. 1970; Figure 3.28). Various small mammals have $T_{exp}$

slightly higher than $T_a$ (Schmid 1976); the giraffe (*Giraffa camelopardalis*) has $T_{exp}$ higher than $T_a$ but below $T_b$ (Langman et al. 1979); and humans have $T_{exp}$ nearly equal to $T_b$.

It is thought that turbinates and endothermy co-evolved (Owerkowicz et al. 2015). For a mammal breathing dry air, about 10% of the heat generated by the ventilation of 1 extra litre of air would be lost by REWL if air were expired at lung temperature (assuming that the lungs are at 37°C where saturation wvp is 45 mg $l^{-1}$, that oxygen extraction is 30%, that metabolic heat is generated by oxidative metabolism at 20.08 kJ $l^{-1}$ of oxygen, and that the latent heat of evaporation of water is 2.4 kJ $l^{-1}$). However, turbinates play an important role in reducing the energetic costs of maintaining a high and stable body temperature, by reducing respiratory heat loss. Not surprisingly, there is a strong allometry of respiratory turbinate surface area in mammals and birds ($\alpha\ M^{0.73}$), but the elevation is about three times higher for mammals (Figure 3.29). The tracheal surface area scales similarly for mammals ($\alpha\ M^{0.76}$) and birds ($\alpha\ M^{0.70}$), but the elevation is about three times higher for birds. The greater reliance by mammals on respiratory turbinates for the conditioning of inspired air, and by birds on tracheal surface area, presumably reflects the shorter neck and larger head of mammals.

The partitioning of total evaporative water loss ($EWL_{tot}$) into its respiratory ($EWL_{resp}$) and cutaneous ($EWL_{cut}$) components varies in different species and at differing temperatures. In thermoneutrality (about 25–30°C), $EWL_{resp}$ varies from as low as 22% of $EWL_{tot}$ in the pallid bat (*Antrozous pallidus*; Chew & White 1960),

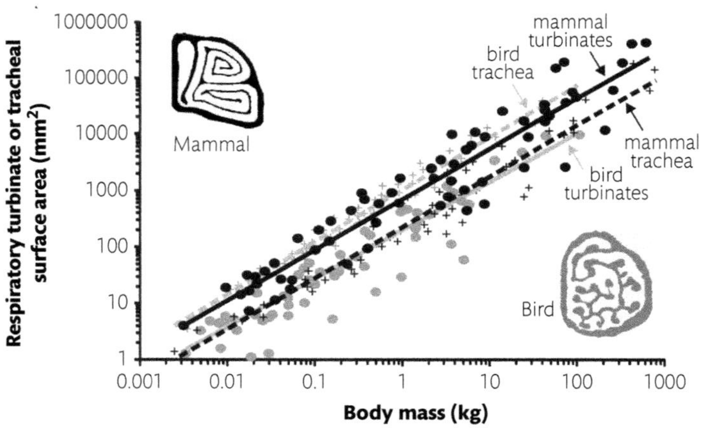


**Figure 3.29** Allometric scaling for respiratory turbinate surface area of mammals (black circles) and birds (grey circles), and tracheal surface area of mammals (black crosses) and birds (grey crosses), with corresponding regression lines. Images are cross-sectional representations of the respiratory turbinates for a mammal (star-nosed mole, *Parascalops breweri*) and a bird (emu, *Dromaius novaehollandiae*). Adapted from Owerkowicz et al. (2015).

to 38–54% in Merriam's kangaroo rat (*Dipodomys merriami*; Chew & Dammann 1961; Tracy & Walsberg 2000), 40% in the hopping mouse (Withers et al. 1979b), 59–60% in house mice and deer mice (Edwards & Haines 1978), 62% in monito del monte (Withers et al. 2012), 65% in the numbat (Cooper & Withers 2004b), to as high as 88% in deer mice (Chew 1955). These species differences could reflect differing experimental methodologies (invasive and potentially disturbed or stressed vs non-invasive and basal conditions), or species differences related to morphological or environmental selective pressures on $EWL_{resp}$ and $EWL_{cut}$ (Withers et al. 2012). For example, both the kangaroo rat and the hopping mouse inhabit arid environments, and a low $EWL_{resp}$ by enhanced countercurrent heat and water exchange might contribute to their high percentage $EWL_{cut}$. The large, naked surface area of the wing membranes of bats presumably contributes to their relatively low percentage $EWL_{resp}$ and high percentage $EWL_{cut}$.

Despite the general assumption that insensible EWL is determined primarily by biophysical properties of the animal and ambient air that drive EWL along a wvp gradient, there is experimental evidence that at least some mammals (both marsupial and placental) can modify their EWL from that expected from biophysical models at ambient temperatures below thermoneutrality (where there is no enhanced EWL for cooling and EWL is generally assumed to be passive). Brushtail possums (*Trichosurus vulpecula*) maintained a constant EWL at relative humidities $\leq 60°C$ at $T_a = 25°C$ (Cooper & Withers 2008), as did the little red kaluta at $T_a = 25$, 30, and 35°C (Withers & Cooper 2014). Perturbing the evaporative environment of the rodent *Pseudomys albocinereus* using helox (21% $O_2$ in helium) also failed to impact on rates of EWL (Cooper & Withers 2014a). These results suggest that mammals may have a previously unappreciated ability to acutely regulate $EWL_{tot}$ under perturbing environmental conditions. Such an ability is likely more widespread than these three species. Re-evaluation of published EWL data for a number of mammals and birds suggested that EWL is not as dependent on wvp deficit as previously believed, and that regulation of insensible EWL may be widespread amongst endotherms (Withers & Cooper 2014). The mechanism(s) of insensible EWL regulation are not yet understood, but for *P. albocinereus*, changes in cutaneous EWL most likely contribute to the observed EWL constancy in a helox environment (Cooper & Withers 2014a). The consequences of regulation of insensible EWL may be water conservation, or may be thermoregulatory; maintaining a constant EWL means that $T_b$, MR, and/or dry thermal conductance are not impacted by variation in humidity or other factors impacting on EWL.

### 3.6.3 The Kidney

For mammals, excretion plays a role in the regulation of both water and solute balance, and the kidney is the primary organ responsible for excretion. Mammals do not have significant extra-renal mechanisms for the excretion of either water or solutes (e.g. they lack salt glands; cf. some reptiles and birds; Bicudo et al. 2010), and their other secretory organs have different roles from water or solute regulation per se (e.g. sweat glands are involved in thermoregulation in the heat in many species, although they can compromise both water and solute balance under extreme thermal challenge). The gut has a significant effect on water balance through its control of faecal water loss and the retention of water during hygric challenge, but this is technically not excretion but elimination. So, the only significant excretory organ of mammals is the kidney.

The mammalian kidney is a paired organ, with each kidney located outside the abdominal cavity/peritoneum, on the dorsal body wall. Each kidney consists of an outer cortex, an inner medulla, and a central renal pelvis where urine drains into the ureter (Figure 3.30A), and then the bladder, where urine is stored prior to urination. The cortex is essentially the typical vertebrate 'kidney', responsible for water and solute balance, and nephrons there cannot produce a urine more concentrated than the blood. The medulla, however, is responsible for urine concentration, and nephrons there can produce a urine more concentrated than the blood. Mammals, and to a lesser extent birds, are the only vertebrates that have this urine-concentrating capacity (Withers 1992).

The functional unit of the kidney is the nephron, or excretory tubule, of which there are many thousands in each kidney. Each mammalian nephron is associated with a glomerulus, a knot of capillaries with afferent and efferent arterioles, where plasma is filtered from the blood. The glomerulus is surrounded by Bowman's capsule, a double-walled invagination of the nephron tubule that collects the fluid

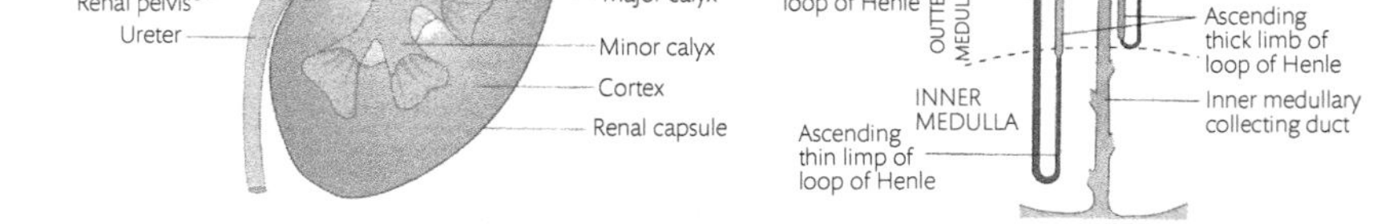

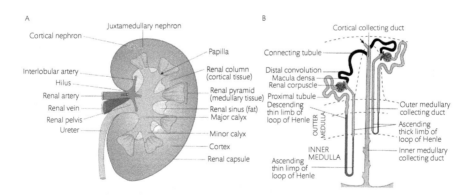


**Figure 3.30** A. Schematic structure of a kidney, showing the primary anatomical features: outer cortex, medullary pyramids, renal pelvis, and arrangement of cortical and juxtamedullary nephrons. B. Schematic of cortical (short-looped; top right) and juxtamedullary (long-looped; central left) nephrons, showing the arrangement of their loops of Henle. From Pocock et al. (2013). Reproduced with permission of Oxford University Press.

filtered from the glomerular capillaries. As such, urine begins as a filtrate of blood, lacking cells and large proteins. The proximal convoluted tubule extends from Bowman's capsule to the loop of Henle—a long, U-shaped loop with a thin descending limb and then an ascending limb with thin then thick sections. From the thick ascending limb, the fluid passes into the distal convoluted tubule, and finally a collecting duct that drains into a renal pelvis and then via the ureter to the bladder (Withers 1992; Guyton & Hall 1996; Sherwood et al. 2005).

The glomerulo-tubular functions of the mammalian kidney include filtration, reabsorption, and secretion, as well as osmo-concentration. The vascular system of the kidney consists of a renal artery supply and venous drainage; there is no renal portal system (unlike in reptiles and birds). The afferent and efferent arterioles carry blood to and from the glomerulus, respectively. The distal convoluted tubule passes through and contacts the fork that is formed by the arterioles. Here, the convoluted tubule and the afferent arteriole form the juxtaglomerular apparatus, which has an important role in regulating the flow of blood in the afferent arteriole. Peritubular capillaries arising from the efferent arterioles surround the length of the nephron tubules and form the vasa recta, made up of hairpin capillary loops that are intimately associated with the loop of Henle. Bird kidneys have a similar arrangement of renal tubules that form a cortex and glomerulus, but the gross structure of the kidneys is highly lobulated, somewhat like the lobulated kidneys of large mammals (Braun & Dantzler 1972).

There are two discrete types of mammalian nephrons. Cortical nephrons have short loops of Henle and are located entirely within the cortical region (except for their collecting ducts; Figure 3.30B). Juxtamedullary nephrons have long loops of Henle, and while their glomerulus, Bowman's capsule, and proximal and distal convoluted tubules are also in the cortex, their loops of Henle extend deep into the medulla. Cortical nephrons carry out functions of filtration, reabsorption, and secretion, but it is the juxtamedullary nephrons that are involved in the countercurrent multiplication system that enables the mammalian kidney to concentrate urine.

The process of urine formation begins with ultrafiltration in the glomerulus, where arterial blood pressure causes fluid to move through the semipermeable membrane that forms the walls of the narrow capillaries. Glomerular filtration rates for mammals are dependent on body mass (Calder & Braun 1983). They range from 10 to 600 ml $kg^{-1}$ $hr^{-1}$ (Yokota et al. 1985) and are most likely determined by metabolic rate (Singer 2001).

The ultrafiltrate collects in Bowman's capsule, and flows into the proximal convoluted tubule, where the reabsorption of water and useful solutes such as glucose, amino acids, and lipids returns them to the venous system via the peritubular capillaries. Typically, more than 95% of the water and total osmolytes that are filtered are reabsorbed. Tubular secretion then provides for the selective excretion of substances such as some organic ions from the peritubular capillary blood into the tubular lumen. The filtrate then flows into the thin descending limb of the loop of Henle. The thin descending limb is permeable to water but relatively impermeable to ions, so water moves out into the interstitial space by osmosis (due to the concentration gradient established in the kidney interstitium by the ascending limb of the loop of Henle; see later), resulting in an increasingly concentrated filtrate as it flows down the descending limb.

In the thick ascending limb of the loop of Henle, which is in the outer medulla, a cotransporter (NKCC2) in the wall of the tubule actively transports $Na^+$, $Cl^-$, and $K^+$ out of the tubule, into the kidney interstitium (Mount 2014). The mechanism of the so-called loop diuretics, like furosemide, is inhibition of this transporter. It is this removal of ions from the increasingly dilute fluid ascending the thick limb that generates a concentration gradient in the outer medullary interstitium. The dilute tubular fluid then passes through the distal convoluted tubule (where further reabsorption occurs) before descending through the collecting duct, which is relatively impermeable to $Na^+$ and $Cl^-$.

The interstitial fluid outside the duct can become successively more concentrated as it flows towards the papillary tip through the interstitial concentration gradient previously established by movement of solutes out of the filtrate in the loop of Henle. The concentrated interstitial fluid creates an osmotic gradient between the interstitial fluid and the filtrate passing down the collecting duct. When water excretion is required, the collecting duct becomes relatively impermeable to water and a dilute urine exits the collecting duct, but when water retention is required, the collecting duct becomes permeable and water flows down its concentration gradient from the duct to the interstitium.

The permeability of the collecting duct depends on the presence of aquaporin, a water channel that allows the movement of water but not ions (Engel et al. 2000). The number of aquaporin water channels in the collecting duct is increased by AVP (also called ADH), which is released from the pituitary in response to an increase in the osmolarity of body fluid. When ADH concentration increases, more aquaporin water channels are placed in the walls of the collecting duct, water is removed from the filtrate, and a more concentrated urine is produced. The maximal urine concentration that can be achieved is isosmotic with the highest papillary interstitial concentration. Urinary fluid can be made even more dilute than the distal tubular concentration by further reabsorption of solutes during passage along the collecting ducts (e.g. nectar feeders that produce very dilute urine; Herrera et al. 2015).

The mammalian kidney concentrates urine by countercurrent multiplication, reflecting the countercurrent arrangement of the descending and ascending segments of the loop of Henle in the renal medulla, where fluid flows through the descending and ascending limbs in opposite directions. This countercurrent arrangement produces a multiplication of the slight osmotic gradient that can be formed between the lumen and the outside of the tubule across any section of the loop of Henle (200 mOsm $l^{-1}$ at most), along its entire length. This results in a considerable osmotic gradient along the loop of Henle, and potentially a considerable osmotic concentration of the urine as tubular fluid passes along the collecting duct, through the renal medullary osmotic gradient. This passive equilibrium hypothesis originally proposed by Kokko and Rector (1972) and Stephenson (1972) remains the most widely accepted hypothesis to explain the maintenance of an osmotic concentration gradient in the inner medulla, despite subsequent advances in molecular explanations for urea transport (Sands 2007).

In addition to the passive countercurrent hypothesis of urine concentration, the Berliner hypothesis (Berliner et al. 1958) provides an explanation as to why increasing urea concentrations along the length of the collecting duct do not result in urea-induced osmotic diuresis. Urea becomes increasingly concentrated down the length of the collecting duct as water is osmotically withdrawn to the interstitial fluid. The high permeability of the terminal region of the collecting duct to urea, especially in the presence of ADH, means that urea accumulates to very high levels in the interstitial fluid of the inner medulla. The countercurrent arrangement of the vasa recta maintains this high interstitial urea concentration, keeping it almost isosmotic with the collecting duct fluid and preventing osmotic movement of water into the collecting duct. The Berliner hypothesis also accounts for the observation by Gamble et al. (1934) that there is 'an economy of water in renal function referable to urea' (Berliner 1976, p. 216). The water required for rats on a low protein diet to excrete various solutes was always additive except if urea was one of the solutes, in which case the water required was only that needed for the other solutes; the urea could be excreted without any additional water. This is due to the urea accumulating in the interstitial fluid of the inner medulla, which osmotically balances the fluid in the collecting duct, so that the salts accumulated in the medullary tissues need only balance the non-urea solutes of the filtrate (Berliner 1976).

Subsequent observations of renal function related to urea provide support for Gamble (1934) and Berliner (1976), but not the passive countercurrent hypothesis, although Sands (2007) noted that their hypotheses are not mutually exclusive, and multiple mechanisms may contribute to urinary function. The osmolarity of the inner medulla continues to increase towards the papilla tip, despite an absence of ion pumps in the thin ascending limb of the loop of Henle. The mechanism for this osmotic increase along the inner medulla is not well understood and remains controversial (Gottschalk 1987; Sands 2007). The ability to concentrate urine is compromised in mammals deprived of protein, and is restored following urea infusion (Sands 2007), suggesting that urea is critically important to the formation of the inner medullary concentration gradient.

Two urea transport genes have been identified in the renal medulla tissues: UT-A is expressed in the inner medullary collecting duct cells and in the lower portion of the descending limbs of both short and long loops of Henle under chronic antidiuretic conditions, and UT-B is found in the descending vasa recta endothelial cells

(Fenton & Knepper 2007). In murine models that have had these urea transporters genetically 'knocked out', the urine concentrating ability is inhibited. Inner medullary urea content was reduced after water restriction in knockout mice, and there was no measurable difference in inner-medullary NaCl between water-restricted 'knockout' and wild-type mice (Fenton et al. 2004, 2005), inconsistent with the passive countercurrent model but supporting Berliner's and Gamble's hypotheses. However, Sands (2007) recognized that methodological issues (measurement of whole papilla) may have hampered identification of the expected NaCl gradients in the papilla tip and masked support for the passive hypothesis. Currently, the details of the mechanism by which urine is concentrated in the inner medulla remain controversial, and although the passive multiplication hypothesis remains the most widely accepted model for urine concentration in the mammalian kidney, there is also support for the contribution of urea transport, and this should be included in future investigation of kidney function (Sands 2007; Sands et al. 2011)

The length of the loops of Henle in the medulla determines the concentrating ability of the mammalian kidney. Therefore, kidneys with a relatively large medulla are capable of producing more concentrated urine than kidneys with a small medulla (Sperber 1944; Schmidt-Nielsen & O'Dell 1961; Heisinger & Breitenbach 1969; Brownfield & Wunder 1976; Beuchat 1990; Figure 3.31). Two renal indices that are often used to anatomically reflect the maximal urine-concentrating capacity are relative medullary thickness (RMT) and relative medullary area (RMA). Body mass is also a major determinant of urine concentrating ability.

Although large mammals have longer loops of Henle than smaller mammals, urine concentrating ability is negatively related to body mass, presumably reflecting a higher mass-specific metabolic rate in small species, and thus higher

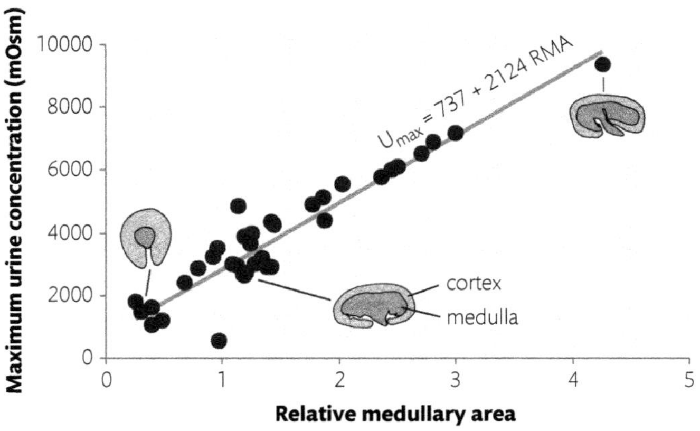


**Figure 3.31** Relationship between relative medullary area and maximum urine-concentrating capacity for mammals. Modified from Withers (1992).

metabolic activity within the kidney, leading to enhanced active transport of the ions necessary for the countercurrent multiplier system (Greenwald & Stetson 1988; Beuchat 1990, 1996). When the effect of body mass is controlled for, kidney morphology, and thus urine concentrating ability, is related to habitat and diet, and reflects adaptive selection for urinary water conservation in species with limited water availability. For example, rodents from arid habitats have relatively larger kidneys with a greater relative medullary thickness than rodents from other habitats, independent of phylogenetic history and body mass (Al-kahtani et al. 2004). Bozinovic et al. (1995) reviewed studies of renal performance and maximal urine osmolality amongst rodent species inhabiting semi-xeric and xeric regions of South America, the southwestern United States, and Australia. The mean value for maximal urine osmolality in the species inhabiting semi-xeric habitats in South America was 4,138 mOsm $kg^{-1}$, similar to that found for nine western North American rodent species (3,930 mOsm $kg^{-1}$). Rodents lacking an inner medulla, like beavers that inhabit rivers and ponds, are unable to form hypertonic urine (Kriz & Kaissling 2000).

Indices examining relative medullary size do not take into account the architectural organization of tubules and blood vessels inside the medulla. For example, the kidneys of cats and dogs have been described as having long looped nephrons, yet they have only an average capacity for urine concentration. Therefore, medullary thickness and the number of long looped nephrons are not always related to the generation of highly hypertonic urine. Variation in the organization of the tubules and vascular bundles inside the renal medulla, especially within the inner stripe of the outer medulla, may also impact urinary concentrating ability. The organization can be simple or complex. In lagomorphs, the structure is simple: the medullary rays run into the interbundle regions and receive a blood supply from a capillary plexus; therefore the tubular and vascular structures are physically separated. However, in a more complex medulla, the thin descending limbs of short-looped nephrons penetrate, in variable degrees, the vascular bundles (Kriz 1981). Therefore, the vascular bundles composed of ascending and descending vasa recta are now formed by the blood vessels and thin descending limbs of short-looped nephrons. The numerical relationship between these elements is as follows: one ascending vasa recta is packed with one descending vasa recta and one short descending loop. Therefore, the typical countercurrent arrangement between ascending and descending vasa recta is replaced by complex bundles, or by giant bundles, in the inner stripe of outer medulla. From a physiological standpoint, the complex bundles may be a site of exchange of solutes from deeper parts of the medulla, in the ascending vasa recta capillaries, towards the short descending limbs present in the complex bundles; this mechanism would add efficiency to the recirculation of solutes that are important for medullary hyper-concentration, such as urea.

### 3.6.3.1 Vasopressin

The concentration of urine can vary from more dilute than plasma to many times higher than the plasma concentration. The concentration of urine is related to the circulating concentration of antidiuretic hormone (ADH). Argenine vasotocin (AVT) is the primary antidiuretic hormone of most vertebrates, but placental mammals and probably monotremes have substituted the chemically similar arginine vasopressin (AVP), while marsupials have a variety of ADHs, including lysopressin and phenypressin (Cooper 2016). Whatever its form, ADH is the most important hormone in the regulation of water balance in mammals. This small peptide is secreted by the neurohypophysis in response to increases in plasma osmolality as small as 2%. The effects of ADH on the renal tubule are mediated by the binding of ADH to $V_2$ type basolateral receptors (Birnbaumer et al. 1992). The functional hydro-osmotic outcomes of the hormone are mediated by an increase in the osmotic water permeability of connecting cells, principal cells, and inner medullary collecting duct cells, via the activity of AQP channels (see later).

Studies of isolated and microperfused inner medullary collecting tubules from rat kidney showed that AVP induces a six-fold increase in osmotic water permeability (Nielsen et al. 1993), while the stimulation of urea and NaCl reabsorption also contributes to the medullary hyper-concentration that constitutes the osmotic gradient for AVP-dependent water reabsorption. In inner medullary collecting duct cells, where urea reabsorption is mediated by the facilitated urea transporter (UT1), AVP stimulates urea reabsorption by phosphorylation of the UT1 transporter, increasing its activity and therefore the rate of urea reabsorption (Sands 1999), as well as by increasing apical plasma membrane accumulation of UT1 (Sands et al. 2011). The reabsorption of NaCl is also stimulated by ADH in the thick ascending limb of Henle. AVP increases the activity and density of the apical $Na^+$-$K^+$-$2Cl^-$ cotransporter (NKCC2; Knepper et al. 1999; Kim et al. 1999); this action has been demonstrated in mouse, but not rat, thick ascending limb; there are no studies of this AVP action in desert-dwelling rodents.

### 3.6.3.2 Aquaporins

Aquaporins (AQPs) are integral membrane proteins that facilitate water transport across the plasma membrane in many cells (Agre et al. 2002). The AQP family is composed of ten members; some isoforms are strict water transporters (aquaporins), while others allow the passage of water as well as small organic molecules (aquaglyceroporins). Prior to the discovery of aquaporin, it was assumed that water simply leaked through the lipid bilayer of cells, but some cells permit rapid water transport, too rapid to be accounted for by simple diffusion. Peter Agre (who won the Nobel Prize for chemistry in 2003 for the discovery of aquaporins) and co-workers were studying the Rh-antigens on red blood cells, and a 28 kD protein kept showing up in the erythrocyte membrane. The protein was expressed in many tissues, and was abundant in erythrocytes and kidney tubules. The molecule was identified as aquaporin and its role in water transport was later established. Studies carried out in aquaporin-null transgenic mice showed that several AQPs play a physiological role in transepithelial water transport; in most epithelia, they are the molecular entities that facilitate water transport (Verkman et al. 1996).

At least seven aquaporins (e.g. AQP-1, AQP-2, etc.) are found within the kidney (Nielsen et al. 2002). AQP-1 is plentiful in the apical and basolateral membranes of proximal tubules and the thin descending loops of Henle. In AQP-1-null laboratory mice, the osmotic permeability of the proximal tubules and thin descending loops of Henle is drastically reduced; these mice are unable to concentrate urine at more than 700 mOsmol $kg^{-1}$ and are polyuric. Therefore, AQP-1 plays an essential role in near iso-osmotic fluid reabsorption and in the countercurrent mechanism allowing the tubular fluid to become hypertonic as it flows through thin descending limbs (Ma et al. 1998). AQP-2 occurs exclusively in the connecting tubule and principal cells of the collecting duct, and is the main AQP on which AVP acts to allow for renal regulation of body water balance. AQP-3 and AQP-4 are also located in the principal cells of the collecting duct, and presumably function to allow water that enters via AQP-2 to exit. AQP-6 is a unique intracellular AQP that is not associated with a cell plasma membrane; its function is unknown, although it may play a role in vesicle acidification. AQP-7 occurs in the proximal tubule's brush border, and also has an unknown physiological function. AQP-8 through AQP-12 occurs at low abundance within the kidney, also with unknown function (Nielsen et al. 2002).

There is little information concerning the role and regulation of renal aquaporins in wild mammals (Bozinovic & Gallardo 2006). For instance, in the kangaroo rat, the complete absence of AQP-4 mRNA and protein in medullary collecting ducts has been reported. Field studies of a South American rodent (*Octodon degus*) during summer and winter showed that it could maintain water balance in both seasons, even though rainfall was near zero during the summer. In summer, the urine osmolality was 3,200 mOsm kg $H_2O^{-1}$, compared to 1,123 mOsm kg $H_2O^{-1}$ in winter. Immunocytochemistry for AQP-2 revealed a marked redistribution of the protein from the cytosol to the apical membrane during summer, and a lower amount of the protein in winter (Bozinovic et al. 2003; Gallardo et al. 2005). No change in immunoreactivity was observed for AQP-1. Thus, the increase in urine osmolality was sustained in part by an increase in water reabsorption in the collecting duct, which was related to the observed changes in AQP-2. The regulation of AQP-2, and possibly of other transporters, adds a great degree of plasticity to the kidney and water balance.

Gallardo et al. (2005) examined the phenotypic flexibility of urine osmolality in response to seasonal rainfall, and the experimental expression of renal AQPs, in the leaf-eared mouse (*Phyllotis darwini*), a South American desert-dwelling rodent, in the field. Urine osmolality was higher in summer than winter. During a rainy year, the urine osmolality in winter was nearly 2,100 kg $H_2O^{-1}$, while in a dry year the value was 2,600 mOsm $kg^{-1}$. Correspondingly, during the rainy year, summer mean field urine osmolality was 3,321 mOsm kg $H_2O^{-1}$, while during the dry year the value was 3,600 mOsm kg $H_2O^{-1}$. Further, dehydration induced an increase in AQP-2 protein compared to the control and water-loaded condition, demonstrating that even though some of the flexibility was expressed to deal with the relative water restrictions in the field, there remained capacity that was expressed when the animals became dehydrated.

## 3.6.4 Nitrogenous Wastes

The metabolism of nitrogen-containing organic compounds (e.g. protein, nucleic acids) and the scavenging of their carbon skeletons for metabolic energy require excretion of the remaining N to prevent the accumulation of potentially toxic nitrogenous wastes (e.g. $NH_4^+$). The excretion of nitrogenous wastes is an important role of the mammalian kidney. Other components of organic compounds also need to be excreted to a lesser extent (e.g. protein sulphur is excreted as $SO_4^{2-}$), but this is a relatively minor role of excretion.

The initial N-containing waste product from protein metabolism is ammonia (Figure 3.32), but for terrestrial amniotes its high toxicity precludes its use as the nitrogenous waste product. Generally, in mammals the ammonia is converted to urea, which is less toxic but also less soluble. In contrast, for terrestrial reptiles and birds the general, N waste product is uric acid. Metabolism of the purine and pyrimidine units of nucleic acids (DNA, RNA) involves conversion to xanthine, followed by a degradation pathway that starts with uric acid. For most reptiles and birds, that pathway ends there at uric acid. For some mammals, that pathway also ends there, but some continue and form various other nitrogenous waste products. Great apes (including humans) and Dalmatian dogs stop at uric acid, but most mammals continue to allantoin; various fishes and marine invertebrates continue to allantoic acid, urea, or ultimately ammonia. In a few specialized mammals, ammonia ($NH_3$) or $NH_4^+$ plays an excretory role in particular dietary or environmental circumstances (see 3.6.4.1).

### 3.6.4.1 Ammonia

Ammonia is an unsuitable waste product for amino-acid nitrogen in mammals because it is extremely soluble in water and because it is a weak base, reacting with water to form ammonium ions ($NH_4^+$). At physiological pH, the equilibrium greatly favours ammonium (Withers 1992, 1998), so it is extremely difficult to excrete ammonia gas across the respiratory or cutaneous surface. Even at high $NH_4^+$ concentrations, the $pNH_3$ is quite low and so is the driving force for its excretion (Table 3.10). Because ammonia/ammonium is so toxic for most animals, including mammals ($LD_{50}$ is about 5 mM of ammonium salts for mammals; Withers 1998a), it isn't possible to achieve a sufficiently high $pNH_3$ to excrete a significant amount of

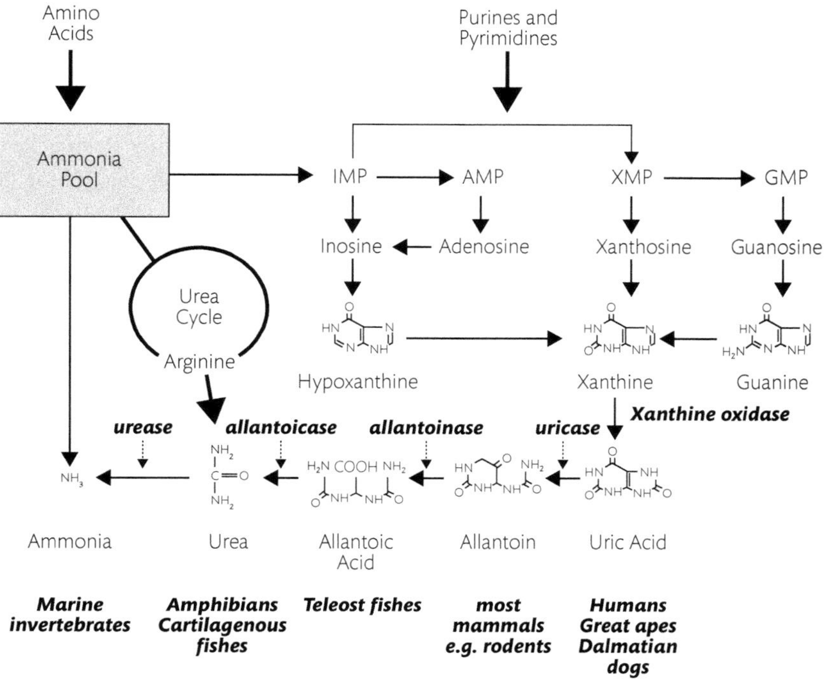


**Figure 3.32** Nitrogenous waste pathways in various vertebrate animals for N degradation from protein (amino acids) and purines (IMP, AMP, XMP, GMP) N degradation. Modified from Balinsky (1972).

**Table 3.10** Relationship between ammonia partial pressure in air and the ammonia and ammonia + ammonium concentration dissolved in water. From Withers (1998a).

| $pNH_3$ (kPa) | $[NH_3]$ (mM) | $[NH_3 + NH_4^+]$ (mM) |
|---|---|---|
| 0.006 | 2 | 89 |
| 0.017 | 5 | 223 |
| 0.033 | 10 | 447 |
| 0.066 | 20 | 893 |
| 0.165 | 50 | 2234 |

$NH_3$, so it must be converted to a less toxic waste product. Mammals are also very sensitive to inspired ammonia; 500 min of exposure to 500 ppm $NH_3$ is lethal to humans, and 40 min is lethal for laboratory rats at 5,000 ppm (Studier et al. 1967).

However, a few mammals such as the Brazilian free-tail bat (*Tadarida brasiliensis*), inhabit environments containing up to about 2,000 ppm $NH_3$

(0.2 kPa partial pressure) emanating from guano decomposition, and are only affected by 3,000 or more ppm (Campbell 1991). They can tolerate breathing 5,000 ppm $NH_3$ for over 5,760 min (4 days) and high blood ammonium concentrations of about 40 mM (Studier 1969; Studier et al. 1967). About half of the inspired $NH_3$ is eliminated via respiratory mucus. Bats apparently retain high levels of $CO_2$, which neutralizes the blood $NH_3$, and the $NH_3$ and $CO_2$ are dumped when the bats return to normal atmospheric air. The nectarivorous bat *Glossophaga soricina* has a diet naturally low in protein, and not surprisingly the bat has a low-maintenance N requirement (Herrera et al. 2006). Although most individuals remain ureotelic when exposed to a low N diet, some become mixed ammono-ureotelic and some become ammonotelic; urea might be recycled in the gut. Gut recycling of urea N is common in many herbivorous and some omnivorous mammals (Gibson & Hume 2002; Singer 2002).

### 3.6.4.2 Urea

Most N derived from amino-acid metabolism is converted by mammals to urea in their liver; urea is a less toxic (but also less soluble) nitrogenous waste product than ammonia. This conversion is accomplished by the urea cycle, a biochemical cycle that essentially combines two $NH_4^+$ with a $CO_2$ to form urea;

$$2NH_3 + CO_2 \rightarrow CO(NH_2)_2 + H_2O$$

(Withers 1992, 1998a; Figure 3.32). The first step in the urea cycle, where $CO_2$ is condensed with $NH_4^+$ to form carbamyl phosphate, is irreversible, because it requires 2 ATP, and the concentration of the catalyzing enzyme, carbamoyl phosphate synthetase, is at a high enough concentration in the mitochondrial matrix to maintain a low $NH_3$ concentration (McGilvery & Goldstein 1983). Urea synthesized in the liver is transported to the kidney, where it is filtered and excreted in the urine at often high concentration.

However, urea is not simply a nitrogenous waste product; it actually has a number of potentially useful roles in various animals, including mammals (Withers 1998a). One of these is establishment of the renal osmotic gradient for urine concentration in the renal medulla. This gradient is established at least in part by a substantial accumulation of urea, in the inner renal medulla, especially (see 3.6.3). The urea is accompanied by a suite of organic solutes, such as betaine and glycerophosphorylcholine, that have a counteracting role (to alleviate the inimical effects of urea on protein structure and function), and also polyols, such as inositol and sorbitol (Yancey 1988). Urea also provides a non-toxic mechanism for ruminant and pseudoruminant mammals to transport N to their symbiotic gut microorganisms. Urea diffuses from the blood supply to the stomach, and is present in saliva that is swallowed into the stomach, where it is metabolized by microorganismal flora and fauna into protein (Nolan 1993).

### 3.6.4.3 Purines

The metabolism of DNA and RNA requires the excretion of their N purine and pyrimidine subunits (Figure 3.32). When pyrimidine nucleotides (thymine, cytosine, uracil) are degraded, the N is converted directly to ammonia, which is converted to urea via the urea cycle (as per protein N degradation). When purine nucleosides (adenine, guanine) are degraded, the N is transferred (through IMP, AMP, XMP, GMP) to xanthine, which is then further degraded through a series of intermediary nitrogenous waste compounds. Marine invertebrates excrete ammonia, which is the last nitrogenous waste compound in this degradation sequence, and it is readily excreted into an aquatic environment. However, the hydrolysis of urea to $CO_2$ and ammonia requires the enzyme urease, and that enzyme was lost in the invertebrate–vertebrate transition, making urea the last possible stage in purine degradation in vertebrates (e.g. cartilaginous fishes and amphibians; Campbell 1991). The enzymes allantoinase and allantoicase were deleted in the amphibian–cotylosaur transition, meaning that mammals can only degrade purines to allantoin (formed from uric acid by uricase activity). Most mammals do indeed excrete allantoin as the purine N waste product. For example, many arid-adapted cricetid rodents excrete sufficiently high concentrations of allantoin in the urine that it crystallizes out of solution, providing a urinary water savings (Buffenstein et al. 1985).

Uricase activity was lost in the reptilian lineage leading to crocodilians and birds at the transcriptional level, not the gene level; this is unsurprising considering that this lineage converts urea to uric acid for the excretion of metabolized nitrogen (Keebaugh & Thomas 2010). Uricase has also been lost in great apes, which consequently excrete uric acid. This loss of uricase activity is related to inactivating mutations of the gene (i.e. it is a pseudogene that undergoes neutral mutation) and the deterioration of immediate downstream genes. This does not seem to be the case for reptiles and birds, in which the uricase gene does not have inactivating mutations; rather it seems to still be encoding a protein that has lost its purine catalytic capacity. In reptiles and birds, the advantage of loss of uricase activity is related to oviparity and water conservation. Although the loss of uricase activity in great apes could be interpreted as an 'evolutionary accident' with unfortunate side effects (e.g. gout in humans), there are possible evolutionary advantages, including increased cognition, an effective antioxidant role, and better regulation of blood pressure (Keebaugh & Thomas 2010). Dalmatian dogs also excrete uric acid even though they have functional uricase, so other biochemical metabolic disorders must be responsible for this.

## 3.7 Neurobiology

The central nervous system (CNS) provides an interface between various modes of sensory information and the control of several motor systems. It receives an enormous amount of sensory information and processes and integrates that information before providing motor control of peripheral organ systems, including skeletal, cardiac and smooth muscle, glands, and neurosecretions.

### 3.7.1 Central Nervous System

The basic structures and functions of the mammalian CNS (Figure 3.33) are quite conservative. The primary sensory inputs and motor outputs are from the peripheral nervous system via the cranial nerves and spinal cord (Kardong 2009). The sensory cranial nerves are as follows: I olfactory (smell); II optic (vision); V trigeminal (facial touch, temperature, pain); VII facial (taste); VIII vestibulocochlear (hearing and equilibrium); IX glossopharyngeal (taste, touch, pressure, pain, temperature, blood pressure, respiration); X vagus (taste, hunger, satiation, gastrointestinal comfort). Various spinal cord nerve tracts have sensory functions, including peripheral sensations, pain, and proprioceptive feedback. The motor cranial nerves are as follows: III oculomotor (eye movements, pupil, focussing); IV trochlear (eye movements); V trigeminal (mastication); VI abducens (eye movement); VII facial (facial expression, tears, salivation, oro-nasal mucosa); IX glossopharyngeal (salivation, swallowing, gagging); X vagus (swallowing, speech, heart rate, bronchoconstriction, gastrointestinal secretion and motility); XI accessory (swallowing; head, neck, and shoulder movement); XII hypoglossal (tongue, food manipulation, swallowing). Various spinal cord nerve tracts have motor functions, including balance and posture, control of head and limb positions, reflexive motor movements, and fine control of proprioceptive limb positioning.

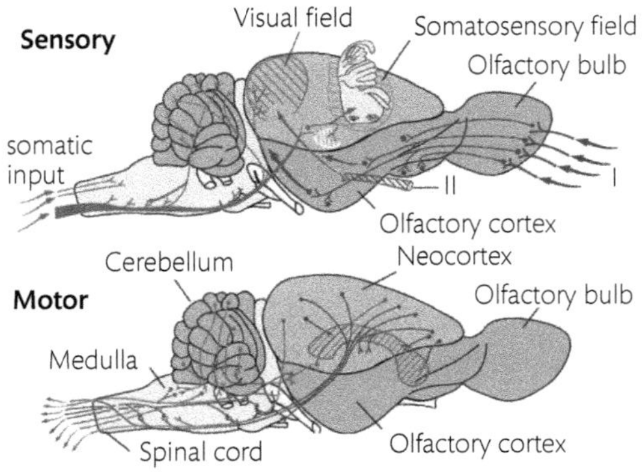


**Figure 3.33** Schematic of the main sensory and motor structures and circuitry of the mammalian (*Didelphis*) brain; Sensory: olfactory nerve (I) input to the olfactory bulb, thence to midbrain; optic nerve (II) input to the visual field; trigeminal (VII) taste sense, and somatic input (spinal cord) to the cerebellum and somatosensory cortex. Motor: motor output to the body via the cranial and spinal cord motor nerves. Modified from Rowe et al. (2011), after Nieuwenhuys et al. (1998). Reproduced with permission from Science (AAAS).

The brain is divided into three major sections: the brainstem, the cerebellum, and the cerebrum (Kardong 2009; Pocock et al. 2013). The posterior brainstem differentiates into the myencephalon (medulla oblongata) and the metencephalon (pons and cerebellum). The medulla has ascending and descending pathways, and is involved with basic sensory and motor functions (sense of touch, pressure, temperature, taste, pain; chewing, swallowing, gagging, vomiting, respiration, vocalization, sneezing, sweating, cardiovascular control, and gastrointestinal control, as well as head, neck, and shoulder movements). The pons has sensory and motor cranial nerve functions, and is involved with respiration, posture, and sleep. The cerebellum is a major centre that serves sensory interpretation and motor coordination. The midbrain has many roles, including visual functions such as eye-tracking, eye reflexes, and eye movements, equilibrium, and balance.

The forebrain consists of the diencephalon (epithalamus, thalamus, and hypothalamus) and the telencephalon (cerebrum). The thalamus is the 'sensory gateway' to the cerebrum; the hypothalamus is a major control centre for the autonomic nervous system and endocrine systems, as well as control of thermoregulation, food and water intake, sleep and circadian rhythms, memory, and emotions. The cerebrum includes the basal nuclei, limbic system, and cerebral cortex. The basal nuclei receive input from, and return output to, the midbrain and cerebral cortex, for motor control. The limbic system (cingulate gyrus, hippocampus, amygdala) has a role in smell, but is a major centre for emotion (pleasure, gratification, aversion, fear, sorrow) and learning. The cerebral cortex is a highly folded sheet of neural tissue responsible for complex integration via association fibres of sensory information into motor actions (via sensory projection fibres).

The monotreme brain is typically mammalian (Griffiths 1989a). The cerebellum is large and divided/folded. The midbrain roof has four protrusions (corpora quadrigemina), in contrast to the two of sauropsids. The very large trigeminal nerve, a monotreme feature, is largely sensory, relaying tactile information to the cortex via the thalamus. The forebrain has large olfactory lobes, a pyriform cortex, a hippocampal cortex, and a neocortex. The neocortex lacks the left-right connecting corpus callosum (like marsupials, see later, and unlike placentals). The echidna has a very folded cortex, unlike the platypus and most marsupials. As in other mammals, the neocortex has sensory and motor areas; the somatosensory, visual, and auditory cortical areas are caudal to the motor cortex. The sensorimotor cortex has a considerable representation of the snout and tongue in the echidna, and especially the muzzle in the platypus. Monotreme brains have direct connections between the spinal cord and the motor cortex (corticospinal and pyrimidal tracts), as do the other mammals. The short-beaked echidna has considerable learning capacity, being the equal of white laboratory rats in position-habit and habit-reversal learning (Saunders et al. 1971a,b).

The marsupial spinal cord and brain generally closely resemble those of placentals. Marsupial and placental brains do differ in their interspheric interconnection (Dawson et al. 1989). Marsupials do not have the important corpus callosum of placentals, which connects the left and right cerebral hemispheres, but rely on the monotreme-grade inter-hemisphere hippocampal and anterior commisures. Diprotodont marsupials also have a fasciculus aberrans commisure, which connects the dorsal neocortices. Marsupials, like placentals, show a range of cortex development, from simple (caenolestids, *Notoryctes*) to most complex in macropods and wombats (cortex complexity is generally related to body mass), but no marsupial brain attains the cortical complexity of placentals, which reaches the most complex forms seen in primates and cetaceans (Dawson et al. 1989) (although see Manger, 2006, for an alternate view of cetacean brain complexity). There is a positive correlation for both brain size and maximum reproductive rate with BMR in marsupials, consistent with a general energetic correlate of brain size evolution in mammals (Isler 2011). There are intergeneric differences for small mammal, and primate, families in brain size with diet and ecology (Harvey et al. 1980). Relative brain size of small mammals is related to diet (lowest for folivores; intermediate for generalists; and highest for frugivores, insectivores, and granivores), stratification (lowest for terrestrial and highest for arboreal species), activity (lowest for diurnal and highest for nocturnal species), habitat (lowest for desert grassland and highest for woodland forest). Patterns are the same for primates, except for stratification (lowest for arboreal and highest for terrestrial species) and activity (lowest for nocturnal and highest for diurnal), and there is a relationship for breeding system (lowest for monogamous and highest for polygynous species).

### 3.7.2 Sensory Systems

It is essential for animals to be able to detect and respond to many aspects of their external environment (to eat, to avoid being eaten, to find shelter and mates), as well as their internal environment (to monitor body position and many internal physiological functions). Sensory receptor cells are specialized nerve or epithelial cells that detect particular stimuli and relay information about those stimuli to the receptive and interpretive parts of the peripheral and central nervous systems. Sensory receptive systems can be defined by the nature of the signals they detect (Table 3.11), such as chemicals, mechanical deformation, temperature, fluid movement and derived senses (based on hair cells), light (visible spectrum in mammals, but including infrared heat detection in some snakes). The lateral line system of vertebrates is a particularly important sensory system in terms of its diversification through vertebrate evolution; it is a highly variable and extensively modified sensory system. Its basic form is a series (line) of mechanoreceptive structures in the skin that detect movement (originally of fluid through subdermal tubes). Its basic sensor units, hair cells, have been modified for other sensory modalities, including electroreception. In mammals, the original hair cell sensory systems (i.e. lateral line) are entirely absent, but hair cells are retained and play a role in hearing and equilibrium (and electroreception in the Guiana dolphin).

**Table 3.11** Categories of sensory signals, sensory structures, and sensations in mammals.

| Sensory signal | Structure | Sense |
|---|---|---|
| Chemicals | Olfactory mucosa | Olfaction |
| | Jacobson's organ | Chemoreception |
| | Taste buds | Taste |
| Cutaneous exteroceptors | Free nerve endings | Vibration, touch, injury, pain, temperature |
| | Hairs, vibrissae | Movement |
| | Disks, corpuscles, endings | Touch and pressure |
| | Eimer's organs | Touch |
| Internal proprioceptors | Tendon and joint organs | Tension |
| | Muscle spindle | Muscle stretch |
| Lateral line derivatives | Hair cell | Hearing |
| | | Static equilibrium |
| | | Dynamic equilibrium |
| Atmospheric pressure | Paratympanic organ? | Baroreception |
| Electric field | Skin glands, vibrissae crypts | Electroreception |
| Magnetic field | Magnetite? cryptochromes? | Magnetoreception |
| Photons of light | Retina | Vision |

### 3.7.2.1 Chemoreception

The two main chemical sensory systems, at least in mammals, are olfaction (smell) and gustation (taste). The vomeronasal organ has an olfaction-like role in many mammals.

The sense of smell is generally quite acute in mammals, with the olfactory mucosal area increased by the extensive turbinate bony scrolls, but it is less important in arboreal and flying mammals, probably because of the poorer formation of scent trails in air. The elephant has more intact olfactory receptor genes (1,948) than rats (1,207), dogs (811), chimpanzees (380), and humans (396), suggesting that they might have the best developed sense of smell of these (and nine other) mammals investigated (Niimura et al. 2014). The presence of turbinates in early mammals (e.g. *Morganucodon*) suggests they had an acute sense of smell, related to their nocturnal habit. The axons of the olfactory mucosa project to the olfactory bulb. Mammals, being primarily terrestrial, move air over the olfactory mucosa, so it is not surprising that aquatic mammals close their nostrils while diving and do not rely much on olfaction; olfactory structures are vestigial in cetaceans.

The echidna has well-developed olfactory organs, unlike the semi-aquatic platypus, with a very large olfactory epithelium over endoturbinals, ectoturbinals, nasoturbinals, and maxilloturbinals (Griffiths 1989a). It is likely that echidnas have an acute sense of smell for locating conspecifics in the breeding season, detecting their ant and termite prey, and allowing the newborn to locate the milk areolar area in the pouch. Platypus close their nares while diving, so a sense of smell is probably not important for foraging but might have roles in courtship and other social interactions (Grant 1989). Marsupials have a complex array of cutaneous odoriferous glands, which are important for social communication, so it is not surprising that they also have a well-developed olfactory system (Dawson et al. 1989). All jawed vertebrates have a small terminal nerve (cranial nerve '0') associated with the surface of the olfactory nerve (Liem et al. 2001), but this nerve is large in some marine mammals. It contains gonadotropin-releasing hormone (GnRH), which suggests that it is involved with olfactory control of reproductive activities.

Jacobson's organ (vomeronasal organ) is a pair of medioventral olfactory organs that open through the nasopalatine duct to the anterior of the oral cavity, the nasal cavity, or both. It is absent in aquatic mammals, most bats, and primates (although there are embryonic vestiges). The vomeronasal organ resembles olfactory epithelia in many ways, but its chemoreceptors have microvilli rather than cilia for chemical detection, and their axons project to the accessory olfactory bulb, thence to different parts of the brain than does the olfactory mucosa. Many mammals use the vomeronasal organ to detect pheromones, related to social and sexual interactions; the Flehmen response conveys air preferentially to the vomeronasal organ.

Gustation (taste) is a chemosense of taste buds. Clusters of endodermal cells form a barrel-shaped structure with an opening to the exterior (oral cavity or pharynx) for inward diffusion of chemicals. Some aquatic vertebrates (e.g. fishes, amphibians) have taste buds on their skin as well as oral and pharyngeal mucosa, but in mammals (and other amniotes), taste buds are limited mainly to the tongue, but also the palate, pharynx, and epiglottis.

Taste is generally important for sensing food, and there are many fewer 'tastes' than smells, with five primary taste sensations: salty, sweet, sour, bitter, and umami (meaty taste produced by amino acids; Yarmolinsky et al. 2009). The capsaicin (hot pepper) taste is actually mediated by TRPV1 (transient receptor potential vanilloid) nociceptive receptors (Simon & de Araujo 2005). The sense of taste is conveyed to the brain via the facial nerve (cranial nerve VII) for oral taste buds and the glossopharyngeal (IX) and vagal (X) nerves for pharyngeal taste buds. Echidnas and platypus have a well-developed area of circumvallate papillae with taste buds. In echidnas, these are located on the posterior of the dental pad of the tongue and are presumably well positioned for tasting prey as they are crushed by the dental pad against the palate. The tongue of *Didelphis* has sharp filiform papillae scattered across the entire tongue, conical filiform and scattered fungiform papillae on the lingual body and tongue tip, and three vallate papillae at the root of the tongue (Mançanares et al. 2012).

### 3.7.2.2 Mechanoreception

Mammalian skin has various touch and pressure receptors (Liem et al. 2001). Some are free nerve endings, whereas others are associated with receptor cells or layers of connective tissue. The latter generally adapt rapidly to stimulation, hence are phasic receptors, whereas the other types are generally responsive to sustained stimuli, and are tonic receptors. Merkel's disks lack laminae and are slow-adapting to touch, whereas Meissner's corpuscles are laminated and rapidly adapt; both of these are located near the epidermal surface, and have narrow receptive fields. Ruffini endings are not laminated and are slow-adapting, whereas Pacinian corpuscles are laminated and rapidly adapting; they are located deeper in the epidermis and have wider, more diffuse receptive fields.

Hairs of mammals have nerve endings around their follicles, hence are effective mechanoreceptors. Even slight movements of hair tips are amplified at the follicle and detected as movement. Vibrissae (whiskers) are highly specialized hairs, generally located on the snout of mammals, which are extremely sensitive to movement, even detecting air or water currents. The hair base is surrounded by a vascular sinus that, along with other structures, provides great sensitivity to movement. These are modified in the Guiana dolphin (*Sotalia guianensis*) as electroreceptors (see later). Eimer's organs are highly specialized touch receptors, found at a very high density on the oral tentacles of star-nosed moles. The 25,000 or so Eimer's organs provide more than five times the touch receptors as are on the human hand, and an exquisitely sensitive tactile sense for the moles' food detection and handling (the moles have a very reduced visual system, as expected for a subterranean mammal).

Monitoring of the position and movement of the internal skeleton is a role of proprioceptors, located in various tendons, muscles, and joints (Moore 1984). Tendon organs are sensory nerve endings encapsulated in connective tissue that monitor contractile forces and avoid damaging, excessive stretch. Muscle spindles are specialized muscle fibres that provide precise monitoring and control of the degree of muscle length and contraction.

### 3.7.2.3 Thermoreception

Thermoreceptors respond to heat and cold. The functional unit for both pain and temperature perception is unmyelinated dendrites of sensory neurons in the skin and deeper tissue. The skin of mammals contains small discrete cold-sensitive and warm-sensitive areas, with five to ten times as many cold spots as warm. In primates, warm receptors begin firing action potentials at about 30°C and the firing rate increases to about 46°C. Low-threshold cold receptors (LCRs) are inactive above about 40°C; their firing rate increases down to about $25-30°C$, and then decreases until the temperature is lower than 10°C, when the low temperature is an effective anaesthetic and neuronal firing ceases (LaMotte & Thalhammer 1982). A separate set of sensory neurons mediates the response to noxious cold and heat and operates below and above, respectively, the range of cold and warm sensors.

The mechanism of temperature transduction appears to involve temperature-sensitive ion channels in the free nerve endings of these dendrites, and while the discovery of the transient receptor potential (TRP) family of receptors has added to our knowledge of temperature transduction, the genetic knockout of most of the candidates in rodent models has had no effect on the normal thermoregulation in the animals, suggesting that other mechanisms are involved. The TRP-V1 channel certainly seems to be involved in the sensation of extreme heat, and knockout of these channels makes rodents much less sensitive to extreme heat, while the TRPM8 receptor seems to mediate the response to extreme cold (Pogorzala et al. 2013).

### 3.7.2.4 Pain Reception

Pain (nociception) is a complex mix of unpleasant sensory, emotional, and cognitive sensations that represent potential physical damage to tissues or psychological damage (Dubin & Patapoutian 2010). Nociceptors (pain receptors) respond to tissue injury. These unpleasant sensations are usually elicited only by extreme temperatures or pressures, inflammation reactions, and toxic chemicals. Like other sensations, nociception is detected, encoded, and transmitted to the central nervous system from cutaneous as well as visceral, joint, and muscle receptors. There are various classes of cutaneous nociceptors (temperature, pressure, chemical), but their stimulation does not result directly in pain sensation. Rather, pain is a higher-level sensation dependent on these sensory inputs as well as their intensity and temporal summation. C-fibre nociceptors have small-diameter, unmyelinated axons; they respond to mechanical stimulation (M), heat (H), or cold (C), or combinations (e.g. MH, MC, MHC). A-fibre nociceptors are myelinated and are responsible for fast-onset pain in response to touch, stretch, vibration, and hair movement.

### 3.7.2.5 Audition

Hearing (or audition) is a role of the inner ear, a structure of very similar derivation as the lateral line system, collectively often called the octavolateralis system (Liem et al. 2001). The inner ear of jawed vertebrates consists of three semicircular canals connected to a chamber (utriculus); these canals lie in the three-dimensional (X, Y, and Z) planes, and are responsible for sensing dynamic equilibrium. The utriculus is connected to a ventral membranous sac, the sacculus, which develops a caudoventral lagena that is elongated in diapsid reptiles and mammals to form the cochlear duct (which is coiled in marsupials and placentals). Static equilibrium is sensed by hair cells that detect the position of dense statoconia (otoliths), particularly in the sacculus, in response to orientation with respect to the gravitational force. Dynamic equilibrium is detected by hair cells in the semicircular canals, responding to turning of the head.

Hearing in mammals (and other tetrapods) is a combined function of various structures: the external ear (pinna), which directs sound vibrations into the external auditory canal; the thin tympanic membrane, which vibrates in response to these vibrations; the middle ear ossicles (malleus, incus, and stapes), which transmit tympanic membrane vibrations to the inner ear; and hair cells, which run along the cochlear duct and detect vibrations in particular parts of the cochlear canal. High frequencies are detected near the base of the cochlea, and low frequencies near the apex.

Most mammals detect frequencies in a range of 5–20 kHz, but many echolocating mammals can detect much higher frequencies (see later). The malleus and stapes have small muscles that can dampen their vibration, and protect the inner ear from excessive vibrations from loud noises (including sounds made by the animals themselves for echolocation; see later). The echidna has unusual ear structures (for a mammal). The ear ossicles have the incus firmly attached to the malleus, which is strongly attached to the petrosal bone, suggesting sensitivity to sound conducted by the body (especially the snout) to the middle ear (Griffiths 1989a). The platypus ear seems less sensitive to sound than the echidna ear. The cochlea of echidnas and platypus is not coiled (like reptiles and early mammals), whereas that of marsupials and placentals is highly coiled.

Echolocation is a sophisticated form of detection of objects for orientation and feeding whereby an animal emits sound and detects echoes via hearing. A few birds, some tenrecs, and some shrews have a relatively unsophisticated system of echolocation for orientation (but not feeding) in their environment. Madagascan tenrecs (*Hemicentetes, Microgale, Echinops*) echolocate using noisy, audible clicks (5–17 kHz, 0.1–0.36 s duration) produced by the tongue and lips; *Hemicentetes* and *Centetes* also have a stridulating organ (dorsal spines) that produce 2–70 Hz noises, presumably for communication (Gould 1965). Some species of shrews (*Sorex, Crocidura, Blarina*) echolocate using high-pitched laryngeal 'twitters'; frequency, 4–8 kHz; duration, 8–16 ms (Buchler 1976; Tomasi 1979; Siemers et al. 2009). Bats have a sophisticated echolocation system for aerial feeding and navigation in the dark (see 4.5.4), and cetaceans have a sophisticated echolocation system for orientation, communication, and feeding underwater (see 4.5).

### 3.7.2.6 Electroreception

Electroreception evolved early in the vertebrate lineage from the lateral line sensory system, and has been lost and re-evolved many times in fishes and amphibians, reflecting its sensory importance in aquatic environments. Their lateral-line-derived electrosensory mode was lost in amniote vertebrates (Baker et al. 2013), so electrosensation is rare for terrestrial vertebrates (Proske et al. 1998; Czech-Damal et al. 2012a) and has evolved independently from fishes and amphibians. Amongst mammals, platypus are responsive to small electrical fields in water (e.g. they are attracted to small batteries underwater; Scheich et al. 1986). They have two types of electroceptors (mucous gland and serous gland) spread densely over the bill (along with push-rod mechanoreceptors). Short-beaked echidnas also have electroreceptors on their snout (Gregory et al. 1989). Monotreme electroreceptors appear to have been derived from skin glands, and are innervated by the trigeminal nerve (which is very large, reflecting this important sensory innervation). The Guiana dolphin (*Sotalia guianensis*) has cutaneous electroreceptors, based on crypts of hairless vibrissae on the beak (Czech-Damal et al. 2012b). Toothed whales often locate food by echolocation, rather than using mechanosensory vibrissae (like seals). It appears that Guiana dolphins have evolved electroreceptors from their vibrissal mechanoreceptor system, and perhaps other toothed whales also have these electroreceptors. As in monotremes, the vibrissal crypt electroreceptors of Guiana dolphins are innervated by the trigeminal nerve.

### 3.7.2.7 Magnetoreception

Despite electroreception being restricted to a few mammals, magnetoreception—or at least the potential for magnetoreception—seems more widespread in mammals. The Earth's magnetic field potentially provides animals with several pieces of information: a north-south orientation using the angle of the magnetic field lines to the horizontal (0° at the equator and 90° at the magnetic poles), and intensity of the field (strongest at the poles). The Earth's magnetic poles move, albeit very slowly (e.g. the magnetic north pole moves about 175 m per day) and experiences sudden reversals about every 300,000 years (Gould 2010).

There is considerable evidence, mainly behavioural but also physiological, for magnetic field detection by many animals, including mammals (Lohmann 2010; O'Neill 2013). Magnetic homing and migration are well studied, especially in birds, but the lesser-studied magnetic alignment that also occurs in many animals provides additional evidence for widespread magneto-alignment. The potential roles that sensation of magnetic alignment could play include providing directionality when moving and synchronizing movements of herd animals, coordinating escape from predators, migrating, orientating to learned landmarks, and estimating distance (e.g. for prey capture; Begall et al. 2013). Amongst mammals, magnetic alignment has been reported for rodents, bats, cattle, two species of deer, and red foxes (Begall et al. 2013). Mole rats (*Cryptomys*) preferentially build nests in a south-east orientation, based on the external magnetic field (Burda et al. 1990a). Bats have a magnetic compass that they calibrate against the sun (Holland et al. 2010).

However, mechanisms for magnetoreception are poorly understood (Lohmann 2010). Elasmobranch fishes might use electromagnetic induction resulting from movement through a magnetic field in a conductive (saline) environment to induce small electrical voltages that are detected by their very sensitive cutaneous electroreceptors, but this mechanism is unlikely for terrestrial animals, such as most mammals. Rather, crystals of magnetite ($Fe_3O_4$) might provide a physical basis for the detection of magnetic fields; the magnetite crystals might be aligned by the magnetic field and stimulate mechanoreceptors. A putative magnetite-based electroreception role has been proposed for the olfactory epithelium of trout and the upper beak skin of birds, associated with the trigeminal nerve; inner ear lagena otoliths have also been suggested as possible magnetoreceptors for birds. In bats, magnetite has been suggested to provide magnetoreception (Holland et al. 2008), but the mammalian inner ear lacks the avian lagena structure (O'Neill 2013). For mole rats, the optic tectum has been implicated in magnetoreception. Magnetic tissues have been detected in humans, but no specific magnetoreceptive structure has yet been described (O'Neill 2013).

Alternatively, the magnetic field might influence magneto-sensitive biochemical reactions. Certain molecules form two radicals with correlated electron spins when they absorb light, and whether these radicals form a singlet or triplet state depends on the local magnetic field; it is possible that a chemical reaction rate could vary with singlet or triplet state, providing a mechanism for magnetic field detection (Lohmann 2010). Cryptochromes are highly conserved photoreceptive proteins involved in circadian rhythmicity that are possible candidates for chemo-magnetoreception, thus implicating the retina as a possible magnetoreceptive organ in many diverse animals, including mammals (Foley et al. 2011; Long et al. 2015). Qin et al. (2015) have described a multimeric rod-shaped protein/iron complex, MagR, which has structural, biophysical, and biological attributes of a magnetosensor.

#### 3.7.2.8 Baroreception

Baroreception, detection of changes in atmospheric pressure, has been reported for birds and mammals, as well as other vertebrates. In birds, Vitali's (or paratympanic) organ is a mechanosensory pouch located near the tympanic membrane and is a putative baroreceptor (O'Neill 2013). This structure is also present in a bat (*Vesperugo pipistrellus*) and some other vertebrates, and is likely homologous with the spiracular organ of fishes that detects jaw movements. In rats, the inner ear mediates behavioural responses to changes in barometric pressure (Funakubo et al. 2010). One ecological role of baroreception would be the detection of weather patterns, and movement to regions with recent rainfall and fresh vegetation. Red kangaroos often follow the paths of storms to escape drought (Newsome 1971), perhaps using pressure changes associated with storms as a cue.

#### 3.7.2.9 Humidity

Humidity preference has been documented in a variety of mammals, and more widely in various other vertebrates and invertebrates. For example, most North American vespertilionid bats select caves with temperatures between 2–10°C and 60–100% RH (Perry 2013). Bligh (1963) postulated that sheep have humidity receptors in their naso-buccal region, but Jessen and Pongratz (1979) suggested that the effect of humidity on respiration in goats was mediated by changes in hypothalamic temperature via the carotid rete, arising because evaporation occurs more readily at lower humidity. Toads (*Bufo*) can detect changes in barometric pressure and humidity, enabling them to search for water sources (Willmer et al. 2009), and many terrestrial invertebrates have hygroreceptors (e.g. millipedes, spiders, ticks, mosquitoes, cockroaches, beetles). However, very little is known of the mechanisms for hygrosensation and in particular the structure of hygroreceptors and the molecules involved in sensing changes in humidity (Montell 2008).

Fruit flies detect humidity gradients (Perttunen & Erkkila 1952) and use transient receptor potential (TRP) channels to detect both dry and moist air (Liu et al. 2007). TRP channels are common detectors for various sensory modalities, including hot and cold temperatures, light, tastants, pheromones, and touch (Venkatachalam & Montell 2007). The roles of both mechano- and thermosensory pathways have been identified in humidity detection for the nematode *Caenorhabditis elegans* (Russell et al. 2014). It appears that the degree of mechanical stretch of sensory branches of subcuticular mechanosensitive multidendritic neurons—together with thermal cues resulting from evaporative cooling, and thus thermal flux, detected by thermosensitive cGMP-gated neurons—allow *C. elegans* to respond to even very small humidity gradients. Similar humidity detection may occur in a wide variety of species, including mammals. The mechanisms and responses associated with humidity detection are a fruitful area of future research, especially in light of recent experimental evidence that mammals can regulate their EWL constant under conditions of variable environmental relative humidity (Cooper & Withers 2014a; Withers & Cooper 2014).

The ecological significance of humidity detection includes selection of suitable microhabitats by small mammals, and long-distance movements of large mammals in response to humidity. For example, migration of black-tail deer (*Odocoileus hemionus*) is closely correlated with seasonal changes in relative humidity, rather than food, rainfall, or absolute humidity (McCullough 1964).

### 3.7.2.10 Vision

Mammalian eyes, like those of other terrestrial vertebrates, require protection against desiccation by eyelids, glandular secretions (lacrimal, Harderian, tear glands), and often a nictitating membrane (Liem et al. 2001). The echidna has good visual discrimination, and is good at learning using visual cues, equivalent to a laboratory rat (Gates 1978). The platypus also seems quite reactive to visual stimuli when on land, but closes its eyes when foraging underwater (Grant & Fanning 2007). The platypus eye has a unique cartilaginous cup structure that supports the eye, and a nictitating membrane. Its relatively flat lens seems more suited for vision underwater than in air. Marsupials generally have a large and prominent eye, but it is reduced in caenolestids and vestigial in marsupial moles (which lack any remnant of retinal photoreceptor and ganglion cells, optic nerves, and cranial nerves;

Sweet 1906). The eyes of placental mammals vary considerably in size, from large and prominent (especially primates) to vestigial (various convergent subterranean species). Eye size (length) scales allometrically with mass, with a slope of 0.196 for vertebrates and 0.225 for mammals (Howland et al. 2004). Eye mass (g) scales $M^{0.74}$ for mammals and birds, and visual acuity scales with eye length$^{1.64}$, so visual acuity is approximately independent of mass ($M^{1.27}$) (Kiltie 2000). Overall, mammals have 15% larger eyes than vertebrates as a whole; primates (and birds) have a relatively large eye (35% larger than vertebrates overall), and other mammals have a smaller eye, especially rodents (39% lower than vertebrates overall). Presumably these differences in eye size are reflected by differences in visual acuity.

The adaptive design of the visual system in mammals has been reviewed in terms of adaptation to different life forms and habitats (Ahnelt & Kolb 2000). Mammals have a 'duplex' retina with rod photoreceptors for scotopic (night) vision and cone photoreceptors for photopic (daylight) and colour vision. The proportion of rods and cones varies considerably between species and correlates with the predominant visual lifestyle—diurnal, crepuscular, or nocturnal (Ahnelt & Kolb 2000). The typical mammalian retina contains two spectral cone types, a majority of middle-to-longwave-sensitive (L-) cones and a minority of short-wave-sensitive (S-) cones (Jacobs 1993). Depending on species, the L-cones have their peak sensitivity in the green to yellow part of the spectrum, and the S-cones in the blue to ultraviolet part.

Nevertheless, amongst mammals, the ratio and the retinal distribution of rod and cone photoreceptors varies considerably, depending on habitat and lifestyle (Jacobs 1993; Ahnelt & Kolb 2000; Peichl et al. 2000). When underwater, and to cope with variations in luminosity, iris muscle contractions create species-specific pupil shapes that regulate the amount of light entering the pupil. Interestingly, the retina of aquatic mammals is similar to that of nocturnal terrestrial mammals. Indeed, it contains mainly rod and a low number of cone photoreceptors. In general, the visual system of aquatic mammals demonstrates a high degree of development and several specific features associated with adaptation for vision in both the aquatic and aerial environments.

Most rodent species are nocturnal and they have a retina dominated by rods, with only a small proportion of cones. But even in these species, the cones comprise two spectral types and thus provide for dichromatic colour vision, the most common form of mammalian colour vision (Jacobs 1993). The two spectral cone types are termed short-wave sensitive (S-) cones and middle-to-long-wave sensitive (M-) cones. In the majority of mammalian taxa, the M-cones are green-sensitive and the S-cones are blue-sensitive. In contrast, a number of rodents have been shown to have ultraviolet (UV)-sensitive S-cones, while their M-cones are conventional (Jacobs et al. 1991). The marsupial honey possum is unusual for marsupials (and most mammals) in being trichromatic, and this may be adaptive for assessing the maturity of its main food source, *Banksia attenuata* flowers (Sumner et al. 2005). The numbat is an unusual marsupial, being diurnal and termitivorous.

Some aspects of its visual organization, such as basic retinal organization, lack of a retinal visual streak and restricted field of view, seem to reflect its dasyurid ancestry rather than its specialized ecology (Arrese et al. 2000).

The photoreceptors of subterranean rodents show some interesting, potentially diagnostic deviations from the typical mammalian pattern. Subterranean species have evolved independently amongst marsupials, insectivores and, most notably, amongst rodents (Nevo 1999; Lacey et al. 2000; see 4.3). They are rarely, if ever, exposed to light. Many subterranean mammals show a number of parallel sensory adaptations that include reduced eyes and poor or absent vision (Burda et al. 1990b; Nevo 1999). This has led to the general assumption that the visual systems of subterranean mammals have undergone extensive convergent regression. The model species that supports such claims is the muroid blind mole rat *Spalax ehrenbergi*, which has atrophied subcutaneous eyes and retinal deficits (Sanyal et al. 1990; Cooper et al. 1993). On the other hand, recent studies on subterranean African mole rats (bathyergid rodents) have reported small but well-developed eyes with photoreceptor properties that suggest visual capabilities at higher light levels (Cernuda-Cernuda et al. 2003; Peichl et al. 2004). Nevertheless, Peichl et al. (2005) examined vision of the subterranean rodent coruro (*Spalacopus cyanus*) by analysing the optical properties of the eye, the presence and distribution of rod and cone photoreceptors, and their spectral sensitivities. Coruro eye size is normal for rodents of similar body size; the cornea and lens are transparent for red to near-UV light; and the retina is well structured. Electroretinography suggests dichromatic colour vision. Immunocytochemistry with opsin-specific antibodies confirms the presence of rods, L-cones, and S-cones. An unexpectedly high percentage (10%) of the photoreceptors are cones, of which S-cones constitute a regionally varying proportion from 2% in the dorsal retina to 20% in the ventral retina. The high cone proportion suggests adaptation to visual demands during the short phases of diurnal surface activity, rather than to the lightless subterranean environment, challenging the general view of convergent adaptive eye reduction and blindness in subterranean mammals.

Although the basic pattern for eutherian mammals is cone dichromacy with two types of cone visual pigment, there is evidence that some Australian marsupials have three spectral cone types and are potential trichromats (Arrese et al. 2002; Cowing et al. 2008). There are only a few studies of the photoreceptors of American marsupials. Walls (1939) provided the first description for the North American opossum (*Didelphis virginiana*) and the mouse opossum (*Marmosa mexicana*). Kolb and Wang (1985) quantified rod and cone densities in the opossum by conventional histology, and Ahnelt et al. (1995) analysed the distribution of photoreceptors in the South American opossum (*Didelphis marsupialis aurita*) with opsin-specific antibodies. Hunt et al. (2009) showed that two nocturnal American opossum species (*Monodelphis domestica* and *Didelphis aurita*) have SWS1 and LWS opsins with $\lambda_{max}$ around 360 nm (UV) and 550 nm, respectively. In addition, the *Monodelphis* genome possesses a single rod opsin gene. In contrast, therefore, to Australian marsupials, where a second rod opsin gene has been found that provides trichromacy, this is not the case for South American marsupials; it is therefore expected that they are dichromats.

## 3.8 Reproduction

Reproductive strategy—in particular, the structure and function of the female reproductive system—has long been used to classify the three major mammalian lineages: the monotremes, the marsupials, and the placentals. The term Monotremata comes from the Greek words for single (mono) and hole (tremos), referring to their single urogenital opening through which urine and faeces are voided, and via which the egg is laid. Monotremes have a very reptile/bird-like reproductive organization (Figure 3.34), hence the early term Ornithodelphia that was used to classify monotremes. In monotremes, the ovaries, oviducts, and uteri are separate, with the ureters opening separately into the urogenital sinus. Marsupials, like monotremes, have a urogenital sinus with a single opening through which the urine and faeces are voided, and the young are born. Placentals have separate urinary, digestive, and reproductive openings. Thus, urogenital anatomy provides an unequivocal definition of monotremes, marsupials, and placentals, and importantly indicates that the female reproductive arrangement of marsupials is not 'intermediate' between that of monotremes and placentals.

The different urogenital anatomy of the three major mammalian groups largely reflects a developmental difference. In fetal mammals, the kidney ducts (ureters, draining the metanephric kidney) and the two genital ducts (Mullerian duct from the gonad and Wollfian duct from the mesonephric kidney) connect to the dorsal surface of the urogenital sinus, and the bladder connects to the ventral surface (Tyndale-Biscoe 2005). This remains the arrangement in adult monotremes; urine

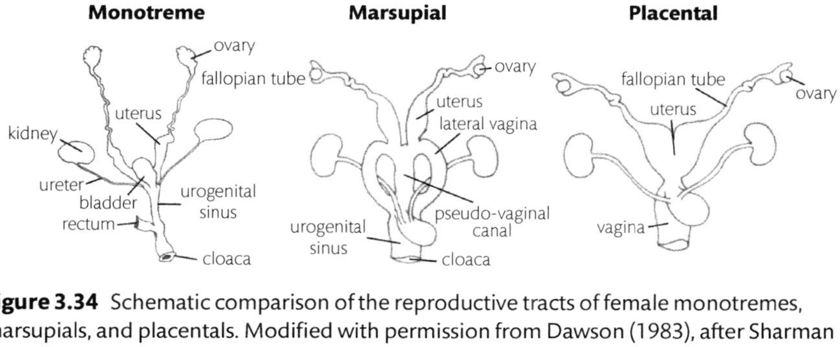


**Figure 3.34** Schematic comparison of the reproductive tracts of female monotremes, marsupials, and placentals. Modified with permission from Dawson (1983), after Sharman (1970) and Hughes and Carrick (1978).

first drains by gravity and ureteral peristalsis into the urogenital sinus, then is presumably moved into the bladder by cloacal peristalsis (as in amphibians; Martin & Hillman 2009). In marsupials and placentals, the ureters migrate to connect directly to the bladder during development; this is presumably a more effective way to convey urine directly to the bladder, by gravity and ureteral peristalsis (Kiil 1973), for storage. However, in placentals, the ureters migrate laterally then ventrally to connect to the bladder, whereas in marsupials they migrate dorsally inside to connect to the bladder (Tyndale-Biscoe 2005; Figure 3.34).

This small developmental difference has a profound consequence for further development of the female reproductive tract (derived from the Mullerian duct). In placentals, the left and right oviducts fuse, and in most species form a single uterus, and fuse in all species to form a single vagina. In marsupials, the two oviducts cannot fuse as the ureters lie between them, so two lateral vaginae develop. In males, there is no developmental consequence of the direction of ureter migration to the bladder, as the Wolffian ducts remain separate as the left and right vas deferentia. When the testes descend, the ureter is dorsal to the vas deferens in marsupials but ventral to the vas deferens in placentals.

The embryological development and structure of the placenta, via which the developing embryo is nourished *in utero* via the maternal blood supply, provides another major distinction between monotremes, marsupials, and placentals; placental structure provides insights into the evolutionary history of these groups (Wildman et al. 2006; Smith 2015; Figure 3.35). The extent of development at birth and nature of the fetal-maternal interface, type of intrauterine nutrient provisioning, nature of interdigitation of maternal and fetal membranes, shape of the placenta, fusion of chorional cells, and membrane type can all vary. The fetal-maternal interface can be epitheliochorial (uterine epithelium is in contact with the fetus, with no epithelial layers absent), synepitheliochorial (syncytium of maternal-fetal cells), endotherliochorial (chorion in contact with the maternal blood vessel epithelium after erosion of the uterine lining), or haemochorial (chorion in direct

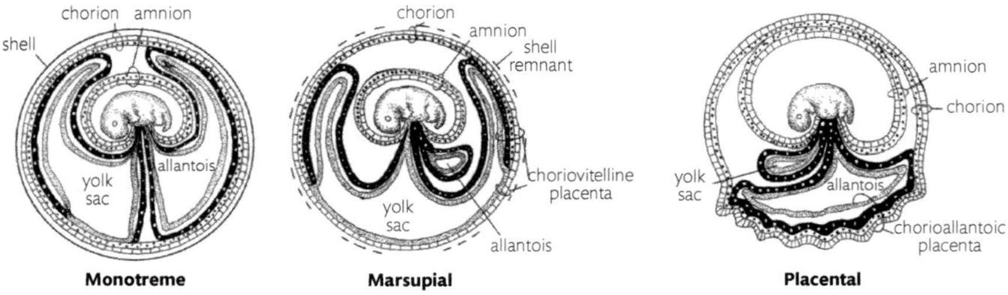


**Figure 3.35** Comparative structures of the extra-embryonic membranes and placenta in marsupial and placental mammals. Modified from Ferner and Mess (2011). Reproduced with permission of Elsevier.

contact with maternal blood). The type of intrauterine nutrient provisioning can be histotrophic, with nutrient uptake only from secretions of the uterine lining, or haemotrophic, with nutrient uptake from the maternal placental blood. The interdigitation of maternal and fetal membranes can be villous (least surface area for exchange between fetus and placenta), folded trabecular (more surface area), lamellar trabecular (even more surface area), or labyrinthine (most surface area). The shape of the placenta can be diffuse (area of fetal-maternal exchange spread over the chorion), limited to cotyledon-shaped areas, zonary (a central zone of fetal-placental exchange), or a distinct disk (discoid) of contact. There can be syncytial fusion of chorional cells. The placental exchange can occur across the yolk sac (choriovitelline) or allantoic membrane (chorioallantoic).

The monotreme egg has a large yolk (huge compared to a marsupial or placental egg), but, unlike most reptile eggs, monotreme eggs have histotrophic nutritional gain before the egg is laid (Griffiths 1989a; Temple-Smith & Grant 2001). There is a marked increase in volume due to absorption of nutrients from the uterus before the tertiary shell membranes, including the thick protective egg shell, are added by the fallopian tube/uterus. This increase in volume is also seen in marsupial and placental eggs (but is not typical of reptiles).

In marsupials and placentals, there is a developmental progression in the absorption of uterine nutrients: the initial autonomous phase is the reliance on the ovarian egg lipids; the absorptive phase is the expansion of the egg by absorption of uterine secretions; the respiratory phase involves exchange of gases and nutrients via intimate contact of the egg with the uterine wall where the allantois forms the primary respiratory surface for the embryo (as in reptiles and birds; Tyndale-Biscoe 2005). Marsupials in general have histotrophic nutritional gain like monotremes, although some lineages develop haemotrophic nutritional gain (exchange via a placenta) (e.g. some didelphids, dasyurids, and peramelids). In most marsupials, the yolk sac forms the placenta, which consists of a two-layered, non-vascularized section where the absorption of uterine fluid nutrients occurs, and a three-layered, vascularized section where respiratory exchange occurs (Figure 3.35; Tyndale-Biscoe 2005). The allantoic sac remains small, with a modest blood supply; its main role is urine storage towards the end of gestation.

In a few marsupials, such as the koala, the allantois is larger but remains non-invasive, whereas in bandicoots, the allantois becomes highly vascular and fuses with the uterine wall to form an invasive allantoic placenta just before birth. It is the more altricial marsupial groups (dasyuromorphs, didelphimorphs, peramelemorphs) that have some invasion by fetal tissues of the uterine lining, syncytial fusion of fetal chorional cells, and haemotrophic nutrition, somewhat like the eutherian condition. Surprisingly, the most altricial group (macropodids) is least like the eutherian condition (Smith 2015). While marsupials generally have an epitheliochorial (relatively uninvasive) and folded placenta, primitive placentals had a haemochorial (invasive) and labyrinthine placenta; both had a discoidal-shaped placenta. Marsupial young are born at a remarkably early stage of development, at least in comparison with placental mammals, and their postnatal survival and development, often in a marsupium or pouch, is equally remarkable.

Within placentals, there is considerable variation in placentation pattern. In general, the allantois retains the primary respiratory exchange role, and assumes a role in nutritive uptake by the embryo, as the allantoic placenta (but some placentals, such as the rabbit, have a yolk sac placenta). The placental structure of placental mammals has evolved and reverted a number of times, within the basic evolutionary pattern of four clades (Figure 3.36; Smith 2015). Phylogenetic reconstructions suggest that the epitheliochorial placenta is the derived condition even though it is the least invasive placentation pattern, with the haemochorial (or endotheliochorial) placenta as the primitive condition. Similarly, most phylogenies suggest that the labyrinthine discoid placenta is the primitive condition. As for marsupials, it appears that less-invasive placentation is the derived condition, with greater reliance on histotrophic nutritional exchange in more precocial young (Smith 2015).

These complex and somewhat surprising patterns of placentation in both marsupial and eutherian lineages might reflect consequences of maternal-fetal conflict, with the fetus (having both maternal and paternal genes) being in potential evolutionary conflict with the mother (e.g. Haig 1993; Smith 2015). Extended intrauterine gestation could result in maternal-fetal conflict for resources, with

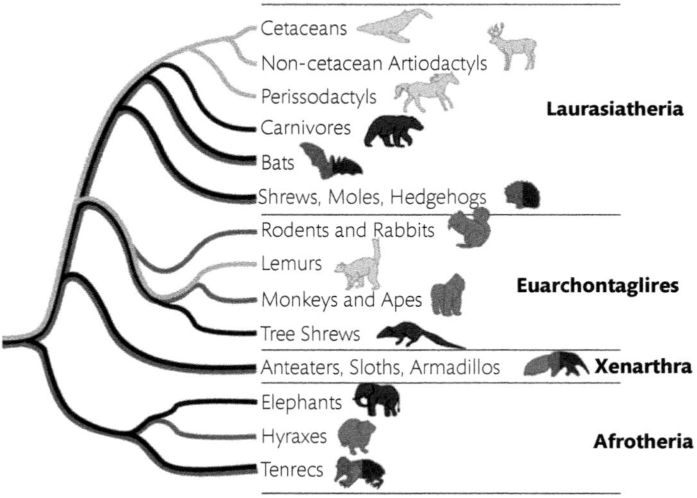


**Figure 3.36** Placental patterns for placental mammals; shades of grey reflect the three basic placenta types: light grey, epitheliochorial placenta; medium grey, haemochorial placenta; dark grey, endotheliochorial placenta. From Smith (2015). Reproduced with permission from Great Transformations in Vertebrate Evolution (Eds. Dial, K. P., Shubin, N. & Brainerd, E. L.), © 2015 The University of Chicago Press.

respect to whether the mother (e.g. in short-gestation marsupials) or fetus (e.g. in long-gestation eutherians) controls the use of maternal resources for fetal development. It could also seriously compromise or kill the mother if there are problems during development or birth. In contrast, marsupial mothers are able to abort their developing fetuses without such serious consequences; for example, in response to environmental problems such as low food or water resources, just as they are able to eject young from the pouch and abandon them when threatened by predators (Armati et al. 2006).

A pouch (marsupium) is often considered to be a general defining trait for marsupial mammals, but not all marsupials have a pouch, and marsupials are not the only mammalian group to have a pouch. Female (and also male) echidnas have a pouch, but the platypus does not (see 3.8.1). Although monotremes, marsupials, and placentals are sometimes considered to represent a graded series from 'primitive' to 'advanced', marsupials do not represent an intermediate stage between monotremes and placentals with respect to reproduction, but are best interpreted as a sister group to placentals that diverged with a differing reproductive strategy based on anatomical constraints (i.e. the dorsal migration of the ureters that prevents the formation of a single uterus). This strategy has proved to be a preadaptation that allows successful reproduction in adverse conditions, but may also be a limitation to reproductive efficiency (Edwards & Deakin 2013).

Indeed there are relative advantages and disadvantages to each of the three reproductive strategies of monotremes (egg-laying, altricial young, pouch and extended lactation), marsupials (live birth, altricial young, pouch and extended lactation), and placentals (live birth, altricial to precocial young, shortened lactation). It could be argued that each strategy 'works' in particular niches and environments, since all three have persisted, but the paucity of extant monotremes and the apparent competitive edge of placentals over marsupials where they compete suggests otherwise.

### 3.8.1 The Monotreme 'Strategy': Egg-Laying

The reproductive system of male monotremes has distinct differences from that of other mammals, although study of the males has received less attention compared to the highly unusual reproductive mode of females (Carrick & Hughes 1978). The relatively simple male reproductive tract of monotremes starts with abdominal testes, situated just dorsal and caudal to the kidneys, large epididymes, and ductus deferentia that join the urethra just cranial to the ureters. Glandular ampullae, seminal vesicles, coagulating glands, and prostate glands are absent, but there is some diffuse glandular tissue around the cranial urethra, and there are well-developed bulbo-urethral glands. The urethra ends in a large, elaborate penis. Abdominal testes are relatively unusual in mammals because their high $T_b$ is detrimental to spermatogenesis, but monotremes have a relatively low $T_b$ (see 3.2.1). The monotreme sperm is very reptilian, being filiform and forming groups of up to

20 spermatids per Sertoli cell (Griffiths 1989a); Griffiths (1984) commented that the monotreme testis is 'as reptilian in character as the interclavicle of the pectoral girdle'. Male platypus and echidnas have a keratinous spur on the inner side of the hind limbs, with a venom gland located further up the leg. Juvenile echidnas have a spur, but it is shed in adult females.

The reproductive system of female monotremes consists of ovaries (the right is rudimentary in platypus), well-developed infundibular funnels, and long fallopian tubes, then paired uteri that open into the urogenital sinus; there is no vagina (Figure 3.34; Hughes & Carrick 1978; Griffiths 1989a; Grant 1989; Temple-Smith & Grant 2001). The ureters open separately into the urogenital sinus near the bladder neck (as occurs in embryonic marsupials and monotremes, but their ureters become incorporated into the bladder neck). The urogenital sinus opens into the cloaca, where eggs, urine, and faeces are voided through the cloacal sphincter.

The reproductive cycle of female monotremes is based on that of reptiles, and is a forerunner to that of marsupials and placentals. When breeding, the ovary surface shows uneven protrusion of yellow-pigmented follicles, much like sauropsid reptiles (which suggested to early mammalogists that monotremes were oviparous). After ovulation, the burst follicle forms a functional corpus luteum, equivalent to that of other mammals, which regulates the secretory and nutritive roles of the uterus during pregnancy. Gestation is about 27 days for both echidna and platypus. Caldwell and Haake independently confirmed in 1884 the supposition that monotremes lay a cleidoic egg (Caldwell 1887). The parchment-covered egg, when laid, contains an embryo equivalent to about a 36-hr chick embryo. Echidnas usually lay one egg (but up to three) and platypus usually lay two (up to three). The eggs hatch after about 10–12 days; the young have a pronounced development of the forelimbs and closely resemble newborn marsupials.

Female and male echidnas both have a ventral pouch area, a relatively hairless concave area, with two antero-lateral hairy areolar areas where milk is secreted by females (Griffiths 1989a; Figure 3.37). During pregnancy, the female's pouch develops thick, tumescent lips reflecting the growth of the underlying mammary glands. When the egg is laid, it appears to be deposited into the pouch by a combination of eversion of the cloaca and contortion of the body. The incubation period is 10–12 days. Just before hatching, an egg tooth forms that assists the neonate to exit the shell. The hatchling moves to one of the areolar areas (which forms a nipple-like protuberance), where it clings to some hairs, and begins suckling milk from the skin surface. The young remain in the pouch for 40–63 days, until the spines start to develop, but the female continues to suckle them in a burrow for about 200 days. The milk resembles that of other mammals. It is initially dilute (about 12% solids, 1.25% fat, 7.85% protein, with a balance of carbohydrate and minerals) compared to mature milk (48.9% solids, 31.0% fat, 12.4% protein, 1.6% hexose, 0.5% minerals; Griffiths et al. 1984).

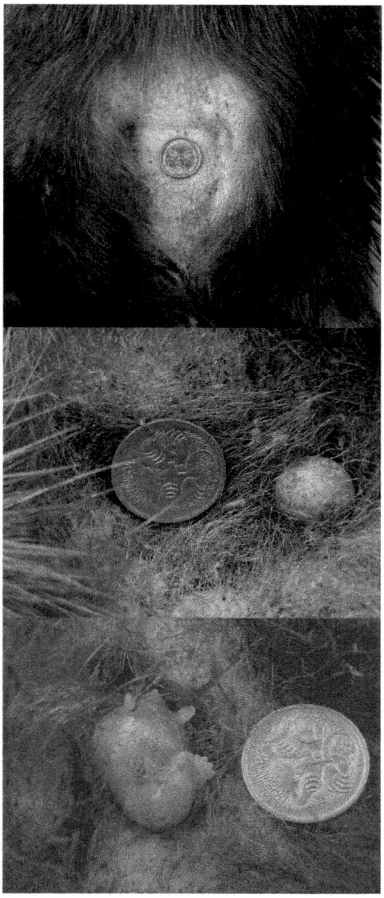

**Figure 3.37** Short-beaked echidna (*Tachyglossus aculeatus* setosus) pouch (top), egg (centre), and pouch young (bottom). Diameter of Australian 5-cent coin is 19.3 mm. Photographs from S.C. Nicol.

Female (and male) platypus lack a pouch, presumably reflecting their semi-aquatic lifestyle (Grant 1989), because it would be impractical for females to forage underwater with an egg in the pouch. Gestation is about 27 days and the eggs are deposited into a breeding burrow, where they hatch after about 10 days; then the young are suckled for 90–120 days. The mature milk is high in solids, like that of echidnas (39.1% solids, 22.2% fat, 8.2% protein, 3.3% hexose; Griffiths et al. 1984). Reproductive differences between platypus and echidnas presumably reflect the semi-aquatic habit of the platypus, so the reproductive patterns of echidnas presumably better reflect that of early mammals, and hence the ancestral pattern for both marsupials and placentals.

Breeding commences in about June for short-beaked echidnas in south-eastern Australia, but there is considerable geographic variation (Griffiths 1989a), presumably related to variation in local environmental conditions. Reproduction and hibernation are incompatible for many mammals, but in Tasmania more than one-third of female echidnas in mating groups are torpid, and can re-enter deep torpor after fertilization; the intense competition between promiscuous males presumably drives this overlap of reproduction and hibernation (Morrow & Nicol 2009). Platypus breed in spring, but again with regional and environmental variation (e.g. earlier onset of breeding in northern Australia; Grant 1989). Echidnas and platypus do not breed every year. Little is known of the breeding biology of long-beaked echidnas (*Zaglossus* spp.). Higgins et al. (2004) measured faecal reproductive steroids of *Zaglossus bruijnii*; they had a lower faecal progestin concentration (420 ng $g^{-1}$) than short-beaked echidnas (860 ng $g^{-1}$).

It is difficult to envisage advantages of the monotreme egg-laying strategy over the marsupial or placental strategies. The relatively large size of the monotreme egg at the time of fertilization (about 15–25 times bigger than ovulated eggs of marsupials and placentals) means that these resources are wasted if fertilization does not occur, or the embryo is not viable (Dawson 1983). The short period of egg incubation, before hatching, and carrying it in a pouch (echidnas) or depositing it in a nest (platypus) mean that if there are any advantages to having the developing young still protected within an egg, they are short-lived. The period of egg development *in utero* (27 days) is roughly equivalent to the gestation period of a similar-size marsupial, and the extra period of egg incubation before hatching is only about 10 days, so the period from ovulation to hatching is not very different from the timing from ovulation to birth in marsupials.

The size of newborn monotremes at hatching (about 10 days older than equivalent neonate marsupials) is 71–96% of that expected for a marsupial, based on maternal mass (calculated from Griffiths et al. 1969; Russell 1982; Rismiller & McKelvy 2003). The egg can apparently be directly deposited into the pouch (echidnas) or in a nest (platypus), so newborn monotremes do not require the pronounced development of the forelimbs that is seen in marsupials and that facilitates their climb from the birth canal to the pouch, but they do require sufficient forelimb development for suckling. The extensive period of lactation, with changing milk composition, is similar in monotremes and marsupials. So, there do not appear to be particular advantages to the monotreme strategy of laying eggs and having a short period of egg incubation, particularly in comparison with the otherwise quite similar marsupial strategy.

### 3.8.2 The Marsupial 'Strategy': Short Gestation

In male marsupials, the testes and epididymis are external to the abdominal cavity, in a scrotal sac, and connect via vasa deferentia to the prostatic urethra, thence to the bladder neck (Dawson et al. 1989). The accessory glands include the prostate and Cowper's (bulbo-uretheral) glands, but not seminal vesicles, ampullae, or coagulating glands (except in caenolestids and mountain pygmy possums, which have an ampulla-like secretory segment of the vasa deferentia). In most marsupials, a pendulous scrotum develops postnatally and testicular descent is complete by about 70–80 days. The scrotum is located anterior to the penis, which is extruded from the urogenital sinus. The penis has a bifid tip, allowing fertilization via both lateral vaginae. The marsupial mole and wombats have sub-integumental, extra-abdominal testes, presumably reflecting their underground lifestyle (Wislocki 1933; Williams & Hutson 1991) and low $T_b$ (Withers et al. 2000). The spermatic cord of scrotal marsupials (and placentals) is enclosed in a thick cremaster muscle (Dawson et al. 1989). Scrotal temperature is controlled by the extent of contraction of the cremaster muscle and vascular heat exchange in a rete mirabile; testicular temperature varies from 1.4°C below core $T_b$ in *Didelphis* to about 6.5°C in *Trichosurus* (see 3.2.4.1). At high $T_a$, testicular temperature can be up to 14°C lower than $T_b$, due to scrotal licking. In the Tasmanian tiger (*Thylacinus*) and the semi-aquatic water opossum (*Chironectes*), the scrotum is located inside a pouch-like skin fold; *Chironectes* can apparently enclose the testes within the 'pouch' while diving, presumably to keep the testes warm for spermatogenesis (Enders 1937).

The reproductive system of female marsupials consists of two ovaries, two oviducts, and two uteri that connect via separate cervices to two lateral vaginae, which open separately into a urogenital sinus (Figure 3.34; Dawson 1983; Dawson et al. 1989; Tyndale-Biscoe 2005). The lateral vaginae open anteriorly into a vaginal cul-de-sac, into which the uterine cervices open separately. More posteriorly, the vaginae, ureters, and urethra are embedded in connective tissue and open into the urogenital sinus. The unusual female reproductive tract of marsupials (compared to placentals) has important consequences for reproduction, gestation, and more general life history strategies. The lateral vaginae and cul-de-sac form the pathway for sperm movement to the ovulated egg, but not the birth canal. In most marsupials, a pseudo-vaginal birth canal forms in the connective tissue strand from the cul-de-sac to the urogenital sinus, but after birth it degenerates, and has to reform for subsequent births. However, in most macropods (Macropodidae) and the honey possum (*Tarsipes*), the medial birth canal remains permanently after the first birth.

The reproductive cycle of female marsupial mammals is a further development from that of monotremes. Development of dormant ova is stimulated by follicle stimulating hormone (FSH) secreted from the pituitary at the start of the oestrous cycle (Tyndale-Biscoe 2005). The follicle cells secrete oestrogen, and fluid that forms a space around the ovum forms a Graafian follicle. From only a few (e.g. kangaroos, brushtail possums) to 50 or more (e.g. Tasmanian devil, Virginia opossum, *Didelphis marsupialis*), follicles begin development during an oestrous cycle. They are ovulated when the Graafian follicle ruptures and releases the ovum into the oviduct. Ovulation is initiated by the release of luteinizing hormone (LH) from the anterior pituitary. The follicle cells persist after ovulation and form the corpus luteum, which secretes progesterone, which prepares the reproductive tract for pregnancy, and relaxin, which peaks in concentration from the middle to the end of gestation. In a few species (e.g. *Antechinus*), there is a protracted period of a few weeks preceding ovulation, when the female mates with males and sperm are stored in oviductal crypts, to fertilize eggs when they are ovulated. The luteal phase then lasts until the end of gestation when the young are born. If there is no pregnancy, the female is anoestrous for the remainder of the year, until the next breeding season.

In most marsupials, however, there can be multiple oestrous cycles in a year; the length of the oestrous cycle varies, from about 21 days (e.g. bandicoots) to 45 days (e.g. kangaroos). Gestation length is shorter or equal to the oestrous cycle, except for the swamp wallaby (*Wallabia bicolor*) and the grey kangaroo (*Macropus giganteus*; Low 1978). Gestation length ranges from 12 to 46 days in marsupials; it is similar for small marsupials and placentals, but is longer for larger (> 1,000 g) placentals than marsupials (Lilligraven 1987). In many species, pregnancy occurs during the luteal phase, which is about twice as long as the pre-ovulation phase, and suckling by the young inhibits the following oestrous cycle. In rat kangaroos and kangaroos, pregnancy lasts for about as long as the oestrous cycle, so the female can mate soon after birth and become pregnant again. However, the pregnancy is put on hold by developmental arrest (embryonic diapause) at the blastula stage if the newborn successfully finds a teat and suckles. Embryonic diapause occurs in pygmy possums, feathertail gliders, and the honey possum, as well as rat kangaroos and nearly all kangaroos and wallabies (and also some placentals). In marsupials, removing the pouch young (an easy experimental manipulation) initiates resumption of embryonic development in the arrested blastula, which either successfully develops to birth or fails to develop properly and dies.

Most but not all female marsupials develop a pouch. The pouch is ventral, and varies from shallow to deep, with a variable number of teats (Russell 1982; Tyndale-Biscoe 2005; Edwards & Deakin 2013). Six types of pouch are the following: Type 1: skin folds develop but do not cover the mammary area; Type 2: skin folds partially cover the mammary area; Type 3: the mammary area, with a circular array of numerous teats, is covered by the pouch; Type 4: the pouch covers teats, located in two pockets; Type 5: the teats are covered by a pouch that opens posteriorly; Type 6: the pouch is covered by an anteriorly opening pouch (Tyndale-Biscoe & Renfree 1987). Some species have a small pouch but a large litter (Figure 3.38); some have fewer young and a well-developed pouch; while some have typically one young in a large pouch. The pouch is ventral in all marsupials, but its morphology varies markedly (Edwards & Deakin 2013). They can be shallow (e.g. many dasyurids) or deep (e.g. macropods), and contain a variable number of teats. A backward-facing pouch is useful for fossorial and semi-fossorial species (e.g. wombats, bandicoots, marsupial moles). The arboreal koala also has a backward-facing pouch to facilitate the young ingesting pap (specialized maternal faeces) for acquisition of fermenting microorganisms.

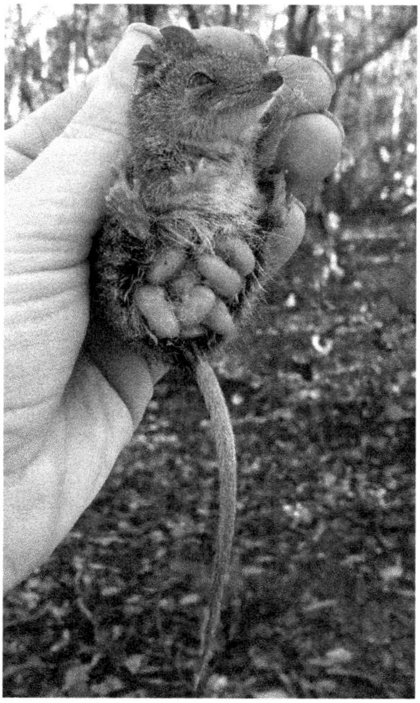

**Figure 3.38** Female antechinus (*Antechinus flavipes*) with pouch young, showing pouch Type 1, with a small pouch and many young. Photograph by C. Cooper.

In marsupials, there is a strong correlation of reproductive life history characteristics (e.g. neonate mass, litter size, developmental stages, lactation duration, pouch type) with maternal body mass and pattern of care, at least within families (Russell 1982; Tyndale-Biscoe 2005; Edwards & Deakin 2013). Species with small pouches and many young (Type A; e.g. dasyurids, didelphids) leave their young in a nest at an earlier stage of development (prior to the eyes opening) and have a shorter lactation period, compared to macropods (Type C), which carry their young (generally just one) for an extensive period in the large pouch. Type B species are intermediate: they have a larger pouch and fewer young than Type A. Small species also make a proportionally larger investment in their young: small dasyurids invest over 300% of the maternal mass in the weight of their litter at weaning, compared to less than 30% for large kangaroos, wombats, and the koala. Total litter mass correlates strongly with maternal body mass, rather than the mass of individual young, because of the co-correlation of litter size.

There appear to be advantages and disadvantages of the marsupial reproductive strategy, compared to placentals. When a pouch is present, it provides many advantages for marsupials: physical protection from the external environment; control of the external environment, including humidity and temperature; and making the young inconspicuous to predators (Edwards & Deakin 2013). Other advantages include the capacity to more easily abort the reproductive investment under adverse conditions (pre- and post-birth), low cost of uterine development for gestation, and lower energetic costs of development for small, ectothermic young (Edwards & Deakin 2013). Marsupials make their maternal investment in the litter over a relatively longer period of time (about 250 days $kg^{-0.25}$ since conception) than do placentals (about 100 days $kg^{-0.25}$), although the total investment is similar, at about 19 MJ $kg^{-1}$ (koala), 21 MJ $kg^{-1}$ (tammar wallaby), and 22 MJ $kg^{-1}$ (ringtail possum, *Pseudocheirus peregrinus*), compared to 23 MJ $kg^{-1}$ for sheep and cattle (Tyndale-Biscoe 2005). This marsupial pattern of relatively low investment in the early stages of pregnancy and lactation means that they can terminate reproduction in response to predation (females can eject their pouch young if threatened) or unfavourable environmental conditions (e.g. extended drought). Female placentals are less able to terminate reproduction with low investment cost, and there are potential costs to the health of the female.

Embryonic diapause gives desert kangaroos a very successful opportunistic strategy for reproduction in their uncertain and unpredictable environment, while sheep experience severe losses during drought in similar environments. However, there are disadvantages to the marsupial reproductive strategy. Their extremely atricial young are required to make their way into the pouch, and this requires precocial development of the forelimbs at birth. This developmental constraint could limit the evolutionary development of specialized forelimbs, such as wings for flight or flippers for swimming (Kelly & Sears 2011a,b). Another potential negative consequence of the marsupial reproductive strategy is a relative inefficiency in provision of nutrients to the young by lactation compared to direct transfer of nutrients across an extensive chorio-allantoic placenta (i.e. placental mammals). Nevertheless, McNab (1986) has suggested that marsupials and placentals are competitive equivalents in environments where resources are limited, although he suggests that placentals are superior competitors over marsupials in resource-rich environments that can support high rates of energy use.

### 3.8.3 The Placental 'Strategy': Prolonged Gestation

In male placentals, the testes and epididymis are generally external to the abdominal cavity, in a scrotal sac, which is posterior to the penis. The penis is not bifid, as it is in marsupials. The spermatic cord of scrotal placentals is enclosed in a thick cremaster muscle (like marsupials). For a number of placental mammal groups, the testes only descend into the scrotum during reproductive activity (e.g. many rodents), while testicond placental mammals have testes permanently located in the abdominal cavity (e.g. hyrax and elephants; Williams & Hutson 1991; Werdelin & Nilsonne 1999), others have partially descended but intra-abdominal testes (cetaceans) or inguinal, testes (prairie dog, *Cynomys* spp., hedgehogs, moles, shrews, sloths, armadillos), emergent just outside the inguinal canal (chinchilla, *Chinchilla lanigera*), or external inguinal testes (elephant seals, tree shrews).

Testicular temperature of scrotal mammals is typically lower than core $T_b$, and tends to decrease at lower $T_a$ (see 3.2.4.1). In elephants, which are testicond, testicular temperature is the same as core body temperature (about 36°C), but other testicond mammals tend to have a low $T_b$, hence testis temperature. If the scrotum evolved prior to the divergence of marsupials and placentals, then presumably it was not to cool the testes, as core $T_b$ was probably not sufficiently high yet to endanger spermatogenesis (Werdelin & Nilsonne 1999). If the scrotum evolved independently in marsupials (pre-penial) and placentals (post-penial), then it is more likely that the selection pressure was to cool the testes to avoid damaging spermatogenesis.

In female placentals, the progenitor lateral vaginae fuse into a single vagina, and in many groups the uteri also fuse into a single medium uterus. This has profound implications for reproductive and life history strategies, centred on their capacity for longer gestation, more complex placentation, more extensive development of young before birth, and a less extensive lactation period prior to weaning compared to marsupials (Capellini et al. 2011).

In placentals, like marsupials, there is a strong correlation between reproductive life history characteristics (e.g. neonate mass, litter size, developmental stages, lactation duration) and maternal body mass. The marked interspecific variation in placental structure and function also has implications for patterns in reproduction. The extent of placental invasiveness and particularly the amount of interdigitation of maternal and fetal tissues influences the gestation period: gestation length is reduced to 44% for species with highly interdigitated labyrinthine placentas

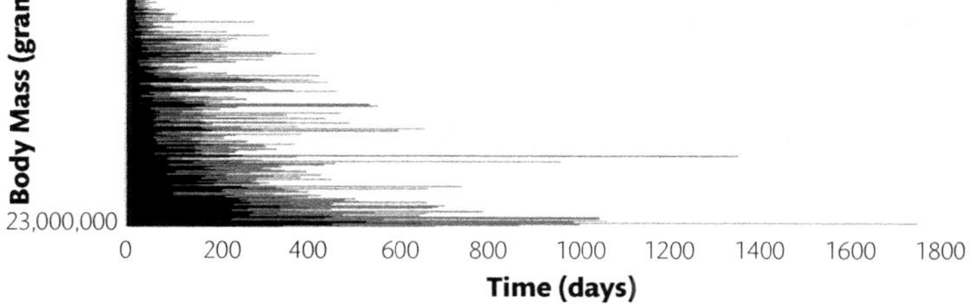

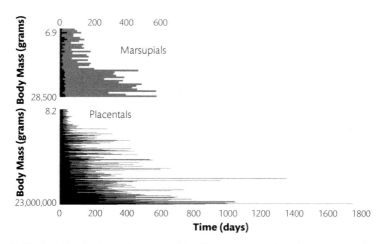


**Figure 3.39** Duration in days of gestation (black) and weaning (grey) for marsupials and placentals, arranged in order of increasing female body mass (from top to bottom). The mass axis for marsupials corresponds approximately with that for placentals. Data from Bielby et al. (2007).

compared to that of species with less interdigitated villous and trabecular placentas (Capellini et al. 2011). The reproductive cycle of female placental mammals is similar to that of marsupials. Gestation length increases with body mass, as does time to weaning (Figure 3.39). Small placentals have similar gestation lengths to marsupials, but gestation is longer for larger (> 1,000 g) placentals than marsupials (Lilligraven 1987). Lactation is generally longer in marsupials than placentals.

There appears to be a 'fast-slow' continuum in mammalian life history strategies, with two contributing factors (or axes; Bielby et al. 2007; Table 3.12). For mammals in general, the first factor (explaining 37.7% of the variation in the data set) includes positive effects of inter-birth interval, weaning age, and age at sexual maturity; collectively, this factor reflects variation in reproductive timing. The second factor (explaining 31.6% of the variation) includes positive effects of gestation length and neonatal mass; this factor collectively reflects a trade-off between offspring size and number. For placentals (Table 3.12), the same life history variables contribute to Factors 1 and 2 as was observed for all mammals (presumably reflecting the high numbers of placentals (227 of 267 species) in the data set, and more of the variance is explained (40.0 and 37.5% of variance, respectively). Litter size also contributes negatively to Factor 2, the offspring size-number trade-off (i.e. long gestation, high neonatal mass, and small litter size).

There are, however, some differences between marsupials and placentals that reflect the fundamental differences in their reproductive strategies. For marsupials (40 species), nearly as much variance is explained (41.4 and 25.7%, respectively), but gestation length contributes positively to Factor 1 (reproductive timing) rather than Factor 2 (offspring size-number trade-off), and age at sexual maturity does

**Table 3.12** Factor analysis of life history traits yields two factor loadings (Factor 1: reproductive timing; Factor 2: offspring size-number trade-off) for life history traits of placental and marsupial mammals. Life history traits contribute slightly differently to Factors 1 and 2. Values in bold indicate more significant loadings (> 0.6). Modified from Bielby et al. (2007).

| | Placentals | | Marsupials | |
|---|---|---|---|---|
| | Factor 1 Reproductive timing | Factor 2 Size-Number trade-off | Factor 1 Reproductive timing | Factor 2 Size-Number trade-off |
| Gestation length | 0.478 | **0.740** | **0.788** | 0.213 |
| Neonatal mass | -0.259 | **0.895** | -0.261 | **0.842** |
| Litter size | -0.288 | **-0.864** | -0.337 | **-0.821** |
| Inter-birth interval | **0.897** | -0.075 | **0.853** | -0.309 |
| Weaning age | **0.727** | 0.369 | **0.951** | 0.133 |
| Age at sexual maturity | **0.830** | 0.106 | 0.225 | -0.014 |

not contribute to Factor 1. Marsupials invest relatively little energy in gestation compared to eutherians, so gestation length is not as suitable a surrogate for life history speed as it is for placentals.

There appear to be advantages and disadvantages of the placental reproductive strategy, compared to marsupials. One advantage is the presumably higher efficiency of the provision of nutrients to the young by direct transfer of nutrients across an extensive chorio-allantoic placenta rather than lactation. McNab (1986) suggested that placentals are superior competitors over marsupials in resource-rich environments that can support high rates of energy use (although marsupials and placentals are competitive equivalents in environments where resources are limited). However, female placentals are less able to terminate reproduction with low investment cost, and with potential detriment to the health of the female.

# 4

# Physiological Adaptations to Extreme Environments

The general bauplan of mammals, in particular their morphological and physiological characteristics, has proven to be extremely flexible. From a small, nocturnal, insectivorous ancestor, an amazing array of sizes, shapes, and physiologies have evolved, including species able to exploit some of the most extreme environments on Earth. Mammals (and birds) inhabit the coldest environments, including polar regions and high altitudes; hot and dry environments, including the harshest deserts; and live underground in fossorial environments. Many mammals are semi-aquatic or aquatic, and some, at least temporarily, occupy very deep underwater depths. Some mammals occupy the aerial environment; some as gliders and bats as powered fliers. Survival in these harshest of environments requires remarkable adaptations.

## 4.1 Cold Environments

The lowest recorded air temperature ($T_a$) at the Earth's surface is $-89.2°C$, measured at Vostok station in Antarctica during the Austral winter of 1983 (Turner et al. 2009). Although this very low $T_a$ occurred in a particularly extreme environment in a location with presumably no natural mammalian inhabitants, sub-zero temperatures occur frequently in a wide variety of habitats, even usually hot desert environments. Polar regions and high altitude locations typically experience winter temperatures $< -35°C$, and mean monthly winter temperatures of $-50$ to $-60°C$ have been recorded for some areas of eastern Siberia and north-western Canada (Scholander et al. 1950a). These regions may only rarely reach temperatures above freezing during summer, but mammals are a conspicuous element of the fauna of these very cold regions year-round. The endothermic bauplan of mammals provides the mechanism to maintain a high and stable body temperature ($T_b$), but at the energetic cost of sustaining a high metabolic heat production. For many mammals in moderately to very cold environments, their suite of physiological

*Ecological and Environmental Physiology of Mammals*. Philip C. Withers, Christine E. Cooper, Shane K. Maloney, Francisco Bozinovic, & Ariovaldo P. Cruz-Neto.  Published 2016 by Oxford University Press.
DOI 10.1093/acprof:oso/9780199642717.001.0001

and morphological characteristics, along with their ecological niche, enables them to remain active and endothermic at a sustainable metabolic cost year-round. However, for some mammals, this energetic cost can become temporarily too difficult to sustain if the required rate of heat production exceeds the maximal sustainable metabolic rate (e.g. at very low $T_a$ or during periods of low food availability). The solution for non-migratory mammals under such circumstances is to become inactive in a more moderate refugia and/or enter periods of prolonged hibernation.

### 4.1.1 Endurers

Many mammals remain active during winter, even in extremely cold environments. Cold winter temperatures in high latitude and altitude habitats require enhanced metabolic heat production, at a time when food accessibility may be reduced. Although challenging for all endothermic homeotherms, small non-hibernating mammals can be most impacted by extreme winters, as small mammals have a large surface area to volume ratio, limited scope to increase fur thickness and therefore insulation, and high mass-specific energy requirements. However, a number of medium and large, as well as small mammals—such as voles, hamsters (Cricetidae), and even shrews (Soricinae; with body mass of < 6 g)—remain active at air temperatures < −30°C. Although small and large mammals have differing approaches to the seasonal acclimatization required for them to remain active in extreme cold environments, both are based on two general responses: increased thermogenesis and decreased heat loss (Heldmaier 1989).

Some small non-social mammals, such as voles (*Clethrionomys* and *Microtus* spp.), shrews (*Cryptotis pava*) and mice (*Peromyscus leucopus* and *P. maniculatus*), form social aggregations during winter. They inhabit communal subnivean nests that can be substantially warmer than ambient conditions and gain thermal and energetic advantages from huddling, reducing heat loss by decreasing their collective surface area to volume ratio (Marchand 2014). For example, nests of taiga voles (*Microtus xanthognathus*) containing five to ten individuals can be as much as 12°C warmer than the surrounding soil (−3 to −5°C) and 25°C warmer than surface air temperature (mean −5 to −23°C) in the field.

In the laboratory, the mean nest temperature of individual voles was lower than that of groups of voles. Voles in both the field and laboratory never all simultaneously vacate the nest, ensuring that nest temperature always remains above ground temperature (Wolf & Lidicker 1981). Winter sociality may also provide opportunities for cooperative defence against predation and guarding of food stores, but can be associated with costs such as increased conspicuousness to predators and higher parasite loads and disease (Wolf & Lidicker 1981; Marchand 2014). Seasonally social mammals may undergo substantial reductions in winter body mass, which presumably reduces total food requirements during winter and limits intraspecific competition. Added insulation from nest building and from restricting activity to the subnivean microclimate also reduces energetic requirements (Chappell 1980). For example, voles build nests from stems and leaves, and the insulation provided by this nesting material reduces heat loss to the environment; a colony of some 19 Brandt's voles accumulate approximately 940 g of winter nesting material (Benedict & Benedict 2001; Zhong et al. 2007). Remaining on the surface can increase energetic costs by 15–25% compared to remaining under the snow, and on cold, clear nights this cost can increase to 40–50%, as temperatures are lower, wind speeds higher, and radiative heat loss is enhanced on the surface compared to the subnivean environment (Chappell 1980).

Small mammals may hoard food to provide an energy supply during cold winter periods when food sources may be limited and ambient conditions become unfavourable for foraging (Marchand 2014). Food, consisting of leaves, grasses, stems, roots, rhizomes, tubers, fungi, and lichens, is stored in piles on the ground underneath snow cover, in underground chambers, amongst rocks or boulders, in tree cavities, and in birds' nests (Benedict & Benedict 2001). For example, Brandt's vole *(Lasiopodomys brandtii)*, which inhabits grasslands and steppes of Mongolia, China, and Inner Mongolia, in regions where winter temperatures may be as low as −40°C, do not hibernate, but store food in the form of plant matter (primarily the prairie sagewort, *Artemisia frigida*) in underground chambers. Groups with a mean number of 16–22 individuals occupy a territory of about 25 m diameter, storing 4.4 to 6.3 kg of food during autumn, with larger groups storing more food (Zhong et al. 2007). A combination of reduced thermoregulatory demands from communal nesting and considerable accumulation of stored food limits winter foraging requirements. Energetic calculations for Benedict's montane voles (*Microtus montanus*) indicate that a group accumulates sufficient food in 3 weeks to support one individual for > 22 days without any additional foraging (Marchand 2014).

Medium and large mammals also have behavioural adaptations that impact on their physiological response to cold and enhance their ability to withstand extremely cold environmental conditions. Amongst primates, huddling, basking, sheltering in caves or snow-covered trees, and postural adjustments such as pressing the furred limbs to the ventral surface of the body, have all been observed (Hori et al. 1977; Zhang et al. 2007). Japanese macaques (*Macaca fuscata*) are the most northerly distributed monkey species, surviving in snow-covered mountains and forests during winter, with some areas of their distribution reaching $T_a$ as low as −20°C (Hori et al. 1977). They have a thick winter pelt that provides sufficient insulation to maintain a sub-zero lower critical temperature when acclimatized to winter conditions. Famously, some Japanese macaques inhabiting the Jigokudani Monkey Park bathe in a man-made, hot-spring-fed pool with a water temperature of 38 – 40°C. Bathing in the hot pool is more frequent in winter than summer (Zang et al. 2007; Figure 4.1), when the monkeys are not fed, and at lower $T_a$, evidence that this behaviour is indeed thermoregulatory. Dominant females and

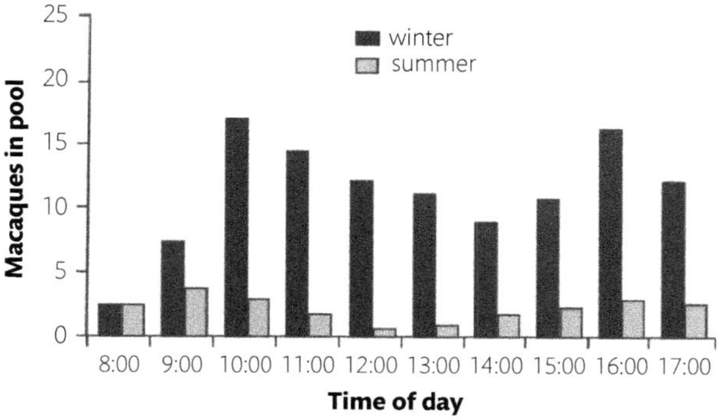


**Figure 4.1** Use of hot pools by a population of Japanese macaques (*Macaca fuscata*) is greater in winter (black bars) than in summer (grey bars). Modified from Zhang et al. (2007).

their offspring bathe more frequently than subordinate females and their offspring, suggesting that hot water is a valued resource. Females are more likely to bathe than males and adolescents; females potentially obtain a greater benefit as they have a lower $T_b$, but also have a greater opportunity to learn this behaviour as they (but not males) maintain site fidelity (Zhang et al. 2007).

Some medium and large mammals are extraordinarily cold tolerant; they can reduce their lower critical temperatures to $<-30°C$ during winter (Scholander et al. 1950b). For example, Norwegian and Svalbard reindeer (*Rangifer tarandus tarandus* and *R. t. platyrhynchus*) reduced their lower critical temperatures from 0 to −30°C and −15 to −50°C from summer to winter, respectively (Nilssen et al. 1984). Much of this thermal acclimation is due to an increase in the insulation of their pelt and associated reduction in whole-body thermal conductance. The fur of large mammals provides excellent insulation in the cold. For example, muskox (*Ovibos moschatus*) in winter have surface temperatures approaching ambient (as low as −40°C), except for small areas around the eyes, snout, and hooves (Figure 4.2). Hart (1956) measured seasonal changes in pelt insulation of temperate and northern Arctic mammals of 12–52% of winter values.

Moulting of the summer pelt occurs in autumn; it is replaced by a winter pelt that has different biophysical properties. The winter underpelt is denser, due to the reactivation of secondary derived hair follicles that regress during summer; more underhairs are therefore produced from shared canals during winter. Winter underfur may also be more crimped than that of a summer pelt, more effectively trapping a layer of insulating still air. Winter guard hairs tend to be finer, longer, and contain medullary air spaces that enhance the pelt's insulative properties (Marchand 2014). Some species, such as the least weasel (*Mustela nivalis*), varying lemming (*Dicrostonyx torquatus*), and Arctic fox (*Alopex lagopus*), also change colour from a darker summer pelt to a white winter pelt, although the functional

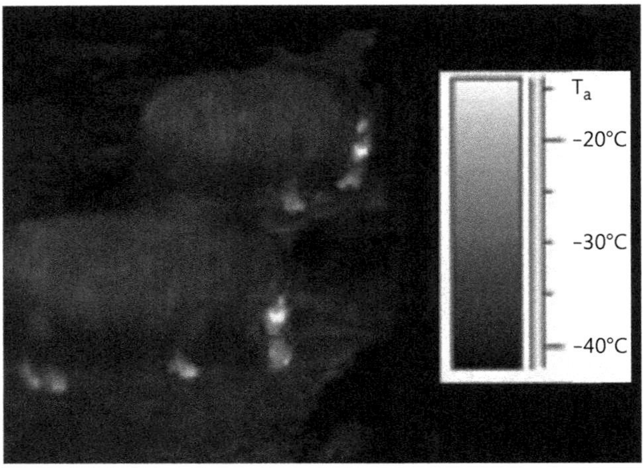


**Figure 4.2** Thermal image of muskox (*Ovibos moschatus*) in winter, showing near-ambient surface temperatures, reflecting effective thermal insulation, except around the eyes, snout, and hooves. Photograph provided by A. J. Munn.

significance of this colour change is likely related to crypsis rather than thermoregulation (Chappell 1980; see 3.2.3.4). For very thick, well-insulating fur, solar heat gain at the level of the skin is limited and colour has minimal impact on thermal balance (Dawson et al. 2013). Small mammals have a limited capacity to increase the insulation of their pelt, as pelt insulation is linearly related to pelt thickness (see 3.2.3.4) and there is a physical limit to how thick a small mammal's pelt can be. Small Arctic mammals therefore have winter pelt insulation similar to that of tropical mammals. However, for Arctic mammals ranging from the size of an Arctic fox (approximately 5 kg) to that of a moose (*Alces alces*; approximately 550 kg), winter pelt insulation is independent of body mass. These medium-to-large mammals also have a much more dramatic seasonal increase in insulation (Scholander et al. 1950a; Hart 1956).

In water, fur pelts of many mammals lose most of their insulative properties, as water replaces air between the hairs. For example, heat loss through polar bear (*Ursus maritimus*) fur increases 20–25 times in ice water compared to air; this increases to 45–50 times for moving water (Scholander et al. 1950a). Despite this, many temperate and Arctic mammals spend all or considerable portions of their time in water with a temperature of $\leq 0°C$. Polar bears use temporal and regional heterothermy, as well as a large body mass (and thus high thermal inertia) and considerable metabolic heat generation to withstand prolonged periods of swimming in icy water. Their pelt provides limited insulation in water and they don't use adipose tissue for insulation; enhanced fat accumulation appears to be only for the purposes of energy storage (Pond et al. 1992; Whiteman et al. 2015).

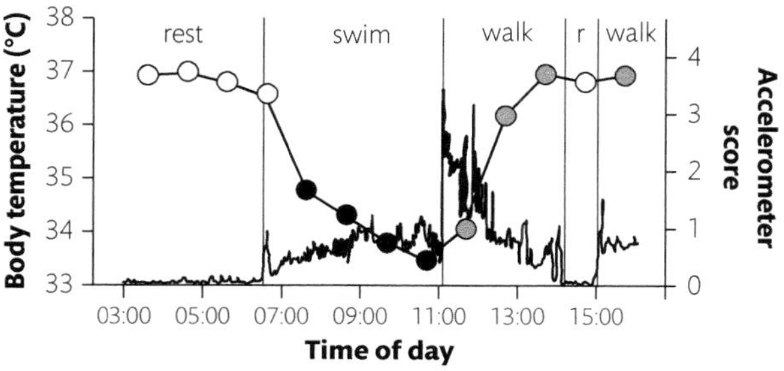


**Figure 4.3** The core body temperature of a wild, free-living polar bear (*Ursus maritimus*) decreases from resting (r, white symbols) during swimming in cold water (black symbols). The bear rewarms during walking (grey symbols) to return to normal $T_b$. Activity level is indicated by the accelerometer score. Redrawn from Whiteman et al. (2015).

This presumably reflects their relatively recent exploitation of marine habitats (0.5 MYBP; Liwanag et al. 2012). During periods of swimming, $T_b$ of free-living polar bears has been observed to fall to $< 35°C$, presumably an active process achieved via vasoconstriction that provides a cool insulating shell for parts of the core (Whiteman et al. 2015; Figure 4.3).

Beaver (*Castor canadensis*), muskrat (*Ondatra zibethicus*), and North American mink (*Mustela vision*) are medium-sized, semiaquatic mammals that forage in water, even during temperate and subarctic winters. Beaver and muskrat construct lodges that provide a favourable micro-environment, and beaver store food, and fat in their tails, to reduce foraging demands (Marchand 2014). All lose considerable heat in water. The thermal conductance of a mink carcass in water is up to 7.9 times greater than in air, and submerged beaver fur has a 10 times higher thermal conductance compared with that in air (Scholander et al. 1950a; Williams 1986). For beavers, the pelt only accounts for 23.5% of whole animal insulation in cold water. Regional heterothermy, facilitated by countercurrent heat exchange in hind foot and caudal *rete mirabilia* (see 3.2.4.1) and in opposing veins and arteries in the thoracic and hind leg regions, reduce heat loss, although beavers still lose heat and experience a decline in $T_b$ at a rate of $0.06°C\ min^{-1}$ when in water at $1-4°C$. This limits their aquatic foraging time; free-living beavers reduce core $T_b$ fluctuations to within 2°C by behaviourally restricting time spent in water (MacArthur & Dyck 1990).

Muskrats and mink also experience declines in $T_b$ during immersion in cold water. For muskrats, dives of only 0.5 to 4 min resulted in a steady decline in $T_b$, with the degree of $T_b$ reduction, post-dive metabolic rate, and recovery time all positively correlated with submersion time, and negatively correlated with water temperature (MacArthur 1984). For mink, heat loss exceeds heat production within 5 min of swimming, heat storage becomes negative ($-8.72W\ kg^{-1}$ at intermediate swimming speeds of $0.36\ m\ s^{-1}$), and $T_b$ declines, even at a moderate water temperature of 20°C. Winter pelages, however, provide more effective insulation in the water than summer pelages (heat storage $-14.28\,W\,kg^{-1}$), more efficiently trapping some insulating air within the pelt. Removing this air resulted in more rapid cooling. This insulation, together with regional heterothermy, particularly from vasoconstriction of blood vessels to the paws, and enhanced metabolic rate, permits mink to forage for short periods in cold water (Williams 1986).

Like mink, river otters spend time on land and in the water, although they forage exclusively in the water. The pelt of river otters is very dense, approximately four times that of a muskrat. It is characterized by guard hairs with a well-developed medulla and interlocking shaft scales, along with a dense and crimped underfur, that trap air. Sebaceous gland secretions coating the hair further limit water penetration to the skin. In addition, their high mustelid basal metabolic rate (BMR) presumably reflects a high capacity for metabolic heat production, and countercurrent exchange in the limbs, peripheral vasoconstriction, and a decrease in heart rate limits heat loss during diving (Marchand 2014).

Sea otters (*Enhydra lutris*) are the smallest marine mammals. Like semi-aquatic mammals, these exclusively marine mustelids rely on a combination of an insulating fur pelt and high metabolism to maintain $T_b$ while almost permanently immersed in cold water (Costa & Kooyman 1982). Sea otters have a high metabolic rate (2.5–3 times allometric predictions; Yeates et al. 2007) and therefore high thermogenic capacity, even compared with other otter species (Kruuk & Balharry 1990). Use of fur rather than blubber for insulation reflects their relatively recent invasion of a marine environment (1.6 MYBP) and their small body size, which would make insulation via blubber impractical. Sea otter fur is exceptionally dense (737–2,465 hair bundles $cm^{-2}$, with 19–91 hairs per bundle), twice that of river otters, and is coated by oily secretions of squalene ($C_{30}H_{50}$) from sebaceous glands that prevent water penetrating the pelt to the level of the skin and maintain an insulating layer of air within the pelt. Sea otter pelts have sparse guard hairs that have interlocking scales at the base, aiding in maintenance of the pelage structure. Underfur hairs also interlock, facilitated by their wavy form and scales, to trap still air. The sloping angle of the hairs (61.9 – 84.3°) also apparently contributes to keeping water from the pelt (Williams et al. 1992).

Maintenance of the pelt requires considerable grooming, > 2 h $day^{-1}$ (Yeates et al. 2007) to remove debris, restore pelt structure, and disperse glandular secretions throughout the pelt. Grooming consists of two phases: washing the fur and then drying it. At the end of a grooming session, sea otters blow air back into the fur or aerate it by churning the water with their forepaws (Kenyon 1969). Contamination of the pelt can lead to greatly increased thermoregulatory costs and hypothermia (Costa & Kooyman 1982). Interestingly, the pelts of European otters (*Lutra lutra*) lose their ability to trap insulating air after several sessions swimming in saltwater, and the otters become prone to hypothermia unless they are able to wash in freshwater and restore the insulating properties of the pelt (Kruuk &

Balharry 1990). This presumably occurs due to salt crystals forming in the fur and disrupting its structure, or the saltwater interfering with the secretions of sebaceous glands and their distribution, or both.

Sea otters rarely leave the water; therefore they are unlikely to have salt crystals form in drying pelts, and they spend considerably longer grooming per day than European or river otters; European otters groom only on land, not in the water (Kruuk & Balharry 1990; Marchand 2014). Use of fur rather than blubber confers several costs for sea otters: air trapped in the fur, although necessary to maintain the pelt's insulative properties, increases buoyancy and therefore contributes to a high cost of diving compared to other marine mammals. Considerable energy—2.4 mJ day$^{-1}$, or 15% of an otter's daily energy expenditure—is also expended on the grooming necessary to maintain the insulative function of the pelt; the metabolic cost of grooming (29.4 ml $O_2$ kg$^{-1}$ min$^{-1}$) is higher than that of feeding and is equivalent to that of swimming (Figure 4.4). Grooming also requires 9.1% of the otters' time budget, but as sea otters spend approximately 40% of their time resting, energetic rather than time costs of grooming are probably more important (Yeates et al. 2007). Sea otter pups can neither swim nor groom despite being born and raised at sea, and so are reliant on their mothers for both floatation and grooming to prevent drowning and hypothermia. Female sea otters spend approximately

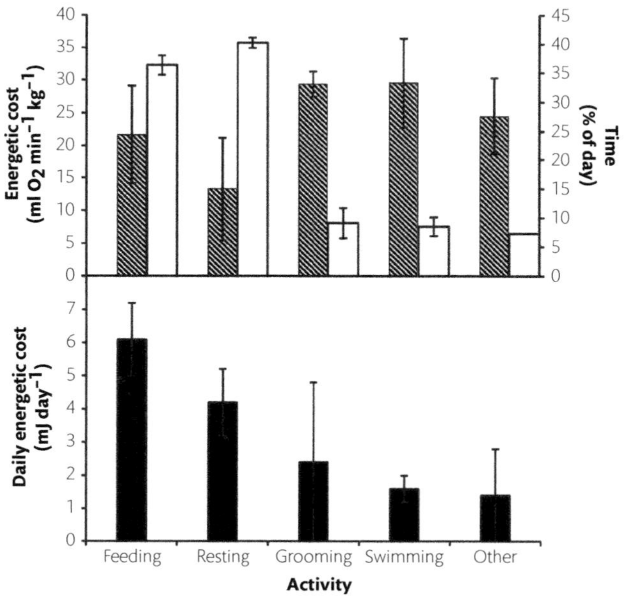


**Figure 4.4** Energetic cost (cross-hatched bars) of various activities for sea otters (*Enhydra lutris*), time spent engaged in each activity (white bars), and overall metabolic cost of each activity (black bars). Data from Yeates et al. (2007).

13% of their time grooming their pups to ensure a waterproof pelt. The layer of air trapped in the pup's pelt is also essential for the pup to maintain buoyancy while the female dives to forage; it is left floating on the ocean surface for these brief periods despite an inability to swim (Cortez et al. 2015).

Pinnipeds, sirenians, and cetaceans spend considerable or all of their time in water. For harbour seals (*Phoca vitulina*), immersion in cold water increases the lower critical temperature from $< -10$ to $+10°C$ (Irving & Hart 1957), but core $T_b$ of seals declines very little during diving at a range of water temperatures (e.g. Gallivan & Ronald 1979; Ponganis et al. 1993), and cetaceans obviously indefinitely maintain core $T_b$ in an aquatic environment. This maintenance of $T_b$ occurs despite restrictions on thermoregulatory heat production that occur due to metabolic adaption to oxygen limitations associated with diving (Boyd 2000). Seals are unusual amongst mammals in using a combination of both fur and blubber for insulation, presumably reflecting their intermediate evolutionary history of 29–23 MYBP in a marine environment, while more recently marine mammals (polar bears and sea otters, < 1.6 MYBP) rely on fur, and exclusively and long-term marine sirenians and cetaceans (50 MYBP) rely on blubber (Liwanag et al. 2012).

Blubber consists of a 'continuous, subcutaneous layer of adipose tissue, reinforced by collagen and elastic fibres' (Liwanag et al. 2012). It is a more effective insulator for deep and/or long-term submersion than fur, when the hydrostatic pressure associated with depth will force air out of the pelt. For example, the ringed seal (*Phoca hispida*) has a very thin pelt, which becomes completely saturated when immersed—but a thick layer of poorly vascularized blubber. Although a poor insulator in air, blubber is much more effective than wet fur in water; the insulation of a seal carcass may be only 5% less in ice water than in 0°C air, but a relatively greater thickness of blubber is required compared with fur (Scholander et al. 1950a; see Figure 3.11). The insulation of blubber can, however, be more readily altered than that of fur, by changing the degree of blood perfusion, and it has the added functions of providing an energy store, aiding buoyancy control, and contributing to attaining a streamlined body form. It has independently evolved in cetaceans, sirenians, and pinnipeds (Liwanag et al. 2012). Amongst pinnipeds, there are allometric, phylogenetic, and ecological patterns in the morphology, biochemistry, and insulative properties of blubber. Fur seals rely more on fur for insulation than sea lions, phocids, and walrus, and this is reflected in differences in fur density and blubber thickness and biochemistry. The various groups of pinnipeds also differ in the composition of the inner and outer layers of blubber: fur seals use their blubber more for storing energy, while phocids and sea lions use the inner layer of blubber for energy storage and the outer layer for thermoregulation, as indicated by fatty acid composition (Liwanag et al. 2012).

Overall, regional heterothermy across the body tissues, facilitated by blubber and vasomotor control, is probably the most important mechanism for maintaining $T_b$

for highly aquatic mammals (Boyd 2000); it can achieve levels of insulation similar to that of the pelt of large terrestrial mammals (Irving & Hart 1957). Indeed, for harbour and harp (*Phoca groenlandica*) seals in cold water, low skin temperatures and steep temperature gradients established over about 6 cm in the body tissues negate any thermoregulatory increase in metabolic rate. Heat loss across the skin of free-ranging Antarctic fur seals in water of $1.4 - 4°C$ ranged from 30 W $m^{-1}$ during diving to more than 600 W $m^{-1}$ while swimming at the surface, the mean being more than twice that recorded for dolphins in warmer tropical water. The skin-to-water thermal gradient of seals was highly variable during their time at sea, ranging from almost 0 to $> 20°C$, and was labile over short time periods of $< 1$ h. This indicates that seals and presumably other largely aquatic mammals use skin temperature as an important thermoregulatory mechanism in aquatic environments (Boyd 2000).

Regional heterothermy that results in cooling of the extremities is common in medium and large cold-climate mammals (and birds) in winter. This reduces heat loss from appendages that have a high surface area to volume ratio. It is generally achieved by countercurrent heat exchange between warm arterial blood and cool venous blood, with anatomical specializations, such as closely opposed vessesls or rete, enhancing the effect. For example, Irving and Hart (1957) observed regional heterothermy in the flippers of harbour seals in water at 6°C. For seals with a core body temperature of 37°C, the tissue temperature amongst the metacarpals was 22°C at a distance of 2.5 cm from the body and $9 - 15°C$ at 5 cm; the subcutaneous temperature of the flipper web was $6 - 7.8°C$ (Irving & Hart 1957). It is essential, however, that this regional heterothermy is controlled, and the temperature of the extremities is regulated above the freezing point of tissues, even when the surface that the animal is contacting may be $> 50°C$ colder. Tropical and non-acclimated temperate mammals generally use vasoconstriction to limit heat loss from the extremities, and undergo pulsatile increases in blood flow to these regions to delay freezing of the tissues.

Cold-acclimatized temperate mammals, such as wolves (*Canis lupus*) and Arctic foxes, use continuous proportional temperature control via regulated vasodilation that allows for warm blood to maintain the appropriate foot temperature (Henshaw et al. 1972). Anatomical specialization of vascular system of the foot allows warm blood to flow directly to the footpad surface, and differential flow of this blood regulates footpad temperature to prevent freezing. For example, adult wolves and Arctic foxes regulate footpad temperature at 3.9°C (variation $\leq 0.7°C$) at temperatures of $-38°C$ for periods of 7.5 h (Henshaw et al. 1972). This may also allow for enhanced heat loss from highly insulated animals during exercise. Indeed, infrared thermography of Arctic, red (*Vulpes vulpes*), and kit (*Vulpes macrotis*) foxes indicated that the lower legs and paws function as effective thermoregulatory surfaces (Klir & Heath 1992).

Moving through snow dramatically increases the cost of transport (COT) for mammals. The deeper and denser the snow, and the faster the animal moves, the greater is the increase in COT. Deep snow may necessitate a change of gait to a bound that requires considerable energy for the complete vertical displacement of the body, and dense snow increases foot drag and prevents wading, requiring the animal to lift its legs free of the snow with each stride (Parker et al. 1984; Marchand 2014). Therefore, considerable energetic costs are associated with foraging for winter-active mammals. For example, energetic costs of simply standing compared to lying down for most ungulates increases metabolic rate by about 21–37%. COT is roughly linearly related to velocity, and increases for uphill and decreases for downhill locomotion. Relative increases of COT increase exponentially with snow depth for mule deer (*Odocoileus hemionus*) and elk (*Cervus elaphus*) walking in snow. In powdery snow with a density of 0.2 g $cm^{-3}$, the relative increase in COT is approximately 40% at a sinking depth of 30% of brisket height, increasing to approximately 375% at a sinking depth of 90% of brisket height. These costs are further increased for snow of higher density—approximately 600% increase in COT at a sinking depth of 80% for wet snow with a density of 0.4 g $cm^{-3}$ (Parker et al. 1984). An animal's foot loading, leg length, and movement velocity also determine how much it will sink in snow and therefore impact on the energetics of locomotion. For example, powdery snow results in greater increases in COT for mule deer compared with elk, as deer have shorter legs, so sinking to the ground means they sink relatively deeper. However, in dense snow, elk have a greater COT increase compared with deer, as deer have proportionally longer legs and lower foot loading, reducing relative sinking depth (Parker et al. 1984). Mammals can also behaviourally mitigate some of the energetic costs of locomotion in snow by using preformed trails, selecting areas of packed snow, or modifying their gait. For example, coyotes in south-eastern Quebec select for shallower and harder snow to reduce their sinking depth, and during conditions of heavy snow preferentially travel on trails of artificially packed snow. These behavioural responses were calculated to reduce heart rate by approximately 5% and therefore presumably to result in energy savings (Crête & Larivière 2003).

Telfer and Kelsall (1984) developed morphological, behavioural, and combined snow-coping indices for a range of large North American temperature and sub-Arctic mammals (Table 4.1). Caribou and moose have the highest morphological snow-coping indices, followed by the predatory wolves, wolverines, and coyotes. Bison (*Bison bison*) and pronghorn antelope (*Antilocapra americana*) had the least favourable indices. Amongst ungulates, behavioural indices were similarly ranked (with the exception of white-tailed deer, *Odocoileus virginianus*, which appears to behaviourally compensate for limited morphological adaption to snow), and therefore so were overall snow-coping indices, with caribou the best snow-adapted species examined. These indices correlate with the severity of snow cover within the species' distributions (Telfer & Kelsall 1984).

**Table 4.1** Morphological ((chest height + foot loading) / 200), behavioural (score of 0–5 for each of six behavioural traits/30), and overall mean snow-coping indices for large North American temperate and sub-arctic mammals. Data from Telfer and Kelsall (1984).

| Category | Species | Morphological Index | Behavioural Index | Mean Overall Snow-coping Index |
|---|---|---|---|---|
| Ungulates | Caribou (*Rangifer tarandus*) | 154 | 26 | 0.82 |
| | Moose (*Alces alces*) | 140 | 19 | 0.67 |
| | White-tailed Deer (*Odocoileus virginianus*) | 112 | 21 | 0.63 |
| | Wapiti (*Cervus canadensis*) | 118 | 18 | 0.60 |
| | Dall Sheep (*Ovis dalli*) | 121 | 17 | 0.59 |
| | Bighorn Sheep (*Ovis canadensis*) | 114 | 16 | 0.55 |
| | Bison (*Bison bison*) | 95 | 15 | 0.49 |
| | Pronghorn Antelope (*Antilocapra americana*) | 81 | 13 | 0.42 |
| Carnivores | Coyote (*Canis latrans*) | 133 | | |
| | Wolf (*Canis lupus*) | 135 | | |
| | Wolverine (*Gulo gulo luscus*) | 135 | | |

Snow (and ice) provide problems in addition to locomotion for mammals in extremely cold environments. Freezing of freshwater sources means that some mammals must ingest frozen water in the form of snow or ice to drink. Ingested frozen water must first be melted, which requires 334 J $g^{-1}$ (latent heat of fusion) and then warmed to $T_b$, another 4.19 J $g^{-1}$ $°C^{-1}$ (specific heat capacity of water; Withers 1992). For small mammals active in subnivean environments, energetic costs associated with consuming frozen water range from 2% (red-backed voles; *Clethrionomys rutilus*) to 12.9% (meadow voles; *Microtus pennsylvanicus*) of their daily energy expenditure during winter (Whitney 1977; Holleman et al. 1982; Berteaux 2000). These energetic costs can be met for winter-active mammals by consuming additional food. However, the necessity to arouse and drink cold or frozen water for the hibernating mountain pygmy possum (*Burramys parvus*) places limitations on hibernation duration and therefore may influence overwinter survival. Ingesting 5% of body mass of cold (2°C) water and warming it to 35°C requires as much energy as 13 h of torpor, and ingesting frozen water requires the energetic equivalent of 45 h (Cooper & Withers 2014b). Over the entire hibernation season, drinking cold water may reduce the potential hibernation period by 11.5 days and eating snow by 30 days, and costs are even higher for juvenile pygmy possums. Consuming cold or frozen food and warming it to $T_b$ also has an energetic cost, which is dependent on food water content, energy content, and digestibility, as well as the $T_a$. However, these costs are generally low compared to drinking cold or frozen water.

Predators rarely experience cold-associated energetic costs of feeding, as freshly caught prey have a $T_b$ similar to their own, and insectivores typically consume high-energy, readily digested food. Herbivores, however, may feed during the winter months on poor-quality, sometimes frozen plant matter that must be melted and warmed (Chappell 1980). These herbivores may preferentially select for plant tissues that avoid freezing at low $T_a$ by accumulation of solutes and therefore avoid the high costs of melting frozen water. For dry plant matter, the energetic costs of warming cold food is relatively low, as cellulose has a specific heat capacity only about one-third that of water (1.3 J $g^{-1}$ °$C^{-1}$) and there is no latent heat of fusion (Berteaux 2000).

Most small, non-hibernating mammals decrease body mass during winter, and some actually reduce body size (a phenomenon known as the Dehnel effect; see 3.2.7), which functions to reduce overall energetic requirements by reducing absolute metabolic rates and therefore food requirements (Heldmaier 1989; Lovegrove 2005). Small species appear to have the most extreme seasonal mass changes; the smallest species may decrease mass by as much as 50% (Lovegrove 2005). Larger species generally increase winter body mass as a consequence of fat storage that may be important for withstanding winter food shortages (Heldmaier1989), although Lovegrove (2005) found no effects on body size or latitude associated with the magnitude of seasonal body mass change for large mammals in general. There are, however, exceptions: body mass of reindeer decreased 8.6 and 3.8% in winter compared to summer (Nilssen et al. 1984). Mass changes may relate to the severity of the local winter conditions. For example, winter body mass reductions were most extreme for the common shrew (*Sorex araneus*) in Finland, were intermediate in Poland, and were smallest in the Rhine valley. Similar patterns are evident for voles (Heldmaier 1989). Lovegrove (2005) found a significant correlation of body mass changes with latitude for mammals after accounting for body size and phylogenetic history.

Many less-well insulated mammals, particularly smaller species, do have to increase their metabolic heat production during winter months. For placental mammals, an increase in brown fat enhances their capacity for non-shivering thermogenesis, and higher myoglobin concentration in skeletal muscle also increases the capacity for oxygen transfer and storage, and thus shivering thermogenesis (Heldmaier 1989; Marchand 2014). Despite these increases in thermogenic capacity, significant decreases in body mass generally result in reduced absolute BMR for small mammals during winter (Lovegrove 2005). Other species decrease their metabolic rate, presumably to reduce food requirements when feed may be limited during winter. For example, the resting metabolic rate of Svalbard and Norwegian reindeer decreased from 2.15 to 1.55 and 2.95 to 2.05 W $kg^{-1}$, summer to winter, respectively. This was accompanied by a decrease in food intake of 57 and 55%, respectively (Nilssen et al. 1984).

## 4.1.2 Avoiders

Some mammals withstand cold winter conditions and associated food limitations by avoiding them. Some migrate to more equable climates (see 4.6.3), but others use long-term hibernation and remain in a secure hibernaculum and reduce $T_b$ and metabolic rate—often to as low as 0°C and 1% of BMR—and await milder spring or summer conditions, when they typically arouse to reproduce. Hibernacula often have more favourable microclimates than the extreme external climatic conditions. For example, the winter subnivian hibernaculum of the mountain pygmy possum remains at a constant temperature of 1.5 – 2.5°C, despite ambient air temperature fluctuating from –8 to 20°C. This constancy occurs due to the insulating effects of overlying rocks, soil, and in particular snow, and is essential for successful overwintering of the possums (Körtner & Geiser 1998). Early snowmelt eliminates the insulative effects of snow and reduces hibernacula temperatures. This results in possums expending more energy during torpor, undergoing more frequent and longer periodic arousals, and undertaking their final arousal before the availability of their spring food, all of which can result in substantial population declines (Smith & Broome 1992; Broome 2001).

Some monotremes, marsupials, and placentals hibernate seasonally for extended periods (although all long-term, cold-climate hibernators undergo periodic arousals during this hibernation period; see 3.2.4.2). Short-beaked echidnas (*Tachyglossus aculeatus*) in cold regions such as the Australian Alps and Tasmania hibernate for up to seven months (Grigg et al. 1992; Nicol & Andersen 2002). Generally, they enter hibernation 68 days after the summer solstice; males arouse about 5 days before the winter solstice; reproductive females arouse about 30 days later; and non-reproductive females may continue to hibernate for another 2 months (Nicol & Andersen 2002). During hibernation, $T_b$ declines considerably (typically < 10°C), but there are periodic arousals throughout the hibernation period (see 3.2.4.2).

Amongst marsupials, the Australian pygmy possums (Burramyidae) and feathertail glider (*Acrobates pygmaeus*; Acrobatidae), as well as the South American monito del monte (*Dromiciops gliroides*; Microbiotheria) undergo long-term seasonal hibernation to avoid cold winter conditions (Bozinovic et al. 2004; Riek & Geiser 2014). The marsupial mountain pygmy possum is a typical hibernator, hibernating for up to 7 months, beginning 130 days after the summer solstice and arousing 100 days after the winter solstice (Nicol & Anderson 2002). The longest recorded hibernation period for any mammal is for the eastern pygmy possum (*Cercartetus nanus*) in the laboratory. Pygmy possums were able to extend their hibernation period to last on average 310 days without food, with one individual lasting 367 days, the only report to date of a mammal extending hibernation for more than 1 year (Geiser 2007).

Placental mammals of the orders Insectivora, Chiroptera, and Rodentia (particularly the families Sciuridae, Cricetidae, and Gliridae) also use seasonal hibernation in cold environments (Ruf & Geiser 2015). Indeed, hibernation was first described by Aristotle (384–322 BC) for the edible dormouse (*Glis glis*; Cooper & Withers 2010). Placental seasonal hibernators typically hibernate for some 7 or so months; sciurids and marmots enter hibernation between 62–72 days after the summer solstice and finally arouse 112–130 days after the winter solstice (Kenagy et al. 1990; Armitage 1998). The longest recorded hibernation season observed for a free-living mammal was for dormice in Austria, in a year of European beech (*Fagus sylvatica*) mast failure. Individuals with a large body mass entered hibernation early, in late July, forgoing reproduction to extend the hibernation season up to 11.4 months (Hoelzl et al. 2015).

There is evidence that during the prolonged hibernation season, circadian cycles cease. Despite some indication of minor circadian cycles for mammals undertaking hibernation in the laboratory, circadian cycles were not apparent for free-living Arctic ground squirrel (*Urocitellus parryii*) $T_b$ or, for males, arousals during or immediately after the hibernation period, but commenced once the squirrels emerged from their hibernacula and were exposed to daylight (Williams et al. 2012; Figure 4.5). Circadian cycles of the clock genes *Per1, Per2*, and *Bmal1* and the clock-controlled gene *arginine vasopressin* are abolished during hibernation in European hamsters (*Cricetus cricetus*), and levels of mRMA for the melatonin rhythm-generating enzyme arylalkylamine N-acetyltransferase remain constant over a 24-h period (Revel et al. 2007). These observations support the hypothesis that hibernation stems from extension of the circadian clock, so that the hibernation period can be considered a single long circadian period (e.g. day). There is a potential molecular explanation for this: normal circadian cycles are driven by transcription of cryptochrome and period (per) genes by CLOCK and BMAL1 transcription factors, which then inhibit *clock* and *bmal1* gene expression. Thus the circadian cycles are driven by oscillations that are likely to be eliminated when the transcription, translation, and mRNA and protein degradation processes they depend on are inhibited by hibernation (van Breukelen & Martin 2015).

Long-term cold-climate hibernators show predictable circannual hibernation patterns (and other seasonal responses, such as activity, body mass, and reproduction) that are intrinsically entrained, and do not rely on external environmental cues. For example, it has been well established that ground squirrels maintain an annual cycle of activity and hibernation even when maintained at constant $T_a$ and photoperiod (Pengelley & Fisher 1963). This level of circannual entrainment reflects their seasonally predictable environments and allows them to still

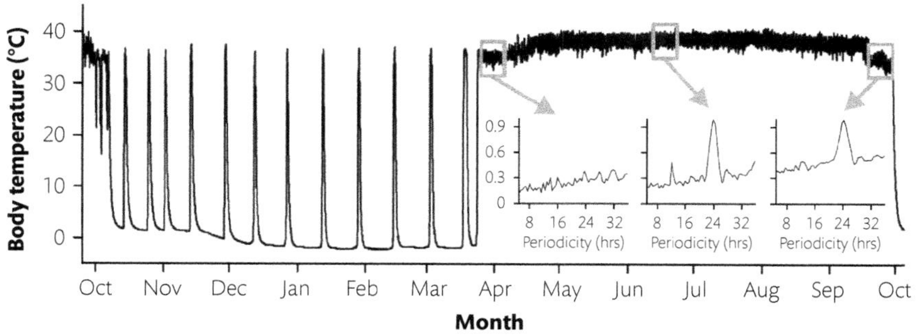


**Figure 4.5** Body temperature during the hibernation period of a male Arctic ground squirrel (*Urocitellus parryii*). Insets show absence of periodicity of body temperature cycles in darkness at the end of hibernation but 24-h entrained rhythms after the squirrel leaves the hibernaculum and is exposed to daylight. Redrawn from Williams et al. (2012). Reproduced with permission of Biology Letters, The Royal Society.

undertake required activities, even when environmental cues are unavailable (e.g. within sealed, underground hibernacula; Körtner & Geiser 2000).

Unlike circadian clocks, the location and underlying mechanisms that determine the function of the circannual clock are still unknown, which hinders understanding of its role in hibernation (Williams et al. 2014a). Despite evidence for intrinsic entrainment of circannual hibernation patterns, changes in the environment require some degree of entrainment to external cues to ensure the circannual cycle remains synchronized to environmental conditions. For example, ground squirrels and marmots hibernate irrespective of changing photoperiod or pinealectomy prior to hibernation, but long-term changes to photoperiod, such as a change in hemisphere or pinealectomy well before hibernation, do interfere with the timing of hibernation, suggesting there is a short period in summer during which the circannual clock of these hibernators is sensitive to photoperiod (Körtner & Geiser 2000). Observations that different populations of ground squirrels have varying hibernation characteristics related to environmental variables, such as altitude, aspect, and snow cover, are further evidence of some plasticity or adaptability in the timing of entrained hibernation (Williams et al. 2014a). $T_a$ appears to have some impact on the circannual cycle of hibernators, with excessively cold spring and autumn temperatures influencing the length of the hibernation period. This presumably has important ecological consequences, allowing hibernating mammals to time their emergence to best exploit or avoid variations in food abundance (Körtner & Geiser 2000).

Hibernators generally rely on fat stores to sustain their reduced metabolic rate during hibernation (although some such as chipmunks, *Tamias*, feed on cached food during interbout arousals), and many will not hibernate or reduce their hibernation period if they have not attained a suitable body mass (Geiser 2013).

For example, only dormice with high body fat extended their hibernation period by entering hibernation early in years of beechmast failure (Hoelzl et al. 2015). A reduction in respiratory quotient (RQ) of hibernators from 1 (carbohydrate metabolism) to 0.7 (fat metabolism) is consistent with hibernators metabolizing fat rather than carbohydrate during hibernation, and pre-hibernation fattening is accompanied by changes in circulating leptin concentrations and leptin sensitivity (Kronfeld-Schor et al. 2000). Age, sex, and reproductive status also impact hibernation; for example, adult male echidnas arise from hibernation much earlier than females, and seek out and mate with females still torpid. Non-reproductive females may hibernate for 2–3 months longer (Nicol & Anderson 2002; Morrow & Nicol 2009). Male ground squirrels also arouse from hibernation and exit their hibernacula earlier than females (Williams et al. 2012), presumably in response to hypothalamus-pituitary-gonadal axis endocrine changes that precede the reproductive season. Social hibernators such as marmots are much more synchronized (Williams et al. 2014a).

Extreme seasonal hibernation requires mammals to undergo a suite of physiological, biochemical, and morphological changes that would prove fatal for other mammals, but the cellular and molecular basis for these changes are not well understood. For example, there is still no clear understanding of the molecular basis for initiating and terminating hibernation, despite extensive research (Frerichs et al. 1998; Carey et al. 2003), although there is some evidence of a role for thermoregulatory neurotransmitters such as serotonin, histamine, and opioids on central nervous system control of hibernation (Sallman et al. 1999).

Examples of seasonal hibernators are found in all three mammalian linages, and heterothermia is believed to be a plesiomorphic trait amongst mammals, so the ability to hibernate is most likely due to patterns in gene expression, rather than to the presence or absence of particular genes (Srere et al. 1992; Carey et al. 2003; Xu et al. 2013). For example, mRNA for genes encoding for thyroxine-binding globulin, apolipoprotein A1, cathepsin H, CIRBP, and $\alpha_2$-macroglobin are upregulated in ground squirrels (*Spermophilus*) during the hibernation season, presumably to enhance function during hibernation; $\alpha_2$-macroglobin, for instance, plays a role in the increased blood clotting times of hibernators (Srere et al. 1992).

The molecular response during hibernation is complex, involving a suite of modifications in gene expression occurring in a variety of tissues and organs, and is probably regulated by metabolic status (van Breukelen & Martin 2015). Some recognized molecular changes during hibernation include differential gene expression that regulates switching from carbohydrate to fatty acid metabolism, with downregulation of mRNA and associated proteins of glyceraldehyde-3-phosphate dehydrogenase and acetyly CoA carboxylase, and upregulation of pyruvate dehydrogenase kinase isoenzyme 4, pancreatic triacylglycerol lipase, hormone-sensitive lipase, and transcription factor PPAR.

Changes in heat production capacity associated with the necessity for periodic arousals are reflected in upregulation of mRMA for uncoupling proteins 2 and 3, and heart- and adipose-type fatty acid binding proteins in tissues including white and brown adipose tissue, skeletal muscle, heart, kidneys, and liver, and in increased myoglobin proteins in skeletal muscle (Carey et al. 2003; Xu et al. 2013). Downregulated mRNAs include those associated with activity, such as prostaglandin $D_2$ in the brain and glycerabdehyde-3 phosphate dehydrogenase in muscle. Other changes in gene expression that occur during hibernation include upregulation of nRMA and proteins associated with protecting the body of hibernators from negative impacts of long-term cold exposure, such as increased risk of ischaemia leading to redox shifts and oxidative stress; mitochondrial proteins such as cytochrome-c oxidase subunit1, ATP synthase 6/8, transcription factor nuclear factor-kB, and glucose-regulated protein 75 (Carey et al. 2000; Hittel & Storey 2002). Despite these reported changes in gene expression, mRNA and proteins associated with the majority of genes remain at normothermic levels during the hibernation season, presumably to facilitate function during periodic arousals (Carey et al. 2003).

As for many physiological processes, numerous cellular and molecular processes are slowed, or cease altogether, during long-term seasonal hibernation (van Breukelen & Martin 2015). RNA transcription is an energetically expensive process; it is reduced during hibernation via reductions or cessation of initiation and elongation processes, consistent with a $Q_{10}$ effect of reduced $T_b$ (van Breukelen & Martin 2002). Translation of mRNA into protein is also affected by hibernation. Protein synthesis is reduced in hibernating compared to normothermic animals, and there are significant losses of polyribosomes in numerous organs of hibernators. Unlike transcription, there is evidence of active suppression of this even more energetically costly process beyond temperature effects (Carey et al. 2003). Brain extracts from hibernating individuals have significantly less translational activity than those from normothermic individuals, but placing polyribosomes from hibernating animals in extracts from normothermic individuals overcomes this suppression (Frerichs et al. 1998). DNA synthesis and cell division are also impacted during hibernation, with DNA synthesis occurring at about 4% of the rate of normothermic animals and mitosis ceasing altogether; it remains unclear if there is regulated suppression of these processes, or if these responses are entirely a consequence of low temperature. They do, however, return to or even exceed pre-hibernation rates during interbout arousals (Carey et al. 2003). Changes in the ultrastructure of cells and mitochondrial respiration also reflect cellular and molecular alternations during torpor.

Seasonal hibernators tend to increase the proportion of monounsaturated and polyunsaturated fatty acids (MUFAs and PUFAs, respectively) in their tissues, which have relatively low melting points compared to saturated fatty acids (SFAs) and may play an important role in maintaining fluidity of body lipids. Lipids can be accessed for metabolism only if they are in a liquid state. Increases in the proportion of PUFAs in cell membrane phospholipids are also observed in deep hibernators, presumably to aid in maintenance of normal membrane function at low $T_b$ by maintaining the usual liquid-crystalline state (Geiser 1993; Florant 1998; Munro & Thomas 2004). Indeed, golden-mantled ground squirrels on a high PUFA diet in the laboratory had a higher hibernation frequency, higher survival, and attained lower $T_b$ than those on diets with lower PUFA concentrations (Frank 1992). Mammals are unable to synthesize PUFAs and therefore must obtain them from their diet; these dietary PUFAs then influence the fatty acid composition of the tissues (Geiser 1990).

Meta-analysis of the impacts of experimental and natural diets on hibernation reveals that low PUFA diets, with about 11% PUFA concentrations, do limit hibernation, with mammals maintaining higher minimum $T_b$ and torpor metabolic rates and having shorter hibernation durations. Providing higher PUFA diets enables mammals to reduce minimum $T_b$ and torpor metabolic rate by incorporating more PUFA into their lipids. However these changes are small (about 1 – 1.2°C for a change in PUFA concentration from 11.2 to 54.4%) and occur at moderate as well as low torpor $T_b$s, suggesting that factors other than lipid state impact the interaction between PUFA availability and hibernation characteristics, possibly by influencing the $T_b$ setpoint or temperature perception (Geiser 1993; Munro & Thomas 2004).

Some mammals show little or no selection for dietary PUFAs, and for some species, diets high in UFAs have no significant impact on hibernation characteristics (e.g. monito del monte; Contreras et al. 2014). Species such as echidnas and other insectivores, which have low concentrations of PUFAs in their natural diet, may substitute MUFAs such as oleic acid in their adipose tissue and membranes, retain and selectively ingest the limited available dietary PUFAs, and employ cholesterol to aid in maintaining lipid fluidity (Schalk & Brigham 1995; Falkenstein et al. 2001). Despite their apparent benefit to lipid and membrane function at low $T_b$, PUFAs are vulnerable to auto-oxidation. During hibernation, mammals are already exposed to enhanced oxidative stress, and therefore the ideal PUFA concentration for a hibernator may involve a trade-off between minimizing potential auto-oxidative effects and maintaining lipid and membrane fluidity. Indeed, evidence suggests that mammals with access to high PUFA diets optimize rather than maximize their PUFA intake; if PUFA levels are too high, hibernation may be limited to restrict negative oxidate effects (Munro & Thomas 2004). For example, the PUFA content of pre-hibernation diets of free-living Arctic ground squirrels vary more than threefold, although no individuals appear to select low PUFA diets. Those ground squirrels that consume an intermediate level of PUFAs (33–74 mg $g^{-1}$) have higher overwinter persistence, longer torpor bouts, fewer arousals, and more torpid days during the hibernation season than those with high PUFA diets (> 74 mg $g^{-1}$), demonstrating the impact of PUFAs on hibernation for wild herbivores (Frank et al. 2008).

The changes in $T_b$ accompanying torpor and hibernation have an impact on acid–base status, because of the effect of temperature on dissociation of water (see 2.5.6), the increased $CO_2$ solubility at low $T_a$, and physiological control of ventilation relative to metabolic rate (MR). For example, daily torpor of the little pocket mouse (*Perognathus longimembris*) is associated with a decrease in MR to 0.05 ml $O_2$ $g^{-1}$ $h^{-1}$ at $T_a = 10°C$ (from 7.04 ml $O_2$ $g^{-1}$ $h^{-1}$ for normothermia), a decrease in minute ventilation ($V_I$) to 6 ml air $g^{-1}$ $h^{-1}$ (from 329), and an increase in $V_I$/MR to 120 ml air ml $O_2^{-1}$ (from 47). This change in $V_I$/MR is reflected in a decreased blood $pCO_2$ during hibernation to 1.9 kPa (from 4.8) and increased pH of 7.51 (from 7.28); this is a ΔpH/ΔT change of about $-0.0085U°C^{-1}$, which is intermediate between that expected from the ionization of water (α-stat hypothesis, −0.017) and constancy of pH (pH-stat hypothesis, 0). The blood $[HCO_3^-]$ did not change from 17.3 to 18.8 mmol $l^{-1}$. For hibernating hamsters (*Cricetus cricetus*), $pCO_2$ also declined (6.0 to 4.4 kPa), but $[HCO_3^-]$ increased (28.2 to 53.8 mmol $l^{-1}$), and pH increased (7.40 to 7.57), with $\Delta pH / \Delta T = -0.006$; for hibernating marmots (*Marmota marmota*), $pCO_2$ declined (5.5 to 4.9 kPa), $[HCO_3^-]$ increased (29 to 52.8 mmol $l^{-1}$), and pH increased (7.45 to 7.57), with $\Delta pH / \Delta T = -0.004$ (Malan et al. 1973). Kreienbühl et al. (1976) reported similar changes for hibernating dormice (*Glis glis*) in $pCO_2$ (5.1 to 3.7 kPa) and pH (7.24 to 7.44), with $\Delta pH / \Delta T = -0.006$.

These results for hibernators are similar to those for the daily torpidator, although blood $[HCO_3^-]$ increased during hibernation, indicating a slightly different acid–base balance, perhaps reflecting the longer temporal scale for hibernators. The reduction of MR during torpor and hibernation is caused by inhibition of cold-induced thermogenesis, the $Q_{10}$ effect on MR due to the decreased $T_b$, and possibly a further metabolic depression caused by $CO_2$ retention and tissue acidosis (Malan 1982, 1988, 2014; see Figure 3.16). Hyperventilation during arousal from hibernation provides further evidence that acidosis might have a metabolic depression effect that needs to be removed for arousal. Geiser (2004) calculated that hibernating mammals have a considerably higher $Q_{10}$ for MR (3–6) than daily heterotherms (1.5–3.5) (cf. typical $Q_{10}$ values of about 2.5 for mammals; Guppy & Withers 1999).

The $T_b$ of at least one mammalian hibernator, the Arctic ground squirrel, drops below 0°C during hibernation, and therefore there is a risk of body tissues freezing (Barnes 1989). They have a $T_b$ as low as −2.9°C while in their hibernacula surrounded by soil temperatures of −6°C, but they do not freeze. Plasma solute concentrations of ground squirrels can account for depressing of the freezing point to −0.6°C, but below this, the ground squirrels supercool. Supercooling occurs when an absence of a nucleating agent allows for cooling below the freezing point without freezing. There is as yet no evidence of antifreeze proteins in ground squirrel blood. Supercooling to $T_b$s as low as −5°C, with subsequent rewarming and survival, has been achieved for small mammals in the laboratory, but only for periods of 50 to 70 min (Kenyon 1961). Nevertheless, Arctic ground squirrels are able to remain in a supercooled state for periods of approximately 3 weeks, between interbout arousals.

Bears (*Ursus*) are an interesting example of mammals that avoid extreme cold winter conditions by becoming inactive, as their physiological characteristics during this seasonal period of dormancy differ considerably from those of other hibernating mammals. This has led to a long-running debate concerning the nature of their winter quiescence: do they hibernate like other smaller mammals, or do they just enter a state of extreme inactivity and prolonged fasting (Hellgren 1995)? Bears fatten considerably in the months preceding the winter hibernation season, and then den for up to 7 months during which they do not eat, urinate, or defaecate. They do, however, maintain some limited activity during this time—standing, drinking, and arranging their nesting material every 1–2 days, for an average of 24 min per day (Tøien et al. 2011; Robbins et al. 2012; but see Folk et al. 1976), and can be easily roused to an active state (Nelson et al. 1983).

Unlike other hibernators that drop their $T_b$ setpoint during hibernation to close to $T_a$, often approaching (or even dropping below) 0°C, bears maintain a relatively high $T_b$, only a few degrees below their normothermic $T_b$ of 37 – 38°C. For example, black bears (*Ursus americanus*) overwintering in dens with a $T_a$ of 0 to –20°C maintained mean $T_b$s of 31.7 – 34°C during mid-hibernation, and as high as 36 – 37°C towards the end of the denning period (Tøien et al. 2011; Figure 4.6). This high $T_b$ is likely a consequence of a large body mass, which limits cooling rates and makes the energetic costs and rate of arousal from very low $T_b$ prohibitive. Bears do not undergo the periodic arousals to normothermia characteristic of all other cold-climate seasonal hibernators, presumably because their relatively high $T_b$ does not result in the same degree of perturbation of homeostasis as the very low $T_b$ of other hibernators, as observed for the tropical hibernating tenrec (*Tenrec ecaudatus*; Lovegrove et al. 2015). However Tøien et al. (2011) did observe

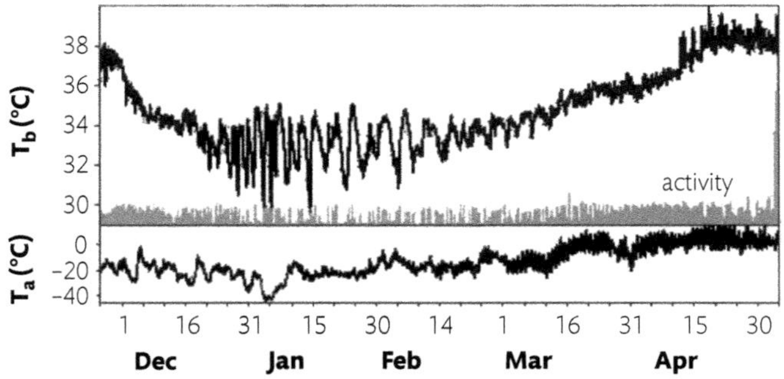


**Figure 4.6** Body and ambient temperatures during overwinter denning by black bears (*Ursus americanus*). Redrawn from Tøien et al. (2011).

a 1.6–7.3 day cycle of $T_b$, with an amplitude of $2-6°C$ that was accompanied by shivering and increased MR, that may reflect periodic arousals, although the circadian cycle appears to have been abolished, as for smaller hibernators.

Metabolic rate of denning bears over winter is reduced to 30–75% of their BMR (Hellgren 1995). For polar bears, the metabolic rate is 1.7–2.2 times higher for fasting but active bears than it is for denning individuals (Robins et al. 2012), which together with observations for black bears of MR reduced well below that expected from the $Q_{10}$ effect of reduced $T_b$, and maintenance of low BMR even after spring exit from the den at normothermic $T_b$, suggests that bears may indeed use metabolic depression to enter a form of hibernation (Tøien et al. 2011). Denning winter bears also have reduced heart rates, from a normal resting rate of 40–60 beats $min^{-1}$ to as low as 8–27 beats $min^{-1}$ (Folk et al. 1970). Denning bears generally lose 15–37% of their body mass during the hibernation period. Their overwinter MR is supported by metabolism of adipose stores rather than lean body mass, evidenced by observations of RQs of $< 0.73$. For example, 74–99% of the body energy content lost by polar bears was from adipose tissue. However, the quantity of adipose stores at the initiation of fasting appears to determine the ability of polar bears to conserve lean body mass (Atkinson et al. 1996).

Amongst hibernating black and grizzly bears, there is no evidence of development of ketosis, as would be expected for a mammal undergoing severe fasting. Fasting but active polar bears in summer have similar blood chemistry to denning black bears in winter (Nelson et al. 1983). Bears suppress urination during the denning period, but a post-hibernating grizzly bear produced only 181 ml of urine after 4.5 months, containing 0.98 g of urea and 72 mg of ammonia, compared to a control sample collected in the bladder over 24 hours, which had a volume of 2,080 ml and contained 55.05 g of urea and 1,785 mg of ammonia (Folk et al. 1976). Remarkably for a mammal, denning winter bears produced urine daily, but the water and solutes were reabsorbed through the bladder wall (Nelson et al. 1975). There were no or very small net changes in blood amino acids, protein, urea, uric acid, or ammonia due to effective protein anabolism rather than due to starvation or cessation of urea production. Regulation of urea balance appears essential for maintenance of hibernation; experimental disruption results in disruption of hibernation (Nelson et al. 1975). Protein metabolism of denning bears increases three to five times, allowing for amino acids to be incorporated into proteins rather than forming urea, and possibly contributing to thermogenesis and the relatively high $T_b$ of overwintering bears (Hellgren 1995).

After spring emergence, bears generally continue to fast for several weeks and their MR remains depressed below BMR, even if food is available (Nelson et al. 1983; Tøien et al. 2011), despite a return to normothermic $T_b$. Bears will not undertake hibernation during the summer months when they are usually active, even when subjected to simulated winter environmental conditions and fasted. Fasting bears during this time use muscle as an energy source and become dehydrated and uremic without access to water (Nelson et al. 1975). In autumn, the energy intake of bears increases more than threefold, and free-living individuals may feed as much as 20 h $day^{-1}$ to build up the adipose supplies necessary to sustain them for the next hibernation period.

Clearly, bears share some aspects of winter hibernation with smaller mammals that undergo classical seasonal hibernation: seasonally restricted dormancy with pre-dormancy fattening, abolishment of circadian patterns, reduced MR and heart rate, and long-term survival without food. However, their winter inactivity is characterized by some profound differences, including a smaller reduction in $T_b$; maintenance of limited activity and responsiveness; no periodic interbout arousals to normothermia; and unique protein and urea management, with an absence of urination for the hibernation period.

## 4.2 Hot Environments

Deserts are traditionally considered to be hot and dry environments. The record air temperature, measured by a standard meteorological weather station (Stevenson screen) is 58.0°C, measured in El Aziza (Libya), with Death Valley (USA) a close second at 56.7°C (Mildrexler et al. 2006). However, the hottest land surface temperature (remotely measured radiometric temperature) is considerably higher because of direct solar radiation and the relatively low conductivity of air; the hottest records are 70.7°C in the Lut Desert (Iran) and 69.3°C in Queensland (Australia). Deserts can not only be hot; they can also be extremely dry. Arica (Atacama Desert, Chile) is reputably the driest place on Earth, with an average annual precipitation of 0.8 mm, and no recorded rainfall for 14 consecutive years (Krause & Flood 1997). Arid zones, which we traditionally synonymize with deserts, are generally defined by a combination of temperature and rainfall characteristics. For example, Köppen's classification scheme defines climate zones in part by the mean annual precipitation ($P_{ann}$; mm), mean annual $T_a$ measured at 2 m height ($T_{ann}$; °$C^{-1}$), and dryness threshold ($P_{th}$; mm); $P_{th}$ is $2\,T_{ann}$ if two-thirds of the annual precipitation occurs in winter, $(2\,T_{ann}) + 28$ if ⅔ of the annual precipitation occurs in summer, and $(2\,T_{ann}) + 14$ otherwise (Kottek et al. 2006). The arid zone (labelled type B) has $P_{ann} < 10P_{th}$, and is subdivided into steppe (BS) if $P_{ann} > 5P_{th}$ and desert (BW) if $P_{ann} \leq P_{th}$. Both BS and BW are divided into hot (h, $T_{ann} \geq 18$°C) and cold (c, $T_{ann} < 18$°C) subregions (e.g. hot desert is BWh and cold desert is BWc). We focus here on the physiological ecology of mammals in hot deserts.

The physiology of mammals provides a number of mechanisms to maintain a stable $T_b$ at high $T_a$ in hot deserts. In moderately warm environments, their suite of physiological and morphological characteristics enables them to dissipate their metabolic heat load, which is substantial given their endothermic bauplan.

Essentially two mechanisms are available to dissipate their endogenous heat load: increasing thermal conductance to facilitate heat loss by conduction and convection, and evaporative heat loss. In very warm environments, mammals have to not only dissipate their metabolic heat load; they must also dissipate any additional environmental heat load. Heat storage is important, and evaporative heat loss becomes the only mechanism for dissipating a metabolic plus environmental heat load, but at the cost of sustaining a high rate of water loss. These physiological mechanisms, along with how mammals exploit their ecological niche, enables them to remain active and endothermic at a sustainable water cost year-round.

There are some mammals, particularly small species, for which the water cost of maintaining activity in the heat can become temporarily too difficult to sustain if the required rate of heat loss taxes their body water balance (e.g. at very high $T_a$ for periods of low water availability). One solution for non-migratory mammals under such circumstances is to avoid the hot conditions; they become inactive and seek refuge in more equable burrows (for small mammals) or caves, or limit their exposure to high ambient temperatures by being active nocturnally when environmental temperatures are usually lower. Many large mammals are unable to escape hot environmental conditions—they must endure them. This is facilitated by their large size, low thermal conductance, considerable thermal inertia, and high dehydration tolerance (Cain et al. 2006). Some large mammals decrease their daytime activity at the hottest time of year; some compensate with increased nocturnal feeding activity (Maloney et al. 2005; Hetem et al. 2012).

## 4.2.1 Endurers

Large mammals in particular must be able to endure the harsh climate of hot deserts, because they are generally unable to escape; shade under large trees is probably their main refuge, if available. Their maintenance of homeostasis in hot environments depends on being able to achieve heat balance or being able to sustain heat storage during the day. Small mammals (e.g. ground squirrels) are able to endure the harsh climate for short periods, but must periodically seek refuge to avoid overheating (see 4.2.2).

When environmental temperature is lower than $T_b$, the temperature differential ($T_b - T_a$) can contribute to heat loss by conduction, convection, and radiation, but is not necessarily sufficient to dissipate all the metabolic heat production, and there is usually a requirement for some enhanced evaporative heat loss (EHL) to dissipate some of the metabolic heat. When environmental temperature exceeds $T_b$, the temperature differential causes heat gain by conduction, convection, and radiation from the environment to the mammal (see 2.4), so heat storage is important and EHL is the only possible mechanism for heat loss; EHL must be sufficient to dissipate metabolic heat plus any environmental heat gain for $T_b$ to remain in steady state. Consequently, the various avenues for evaporation, hence EHL, becomes paramount for endurers: salivation, sweating, and panting.

Many mammals increase saliva production when they are exposed to heat, and spread it onto the body surface, generally by licking their fur or naked patches of skin, to facilitate EHL when the saliva evaporates. Because salivation is a common thermoregulatory response of many mammals once considered to be 'primitive', such as opossums (Higginbotham & Koon 1955), it has often been considered to be a 'primitive' thermoregulatory response. However, it is clearly a controlled and quite a sophisticated thermoregulatory response of many mammals, particularly rodents, which lack sweat glands. It involves the coordinated physiological and behavioural responses of an increase in the production of a protein-poor saliva, the behavioural response of licking to spread the saliva, and often an increase in blood flow to the skin beneath the licked surface. Rodents lack the ability to pant or sweat, and laboratory rats increase their salivation when exposed to temperatures higher than $T_b$ (Toth 1973). Rats acclimated to high $T_a$ produce more saliva at a lower core temperature, and the ligation of the salivary glands makes the animals vulnerable to rapid hyperthermia (Horowitz et al. 1983). Kangaroos (*Macropus*) will salivate in the heat (unless they are dehydrated) and lick their forearms to evaporatively cool (Dawson 1973b; Figure 4.7). Their mandibular salivary glands may preferentially contribute to saliva production for thermoregulation compared to the parotid salivary glands (see 4.8.2). Vascular casting of the forearms reveals a dense superficial network of fine blood vessels underlying the forelimb skin (Needham et al. 1974). In a marriage of physiology, morphology, and behaviour, an increase in the core temperature of kangaroos results in a 3–4 times increase in blood flow to the forearms and consequent loss of heat by evaporative cooling of saliva (Needham et al. 1974).

For many large mammals (e.g. humans, horses, some antelope), sweating is the primary means of evaporative heat loss, but small mammals tend to rely more on salivation (see earlier) or panting (see later). Many rodents and lagomorphs lack sweat glands, so sweating is not an option for them. Aquatic mammals such as

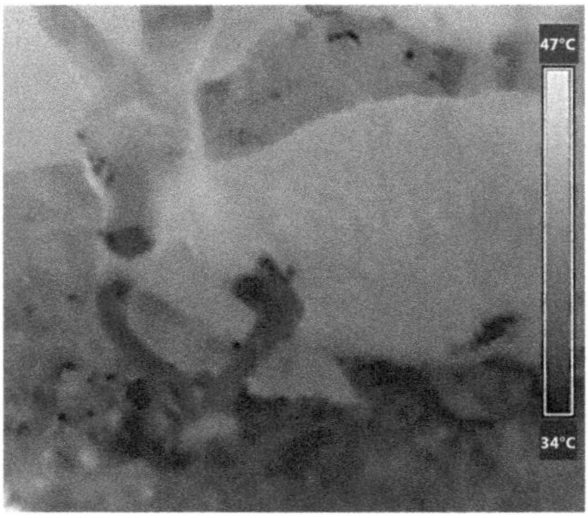


**Figure 4.7** Thermal image of an eastern grey kangaroo (*Macropus giganteus*) in the heat, showing the forelimbs cooled by licking. Photograph provided by A. J. Munn.

cetaceans and hippopotamus also lack sweat glands, presumably as EHL is redundant in an aquatic environment. The total sweat gland number can range from as few as 20–30 $cm^{-2}$ for pigs to more than 2,000 $cm^{-2}$ for zebu cattle. African dik-dik (*Rhynchotragus*) are relatively small African ungulates (about 2–8 kg) that have a moderate density of sweat glands, about 190 $cm^{-2}$, and a relatively low sweating rate when heat challenged, about 19 $g\ m^{-2}\ h^{-1}$; they rely more on panting to dissipate heat, like larger African ungulates (Maloiy 1973). Sweat glands can also have functions other than thermoregulation, including reducing friction (palms and eyelids), excreting wastes, providing antibacterial protection (e.g. the antibacterial peptide dermicidin in human sweat; Schitteck et al. 2001), and for interspecific communication (Jenkinson 1973; see 2.2.4.1).

Sweat is produced by specialized glands in the skin (see 2.2.4.1). In some taxa, the glands occur only in specialized regions (e.g. the footpads of felids and canids), while in other taxa they are distributed all over the body surface. Atrichial (eccrine) sweat glands are not associated with hair follicles; they are found in humans, moles, and those species with plantar glands. Most other mammals have epitrichial (apocrine) sweat glands, associated with hair follicles, and some taxa, such as primates, have both atrichial and epitrichial sweat glands. Sweat glands consist of a fundus and a duct. The fundus varies from a simple sac-like structure to a highly coiled tube, and is formed by two layers of epithelium: an inner secretory layer and an outer myoepithelium. Sweat gland ducts are usually fairly straight, with those of atrichial glands opening directly onto the skin surface, and those of epitrichial glands opening in the pilosebaceous canal, allowing for mixing of sweat and sebum (Jenkinson 1973). Sweat begins as a filtrate of plasma in the sweat gland, the filtrate being modified as it travels down the sweat duct to the skin. Sweating is stimulated by catecholamines in bovids (Robertshaw 1975), while in primates sweating is cholinergically stimulated; cholinergic postganglionic sympathetic nerves lie close to the sweat gland. Marsupials, at least kangaroos, sweat profusely when they exercise but do not sweat during heat exposure (Dawson et al. 1974). Sweating is presumably a more appropriate mechanism to dissipate internal heat, because in a hot environment a consequent reduction in skin temperature would promote heat gain, whereas panting is advantageous to dissipate an external heat load because it dissipates core body heat and does not cool the skin.

Panting can markedly increase respiratory evaporative water loss, and is an important mechanism for heat dissipation in many small and large mammals. It involves a coordinated increase in ventilation combined with an increase in blood flow to the upper respiratory tract; the upper respiratory tract is kept wet, and air flow across its surface is increased by an elevated ventilation rate (Vesterdorf et al. 2011) while the respiratory mucosa is kept cool by limiting its blood flow (Jessen 2001a). The specialized respiratory turbinates that are present in the nose of mammals (and birds) are thought to have evolved for precisely the opposite reason—to minimize evaporative water loss by recovering water vapour from the expired air (Hillenius 1992; see 3.6.2). So, when enhanced respiratory evaporative water loss is required, as in panting, the process of recuperative heat exchange is short-circuited by a large increase in arterial blood flow to the mucosa to prevent condensation and cooling of the expired air (Hales 1973). Recuperative heat exchange is sometimes further reduced by expiration via the oral cavity rather than the respiratory chamber (Schmidt-Nielsen et al. 1970b). For example, reindeer (*Rangifer tarandus*) alternate abruptly between closed- to open-mouth panting, with more open-mouth panting at higher $T_a$ (Aas-Hansen et al. 2000). The heat gained by the air on inspiration is thereby exhaled, and the cooled venous blood draining the mucosa cools the general body.

The increase in ventilation required to achieve respiratory heat loss is not without physical consequence, because ventilation rate is linked to gas exchange and the control of pH, $pO_2$, and $pCO_2$ (see 2.5.6). An increase in alveolar ventilation ($V_A$) results in a washout of $CO_2$ and a decrease in arterial $pCO_2$ ($p_aCO_2$), as shown by the alveolar gas equation: $p_aCO_2 = VCO_2K / V_A$, where $\dot{V}CO_2$ is the $CO_2$ production rate and K is the proportionality constant. During mild hyperthermia, increased air flow in the upper respiratory tract occurs without any major increase in alveolar ventilation; respiratory rate ($f_R$) increases but there is a decrease in tidal volume ($V_T$), so the increased ventilation is of the respiratory dead space rather than alveolar ventilation ($V_A$; Hales 1976). This ventilatory pattern is Type I panting. Under more severe heat stress, mammals switch to Type II panting: $V_T$ increases, $V_A$ increases, and $P_aCO_2$ decreases, leading to acid–base disturbances (Hales 1976). Panting involves extra work by respiratory muscles, so the added heat production associated with panting can also become important; the laboured breathing associated with Type II panting has been estimated to be responsible for 11% of the increased heat production in severe heat (Hales & Findlay 1968).

Overall, salivation, sweating, and panting are the main evaporative responses to heat for different mammals. However, the mechanism for evaporation varies from species to species, and also with acclimation. Important in these considerations is the maximum evaporative heat loss that the mammal can achieve, because a higher heat loss ability will allow thermal equilibrium at larger heat loads. In general, the magnitude of heat loss possible by sweating is greater than by panting, and more energetic and larger species tend to depend on sweating (for example, humans and horses).

Taylor and Lyman (1972) reported that a large running antelope (Thomson's gazelle *Gazella thomsoni*) produced large amounts of internal heat when running and $T_b$ increased considerably (+4.6°C), but brain temperature ($T_{br}$) increased more slowly than $T_b$ and could be up to 2.7°C lower. This selective brain cooling was thought to be an important adaptation to heat stress, protecting a thermally vulnerable brain from heat damage. Laboratory studies showed that selective brain cooling was activated above a threshold core $T_b$ via the control of venous blood perfusing the cavernous sinus where there is heat exchange in the carotid rete between warm arterial blood going to the brain and cool venous blood returning from the nasal mucosa (Figure 4.8; see 3.4.4). While some early studies reported magnitudes of selective brain cooling (the extent to which $T_{br}$ was lower than arterial blood temperature) of nearly 4°C, later studies showed that this large magnitude was probably an artefact of rapidly changing $T_b$ during exercise (Maloney et al. 2009); the maximum magnitude of selective brain cooling in steady state is 1 – 1.5°C (Fuller et al. 2007; Hetem et al. 2012). It is worth noting that an increase in core $T_b$ of 1°C is sufficient to considerably elevate respiratory evaporative loss (e.g. by about six times in the goat; Kuhnen & Jessen 1994).

Selective brain cooling is activated above a threshold core $T_b$ when cool blood is diverted to the cavernous sinus; at low core $T_b$, venous blood bypasses the sinus and no heat exchange occurs in the rete (Kuhnen & Jessen 1994). However, the first measurements of $T_b$ and $T_{br}$ for a free-living mammal, the black wildebeest (*Connochaetes gnou*), indicated that selective brain cooling was not always activated when $T_b$ increased, and that during exercise, when $T_{br}$ was the highest (41 – 42°C), the wildebeest did not use selective brain cooling at all, casting doubt on its protective role (Jessen et al. 1994). Further, the control of selective brain cooling was much more varied than had been found in laboratory studies, with a large range of core $T_b$ where selective brain cooling may or may not be activated.

An alternative role proposed by Jessen (1998) is that the adaptive significance of selective brain cooling is to modulate the use of water for thermoregulation, since it reduces the temperature of the hypothalamus, where the neural control centres

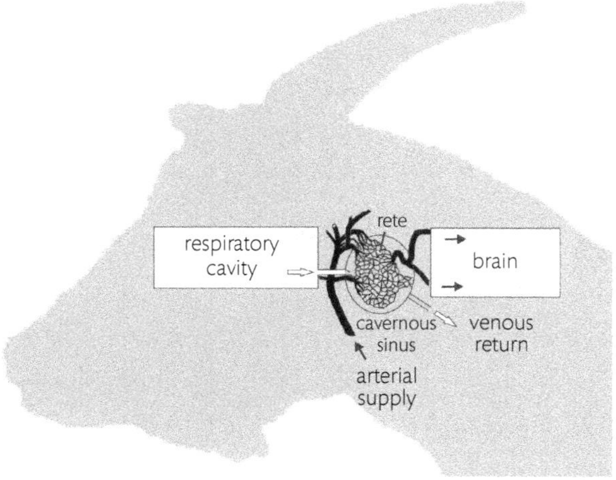


**Figure 4.8** Schematic of heat exchange in the carotid rete of a goat, between warm arterial blood flowing through the cavernous sinus to the brain, and cool blood returning from the nasal cavity to the body. Modified from Daniel et al. (1953) and Taylor and Lyman (1972).

for heat defence, including evaporative water loss (EWL), are located. Kuhnen (1997) had shown that manipulation of selective brain cooling in goats altered their EWL during heat exposure. Two studies offer solid support for the role proposed by Jessen (1998). First, if selective brain cooling does modulate water use, then an osmotically stressed mammal could reduce water use by augmenting selective brain cooling. Sheep that are exposed to heat and deprived of drinking water for five days increase selective brain cooling (Fuller et al. 2007). Later, Strauss et al. (2015) showed that the individual sheep that used selective brain cooling more during dehydration and heat exposure used less water than those individuals that used selective brain cooling less, concluding that the activation of selective brain cooling leads to a reduction of water use for evaporative cooling during heat exposure. It has been argued that the evolution of the carotid rete and the ability of artiodactyls to use selective brain cooling and modulate respiratory water loss is what facilitated the increased diversity of artiodactyls since the Eocene in both hot and cold environments, in contrast to the contraction in diversity of perissodactyls, which do not have a carotid rete (Mitchell & Lust 2008).

Another mechanism for mammals to reduce water requirements in the heat is to allow $T_b$ to increase, storing the heat that would normally need to be dissipated by evaporation, and also increasing the temperature differential ($T_b - T_a$) for heat loss if $T_a$ is slightly lower than $T_b$, or reducing the gradient for heat gain from a hot environment when $T_b < T_a$. In a landmark study in the 1950s, Schmidt-Nielsen and colleagues showed that this is exactly what happens for camels when they are water-deprived and exposed to high environmental temperatures (Figure 4.9; Schmidt-Nielsen et al. 1957). The amplitude of the daily rhythm of core $T_b$ increased from the normal 2°C to more than 6°C when camels were deprived of drinking water and exposed to $T_a$ exceeding 40°C. When hydrated, the considerable EHL and minor heat storage balance metabolic heat production (MHP) and

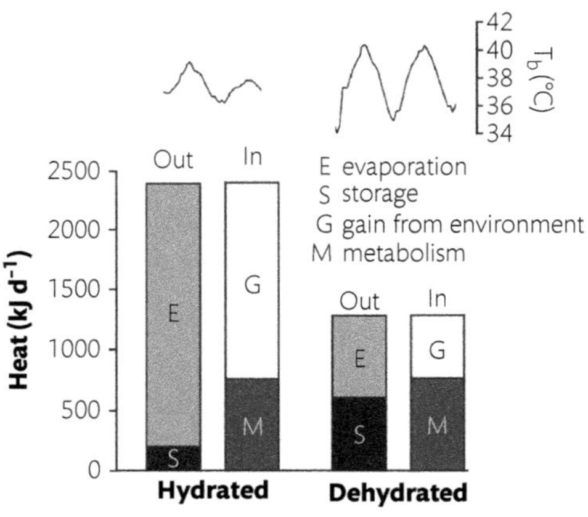


**Figure 4.9** Heat balance of camels (*Camelus dromedarius*) when hydrated and dehydrated, and daily $T_b$ amplitude. Modified from Schmidt-Nielsen et al. (1957).

heat gain from the environment. When dehydrated, the reduced evaporative heat loss and increased heat storage balance MHP and heat gain from the environment; note that the total heat gain is considerably reduced for dehydrated camels. Later, Taylor (1968) reported that the amplitude of the daily rhythm of core $T_b$ for several species of large African ungulates increased when they were exposed to heat, and more so when they were deprived of drinking water.

These seminal studies became the basis for the widely held view that the survival of large mammals in hot environments depends on the relaxation of homeothermy and the adoption of a more heterothermic pattern of core $T_b$. However, studies using bio-logging of mammals free-ranging in their natural habitat consistently find that the amplitude of the daily rhythm of core $T_b$ is independent of environmental temperature, as long as the animals have access to drinking water (Fuller et al. 2014). It seems that under normal conditions, large mammals activate EHL in the face of environmental heat load, and that heat loss is sufficient to maintain core $T_b$ (but see later). When water intake is limited, osmotic stress causes changes in thermoregulation that may be due to the integration of signals from osmosensitive and thermally sensitive neurons in the hypothalamus, or may be due to the activation of selective brain cooling that reduces hypothalamic temperature and suppresses the activation of EHL, leading to a larger amplitude of core $T_b$ (Fuller et al. 2014).

In hot, humid environments, sweating and panting might be stimulated, but the high humidity is not conducive for evaporation. In humid tropical environments, mammals generally have a very high heat loss capacity and usually other adaptations that facilitate heat balance. For example, the predominant cattle breeds used for meat and milk production are based on *Bos taurus taurus*, but these breeds generally do not cope well with the tropics; breeds based on *Bos taurus indicus* cope better with tropical conditions. The essential characteristics of the *indicus* breed that makes it better in the tropics are higher sweating capacity (more and larger sweat glands) and a lower metabolic heat production (Beatty et al. 2006). The lower metabolic heat production is an advantage in the heat, but conflicts with the intention of animal production, which is to produce more animals. Attempts to meld the productivity of *taurus* with the heat tolerance of *indicus* have not generally been successful; any increase in productivity is usually associated with a decrease in heat tolerance (Cunningham 1991).

A reduction in metabolic rate can also be achieved by a reduction in food intake, which should assist with heat balance in hot environments. Many mammals do decrease feed intake in the heat. Indeed, rats forced to maintain the same level of food intake in the heat suffer mortality (Hamilton 1976). For some grazing species, reduced food intake may be an indirect effect of shade-seeking in hot conditions, but even in the absence of radiant heat, an effect of heat load on food intake is well described (Yousef et al. 1968). Metabolic heat production can also be reduced during prolonged exposure to heat by lowered thyroid activity (Yousef et al. 1967).

Thyroid activity decreases when food intake is restricted (Blincoe & Brody 1955) and so it is logical to reason that the reduction in food intake causes a decrease in circulating thyroid hormone, with a subsequent decrease in metabolism. However, several studies have shown that thyroid activity and heat production decrease even if animals are force-fed to maintain the thermoneutral level of feed intake during heat exposure (Yousef & Johnson 1966; Kibler et al. 1970), so it seems likely that $T_b$ directly affects thyroid hormone release and metabolism—evidence for a direct effect of $T_b$ is equivocal. Andersson (1963) suppressed feed intake in goats by heating their hypothalamus, but Spector et al. (1968) and Hamilton and Ciaccia (1971) showed that brain heating increased feed intake in rats. Effects of $T_b$ on rumen function might also contribute to reduced appetite of ruminant mammals (Collier et al. 1982).

### 4.2.2 Avoiders

Small mammals, in particular, are able to seek refuge from hot environmental conditions. Many are nocturnal, avoiding harsh daytime temperatures, but some are diurnal and some are flexible with respect to timing of activity (MacMillen 1972; Walsberg 2000). For example, two species of spiny mice coexist in Israeli desert areas; the common spiny mouse (*Acomys carinus*) is nocturnal whereas the golden spiny mouse (*Acomys russatus*) is more behaviourally flexible and is frequently diurnal (Shkolnik 1971). Diurnally active desert mammals generally avoid exposure to direct sunlight, but can resort to physiological mechanisms to cope with solar heat loads. The white-tailed antelope ground squirrel (*Ammospermophilus leucurus*) has an exceptionally labile $T_b$ when diurnally active, and especially in summer they rely on periodic bouts of hyperthermia to store heat when surface-foraging, then they dump the stored heat by thigmothermy or retreat to their burrow to dump heat to cool soil (Figure 4.10). For these small diurnal desert mammals, hyperthermia is a short-term strategy of heat storage then dumping, compared to the day-long strategy of large mammals (such as camels; see Figure 4.9), because of their much higher surface area to volume ratio and lower thermal inertia, which means they heat and cool much faster.

Aestivation, or 'summer dormancy', is a common response of many vertebrate and invertebrate animals during which metabolic rate and water loss are depressed (Withers & Cooper 2010). For ectothermic vertebrates, the metabolic depression associated with aestivation is intrinsic, apparently unrelated to body temperature, $pO_2$, or body water changes (Withers & Cooper 2010). For endothermic mammals (and a few birds), aestivation is essentially the ecological equivalent of winter torpor and hibernation (see 4.1.2), but in hot and dry conditions. Aestivation is phylogenetically widespread amongst mammals (Geiser 2010). Some mammalian aestivators, such as the cactus mouse (*Peromyscus eremicus*; MacMillen 1965) aestivate on a daily basis, whereas others such as the Mohave ground squirrel (*Citellus mohavensis*) aestivate for weeks to months (Bartholomew & Hudson 1960).

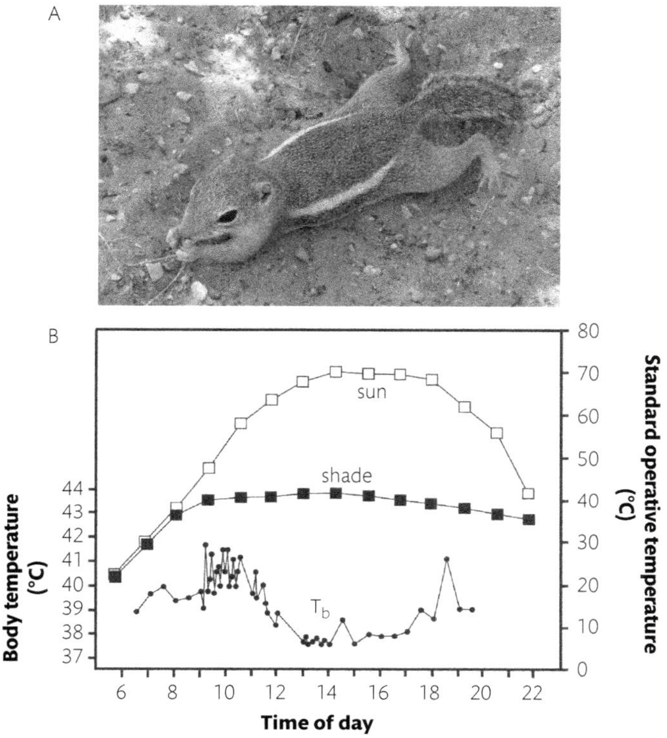


**Figure 4.10** A: An antelope ground squirrel (*Ammospermophilus leucurus*) dumping heat to the substrate by thigmothermy. Photograph by C. E. Cooper. B: Body temperature of an antelope ground squirrel during a summer day, compared with environmental temperature (standard operative temperature, $T_{es}$) in sun and shade. Modified from Chappell and Bartholomew (1981b).

The pattern of change in body temperature and metabolic rate with $T_a$ is the same for aestivation as for torpor and hibernation. In fact, we cannot discriminate on physiological grounds between single-day aestivation, multi-day aestivation, single-day winter torpor, and multi-day hibernation (Wilz & Heldmaier 2000; Van Breukelen & Martin 2015) other than aestivation occurs in the summer, at higher ambient temperatures, under dryer conditions. So, summer aestivation presents different challenges than winter torpor and hibernation, with high ambient temperatures and low water availability being as or even more challenging than lack of food (Geiser 2010). For aestivating cactus mice, $T_b$ declines with $T_a$, becoming closer to $T_a$ at lower temperatures; for example, $(T_b - T_a)$ is $< 1°C$ at $T_a = 10°C$, and about 3°C at $T_a$ of 20–30°C (MacMillen 1965; Figure 4.11). Since aestivation commonly occurs at relatively high $T_a$, there is presumably a lesser reduction

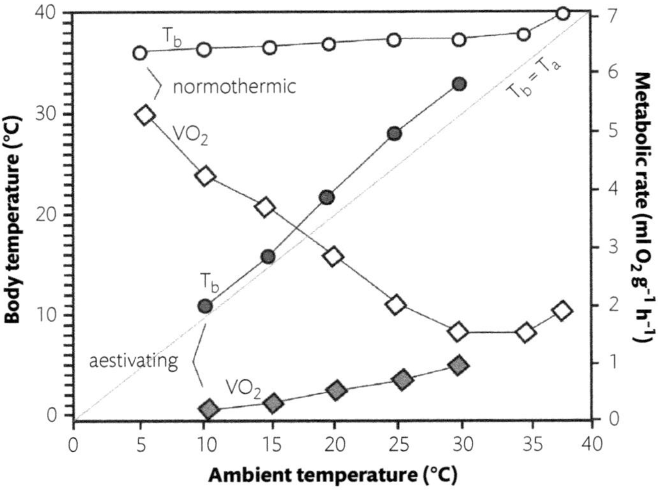


**Figure 4.11** Body temperature and metabolic rate of cactus mice (*Peromyscus eremicus*) when normothermic and aestivating. Modified from MacMillen (1965).

of metabolic rate and water loss rate than torpor/hibernation at lower $T_a$, as $T_b$ must remain above $T_a$. Furthermore, aestivation presumably results in a poorer relative water economy (RWE = MWP / EWL) than for normothermy (as for torpor; see 3.2.4.2), because aestivation reduces MR more than EWL, and MWP is likely less than EWL (i.e. RWE < 1).

Aestivation can be induced by food restriction, water restriction, or both. For example, some cactus mice enter aestivation in response to water restriction, but some individuals do not (MacMillen 1965). Some (20%) least gerbils (*Gerbillus pusillus*) enter aestivation when water-deprived, but more (88%) do so when food- and water-deprived (Buffenstein 1985). The stripe-faced dunnart (*Sminthopsis macroura*) does not increase its use of torpor when water-deprived if it has access to moist food, but does when only dry food is provided (Song & Geiser 1997). The tendency of mammals to enter aestivation in response to water restriction seems somewhat equivocal, and their response to water deprivation with food available depends on the potential of the food to provide sufficient water. Many species restrict food intake in response to water deprivation. Nevertheless, food and water restriction are powerful stressors than induce aestivation.

Little is known of the physiological role of aestivation in the field. Two rodents found on rocky outcrops in the Namib Desert (pygmy rock mouse, *Petromyscus collinus*; Namaqua rock rat, *Aethomys namaquensis*) normally have a $T_b$ of 33.6 and 34.0°C, respectively, but during aestivation at $T_a$ of 18–22°C, their $T_b$ is reduced to 18.0–23.6°C (Withers et al. 1980). In the field, both species have remarkably low water turnover rates, of 0.8 ± 0.1 and 2.2 ± 0.2 ml $day^{-1}$ respectively, presumably reflecting the use of aestivation. Interestingly, their water turnover rates increased in the field after an advective fog, to 1.4 ± 0.2 and 3.2 ± 0.2 ml $day^{-1}$, respectively.

## 4.3 Underground Environments

A fossorial (subterranean) existence has developed in several mammalian taxonomic orders, including marsupials, rodents, insectivores, and edentates, and includes species that live almost exclusively underground and only rarely come to the surface. The convergent evolution of fossorial mammals is a fascinating and puzzling evolutionary phenomenon. Convergent morphological features for burrowing include compact bodies, short tails and necks, microphthalmic eyes, and large and powerful forefeet, pectoral girdles, and associated muscles (McNab 1966; Nevo 1999; Warburton et al. 2003; Warburton 2006). The abiotic microenvironment inhabited by fossorial mammals is relatively humid, hypoxic and hypercapnic, and thermally constant (McNab 1966; Arieli 1979; Cooper & Withers 2005). Living underground has resulted in a number of physiological adaptations to these conditions, including respiratory responses to hypoxia and hypercapnia, mechanisms for maintaining heat balance in warm and humid environments, and sustaining the metabolic cost of digging burrows.

Many mammals live both underground and above ground, often on a daily cycle. These semi-fossorial species are generally not so specialized for an underground existence with respect to their morphology or physiology, because they are also active above ground. Nevertheless, many semi-fossorial species show similar adaptations to an underground existence as fully fossorial species, albeit often less extreme.

### 4.3.1 Hypercapnic Hypoxia

An important constraint faced by fossorial mammals is potentially low $O_2$ availability (hypoxia) and excess $CO_2$ (hypercapnia) underground (Arieli 1979, 1990). The depletion of $O_2$ in burrow environments due to animal (or soil microbe) metabolism is directly correlated with an increase in $CO_2$, so this form of hypoxia is termed 'hypercapnic hypoxic' (cf. 'hypoxic hypoxia' with altitude; 4.4.1). For example, levels of $CO_2$ as high as 6.1% and $O_2$ as low as 7.2% were recorded in the breeding mounds of a blind mole rat (*S. carmeli*) in a flooded, poorly drained field of heavy clay soil with very high volumetric water content (Shams et al. 2005). Gaseous interchange between burrows and the atmosphere depends on the gas permeability properties of the soil (Wilson & Kilgore 1978; Withers 1978; Arieli 1979), and some air ventilation caused primarily by animal movements (Buffenstein 2000) or wind-induced convection (Vogel et al. 1973). These factors mean that burrow gas composition can differ considerably from atmospheric air, and any activity of the inhabitants increases such differences. Models of diffusive gas exchange (Withers 1978) and experimental data show that, unless the soil is completely devoid of biotic substances, burrow atmospheres will always be hypoxic and hypercapnic relative to the surface atmosphere. Faced with low $pO_2$ and the potential $CO_2$ perturbation of their acid–base balance in the burrow atmosphere, burrow-dwelling mammals would be expected to have physiological mechanisms to avoid excessive energy expenditure.

Tomasco et al. (2010) reviewed studies of the respiratory responses of mammals to underground environments. The primary respiratory driver for mammals is $CO_2$ rather than $O_2$ (see 3.3.3), so hypoxia is often less significant than hypercapnia for non-fossorial mammals. An attenuated ventilatory response to hypoxia is not a general characteristic of semi-fossorial or fossorial mammals. Many semi-fossorial species, such as the Syrian hamster (*Mesocricetus auratus*), woodchuck (*Marmota monax*), golden-mantled (*Spermophilus lateralis*) and Columbian (*S. columbianus*) ground squirrels, do not differ in their response to hypoxia to similar-sized non-fossorial species.

However, some burrowing species do have an attenuated sensitivity to hypoxia; the hyperventilatory response to hypoxia of the echidna, armadillo (*Dasypus novemcincus*), and hairy-nosed wombat (*Lasiorhinus latifrons*) is depressed (Frappell et al. 2002). The Chilean fossorial coruro (*Spalacopus cyanus*) and semi-fossorial degu (*Octodon degus*) both have a low sensitivity to hypoxia, but the coruro has a more acute response to hypoxia than the degu (Tomasco et al. 2010). These findings support the conclusion that persistent changes in the neural control system for respiratory ventilation are generated based on prior experience (Mortola 2004). Chronic sustained hypoxia ($pO_2$ 6.7 – 9.3 kPa) elicits plasticity in the carotid body chemoreceptors, with delayed effects on the central neural integration of carotid chemo-afferent neurons that become more prominent as the duration of hypoxia is extended. In the case of fossorial coruros, hypoxia may last for long periods inside closed burrows, and their enhanced ventilatory response is the result of the potentiation of the carotid chemoreflex to hypoxia. Degus, however, might tolerate intermittent hypoxia, experiencing hypoxia only while resting at night; this may require plasticity via central neural mechanisms of respiratory control (Ling et al. 2001).

Burrowing mammals often encounter significant hypercapnia, and generally have a reduced sensitivity and response to $CO_2$ compared to non-fossorial (and non-diving) species (Boggs et al. 1984; Tomasco et al. 2010). The most $CO_2$-insensitive species are the fossorial pocket gopher (*Thomomys bottae*; Darden 1972) and Middle East blind mole rat (*Spalax ehrenbergi*; Arieli & Ar 1979). Unlike hypoxia, which seems to result in adaptations via sensory input, severe hypercapnia is often associated with long-lasting depression of respiratory motor output. During hypercapnia, respiratory activity initially increases, but then decreases

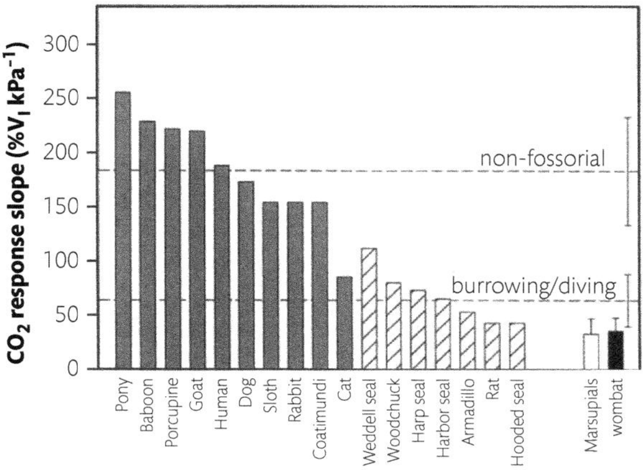


**Figure 4.12** Respiratory responses to hypercapnia (expressed as percentage change in minute volume per kPa partial pressure of $CO_2$) for non-fossorial placental mammals, fossorial or diving placental mammals, marsupials in general, and a semi-fossorial marsupial (wombat). Dashed lines are averages ± standard deviation for non-fossorial, and burrowing and diving placental mammals. Modified from Frappell et al. (2002). Reproduced with permission of the University of Chicago Press.

progressively when the hypercapnia is sustained. Interestingly, marsupials have a low $CO_2$ sensitivity, and there would appear to be little difference between aboveground and burrowing species (Frappell & Baudinette 1995; Frappell et al. 2002; Figure 4.12). A reduction in the gain of the ventilatory control system in marsupials to a level akin to that of burrowing placentals could reflect potential exposure to hypercapnia during development—burrowing animals within the burrow and marsupials within the pouch. In fact, exposure to hypercapnia during development causes long-lasting attenuation of the acute hypercapnic ventilatory response so hypercapnia-induced developmental plasticity may play a role in the reduced hypercapnic ventilatory responses commonly observed in fossorial mammals.

## 4.3.2 Temperature and Energetics

Burrowing mammals can have a reduced capacity for convective heat loss, and evaporative heat loss can be low, because of the relatively high $T_a$ and ambient humidity in burrows (McNab 1966, 1979). Many fossorial mammals have a slightly lower-than-expected $T_b$ (35 – 37°C), which would reduce their metabolic heat production and reduce the risk of overheating. The naked mole rat (*Heterocephalus glaber*) is exceptional in this regard, having a $T_b$ of only 32°C, which is close to soil temperature (about 31°C), and being essentially ectothermic even when huddling in groups (McNab 1966; Withers & Jarvis 1980; Buffenstein & Yahav 1991).

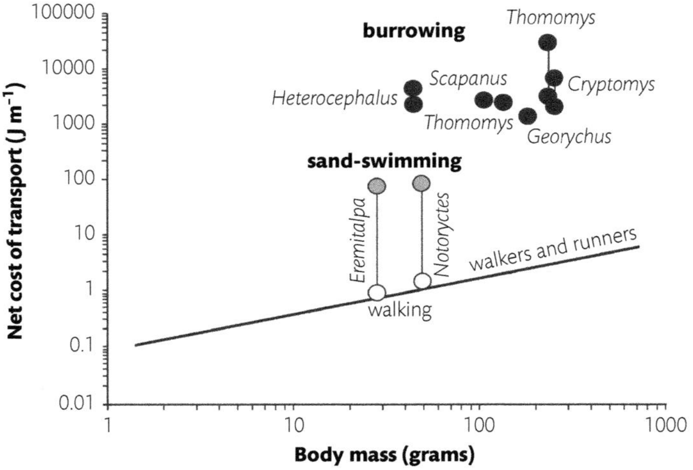


**Figure 4.13** Net cost of transport for burrowing mammals (black symbols) and sand-swimming mammals in sand (grey symbols) and walking (open symbols), and walking and running mammals (regression line). Data from Vleck (1979), Du Toit et al. (1985), Lovegrove (1989), Seymour et al. (1998), and Withers et al. (2000). Modified from Withers et al. (2000) and Xu et al. (2014).

Fossorial mammals also often have a high capacity for non-evaporative heat dissipation, reflecting the limited capacity for evaporation in a humid burrow. For example, the Talas tuco-tuco (*Ctenomys talarum*), which shelters in a sealed burrow but forages above ground (Baldo et al. 2015), has a considerable capacity to increase dry thermal conductance (by about fourfold) to facilitate non-evaporative heat dissipation at $T_a = 35°C$ compared to lower $T_a$s, compared with only about a 1.3 times increase in EHL.

Fossorial mammals generally have to dig their own burrow, so they must cope with the high energy costs of digging as well as foraging (Ebensperger & Bozinovic 2000). Digging can be an energetically demanding process, and requires some anatomical specializations of the limbs or incisors. Loosened soil must also be transported back along the burrow and disposed of, often via a lateral tunnel to the surface (Vleck 1979). The metabolic cost of digging is therefore high relative to other costs of transport for mammals, and varies with soil density, cohesiveness, and burrow size and structure. Digging can be 360 to 3,400 times more energy-consuming than moving the same distance by walking (Vleck 1979). The cost of transport has been found to be similarly high for a range of burrowing mammals (Figure 4.13).

A different 'burrowing' strategy is sand-swimming by small fossorial mammals that live in loose, aeolian sand dunes. Sand-swimming is 'swimming' through the loose sand using limb movements and an undulatory body motion to push the sand away from the front of the animal; there is no need to move spoil, as must a burrowing mammal, because loose sand fills the space left behind the sand-swimmer. The Namib Desert golden mole (*Eremitalpa granti namibensis*) is an adept sand-swimmer (Holm 1969). It is nocturnally active on the dune face, foraging and periodically 'dipping' its head into the sand (to sense food); it rests under the sand during the day. Its metabolic cost of sand-swimming is considerably less than that of burrowers in compacted soil, but nevertheless substantially higher than when walking on the surface (Seymour et al. 1998; Figure 4.13). The Australian marsupial mole (*Notoryctes caurinus*) is a remarkably convergent sand-swimmer that, when sand-swimming or walking, has a similar net cost of transport as the Namib Desert golden mole (Withers et al. 2000).

Rates of metabolism have been linked to different biotic and abiotic factors as evidence of metabolic adaptation to environmental conditions and geographic distribution (Bozinovic 1992; Spicer & Gaston 1999; McNab 2002). The observation that fossorial mammals have lower than allometrically expected BMR (McNab 1979; Contreras 1983; Lovegrove 1986) has been explained as an adaptation to hypoxic conditions within the burrow and the energetic costs of digging. Several competing hypotheses have also been suggested to explain how physical microenvironmental conditions and underground life affect the energetics of fossorial mammals. Two of these are the thermal-stress hypothesis and the cost-of-burrowing hypothesis (but see Lovegrove and Wissell, 1988, and Lovegrove, 1989, for alternative ecological hypotheses, such as the aridity-food distribution hypothesis).

The thermal-stress hypothesis posits that a lower mass-independent BMR reduces overheating in burrows where the capacity for convective and evaporative heat loss is low (McNab 1966, 1979). The cost-of-burrowing hypothesis states that a lower mass-independent BMR compensates for the extremely high energetic cost of digging during foraging activity (Vleck 1979, 1981; see 4.3.3). Using phylogenetic and conventional allometric analyses, White (2003) examined both hypotheses for approximately 100 species of fossorial and semi-fossorial species in a biogeographic scenario. He concluded that mammalian species from mesic habitats support the thermal-stress hypothesis, but species from arid habitats support the cost-of-burrowing hypothesis. BMRs of large ($> 77$ g) fossorial (i.e. truly subterranean) mammals from mesic and arid habitats are not different from BMRs of their semi-fossorial (i.e. less adapted to digging) counterparts despite expected differences in their foraging costs, a result consistent with the thermal-stress hypothesis (White 2003). On the other hand, small ($< 77$ g) fossorial mammals from arid habitats have lower BMRs than their similarly sized but semi-fossorial counterparts, a result consistent with the cost-of-burrowing hypothesis (White 2003). These results led to the conclusion that the two hypotheses are not mutually exclusive. Bozinovic et al. (2005) tested the thermal-stress and cost-of- burrowing hypotheses at an intraspecific level. They compared seven populations of the coruro from different geographic localities with contrasting habitat conditions. Their results did not support the thermal stress or the cost-of-burrowing hypotheses. Coruros from habitats with contrasting climatic and soil conditions had similar basal and digging metabolic rates when measured under similar semi-natural conditions. It is possible that *S. cyanus* originated in Andean locations where it adapted to relatively hard soils, and later, when populations dispersed into coastal areas characterized by softer soils, they may have retained the original adaptation without further phenotypic changes.

The thermal-stress hypothesis neglects potential behavioural thermoregulatory adjustments. There are daily and seasonal patterns in the timing of daily activity for small ground-dwelling mammals (e.g. Chappell & Bartholomew 1981a,b). Coruros, for instance, adjust their surface activity according to diurnal and seasonal changes in environmental temperature (Rezende et al. 2003). They decrease their surface activity (digging, foraging, and vigilance) during the warmest time of day (from midday to early afternoon) in summer, but not during winter (Rezende et al. 2003). Thus, coruros might cope with thermal constraints inside burrows by shifting their activity according to changes in environmental temperature in time and space.

Low metabolic rates of fossorial small mammals may also be interpreted as adaptation to the low $pO_2$ of underground environments (see Nevo 1999 and references therein), which would predict a similar lower BMR at high altitude (although this may be confounded by adaptation to lower $T_a$ at higher altitude). Future study of *Ctenomys* species might provide a suitable test of this hypothesis, as fossorial ctenomyids include Andean species living above 3,500 m. Indeed, there are some parallels in the physiology of fossorial and high altitude species with respect to haematological adaptations to low $pO_2$ (see 4.4.1). It has been postulated small erythrocytes have a larger relative surface area and hence faster diffusional exchange. For example, the fossorial Palestine mole rat (*Spalax ehrenbergi*), which is very tolerant of hypoxia and hypercapnia (Arieli et al. 1977), has small erythrocytes (Shams et al. 2005). This allows a reduced capillary diameter, which in turn allows a high tissue capillary density and mitochondrial volume. Its erythrocyte count and lung diffusion capacity are high, and it has functional specializations of its haemoglobin and myoglobin, all of which presumably facilitate gas exchange under hypoxic conditions.

## 4.4 High Altitude Environments

Mt Everest (8,848 m; Himalayas, Tibet/Nepal) is the highest mountain on Earth while Mt Aconcagua (6,961 m; Andes, Argentina-Chile) is the highest mountain outside Asia. The peaks of these and other high mountains are above the highest elevations inhabited by mammals (see McNab 2002). The yak (*Bos grunniens*) is the mammal that inhabits the highest elevations, living at up to 5,500 m, but domesticated pack yaks can work intermittently at up to 7,200 m (Wiener et al. 2006). They, along with ibex (*Capra [ibex] sibirica*), Tibetan gazelle (*Procapra picticaudata*), and vicuna (*Vicugna vicugna*), are the most eurybaric mammals, tolerant of low barometric pressures. Many stenobaric mammals (less tolerant of low pressure), such as rabbits, pikas, felids, and mountain goats and sheep, occur in less extreme but nevertheless high altitude environments (an elevation of 2,500 m defines the lower limit of 'high altitude').

High altitude habitats are characterized by low $pO_2$, cold and significantly variable $T_a$, low humidity, high solar radiant energy, high winds, and low primary productivity. As such, high altitude is a multiple stress environment for mammals (Monge & León-Velarde 1991). Although the identification of specific physiological traits has been of evident value, it is the overall metabolic response to hypoxia that gives an integrated measure of the extent of adaptation to high altitudes (Rosenmann & Morrison 1975). In this vein, adaptation to high altitude involves many steps in the $O_2$ cascade (see 2.5.1) from the atmosphere to its final utilization in the cell (Dill 1938; Morrison 1964). Mammals can function to some extent during hypoxia with no change in their MR, but at some threshold $pO_2$ their MR decreases. Classical studies in comparative physiology demonstrated that this critical pressure ($P_C$) is a measure of the hypoxic sensitivity of the species, but is also related to the level of oxygen demand by the individual (Rosenmann & Morrison 1974b). As $pO_2$ is lowered below $P_C$, the MR decreases to a lethal level.

## 4.4.1 Hypoxic Hypoxia

The most important factor that may limit the overall altitudinal range of mammals is a lack of oxygen, a consequence of the decrease in barometric pressure ($P_b$) with increased elevation (Dill et al. 1964). Altitudinal hypoxia (or 'hypoxic hypoxia') is low $O_2$ partial pressure, resulting from the altitudinal decline in $P_b$; the partial pressure of $CO_2$ is also reduced because of the lowered $P_b$ (whereas $pCO_2$ is elevated in the 'hypercapnic hypoxia' of burrows).

Barometric pressure declines with increasing elevation (Table 4.2), from 'standard atmospheric pressure' of 101.3 kPa at sea level to 55 kPa at the highest human habitation (La Rinconada, Peru, 5,100 m), 43 kPa at 7,010 m (the usual air-breathing ceiling for humans, where arterial saturation is only 50%), 15 kPa at 14,326 m (the ceiling for humans breathing pure $O_2$, where arterial saturation is again only 50%), to 6.3 kPa at 18,190 m, which is the saturation water vapour pressure for water at 37°C, meaning that water boils at 37°C at elevations above 18,190 m. The physiological effect of this decrease in $P_b$ with elevation arises from the concomitant decline in the partial pressure of oxygen; $pO_2 = FO_2 P_b$, where $FO_2$ is the fractional content of air that is $O_2$. At sea level, $FO_2$ is 0.2095, and this value does not change significantly with elevation, so $pO_2 = 0.2095\ P_b$; therefore $pO_2$ at sea level is 21.2 kPa $(0.2095 \times 101.3)$ and at La Rinconada (5,100 m) is 11.5 kPa $(0.2095 \times 55)$. The decline in ambient $pO_2$ with elevation causes a decline in alveolar $pO_2$, and other levels through the $O_2$ cascade to the mitochondria, and compromises $O_2$ delivery.

Humans lose consciousness if their arterial percentage $O_2$ saturation is less than about 50%, which occurs at an elevation of 7,010 m, where $P_b$ is 43 kPa and $pO_2$ is 9.0 kPa; this defines the elevation 'ceiling' above which a human breathing air will lose consciousness (Table 4.2). This ceiling is below the top of Mt Everest, so in theory mountaineers should not be able to summit Mt Everest if they breathe ambient air. However, breathing pure $O_2$ raises the ceiling to 14,326 m $(P_b = 15\ \text{kPa}, pO_2 = 3.1\ \text{kPa})$, which facilitated the first summit of Mt Everest by Tenzing Norgay and Edmund Hillary, aided by the physiological research of Griffith Pugh, who demonstrated that the weight of oxygen tanks was more than compensated for by the extra aerobic capacity that oxygen breathing afforded. These limitations apply to un-acclimated humans; two highly trained

**Table 4.2** Effect of elevation on barometric pressure and partial pressure of oxygen.

| Location | Elevation (m) | Barometric pressure ($P_b$, kPa) | Oxygen partial pressure* ($pO_2$, kPa) |
|---|---|---|---|
| Sea level | 0 | 101.3 | 21.2 |
| Mt Kosciuszko, Australia | 2,228 | 78 | 16.3 |
| Lower limit of 'high altitude' | 2,500 | 76 | 15.9 |
| Lower limit of 'very high altitude' | 3,658 | 66 | 13.8 |
| Mt Blanc, France | 4,810 | 57 | 11.9 |
| Mt Kilimanjaro | 4,877 | 57 | 11.9 |
| La Rinconada, Peru (highest human habitation) | 5,100 | 55 | 11.5 |
| Lower limit of 'extremely high altitude' | 5,500 | 53 | 11.1 |
| Mt Elbrus | 5,642 | 52 | 10.9 |
| Mt McKinley | 6,190 | 48 | 10.1 |
| Mt Aconcagua | 6,961 | 43 | 9.0 |
| Human air 'ceiling' + | 7,010 | 43 | 9.0 |
| Mt Everest | 8,848 | 34 | 7.1 |
| Human pure $O_2$ 'ceiling' $+^{+}$ | 14,326 | 15 | 3.1 |
| Water boils (at 37°C) | 18,190 | 6.3 | 1.3 |

* $pO2 = 0.2095\ P_b$;

\+ limit of consciousness when breathing air;

$+^{+}$ limit of consciousness when breathing pure $O_2$

and altitude-acclimated mountaineers (Reinhold Messner and Peter Haberler) proved the fallacy of this widely held dogma concerning altitudinal limits when they ascended Mt Everest without breathing pure $O_2$ (Oelz et al. 1986).

There are two basic types of physiological responses of animals to reduced $pO_2$: (1) for oxyconformers, metabolism is linearly dependent on ambient $pO_2$; (2) for $O_2$-regulators, such as mammals, metabolism is constant with decreasing $pO_2$ to the $P_c$, below which MR decreases linearly with $pO_2$. Their $P_c$ is modified by several exogenous and endogenous factors, including $T_a$ (Rosenmann 1987). Thus, mammals that live or evolved in environments with lower $pO_2$ have a lower $VO_2$ and lower $P_c$ than mammals living at higher $pO_2$ (Rosenmann & Morrison 1975). Consequently, $P_c$ can be used as a measure of sensitivity to hypoxia in different mammal species inhabiting high and low altitudes (Rosenmann & Morrison 1974).

Overall, mammals are much more sensitive to and dependent on environmental $pO_2$ than birds, and consequently their altitudinal distribution is more restricted. So, although birds live and nest at high altitudes (4,000 – 6,500 m), migrate to even higher altitudes (Bouverot 1985), and can fly still higher (a griffin vulture collided with an aircraft at 11,300 m; Laybourne 1974), only a few mammals can attain 6,000 m (see McNab 2002). Unlike birds (which have highly efficient cross-current lungs with unidirectional airflow; Bicudo et al. 2010), mammals have an alveolar lung with a tidal (inspiration-expiration flow change). In functional terms, this and the alveolar dead space limit the maximum alveolar $pO_2$, which can only be increased by increased respiratory ventilation. Indeed, the first response of mammals faced with experimental hypoxia is increased respiratory frequency ($f_R$; Rosenmann & Morrison 1974; Nice et al. 1980). Rodent species that usually live at low altitudes generally have a much greater increase in $f_R$ in response to hypoxia than high altitude species (Rosenmann & Morrison 1975). A similar pattern can be seen with $T_b$, which decreases more markedly in lowland than high altitude species exposed to similar levels of hypoxia. Thus, mammals generally have differential tolerance to hypoxia, which is closely related to the altitude at which they live. Accordingly, the metabolic response to hypoxia ($P_c$) of mammals depends on the altitude of their evolutionary point of origin (Novoa et al. 2002). Data for 27 rodent species from high and low altitudes show that metabolic demand greatly influences the magnitude of $P_c$ (Table 4.3). Overall, native high altitude species have a significantly lower $P_c$ (14.9 kPa) than native low altitude species (16.2 kPa).

Mammals have a reduced maximal metabolic rate (MMR) at low $P_b$ (< 75 kPa; Rosenmann & Morrison 1975), and this impairs their capacity for sustained activity and thermoregulation in the cold. Consequently, some native high altitude mammals have specific adaptations that maximize gas exchange and thermogenesis. There is strong evidence that evolutionary adaptation of hypoxia tolerance is associated with the $O_2$ content of blood leaving the lungs, which in turn is highly dependent on haemoglobin (Hb), the number and size of red blood corpuscles, haematocrit (Hct), and Hb-$O_2$ affinity ($P_{50}$). For example, high altitude adaptations of yaks

**Table 4.3** Critical pressure of oxygen ($P_c$) in rodents from low and high altitude. Table modified from Novoa et al. (2002); data from Rosenmann and Morrison (1975) and Novoa et al. (2002).

| Species | Body mass (g) | Critical $pO_2$ $P_c^*$ (kPa) |
|---|---|---|
| Low altitude species | | |
| *Phyllotis darwini limatus* | 56 | 18.1 |
| *Octodon degus* | 181 | 18.5 |
| *Acomys cahirinus* | 49 | 17.5 |
| *Citellus undulatus* | 472 | 17.5 |
| *Glaucomys volans* | 67 | 17.1 |
| *Baiomys taylori* | 8 | 16.9 |
| *Microtus oeconomus* | 36 | 16.4 |
| *Meriones unguiculatus* | 48 | 16.4 |
| *Dicrostonyx g.stevensoni* | 52 | 16.4 |
| *Clethrionomys rutilus* | 33 | 16.3 |
| *M. musculus* (white) | 35 | 15.5 |
| *M. musculus* (feral) | 17 | 14.9 |
| *Calomys callosus* | 48 | 14.8 |
| *Peromyscus m. bairdi* | 21 | 14.5 |
| *Akodon olivaceus* | 27 | 14.1 |
| *Microtus pennsylvanicus* | 29 | 12.9 |
| High altitude species | | |
| *Phyllotis darwini chilensis* | 29 | 15.9 |
| *Eligmodontia puerulus* | 21 | 15.9 |
| *Akodon andinus* (Farellones) | 29 | 15.9 |
| *Cavia porcellus* | 481 | 15.6 |
| *Mus musculus* (feral) | 19 | 15.5 |
| *Akodon boliviensis* | 26 | 15.1 |
| *Akodon andinus* (Parinacota) | 27 | 14.5 |
| *Calomys ducilla* | 18 | 14.4 |
| *Phyllotis darwini posticalis* | 78 | 14.3 |
| *Ochotona rufescens* | 176 | 13.3 |
| *Auliscomys boliviensis* | 87 | 13.3 |

*$P_c$ corrected for different metabolic loads according to $VO_2 = 3.8\,M^{-0.27}$, and an increase in $P_c$ of 1.76 kPa per metabolic load (after Rosenmann & Morrison 1975).

include increased red corpuscle count and decreased mean corpuscular volume (Ding et al. 2014). Yaks also have large lungs and heart relative to their mass, and their haemoglobin has a high $O_2$ affinity (Wiener et al. 2006). Juveniles of two South American Andean rodents (*Abrothrix* and *Phyllotis*) born in captivity at sea level retain the high Hb and Hct of adults, maintaining their functional architecture, and suggesting evolutionary adaptation to high altitude (Ruiz et al. 2005).

The affinity of Hb for oxygen and its relationship with altitude and hypoxia have been widely examined as physiological adaptations, as have the size and number of erythrocytes. One solution to low $pO_2$ at altitude is a high Hb-$O_2$ affinity to facilitate $O_2$ loading in the lungs. However, a high affinity would not be beneficial to overcome the problem of delivery of $O_2$ to tissues (see Figure 3.21). It seems that in hypoxic environments it is $O_2$ loading that is paramount as without adequate arterial $O_2$ saturation, facilitating $O_2$ delivery to tissues becomes less relevant. Physiologists initially considered the decreased Hb-$O_2$ affinity and high Hb and Hct of humans to be beneficial adaptations to high altitude, whereas they now acknowledge that these responses may not be generally adaptive for high altitude mammals. It is now generally acknowledged that high Hb-$O_2$ affinity is characteristic of hypoxia-tolerant mammals, and this pattern has been observed for widely divergent vertebrate species. Nevertheless, it is important to interpret 'adaptations' to altitude hypoxia in an evolutionary perspective. For instance, the high Hb-$O_2$ affinity (low $P_{50}$) of South American camelids was initially considered to be an adaptation to high altitudes. Studies of Old World camels and dromedaries now indicate that high Hb-$O_2$ affinity is a common trait in the family, being present in the lineage before the colonization of the Andes, suggesting that camelids in South America exploited their low haemoglobin $P_{50}$ as an exaptation to life at high altitude (see Rezende et al. 2005).

The respiratory functions of Hb are a product of both its intrinsic $O_2$-binding affinity and interactions with allosteric effectors such as $H^+$, $Cl^-$, $CO_2$, and organic phosphates (e.g. 2,3-diphosphoglycerate, DPG). For many high altitude mammals, it is possible to identify specific mechanisms of Hb adaptation to hypoxia. Indeed, functional studies of human Hb mutants also suggest that there is ample scope for evolutionary adjustments in Hb-$O_2$ affinity.

At a molecular level, Storz et al. (2009) showed that adaptive modifications of heteromeric proteins (such as haemoglobin) can involve genetically based changes in single subunit polypeptides or parallel changes in multiple genes that encode distinct, interacting subunits. Their evolutionary and functional analysis of duplicated globin genes showed that natural populations of deer mice are adapted to different elevational areas. The adaptation of the haemoglobin of *P. maniculatus* to high altitude involves parallel functional differentiation at multiple unlinked gene duplicates, two $\alpha$-globin paralogs, and two $\beta$-globin paralogs. Differences in the $O_2$-binding affinity of the alternative $\beta$-chain haemoglobin isoforms are entirely attributable to allelic differences in sensitivity to DPG, a classical allosteric cofactor that stabilizes the low-affinity, deoxygenated conformation of the haemoglobin tetramer. The two-locus β-globin haplotype that predominates at high altitude is associated with suppressed DPG sensitivity and therefore increased haemoglobin-$O_2$ affinity and enhanced pulmonary $O_2$ loading under hypoxia.

Deer mice (*Peromyscus maniculatus*) have an array of Hb polymorphisms, which are inherited as two different haplotypes—similar to what is expected with Mendelian inheritance of a single locus with two alleles—and haplotype frequencies are correlated with altitude (Snyder et al. 1988). Individuals with the high altitude haplotype had increased Hb-$O_2$ affinities and MMR during cold exposure when measured at high altitudes, whereas the opposite was observed for the low altitude haplotype, suggesting that the correlation between allelic frequencies and altitude may reflect local adaptation.

For humans, the evolutionary adaptation of Tibetans to hypoxic high altitudes on the Tibetan plateau is remarkable, as it provides evidence of incorporation of genetic material from another species into the human genome (Bigman 2010; Yi et al. 2010). The hypoxia pathway gene (*EPAS1*) was identified as having the most extreme signature of positive selection, being associated with phenotypic differences in Hb concentration at high altitude. This gene has an extraordinary haplotype structure that can only be explained by introgression of DNA from Denisovan-related individuals (fossil hominins) into humans (Huerta-Sánchez et al. 2014), with the selected haplotype found only in Denisovans and Tibetans, suggesting that admixture with other hominin species has provided genetic variation that facilitated humans adapting to new environments.

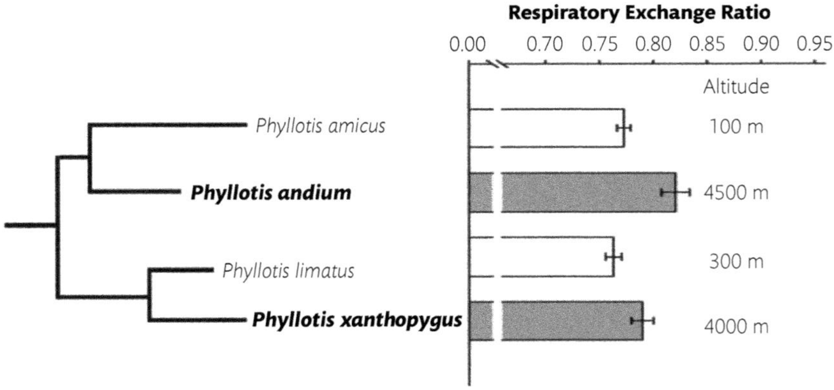


**Figure 4.14** Comparison of carbohydrate metabolism at 75% of maximal metabolic rate, showing a higher respiratory exchange ratio for high altitude Andean rodents (bold font) compared to low altitude species, suggesting the independent evolution of high carbohydrate substrate use (high RER) as an adaptation to hypoxia. Modified from Schippers et al. (2012).

A very different metabolic strategy for adapting to altitude hypoxia is to minimize the amount of $O_2$ required for metabolism by modifying metabolic pathways. Hochachka (1985) suggested that increased reliance on carbohydrates as the substrate for metabolism requires less $O_2$, since the chemical stoichiometry of carbohydrate metabolism requires less $O_2$ per kJ than lipid or protein metabolism (see Table 2.5). Schippers et al. (2012) found that two species of Andean rodent (*Phyllotis andium* and *P. xanthopygus*) use more carbohydrate than low altitude counterparts (*P. amicus* and *P. limatus*) during activity, suggesting that the high altitude species have indeed adapted their fundamental metabolic physiology in response to hypoxia (Figure 4.14). Cardiac muscle of the high altitude species also had enhanced oxidative capacity, facilitated by higher citrate synthase and isocitrate dehydrogenase (although skeletal muscle had no striking differences in enzyme levels).

The nature of altitudinal adaptations of mammals reflects to some extent whether they are preadapted, evolutionarily (genetically) adapted, developmentally adapted individuals, or short-term phenotypically acclimated individuals (Monge & León-Verlarde 1991). Camelids, for example, are considered to be pre-adapted to high altitude (Rezende et al. 2005; Weber 2007). Sometimes the genetic adaptations of mammals native to high altitude are different or even opposite to the phenotypic adaptations of short-term acclimated individuals (see Monge & León-Verlarde 1991). Native high altitude mammals generally have a normal haematocrit and slight pulmonary vasoconstriction and hypertension, and a high $P_{50}$ and low tissue $pO_2$; changes in blood flow are important in exercise and short-term hypoxia. In contrast, many mammals acclimated to altitude (e.g. humans) have a more pronounced pulmonary vasoconstriction and hypertension, and elevated haematocrit (polycythaemia), which puts a greater load on the right side of the heart, and a lowered Hb $P_{50}$ (e.g. related to increased blood DPG levels), which at least partially maintains high tissue $pO_2$. Developmental effects of hypoxia include reduced placental diffusion distance and increased blood flow, and increased lung volume. Nevertheless, there is evidence of growth retardation for humans and other non-native mammals born at high altitude. Finally, non-native mammals adapted to high altitude often lose their phenotypic adaptations when they return to low elevations, whereas native high altitude mammals do not, and are generally as adept at survival in low as well as high altitudes. The guanaco, which is a typical high-elevation native mammal in, for example, Peru and Bolivia, is also a typical low altitude mammal, for example in Argentina.

### 4.4.2 Thermal Balance

Cold stress is the second major physiological challenge at high altitude. The typical response of mammals to cold stress is increased thermogenic capacity (see 4.1), but the hypoxia associated with altitude compromises gas exchange, hence metabolism and thermogenesis. Increased insulation is consequently an especially viable strategy for counteracting the cold at high altitude. For yaks, temperature is the most important factor determining their survival and distribution (Wiener et al. 2006). They prefer an average ambient $T_a$ of about 5°C, and survive well at even −40°C. Their primary thermal strategy is to conserve heat, not thermogenesis. They have a thick fur coat, a thick layer of subcutaneous fat, and thick skin, all of which aid heat conservation. They almost completely lack functional sweat glands.

Thermogenesis is nevertheless important for some high altitude mammals, especially small species for which increased insulation is not a viable option. The plateau pika (*Ochotona curzoniae*; 3,000–5,000 m) has an unusually high resting metabolic rate and capacity for non-shivering thermogenesis (Li et al. 2001). Hayes and O'Connor (1999) found some evidence for natural selection of maximum thermogenic capacity ($VO_{2,max}$) amongst high altitude (3,800 m) deer mice in the field, which should facilitate thermoregulation in the cold and also allow sustained maximal activity. In one of two years, there was strong selection for high $VO_{2,max}$ in the deer mice, suggesting that a high thermogenic capacity is important in the cold (see 4.1); there was also weak selection for reduced body mass (counter to predictions from Bergmann's rule).

## 4.5 Aquatic Environments

Diving mammals have a suite of adaptations that range from the molecular and cellular level, through morphology and physiology, to behaviour and ecology. The adaptive consequence of many of these is to facilitate their swimming and to extend their breath-hold time, hence depth of diving and foraging duration (aerobic dive limit, ADL). However, breath-hold time and foraging duration are only some of the problems faced by aquatic mammals; thermal balance can also be problematic for diving mammals. In cold water, considerable heat is lost to water, so heat retention is important; in warm water, they can experience the opposite problem of having sufficient heat loss to maintain constant $T_b$. For marine mammals, water and solute balance can also be problematic, especially if they are consuming food that is isosmotic with seawater (e.g. most marine invertebrates, particularly very hydrated ones, such as jellyfish). Seawater is generally their only source of drinking water, so this exacerbates their salt load if they do indeed drink. Since mammals lack specialized salt glands (unlike marine birds; Bicudo et al. 2010), they rely on kidneys for maintenance of body water volume and solute balance. Finally, deep-diving marine mammals also experience considerable hydrostatic pressures (pressure increases by 1 atm every 10 m of water depth). For the deepest-diving mammals (e.g. Cuvier's beaked whale, *Ziphius cavirostris*; 2,992 m), this corresponds to a hyperbaria of about 29,652 kPa. Even though water is essentially incompressible, such high pressures can have physical and biochemical effects on cells, as well as extreme effects on air spaces (e.g. lung volume) and the dynamics of gas exchange (hence problems such as the 'bends').

Semi-aquatic mammals, such as beaver, muskrat, and sea otters, have some adaptations for swimming and extending their ADL (Williams & Worthy 2009; Berta et al. 2015; Davis 2014). They have webbed feet for swimming and typically have a high buoyancy, which is related to the volume of air trapped in their fur. For example, the pelage air layer of muskrat is 21% of body volume, which reduces its density to 0.79. Although such a high buoyancy is good for floating at the surface and contributes greatly to pelt insulation (see 4.1.1), it increases the energetic cost of diving for these semi-aquatic mammals (Williams et al. 1992; Fish 2000; Fish et al. 2002), and they do reach considerable depths; sea otters dive up to 100 m (Berta et al. 2015).

Sirenians (dugong, *Dugong dugon* and manatees, *Trichechus* spp.) are permanently aquatic and forage on sea grasses in shallow water. They have high-density bone that facilitates neutral buoyancy while foraging in shallow water. Nevertheless, sirenians have been reported to dive for 6 min to 600 m for manatee and 8 min to 400 m for dugong (Berta et al. 2015). However, it is the deep-diving marine mammals, such as many cetaceans and seals, that rely in extremis on many levels of adaptation to achieve their remarkable breath-hold time and capacity for deep diving. For example, phocid seals dive up to 1,530 m and for up to 120 min; otariid seals, 482 m and 18 min; mysticete cetaceans, 500 m and 80 min; and odontocete cetaceans, 3,000 m and 138 min (Berta et al. 2015). Cuvier's beaked whale dives up to 2,992 m depth and for up to 137.5 min; deep dives on average were to 1,401 m depth and 67.4 min duration (Schorr et al. 2014).

An obvious morphological adaptation of diving mammals is their streamlined profile, with limbs modified for swimming (flippers of seals, flukes of cetaceans). Streamlining provides a marked energetic advantage for swimming (Davis 2014). A fineness ratio (body length/diameter) of about 4.5 is optimal for drag reduction; pressure drag increases at lower ratios and frictional drag increases at higher ratios (Vogel 1994). The high aspect ratio ($span^2$/area) of flippers and flukes aids thrust production and swimming efficiency (Fish 1998). Surface swimming increases the energy cost of swimming by up to five times because of the formation of a swimming wave, but swimming three body diameters below the surface eliminates the wave energy dissipation (Fish 2000). Not surprisingly, most marine mammals minimize their time swimming at the surface, but they do have to occasionally breathe. Deep-diving marine mammals have little hair, so pelage buoyancy is not an issue; rather, they have blubber (20–30% of body mass), which provides some buoyancy as well as thermal insulation, enhances streamlining, provides energy reserves, and has lower maintenance costs (Fish 2000; see also 4.1.1). Changes in dynamics of buoyancy of these mammals with diving are related more to lung volume and lung compression at depth. For them, changes in swimming behaviour seem to be the main compensation for changes in buoyancy during diving.

Hydrodynamic lift from the flippers or flukes can counterbalance the reduced buoyancy at depth due to compression of the air spaces. Deep-diving mammals typically exhale upon diving to facilitate negative buoyancy and to limit exposure to high partial pressures of gases while submerged. Their negative buoyancy at depth conserves $O_2$ stores and facilitates extended feeding durations. They can also exploit intermittent swimming to reduce locomotory costs by gliding during decent (Skrovan et al. 1999; Fish 2000).

### 4.5.1 Diving Response

The obvious problem for diving mammals of extending breath-hold time (hence foraging duration) meant that the physiology of breath-holding (apnoea), or the diving response, was one of the first adaptations investigated for diving. The concept of the diving response was based on early observations by Bert and Richet (see Castellini 2012 and Davis & Wiliams 2012 for historical reviews). The physiology of the diving response was examined in the 1930s by Scholander and Irving using simple (if somewhat crude) laboratory experiments involving forced submersion of various mammals and birds, to examine the hypothesis that during diving there were circulatory adjustments that confined $O_2$ delivery to critical tissues such as heart and brain. These early experiments clearly established the cardiovascular sequelae of diving apnoea. There is a rapid decrease in heart rate (bradycardia) upon cessation of breathing (Figure 4.15A) that is associated with reduced distribution of blood flow to many organs (e.g. muscle, gut, liver, kidney, skin) but not the more essential aerobic organs (e.g. heart, brain, lungs, adrenals). The decrease in heart rate combined with the peripheral vasoconstriction maintains central blood pressure (Figure 4.15B). In fact, this cardiovascular adjustment is a general and primitive response of mammals (even non-diving species) and other vertebrates to hypoxic challenge (Panneton 2013). Humans also have a pronounced diving response to breath-holding and water contacting the face (Gooden 1994). Scholander (1963) described this dramatic cardiovascular response to apnoea and hypoxia as the 'master switch of life'.

Although these early investigators of the diving response appreciated that there may be physiological differences between forced and voluntary diving, it was not until the 1970s that technology became available to study the physiology of mammals and birds diving voluntarily. Kooyman and Campbell (1972) showed that the diving response was generally less dramatic for voluntary dives. This and other pioneering studies of Kooyman further showed that seals would voluntarily make both shallow and short dives, and deep and long dives, and that there were differences, such as more pronounced bradycardia for the latter (even at the initiation of diving). More recent studies of free-diving cetaceans indicate that the bradycardia response to diving is quite flexible, being modulated by behaviour and the nature of their activity. For example, free-swimming bottlenose dolphins (*Tursiops*

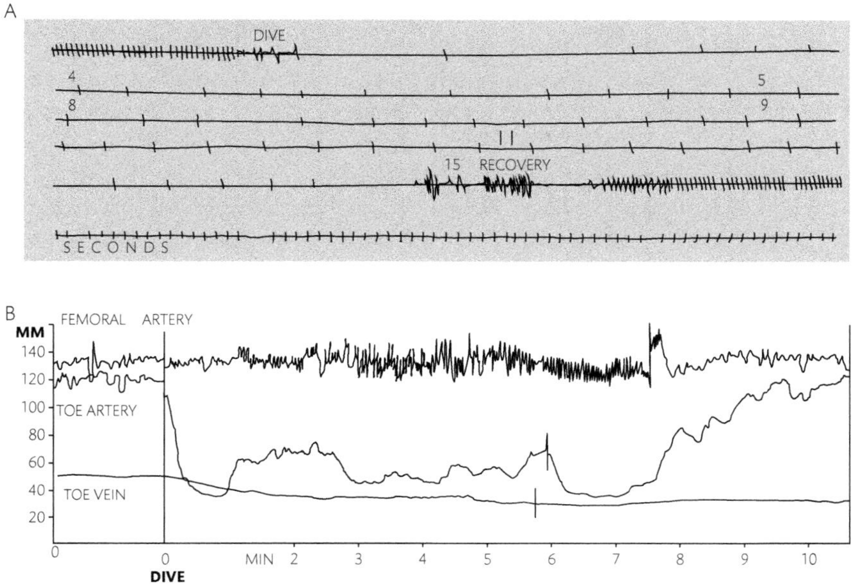


**Figure 4.15** Diving response for a common harbour seal (*Phoca vitulina*) during a forced dive. A: Heart rate pre-dive, at various stages during a 15-min dive (numbers indicates minutes into the dive), and post-dive ('15 RECOVERY' indicates artefact from struggling, not heart rate). B: Central arterial (femoral) and peripheral (toe), and peripheral venous (toe) blood pressure during diving; vertical line indicates the end of the dive for each pressure trace. From Irving et al. (1942). Reproduced with permission of the American Physiological Society.

*truncatus*) have a lower but variable heart rate while underwater compared to when resting at the surface (105 beats $min^{-1}$, compared to their maximum heart rate of 128 $min^{-1}$), submerged resting (40 $min^{-1}$), horizontal swimming (37 $min^{-1}$), and head bobbing (56 $min^{-1}$; Noren et al. 2012). Thus, Scholander's concept of a 'master switch' was in reality more of a graded cardiovascular response ('dimmer switch') that anticipated the extent of the dive.

The next conceptual leap in diving physiology was addressing the question of whether diving metabolism was primarily aerobic or anaerobic (Davis 2014; see 2.3). Diving mammals have cardiac and microvascular adaptations that conserve essential nutrients (especially oxygen), and isolate lactate that is produced in active muscle while diving underwater (Irving et al. 1942; Zapol et al. 1979). In a forced dive, the extreme diving bradycardia is associated with a major redistribution of blood flow away from many peripheral tissues (gut, kidneys, muscle) that helps to conserve oxygen for hypoxia-sensitive tissues such as the heart and brain.

The hypoxic tissues will rely more on anaerobic metabolism during the dive, and accumulate sometimes very high levels of lactate.

In natural dives, the bradycardia and redistribution of blood flow is less pronounced, and whole-body metabolism remains largely aerobic (Kooyman et al. 1980; Fedak et al. 1988). Here, the primary role of the diving response seems to be to regulate the level of hypoxia in the skeletal muscle for more efficient use of blood and tissue oxygen stores and maximize aerobic dive duration (Davis 2014; see 4.5.2). Indeed, because many diving mammals rely on their muscle myoglobin $O_2$ stores, a low intramuscular $pO_2$ (less than about 1.3 kPa) is required to unload $O_2$ from the myoglobin. The redistribution of blood flow and reduced flow to particular organs and tissues due to the diving response is an important means of making these tissues hypoxic but not anaerobic, to facilitate $O_2$ unloading from the myoglobin stores. As the dive is extended, the overall reduction in arterial $pO_2$ (hypoxic hypoxia) also facilitates this myoglobin $O_2$ unloading. The intensity of the dive response is adjusted to optimize the use of $O_2$ stores, and maximize the ADL. For $O_2$-sensitive tissues without myoglobin $O_2$ stores (e.g. heart, brain, kidneys), there is a minimum heart rate and cardiovascular transport for sustaining aerobic metabolism. For Weddell seals, a minimum heart rate of 12 $min^{-1}$ is predicted for sustaining aerobic metabolism in these organs; this is essentially the same as that observed (11 $min^{-1}$) for free-diving, drifting seals at depth (Davis & Williams 2012).

The aerobic organs of diving mammals also have their own particular adaptations (Davis 2014). The heart has a high myoglobin concentration and lactate dehydrogenase (LDH) activity, and a high glycogen content, and the kidneys can have a high mitochondrial volume density. The splanchnic organs (liver, stomach, intestine) have high mitochondrial volume density and LDH levels. If the ADL is not exceeded, then kidney and splanchnic blood flow are similar to resting levels, and function is sustained. If the ADL is exceeded, blood flow and organ function is dramatically reduced, but is rapidly re-established after $O_2$ availability is restored.

Diving-induced ischaemia/reperfusion produces reactive oxygen species (ROS), and ROS production is higher for long- than short-duration divers (Zenteno-Savín et al. 2012). Oxidative damage is not limited by restricting ROS production, but is limited by the synergistic effects of higher tissue activities of antioxidant enzymes and non-enzymatic compared to tissues of non-diving mammals. High levels of antioxidant enzymes have been reported in tissues of marine mammals; for example, superoxide dismutase (SOD), glutathione peroxidase (GPx), glutathione reductase (GR), and glucose-6-phosphate dehydrogenase (G6PDH). High levels of a non-enzymatic antioxidant, γ-L-glutamyl-L-cysteinyl glycine (GSH), have been reported for various terrestrial animals that routinely experience hypoxia, including hibernation and aestivation, environmental hypoxia and diving (Hermes-Lima & Zenteno-Savín 2002; Zenteno-Savín et al. 2012). The antioxidant systems of diving mammals and birds are effective in that their tissues do not have higher oxidative damage than non-divers, and GSH seems to have an important role in both, but the details of antioxidant patterns are different between seals and penguins.

## 4.5.2 $O_2$ Stores and Aerobic Dive Limit

Diving mammals are isolated from an air supply while underwater, hence any use of $O_2$ for energy generation while diving must rely on $O_2$ stored within the body. There are three main potential $O_2$ storage sites during diving: lung $O_2$ stores, blood $O_2$ stores, and muscle myoglobin $O_2$ stores (Table 4.4).

**Table 4.4** Lung, blood, and muscle oxygen store parameters for various diving mammals, and cattle as a terrestrial comparison. DLV, diving lung volume; BV, blood volume; Hb, haemoglobin concentration; Mb, myoglobin concentration. Data from Ponganis (2011), Gerlinsky et al. (2013) and Davis (2014).

| $O_2$ Store | Lung Stores | Blood Stores | | Muscle Stores | |
|---|---|---|---|---|---|
| Species | LV (ml $kg^{-1}$) | BV ml ($kg^{-1}$) | Hb (g $dl^{-1}$) | Mb (g $kg^{-1}$) | Muscle (g $kg^{-1}$) |
| Cattle | 66 | 71 | 12 | 0.4 | 330 |
| Bottlenose dolphin | 90 | 71 | 14 | 3.3 | |
| Manatee | 37 | 80 | 15 | 0.4 | 350 |
| Sea otter | 207 | 91 | 17 | 2.6 | 30 |
| Steller sea lion | 55 | 97 | 120 | 2.7 | |
| Hooded seal | 40 | 106 | 23 | 9.5 | 270 |
| Walrus | 58 | 106 | 16 | 3.0 | 300 |
| Pacific white-sided dolphin | | 108 | 17 | 3.5 | |
| Northern fur seal | 72.5 | 109 | 17 | 3.5 | 300 |
| California sea lion | 58 | 120 | 18 | 5.4 | 370 |
| Beluga whale | 57.8 | 128 | 21 | 3.4 | 300 |
| Harbour seal | 39 | 132 | 21 | 5.5 | 300 |
| Dall porpoise | | 143 | 20 | | |
| Harp seal | | 168 | 23 | 8.6 | 250 |
| Baikal seal | | 177 | 27 | 6.0 | 300 |
| Australian sea lion | | 178 | 19 | 2.7 | |
| Sperm whale | 28 | 200 | 22 | 5.4 | 340 |
| Weddell seal | 27 | 210 | 26 | 5.4 | 350 |
| Elephant seal | 20 | 216 | 25 | 6.5 | 280 |

Lung $O_2$ stores depend on the volume of air in the lungs (where gas exchange occurs) on commencement of diving. For near-surface swimmers, the inspiratory volume is assumed to be available as an $O_2$ store for manatees, 60% of lung volume for sea otters and 50% for pinnipeds, compared to 100% of lung volume for cetaceans (Ponganis et al. 2011). Further, an extraction of 15% lung $O_2$ is assumed, to calculate the effective lung $O_2$ store. The relatively large lung size of delphinid and phocoenid dolphins, which are rapid breathing, short-duration, shallow divers, may enable their lung to function as a site of respiratory gas exchange throughout a dive (Piscitelli et al. 2010). For blood $O_2$ stores, essentially all the $O_2$ is bound to haemoglobin; an insignificant amount is dissolved in the plasma. Consequently, the $O_2$ store is proportional to the blood haemoglobin concentration, saturation levels of arterial/venous blood, and the total blood volume.

Not surprisingly, marine mammals typically have much higher blood haemoglobin concentrations than terrestrial mammals (see 3.4.1); there is no scope for adaptive variation in how much $O_2$ can be bound to haemoglobin, as all vertebrate haemoglobins typically bind 1.34 ml $O_2$ per gram (Withers 1992). It is usually assumed that one-third of the blood is arterial and two-thirds is venous, initial arterial blood is 95% saturated and end arterial saturation is 20%, initial venous content is 5 ml $dl^{-1}$ lower than 95% arterial blood saturation, and end venous blood is 0% saturated (Ponganis et al. 2011). Muscle $O_2$ stores rely on intracellular myoglobin, which is essentially a single subunit of tetrameric haemoglobin, and typically binds 1.34 ml $O_2$ per gram (like haemoglobin). Myoglobin stores are assumed to be 100% saturated prior to diving, and as low as 0% saturated at the end of a dive. Not surprisingly, marine mammals typically have much higher muscle myoglobin concentrations than terrestrial mammals (see later).

The suite of physiological adjustments of diving mammals that increase total body $O_2$ stores, and morphological and locomotory adaptations that maximize the energetic efficiency of diving (e.g. streamlined shape, prolonged gliding), extend their ADL (Kooyman et al. 1980; Williams et al. 2000; Ponganis et al. 2011). Although the ADL can be measured experimentally by measuring blood lactate levels, it is most often determined as the ratio of calculated $O_2$ stores to diving metabolic rate. Most marine mammals are thought to dive within their ADL, but the specialized deep and long-diving beaked whales (e.g. *Ziphius cavirostris*) and high-speed sprinting fin whales (e.g. *Globiocephala macrorhynchus*) might challenge this point of view. Both groups of whales have high myoglobin, as expected, and muscle fibre specializations that extend their ADL (Velten et al. 2013). Beaked whales have mainly fast-twitch fibres that decrease the metabolic cost of diving, and fin whales have some rare slow-twitch oxidative fibres to support high swimming activity. Calculated ADLs suggest that both groups of whales have sufficient $O_2$ stores to sustain aerobic dive limits.

Muscle myoglobin concentration ([Mb]) has increased independently in a variety of mammalian divers. Mirceta et al. (2013) reconstructed the evolutionary

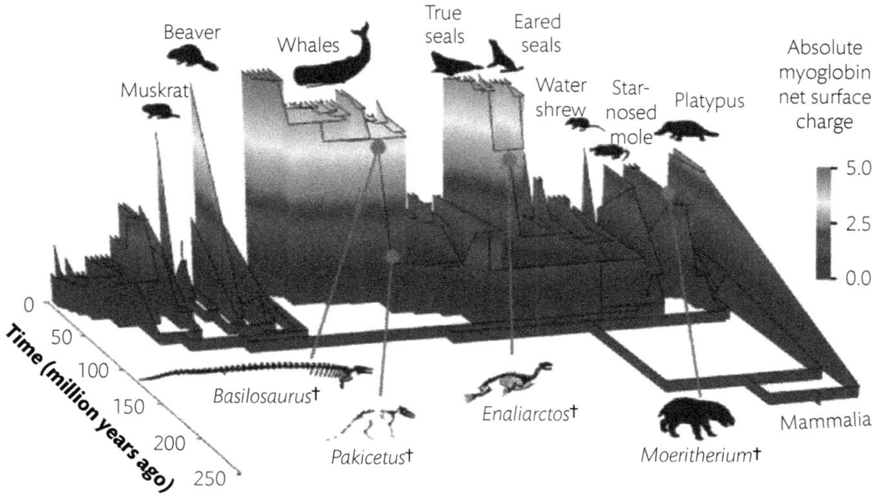


**Figure 4.16** Evolutionary reconstruction of myoglobin net surface charge (a proxy for myoglobin concentration) for terrestrial and aquatic mammals. The myoglobin net surface charge increases in all lineages of mammalian divers with an extended aquatic component, and is used to infer the diving capacity of extinct species representing stages during mammalian land-to-water transitions. From Mirceta et al. (2013). Reproduced with permission from Science (AAAS).

history of [Mb], using the net surface charge density of myoglobin as a proxy for maximal [Mb]. Their findings not only show the pattern of independent evolution of high Mb levels in various extant mammalian diving lineages (Figure 4.16), but imply an aquatic ancestry of related but now non-aquatic mammals, such as echidnas (cf. platypus), talpid moles (cf. star-nosed moles), and hyrax and elephants (cf. sirenians).

For shallow semi-aquatic divers (and diving birds), and deep divers prior to lung collapse, the $O_2$ in the lung air is a substantial store (Figure 4.17). However, the lung $O_2$ store of deep divers is compromised at depth because of the hyperbaric compression of particularly their lung (Ponganis 2011; see 4.5.3), so they typically have a very low lung $O_2$ store. For most divers, blood $O_2$ is the main store. Muscle $O_2$ stores can range from being a relatively small store to the main store, depending on the muscle myoglobin content of particular species.

### 4.5.3 Hyperbaria: Diving under Pressure

Three physiological syndromes are experienced by human divers when breathing air at considerable depth, hence pressure: decompression sickness, $N_2$ narcosis, and $O_2$ toxicity. These are all consequences of the fact that pressure increases

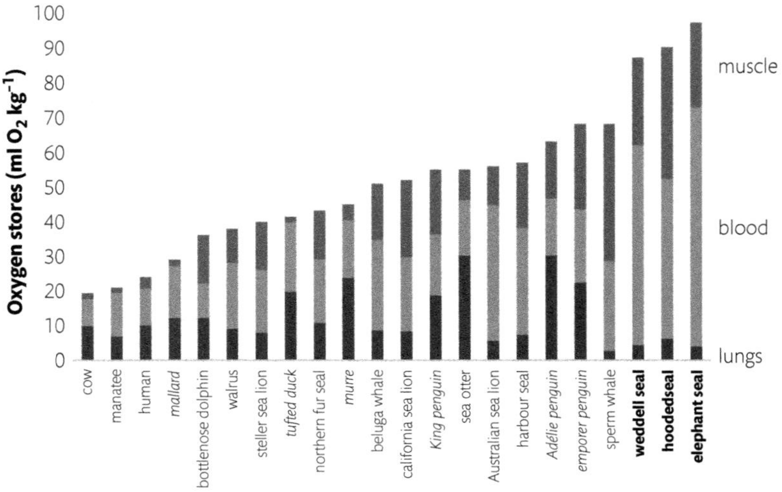


**Figure 4.17** Lung, blood, and muscle $O_2$ stores in non-diving mammals (human, cow), shallow-diving marine mammals (manatee, sea otter), various seals and cetaceans, and some diving birds (italics) for comparison. Deep-diving marine mammals (in bold, far right) typically have low lung $O_2$ stores, especially compared to diving birds and the surface-diving sea otter. Redrawn using data from Ponganis (2011), Ponganis et al. (2011), Davis (2014), Stephenson et al. (1989), and Croll et al. (1992)

dramatically with depth underwater, by about 1 atmosphere pressure (101.3 kPa) per 10 m depth (Figure 4.18). For an air space underwater (e.g. the respiratory system), the increase in pressure collapses the volume of air (unless there is sufficient physical strength of the vessel wall to resist collapse; e.g. the hull of a submarine), and the partial pressures of all gases are increased in proportion to the pressure. For example, at 10 m depth, the pressure is 202.6 kPa (2 atm), and for normal air the $pO_2$ is 42.4 kPa (compared to 21.2 at 1 atm) and the $pN_2$ is 160.2 kPa (compared to 80.1).

Breathing high partial pressures of gases increases the amount of those gases dissolved in the body fluid and fat (Henry's law; see 2.5.4). Over an extended time at even moderate depth (e.g. 10 m), more and more $O_2$ and $N_2$ dissolve into the body fluids and fat. When the diver returns to the surface, these gases become supersaturated and can bubble out of solution; such bubbles can cause pain, tissue damage, and even death; this is decompression sickness (the 'bends', or caisson disease; Ponganis 2011; Pocock et al. 2013). Consequently, divers must ascend slowly, adhering to strict decompression schedule tables. Otherwise, they must be re-compressed at the surface (or by returning underwater) for long enough that gases in the body fluids and fat are 'blown off'.

Free-diving (without compressed air) occurs to relatively limited depth, for short periods, so insufficient gases dissolve into the body fluids and fat to cause

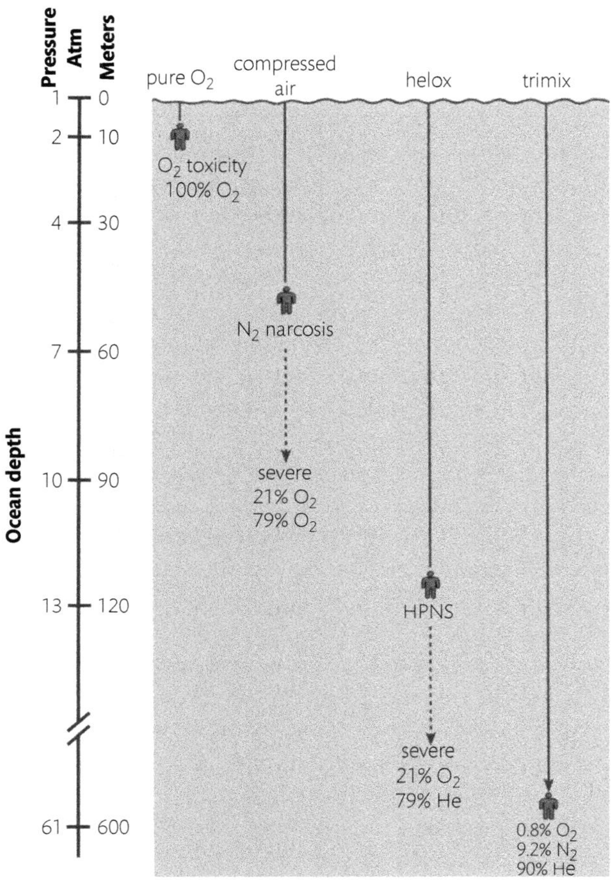


**Figure 4.18** Effect of water depth on barometric pressure, and use of various gases in diving to counteract decompression sickness, nitrogen narcosis, and high pressure nervous syndrome (HPNS). Modified from Pocock et al. (2013). Reproduced with permission of Oxford University Press.

decompression sickness, although continuous intense free-diving (e.g. pearl and sponge divers) can apparently rarely experience mild decompression sickness (Wong 2000). High $pO_2$ has direct physiological effects, resulting in $O_2$ toxicity, or high pressure nervous syndrome (HPNS; Ponganis 2011; Pocock et al. 2013). Even breathing pure $O_2$ at 1 atm pressure has physiological sequelae (pulmonary congestion, impaired mental activity), and above 2 atm leads to $O_2$ toxicity (nausea, dizziness, feeling of intoxication, tremors, convulsions). A rapid rate of increase in pressure exacerbates the extent of HPNS; humans typically experience HPNS at about 190 m depth. High $pN_2$ also has physiological effects at depths below about 50 m when breathing normal air, resulting in nitrogen narcosis. Symptoms are mental confusion and poor motor coordination, accompanied by a sense of euphoria ('raptures of the deep').

These various syndromes can be alleviated by breathing special gas mixtures (Figure 4.18). Breathing helox (or heliox; about 21% $O_2$ and 79% He) largely overcomes the problems of nitrogen narcosis, and its lower solubility than $N_2$ reduces the problems of decompression sickness. However, the high thermal conductivity of helium can lead to thermal balance problems, and its lower density makes vocal communication difficult (the voice becomes very high pitched). Trimix, a mixture of $O_2$ (0.8%), $N_2$ (9.2%), and He (90%), uses the narcotic effect of $N_2$ to reduce $O_2$ toxicity. Diving mammals might be equivalent to free-diving humans, but their extended dive times and considerably greater depths and pressures potentially make them as, or even more, susceptible to these syndromes (Ponganis 2011). Many cetaceans and pinnipeds exceed the 190 m HPNS onset depth experienced by humans, and should potentially experience nitrogen narcosis and decompression sickness, so they have particular adaptations to avoid these problems.

First, diving to considerable depth exposes the mammal to physical barotrauma, resulting from compression of air-filled spaces (Ponganis 2011). In some diving mammals, the aortic arch is relatively compliant whereas the aorta is relatively incompliant (Lillie et al. 2013). The compliant arch section acts as a windkessel, maintaining arterial blood pressure during the extended bradycardia. Extreme differences in stiffness of the arch and the aorta in species such as fin whales may be an adaptation to avoid haemodynamic consequences of rapid depth-related changes in pressure across the aortic wall, between intravascular blood and the thoracic cavity. Remarkable collapsibility of the chest wall and lung of various seals can provide nearly unlimited compensation for high pressures. Diving sea lions experience lung collapse at about 225 m, although the depth for collapse depends on the maximum dive depth; this suggests that sea lions inhale more deeply for anticipated deeper dives (McDonald & Ponganis 2012). Scholander (1940) noted that the excised lung of fin whales was able to collapse nearly completely with hyperbaria, with little remaining residual volume, and speculated that lung collapse avoided gas exchange at depth, hence avoided decompression sickness (Kooyman 2015).

Lung collapse not only reduces the uptake of $N_2$ at depth (and therefore reduces the risk of decompression sickness and nitrogen narcosis) but also preserves the lung $O_2$ store for use on ascent. Beaked whales, which dive deeply more regularly than other cetaceans, might be more susceptible to decompression sickness, and have additional adaptations that reduce decompression bubble nucleation such as specialized endothelial structures or elevated levels of nitric oxide, together with behavioural management of acent rates and recompression with subsequent dives (Hooker et al. 2009). Indeed, the adaptations of deep-diving marine mammals might be more to manage the $N_2$ load during ascent rather than to minimize $N_2$ loading during diving (Hooker et al. 2011). It has been suggested that human activities (e.g. sonar) might increase the risk of decompression sickness and stranding, by modifying acent behaviour (e.g. causing whales to acend more rapidly) or sonar directly interacting with tissues supersaturated with nitrogen (e.g. Weilgart 2007) although this idea remains controversial (Kvadsheim et al. 2012).

### 4.5.4 Vision and Echolocation

Light does not penetrate water, even clear water, very well. For turbid water, light penetration can be almost zero. Consequently, much of the aquatic environment is dark even during the daytime, so vision becomes difficult. Nevertheless, vision remains an important sense for marine mammals, and most have relatively large eyes, although walrus and sirenians have relatively small eyes (Berta et al. 2015). Marine mammals generally have a high visual acuity, facilitated by their nearly circular lens, thick retina with a high proportion of rod photoreceptors, and a well-developed tapetum lucidum to reflect light from the back of the retina back to the photoreceptors. Not surprisingly, sea otters and polar bears have more terrestrial-type eyes, relying more on aerial vision and other senses (e.g. chemoreception).

Marine mammals (dolphins, seals, manatee) have a higher carbohydrate concentration in their tears than terrestrial mammals (humans, camelids; Davis & Argüeso 2014), presumably to help maintain the hydration of the cornea in seawater (Tarpley & Ridgway 1991). The tear film of dolphins (Tarpley & Ridgway 1991) and seals (Davis et al. 2013) lacks lipids, unlike typical terrestrial mammals, which have a reduced ocular EWL because of the lipid component of their tear film (see 3.6.2). Cetaceans lack a nictitating membrane, so a lubricating role of the tear film is less important than for mammals that have a nictitating membrane (e.g. seals, polar bear).

Deep-diving marine mammals must locate their prey in a dark environment, unless they have echolocation (see later), although the Guiana dolphin (*Sotalia guianensis*) has cutaneous electroreceptors for feeding underwater (Czech-Damal et al. 2012b; see 3.7.2.6), and the platypus (*Ornithorhynchus anatinus*) has electroreceptors on its bill for feeding underwater (with its eyes closed; Grant 1989). The vision of many marine mammals is adapted to low intensity light with a peak sensitivity at 485 nm, which matches the wavelength of bioluminescence produced by a large range of marine organisms, including myctophid fish (Vacquié-Garcia et al. 2012). Bioluminescence thus likely is a key to predator-prey interactions in deep, dark marine environments.

Hearing is an important sense for marine mammals (Berta et al. 2015). Polar bears and sea otters vocalize and hear much like terrestrial mammals, but pinnipeds have a number of specializations for underwater and terrestrial hearing. Many pinnipeds vocalize considerably on land, for various social interactions, including mother-pup identification calls that are important for re-establishing contact when the mother returns from feeding (other senses such as olfaction, vision, and spatial awareness are also important). Various species also have underwater vocalizations for general social interactions. Male walrus emit strange 'knocking' sounds during the breeding season; these 'songs' seem to reinforce dominance status. Some pinnipeds (e.g. harbour seal, *Phoca vitulina*; ringed seal, *Pusa hispida*; harp seal, *Pagophilus groenlandicus*; grey seal, *Halichoerus grypus*; hooded seal, *Cystophora cristata*) emit underwater clicks, but it is unclear whether these are used for echolocation. The ear of pinnipeds is generally similar to that of terrestrial mammals, although some modifications amplify auditory signals. The cavernous venous sinuses in tissues of the middle and outer ear may enhance sound transmission to the inner ear and also provide volume compensation for the hyperbaric effect of deep diving (Nummela 2008).

Cetaceans have a sophisticated, high-frequency echolocation system (sonar) for orientating and capturing prey underwater (Berta et al. 2015). Sound travels about five times faster in water (about 1,450–1,559 m $s^{-1}$) than air (about 340 m $s^{-1}$), and acoustic signals are useful, especially in turbid water or darkness, for feeding, sensing environmental features, and often remarkably sophisticated communication. Cetaceans emit both high- and low-frequency acoustic signals, with different roles. The source of the acoustic signals for small odontocetes is the complex nasal sac system, located just inside the blow hole. This is coupled to the role of the melon, a low-density, lipid-filled structure that sits above the upper jaw. The melon functions as an acoustic lens that creates a focused and highly directional acoustic beam emitted in front of the animal.

Toothed whales are a morphologically and ecologically diverse group of predators; sperm whales (*Physeter macrocephalus*) are deep-sea squid predators, dusky dolphins (*Lagenorhynchus obscurus*) prey on oceanic fish schools, and shallow water river dolphins (Platanistoidea) feed on individual prawns and fish (Jensen et al. 2013). Their acoustic calls are equally diverse, of four general types. The echolocation clicks of most odontocetes are short and relatively broadband, about 10 – 150 kHz, with a range of about 100 m (Berta et al. 2015). Sperm whale echolocations are highly directional, low frequency (3 – 15 kHz), and high intensity (235 dB). Killer whales (*Orcinus orca*) produce low- (80 Hz – 10 kHz), mid- (10 kHz) and high- (100 – 160 kHz) frequency signals, for object detection (low frequencies) and prey discrimination and compass bearing direction (high frequencies). They also have group 'dialects' for pod recognition. Some odontocetes use high-intensity sounds to stun their prey. Beaked whales have low-intensity, frequency-modulated clicks (about 45 kHz). Whistling dolphins have higher frequency (60 – 80 kHz), short, loud (220 dB) broadband clicks. Dolphins also produce a more harmonic and narrow frequency band whistle (or squeal). These are used for individual or group recognition. A range of porpoises, pygmy sperm whales, and other unrelated delphinids have narrowband clicks (about 130 kHz). River dolphins use a high repetition of relatively low frequency (61 – 92 kHz) and low intensity (180–195 dB) clicks compared to similar-size marine delphinids, perhaps reflecting their lesser need for long-distance detection.

The inverse scaling of echolocation frequency with body mass seems to be a major driver of echolocation call diversity in cetaceans (compared to habitat and diet variation, which drives much of the acoustic diversity in bats; see 4.7.4). Standard allometry of echolocation frequency and body mass for whales is strong ($R^2 = 0.86 - 0.93$), suggesting an allometric constraint on frequency, but there is a strong phylogenetic component to this pattern, as it is considerably weaker ($R^2 = 0.27$) after phylogenetic correction (May-Collardo et al. 2007). Mysticetes are again less understood than odontocetes with respect to their uses of acoustic signals. Many emit broadband clicks that may be used for echolocation and for communication; for instance, the complex, low-frequency 'songs' of humpback whales (*Megaptera novaeangliae*) are used for long-distance communication, over hundreds or even thousands of kilometers.

The cetacean auditory system is considerably modified for underwater hearing (Berta et al. 2015). The external auditory canal is very narrow, and in mysticetes is plugged and may not be functional. For odontocetes, echoes of sound reflected from objects in the environment are detected primarily by the dense bone of the lower jaw ('jaw hearing'; Norris 1964) and fat bodies that channel sound to the middle ear. The two bony parts of the ear, the tympanic and periotic bones, are very dense bone; they differ markedly from the typical terrestrial mammal ear. Inner ear cochlear structure appears modified for high-frequency hearing (e.g. the basilar membrane is relatively narrow and thick). The acoustic system of mysticete cetaceans is less understood but likely to be generally similar to odontocetes. They lack vocal cords but have a U-fold in the larynx that seems homologous in function to vocal cords, and distribution of sound into the environment seems to involve cranial sinuses (Berta et al. 2015).

Less is known of vocalization and auditory reception by sirenians (Berta et al. 2015). They emit 'chirp-squeaks' presumably from vocal folds in the larynx, and may be able to use their nasal cavity and fat pads to direct these sounds (like their proboscidean relatives; Landrau-Giovanetti et al. 2014). The chirp-squeaks are short, frequency-modulated sounds ($1 - 18$ kHz) used for communication (e.g. female-young bonding, individual identity, aggression, and perhaps the size of the signaller). Dugongs and manatees lack external ears and have a much-reduced auditory meatus and ear canal leading to the highly ossified tympano-periotic complex (like cetaceans).

## 4.6 Extreme Terrestrial Locomotion

Terrestrial locomotion is very diverse in mammals, reflecting both size and ecological niche, from slow ambulatory mammals, such as echidnas, pangolins (*Manis* spp.), aardvarks (*Orycteropus afer*), and elephants, to super-fast predators (e.g. cheetah, wolves), to their nearly-as-fast prey (e.g. various antelopes and pronghorn). High speed cursoriality is the extreme, most metabolically demanding, example of terrestrial locomotion, involving many anatomical and physiological adaptations. Brachiation ('arm swinging') through trees, similarly requires anatomical and physiological specializations. Migrating, often for considerable distances, is a different kind of extreme terrestrial locomotion. Each of these aspects of extreme terrestrial locomotion is examined later. Other extreme forms of mammalian locomotion include burrowing underground, which can be extremely metabolically costly (see 4.3.3); flight, which is costly in terms of the immediate metabolic cost but relatively economical in terms of cost of transport (see 2.8.2, 4.6.1); and diving, which is not so metabolically costly but entails many other physiological challenges (see 2.8.3, 4.5).

## 4.6.1 Cursorial Locomotion

The running speed of terrestrial mammals depends largely on body mass; cursorial mammals are those that run fast (typically artiodactyls, perissodactyls, and carnivores), with associated morphological adaptations such as long distal limb segments, shortened proximal segments, and change in foot stance from plantigrade to digitigrade or unguligrade (Garland & Janis 1993). Speed relative to body length (BL) ranges from $< 5$ BL $s^{-1}$ in elephants (*Loxodonta africana, Elephas maximus*), hippopotamus (*Hippopotamus amphibious*), rhinoceros (*Ceratotherium simum, Diceros bicornis*), camels (*Camelus dromedarius*), kouprey (*Bos sauveli*), giraffe (*Giraffa camelopardalis*), and polar bear to $> 40$ BL $s^{-1}$ for small marsupials, rabbits and hares, and various rodents (Iriarte-Díaz 2002). However, the allometric scaling exponent varies with mass; for small mammals ($M < 10$ kg), BL $s^{-1} \alpha M^{-0.09}$, for intermediate mass (10 kg $< M <$ 100 kg), BL $s^{-1} \alpha M^{-0.34}$, and for large mammals ($>$ 100 kg), BL $s^{-1} \alpha M^{-0.51}$. These differing scaling exponents seem to reflect differences in the allometry of bone-bending stress scope ($\alpha M^{-0.27}$); larger mammals have less scope for bending stress (a lower safety factor).

In terms of absolute speed, many small marsupials and rodents are slow ($< 5$ m $s^{-1}$), whereas the cheetah (*Acinonyx jubatus*) is the fastest (30 m $s^{-1}$), closely followed by the pronghorn (*Antilocapra americana*; 28 m $s^{-1}$) and numerous African artiodactyls (25–28 m $s^{-1}$). Lovegrove (2004) showed that maximum running speed (MRS) of mammals differed systematically with locomotor limb anatomy; speeds increased with mass for plantigrade (m $s^{-1} \alpha M^{0.124}$), digitigrade (m $s^{-1} \alpha M^{0.194}$), and lagomorph (hopping; m $s^{-1} \alpha M^{0.319}$) mammals, but decreased with mass for unguligrade mammals (m $s^{-1} \alpha M^{-0.115}$; Figure 4.19). Lovegrove and Mowoe (2014) described a micro-cursorial digitigrade locomotor mode for elephant shrews (*Elephantulus*) based on their very high metatarsal:femur ratio (of 1.07) that is typical of unguligrade cursors; their MRS is at the high end of the range

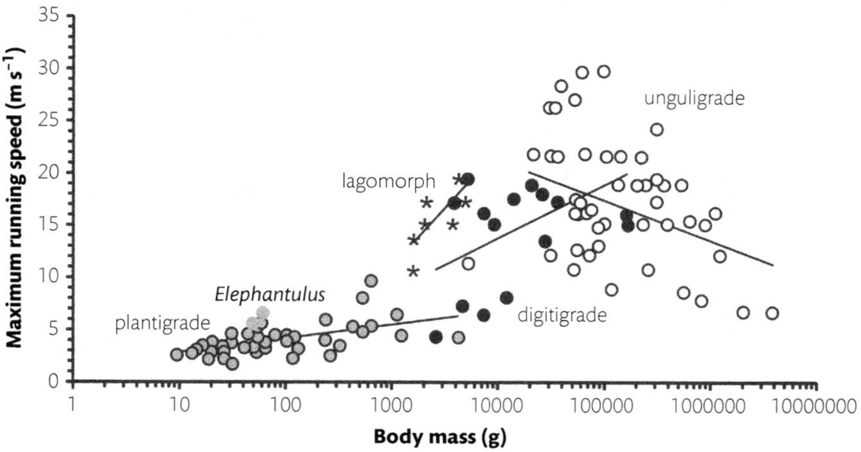


**Figure 4.19** Relationship between maximum running speed and body mass, showing the different allometries for five anatomical locomotor types: plantigrade, micro-cursorial (*Elephantulus*), digitigrade, lagomorph hopping, and unguligrade. Modified from Iriarte-Díaz (2002), Lovegrove (2004), and Lovegrove and Mowoe (2014).

for unguligrade mammals, relative to their body mass. The low speeds of plantigrade mammals may constrain the evolution of large body mass because of risks and costs of predation, unless they evolve protective armour (Lovegrove 2001).

Whether fast cursorial mammals achieve their extreme performance by having an exceptional aerobic capacity and high locomotor muscle mass, or anatomical adaptations that makes running more efficient, differs for different cursors. Pronghorn antelope, which have a maximal running speed of about 28 m $s^{-1}$ (100 km $h^{-1}$) and exceptional endurance (e.g. running 11 km in 10 min), have an exceptionally high aerobic capacity (Lindstedt et al. 1991). Pronghorns running at about 10 m $s^{-1}$ up an 11% incline had maximal $O_2$ uptake of about 18 ml $O_2$ $g^{-1}$ $h^{-1}$, which is over three times the maximum predicted for a similar mass mammal (and about that for a 10-g mouse compared to 32 kg for the pronghorn).

In contrast to pronghorn, running cheetah (the fastest land mammal) have a lower metabolic rate at 5 m $s^{-1}$ of about 3.2 ml $O_2$ $g^{-1}$ $h^{-1}$, and the expected net cost of transport of 0.14 ml $O_2$ $g^{-1}$ $km^{-1}$ for a running mammal (Taylor et al. 1974). These values are similar to those of one of their prey (*Gazella gazella*) of 3.4 ml $O_2$ $g^{-1}$ $h^{-1}$ and 0.16 ml $O_2$ $g^{-1}$ $km^{-1}$. The remarkable running speed of cheetah is due in part to their considerable spinal flexion that increases stride length (Figure 4.20); this probably contributes about 1.3 m $s^{-1}$ at 27 m $s^{-1}$ (Hildebrand 1959). Wilson et al. (2013a,b) have recently demonstrated elegant variation of the running strategy of cheetah during a hunt, using miniature data loggers that

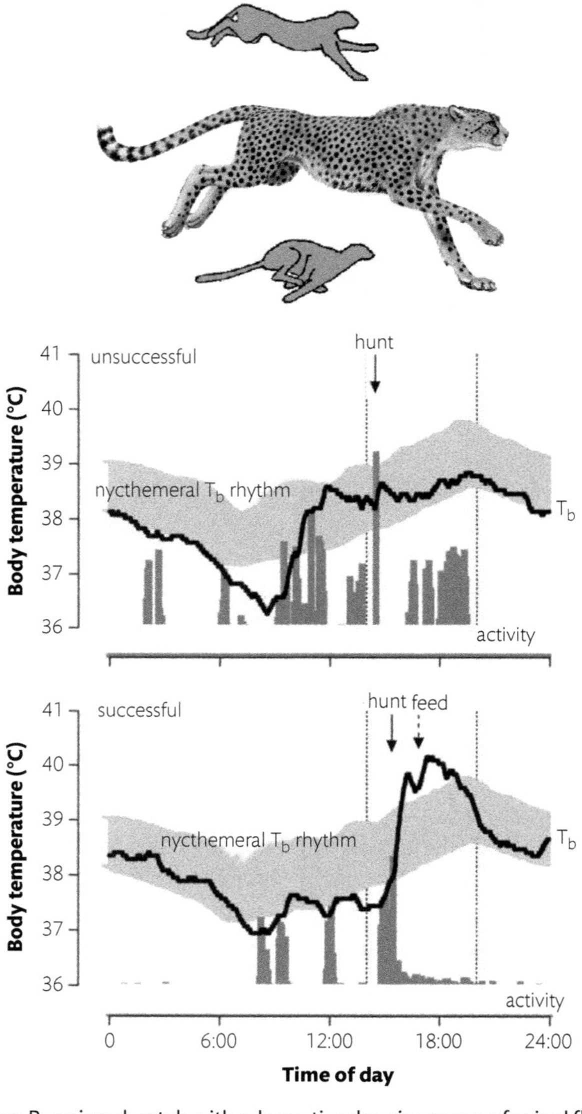


**Figure 4.20** Top: Running cheetah with schematics showing range of spinal flexion. Adapted from Hildebrand (1959) and McDonald (2010). Bottom: Body temperature of free-ranging cheetah for a successful hunt, showing the initial small hyperthermia from metabolic heat accumulation when running, followed by a stress hyperthermia, compared to the minimal metabolic hyperthermia of an unsuccessful hunt. Modified from Hetem et al. (2013).

recorded speed and acceleration (see 5.4). Top speed was recorded as 25.9 m $s^{-1}$ but most hunts were at more moderate speeds (Wilson et al. 2013a). Cheetah accelerated at up to 7.5 m $s^{-1}$ (cf. gravitational acceleration is 9.8 m $s^{-1}$) and reached hunting speeds up to 18.94 m $s^{-1}$ at the start of a hunt, but reduced their speed about 5–8 s before the end, to make rapid turns in response to their prey's evasion tactics (Wilson et al. 2013b).

Lions (*Panthera leo*), in contrast to cheetah, have a strong social structure and often do not hunt alone. They have a high net cost of transport (0.36 ml $O_2$ $g^{-1}$ $km^{-1}$) compared to predicted (0.11 ml $O_2$ $g^{-1}$ $km^{-1}$), but why is not clear—it is not associated with a peculiar limb morphology (Chassin et al. 1976). Puma (*Felis concolor*) have an entirely different hunting strategy in very different environments from cheetah and lion. They are low-energy-cost, sit-and-wait cryptic predators in diverse, rugged habitats that precisely match their pouncing force to prey size (Williams et al. 2014). Nevertheless, their energetic cost of free-ranging hunting is, like that of other large mammalian carnivores, about 2.3 times higher than that predicted by laboratory studies for routine cost of transport based on mass, and 3.8 times higher than the minimum cost of transport. Hunting is clearly an energetically expensive feeding strategy.

The high metabolic cost of running fast can create a thermal problem from accumulation of metabolic heat. It is impossible for a fast runner to dissipate all its metabolic heat production, so there is a continual storage of heat in the body, causing $T_b$ to rise. For example, cheetahs running at 28 m $s^{-1}$ produce more than 60 times as much heat as at rest, and have to store most of this heat by an increase in $T_b$. Taylor and Rowntree (1973) determined that cheetah running on a treadmill at 5 m $s^{-1}$ stored 90% of their metabolic heat; since cheetah refuse to run if $T_b$ exceeded 40.5°C, they calculated that cheetah could only run for about 1 km, which is about their pursuit distance in nature. Consequently, they speculated that cheetah's sprinting might be limited by their capacity to store heat. However, Hetem et al. (2013) found that free-hunting cheetah had a $T_b$ of only about 38.4°C after hunting, so heat storage did not compromise hunting. This suggests that perhaps sprinting is limited by the cheetah's aerobic capacity, and associated anaeroblic limits and lactate accumulation.There is actually a greater hyperthermia for cheetah after successful hunts ($T_b$ increases by 1.3°C) than unsuccessful hunts ($T_b$ increases by 0.5°C; Figure 4.20), likely reflecting a stress response to the danger of being attacked by other large predators (e.g. leopard, lion) prior to their commencing to feed on the carcass. Amazingly, Hetem et al. (2013) found that a lame male cheetah, which joined his female sibling after her successful hunt but did not participate in the hunt, showed the same stress hyperthermia as the hunter! Cheetahs, like many other predators, experience kleptoparasitism, the theft of their prey by other animals. About 25% theft of prey requires cheetahs to hunt for about an extra 1.1 h per day, increasing their daily energy expenditure by about 12% (Scantlebury et al. 2014). The stress of kleptoparasitism and predation by other predators has a major physiological impact on cheetah, and is presumably a major evolutionary cost.

Kangaroos and wallabies are large bipedal hoppers, and have a remarkable uncoupling of the metabolic cost of transport and speed, reflecting elastic storage of energy in tendons and tensed muscles during hopping. The classic study of Dawson and Taylor (1973) showed that the metabolic rate of red kangaroos (*Macropus rufa*) increased linearly with speed when locomoting pentapedally (using forelimb, hind limbs, and tail) at low speed, but actually decreased at higher speeds after the transition (at about 1.8–1.9 m $s^{-1}$) to bipedal hopping. Stride frequency showed essentially the same pattern with speed, but stride length increased linearly with speed; hopping faster is accomplished by jumping further per hop.

This remarkable conservation in the metabolic cost of locomotion over varying speeds reflects the elastic storage of energy in tendons, ligaments, and active muscle on footfall, and its release on take-off to decrease the metabolic cost of hopping. The tammar wallaby (*Macropus eugenii*) shows the same remarkable pattern for metabolic rate and hopping speed (Baudinette et al. 1992). The smaller potoroo lineage probably diverged from macropods about 20–30 MYBP, but it is unclear whether their pattern of hopping energetics is the same as kangaroos. The metabolic rate of the long-nosed potoroo (*Potorous tridactylus*) increases linearly with speed, but it is primarily quadrupedal, only using bipedal hopping intermittently and for short bursts at high speed (Baudinette et al. 1993). The rat-kangaroo lineage of marsupials is even older than the kangaroos and wallabies (diverged about 26 MYBP), and the rat-kangaroo *Bettongia penicillata* has a linear increase in metabolic rate with hopping speed, albeit at a lower-than-predicted rate of increase (Webster & Dawson 2003). Other hopping mammals, weighing less than 5 kg, have the typical linear relationship between metabolic rate and running speed (Thompson et al. 1980). So, the remarkable independence of metabolic rate on speed currently appears unique to the larger macropods. However, the use of elastic storage in locomotion is not unique to macropods, and is relatively common in many terrestrial runners (e.g. Biewener 1998), but is insufficient to have such an impact of the metabolic cost of transport as is seen for macropods.

## 4.6.2 Brachiation and Climbing

Brachiation, or 'arm swinging', is a form of arboreal locomotion whereby the body is supported by the arms, as for primates such as gibbons (Hylobatidae), siamangs (*Symphalangus syndactylus*), orangutans (*Pongo*), atelines (Cebidae), and suspensory quadrupeds (e.g. loris, Lorisinae; and sloths, *Choloepus didactylus; Bradypus variegatus*). It is associated with anatomical specializations such as long forelimbs, high shoulder joint mobility, and modified finger/elbow flexion musculature (Nyakatura & Andrada 2013). In gibbons, there are two brachiation gaits: (1) continuous contact brachiation is where at least one hand is always in contact with the substrate and is low speed, and (2) ricochetal brachiation has a flight (no-hand contact) phase and is higher speed (Michilsens et al. 2012). Continuous contact brachiation is a continuum of four transition types, which presumably confers efficient determination of locomotor cost at varying speed, like the variable gait of terrestrial mammals such as horses (Hoyt & Taylor 1981) and lemurs (O'Neill 2012). A mechanical cost of brachiating by primates is the substantial dynamic forces resulting from swinging from branches. Gibbon's hindlimbs also show adaptations to bipedalism. Their Achilles tendon has biomechanical properties suitable for storage and release of elastic energy, acting as an elastic 'spring' to reduce the locomotor cost of bipedalism (like humans and kangaroos), whereas the high stiffness of their patellar tendon might enhance leaping performance (Vereecke & Channon 2013).

The energetic cost of brachiating does not appear to be lower than for terrestrial locomotion. The metabolic rate of spider monkeys (*Ateles* sp.) when hanging motionless by the arms is about two times their resting rate, and increases linearly with speed when brachiating; their energetic cost of brachiation is higher than when walking, regardless of speed (Parsons & Taylor 1977). So, the advantage of brachiation is unlikely to be a low energetic cost of transport per se, but there is likely a reduced effective cost of locomotion by being able to move more directly by brachiating through forests rather than climbing up and down trees. In comparison, the energetic cost of climbing is the same as for terrestrial locomotion for small primates ($< 0.5$ kg), which presumably contributes to their ecological success in finely branching arboreal environments, and the success of early primates (Hanna & Schmitt 2011). In contrast, the cost of climbing is about two times terrestrial locomotion for larger primates.

Slow loris (*Nycticebus coucang*) are suspension quadrupeds that essentially walk slowly, upside down; their metabolic rate when suspended from a rope, or standing on top, is about 1.4 times resting, and increases linearly with speed (Parsons & Taylor 1977). The red slender loris (*Loris tardigradus*) has the same metabolic cost of transport when walking and climbing (Hanna & Schmitt 2011). Sloths brachiate slowly, using relatively extended limbs, with 'trot-like' sequences (Nyakatura & Andrada 2012). Their 'slow-motion' movements reduce the risks of breaking contact with the substrate, which is important because they are slow-moving and unlikely to be able to respond quickly to avoid potentially lethal falls—but they lose the dynamic mechanical benefits of primate pendular mechanics. Slow movement is also advantageous for remaining cryptic and avoiding predation, and is consonant with their low metabolic expenditure associated with their use of a low-energy food source (see 4.8.2).

### 4.6.3 Migration

Many mammals migrate seasonally or shorter term. Migration is the regular, two-way movement of animals, generally on an annual cycle, between feeding grounds and breeding sites. Dispersal, in contrast, is a one-way movement, generally by juveniles away from their natal origin to a new home range. The reasons for migration vary, but include avoidance of unsuitable weather, short (or no) photoperiod, and exploitation of heterogeneous food or water resources. Costs of migration include the high energetic cost of moving, potential for exhaustion, and a substantially increased mortality from predation. Migration distance varies, with a strong effect of body mass (Figure 4.21) for both terrestrial and marine migrators, with terrestrial species having lower mass and shorter migration distances. However, small and highly mobile bats have migration distances that rival those of marine mammals weighing six orders of magnitude more.

Various terrestrial mammals migrate, typically seasonally; Harris et al. (2009) list 24 large terrestrial mammal species (all ungulates) that are considered to migrate. Maximum migration distance varies from relatively short (e.g. oryx, *Oryx dammah*, 100 km; American bison, *Bison bison*, 160 km; elk, *Cervus elephas*, 200 km); intermediate (blue wildebeest, *Connochaetes taurinus*, 600 km; chiru, *Pantholops hodgsoni*, 600 km); and very long (Siberian roe deer, *Capreolus pygargus*, 1,000 km; Mongolian gazelle, *Procapra gutterosa*, 1,000 km; saiga, *Saiga tatarica*, 2,400 km; caribou/reindeer, *Rangifer tarandus*, 3,031 km). These migration distances are being adversely limited by human activities such as fencing of agricultural and conservation areas. Various smaller terrestrial mammals migrate over much smaller distances in absolute terms, but similar distances relative to their body length.

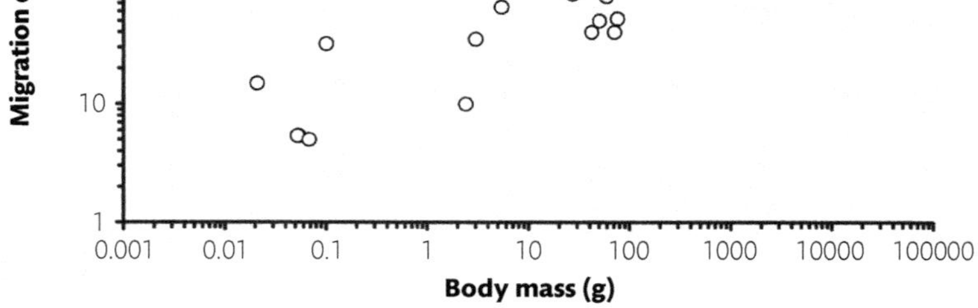

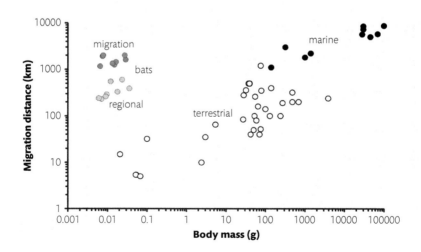


**Figure 4.21** Migration distance for terrestrial (white symbols), marine (black symbols), and flying mammals (bats; light and dark grey symbols), as a function of body mass. Data in part from Fleming and Eby (2003) and Hein et al. (2012).

Most baleen whales migrate annually from productive feeding grounds in the higher latitudes to more tropical breeding grounds. The considerable cost of swimming such long distances indicates that there must be a substantial evolutionary advantage. One of the longest mammalian migrations is that of the humpback whale (*Megaptera novaeangliae*) that moves up to 8,500 km annually from feeding areas off Antarctica to more favourable breeding areas along the Pacific coast of Central America (Rasmussen et al. 2007). Calves that develop in warmer water may have higher survival and grow into larger adults, providing an evolutionary advantage that counters the energy cost of such a long migration. Other possible advantages include a lower metabolic cost of thermoregulation in warmer waters and avoidance of killer whales. Recently, (Silva et al. 2013) have used satellite-linked radio transmitter tags (see 5.4) to track fin (*Balaenoptera physalis*) and blue whales (*B. musculus*) to monitor their annual migration from Greenland/Iceland (feeding grounds) to the Azores (breeding ground). Interestingly, some whales feed mid-latitude during their migration, and even suspend their migration for extended periods. Northern elephant seals (*Mirounga angustirostris*) are a good example of migration by pinnipeds. Satellite tracking (see 5.4) has revealed the extent of their migrations as well as ocean physiographic conditions at various depths. They are at sea for most of the year, and range over the eastern North Pacific Ocean on double annual migrations between Californian breeding grounds and distant foraging grounds (Brillinger & Stewart 1998).

Bats, being smaller but more mobile by virtue of flight than other migrating mammals, move remarkably long distances (Figure 4.21), rivalling some migratory birds. Migration by bats has evolved independently in a number of lineages, particularly for temperate species; less is known for tropical species (Moussy et al. 2013). Many bats move annually, typically between summer feeding grounds to winter hibernacula sites, but some move regionally, over shorter distances. Cryan et al. (2004) used stable H analysis of hair (see 5.6) from live and museum specimens of hoary bats (*Lasiurus cinereus*) to determine their latitudinal movements; the $\delta D_{fur}$ (difference in deuterium and hydrogen isotopes of fur) reflects the $\delta D_{pre}$ of the precipitation where the fur was grown, which varies latitudinally. They showed that hoary bats are long-distance migrants, moving > 2,000 km between Central America and Canada (e.g. one individual migrated from Chihuahua, Mexico, to north of the Canadian border). Tri-coloured bats (*Perimyotis subflavus*) also range from Central America to Canada. Fraser et al. (2012) found that these bats also show regional movements as well as north-south annual migration. Nathusius's pipistrelle (*Pipistrellus nathusii*) migrates annually from north-eastern to south-western Europe; individuals fly more than 2,000 km one way. Bats have a lesser capacity than birds to store energy for their annual migration as fat. Voigt et al. (2012) found that when migrating Nathusius's pipistrelles consume insects, they preferentially oxidize the protein and carbohydrate content for energy, but use the lipid to replenish their (meagre) body fat stores. In contrast, non-migratory bat species oxidize the lipid as well as protein carbohydrate in their diet. Bats accumulate especially $C_{14}$ and $C_{18}$ saturated and unsaturated lipid as a mass-effective energy store, with lipid indices (g lipid per g lean dry mass) of 0.38–0.87 (Blem 1980).

## 4.7 Flying Mammals

Flight is divided into two categories: gliding, an unpowered and passive mode of flight where aerodynamic forces are passively exerted on a membranous structure stretched between a range of body parts (the patagium); and flapping, an active and powered flight mode, where muscular power is used to generate aerodynamic forces (Norberg 1990). Gliding is found in more than 60 species of mammals, from six families of three distantly related orders (Dermoptera, Rodentia, and Diprotodontia) and in four extinct lineages (Byrnes & Spence 2011; Jackson & Schouten 2012). In marsupials, gliding occurs in the feathertail gliders (Acrobatidae, *Acrobates*), gliding possums (Petauridae, *Petaurus*), and the greater glider (Pseudocheiridae, *Petauroides*). Gliding in rodents is observed in flying squirrels (family Sciuridae; e.g. *Glaucomys, Iomys, Petinomys, Hylopeles, Petaurista*) and in scaly tailed squirrels (Family Anomaluridae; e.g. *Anomalurus, Idiurus, Zenkerella*). The quintessential examples of gliding mammals are two dermopterans of the family Cynocephalidae (flying lemurs or colugos; *Cynocephalus* and *Galeopterus*). Byrnes and Spence (2011) reviewed the multiple hypotheses proposed to explain the evolution of gliding in mammals. They suggested that the selective pressure(s) that led to gliding may have been to decrease predation, increase foraging efficiency, and/or control landing forces. Using a phylogenetically based comparative analysis, they argued that gliding evolved in folivorous, frugivorous, or exudivorous groups, suggesting that it was a response to poor-quality (especially protein-deficient) diets that require extensive foraging.

Amongst vertebrates, powered flight independently evolved in three lineages: bats, birds, and the extinct pterosaurs. Bats are the only extant mammals that use powered flight (Figure 4.22). It has been suggested that Old World megachiropteran fruit bats (Pteropodidae) independently evolved flight from microchiropteran insectivorous bats (Pettigrew et al. 1989), but the current consensus is that bats are a monophyletic group, so that powered flight evolved only once in mammals (Teeling et al. 2005; Simmons et al. 2008). The earliest known bat, *Onichonycteris finney*, from the early Eocene period (ca 52 MYBP), was capable of powered flight but, unlike modern bats, its digits had tiny claws, which suggests that it could climb trees (Simmons et al. 2008).

Bats probably evolved powered flight from an arboreal species that could glide (Norberg 1990; Dudley et al. 2007). The transition from gliding to powered flight by bats required extensive changes from the ancestral glider bauplan. Although we lack transitional fossil records to accurately describe the steps involved in this

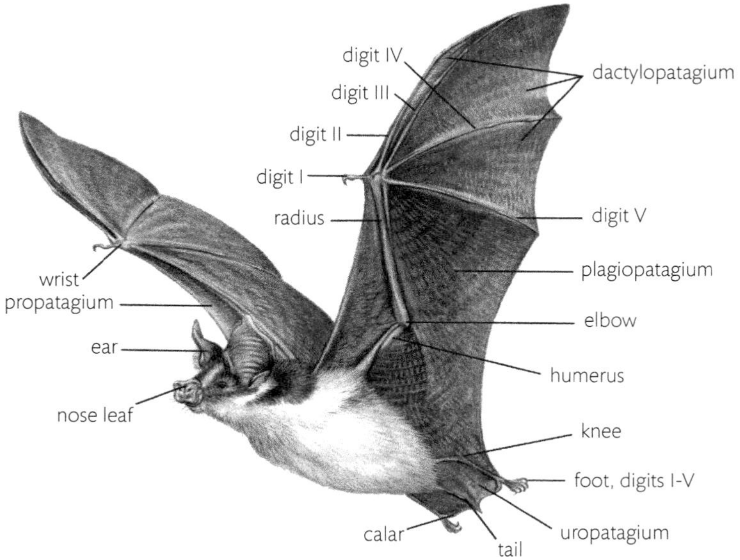


**Figure 4.22** General body plan of a microchiropteran bat (diadem leaf-nosed bat, *Hipposideros diadema*). From Macdonald (2009). Reproduced with permission of Oxford University Press.

transition, several authors analysed the possible steps involved using a combination of aerodynamics, biochemical, and morphological analysis (Norberg 1985; Swartz et al. 1992, 2006; Thewissenn & Babcock 1992; Dudley 2002; Dudley et al. 2007). Bishop (2008) provided an integrative analysis of these results and proposed that the following steps were involved: the evolution of a wing capable of acting as an aerofoil; the evolution of enough power acting on this wing to generate thrust while maintaining lift; liberation of the forelimbs from other functional constraints so that the wings could be fully developed; and, finally, the evolution of physiological machinery able to supply enough fuel and $O_2$ to sustain the metabolic costs of powered flight.

The evolution of flight requires extensive, and intertwined, modifications in virtually all organ systems and biological process, mostly notable in the sensory, locomotory, digestive, circulatory, and respiratory systems (Maina 2000a). Some of these modifications, especially in the digestive and circulatory systems, are remarkably convergent between bats and birds (see 4.7.3). Others, however, are strikingly different (e.g. respiration), reflecting the independent evolution of flight in these two quite different ecologically and divergent vertebrate groups (the ancestral mammals diverged from the ancestral birds about 310 MYBP; Kumar & Hedges 1998). For example, bats, like birds and pterosaurs, evolved a number of adaptations of the skull to become lighter but, at the same time, strong enough to sustain the large aerodynamics forces imposed by flight (Swartz et al. 1992). In bats and birds, the main flight muscle is the pectoral muscle, which in bats has a relatively smaller mass than in birds. However, in birds, only one set of muscles (the supracoracoideus) is involved in flight, whereas in bats the flight musculature is more complex and involves at least 17 muscles that are active during the downstroke and upstroke of the wings (Norberg 1990). Bats, like their putative ancestral mammal, are nocturnal, whereas birds are primarily diurnal. Competition with birds, increased risk of predation, and hyperthermia have been invoked to explain why bats are evolutionary trapped in a nocturnal niche (Rydell & Speakman 1995; Speakman 1995, 2005; Voigt & Lewanzik 2010).

An evolutionary commitment to a nocturnal niche required profound changes in the sensory system of bats. Chief amongst these changes was the evolution of a sophisticated echolocation system (see 4.7.4). Not all bats echolocate (e.g. some members of the more diurnally active and visually oriented members of the Pteropodidae), and those that do can produce broadband (frequency-modulated), narrowband, or long constant-frequency calls. The larynx produces these calls, but brief broadband calls can also be emitted by tongue clicking (Jones & Teeling 2006). Laryngeal echolocation calls are more sophisticated and complex, and are regarded to be the ancestral mode of echolocation in bats (Springer et al. 2001; Jones & Teeling 2006). Echolocation calls are energetically expensive to produce for resting bats, but these costs are negligible during flight (Speakman & Racey 1991; Voigt & Lewanzik 2012) due to the 1:1 synchronization while flying between the respiratory and the flight muscles (Lancaster et al. 1995; see also 4.7.3).

The body mass range and the maximum body mass attainable by bats are much smaller than those reported for birds capable of powered flight. The body mass of bats ranges from 2 g (bumblebee bat, *Craseonycteris thonglongyai*) to 1.6 kg (golden-crowned flying fox, *Acerodon jubatus*), whereas body mass ranges for birds from 2 g (bee hummingbird, *Mellisuga helenae*) to 12–14 kg bustards (*Ardeotis kori*), California condor (*Gymnogypus californinanus*), and mute swan (*Cygnus olor*). There is a complex interplay between the scaling relationship of wingbeat frequency and flight muscle mass, patterns of echolocation calls, wing morphology, and the capacity to delivery fuel and $O_2$ to the muscles during flight, that placed constraints on the maximum body mass that bats are able to achieve (Arita & Fenton 1997; Bullen & McKenzie 2002, 2004; Norberg & Norberg 2012; Ruxton 2014). As for other mammals, body mass is the most important factor affecting life history traits of bats, so the evolution of flight, by imposing constraints on the range of body mass of bats, has also impacted their evolution of reproductive patterns (Barclay & Harder 2003).

In bats, unlike birds, the wings evolved from a gliding membrane, and the ancestral bats had to develop a different range of structural modifications to fully develop and support a wing capable of sustained, powered flight (Sears et al. 2006; Cretekos et al. 2008; Konow et al. 2015). The wingbeat kinematics of bats is different from those of birds, a direct consequence of the evolutionary constraints imposed by the more compliant skin membrane of bats compared to the more rigid feather membrane of birds (Swartz et al. 2006; Hedenström & Johansson 2015). The aerodynamic forces acting on a bat's wing are the same as those for a glider (Figure 2.22), but a bat uses the power of the wing downstroke to maintain horizontal flight (Figure 4.23). The two common measurements that describe the size and shape of wings are wing loading (WL) and aspect ratio (AR; see 4.7.4). WL is the mass of the bat divided by the area of the wing elements; aspect ratio (AR) is the square of the wingspan (distance from the tip of the wing to the central axis of the body, multiplied by two) divided by the area of the wing elements (Norberg & Rayner 1987). There is a positive linear relationship between AR and

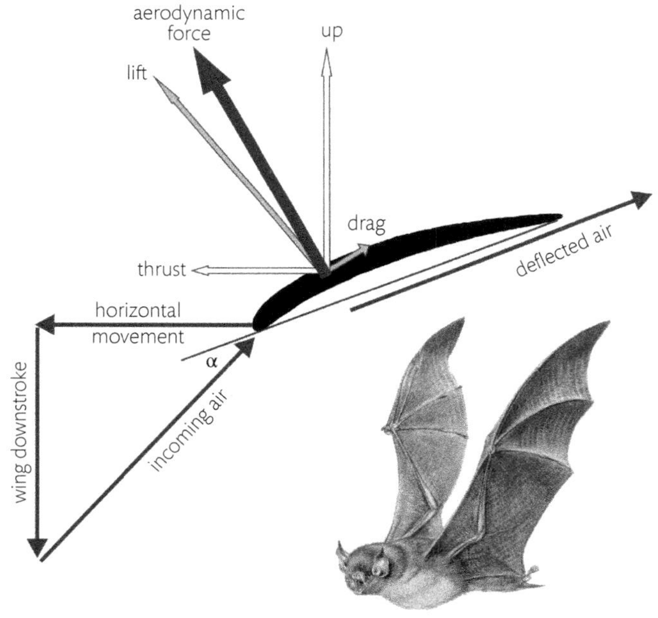


**Figure 4.23** Aerodynamic forces acting on a bat wing in horizontal flight; forward movement in concert with wing downstroke creates the upward- and backward-directed incoming air flow. The aerodynamic force is resolved into lift (perpendicular to incoming airstream) and drag (parallel to incoming airstream), and horizontal thrust to counterbalance the other drag forces acting on the bat, and an upward force to counterbalance body weight; $\alpha$ is the angle of attack. Inset shows a Peter's disk-winged bat *Thyroptera discifera*; reproduced with permission of Oxford University Press.

WL, with the combination of these characteristics determining aerodynamics of bat flight; AR impacts the cost of transport while WL influences agility and manoeuvrability (Fullard et al. 1991). Bats with a high AR and high WL have long, thin wings, resulting in poor manoeuvrability but a low cost of transport, whereas bats with a low AR and low WL have short, thick wings, with good agility but a high cost of transport (Norberg & Rayner 1987). So, in general (as for birds) high WL characterizes species adapted for fast flight but at the expense of a reduced manoeuvrability, while a high AR characterizes species capable of flying with reduced energetic cost. A particular combination of WL and AR can be interpreted as a compromise between speed, endurance, economy, and agility, which, in turn, can be linked to the food habits, habitat use and, for insectivore bats, to the structure of echolocation calls (see 4.7.4; Bogdanowicz et al. 1999; Jennings et al. 2004; Siemers & Schnitzler 2004; Mancina et al. 2012; Voigt & Holderied 2012; Falk et al. 2014; Marinello & Bernard 2014). Wing tip proportions (pointed vs rounded) also impact aerodynamics, particularly manoeuvrability, and can be described by the proportions by which the hand wing and arm wing areas contribute to the total wing area; a rounded wing is characterized by a short, broad hand wing (Aldridge & Rautenbach 1987).

Not surprisingly, wing morphology correlates with the foraging niche of insectivorous bats. Broadly speaking, insectivorous bats forage in three locations: in closed environments (gleaning, within stands of trees or amongst the canopy), beside vegetation (beside or on the edge of stands of trees, just above or under the canopy), and in the open (well above the canopy; Fullard et al. 1991). Bats that primarily forage in open environments tend to have high AR and WL and pointed wing tips; those in closed environments have low AR and WL, with rounded wing tips; while bats foraging beside vegetation have intermediate wing morphologies (Aldridge & Rautenbach 1987; Norberg & Rayner 1987). The aerodynamic principles governing the evolution of the wing morphology for insectivorous bats extend to bats with other diet types. Piscivores and carnivores snatch prey from the ground or water, and therefore require slow flight with a low WL, although a high AR improves flight economics, and long wings are possible in an uncluttered environment. Nectarivores and frugivores often have to travel long distances between food sources, so generally have high AR and WL, although this structure may be somewhat tempered by the requirement to hover and manoeuvre amongst vegetation to access food (Norberg & Rayner 1987). Although some bats are generalists and forage in a variety of habitat types, adaption to the more extreme niches may severely limit the potential to exploit other environments. For example, the long wings of bats that typically forage in open areas mean that they are unable to negotiate the cluttered environment within vegetation; similarly, there are flight constraints on closed-environment bats foraging in the open, particularly in adverse environmental conditions such as high wind speeds (Fullard et al. 1991).

### 4.7.1 Metabolic Cost of Flight

Of the two basic modes of flight, gliding is less costly than powered flight. For birds, the metabolic rate while gliding is only twice the BMR (Baudinette & Schmidt-Nielsen 1974; Sapir et al. 2010). For bats, powered flight is up to 17 times BMR (see later). A gliding mammal uses the potential energy gained by its previous climbing, and, because the power required to control and manoeuvre is negligible, the only costs are those necessary to keep the gliding membrane rigid to resist deformation by the aerodynamic forces (Norberg 1985). In fact, it has been postulated that the energetic economy achieved during gliding was one of the main selective forces responsible for the evolution of this mode of flight (the 'energetic-economy hypothesis'; Dudley et al. 2007; Byrnes & Spence 2011). Mathematical models suggest that the economy achieved by gliding seems to be high at intermediate body mass, but that large gliders must glide a much longer distance than smaller gliders before a substantial economy can be accrued (Dial 2003b; Scheibe et al. 2006). Byrnes et al. (2011) compared the costs of gliding with those of moving horizontally in the canopy for the Malayan colugo (*Galeopterus variegatus*) and concluded that gliding is energetically inexpensive. However, because the initiation of gliding requires climbing to a certain height, the overall cost of gliding (climbing + gliding) was higher than what it would be if the colugo moved horizontally through the canopy. Nevertheless, since colugos spend a small fraction of their time budgets engaged in gliding, the impact of these costs to their daily energy budget was small.

The basic aerodynamic principles for bat flight are the same as for gliders (see 2.8.2, Figure 4.23), except that the powered downstroke of the wing provides sufficient thrust and vertical force to overcome the drag and weight of the bat, enabling it to fly horizontally indefinitely, whereas the glider eventually descends to the ground. The power requirements for powered flight are complex (Norberg 1990; Rayner 1999). In theory, the total aerodynamic power required for flight ($P_{tot}$) is composed of three power requirements, parasite power ($P_{par}$), profile power ($P_{pro}$), and induced power ($P_{in}$); $P_{tot} = P_{par} + P_{pro} + P_{in}$. Parasite power ($P_{par}$) is the power required to overcome the aerodynamic drag of the body; this increases with flight velocity. Profile power ($P_{pro}$), the power required to overcome the aerodynamic drag of the wings, is much lower than $P_{par}$ because the wings are more streamlined than the body; however, $P_{pro}$ also increases with velocity. Induced power ($P_{in}$) is the drag associated with the wings having to deflect oncoming air to generate an aerodynamic force. $P_{in}$ decreases with velocity, because at high velocity relatively little deflection is required to obtain the momentum change needed to counterbalance the body weight; conversely, at low velocity, the slower moving air needs to be deflected downwards more to achieve the same momentum change. Consequently, induced power decreases with speed. The net effect is that the total power required

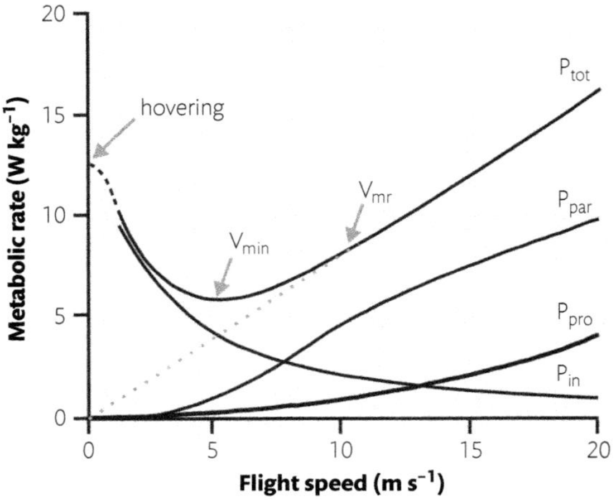


**Figure 4.24** Schematic of the partitioning of total cost of flight ($P_{tot}$) as a function of flight velocity, into parasite power ($P_{par}$), profile power ($P_{pro}$), and induced power ($P_{in}$), with the velocity for minimum cost of flight ($V_{min}$) and maximum range velocity ($V_{mr}$). Modified from Withers (1992).

for flight ($P_{aero} = P_{par} + P_{pro} + P_{in}$) has a U-shaped curve, being high at low and high flight velocities, and lowest at intermediate velocities (Figure 4.24). The flight velocity for maximum range ($V_{mr}$) is higher than the velocity for minimum cost ($V_{min}$), equivalent to the velocity where a line from the origin makes a tangent with the parabolic curve.

Another power requirement is for inertial power ($P_i$), the power necessary to overcome inertial forces (joint friction, internal tissue viscosity) and move the wings up and down. At medium to high speed, $P_i$ is thought to be negligible because the wing inertia is converted into aerodynamic work at the end of each stroke (Norberg 1990). However at very low speed and, especially for stationary (hovering) flight, $P_i$ can be high. In stationary flight, there is no horizontal air flow, so that all air flowing over the wings must be supplied by forces generated by the muscle activity necessary to beat them. This explains the high power requirement of hovering flight compared to horizontal flight. For the nectar-feeding bat *Glossophaga soricina*, $P_{tot} (= P_i + P_{aero})$ was estimated to be 0.34 W for hovering, with $P_i$ accounting for more than 60% of the cost (Norberg et al. 1993). In comparison, the metabolic cost for forward flight near the minimum-cost flight speed (see later) is about 0.14 W.

Each of the individual power requirements in this model embraces a suite of aerodynamic components associated with body and wing morphology, aerodynamic 'cleanliness' attributes, and dynamic factors that must be quantified if the earlier equation is to accurately predict the aerodynamic power required for animal flight. However, measurement of such parameters is challenging (Bullen & McKenzie 2001, 2002, 2008; Norberg & Norberg 2012). This classical model assumes a fixed wing and a steady-state flight (Norberg 1990; Rayner 1999), but kinematic studies show that the compliant membranous wings of bats have more degrees of change than a bird wing, being capable of pronounced deformation under different aerodynamic loads (Swartz et al. 2006; Cheney et al. 2014). Additionally, unsteady effects have been reported for bats during flapping flight, especially at low speeds (Muijres et al. 2008). Furthermore, this model indicates only the mechanical power necessary for flight, but not the metabolic power actually used during flight.

Basic thermodynamic laws dictate that the conversion of chemical energy to mechanical energy is far less than 100%, so the mechanical costs calculated by aerodynamic models represent only a small part of the total metabolic costs. In fact, the estimated mechanical costs derived from aerodynamic models are nearly five times lower than the actual metabolic costs (Norberg et al. 1993; Speakman & Thomas 2003; Bullen et al. 2014). Mechanical efficiency ($\eta$), the effectiveness of converting chemical energy to mechanical work, has been indirectly assessed by comparing measurements of metabolic rate during flight with mechanical power derived from the classical power-speed curve. Under the assumption that $\eta$ is similar at different speeds, these results gave mean $\eta$ values for bats of 20% (range 10–28%; Speakman & Thomas 2003), which are similar to those calculated for birds (range 12–40%; Ward et al. 2001). However, these are probably overestimates. Bullen et al. (2014), using a quasi-steady model, calculated $\eta$ for eight species of bats to vary between 3.8 and 18.2%, and reported that this efficiency varies allometrically and with flight speed (from 4 to 10 m $s^{-1}$). At speeds close to $V_{min}$ (5–8 m $s^{-1}$), $\eta$ varied from 4.3 to 10.8%. von Busse et al. (2014) used a more complete analysis for Seber's short-tailed bat, *Carollia perspicillata*. Combining isotopic measurements of flight costs with simultaneous measurements of mechanical power output, they calculated $\eta$ at 5.6–11.3%, which, like those of Bullen et al. (2014), overlapped with the values calculated for nectar-feeding bats (Voigt & Winter 1999).

The metabolic cost of flight has been measured for 14 species of bat (Speakman & Thomas 2003; von Busse et al. 2013), at different flight speeds, and using different methods (doubly labelled water, $^{13}C$-labelled $NaHCO_3$, mass-balance experiments, and mask respirometry). The U-shaped curve expected for the classical power-speed curve derived from fixed-wing aerodynamic models was observed for Seber's short-tailed bat (Von Busse et al. 2013), which is more in concert with those reported for birds (Ellington 1991; Bundle et al. 2007). Other studies for the greater spear-nosed bat (*Phyllostomus hastatus*) and the black flying fox (*Pteropus gouldi*) suggested a J- or L-shaped curve (Carpenter 1985, 1986; Thomas 1975). Data for a nectar feeding bat (Pallas's long-tongued bat, *Glossophaga soricina*) showed that hovering flight increased metabolic rate by 1.2 times above the cost of forward flight, which suggests a shallow U-shape curve (Winter 1998). Differences in body mass and methodology, by affecting the range of flight velocities used in these studies, probably accounts for these differences. Large bats, such as the greater spear-nosed bat and the black flying fox, cannot fly at low velocities, and the use of masks to measure flight metabolic rate also prevented measurements at high flight speeds for these species. Von Busse et al. (2013) reported that the minimum metabolic cost of flight for Seber's short-tailed bat was at a flight speed (4 m $s^{-1}$) close to the minimum power predicted by the fixed-wing aerodynamic model ($V_{mc} = 5$ m $s^{-1}$), but different from the power requirement at $V_{mr}$.

As with birds and running mammals, the metabolic cost of flight increases with body mass in bats (Figure 4.25). As noted by Speakman and Thomas (2003), this relationship is surprisingly tight for bats ($r^2 > 0.99$) if we consider that the data came from results that used different methodologies and included costs that, expect for hovering flight, do not necessarily reflect maximum costs. The highly significant allometric relationship precludes analysis of residual variation to test for the effects of morphological variables, such as aspect ratio, on the metabolic cost of flight in bats (Speakman & Thomas 2003). When the comparison is restricted to the same mass range, the allometric exponent of this relationship (0.74) is similar to those reported for BMR of bats (0.79; Cruz-Neto & Jones 2006). Flight by bats increases their metabolic rate up to 17 times above BMR, but this scope is probably too high; it is unlikely that the basic physiological processes that determine BMR (e.g. $T_b$) would not increase during flight.

The scaling exponent for bats is also very similar to those reported for running mammals and birds with a similar mass range (0.75–0.79). As expected (Norberg 1990; Bishop 1999), flight is more expensive than running (2 times in bats and 2.8 times in birds). Birds are able to fly faster and more efficiently than bats (Muijres et al. 2012), so we might expect that the cost of transport would be higher for bats than birds. However, Winter and von Helversen (1998) suggested that bats

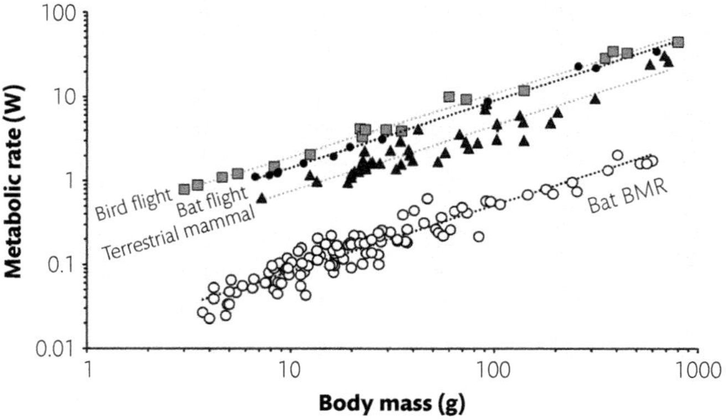


**Figure 4.25** Comparison of metabolic rate for flying bats (0.233 $M^{0.939}$) with flying birds (0.340 $M^{0.754}$), terrestrial mammals (0.121 $M^{0.795}$), and BMR of bats (0.135 $M^{0.790}$). Data from Hails (1979), Bartholomew and Lighton (1986), Suarez et al. (1991, 1997), Speakman and Thomas (2003), Cruz-Neto and Jones (2006), and Dlugosz et al. (2013).

fly more cheaply than birds, and more recent analyses (Speakman & Thomas 2003; Hedenström et al. 2009) suggest that the costs of flight for bats and birds are not significantly different (1.0–1.3). Although the effects of flight speed were not accounted for, these results suggest that aerodynamic considerations and kinematics of flight should be more important in determining the supposed differences in the costs of transport between birds and bats. During hovering, the costs of flight also do not differ between bats and birds (Voigt & Winter 1999), which probably reflects the similarities in aerodynamics and the enzymatic capacity of the flight muscles observed between these two groups (Suarez et al. 1997, 2009).

## 4.7.2 Thermal Balance

Bats are considerably thermolabile mammals. Normothermic resting $T_b$ of bats varies with body mass and ranges from 31 to 38°C (Clarke & Rothery 2008), but the low average $T_b$ of bats of 35.3°C compared to most other placental mammals might reflect methodological difficulties with measuring their normothermic $T_b$ (Willis & Cooper 2009; see 3.2.1; Table 3.3). During hibernation or daily torpor, $T_b$ of bats can drop to $5-25$°C depending on $T_a$ and M (Geiser 2004; McNab 2012).

The capacity of bats to tolerate an increase in $T_a$ to levels close or even higher than their normothermic $T_b$ is quite variable. Early studies suggested that bats become hyperthermic when exposed to higher $T_a$, but survival is only possible for a short period of time (Licht & Leitner 1967a,b). At high $T_a$ ($>35$°C), EHL accounted for more than 50% of MHP for Gould's long-eared bat (*Nyctophilus gouldi*; Morris et al. 1994). For the Angolan free-tailed bat (*Mops condylurus*), a species that thrives in roosts that often exceed 40°C, virtually all their heat production is dissipated by evaporative cooling at $T_a$s approaching roost temperature (Maloney et al. 1999). Bats use behavioural and physiological mechanisms such as wing-fanning to increase evaporative cooling and increase blood flow to the peripheral parts of the body, especially to the wings and to naked and vascularized regions on the side of the body (Ochoa-Acuna & Kunz 1999; Makanya & Mortola 2007; Reichard et al. 2010a).

Total evaporative water loss ($EWL_{tot}$) is made up of both cutaneous and respiratory components. Respiratory evaporative water loss ($EWL_{resp}$) is higher than cutaneous evaporative water loss ($EWL_{cut}$) for bats at high $T_a$ (Minnaar et al. 2014; Muñoz-Garcia et al. 2016). The increase in $EWL_{tot}$ at high $T_a$ places a burden on the capacity of bats to maintain water balance, so we might expect that there are mechanisms to minimize $EWL_{tot}$ and maintain water balance. Different species of bats show differences in their $EWL_{cut}$ and/or $EWL_{resp}$ responses (Muñoz-Garcia et al. 2012a,b, 2016), as well as the capacity of their kidneys to concentrate urine (Studier & Wilson 1983; Schondube et al. 2001; Casotti et al. 2006; Gopal 2013) that are related to body mass, habitat, and diet.

However, these capacities are not markedly different from what would be expected for similar-sized non-flying mammals having the same diet and living in a similar habitat (Webb 1995; Beuchat 1990, 1996; Al-kahtani et al. 2004; Van Sant et al. 2012). Differences in the capacity of different bat species to minimize $EWL_{tot}$ may partially explain why some bats use torpor in hot, desert areas (Muñoz-Garcia et al. 2016), and also the differential susceptibility of bats to heat waves (Welbergen et al. 2008; Bondarenco et al. 2014; Dey et al. 2015).

During flight, MHP of bats increases by up to 17 times above resting MHP (see 4.7.1), which poses challenges for their capacity to maintain thermal and water balance. To maintain a constant $T_b$ during flight, this excess heat must be dissipated; flight-induced air convention, radiative heat loss, $EWL_{cut}$, and $EWL_{resp}$ all play role in dissipation. For the mid-sized spear-nosed bat (*Phyllostomus hastatus*; 90–100 g), MHP and $T_b$ increase with $T_a$ during flight (Thomas et al. 1991). The grey-headed flying fox (*Pteropus poliocephalus*; 700–800 g) becomes hyperthermic when flying at $T_a$ above 25°C (Carpenter 1985, 1986). $EWL_{resp}$ increases during flight but, as for birds (Torres-Bueno 1978; Giladi & Pinshow 1999; Michaeli & Pinshow 2001), the heat dissipated by this route accounts for only a small fraction (10–14%) of the MHP during flight in bats (Carpenter 1985, 1986; Thomas et al. 1991). The 1:1 synchronization between wing beat and pulmonary ventilation might result in an inability to match $EWL_{resp}$ to increasing MHP (Boggs 2002), with the capacity to increase evaporation more limited than metabolism.

Modulation of $EWL_{cut}$ and heat loss by radiation and convection, thus, might be very important for bats during flight. These possibilities were investigated by Reichard et al. (2010a,b) for thermoregulatory adjustments during natural flight for the Brazilian free-tailed bat (*Tadarida brasiliensis*), a bat capable of fast flight and high endurance. Thermographic IR images showed that surface temperature ($T_{surf}$, measured at the centre of the body) was always higher than wing temperature ($T_{wing}$) at $T_a$ 20 – 35°C, but $T_{wing}$ was always lower than $T_a$ within this range. This suggests that the Brazilian free-tailed bat is able to shunt blood away from the wings. Although this might also be viewed as a strategy to avoid an increase in $EWL_{cut}$, such regional hypothermia reduces convective heat loss from the wings during flight. The reduction seems to be partially compensated for by an increase in radiative heat loss and by the presence of thermal windows at the flanks of the body and pelvic region, which are capable of dumping any excess heat generated during flight, especially at high $T_a$. Finally, this bat can fly continuously at high altitudes (up to 2,000 m) for long distances (up to 100 km; McCracken et al. 2008). Birds that fly at high altitudes for long distances may remain in water balance and avoid dehydration (Torres-Bueno 1978; Landys et al. 2000), so it is possible that bats might also be able to maintain water balance during long flights at high elevation.

The possibility that renal function, by controlling sensible water loss, is correlated with flight capacity and, hence, affects flight performance has been investigated by Happold and Happold (1988) for an assemblage of insectivorous bats in Africa. Although they found some correlations between habitat use, flight capacity, and renal function, the causal link underlying these mechanisms remains unclear. Nevertheless, some of the putative mechanisms suggested by Happold and Happold (1988) might explain the broad differences observed between studies of Carpenter (1985, 1986) and Thomas et al. (1991), on one hand, and that of Reichard et al. (2010), on the other. For example, the Brazilian free-tailed bat seems to have adaptations that reduce the amount of water loss during flight without jeopardizing its ability of maintain $T_b$ at constant levels. As also expected from its insectivorous diet, this bat has a low BMR (and hence low $EWL_{tot}$ during rest) and kidneys with a higher relative medullary thickness and a higher capacity for concentrating urine than the frugivorous flying-fox and the omnivorous spear-nosed bat (Geluso 1978; McNab 1992; Casotti et al. 2006). Furthermore, Brazilian free-tailed bats are able to fly fast, high, and for long periods—an ability not shared by the other two frugivorous species. This interplay between renal morphology, flight capacity, and mechanisms to adjust EWL may thus explain the greater capacity of Brazilian free-tailed bats to maintain water and thermal balance during flight.

### 4.7.3 Digestion, Respiration, and Circulation

The gross anatomy of the digestive system of bats does not depart widely from that of other mammals (Stevens & Hume 1995). One exception is that, as for birds, the large intestine tends to be short, with less nominal surface area in bats compared to similar-sized non-flying mammals (Caviedes-Vidal et al. 2007; Price et al. 2015a). The reduction in intestinal volume, and hence mass of food carried, would be advantageous, because the costs of flight increase with load carried, and because takeoff and manoeuvrability are diminished at heavier mass (Kvist et al. 2002). One drawback of this reduction in the large intestine length and surface area is that food retention time is reduced for bats (Price et al. 2015a).

To compensate for reduced large intestine length and surface area, bats show some digestive adaptations that are convergent with those of birds. Unlike non-flying mammals, but similar to birds, bats usually rely on passive, paracellular absorption of water-soluble nutrients, such as glucose and amino acids, to a greater extent than active-transport transcellular uptake (up to 70% paracellular; Tracy et al. 2007; Caviedes-Vidal et al. 2008; Fasulo et al. 2013; Price et al. 2013, 2015b; Brun et al. 2014). Evidence suggests that the high paracellular absorption of bats results from an increased number of tight junctions in the intestine, which have a high permeability to nutrient-sized molecules (Brun et al. 2014; Price et al. 2015a). Compared to other mammals, bats also have more villous area and a higher number of enterocytes per $cm^2$ of nominal surface area (Zhang et al. 2014), which may provide a mechanistic explanation for their high reliance on paracellular absorption. All these differences are related to the costs of flight; a high reliance on paracellular absorption has also been observed for birds (Lavin et al. 2008; Karasov et al. 2012), presumably as a mechanism to overcome the reduction in intestinal length and area.

One functional consequence associated with the high reliance of bats on paracellular absorption is that they are able to meet their high energetic demands of flight using dietary nutrients. Non-flying mammals, which rely more on transcellular absorption, usually meet increased energy demands by mobilizing endogenous substrates such as fats or glycogen. Bats, on the other hand, are capable of channelling the substrates of recently eaten food to meet their energetic requirements (Voigt & Speakman 2007; Welch et al. 2008; Amitai et al. 2010; Voigt et al. 2010, 2012). Again, this is convergent with birds. For example, nectar-feeding birds, such as hummingbirds, also rely heavily on paracellular absorption and are able to meet the extremely high costs of hovering flight using exogenous sources (McWhorther et al. 2006; Welch et al. 2008; Chen & Welch 2014).

The capacity of bats to mobilize exogenous energy sources might also alleviate the costs of long-distance migration (see 4.6.3). Unlike birds, bats cannot mobilize fat stores to support the costs of long-distance migrations, and do not have the equivalent physiological modifications such as fattening, observed for birds prior to migration. Voigt et al. (2012) showed that the migratory Nathusius's pipistrelle (*Pipistrellus nathusii*), in addition to using endogenous fatty acids, is able to oxidize proteins directly from the insects captured en route, and to mobilize the fatty acids of its diet to replenish body reserves. Reliance on exogenous sources may decrease the energetic costs associated with lipogenesis and gluconeogenesis, which, coupled with the energetic savings that accrue during torpor, may reduce the total energetic costs associated with migrations in bats (Guglielmo 2010).

The respiratory system of bats, like their digestive system, does not depart widely from the typical mammalian bauplan, but it does show some refinements associated with flight. Bats have relatively larger lung volume and heavier lungs compared to similar-size non-flying mammals (Jürgens et al. 1981; Canals et al. 2005a, 2011), and compared also to the volume of lung parenchyma in birds (Maina 2000a,b). The allometric relationship between lung volume ($V_L$; ml) and M (g) for nine species of bats, ranging from the 3-g Kalinowski's mastiff bat (*Mormopterus kalinowski*) to the 928-g grey-headed flying fox (Canals et al. 2005a), is $V_L = 0.0714M_b^{0.903\pm0.019}$ ($r^2 = 0.996; P < 0.001$).

Besides having a large lung, bats show modification of some parts of the respiratory system, which is thought to enhance its capability for gas exchange and, to some extent, to reduce the costs of breathing. For example, breathing work depends on the airways resistance of all components of the respiratory system, especially the proximal airway, to air convective flow (Zakynthinos & Roussos 1991). Canals et al. (2005b) showed that the Brazilian free-tailed bat (*Tadarida brasilensis*), when compared to rodents, had extensive modifications in the geometry of its bronchial tree that would reduce the cost of breathing and minimize the energy cost of flight. Other respiratory structural parameters seem also to be refined in bats. For example, the alveoli of bats are smaller (Maina 2000a; Maina et al. 1991), which, combined with their high lung volume, means that the total respiratory surface area available for gas exchange is higher for bats (Canals 2005b; Maina 2000a,b). In fact, the highest mass-specific respiratory surface area reported thus far for a vertebrate (138 $cm^3\ g^{-1}$) was for Wahlberg's epauletted fruit bat (*Epomophorus wahlbergi*; Maina et al. 1991). In addition, the alveoli are highly vascularized (see later) and the thickness of the alveolar-capillary barrier is greatly reduced in bats compared to non-flying mammals (Maina et al. 1991).

Bats rely on the basic mammalian pattern of a bidirectional convective flow to ventilate their large lungs. The highly vascularized flight membrane of bats is a suitable site for the passive diffusion of gases, but under resting conditions it accounts for only 2–10% of the total $O_2$ uptake (Herreid et al. 1968; Makanya & Mortola 2007). It is highly unlikely that cutaneous $O_2$ uptake, even if it increased during flight, would be enough to accommodate the extra metabolic demand for flight. Rather, the respiratory accommodation of the metabolic cost of flight requires adjustment of respiratory minute volume ($V_I$), which is the product of breathing frequency ($f_R$) and tidal volume ($V_T$). At resting, thermoneutral conditions, $V_I$ is usually higher for bats (and birds) than for similar-sized non-flying mammals, due to their larger $V_T$ and $f_R$ (Table 4.5). The amount of $O_2$ extracted per volume of air that reaches the lung ($EO_2$) at rest is available for only three species, and does not allow for meaningful comparisons. The greater spear-nosed bat (*Phyllostomus hastatus*) has an $EO_2$ of 29%, and the other two species (black flying fox *Pteropus gouldii* and lesser bulldog bat *Noctilio albiventris*) of 16 and 18% respectively, all within the range measured for other mammals and also birds.

During flight, the metabolic scope of bats increased dramatically, reaching values that are higher than those observed during thermoregulation in the cold, and similar to those reported for flying birds. To accommodate this increase in $O_2$ demand during flight, bats can either increase $V_I$, via changes in $f_R$ or $V_T$, or both, and/or increase $EO_2$. $V_I$ tends to increase by 10–17 times for bats during steady flight (Thomas 1981; Thomas et al. 1994). For all bats studied to date, $f_R$ increases by up to sixfold (Thomas & Suthers 1972; Thomas 1981; Carpenter 1985, 1986; Thomas et al. 1994).

It is difficult, however, to generalize about the relative contribution of $f_R$ to the increase in $V_I$ during flight, as only two studies concomitantly measured $f_R$, $V_T$, and $V_I$. In the black flying fox (*Pteropus alecto*), $V_T$ and $f_R$ increased in roughly the same proportion (Thomas 1981), while in the greater spear-nosed bat, $f_R$ increased by more than fivefold while $V_T$ increased only 1.8 times above resting levels

**Table 4.5** Body mass (M, g), breathing frequency ($f_R$, $min^{-1}$), tidal volume ($V_T$, ml), minute volume ($V_I$, ml $min^{-1}$), and oxygen extraction ($EO_2$,%) for bats at rest (or near the lower critical limit of the thermoneutral zone). Numbers in parentheses denote the ratio of observed value with those expected from allometric equation for mammals and birds (in bold). All respiratory values are in standard temperature and pressure units.

| Species | M g | $f_R$ $min^{-1}$ | $V_T$ ml | $V_I$ ml $min^{-1}$ | $EO_2$ % |
|---|---|---|---|---|---|
| *Scotorepens balsoni*[1] | 8 | 68.0<br>(0.36; **0.77**) | 0.05<br>(0.99; **0.4**) | 4.5<br>(1.41; **1.36**) | |
| *Plecotus auritus*[2] | 8.5 | 159.2<br>(0.82; **1.85**) | | | |
| *Leptonycteris sanborni*[3] | 22.0 | 30<br>(0.20; **0.48**) | | | |
| *Noctilio albiventris*[4] | 40.0 | 91.5<br>(0.74; **1.80**) | 0.20<br>(0.74; **0.28**) | 17.5<br>(0.61; **0.46**) | 18.3<br>(1.1; **0.88**) |
| *Phyllostomus discolour*[5] | 43.1 | 170.5<br>(1.40; **3.44**) | 0.39<br>(1.33; **0.51**) | 61.05<br>(1.99; **1.52**) | |
| *Phyllostomus hastatus*[6] | 110 | 81.6<br>(0.86; **2.27**) | 0.88<br>(1.14; **0.42**) | 72.05<br>(1.11; **1.89**) | 29<br>(1.81; **1.38**) |
| *Macroderma gigas*[7] | 150 | 55<br>(0.63; **1.69**) | | | |
| *Pteropus dasymullus*[1] | 379 | 56.0<br>(0.81; **2.36**) | 4.81<br>(1.71; **0.60**) | 272.8<br>(1.56; **1.42**) | |
| *Pteropus gouldii*[8] | 777 | 44.8<br>(0.78; **2.42**) | 9.8<br>(1.56; **0.56**) | 436.7<br>(1.41; **1.36**) | 16.0<br>(1.1; **0.84**) |

Allometric equations for mammals from Stahl (1967) and for birds from Frappell et al. (2001; conventional equations). Expected data (and ratio) for $EO_2$ taken directly from the source.

Sources:

[1] Frappell et al. (1992); [2] Webb et al. (1992); [3] Carpenter & Graham (1967); [4] Chappell and Roverud (1990); [5] Walsh et al. (1996); [6] Thomas et al. (1984); [7] Leitner and Nelson (1967); [8] Thomas (1981).

(Thomas et al. 1994). For both species, as observed during resting conditions, the increase in $f_R$ during flight was higher than that observed for birds. This is expected, as bats shows a rigid 1:1 synchronization between wingbeat and respiratory cycle, while only few species of birds display such a rigid match (Carpenter 1985; Boggs 2002). $V_T$ during flight, however, was lower for these two bats compared to birds, but similar to what was observed for active non-flying mammals. This is also expected based on the functional constraints imposed by a typical mammalian lung (Maina 2000a). It is worth noting that $EO_2$ during flight decreased for both species, from 29 to 22% for *P. hastatus* and from 16 to 10% for *P. gouldii* (Thomas 1981; Thomas et al. 1984). Why $EO_2$ differs between these two species is unclear, but the overall decrease in $EO_2$ suggests that bats might hyperventilate their lungs during flight (Thomas 1984).

These morphological refinements of the respiratory systems accommodate the increase in $O_2$ demand during flight (Canals et al. 2011); refinements of the cardiovascular system might also contribute to ensuring an adequate supply of $O_2$ to the tissues in bats. As with birds, bats have a proportionally larger heart compared to similar-sized non-flying mammals (Jürgens et al. 1981; Bishop 1997; Canals et al. 2005a). This is mostly due to an increased mass of the right atrium and ventricle, which presumably accommodates an increased venous return during flight and pumps this blood for oxygenation in the lungs (Kallen 1977). For bats, heart mass ($M_h$; g) scales with M (g) as follows: $M_h = 0.0173M_b^{0.741\pm0.023}$ (data from Table 4.6).

**Table 4.6** Body mass (M, g), heart mass ($M_h$, g), relative heart mass ($RHM = M_h / M$), $M_{h\_e}$ (expected heart mass based on the allometric equation $M_h = 0.0173M^{0.741}$), and the ratio between observed/expected $M_h$ ($M_h/M_{h\_e}$) for various species of bat.

| Species | $M_b$ | $M_h$ | RHM | $M_{h\_e}$ | $M_h/M_{h\_e}$ |
|---|---|---|---|---|---|
| *Mormopterus kalinowski*[3] | 3.10 | 0.057 | 0.018 | 0.040 | 1.423 |
| *Pteronotus quadridens*[1] | 4.30 | 0.010 | 0.002 | 0.051 | 0.19 |
| *Pipistrellus pipistrellus*[2] | 4.85 | 0.061 | 0.013 | 0.056 | 1.09 |
| *Myotis chiloensis*[3] | 6.88 | 0.096 | 0.014 | 0.072 | 1.32 |
| *Lasiurus borealis*[1,3] | 7.87 | 0.069 | 0.009 | 0.080 | 0.86 |
| *Mormoops blainvilli*[1] | 8.30 | 0.015 | 0.002 | 0.083 | 0.18 |
| *Monophyllus redmani*[1] | 8.60 | 0.280 | 0.033 | 0.085 | 3.28 |
| *Eptesicus fuscus*[1] | 9.30 | 0.090 | 0.010 | 0.090 | 0.991 |
| *Histiotus macrotus*[3] | 9.65 | 0.166 | 0.017 | 0.093 | 1.787 |
| *Tadarida brasiliensis*[1,3] | 9.80 | 0.178 | 0.018 | 0.094 | 1.893 |
| *Molossus molossus*[3] | 11.50 | 0.320 | 0.028 | 0.106 | 3.02 |
| *Histiotus montanus*[3] | 12.50 | 0.272 | 0.022 | 0.113 | 2.41 |
| *Lasiurus cinerus*[1] | 12.76 | 0.173 | 0.014 | 0.114 | 1.51 |
| *Erophylla sezerkoni*[1] | 13.20 | 0.038 | 0.003 | 0.117 | 0.32 |
| *Myotis myotis*[2] | 20.60 | 0.202 | 0.010 | 0.163 | 1.23 |
| *Molossus ater*[2] | 38.20 | 0.371 | 0.010 | 0.258 | 1.43 |
| *Artibeus jamaiscensis*[1] | 39.40 | 0.158 | 0.004 | 0.264 | 0.59 |
| *Brachyphylla cavernarum*[1] | 43.50 | 0.109 | 0.003 | 0.284 | 0.38 |
| *Phyllostomus discolor*[2] | 45.20 | 0.425 | 0.009 | 0.292 | 1.45 |
| *Noctilio leporinus*[1] | 56.60 | 0.130 | 0.002 | 0.345 | 0.37 |
| *Rousettus aegyptiacus*[2] | 146.00 | 1.226 | 0.008 | 0.696 | 1.76 |

[1] Rodriguez-Duran and Padilla-Rodriguez (2008);
[2] Jürgens et al. (1981);
[3] Canals et al. (2005a).

The relatively low predictive power of this equation ($r^2 = 0.42; P < 0.01$) means that other variables besides $M_b$ can account for the variability in $M_h$. For example, Rodriguez-Duran and Padilla-Rodriguez (2008) showed that wing loading accounted for much of the variability in $M_h$ for bats. This in turn suggests that factors such as flight style and diet can also explain the variability in $M_h$. For instance, the highest relative $M_h$ (as well as the largest positive deviations from the expected value based on the allometric equation) was for the Greater Antillean long-tongued bat (*Monophyllus redmani*), a bat capable of energetically challenging hovering flight.

Cardiac output (Q) is the product of heart rate ($f_H$) and stroke volume ($V_S$). No data are available for Q or $V_s$ in bats. The $f_H$ of resting bats is comparable to those of similar-sized non-flying mammals (Carpenter 1985), but shows a capacity for rapid and large increases to accommodate changes in $O_2$ demand. For example, bats can change $f_H$ from less than 10 $min^{-1}$ during torpor to more than 800 $min^{-1}$ while thermoregulating at low $T_a$ (Currie et al. 2014). During flight, $f_H$ increases up to sixfold above pre-flight levels (Studier & Howell 1969; Carpenter 1985, 1986), which is higher than that of running mammals. The capacity of bats to vary $f_R$ might be related to their capacity for increased venous return. Venous return occurs through two venae cavae, with the more muscular posterior part helping to store blood while at rest and then quickly releasing this blood in flight (Kallen 1977). $V_s$ is expected to vary as a function of $M_h$ (Bishop 1997), and therefore one might expect it to be higher in bats than non-flying mammals and birds. However, like Q, no studies have actually measured $V_S$ for bats. It is likely that the increase in Q to accommodate the changes in $O_2$ demands during flight is mostly regulated by increases in $f_H$ in bats.

Kallen (1977) presented a detailed overview of the general circulatory pattern in bats. Relative to the increased $O_2$ demands of flight, the most extensive modifications can be seen at the level of blood flow and pressure in capillaries supplying the lungs and pectoral muscles. Although bats have some interesting aspects associated with blood circulation to the wings (Kallen 1977; Davis 1988), these probably have more to do with regulation of heat transfer than with $O_2$ uptake during flight (see also section 4.7.2). Lungs and pectoral muscles in bats are highly vascularized, and a great portion of Q must be diverted to supply these tissues as bats commence flight. Pulmonary blood flow is higher in bats than in non-flying mammals, and the thinner interface between the highly branched capillary bed perfusing the lungs and the alveoli facilitates $O_2$ uptake (Maina 2000). The capillary bed supplying the pectoral muscle is also highly branched (Kallen 1977), and the number of capillaries per muscle fibre—compared to the bat hindlimb or the rat soleus muscle—is also high (Mathieu-Costello et al. 1992; Mathieu-Costello 1993). This facilitates delivery of $O_2$ to the flight muscles at a rate that is comparable to those observed in birds.

A suite of haematological changes further enhances the capacity of bats to deliver $O_2$ to their tissues. Blood volume and blood Hb-$O_2$ affinity ($P_{50}$) of bats are not different from those of other similar-size mammals (Snyder 1976; Jürgens et al. 1981; Boggs et al. 1999). However, erythrocytes of bats are smaller, and the number of erythrocytes per volume of blood is higher for bats than for non-flying mammals (see 3.4.1). The mean Hct of bats (around 60%) is far higher than those reported for non-flying mammals and birds, with some insectivorous species (*Pipistrellus pipistrellus, Molossus sinaloae*, and *Molossus bondae*) reaching 64–65% (Jürgens et al. 1981; Rodriguez-Duran & Padilla-Rodriguez 2008; Schinnerl et al. 2011). Although this high Hct increases blood viscosity, the negative consequences for blood flow might be counterbalanced by the high $M_h$ and $f_H$ observed in bats. In concert with the elevated number of red blood corpuscles, bats also have a high Hb (180–240 g $l^{-1}$), which gives a blood oxygen carrying capacity of up to 30%, values that are higher than for non-flying mammals and birds (Thomas 1987; Jürgens et al. 1981; Canals et al. 2011).

In summary, bats have evolved a myriad of changes (or refinements) of their digestive and cardiorespiratory systems to secure an adequate rate of nutrient acquisition—and supply and delivery of oxygen to their tissues—to meet the high energetic demands associated with flight (Table 4.7). Some of these changes, such as those observed in the respiratory system, were made within the design constraints imposed by the basic mammalian bauplan. Other modifications, such as those observed in the digestive, and especially at the cardiovascular, systems, were

**Table 4.7** Summary of the main digestive, respiratory, and circulatory adaptations of bats as compared to non-flying mammals. After Maina (2000) and Canals et al. (2011).

| Digestive adaptations | Respiratory adaptations | Circulatory adaptations |
|---|---|---|
| Shorter intestine* | Increased lung volume | Increased heart mass* |
| Reduced food transit time* | Small alveoli | Increased heart rate* |
| Increased villus area | Thin blood-gas barrier | Development of right side of the heart |
| Increased number of enterocytes | High $O_2$ diffusing capacity | Increased venous return |
| High paracellular absorption* | High respiratory frequency* | Increased capillary density in muscle* |
| High use of exogenous fuel* | Changes in the proximal airway | High hematocrit* |
| | | High haemoglobin concentration* |
| | | High $O_2$ transport capacity* |

*Denotes traits whose magnitude are thought to be similar to or even higher than in birds.

more extensive; and some of them represent classical cases of convergent evolution with birds.

### 4.7.4 Echolocation

Bats generally have sophisticated echolocation capabilities, for orientation and prey location and even interspecific communication (Kunz & Fenton 2003; Jones & Holderied 2007; Voigt-Heucke et al. 2010). Although many bat calls are quite loud (e.g. 128–133 dB) to maximize the echo strength, 'whispering' bats, which listen for the faint rustling noises of their prey (e.g. scorpions), have lower call intensity (e.g. 82 dB). Frequencies vary from 11 kHz (e.g. *Euderma maculatum*) to 212 kHz (*Cloeotis percivali*), with most insectivorous species 20–60 kHz (Jones & Holderied 2007). Low frequencies are avoided because the echo is weak for wavelengths longer than about an insect-wing length, and very high frequencies are avoided because they suffer from severe atmospheric attenuation.

The often complex acoustic structure of different bat calls has adaptive significance. In general, long narrowband calls are used for long-range detection, whereas short broadband calls are used for accurate localization and precision. For example, calls of *Myotis nattereri* range from 16 to 135 kHz; this enables the bat to discriminate its prey from background clutter. The acoustic characteristics of bats' echolocation calls can, therefore, be directly related to foraging environment and are closely related to wing morphology, mediated by the requirements of their foraging niche, which has placed strong selection pressure on the characteristics of both these important components of bat ecology (Norberg & Rayner 1987). Cluttered vs uncluttered environments require quite different acoustics for successful echolocation, and as a consequence of similar selection pressures, wing morphometrics and echolocation calls are closely correlated (Aldridge & Rautenbach 1987; Norberg & Rayner 1987). Echolocation calls can be characterized by their duration, intensity, peak and minimum frequency, and bandwidth, with peak frequency probably the most important determinant of echolocatory ability, as this appears to correlate most strongly with habitat and wing morphometric characteristics (Fullard et al. 1991).

Bats that forage in closed environments must contend with background acoustic contamination. Short, faint, high-frequency calls with wide bandwidth are most beneficial in these cluttered habitats, but these calls are impacted by atmospheric attenuation. Bats foraging in the open have more intense, longer calls, with lower frequency and narrower bandwidth, as these will attenuate less and provide better long-distance resolution, although they are more impacted by clutter (Aldridge & Rautenbach 1987; Norberg & Rayner 1987; Fullard et al. 1991). Consequently, bats foraging in open habitats have high WL and AR together with low-frequency, low-bandwidth calls, adapted for energetically efficient flight and long-range prey detection. Those in more cluttered environments, although morphologically and

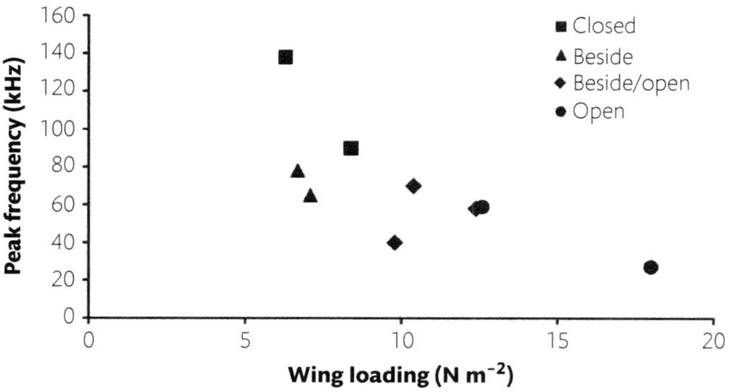


**Figure 4.26** Relationship between wing loading (which is correlated with aspect ratio), peak call frequency, and foraging location relative to vegetation for a guild of bats in Kruger National Park, South Africa. Data from Aldridge and Rautenbach (1987).

physiologically more variable (as a consequence of subtle variations in microhabitat and their associated demands), have low WL and AR combined with higher-frequency, broader-bandwidth calls (Aldridge & Rautenbach 1987; Norberg & Rayner 1987; Figure 4.26).

For example, the fast flying molossids have the highest WL amongst bats (and similar to birds of the same body mass and diet), but their decreased agility associated with a high WL, which elevates the costs of manoeuvrability, restricts these bats to forage in open spaces above the vegetation. These bats use narrowband echolocation calls of long duration, which are ideal for locating targets at long distances in open spaces. On the other hand, Vespertilionid bats, which glean for insects amongst foliage, need to be more manoeuvrable, with wings of lower WL. These bats use short, broadband calls, which are well suited to locate prey at short distances. The need to be more manoeuvrable also explains the comparatively low WL of fruit- and nectar-feeding bats, which have to find their food amidst vegetation. For these bats, however, especially for hovering nectar-feeding bats, selective pressure was probably directed more towards an increase in endurance and economy, so their wings also have a high AR.

## 4.8 Difficult Digestion

The role of digestion is to provide the nutrients and energy required by mammals for maintenance and growth (Barboza et al. 2009). Mammals that ingest easy-to-digest material, such as vertebrate flesh and fruits, generally have a simple digestive tract, consisting of a stomach and short intestine (see 2.6, 3.5). However, many mammals have specialized diets, and consume less digestible foods including proteins, such as keratin (in skin, nails, hooves, horns); bones; lipids, such as waxes (in many marine crustaceans, such as krill); the polysaccharide chitin (a nitrogenous polysaccharide found mainly in insect cuticle); and plant wall materials (glucose polymers, including cellulose, hemicellulose, and lignins). Digesting these materials requires specialized gut morphologies and physiologies.

### 4.8.1 Keratin, Bone, Wax, and Chitin

Keratin, in the form of skin, hooves, and horns, is particularly indigestible, and most carnivores avoid ingesting these components of carcasses. However, hypercarnivores such as hyaenas consume essentially all parts of a carcass, including skin, hooves, horns, and bones, and digest much of it. Nevertheless, hyaenas regurgitate much of the skin, hooves, and horns (Bearder 1977). For example, remains of skin are present in 35% of regurgitated material compared to 9% of scats, intact hooves in 27% of regurgitations compared to 0.4% of scats, and ceratin (horn) remains in 41% of regurgitations and 3% of scats. Hair is present in almost all scats, with 24% of scats having 10–20% hair content, and 15% of scats containing 80–100% hair. It is not clear how much, if any, energy is derived from ingested skin, hooves, horn, or hair.

Mammal bones have a very high mineral content (about 37% of wet mass) relative to their organic (primarily bone marrow; about 31%) and water (about 32%) content (Houston & Copsey 1994). Nevertheless, bones have a higher energy content per wet weight (6.7 kJ $g^{-1}$) than muscle (5.8 kJ $g^{-1}$; Brown 1988 cited by Houston & Copsey 1994), partly reflecting their high lipid content, so osteophagy ('bone-cracking') is an energetically viable, if morphologically demanding, strategy. Bone is generally a relatively small fraction of the diet for mammalian carnivores (e.g. hyaenas), but it is 70–90% of the diet for bearded vultures (*Gyptaeus barbatus*), which have a mean digestibility of 50% for bones (Houston & Copsey 1994). Digestion and mechanical breakdown for passage through the gut is presumably facilitated by the low stomach pH of bearded vultures, and mammalian scavengers also have a lower stomach pH (1.9) than other mammals (3.7), with herbivorous foregut-fermenting mammals having the highest pH (6.1; Beasley et al. 2015).

Consumption and digestion of bones requires the mechanical capacity to crack the bones prior to ingestion, hence a robust dentition and high bite force (Wroe et al. 2005; Tseng 2013). The killing bite of carnivores, using the canines, also requires maximal bite force. Wroe et al. (2005) compare the bite force of different mammalian carnivores from an allometric and phylogenetic perspective. Using a bite force quotient (BFQ), calculated as the residual of the allometric relationship for bite force (i.e. mass-adjusted bite force) normalized to 100, they found that the extant mammal with the highest BFQ was the Tasmanian devil (*Sarcophilus harrissi*; 181), with the highest placental carnivore being the African hunting dog (*Lycaon pictus*; 142). Including fossil species, two extinct marsupial lions had higher BFQs (*Thylacaleo carnifex*, 194; *Priscileo roskellyae*, 196), and the highest placental was the dire wolf (*Canis dirus*, 163). In general, hypercarnivores, which prey on animals larger than themselves, have a higher BFQ (120 ± 8) than other carnivores (86 ± 7). Surprisingly, the osteophagous (bone-cracking) hyaenas did not have particularly high BFQs (*Hyaena hyaena*, 113; *Crocuta crocuta*, 117), despite their very robust dentition, although the Tasmanian devil is also osteophagous. The ability to crack bones is perhaps related more to dynamic than static bite forces (related to BFQ), and to unilateral biting using the carnassial to fracture bones. Not surprisingly, the specialized termitivorous aardwolf (*Proteles cristatus*), which is closely related to hyaenas, has a low BFQ (77).

Mammals are generally able to easily digest most lipids, but digestion of waxes is problematic. Waxes are esters of long-chain fatty acids and mono-hydroxylic alcohols (Stevens & Hume 1995), each of which is readily digestible separately but not when esterified. Waxes are found in plant cuticle and beehives, but are especially predominant in marine ecosystems (as is chitin; see later), in marine invertebrates such as planktonic crustaceans (e.g. krill), fishes, and the spermaceti of whales. Waxes supplement or supplant triglycerides for energy storage. Planktonic crustaceans are very ecologically important in marine food chains, and up to 50% of their lipid synthesis is waxes (Lee et al. 1971).

Terrestrial mammals do not seem to have a particularly effective pancreatic lipase for wax digestion (< 50% digestibility), or particularly high levels of bile salts in the gall bladder (21–358 mM) to emulsify fats, but many seabirds have much higher digestive efficiencies (> 90%) and high bile salt concentration (469–507 mM; Place 1992). Seabirds can also return gastric and duodenal contents to their gizzard for further mechanical processing, and overall have about equivalent utilization of waxes and triglycerides. Unlike terrestrial mammals, Minke whales (*Balaenoptera acutorostrata*) have a high wax digestibility (94%), equivalent to seabirds, as would be expected, because wax esters are 21% of the total energy and 47% of the total lipids in their krill diet (Nordøy 1995). Swaim et al. (2009) estimated from models of ingestion and defaecation that North Atlantic right whales could digest more than 99% of the about 58 kg of wax esters ingested per day (by a 40,000-kg animal).

Chitin is a structural carbohydrate (polymer of n-acetyl-glucosamine) that is present in many animals; it is up to 60–85% of dry mass for arthropods, and is present in fungal cell walls as well as in jaws, chaetae, exoskeletons, and the like of many marine invertebrates (Stevens & Hume 1995). It is broken down to chitobiose by the enzyme chitinase, then to glucosamine by chitobiase. Some mammals produce endogenous chitinase to digest dietary chitin (e.g. rodents, monkeys, pigs, and especially some insectivore-consuming carnivores and bats; Cornelius et al. 1975; Stevens & Hume 1995; Strobel et al. 2013).

There are approximately 22 species of myrmecophagous mammal: highly specialized consumers of ants and/or termites (e.g. placental armadillos, *Cabassous* and *Tolypeutues* spp.; silky anteater, *Cyclopes didactylus*; giant anteater, *Myrmecophaga tridactyla*; tamandua, *Tamandua tetradactyla*; pangolins; aardvark; aardwolf, *Proteles cristatus*; the marsupial numbat, *Myrmecobius fasciatus*; and the echidna; Redford & Dorea 1984; Redford 1987). Despite representing very different evolutionary linages, myrmecophagous mammals are highly convergent and share a suite of anatomical and physiological characteristics that can be interpreted as adaptations and preadaptations to their diet. Typically, these mammals have structurally reduced, peg-like teeth (e.g. aardwolf), no teeth at all (e.g. echidna, giant anteater), and sometimes supernumerary teeth (e.g. numbat; reflecting the lack of selection pressure on its dentition); a long, extendible, vermiform tongue, which is effective in reaching the inner recesses of ant and termite nests; enlarged salivary glands, which produce copious sticky saliva to facilitate the capture of social insects with the tongue; an elongate snout and palate; anomalous stomachs; and digging-adapted forelimbs (Griffiths 1968). Myrmecophagous mammals also share a low-energy physiology, characterized by low $T_b$ and low BMR (McNab 1984; 2000). Although it is unclear if these shared physiological characteristics are really dietary adaptations or have arisen as a consequence of the phylogeny and other characteristics (e.g. semi-fossorial, armoured) of myrmecophages, it is apparent that this low-energy approach is necessary for specialization on a diet of ants and termites (Cooper & Withers 2002). Despite the localized abundance of ants and termites, a myrmecophagous diet has a low energy yield due its temporally and spatially patchiness, short feeding bouts (due to the prey's physical and chemical defences), low energy density (due to the inevitable ingestion of indigestible debris while feeding), and low digestibility (due to the chitin content of the prey; Redford & Dorea 1984; McNab 1984; Cooper & Withers 2004).

Myrmecophages don't appear to digest the chitin of their prey; rather, they digest the tissues and eliminate the chitinous exoskeleton in their faeces (Cooper & Withers 2004). Many species, such as aardvarks, pangolins, anteaters, and armadillos, have specialized stomachs with a large muscular wall, cornifed stratified epithelium, and keratinized 'teeth' that grind termites (and presumably expose the tissues to digestion) and manage the large quantities of abrasive dirt ingested during feeding (Griffiths 1968). However, apparent digestibility does not appear to differ from those species that do not have these digestive specializations, being about 64–81% of dry matter (Cooper & Withers 2004).

Another significant group of chitin-consuming mammals are the baleen (mysticete) whales, which filter zooplankton and small crustaceans (Macdonald 2010). They lack teeth (except as embryonic vestigial buds), but have extensive baleen plates, which are keratinous, fringed filters in the oral cavity that strain the contents of large volumes of seawater. Some mysticetes, like right whales (*Eubalaena*), skim the ocean surface to feed, whereas others, like Sei whales (*Balaenoptera borealis*), suck seawater into their extremely large oral cavity, then squeeze it out though the baleen. The filtered zooplankton and crustaceans are then scraped off the baleen using the tongue, and swallowed. Not surprisingly, these marine mammals have chitinolytic bacteria to digest dietary chitin (Souza et al. 2011). Minke whales, for example, have bacterial chitinase in their aglandular forestomach that dissolves chitinous crustacean exoskeletons (Olsen et al. 1999). The digestibility by Minke whales of chitin is 93% (joule content of krill is 23.8 kJ $g^{-1}$); the crabeater seal, which also consumes krill but has a single-chambered stomach, has a slightly lower digestive efficiency of 84% (Mårtensson et al. 1994).

### 4.8.2 Plant Fermentation

The Earth has experienced profound changes in climate, hence plant diversity, since about 65 MYBP, and these changes have been accompanied by the diversification of mammals, especially the herbivorous mammals (Stevens & Hume 1995). It is these herbivorous mammals that have evolved the greatest specializations for digestion, because their plant food is often high in carbohydrates such as cellulose and hemicellulose that are difficult or impossible for normal vertebrate digestive enzymes to hydrolyze.

In terrestrial ecosystems, the primary structural carbohydrates are plant cell wall components, such as cellulose, hemicellulose, and lignin (Withers 1992; Stevens & Hume 1995). Cellulose is the main constituent of plant cell walls, being 20–40% of the dry matter. Cellulose is straight polymeric chains of glucose (as is animal glycogen), but the glucose units are joined by a different three-dimensional bond arrangement; the linkages between glucose subunits are described as α-1,4 linkages in glycogen and β-1,4 linkages in cellulose. This three-dimensional difference in α and β bonds renders enzymes that can hydrolyze α bonds ineffective for β bonds, and vice versa. Hemicellulose is a branching polymer of polysaccharides (e.g. xylose, glucose, mannose, arabinose), often based on xylose subunits, joined by β-1,4 linkages; it is covalently bound to lignin, making it less water soluble. Lignin is not a polysaccharide; rather it is a phenyl-propane polymer with cutin, tannins, proteins, and silica. This difference in α and β three-dimensional linkages renders cellulose impervious to the action of animal amylases; cellulase is required for cellulose breakdown. In fact, three different cellulases are required for the complete breakdown of cellulose: endo-β-gluconases split β-linkages; exo-β-gluconases split glucose or cellobiose from the end of polysaccharides; and β-glucosidases hydrolyze cellobiose to glucose (Withers 1992). Some invertebrate animals endogenously produce cellulase, and consequently are able to digest cellulose (e.g. some crustaceans, silverfish, snails, and wood-boring beetles), but no vertebrate animals are able to endogenously produce a cellulase enzyme. Nevertheless, many herbivorous mammals are able to digest cellulose and various other plant cell wall structural carbohydrates, via the fermentative action of symbiont microorganisms (see later).

There have been various changes in the digestive anatomy and physiology of herbivores to utilize the ingested plant cell wall biomass (Batzli & Hume 1994; Stevens & Hume 1995). The maximum volume of the digestive tract is up to about 25% of body mass for herbivores (cf. < 10% for carnivores and omnivores), and the retention time for food in the gut is correspondingly increased (Barboza et al. 2009). In general, the gut becomes more complex for more difficult-to-digest foods. The simple digestive tract of carnivores, by contrast, is essentially a simple tube with digesta flowing through it in a pulsatile manner.

Herbivorous mammals adopted essentially four strategies to exploit plant food sources. First, some (e.g. bears, *Ursus americanus*) have an elongated midgut for greater digestion of nutrients, and possibly for accommodating some microorganismal fermentation. Second, some (e.g. rabbits, *Oryctolagus cuniculus*; koalas, *Phascolarctos cinereus*) have an enlarged caecum for microorganismal fermentation, and many use coprophagy or caecophagy to facilitate uptake of the products of this fermentation (see later). Third, perissodactyls (e.g. horses, rhinoceros, tapirs), elephants, and primates rely on an enlarged hindgut for fermentation. Fourth, many diverse groups of herbivores have a complex, often multi-chambered stomach that acts as a fermentation chamber; these are foregut fermenters. Ruminant artiodactyls (e.g. bovids, deer, antelope, camels, llamas, giraffes) have a complex four-chambered stomach, and regurgitate stomach contents into their oral cavity to rechew the digesta; this is rumination and the reason they are called ruminant mammals. Other artiodactyls (pigs, peccaries, and hippopotamus) do not ruminate. Kangaroos, sloths, and langur and colobid monkeys rely on an enlarged and generally compartmentalized stomach. Artiodactyls and perissodactyls largely evolved on separate continents (Africa/Eurasia and North America, respectively) with the global spread of grasslands. The current predominance of artiodactyls over perissodactyls suggests that foregut fermentation has advantages over hindgut fermentation (as does brain cooling; see 3.4.4, 4.2.1).

Ruminant mammals, the artiodactyls, have a characteristically large and multi-chambered stomach (Stevens & Hume 1995; Dijkstra et al. 2005; Figure 4.27). Ingested food is first physically processed in the oral cavity and swallowed into the largest stomach compartment, the rumen, where conditions (temperature, moisture) are ideal for microorganismal digestion (by bacteria, ciliates, flagellates, and fungal sporangia). Digesta are mixed in the rumen, and can be regurgitated into the oral cavity for rechewing; this process is termed rumination (Kennedy 2005). The rumen has a large surface area for absorption of microorganismal digestive products, primarily volatile fatty acids (VFAs; France & Dijkstra 2005). Digesta are then passed from the rumen into the reticulum, which is also a fermentation chamber but with a honeycomb lining; here, material is separated into coarse particles for return to the rumen, and fine particles to be passed to the omasum.

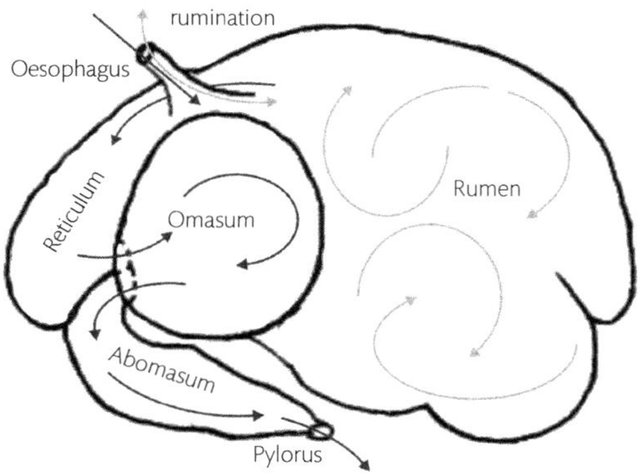


**Figure 4.27** Schematic of the four-chambered stomach of a ruminant mammal (cow), showing the flow of digesta through the rumen (with internal mixing and regurgitation for rechewing), reticulum, omasum, and abomasum, thence to the small intestine.

In the omasum, which has highly folded 'leaves' like book pages, digesta are further physically mixed and ground into smaller particles, before being passed to the abomasum, which is the 'true stomach' with the normal functions of the mammalian stomach, such as $H^+$ secretion and protein digestion. The ruminant system converts food to short-chain VFAs, methane, ammonia, and sometimes lactic acid (Russell & Strobel 2005). For a ruminant mammal such as a sheep digesting alfalfa, about 41% of the ingested energy is lost as faeces; 5% is eructed as methane; 33% is converted to VFAs; 18% is converted to microorganismal growth in the stomach; 18% is absorbed in the abomasum and the small intestine; and 3% is lost as heat from microorganismal metabolism (Withers 1992). Cetaceans are phylogenetically closely associated with artiodactyls, as the Cetartiodactyla (Table 1.1; Appendix). Interestingly, although some cetaceans have a single-compartment stomach others have a stomach with two or three compartments; the forestomach houses bacteria responsible for chitin and wax digestion (Nordøy 1995; Olsen et al. 1999; Swaim et al. 2009). Stones in the forestomach might enable it to function as a grinding gizzard.

Pseudoruminant mammals include a variety of other mammals that have convergently evolved a similar fermentative digestive strategy using the stomach, but do not ruminate (Stevens & Hume 1995). Placental pseudoruminants include the non-ruminant artiodactyls (pigs, peccaries, hippopotamus), sloths, colobid monkeys, and langur monkeys. The domestic pig and warthog have a simple stomach, but some suids (e.g. babyrousa, *Babyrousa celebensis*) have a compartmentalized stomach. This hippopotamus has a very complex, compartmentalized stomach, but no caecum and a very short large intestine.

Sloths are arboreal folivores, and have a complex, three-compartment stomach: the first compartment is large, divided into two sacs; the second is small, with a

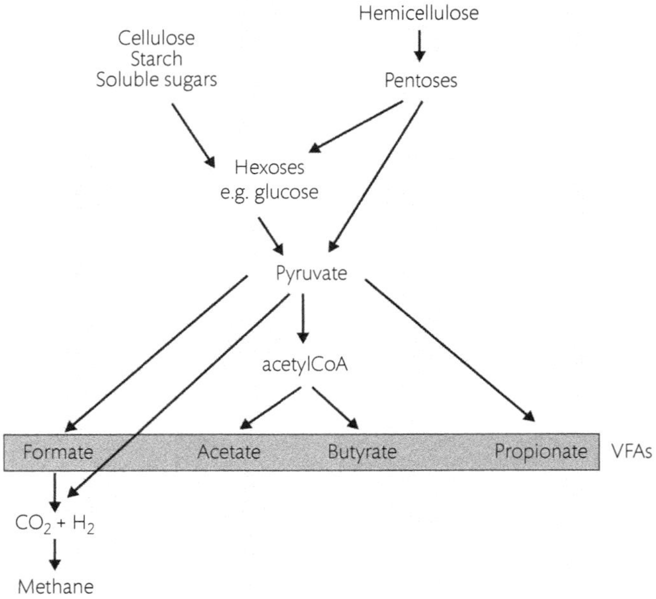


**Figure 4.28** Schematic of microorganismal fermentation of cellulose, starch, soluble sugars, and hemicellulose to pentoses and hexoses (e.g. glucose), then pyruvate, then microorganismal metabolic end products (volatile fatty acids, VFAs), and finally the combination of $CO_2$ with $H_2$ to form methane. Modified from France and Dijkstra (2005).

groove running to the third large and tubular compartment. Langur (*Presbytis*) and colobus (*Colobus*) monkeys also have a complex, three-compartment stomach. Macropod marsupials (e.g. *Macropus, Potorous*) are also pseudoruminants. They have a complex tubular stomach divided into three haustra. The forestomach has a sacciform anterior sac and a tubiform sac, whereas the hindstomach sac is the 'true' mammalian stomach, with the gastric and pyloric mucosa. Macropods chew their food more thoroughly than do ruminants, but they show a similar action of regurgitating and rechewing digesta, termed merycism (Barker et al. 1963; Hume 1982). Even some carnivorous marsupials undertake merycism (Archer 1974).

Sheep and kangaroos in Australia provide an informative comparison of a ruminant and a pseudoruminant mammal. Where sheep and red kangaroos coexist, kangaroos are more selective in their diet (Munn et al. 2009b). Sheep have a higher dry matter intake and a longer retention time than kangaroos (Hume 1982) but achieve fairly similar digestibilities (59 and 52%, respectively), despite their different fermentation strategies; sheep feed in short bouts whereas kangaroos have longer bouts (Munn et al. 2010b). Kangaroos can meet their daily energy requirements at lower dry matter intakes and higher fibre contents than sheep, reflecting their lower energy metabolism (Munn et al. 2009b). Field metabolic rate (FMR) was 16,664 kJ day$^{-1}$ for sheep (mass = 50.2kg) and 4,872 kJ day$^{-1}$ for kangaroos (mass = 23.6kg); when corrected for mass differences, a kangaroo is energetically equivalent to 0.7 of a sheep (Munn et al. 2009b). Interestingly, sheep and kangaroos differ markedly in water requirements: red kangaroos use only 13% of the water that a sheep uses, and have more concentrated urine (1,852 compared to 1,000 mOsm kg$^{-1}$). Since red kangaroos are not so dependent on water, their impact on rangelands is less and more broadly distributed.

For foregut fermenters, digesta encounter the fermenting microorganisms before the mammal's 'normal' digestive tract. Therefore, one disadvantage of foregut fermentation is that the microorganisms not only hydrolyze the structural carbohydrates such as cellulose and hemicellulose to monosaccharides (e.g. glucose), but also anaerobically metabolize these monosaccharides as their energy source, producing VFAs as their metabolic waste product. Although these VFAs are an important energy source for the host, there is little glucose remaining for the host's metabolism (Brockman 2005); for example, soluble sugars disappear from the forestomach of kangaroos, and there are low levels of disaccharases in the small intestine (Hume 1982). This low soluble carbohydrate availability is a serious physiological consequence because blood glucose concentration is low (compared to non-foregut fermenters), but glucose is essential for brain metabolism. To provide glucose to the body, the liver of ruminants and pseudoruminants synthesizes glucose from non-carbohydrate precursors, amino acids, by gluconeogenesis (Ballard et al. 1969). The liver of sheep and kangaroos releases glucose into the blood regardless of whether they have recently fed or are fasted, whereas the liver of a non-foregut fermenter, such as the dog, absorbs glucose after feeding but releases glucose when fasted.

An advantage of pre-peptic fermentation is that the rumen microorganisms can use the N waste of the host (urea) as a nitrogen source, thereby recycling N to the host through microorganismal protein synthesis (Nolan & Dobos 2005). In ruminants, rumen ammonia production can be 17–84% of the dietary N intake and the source for 69–80% of microorganismal protein; in macropods, ammonia is the N source for about 63% of microorganismal protein (Hume 1982). Urea is transferred to the rumen contents via saliva and diffusion from the blood. For kangaroos, the urea concentration of parotid and mandibular saliva (each produced at 0.05–4.5 ml min$^{-1}$) is correlated with flow rate, from 1.04 to 0.92 (relative to plasma) for parotid, and 0.84 to 0.08 for mandibular salivary glands, respectively. Parotid saliva flow is continuous, and has been suggested to be the main route for salivary urea recycling to the stomach, whereas mandibular saliva is likely the primary source of water for evaporative cooling (e.g. via forelimb licking; Beal 1987). Another fermentative strategy of mammals is to use sites after the stomach as the fermentation chamber (i.e. the caecum, hindgut, or both). In general, small mammals (< 5 kg) use the caecum, whereas large species (> 50 kg) use the hindgut; intermediate-size species use either or both (Hume 1982; Hume 1989; Stevens & Hume 1995; Gibson & Hume 2002). In general, hindgut fermenters have a relatively high passage time and a low digestibility; they achieve equivalent daily energy intakes as foregut fermenters by having a higher dry matter intake, particularly on poor-quality diets, so long as food is not limiting (Ilius & Gordon 1992). An advantage of hingut fermentation is that the digesta is subjected to normal digestive processes before fermentation so any digestible materials, such as glucose, are directly available to the mammal. However, hindgut fermenters suffer the disadvantage of having microorganismal fermentation located after the stomach and small intestine; the useful products of fermentation (VFAs, microorganismal production) are not subject to the normal mammalian stomach digestion and small intestine absorption functions.

Caecum fermenters include many rodents and lagomorphs (rabbits, hares, pika). The koala (*Phascolarctos cinereus*) has a relatively enormous caecum and hindgut, and many possums (e.g. brushtail, *Trichosurus vulpecula*) have a large caecum and hindgut. Coprophagy (or caecophagy) involves the elimination of specialized faecal pellets containing the product of microorganismal fermentation. These soft faeces, which are quite different from the normal harder and drier faecal pellets, are ingested to recycle the products of fermentation to the stomach for digestion, rather than simply eliminating these nutrient-rich materials, thereby overcoming the nutritional problems of having fermentation occur after the stomach (Figure 4.29). Coprophagy is best known for lagomorphs (rabbits, hares, pikas), but also occurs in rodents, shrews, and a folivorous prosimian (*Lepilemur*; Kenagy & Hoyt 1980). Coprophagy has been reported for ringtail possums and young koalas, presumably for inoculation with symbionts as well as nutrition. For the chisel-toothed kangaroo rat (*Dipodomys microps*), about one-quarter of the faecal pellets are of caecal origin, and are ingested over about 8 h during the day when not foraging (Kenagy & Hoyt 1980); their caecal pellets have more nitrogen and water than the normal faecal pellets. The herbivorous California vole *Microtus californicus* also ingests about one-quarter of its faeces, but in one-to-several-hour bouts, reflecting their continuous day and night foraging pattern. The diurnal degu rodent (*Octodon degu*) consumes its caecal pellets primarily at night (Kenagy et al. 1999).

A final aspect of microorganismal-based fermentation as an herbivorous strategy for mammals (and other animals) that has global significance is the incidental production of hydrogen, and particularly methane, by fermentation. Methane ($CH_4$) is an important greenhouse gas, and ruminant livestock are the largest single source of methane, about 28% of global production (Klieve 2009). Archaean microbacteria in the rumen and forestomach of other fermentative herbivores, such as kangaroos and sloths, reduce $H_2$ from fermentation to $CH_4$; this reaction is the main sink for $H_2$ in the stomach and is of considerable importance, as $H_2$ must be removed from the rumen for continued efficient fermentation. Methane

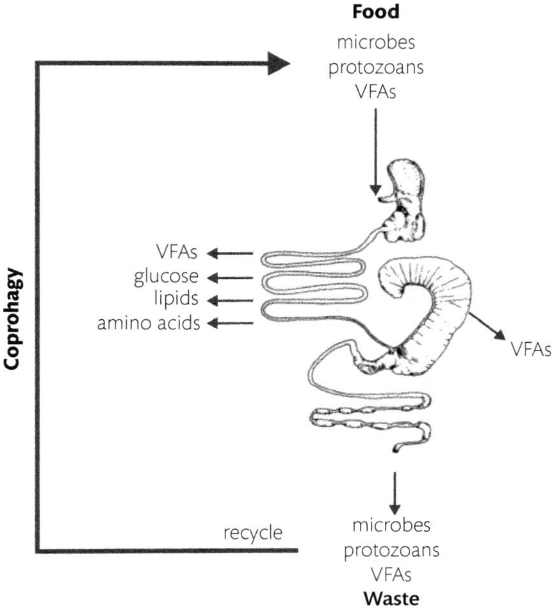


**Figure 4.29** Schematic of the fates of ingested food and reingested faeces for a post-gastric fermenter with coprophagy (rabbit). Gut schematic from Stevens and Hume (1995).

is eliminated via the mouth by eructation. Methane production of ruminants and equids increases isometrically with mass (ml $CH_4$ $day^{-1} = 0.66M^{0.97}$ and $0.18\ M^{0.97}$, respectively), compared to $M^{0.75}$ for metabolic rate, so methanogenic energy loss increases with mass and could limit the maximal mass of ruminants but not equids (Franz et al. 2010; Vendl et al. 2015a; Figure 4.30). Methane production of various non-ruminant herbivores scales the same as for equids (ml $day^{-1} = 0.18M^{0.97}$; Franz et al. 2011), but domesticated pigs are lower (0.07 $M^{0.99}$; Franz et al. 2010). Kangaroos are similar to or lower than non-ruminants (Vendl et al. 2015a); sloths are slightly higher (Vendl et al. 2015b), as are peccary and pygmy hippopotamus (Vendl et al. 2016). Although informative, these allometries of absolute methane production need to be carefully interpreted in terms of methanogenesis per metabolic activity of the herbivore. Methane production of macropod marsupials increases compared to placental mammals when expressed per dry matter intake or metabolic rate, generally falling between that of ruminant and non-ruminant placentals, and especially increases for sloths, to about that of ruminants (Vendl et al. 2015a). Despite the independent evolution of foregut fermentation in various mammals (and birds; e.g. the hoatzin), the convergent role of methanogenesis has resulted in convergence of the microbial communities in these

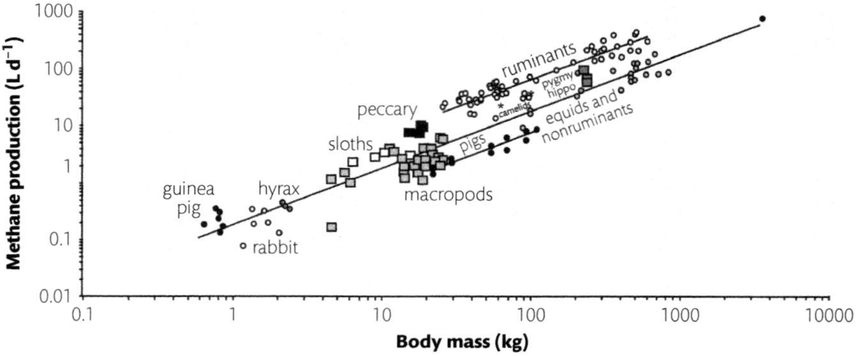


**Figure 4.30** Methane production of herbivorous mammals: ruminants, macropods, sloths, hyrax, rabbit, guinea pigs, peccary, camelids, elephant, equids, and non-ruminants. Data from Franz et al. (2010, 2011), Vendl et al. (2015a,b), and Vendl et al. (2016).

phylogenetically distant herbivores, even to the level of cow and hoatzin rumen/crop communities being more similar than their respective colon/ceca communities (Godoy-Vitorino et al. 2012).

Despite microorganismal-assisted digestion, digestibility of plant matter is generally relatively low compared to other diets. Digestive efficiency (e.g. percentage dry matter digestion) varies with digestive strategy (which varies with body mass) and diet, rather than with the phylogenetic affiliation of species (Table 4.8). As a consequence, digestibility is higher for small herbivores and ruminants than for large hindgut fermenters.

There is a general relationship between the body mass of herbivores and the quality of food consumed (as well as location of the fermentation chamber; see earlier). For example, amongst macropod marsupials, the small musky rat-kangaroo (*Hypsiprymnodon moschatus*; 1.8 kg) consumes fruit and insects; rat-kangaroos (*Potorous, Bettongia*; 3.8 kg) consume fungi, tubers, and insects; tree kangaroos (*Dendrolagus*; 11 kg) consume leaves and fruit; swamp wallabies (*Wallabia bicolor*; 14 kg) browse on shrubs; wallaroos (*Macropus robustus*; 22 kg) are mixed browsers/grazers; and large kangaroos (*Macropus* spp; 25–57 kg) graze on grass and herbs (Hume 2014). Rock wallabies (*Petrogale*; 8.6 kg) and hare-wallabies (*Lagorchestes, Lagostrophus*; 3 kg) are unexpectedly small grazers.

Similar patterns are evident for some assemblages of large African herbivores, but this simple pattern is not universal and is 'blurred' by differences in species' bauplans (Clauss et al. 2013). For example, large herbivores of Kruger National Park vary markedly in diet, as evidenced by the percentage of $C_4$ plants in their diet (Codron et al. 2006), which is related to their $\delta C^{13}$ isotope ratio (see 5.7.2; Figure 4.31). The percentage (%) $C_4$ plants and ‰ $\delta C^{13}$ are remarkably consistent over geography (north and south of the park) and seasons (wet and dry season) for

**Table 4.8** Comparison of dry matter intake and digestibility for various ruminants differing in body mass, diet, and digestive strategy.

| Species | Mass (kg) | Diet | DMI (kg $d^{-1}$) | DM digestibility (%) | Comments |
|---|---|---|---|---|---|
| Pika (*Ochotona princeps*)[1] | 0.14 | High tannin leaves + pelleted food | 0.052 | 60 | Caecum fermentation; coprophagic; high tannin diet |
| Brushtail possum (*Trichosurus vulpecula*)[2] | 2.49 | Eucalyptus leaves | 0.074 | 51 | Caecum + hindgut fermentation; high phenolic diet |
| Koala (*Phascolarctos cinereus*)[3] | 6.56 | Eucalyptus leaves | 0.17 | 54 | Caecum + hindgut fermentation; high phenolic diet |
| Reedbuck (*Redunca redunca*)[4] | 14 | Grass mixture | 0.24 | 49 | Ruminant fermentation |
| Grant's gazelle (*Nanger granti*)[5] | 80 | Free-ranging | 2.5 | 61 | Ruminant fermentation |
| Cape Buffalo (*Syncerus caffer*)[4] | 235 | Grass mixture | 7.2 | 46 | Ruminant fermentation |
| Zebra (*Equus burchellii*)[5] | 270 | Free-ranging | 7.2 | 42 | Hindgut fermentation |
| Hippopotamus (*Hippopotamus amphibius*)[4] | 1,200 | Elephant grass | ≈ 8 | 68 | Non-ruminant artiodactyl |
| Asian elephant (*Elephas maximus*)[6] | 2,427 | Hay + pelleted food | 37.9 | 36 | Hindgut fermentation |
| African elephant (*Loxodonta africana*)[7] | ≈ 4,578 | Free-ranging | 54–67 | 30–45 | Hindgut fermentation |

[1]Dearing (1997);
[2]Foley and Hume (1987);
[3]Cork et al. (1983);
[4]Arman and Field (1973);
[5]Abaturov et al. (1995);
[6]Clauss et al. (2003a);
[7]Meissner et al. (1990).

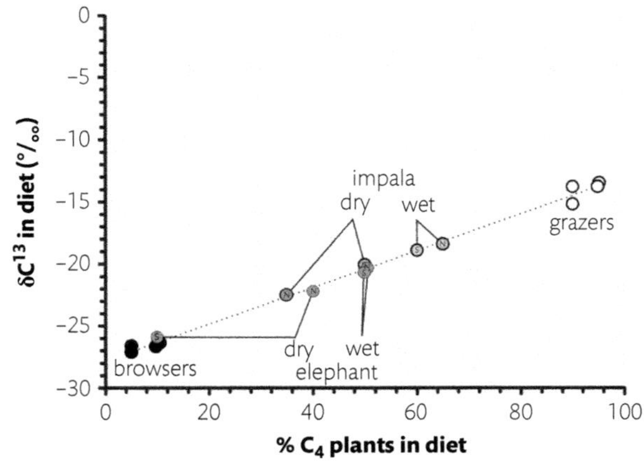


**Figure 4.31** Dietary percentage of $C_4$ plants and ‰ $\delta C^{13}$ content of the diet for African browsers (giraffe, *Giraffa camelopardalis*; greater kudu, *Tragelaphus strepsiceros*), grazers (Burchell's zebra, *Equus birchellii*; African buffalo, *Syncerus caffer*; blue wildebeest, *Connochaetes taurinus*), and mixed feeders (impala, *Aepyceros melampus*; elephant, *Loxodonta africana*), in the north and south of Kruger National Park, in the wet and dry seasons. Data from Codron et al. (2006).

large browsers (about 8% and −27‰) and large grazers (about 93% and −14‰). A mixed browser/grazer (impala; *Aepyceros melampus*) has intermediate %$C_4$ and ‰ $\delta C^{13}$, which is lower in the dry season (and lower in the north of the park) than in the wet season (with no geographic difference). However, the elephant, a very large herbivore, is a browser in the south of the park in the dry season, and a mixed browser/grazer in the north; in the wet season, they are even more mixed browser/grazer, with no geographic difference.

Large body size has been considered to offer nutritional advantages to mammals, because of their lower mass-specific metabolic rates, relatively large gut volume, and longer passage times, but there is a limit to increasing body mass for optimization of fermentative digestion (Clauss et al. 2003b, 2013; Barboza et al. 2009). Large foregut fermenters are designed to slow the passage of ingesta and maximize physical and chemical digestion and fermentation, but this inherently limits their passage rate, hence amount of material potentially available for digestion (e.g. the hippopotamus). In contrast, large hindgut fermenters have a relatively fast passage time and low digestibility and are more adaptable to large body mass. This hypothesis is supported by the fossil record, which suggests that fossil foregut fermenters did not generally exceed the highest body mass of extant foregut fermenters, and the very large fossil mammalian herbivores are thought to have been hindgut fermenters. On various continents, there was an increase in maximum mass after the Cretaceous/Paleogene boundary (66 MYBP) that levelled off at about 10,000 kg, around 40 MYBP (Smith et al. 2010); these largest terrestrial mammals included hindgut-fermenting perissodactyls (*Indricotherium*) and proboscideans (*Deinotherium*). While digestive constraints, such as the allometry of mean retention time, might explain the body mass limit for herbivory, they cannot alone be considered the evolutionary constraint on body mass—other factors such as fasting endurance or running speed may be involved (Clauss et al. 2007, 2013).

There are also various ecological consequences of herbivory, including how herbivores overcome plant defences. The evolutionary 'arms race' between plants and herbivores, like that between prey and predators, has resulted in continuing adaptation and counter-adaptation of physical and chemical 'weapons'. Herbivorous mammals also affect plant diversity: regions where mammal herbivory is more important than insect herbivory tend to have lower plant diversity (Becerra 2015).

Many plants have physical defences against mammalian herbivores, such as spines, stinging hairs, or highly sclerotized leaves. Although spines are effective against some herbivores, others have responded with adaptations to circumvent the deterrence of spines. For example, the white-throated woodrat (*Neotoma albigula*) is an *Opuntia* cactus specialist; it is adept at clipping off the spines before transporting the cactus to its den for consumption, and actually prefers to collect spiny cacti over (artificially) de-spined cacti (Kohl et al. 2014).

Many plants use chemical defences against herbivory, in the form of plant secondary metabolites (PSMs). PSMs are an incredibly diverse assemblage of plant metabolites that are not just waste products, but also defend plants against herbivory and abiotic stress, and they are major contributors to community dynamics with effects that cascade through ecosystems (Iason et al. 2012). Many PSMs have marked effects on diet selection, digestion, and reproductive success of herbivorous mammals through aversive, detrimental, or toxic effects (DeGabriel et al. 2014; Moore et al. 2015). PSMs can impart a noxious taste to foliage; for example, many tannins are very astringent and potentially deter herbivores from their consumption. Some PSMs have adverse effects, such as reducing digestion of specific dietary components; for example, tannins bind to dietary and endogenous proteins and inhibit digestive enzymes (e.g. $\alpha$-amylase and $\alpha$-glucoamylase; Barrett et al. 2013). Nevertheless, many herbivorous mammals ingest PSMs, often in sufficient amounts that they have marked physiological effects.

Phenolics are a common toxic PSM; for example, eucalypt leaves (*Eucalyptus punctata*) have a phenolic content of about 28% dry weight (Cork et al. 1983); *E. urograndis* leaves contain about 0.04% hydroxybenzoic acid (a derivative of benzoic acid; Chapuis-Lardy et al. 2002). Terpines are a diverse group of volatile unsaturated hydrocarbons found in plant essential oils. For example, *E. camaldulensis* leaves contain about 1.1% dry weight cineole, a terpinoid oxide, and 2.05% total terpinoids (Stone & Bacon 1994).

As many PSMs have toxic effects on their consumers, detoxification is the consumers' counter-response. Detoxification is a series of biochemical reactions that convert the PSM into a suitable form for excretion. Herbivores can either hydrolyze, oxidize, or reduce the PSM for excretion (Phase I reactions), or conjugate the PSM with glucuronic acid, glycine, or sulphate for excretion (Phase II reactions), or both (Marsh et al. 2006). Phenolics are detoxified by conjugation with glucuronic acid for urinary excretion; cineole is detoxified by oxidation and conjugation with glucuronic acid; benzoic acid is mainly conjugated with glycine (Marsh et al. 2006). Glucuronic acid excretion accounts for about 40% of urinary energy loss of woodrats consuming creosote (*Larrea tridentata*) resin (about 1.9% of total metabolizable energy; Mangione et al. 2004). For brushtail possums (*Trichosurus vulpecula*), consumption of diets high in benzoate incurred a considerable cost in protein, about 30% of the daily N intake (Au et al. 2013).

Some mammals subsist entirely on the highly defended leaves of plants (e.g. the koala consumes only *Eucalyptus* leaves). Their stomach is small, as is their small intestine, but their colon and caecum are very long and wide (the koala has the longest caecum relative to body size of any mammal). The microorganisms present in a koala's hindgut seem to detoxify essential *Eucalyptus* oils (Eberhard et al. 1975), and glucuronic acid excretion accounts for more than 7% of their urinary energy excretion (but only about 1% of the total metabolizable energy; Cork et al. 1983).

# 5

# Concepts, Approaches, Techniques, and Applications

Advances in ecological and environmental physiology rely on the improvement of current methodologies and approaches, the application of existing technologies in new ways, as well as the development of new methodologies and approaches. For example, the measurement of mammalian metabolism was one of the first techniques developed for the study of mammalian physiology, when in the 1770s Lavoisier developed an ice-calorimeter and correlated the heat production measured by the calorimeter with the carbon dioxide production of a guinea pig (Frankenfield 2010; see 5.3.1). Techniques used now are much more sophisticated and diverse (see 5.3). Methods for radio-tracking and GPS-locating mammals are being revolutionized by the miniaturization of electrical components (largely driven by mobile phone technology), developments in satellite technology, and computer systems (see 5.3.2), and modern electronics is benefitting from rapid advances in the energy density of batteries. The rapid development in techniques and approaches for molecular biology, particularly since the 1990s and early 2000s, has revolutionized all aspects of the basic biological sciences, although the main impetus of much of this development is the biomedical and clinical sciences (see 5.1, 5.7). As a result, we are now better able to study the field biology of mammals, which is becoming particularly critical given the potentially catastrophic effects of global climate change on many mammalian species (section 6.2).

We briefly describe here some of these major developments in concepts, approaches, techniques, and their applications.

## 5.1 The Comparative Method

A basic requirement of comparative studies is to know the phylogenetic relationships of what is being compared. The Mammalia is a monophyletic clade of vertebrates that has had an extensive radiation since its origin, about 250 MYBP. Of the extant mammals, monotremes diverged about 166 MYBP from the sister lineages of marsupials

*Ecological and Environmental Physiology of Mammals*. Philip C. Withers, Christine E. Cooper, Shane K. Maloney, Francisco Bozinovic, & Ariovaldo P. Cruz-Neto. © Philip C. Withers, Christine E. Cooper, Shane K. Maloney, Francisco Bozinovic, & Ariovaldo P. Cruz-Neto 2016. Published 2016 by Oxford University Press.
DOI 10.1093/acprof:oso/9780199642717.001.0001

and placentals, which diverged from each other about 147.7 MYBP (Chapter 1). The mammalian 'super-tree' (Bininda-Emonds et al. 2007) provides a phylogenetic framework for about 4,500 extant mammalian species (see Figure 1.2; section 5.6). It is perhaps obvious that any particular species is more closely related to some species (e.g. those in the same genus) than with others (e.g. species in other families and orders). So, when we compare a physiological trait (or any other type of trait) in a group of mammals, it is important to appreciate that more closely related species are more likely to be similar than more distantly related species. For example, the body mass of a species of shrew (Soricidae) is more likely to be similar to that of other species of shrew than to other species from different families or orders—whales (Cetacea), for instance, or elephants (Proboscidea). Taking a quantitative approach, we would expect the variation in body mass of species of shrews in a particular genus to be smaller than the variance in body mass in different genera of mammals from a particular family, or families of mammals within an order, or orders of mammals within the class. For body mass of adult placental mammals, and for a range of other life history traits, about 60–70% of the total variation occurs at the level of orders within the Class Mammalia, whereas only 2–10% of the total variation is for species of mammals within their genus (Read & Harvey 1989; Harvey & Pagel 1991; Table 5.1).

### 5.1.1 Phylogenies

The construction of phylogenetic trees is an essential aspect of applying the comparative method to the analysis of biological traits, because we need some framework within which to account for the likely effects of relatedness of species when

**Table 5.1** Partitioning of variance in life history traits for placental mammals at different taxonomic levels, species within genera, genera within families, families within orders, and orders within the Class Mammalia. After Read and Harvey (1989); from Harvey and Pagel (1991). Reproduced with permission of Oxford University Press.

| **Among:**<br>**Within:** | **Species**<br>**Genera** | **Genera**<br>**Families** | **Families**<br>**Orders** | **Orders**<br>**Class** |
|---|---|---|---|---|
| Relative variance component[a] | 3 | 7 | 21 | 69 |
| Adult mass | 3 | 5 | 27 | 65 |
| Neonatal mass | 2 | 6 | 21 | 71 |
| Gestation length | 8 | 11 | 19 | 62 |
| Maximum reproductive life | 10 | 10 | 12 | 68 |
| Annual fecundity | 5 | 7 | 14 | 74 |
| Annual biomass production | 6 | 8 | 18 | 68 |

[a] Relative variance component = 100*variance/total variance.

analysing patterns of traits. Currently, molecular data (e.g. DNA sequences) are routinely used to construct phylogenetic trees. Prior to the molecular biology revolution, biologists had access to taxonomic hierarchies (e.g. species, genus, family, order, class levels, as well as possible intermediate levels, such as sub-species, sub-genus, sub-family). The different approaches to taxonomy include pheneticists, cladists, transformed cladists, and evolutionary taxonomists, amongst others (Harvey & Pagel 1991).

Traditional taxonomic hierarchies do not necessarily reflect modern phylogenies. For example, humans have traditionally been classified as Hominidae, whereas orangutans, gorillas, and chimpanzees were ascribed to the family Pongidae. Modern phylogenies, however, show that humans are most closely related to chimpanzees, then to gorillas—excluding humans from the Pongidae makes the family paraphyletic. To resolve this paraphyly, family Hominidae could be redefined as containing gorillas, chimpanzees, and humans, leaving orangutans to Pongidae. Nevertheless, taxonomic information can be used, with care, as a proxy for a phylogenetic tree to analyse physiological or ecological traits in a simple hierarchical pattern, such as the subclasses Prototheria, Metatheria, and Eutheria, or a more complex hierarchical arrangement with varying numbers of levels using nested analysis of variance (ANOVA, see later).

A phylogeny (or phylogenetic tree) is a genealogical representation of a group of species, with the presumed common ancestral species as the root of the tree (Harvey & Pagel 1991; Figure 5.1, top). The phylogenetic tree shows us the dichotomously bifurcating lineages. Ideally, the phylogeny includes knowledge of the absolute time (in MYBP) of the various divergence points (nodes). From the phylogeny and divergence times, we can construct a distance matrix, which summarizes the distance of every species from every other species (Figure 5.1, middle). A nexus file (Figure 5.1, bottom) is another commonly used way of summarizing the information contained in a phylogeny. It is a text file that includes information such as the numeric code used for the species, and a TREE statement that includes branch pattern and length information for the tree. Frequently, tree reconstruction methods provide similarity information for the nodes, rather than an MYBP metric, so similarity needs to be transformed to distance, if a distance matrix or nexus file is required for analysis.

The procedures for the reconstruction of phylogenetic trees have been debated for many years, and the relatively recent addition of molecular data has added an extra level of complexity and new impetus to tree construction (e.g. Harvey & Pagel 1991). Of the many possible trees that can be derived for a data set, the most parsimonious is the tree that breaks a specified criterion rule (e.g. the least number of character state changes) the minimum number of times. Alternatively, a compatibility criterion—such as the tree with the least number of changes for all characters—can be used. The maximum likelihood approach seeks a tree with the highest probability of obtaining the observed data set consistent with the

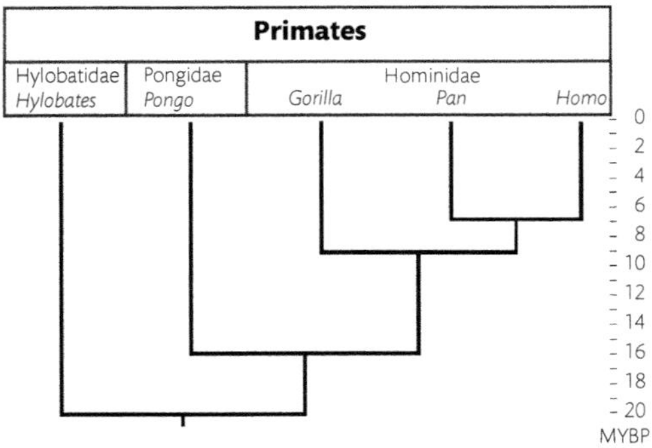


**Distance Matrix**

| | *Hylobates* | *Pongo* | *Gorilla* | *Pan* | *Homo* |
|---|---|---|---|---|---|
| *Hylobates* | 0 | 19.9 | 19.9 | 19.9 | 19.9 |
| *Pongo* | 19.9 | 0 | 15.8 | 15.8 | 15.8 |
| *Gorilla* | 19.9 | 15.8 | 0 | 8.9 | 8.9 |
| *Pan* | 19.9 | 15.8 | 8.9 | 0 | 6.6 |
| *Homo* | 19.9 | 15.8 | 8.9 | 6.6 | 0 |

**Nexus File**

```
begin TREES;

    TRANSLATE
            1 Hylobates,
            2 Pongo,
            3 Gorilla,
            4 Pan,
            5 Homo;

      TREE 'Default symmetrical+' =
(1:19.9, (2:15.8, (3:8.9, (4:6.6, 5:6.6) :2.3) : 6.9) : 4.1);
end;
```

**Figure 5.1** Top: Phylogenetic tree for Primate families (gibbons, orangutan, gorilla, chimpanzee, and humans) showing the possible taxonomic reclassification of Hominidae to avoid Pongidae being paraphyletic, as it would if it included *Pongo*, *Gorilla*, and *Pan*. Divergence times from www.timetree.org (Hedges et al. 2006). Centre: Distance matrix for this phylogenetic tree. Bottom: Nexus file for this phylogenetic tree.

probabilities of character change. It is impossible to be certain that a constructed tree is the correct tree, but by using a variety of methods, the analysis should converge on the correct one.

It is important that a phylogenetic tree has the correct root (i.e. the root of the tree is indeed the common ancestor). This is not necessarily easy to determine, because character states can reverse during evolution. Various approaches can be used to root the tree correctly, such as including an outgroup in the analysis; the outgroup is a taxon that is highly likely to possess the ancestral state (e.g. black bear as an outgroup for canids in Figure 5.2). Other possible approaches to rooting the tree include using the fossil record and making inferences from ontogenetic development. Two final caveats with respect to the correctness of a constructed tree: (1) it is possible that a tree will include polytomies (a single multiple, rather than dichotomous, branching pattern), and (2) there can be uncertainty about the

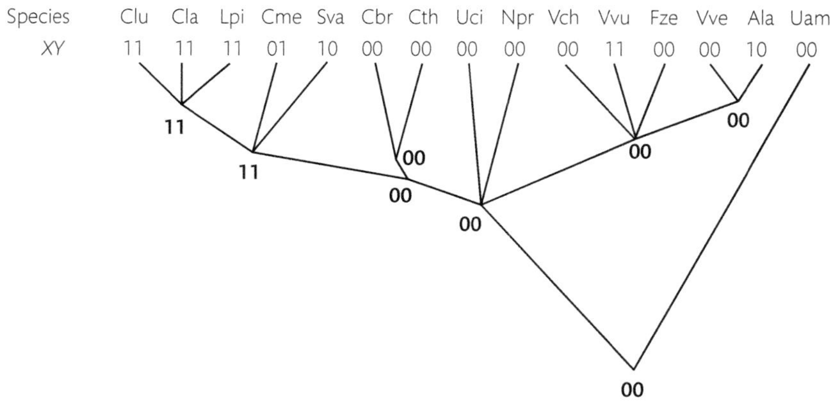


**Figure 5.2** Example of reconstruction of ancestral states for characters X (diet; 0 = omnivory, 1 = carnivory) and Y (parental care; 0 = non-communal, 1 = communal) for 14 canid species, with the black bear as an outgroup. Species are as follows: Clu, *Canis lupus*; Cla, *Canis latrans*, Lpi, *Lycaon pictus*; Cme, *Canis mesomelas*; Sva, *Speothos vanaticus*; Cbr, *Chrysocyon brachyurus*; Cth, *Cerdocyon thous*; Uci, *Urocyon cineroargenteus*; Npr, *Nyctereutes procyonoides*; Vch, *Vulpes chama*; Vvu, *Vulpes vulpes*; Fze, *Fennecus zerda*; Vve, *Vulpes velox*; Ala, *Alopex lagopus*; Uam, *Ursus americanus*. Modified from Harvey and Pagel (1991). Reproduced with permission of Oxford University Press.

actual divergence dates. Fortunately, many comparative approaches are relatively robust with respect to these potential problems, particularly if uncertainty is concentrated towards the tree tips (recent divergences) rather than deep in the tree (early divergences).

Related to the process of constructing a tree is the determination of hypothetical ancestral character states. Unless all of the data are derived from the fossil record, it is unlikely that ancestral states at any of the nodes within a tree are known, so they need to be inferred. Reconstructing ancestral states is aided by knowledge of an appropriate outgroup, and by rules related to parsimony of the most likely pattern of trait evolution (Harvey & Pagel 1991).

Molecular biology has revolutionized the phylogenetic interpretation of the relationships between animals, including the Mammalia (e.g. Springer et al. 2004). The reconstruction of molecular trees, using various possible selections of both the nuclear and mitochondrial genome, is now a common, even routine, procedure. The availability and general utility of molecular trees has had a profound effect on the understanding of mammalian relationships; having ready access to detailed phylogenetic trees has in turn enabled the development and use of complex comparative methods for the analysis of physiological, ecological, and other types of trait data (see 5.1).

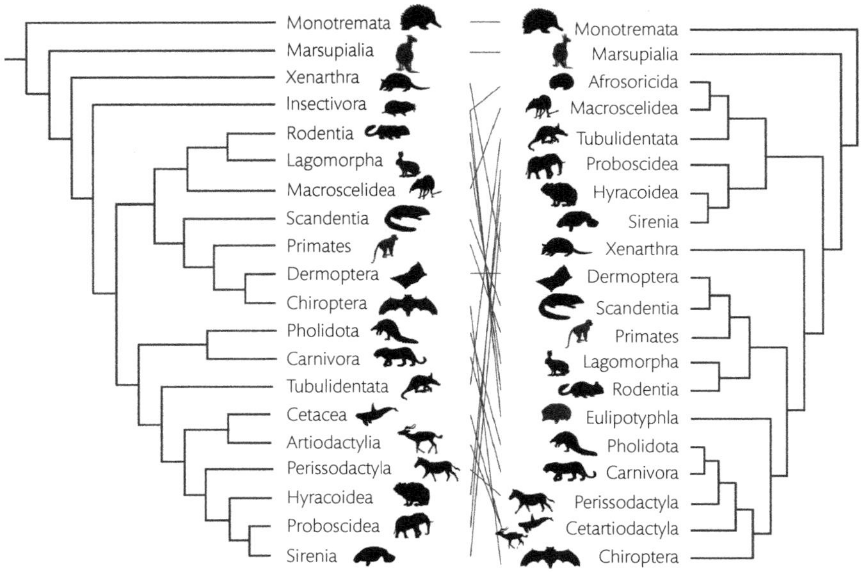


**Figure 5.3** Comparison of a morphological-based (left) and molecular-based (right) phylogenetic tree of mammals, highlighting the major divergences in the two approaches. Three new clades appear in the molecular tree, and there is considerable rearrangement of other lineages (see connecting lines). Modified from Springer et al. (2004).

Molecular phylogenies often provide strong support for conventional trees, but sometimes indicate quite different phylogenetic arrangements than were previously considered. For example, a well-accepted conventional phylogeny of placental mammals (a sister group to marsupials, with monotremes representing an earlier divergence) consists of a xenarthran lineage first, then insectivores, then a lineage consisting of three major clades: the Anagalida, Archonta, and Ungulata (Figure 5.3, left panel; Springer et al. 2004). A molecular tree confirms much of the major divergence sequence for mammals—monotremes diverging first, then placentals and marsupials as sister taxa—but in contrast to the conventional tree, the molecular tree has placentals arranged in four major clades: Afrotheria, Xenarthra, Euarchontoglires, and Laurasiatheria (Figure 5.3, right panel; see also section 1.1.1). This molecular phylogeny includes some fairly radical rearrangements for some taxa (e.g. elephants and hyrax).

Sometimes molecular phylogenies provide unexpected relationships. For example, bats were traditionally arranged into two sister clades, the 'micro' bats (Microchiroptera) and the 'mega' bats (Megachiroptera). These clades differ in many characteristics, including body size and the laryngeal mechanisms that produce echolocating clicks (Springer et al. 2004). However, this traditional phylogenetic arrangement was challenged by data for penis morphology (Smith

& Madkour 1980) and retino-tectal projection pathways (Pettigrew et al. 1989), which allied the megabats with primates rather than microbats. Although subsequent molecular and morphological phylogenies provided support for the original sister-group status of micro- and megabats, molecular studies surprisingly showed that bats from the superfamily Rhinolophoidea were more closely related to megabats than they were to other microbats, thus making the original Microchiroptera paraphyletic (Pumo et al. 1998; Teeling et al. 2005). As pointed out by Teeling et al. (2005), this suggests that the laryngeal echolocation structures either evolved separately in the rhinolophoid and other microbats, or evolved once in the common ancestor to those and the megabats, and was subsequently lost in the megabats. Morphological and fossil evidence favours the latter hypothesis (Springer et al. 2004). The distance-based 'supertree' phylogeny (Bininda-Emonds et al. 2007) for 4,510 (of the then-recognized 4,554) extant mammalian species suggests that the major mammalian orders and superorders were present about 93 MYBP, and that the rate of accumulation of extant lineages has accumulated relatively linearly since the Cretaceous-Tertiary boundary.

## 5.1.2 Phylogenetic Methods

Phylogenetic analysis can be applied to discrete or continuous data. Discrete trait data can take only certain discrete values (e.g. 0 or 1 for a dichotomous trait, or 0, 1, 2, …, n for an n-dimensional trait). For the example in Figure 5.2, diet is dichotomous (0 for omnivory, 1 for carnivory) as is the type of parental care (0 for non-communal, 1 for communal). Continuous trait data are numeric values that can take any value along a continuous scale (e.g. body mass, morphometric measurements of body parts, metabolic rate, water loss, many life history characteristics, running speed). The general assumption that would be made about the data is that closely related species would not be completely independent points (e.g. species within a genus would have similar body mass, whereas families within orders would be more independent, and dissimilar in mass). Therefore, closely related species cannot be assumed to be independent data points (independence of data points is a common assumption in most statistical tests); a sample of, say, 100 mammal species does not necessarily represent a sample of 100 independent data points. The solution is to use statistical methods that, based on the phylogenetic tree, control for the potential non-independence of data points.

Whether there is a phylogenetic signal for a particular trait of interest clearly depends on the nature of the trait in question, the number of species being investigated, and the particular phylogeny. Nested analysis of variance for taxonomic hierarchies (see earlier example) is one such indicator of phylogenetic signal, but the somewhat arbitrary nature of taxonomic hierarchies compared to a phylogeny, and lack of branch length information, makes phylogenetic-based measures more desirable.

There are various approaches to calculating the phylogenetic signal (Blomberg et al. 2003), including the use of autoregression coefficients (Cheverud et al. 1985), Moran's index (Gittleman & Kot 1990), serial independence (Abouheif et al. 1999), Grafen's $\rho$ (Grafen 1989), and Pagel's $\lambda$ (Pagel 1992). Blomberg's K (and K*) parameter is another useful measure of phylogenetic signal; it reflects the ratio of the trait's mean square error when accounting for phylogeny (MSE) relative to the mean square error when phylogeny is ignored ($MSE_0$). If the phylogenetic structure is closely related to the variance-covariance structure of the trait values, then MSE is low and $K = MSE_0 / MSE$ is high, indicating a strong phylogenetic signal ($K \sim 1$); that is, the phylogeny accounts for all the variance in the trait. Alternatively, if the trait values are not closely associated with the phylogenetic structure, then MSE is high and K is low, indicating a weak phylogenetic signal. K* uses the MSE*, calculated for a star phylogeny using the mean trait value (rather than the phylogenetically calculated mean value); it is highly correlated with K, so either K or K* can be used as a measure of phylogenetic signal. K (or K*) $= 0$ indicates no phylogenetic signal; K (or K*) $= 1$ indicates a perfect relationship between the trait value and the phylogeny, and K (or K*) $> 1$ indicates that closely related species are more similar than expected under the Brownian motion model of evolution for the phylogeny. The significance of a K (or K*) value can be estimated using a randomization test. Phylogenetic signal representation (PSR) provides curves that indicate whether traits are evolving slower or faster than expected by Brownian motion; the PSR area under the curve is highly correlated with Blomberg's K (Diniz Filho et al. 2012).

In general, most biological traits have some phylogenetic signal. From analysis of 121 traits of various kinds, from 35 different phylogenies of animals and plants, Blomberg et al. (2003) found that behavioural traits had the lowest K values (0.355), compared to physiology (0.537), life history (0.631), and morphology (0.708); body mass had the strongest phylogenetic signal (0.832). Thus, behavioural and ecological traits generally have smaller phylogenetic signals than morphological and physiological traits, suggesting that the former are more 'plastic' as new species evolve (Blomberg et al. 2003).

One approach to analysing trait data and accounting for the relatedness of species is independent contrasts (IC; Felsenstein 1985; Figure 5.4). Independent contrasts are the differences in trait values between closely related species or ancestral nodes. Contrasts can be calculated for a single trait, but often two traits are analysed to examine the regression relationship between them. For example, in Figure 5.4, we have X and Y trait values for four species, with two sets of two-species sister groups, with a common ancestor. The contrasts (differences $C_1$, $C_2$, and $C_3$) provide independent comparisons of trait values, under a Brownian motion model of evolution. The X contrasts are all positivized, and the sign of the corresponding Y contrast is adjusted if necessary. When graphed, the slope of the regression between the X and Y contrasts is independent of the phylogenetic relationships.

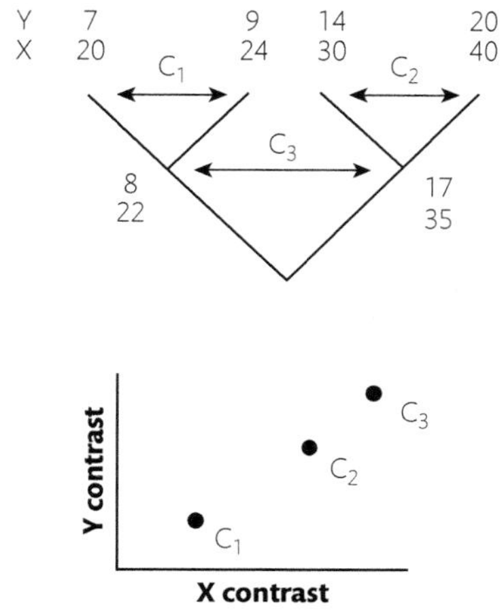


**Figure 5.4** Simplified example of independent-contrasts calculations for two traits (X, Y) for four species consisting of two, two-species sister groups with a common ancestor. Measured trait values for the four species are given, along with the calculated node values. The pattern for calculating contrasts (differences between species of node pairs, C) is shown in the top panel, with values for $C_1$, $C_2$, and $C_3$ for X and Y traits; these contrasts for X and Y traits are plotted in the lower panel. Branch lengths are ignored. Modified from Harvey and Pagel (1991). Reproduced with permission of Oxford University Press.

The regression slope is calculated not using ordinary least squares regression (which assumes no measurement error in the X variable), but a structural relations model, which accounts for error in both the X and Y contrasts; the regression is also forced through the origin, because there is no contrast in Y when there is no contrast in X (Harvey & Pagel 1991).

Independent contrasts are differences between species/nodes, and there is one less contrast than species; consequently, contrasts cannot be examined (e.g. by ANOVA) for patterns. However, species residuals (ε) from the phylogenetically corrected regression can be obtained by fitting a line, with a slope equal to that obtained from the IC regression, to a graph of the raw X and Y data, forcing this line through the mean X and mean Y value, and then calculating the residual for each species from this line (Garland & Ives 2000). The ε values can then be analysed according to other biological traits (e.g. diet) or environmental variables (e.g. mean annual temperature or rainfall) to examine patterns that are independent of phylogenetic relationships.

Many other phylogenetic methods are available (Cheverud & Dow 1985; Harvey & Pagel 1991; Diniz-Filho et al. 1998, 2012; Rohlf 2001), including phylogenetic generalized least squares regression (PGLS), autoregression (AR), phylogenetic vector regression (PVR), and phylogenetic signal-representation (PSR). PGLS is essentially a weighted regression approach for two (X and Y) traits, using a variance-covariance matrix (derived from the distance matrix mentioned earlier) as weights for the regression. Although conceptually quite different from IC, PGLS is numerically equivalent to IC and yields the same estimates for regression slope, species residuals, etc. (Garland & Ives 2000; Rohlf 2001).

Autoregression, like IC, can be used to analyse a single trait for the 'phylogenetically predicted component' of the trait (i.e. the trait value for each species that is correlated with the phylogenetic relationship) and the 'phylogenetically independent component' (i.e. the residual ε for each species of the actual trait value from the phylogenetically predicted value). The basic model for AR is $Y = \rho WY + \varepsilon$, where ρ is a scaling coefficient, obtained by iterative calculation using a predetermined ρ to minimize the εs; W is a matrix of the reciprocals of the distance matrix; and Y is the vector of species trait values. Regression analysis of phylogenetically corrected X and Y values will then provide a slope and a set of residual values that are independent of phylogenetic relationships.

PVR uses the significant eigenvalues and eigenvectors of the distance matrix, transformed by $-0.5d_{ij}^2$, where $d_{ij}$ is the distance between species *i* and *j*, and the eigenvectors (scaled by the eigenvalue$^{0.5}$) are regressed against the trait. The trait values predicted by the multiple regression are the phylogenetically predicted values, and the residuals are the phylogenetically independent values (εs). Regression analysis of X and Y ε values provides a slope and species residual values that are independent of phylogenetic relationships. In PSR, sequential PVR models are obtained by successively increasing the number of eigenvectors. The graph of the regression coefficient of determination ($r^2$) against the number of eigenvalues (PSR curve) is expected to be linear under Brownian evolution, and its shape is expected to change under alternative evolutionary models. Garland et al. (1993) used Monte Carlo computer simulations to incorporate phylogenetic information into ANOVA or analysis of covariance (ANCOVA) analyses.

Different phylogenetic methods have their particular advantages and disadvantages (Rohlf 2001). IC is conceptually elegant, but it refers to differences between species (or ancestral nodes), not to the species themselves. Consequently, it is possible to graph contrasts for Y against contrasts for X, but not species Y values independent of phylogeny against species X values independent of phylogeny. Predicted Y values after phylogenetic correction by IC need to be calculated from the IC regression line fitted to the raw data; it is possible then to calculate species Y residuals, but not species X residuals. PGLS is a very different computational method than IC, but gives the same results, and has the same limitation with obtaining species predicted Y and residuals. Autoregression and PVR, in contrast, enable the calculation of both X and Y for species, independent of phylogeny, and phylogenetically informed residuals for both Y and X.

In general, the allometric slopes for traits resulting from phylogenetically corrected analyses are different from those resulting from non-phylogenetically informed analyses. However, the direction and significance of the difference varies between different clades, and partly depends on the strength of the phylogenetic signal for BMR, measured by λ (Freckleton et al. 2002) or K and K* (Blomberg et al. 2003), as well as on tree topology and especially on estimated branch lengths (Sieg et al. 2009; Capellini et al. 2010; White et al. 2009; White 2011). Although there is criticism of the use of phylogenies and assumptions of various phylogenetic methods, it is clear that some kind of taxonomic- or phylogenetic-informed analysis, even when the information about the tree topology and branch lengths is incomplete or approximate, is better than non-phylogenetic analysis (e.g. Capellini et al. 2010; White 2011; McNab 2012, 2015). It is unlikely that inclusion of new data points will change the conclusions obtained using current analyses, but the development of improved phylogenetically informed analyses using alternative line-fitting methods (such as nonlinear regressions or polynomial relationships) might change these conclusions in future studies.

Many phylogenetic analysis programs are freely available in the R environment (R Development Core Team 2013); examples include APE (Analyses of Phylogenetics and Evolution), PHYTOOLS (Phylogenetic Tools for Comparative Biology), GEIGER (Analysis of Evolutionary Diversification), CAPER, PICANTE (R tools for Integrating Phylogenies and Ecology), and PALAEOTREE (Palaeontological and Phylogenetic Analyses of Evolution). Other freely available programs include PDAP (Phylogenetic Diversity Analysis Programs), PAUP (Phylogenetic Analysis Using Parsimony), PHYLIP, Mesquite, and MacClade. PDAP includes a number of programs for IC analysis (PDTREE), phylogenetic ANOVA (PDANOVA), and Monte Carlo simulation of various evolutionary models (PDSIM).

### 5.1.3 Allometry and Scaling

Even after phylogenetic relationships are accounted for, many physiological (and other) traits are strongly related to body mass (allometry) and other biological traits (scaling). The general allometric form of comparative relationships is $Y = a\,X^b$, which is log-transformed to the linear relationship $\log(Y) = \log(a) + b\log(X)$ (i.e. $Y' = a' + b\,X'$). The linear form is much more amenable to statistical analysis using well-developed regression models (ordinary least squares and structural models; Harvey & Pagel 1991); log-transformation tends to equalize the variances in Y at different X. A widespread use of allometric regression is to calculate residuals from the regression; the residuals can then be analysed for patterns, having removed the effect of body mass. For example, Withers et al. (2006) found similar allometries for the BMR of marsupials using both conventional and autoregression analyses (Figure 5.5), and used regression residuals to show that BMR was related to environmental rainfall variability.

However, a caveat to this approach is that selective forces other than mass could contribute to, and be obscured by, the allometric relationship (Harvey & Pagel 1991). For example, there is a strong allometric relationship between antler size and body mass for deer, but strong sexual selection favouring large antler size could simultaneously increase body mass; the resulting strong allometric pattern might suggest that antler size is determined only by body mass, and the importance of sexual selection would not be evident. Predicted Y values from a regression are used to represent the expected mass effect, but they are biased by the log transformation; anti-log transforming the predictions to the original scale requires correction for

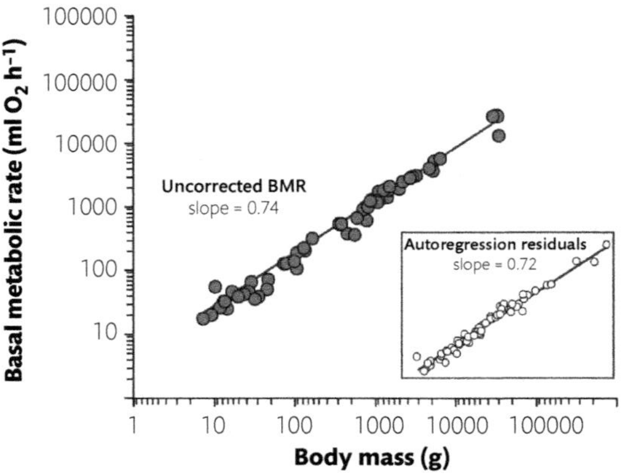


**Figure 5.5** Regression relationship for conventional and autoregression (inset) of $\log_{10}$ basal metabolic rate and $\log_{10}$ body mass for marsupials, with conventional slope of 0.74 and autoregression slope of 0.72. Modified from Withers et al. (2006).

this bias (Hayes & Shonkwiler 2006). The analysis of log-transformed biological data has been criticized as being potentially biased and misleading (Packard & Boardman 2008).

Nevertheless, there are many examples of the informative use of allometric analysis for comparison of physiological and other biological traits. For example, the historical development of the understanding of allometry of mammalian BMR over time shows an increasing sophistication in the recognition of phylogenetic differences and comparative analyses. Kleiber (1932) examined the allometry of BMR for birds and mammals combined, then Kleiber (1947) restricted the data set to mammals only, and extended the sample size to 26 individuals (of 12 species), to find that BMR (kcal $day^{-1}$) $= 67.6\,kg^{0.756}$.

Harvey and Pagel (1991) used IC to examine the allometry of mammals; their structural regression of contrasts for body mass and BMR has a major-axis slope of 0.75, conforming nicely to the typical value commonly observed for mammals. Dawson and Hulbert (1970) recognized further taxonomic differences in BMR (and other physiological variables such as body temperature, $T_b$) for marsupials and placentals, and calculated separate regressions. They also calculated that monotremes (echidna, *Tachyglossus aculeatus*) and marsupials had a similar mass-specific BMR when corrected to a common body temperature ($T_b$) of 38°C, but that it was lower than placentals (and also non-passerine and passerine birds). Dawson et al. (1979) then determined that *Tachyglossus* and the other echidna genus (*Zaglossus*) had BMRs considerably lower than marsupials, whereas the platypus (*Ornithorhynchus anatinus*) had a BMR that was only slightly lower than that of marsupials.

Separate allometries have been proposed for various placental taxa (see Harvey et al. 1990; Withers 1992; Table 3.1) and mammals from different zoogeographic regions (Lovegrove 2000). White et al. (2009) and Capellini et al. (2010) rejected the concept of a single common allometric slope (e.g. about 0.75) for all mammals, finding differences in slope for different families, and for small and large mammals. They also found that taxonomic level (genus and family) did not contribute significantly to variation of BMR contrasts.

## 5.2 Mass, Temperature, and Humidity

Possibly the two most common measurements made by ecophysiologists are the body mass of animals and their temperature. Body mass affects many important physiological and ecological aspects of how animals work in their environment, through isometric or allometric relationships. Similarly, temperature affects many important physiological and ecological aspects of how animals work in their environment through the kinetic effect of temperature on chemical reactions ($Q_{10}$ effect) and the energy cost of maintaining a temperature gradient between the body core and the environment. Consequently, it is important to appreciate how mass and temperature are measured, because both are fundamental to how mammals work.

### 5.2.1 Mass

Body mass (M) is probably the most useful and commonly measured biological property of animals (see 2.1). It is necessary to distinguish between mass (SI unit is kilograms) and weight (W; kilograms force or Newtons), because it is M that is physiologically and ecologically relevant. Mass reflects the amount of material in an animal, whereas W is the force acting on the animal due to gravity; W = M a, where a is the gravitational constant (9.8 m $s^{-2}$). The gravitational force acting on a mass of one kilogram is 9.8 Newtons (= 1 × 9.8). The value of g varies only slightly over the Earth's surface. In zero gravity (e.g. in orbit in outer space), an animal has its normal mass but zero weight; nevertheless its basic physiology remains the same, hence it is mass not weight that determines the basic physiology and ecology of animals (nevertheless, zero gravity does have some physiological effects on animals, such as the function of their circulatory system). Earth-bound ecophysiologists determine the mass of animals by measuring their weight, ignoring the small geographic variations in the gravitational constant.

The measurement of mass is potentially one of the most accurate measurements that can be made on an animal, because of the high accuracy and precision of mechanical and—particularly modern—electronic balances. However, biological variability results in considerably greater inaccuracy and imprecision. For example, animals might not remain completely inactive during mass measurement, resulting in variation in their recorded mass; they may also vary in mass temporally due to short- or long-term changes in their mass balance (e.g. drinking, food consumption, urination, defaecation, evaporation, growth, fasting, or starvation).

Measurement of mass is now routine and accurate in small animals or biological samples of 1 mg or so, up to 100 kg or so. Even smaller animals or samples (e.g. a microgram) can be measured accurately using electronic microbalances.

Measuring larger masses is more problematic, and can be dangerous. Commercial high-load scales are available, and while infrastructure and planning are required, the weighing of large mammals after they are sedated is reasonably easy using a block and tackle to raise and suspend the mammal in a sling suspended from the scale. To facilitate weighing, an A-frame houses the scale, under which a vehicle transporting the animal can be positioned (Figure 5.6). Weighbridges used to measure the load on trucks can be used if the tare mass of the empty truck carrying an animal can be measured accurately. Benedict (1936), for his classic study of the physiology of Asian elephants (*Elephas maximus*), either estimated mass or used scales; the elephant used in his metabolism studies ('Jap') was weighed on the Campgaw (New Jersey) certified municipal scales.

Body mass can also be estimated without handling an animal via photogrammetry, using photographs or 3D images/models of animals, calibrated using animals of known mass (e.g. Haley et al. 1991). Body volume estimated from about 10 photographs per individual, and converted to mass using a density of 1,010 kg $m^{-3}$ for 14 large mammal species (25–4,060 kg), closely predicted body mass ($r^2$ for predicted versus actual mass was 0.91–0.99) for various large mammal species (Postma et al. 2015).

Body mass can also be estimated from osteological measurements, a method that is particularly useful for estimating the body mass of fossil specimens, using the mass and osteological measurements for extant specimens of mammals with similar body proportions. For example, Christiansen (2004) used data from extant elephants to estimate the mass of fossil proboscideans, some of which were much larger than extant species. A simple approach is to estimate mass (M, kg) from shoulder height (SH, cm): $\log(M) = -4.016 + 3.161\log(SH)$, but osteometric measurements (such as long bone length and diaphysis proportions) are better predictors of mass.

**Figure 5.6** African buffalo (*Syncerus caffer*) suspended in a sling below an A-frame by a winch system with a calibrated electronic scale between the sling and the suspension system. The mass of the empty sling and cabling is subtracted to obtain the mass of the animal. Photo courtesy of Arista Botha.

### 5.2.2 Temperature

Thermal environmental conditions hold considerable significance for most hierarchical levels of biological function. Because environmental temperature varies in time and space at different times and scales, organisms are continuously challenged to maintain homeostasis. Thus thermal physiology may be a significant factor underpinning the ecological and evolutionary success of animals. The environmental temperature impacts on most aspects of a mammal's physiology, and an animal's core $T_b$ has considerable biological significance through the $Q_{10}$ effect (see 2.4.1). Environmental air temperature is probably the most useful and commonly measured ambient environmental variable. It is one of many important meteorological measurements, along with relative humidity (RH), barometric pressure, wind speed and direction, and precipitation (Harrison 2015). The SI unit for temperature is the degree Kelvin (°K), but by convention biologists use the Celsius (°C), or Centigrade, scale. There are 100°C between the original reference points of the melting and boiling temperatures of water. The Kelvin is an absolute scale, with $0°K(-273.16°C)$ occurring when molecules have no kinetic energy. A difference of $1°C = 1°K$, and the scales can be interconverted easily, with $°C = °K + 273.16$. It is not meaningful to express °C values as ratios (i.e. 20°C is not really twice as high a temperature as 10°C).

A variety of mechanical and electrical methods are used for temperature measurement; they vary in size, complexity, cost, and application. Bimetallic thermometers consist of two metals, with differing coefficients of thermal expansion; when fixed at one end, the bimetallic strip bends in proportion to the temperature. These are mostly used for ambient temperature measurement. Glass thermometers are fine-bore glass tubes filled with mercury (or alcohol), which expands more than the glass when warmed, thereby rising up the bore. They are very portable and relatively accurate, but fragile, and some contain mercury (and are hence a health hazard). Schultheis cloacal rapid reading mercury thermometers are smaller thermometers with a very fast response time. They were used extensively by physiologists to measure the body temperature of small animals, particularly reptiles (e.g. Pianka 1970).

Thermocouples and thermistors have largely supplanted glass thermometers, as they provide a fine measuring probe, and their electronic meters have become miniaturized, portable, accurate, and economical. Thermocouples are a traditional electrical methodology consisting of two dissimilar metal wires joined in a circuit; a voltage is generated in the circuit that is proportional to the temperature difference between the ends. One end is generally placed at a reference temperature (commonly an ice bath); the voltage is proportional to the temperature at the other end. Copper-constantan (Type T) and chromel-alumel (Type K) are commonly used for the biological temperature range.

Thermistors are temperature-sensitive resistors; their resistance varies directly (PTC) or inversely (NTC) and non-linearly with temperature. Resistance temperature detectors, such as platinum resistance thermometers, are suitable for very precise temperature measurement (e.g. Pt100 sensors have a resistance of 100 Ω at 0°C, and changes by 0.00385 Ω°C$^{-1}$). Temperature-sensitive silicon diodes have a stable thermal coefficient of about 2.3 mV °C$^{-1}$. These various electrical temperature sensors require some electronic circuitry and a meter for display. Temperature-sensitive radio-transmitters and data loggers have been routinely used to remotely record the body temperature of mammals in the laboratory and field (see 5.4).

A common measure of environmental temperature is the shade air temperature (dry bulb $T_a$), as is routinely measured in the standard Stevenson screen weather station. However, although this $T_a$ measure is commonly used and useful, it does not necessarily reflect the thermal environment experienced by an animal because heat exchange depends also on humidity, air movement, and the radiation temperature. The IUPS Thermal Commission (2003) provides a comprehensive glossary of terms for thermal physiology. The wet-bulb temperature ($T_{wb}$) is the temperature measured using a thermometer with a wet cover; it is lower than dry-bulb $T_a$ ($T_{db}$) because of evaporative cooling if relative humidity (RH; see 5.2.3) is < 100%. For example, a sling psychrometer is spun to maximize evaporation from the wet bulb. Standard tables can be used to calculate RH from $T_{db}$ and $\left(T_{db} - T_{wb}\right)$.

The black globe (or black bulb) temperature ($T_{bg}$) is the temperature measured at the centre of a 15-cm-diameter copper sphere painted flat black. The globe thermometer is influenced by the radiation temperature and the wind speed in addition to the air temperature. If radiation temperature is high (e.g. via exposure to solar radiation, which can exceed 1,100 W m$^{-2}$ in dry arid areas) and wind speed is low, the globe thermometer will equilibrate at close to the radiation temperature. With an increase in wind speed, the globe temperature moves closer to the air temperature. Miniglobes (miniature black globe thermometers) can be used to calculate $T_{bg}$; they can be deployed on the collar of large mammals (Hetem et al. 2007). The wet-bulb-globe temperature ($T_{wbg}$) is calculated from $T_{bg}$ and $T_{wb}$, and so accounts for the capacity of the environment to accept evaporation; it is a useful index of the heat stress level for humans and other animals. $T_{wbg}$ can be calculated from a relationship between $T_a$, $T_{wb}$, and $T_{bg}$, or from a complex algorithm including $T_a$, wind speed, dew point, solar irradiance, direct and diffuse solar radiation, and zenith angle (Dimiceli et al. 2011).

The effective operative temperature ($T_e$) of an object (or of an animal that lacks physiological thermoregulatory mechanisms) in a particular thermal environment better reflects the functional temperature for the object than does air temperature. $T_e = T_a + \Delta T_r$ where $\Delta T_r$ reflects the increase (or decrease) in temperature of the object required to dissipate any excess radiative heat gain (or loss) by convection and radiation (Bakken 1992; Dzialowski 2005). $\Delta T_r$ is $(Q_a - A_e \sigma \varepsilon T_a^4) / (C_H + C_R)$, where $Q_a$ is radiative heat load, $A_e$ is effective radiative exchange area, $C_H$ is convective, and $C_R$ is the radiative conductance of the animal (W °C$^{-1}$). $T_e$ can be measured using a taxidermy mount or thermally appropriate model. Standard operative temperature ($T_{es}$) is the effective ambient temperature relative to the body temperature ($T_b$) of an animal, corrected for its heat production that must be dissipated by thermal loss; $T_{es} = T_b - (MPH - EHL) / (C_H + C_R)$, where $(MHP - EWL)$ is metabolic heat production minus evaporative heat loss. Measurements for a small rodent (*Thallomys nigricauda*) of black globe temperature and $T_e$ using copper models suggest that $T_{bg}$ approximates $T_e$ in summer and in winter, although the measures diverged during the day in full sun, and there was

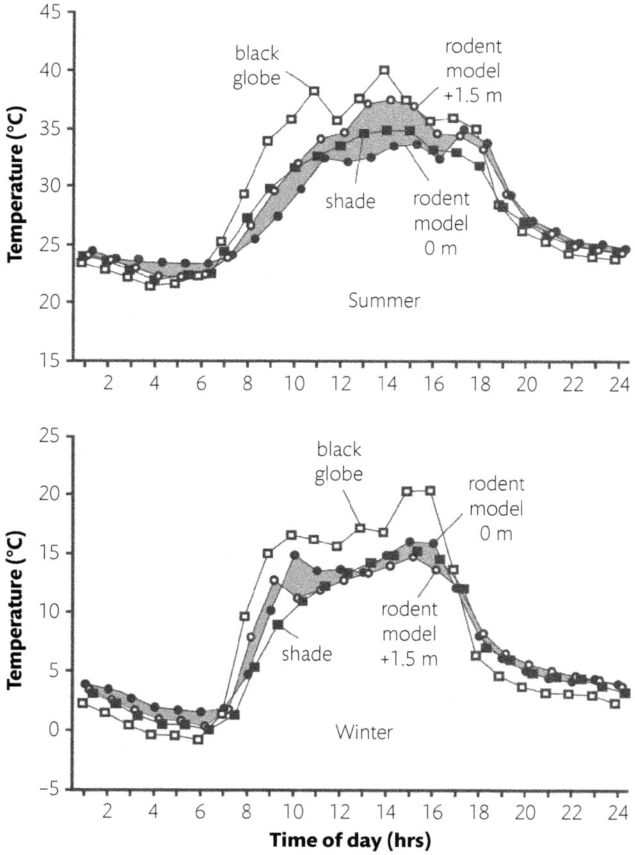


**Figure 5.7** Diurnal cycle in summer (top) and winter (bottom) shade air temperature (black squares), black globe temperature (white squares), and physical copper models (shaded area) of the arboreal rodent *Thallomys nigricauda* located at ground level (black circles) and +1.5 m height (white circles)), at Molopo, South Africa. Modified from Coleman and Downs (2010).

a difference in the $T_e$ measured by models placed at ground level and tree height of 1.5 m over the entire daily cycle (Coleman & Downs 2010; Figure 5.7).

Infrared thermography (IRT) determines the surface temperature of objects based on the physics of electromagnetic radiation emitted by all objects warmer than 0°K. For objects at biologically relevant temperatures of about 300°K (27°C), the peak radiation is in the infrared region. Thermal imaging systems have been used widely in the physical sciences (e.g. climatography, oceanography, geography, meteorology; Kuenzer & Dech 2013). Thermal imaging cameras have also been used to determine the surface temperatures of animals, especially birds and mammals (because they are endotherms) since the 1960s (McCafferty et al. 2011). Such IR images show clear patterns of cold and hot areas on the body surface of mammals, and are useful for graphically interpreting the patterns and sites of radiant heat loss (e.g. the presence and role of thermal windows). In addition, since a considerable (and potentially quantifiable) fraction of a mammals' heat loss is IR radiation, IR images can be used to estimate the heat loss of mammals and hence estimate their heat production (i.e. metabolic rate; McCafferty et al. 2011).

For example, Cross et al. (2016) used infrared imaging to estimate the extra heat production required by wolves in Montana that were infected with mange (a skin

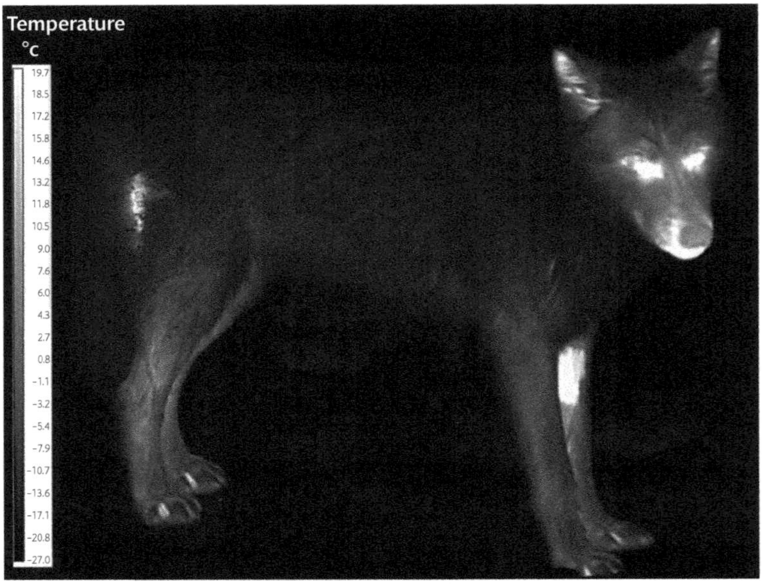


**Figure 5.8** Thermal infrared image of a wolf (*Canis lupus*) showing surface regions with relatively high radiative heat loss from natural thermal windows (eyes, snout, ears, and between the toes) and shaved patches on foreleg and rump. Image provided by Paul Cross and Catherine Hasse (U.S. Geological Survey).

disease caused by parasitic mites that causes hair loss, and thus loss of insulation from mammals). By taking IR pictures of wild wolves during the night (Figure 5.8) and measuring environmental variables (air temperature, radiation temperature, and wind speed), the radiative and convective heat loss from the wolves could be estimated. The method was validated by shaving patches from captive wolves and using IR to estimate the heat loss from those animals. Assuming that the wolves maintained their core temperature, the metabolic heat production was calculated, and severe mange was found to increase metabolic requirements by 5–12 MJ $day^{-1}$, in the context of a daily energy expenditure of 12–22 MJ $day^{-1}$ for a healthy wolf. Based on these estimates, a typical wolf pack of 40 individuals, with 20% of the pack infected with mange, would need to harvest an extra 12 elk per winter, compared to the 300 elk per winter for a pack with no mange-infected individuals. The authors comment on the irony of the cascading effect of a parasite that was originally introduced to eradicate wolves and protect ungulates, but which may in fact be doing the opposite.

Speakman and Krol (2010) suggested that in at least some conditions, a limit on energy processing might be determined by the ability of the animal to deal with the excess heat associated with the inefficiency of energy assimilation and transduction. There was a lack of effect of various manipulations on altering the energy transfer in milk from mouse mothers to pups, and their finding that shaving a patch of fur from the mothers (thereby allowing heat to be dissipated more easily) increased the transfer of energy in milk. Similarly, mouse pups grow more rapidly when pups and mother are housed at 22°C, below the thermoneutral zone, than within the thermoneutral zone at 30°C (Maloney & Sorenson, unpublished data). Speakman and Krol's hypothesis has led to the reinterpretation of many known facts into the context of animals dealing with the heat of inefficiency, trading off thermal limits with productivity. For example, it has long been known that high ambient temperature leads to reduction in reproductive activity in production species (Gwazdauskas 1985), to a reduction in milk output in dairy cattle (West 2003), and also to the abandonment of foraging activity in many ruminant mammals (Maloney et al. 2005).

### 5.2.3 Humidity

Humidity is a measure of the water content of air. Water vapour in air, like other gas components, has a water vapour partial pressure (wvp; kPa), but a more common measure of the water content of air is the absolute humidity (AH; mg $H_2O$ $l_{air}^{-1}$). When saturated with water vapour, the AH and wvp increase exponentially with air temperature. At any given $T_a$, the RH (%) is the wvp (or AH) expressed as a fraction of the saturation wvp (or AH). Relative humidity is a useful measure of the water content of air and the driving force for evaporation at a particular $T_a$, but is problematic at varying $T_a$, because the driving force for evaporation is the difference in wvp (or AH), not RH, between the evaporating surface and the air. For example, the cutaneous evaporative water loss of endothermic mammals and birds theoretically depends on the wvp difference between the skin surface—based on relative humidity (RH) at the skin surface ($RH_{skin}$) being 100% and saturation wvp at $T_{skin}$—and the ambient air, based on ambient RH and saturation wvp at $T_a$; that is, $EWL \propto (RH_{skin}\ wvp_{saturation\ at\ Tskin}) - (RH_{ambient}\ wvp_{saturation\ at\ ambient\ Ta})$, not $(RH_{skin} - RH_{ambient})$.

The diffusion of water vapour in air, or through a biological integument, is determined by the same physical principles as other gases (Monteith & Campbell 1980), with the driving force equal to the difference in wvp (Δwvp) or AH (ΔAH). The rate of diffusion ($J_{H2O}$) is determined by the water vapour permeability (water vapour flux per Δwvp; $P_{H2O}$), or resistance ($R_{H2O}$; s $cm^{-1}$); $J_{H2O} = -\Delta AH\, P_{H2O} = -\Delta AH / R_{H2O}$. The $P_{H2O}$ and $R_{H2O}$ are generally properties of the particular biological system and not the gradient driving evaporation, but hygroscopic materials such as vertebrate skin (and snail epiphragm) vary markedly in water content at different RH, hence $P_{H2O}$ and $R_{H2O}$ (Barnhardt 1983; Blank et al. 1984;Withers 1998b; Lillywhite 2006). The mammalian epidermis generally has a resistance of about 100–300 s $cm^{-1}$, which is a substantially higher resistance than that of a free water surface or the skin of amphibians. The resistance of mammalian skin is similar to that of cocooned and waterproof amphibians, many reptiles and birds, but considerably less than that of xeric-adapted reptiles (see 2.7.2).

There are various methods for measuring humidity. Hair hygrometers have routinely been used in hygrothermographs to measure environmental RH (e.g. in the weather station Stevenson screen); horse (or human) hair changes in length depending upon the relative humidity, and this change in length is typically recorded as an ink-tracing on a rotating mechanical drum. A wet-dry bulb thermometer was also often used to measure environmental RH (see earlier). A dewpoint hygrometer determines the temperature at which water vapour condenses (or freezes), and hygrometric calculations are used to convert dewpoint °C to absolute humidity. This can provide a very accurate measure of RH, depending on the accuracy of the thermal measuring device to determine dewpoint, and the accuracy of the equations used to convert dewpoint and $T_a$ to RH (e.g. Parish & Putnam 1977).

Later, thin-film capacitance hygrometers came into common use for measuring RH; RH is inferred from the change in capacitance between two electrodes separated by a hygroscopic material. Dewpoint and capacitance hygrometers are typically electronic laboratory or hand-held units, but miniature loggers are available (e.g. iButton temperature/RH loggers; Maxim Corp.). The relative humidity of an air space can be controlled by equilibration with a saturated salt solution; the RH depends on the nature of the saturated salt and temperature (Wexler & Hasegawa 1954); saturation RH at a $T_a$ of about 20°C is 33% for LiCl, 56% for $Na_2Cr_2O_7$, 76% for NaCl, 81% for $(NH4)_2SO_4$, and 93% for $KNO_3$.

Ambient humidity has a significant influence on mammals, through its potential effects on EWL and hence water balance (2.7, 3.6.1) and heat balance (2.4). In the laboratory, EWL is often measured by closed or flow-through hygrometry (see 5.3.1).

In flow-through hygrometry, air of known AH is passed through a chamber containing an animal, and the excurrent AH is measured so that the EWL can be calculated from the change in AH, accounting for the air flow rate (cf. flow-through respirometry for measurement of $O_2$ consumption and $CO_2$ production; 5.3.1). One problem with measurement of EWL using flow-through respirometry is that the washout of residual water in the respirometry system is relatively slow (compared to $O_2$ and $CO_2$), so it becomes difficult and time-consuming to determine minimum EWL values.

EWL from the skin surface can be measured using closed capsules, holding a desiccant such as silica gel against the skin, or ventilated capsules held against the skin using a hygrometer to measure the change in AH. For example, Love and Shanks (1962) used a ventilated capsule to measure the sweat rate from the forearm of humans at 35°C of about 12 mg $H_2O$ $min^{-1}$ (equivalent to an R of about 4 s $cm^{-1}$; cf. Table 2.9, non-sweating values). Dunkin et al. (2013) used a ventilated capsule to show that cutaneous EWL of Asian elephants (*Elephas maximus*) increased with the Δwvp, as expected.

Saturation water vapour pressure can be calculated using Teten's formula, $wvp_{sat}(T) = a\,e^{bT/(T+c)}$, where T is the temperature in degrees Celsius (°C), with constants a (0.611 kPa), b (17.502), and c (265.5°C); if $T < 0$°C, then $b = 21.87$ and $c = 265.5$°C (Campbell & Norman 1998). Alternatively, values for $wvp_{sat}$ can be obtained from published tables (e.g. List 1971). The saturation water content (absolute humidity, χ; kg $m^{-3}$) can be calculated from other psychrometric equations; for instance, Parish and Putnam (1977), $\chi(\text{kg m}^{-3}) = k(T+273)^{-1}(T+273)^{a}10^{(c+(b/(T+273)))}$, where T is the temperature (°C), $k = 0.21668$, $a = -4.9283$ ($-0.32286$ over ice), $b = -2937.4$ ($-2705.21$ over ice), and $c = 23.5518$ (11.4816 over ice).

## 5.3 Energetics

The energy balance of animals is a fundamental aspect of their functioning. The chemical energy in food is the basis of the 'Fire of Life' (Kleiber 1975). Energy is required for maintaining the minimal essential functions of organisms when at rest, and to fuel the various activities that animals must undertake. Calorimetry is the study of heat balance; bomb calorimetry is used to measure the combustible energy of biological materials; indirect calorimetry measures the heat production of animals but through indirect measurement of $O_2$ and/or $CO_2$ exchange, and direct calorimetry measures the heat produced by animals (McLean & Tobin 1987; Lighton 2008).

### 5.3.1 Laboratory Energetics

In the 1770s, Antoine Lavoisier pioneered the measurement of metabolic rate in animals along with actually describing the chemical element oxygen (which he termed dephlogisticated air). Unfortunately, the French Revolution and the guillotine brought his scientific career to a premature end, as he was also a tax collector (Bensaude-Vincent 1996). Lavoisier devised methodologies for both direct calorimetry (measurement of heat production) and indirect calorimetry (e.g. measurement of $O_2$ consumption or $CO_2$ production).

As mentioned, direct calorimetry measures the actual heat production of an animal, and so measures both aerobic and anaerobic metabolism. Lavoisier measured it by the rate that an animal's heat melted ice in an ice jacket surrounding the chamber that housed the animal, but this is clearly not a generally practical (or particularly accurate) methodology. Today, direct calorimetry measures heat loss from an animal, accounting for the four possible modes of heat exchange: conduction ($HE_K$), convection ($HE_C$), radiation ($HE_R$), and evaporation ($HE_E$). When there is no storage (S) of heat in the animal (by decrease or increase in body heat content), total heat production is the sum of these four avenues of heat loss (Kaiyala & Ramsay 2011). The sum of $HE_K + HE_C + HE_R$ is termed the 'dry' heat loss; $HE_E$ is the 'evaporative' heat loss.

There are four types of modern direct calorimeter systems. Isothermal direct calorimeters measure the heat loss (e.g. using heat flow sensors) from a calorimeter maintained at a constant temperature. Heat sink calorimeters measure the sensible heat removed using a liquid heat exchanger (e.g. water circulating through a jacket; cf. Lavoisier's ice-jacket calorimeter). A direct convection calorimeter measures the thermal loss via air exiting an insulated calorimeter (e.g. the Snellen human calorimeter; Rearden et al. 2006). A differential calorimeter uses the rate of electrical heat generation in a control (no animal) chamber that matches the temperature increase in an identical chamber housing the animal.

Small-animal calorimeters are typically isothermal, using thermal flux sensors to measure the rate of heat loss across the walls of the calorimetry chamber containing the animal, thermometry to measure the heat lost by an increase in the temperature of the air flowing through the calorimeter, and hygrometry to measure the rate of heat loss by evaporation. For example, Walsberg and Hoffman (2005) used thermopiles (consisting of about 3,000 thermocouple elements) to measure heat flux across each of the six thick aluminium walls of their calorimeter; they also accounted for convective heat loss via the air flow through the chamber, and evaporative heat loss from their kangaroo rats. Burnett and Grobe (2013) describe a similarly designed small-animal calorimeter. Commercially available coolers can be 'reverse-engineered' to convert their Peltier effect cooling element into a heat flux sensor (using the Seebeck effect) and the cooler into a calorimeter (Wesolowski et al. 1985; Lighton 2008; Figure 5.9). Isothermal calorimeters have been scaled up to measure larger mammals, such as rabbits, pigs, and cattle.

Indirect calorimetry (also called respirometry) measures the consumption of reactants (e.g. $O_2$) or the formation of products (e.g. $CO_2$) of the chemical reaction for metabolism

$$C_6H_{12}O_6 + 6O_2 \rightarrow 6CO_2 + 6H_2O + 2{,}874 \text{ kJ mole}^{-1}$$

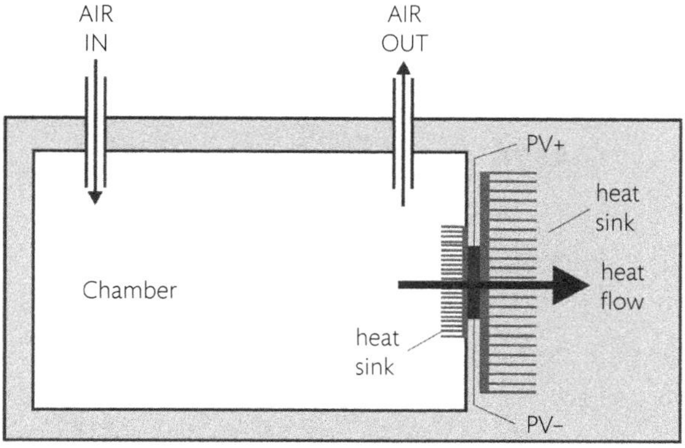


**Figure 5.9** Schematic of an isothermal calorimeter constructed from a thermoelectric cooler. Heat loss from the animal (which would be in the central chamber) passes through the inner, then outer, heat sinks, generating a voltage differential from the Peltier element, proportional to the rate of heat flux; PV+ and PV− are the voltage outputs from the Peltier element. Modified from Lighton (2008).

Relating these to metabolic heat production via the stoichiometry of the reaction. Although theoretically any of the reactants or products could be used to measure metabolic rate, in practice the rate of consumption of $O_2$ ($VO_2$) and/or the production of $CO_2$ ($VCO_2$) are the most feasible indirect measures of metabolism and so are by far the most commonly measured. Ideally, both $O_2$ and $CO_2$ are measured simultaneously, enabling accurate calculation (but see later, discrepancy from Hess's law) of metabolic heat production from the respiratory exchange ratio (RER; ratio of oxygen consumed to carbon dioxide production).

The respiratory quotient (RQ) is the $VCO_2/VO_2$ and varies with the biochemical stoichiometry of the substrates being metabolized; in steady state, RER = RQ, but in non-steady state, RER can be greater than RQ (e.g. hyperventilation) or less than RQ (e.g. a decreased body temperature increases the solubility of $CO_2$ in body fluids). Carbohydrate metabolism produces an RQ of 1, but RQ is 0.7 for lipid metabolism; protein metabolism has an intermediate RQ (about 0.81; Walsberg & Hoffman 2005). It is unusual, but possible, for RQ to be $< 0.7$ or $> 1$ (Kleiber 1975; Walsberg & Hoffman 2005). For example, if an animal is laying down fat, RQ can be greater than 1 (e.g. conversion of glucose to palmitic acid has RQ = 8). Note, however, that RER can be $< 0.7$ or $> 1$, even if RQ is in that range, if there is $CO_2$ retention or washout resulting from non-steady-state physiological effects.

Metabolic heat production can be calculated from $VO_2$ and RQ (or $VCO_2$ and RQ), using conversion coefficients that depend on RQ (Walsberg & Hoffman 2005; see 2.4.3). Joule equivalents and metabolic water production (MWP) can be interpolated for different mixtures of carbohydrate and lipid (Table 5.2). For protein metabolism, the joule equivalents are 19.2 kJ l $O_2^{-1}$ and 23.8 kJ l $CO_2^{-1}$.

**Table 5.2** Joule equivalents and metabolic water production (MWP) for metabolism of lipids (palmitate; RQ = 0.70) and carbohydrate (glucosyl units; RQ = 1.00), and interpolated mixtures of lipid and carbohydrate (see Table 2.5; Withers 1992).

| | RQ | $O_2$ equivalent (J/ml $O_2$) | $CO_2$ equivalent (J/ml $CO_2$) | MWP (mg $H_2O$/ml $O_2$) | MWP (mg $H_2O$/ml $CO_2$) |
|---|---|---|---|---|---|
| Palmitate | 0.70 | 19.5 | 27.9 | 0.565 | 0.808 |
| | 0.71 | 19.6 | 27.6 | 0.569 | 0.801 |
| | 0.72 | 19.6 | 27.3 | 0.572 | 0.794 |
| | 0.73 | 19.7 | 27.0 | 0.575 | 0.788 |
| | 0.74 | 19.8 | 26.7 | 0.579 | 0.782 |
| | 0.75 | 19.8 | 26.4 | 0.582 | 0.776 |
| | 0.76 | 19.9 | 26.2 | 0.585 | 0.770 |
| | 0.78 | 20.0 | 25.7 | 0.592 | 0.759 |
| | 0.79 | 20.1 | 25.4 | 0.595 | 0.753 |
| | 0.80 | 20.1 | 25.2 | 0.598 | 0.748 |
| | 0.81 | 20.2 | 24.9 | 0.601 | 0.742 |
| | 0.82 | 20.3 | 24.7 | 0.605 | 0.738 |
| | 0.83 | 20.3 | 24.5 | 0.608 | 0.733 |
| | 0.84 | 20.4 | 24.3 | 0.611 | 0.727 |
| | 0.85 | 20.5 | 24.1 | 0.614 | 0.722 |
| | 0.87 | 20.6 | 23.7 | 0.621 | 0.714 |
| | 0.88 | 20.6 | 23.5 | 0.624 | 0.709 |
| | 0.89 | 20.7 | 23.3 | 0.628 | 0.706 |
| | 0.90 | 20.8 | 23.1 | 0.631 | 0.701 |
| | 0.91 | 20.8 | 22.9 | 0.634 | 0.697 |
| | 0.92 | 20.9 | 22.7 | 0.637 | 0.692 |
| | 0.93 | 21.0 | 22.5 | 0.641 | 0.689 |
| | 0.94 | 21.0 | 22.4 | 0.644 | 0.685 |
| | 0.95 | 21.1 | 22.2 | 0.647 | 0.681 |
| | 0.96 | 21.1 | 22.0 | 0.65 | 0.677 |
| | 0.98 | 21.3 | 21.7 | 0.657 | 0.670 |
| | 0.99 | 21.3 | 21.6 | 0.66 | 0.667 |
| Glucosyl | 1.00 | 21.4 | 21.4 | 0.663 | 0.663 |

The calculation of metabolic heat production from $VO_2$ alone can result in an error of 9–10%, while calculation from $VCO_2$ alone can result in an error of 24–32%. Given the inadequacy of RQ's being able to reflect an exact chemical stoichiometry for cellular metabolism, potential short-term mismatch between RQ and RER, and possible errors in the conversion factor used to calculate MHP, it is not

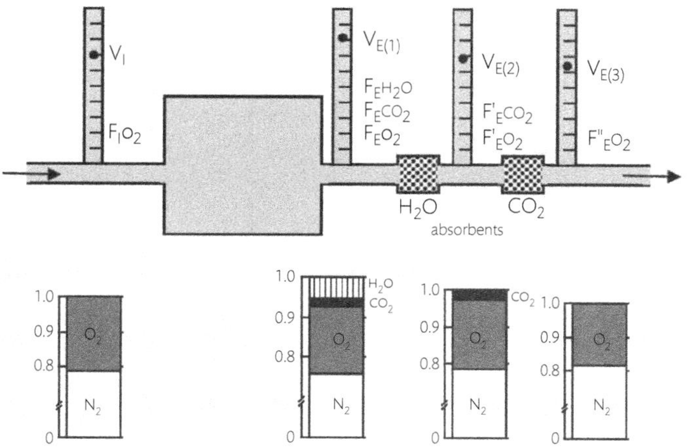


**Figure 5.10** Schematic representation of a typical flow-through respirometry system, showing the varying flow rates through the system, from dry, $CO_2$-free inlet air (V) to $V_{E(I)}$ after the chamber, $V_{E(2)}$ after the $H_2O$ absorbent, and $V_{E(3)}$ after the $CO_2$ absorber. The terminology for the corresponding fractional $H_2O$, $O_2$, and $CO_2$ concentrations are also shown. The gas composition and relative flow rates at these flow locations are shown underneath. Modified from Withers (2001).

unexpected that direct and indirect calorimetry might not always closely agree (e.g. Kleiber 1975; Walsberg & Hoffman 2005; see later, Hess's law).

The measurement of $VO_2$ and/or $VCO_2$ can be made in a sealed chamber (closed-system respirometry) or in a chamber (or with a mask) through which air flows at a known flow rate (open-flow respirometry). Closed-system respirometry is impractical for mammals due to their relatively high metabolic rates, necessitating excessively short measurement durations to maintain chamber air composition at a level that will not have physiological implications for the mammal (generally the change in $O_2$ and $CO_2$ is $< 0.5$ to 1%). Closed-system respirometry also has the problem that it averages metabolic rate over the measurement time interval, so metabolic rates obtained in this way can include different metabolic states (e.g. bouts of activity and rest), that may hamper interpretation of the resulting data.

Open-flow respirometry is by far the most common method of measuring mammalian metabolic rate. The concept of open-flow respirometry is as follows (Withers 2001; Lighton 2008; Chaui-Berlinck & Bicudo 2006; Figure 5.10). Air with a known oxygen fractional content ($F_{IO2}$) is passed at a known flow rate ($V_I$) through a chamber containing an animal. Air exits the chamber with a reduced oxygen content ($F_EO_2$), and the difference between the incurrent and excurrent fractional $O_2$ contents ($F_IO_2 - F_EO_2$) reflects the mammal's metabolic (oxygen consumption) rate. In many respirometric studies, the incurrent and excurrent $CO_2$ concentrations ($F_ICO_2$, $F_ECO_2$) are also measured, enabling the calculation of $VCO_2$ (and RER) from $F_ECO_2 - F_ICO_2$. In some respirometric studies (but not enough), the incurrent and excurrent water content is measured as well, to enable calculation of evaporative water loss (EWL) from ($F_EH_2O - F_IH_2O$). Water content is typically measured as relative humidity (RH;%), which can be converted to AH ($\chi$; mg $l^{-1}$) or wvp (kPa). These conversions from RH to $\chi$ and wvp can be accomplished using derived equations (e.g. Parish & Putnam 1977), and $\chi$ can be converted to a fractional water content using the gas law.

While the concept of open-flow respirometry is straightforward, the calculations for $VO_2$, $VCO_2$, and EWL from respirometry are complex, because of the varying flow rate in different parts of the respirometry system (Figure 5.10). The addition or removal of $O_2$, $CO_2$, and $H_2O$ not only changes the concentrations of those gases in the metabolic chamber, but also changes the rate of air flow, because generally the volume of the gases added does not equal to the volume of the gases removed. In a respirometry system in which only $VO_2$ is measured, the simplest approach is to remove $H_2O$ and $CO_2$ from the incurrent and the excurrent air, so that the $F_IO_2$ is measured for dry, $CO_2$-free inlet air (flow rate $V_I$) and dry, $CO_2$-free excurrent air ($F_E''O_2$) (Figure 5.10). The excurrent air flow rate ($V_E''$) equals $V_I - VO_2$. By mass balance, $VO_2 = (V_I\ F_IO_2) - (V_E''F_E''O_2)$, and substitution of $V_E'' = V_I - VO_2$ yields the basic respirometric equation of Depocas and Hart (1957) for this simple respirometric system: $VO_2 = V_I(F_IO_2 - F_E''O_2) / (1 - F_E''O_2)$.

For more complex respirometric systems, where $CO_2$, $H_2O$, or both are retained in the excurrent airstream, there are additional excurrent flow rates and fractional gas compositions, and more complex equations are required to calculate $VO_2$, $VCO_2$, and EWL (see Withers 2001; Lighton 2008). If air flow is measured in the excurrent airstream (e.g. when using a mask), then different equations are also required to account for $CO_2$ and $H_2O$ in the incurrent air (e.g. Withers 1977a).

The gas levels measured by the various analysers in respirometry systems do not instantaneously change as the metabolic exchange and water loss of the animal change. Rather, changes are delayed due to mixing of air in the metabolic chamber. The washout time ($\tau$) depends on the flow rate ($V_{FR}$) and chamber volume, ($C_{Vol}$); $\tau = 2.3\ C_{Vol} / V_{FR} \log(1/(1-z))$, where $z$ is the percent of equilibrium to be attained (Lasiewski et al. 1966). For 99% equilibrium, $\tau = 4.6\ C_{Vol} / V_{FR}$. Washout time is reduced by increasing the flow rate or decreasing the chamber volume. If the washout characteristics of the respirometry system are determined, then it is possible to apply an 'instantaneous correction' to adjust for washout and estimate the instantaneous changes in $VO_2$ and $VCO_2$ (Bartholomew et al. 1981). Washout of water vapour is more difficult to deal with, because water vapour seems 'stickier' than $O_2$ and $CO_2$, and washes out considerably slower than expected.

Respirometric systems can be modified to enable the calculation of respiratory ventilation variables—respiratory rate ($r_f$), tidal volume ($V_T$), and minute volume ($V_I = r_f\ V_T$) —by plethysmography, measuring the small pressure changes that occur in the metabolic chamber due to the warming and humidification of inspired air (Malan 1973; Withers 1977b). Szewczack and Powell (2003) have described a useful correction to such respirometric calculations to mathematically convert the ventilatory and calibration pressure changes from the open-flow respiratory system to a closed system, thereby improving the accuracy of calibrating the pressure changes. The calculation of the volume of air inspired that is required to obtain a given pressure change requires knowing the temperature that the air attains, usually assumed to be lung temperature, which is assumed to be equal to core $T_b$. The availability of small temperature transmitters and passive integrated transponders (PITs), which can be implanted into the abdominal cavity of a mammal, enables the continuous recording of $T_b$ throughout the respirometry experiment. This facilitates the interpretation of metabolic patterns (e.g. during torpor). Such technology also enables the continuous calculation of wet thermal conductance ($C_{wet}$) and dry thermal conductance ($C_{dry}$), which are useful indices of the heat transfer capacity of the mammal, including and excluding evaporative heat loss (Burton 1934; McNab 1980). $C_{wet} = MHP / (T_b - T_a)$ and $C_{dry} = (MHP - EHL) / (T_b - T_a)$, where MHP is metabolic heat production and EHL is evaporative heat loss.

The measurement of the basal metabolic rate by open-flow respirometry requires strictly standardized conditions, allowing for highly repeatable measurements and direct comparisons between different species. Therefore, it is essential that measurement protocols for BMR are stringently adhered to. BMR is generally measured for adult, post-absorptive, non-reproductive homeothermic mammals at rest in a thermoneutral environment during their inactive phase (Dubois 1924, 1930; Kleiber 1932, 1975; Aschoff & Pohl 1970; McNab 1997; Cooper & Withers 2009; Page et al. 2011).

Several aspects of open-flow respirometry need to be carefully considered when measuring BMR (and other physiological states). Sufficient measurement duration to attain basal values is an essential requirement; generally, a minimum of 4–5 h might be required, but often 8–9 h or even longer is necessary (Cooper & Withers 2009; Connolly & Cooper 2014). The use of switching systems to enable the measurement of multiple animals using a single set of analysers is useful for maximizing the rate of data acquisition, but it is important to ensure that this does not compromise the capacity to determine the actual minimal metabolic rate, because each individual animal is measured for only a fraction of the time as the system is switched between individuals (Cooper & Withers 2010). The measurement of BMR also requires measurement of the minimal rates at the appropriate part of the circadian cycle (Aschoff & Pohl 1970; Connolly & Cooper 2014). The repeated measurement of BMR for individuals can also lead to a declining BMR with successive measurements as the animal becomes accustomed to the procedures (Jacobs & McKechnie 2014). The accurate calibration of instrumentation, including gas analysers and flow meters, is also clearly a fundamental requirement for accurate measurement. Not surprisingly, given the strict criteria and protocols required to achieve BMR measurement, some published measurements of 'BMR' are likely to be over-estimates because they did not achieve the stringent requirements. The interpretation of a given species' BMR, comparisons across species, and allometric predictions all require appropriate data (McKechnie & Wolf 2004).

Hess's law of constant sums—that the heat of a chemical reaction is independent of the pathway—predicts that direct and indirect calorimetry should measure the same rate of heat production, provided no external work is done by the animal (Kleiber 1975). Although direct and indirect calorimetry generally provide similar estimates of metabolic rate (in the absence of significant anaerobic metabolism), studies have suggested that there can be discrepancies between the two measurements.

Walsberg and Hoffman (2005, 2006) showed similar patterns in direct and indirect calorimetric measures of metabolic rate for both endotherms (including the kangaroo rat) and an ectotherm (python). Both methodologies were very similar (within 1%) for the ectotherm, but there were some discrepancies for the endotherms. Although the patterns over time for $VO_2$, $VCO_2$, and heat production were similar, the RER declined from about 0.95 to 0.80 (Figure 5.11), reflecting a change in metabolic substrate from mainly carbohydrate (1.00) to protein (0.81) or a mixture of carbohydrate and lipid (0.71). The discrepancy between direct and indirect measurements was apparent from the calculated conversion coefficients for both $VO_2$ and $VCO_2$ with direct heat production. For $VO_2$, the expected ratios vary from 19.2 (protein) to 19.8 (lipid) and 21.1 (carbohydrate), but the calculated ratios for kangaroo rats were generally below these, except for the last few hours. For $VCO_2$, the expected ratios vary from 21.1 (carbohydrate) to 23.8 (protein) and 27.8 (lipid), and calculated ratios for kangaroo rats were generally within these, except for the first few hours.

Burnett and Grobe (2013, 2014) compared direct and indirect measurements of resting metabolic rate for laboratory mice on a high-fat diet. They found no discrepancy between the two methodologies on the high-fat diet, but respirometric values were about 6% lower than direct heat production for the control laboratory chow diet. The reasons for these discrepancies are not readily apparent, but they cannot be ascribed to errors in either indirect or direct methodologies. We simply do not have sufficient understanding of the biochemical bases for these complex metabolic processes, so further study is needed. However, we should not abandon now—or possibly ever—respirometry (Speakman 2014), as it remains the most practical 'gold standard' for routine metabolic measurement in the laboratory and for comparative analyses.

## 5.3.2 Field Energetics

A major thrust of studies of mammalian energetics was the extension from laboratory-based studies to the measurement of energy and water use in free-living mammals in their semi-natural or natural environments. Variables such as mass,

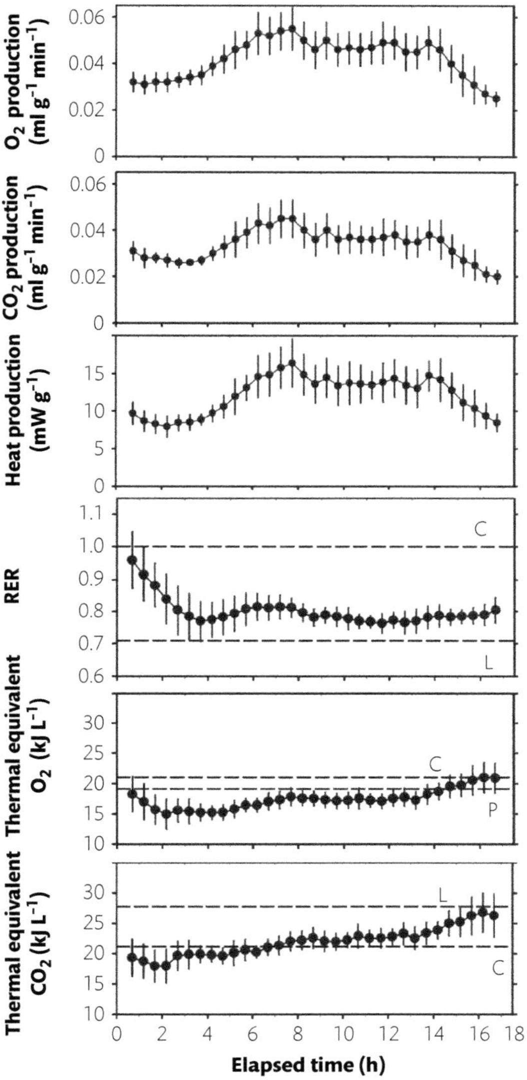


**Figure 5.11** Comparison of direct and indirect calorimetry for a kangaroo rat (*Dipodomys merriami*), comparing $O_2$ consumption, $CO_2$ and heat production, and respiratory exchange ratio (RER), and the calculated thermal equivalents for $O_2$ and $CO_2$ compared with theoretical upper and lower bounds (C, carbohydrate; P, protein; L, lipid). Error bars are 95% confidence limits. Modified from Walsberg and Hoffman (2005).

phylogeny, environmental conditions, activity, torpor (including hibernation and aestivation), and food availability all cause changes in energy and water use. In the field, the complexity of the interactions between all of these variables, in addition to biotic interactions (competition, predation, etc.) and abiotic interactions (weather, etc.), increases the complexity of energy and water use. It can be difficult to separate these various effects; for example, winter cold can increase, but use of torpor will decrease, metabolic requirements. Nevertheless, despite these complexities, it is essential to understand the field metabolic rate (FMR) and water turnover rate (WTR) of free-living mammals to fully understand how they survive in their natural environment. The overall measurement of FMR, and how it is allocated for various activities, is central to understanding the physiological, behavioural, and evolutionary ecology of mammals (Butler et al. 2008).

While it is now possible to measure the energy and water used by free-living mammals (like other free-living animals), such measurements were not initially feasible, and so various proxy measurements have been used to estimate energy balance. The use of time-energy budgets was one early approach to estimating field metabolic rate. Belovsky and Jordan (1978) analysed the diet and food consumption of moose (*Alces alces*) to estimate their FMR; FMR was calculated to be 60,600–65,300 kJ day$^{-1}$ based on their consumption of different food classes, the gross caloric content of each food type, the dry matter digestibility of each food type, and the locomotory cost of foraging.

Kenagy and Hoyt (1989) used laboratory-based measures of the energetics of locomotion and field observations of locomotor patterns to estimate a movement distance of 5 km day$^{-1}$ walking and 3.5 km day$^{-1}$ running for golden-mantled ground squirrels (*Citellus lateralis*), with corresponding net costs of transport calculated to be 0.076 and 0.049 kJ g$^{-1}$ day$^{-1}$, respectively. These are 8 and 5%, respectively, of the total field metabolic rate (measured by doubly labelled water; see later) of 0.99 kJ g$^{-1}$ day$^{-1}$ (Kenagy et al. 1989). The sum of the net costs of transport (0.125) and BMR (0.436 kJ g$^{-1}$ day$^{-1}$) are, nevertheless, only about 57% of FMR, indicating significant other components of FMR (e.g. costs of feeding, digestion, thermoregulation, etc.). Clearly, such time-energy budget approaches provide a useful indication of field energetics, but will always be open to criticism. Methods that measure the energy use directly remove all of that uncertainty.

Another indirect proxy of FMR that has been used extensively is heart rate (Butler et al. 2008; Green 2011; Tomlinson et al. 2014). At least for moderate to large animals (about 40 g or more), it is feasible to implant or externally fit heart rate loggers or transmitters. Laboratory 'calibration' of heart rate with metabolic rate provides a conversion factor (or factors) to calculate FMR from daily heart rate measurements. Generally for mammals, metabolic rate (MR; ml $O_2$ kg$^{-1}$ h$^{-1}$) $= 6.38\ f_h + 38.2$, where $f_h$ is the heart rate in beats per minute (for birds, MR $= 9.47\ f_h + 232.8$; Tomlinson et al. 2014).

A conceptually similar proxy for MR is muscle activity. Coordinated muscle activity such as shivering correlates with MR in the cold in mammals, though not as strongly as it does for birds, probably because birds do not develop non-shivering thermogenesis with cold acclimation, whereas mammals do (West 1965, see 4.1.1). An alternative, more general approach to monitoring muscle activity as a proxy of metabolic rate is the measurement of overall dynamic body acceleration (ODBA) using accelerometers (Wilson et al. 2006; Gleiss et al. 2011; Lush et al. 2015). ODBA is calculated as the vector sum of the three orthogonal (X, Y, Z) accelerometer axes. For example, for humans on a treadmill, MR (ml $O_2 min^{-1}$) = 1,132 ODBA + 615 (Qasem et al. 2012). Williams et al. (2014b) used the laboratory-measured ODBA-$VO_2$ relationship for puma (*Puma concolor*) to document the energetics of their hunting in the field. There was marked individual variation for pumas in their foraging strategy, from sit-and-wait to active foraging, for which the average energetic cost was 2.3 times that predicted. Fish et al. (2014) used bubble digital particle image velocimetry (DPIV) to estimate the thrust production of swimming dolphins as up to 1,468 N, by visualizing their wake.

A more direct measurement of FMR is provided by the use of doubly labelled water (DLW)—water containing of a mixture of isotopes of oxygen ($^{18}O$, a stable isotope) and hydrogen ($^2H$, deuterium, a stable isotope; or $^3H$, tritium, a radioactive isotope; see 5.7.1). The rationale for this methodology, pioneered by Lifson and McClintock (1966), is that $^2H$ (or $^3H$) is lost from a mammal only as water ($^2H_2O$), but $^{18}O$ is lost as both water ($H_2{}^{18}O$) and carbon dioxide ($C^{18}O_2$) because of the reversible reaction of $CO_2$ with $H_2O$ to form $H_2CO_3$. The difference in H turnover and O turnover is therefore a measure of $CO_2$ turnover, which can be stoichiometrically related to $O_2$ and energy turnover, both being measures of FMR. Speakman (1997) described the DLW methodology in considerable detail, including the various assumptions and calculations required; Munn et al. (2013b) described the calculations for FMR of red kangaroos. A large number of reviews (e.g. Nagy 2005) have documented the FMR of many mammals (and other vertebrates).

Various other isotopes have been used to measure FMR (Odum & Golley 1963; Baker & Dunaway 1975). Sodium ($^{22}Na$) turnover is a useful technique for estimating food consumption, hence FMR, for carnivores in particular (e.g. dingoes) because of the relatively constant Na content of their food. However, the variability in dietary Na for plant components of herbivores' diet would require detailed knowledge of the exact dietary composition to estimate food consumption and FMR (Green 1978). Rubidium ($^{86}Rb$) turnover is a useful proxy for FMR in mammals (Tomlinson et al. 2013a) and other vertebrates, and has some advantages over the DLW technique, including ease of whole-body $^{86}Rb$ measurement.

Another powerful isotopic method to measure rates of energy expenditure in free-ranging mammals is the $^{13}C$ labelled bicarbonate technique. Originally, this technique used the radioactive isotope $^{14}C$, and the labelled solution was administered by continuous infusion until it reached an equilibrium with the body bicarbonate pool (Corbett et al. 1971). The heavy isotope was then predominantly eliminated in the expired $CO_2$ (98%), which allow researchers to estimate $VCO_2$ from the exhaled air. Although this method was successfully used in large, agriculture mammals (Corbett et al. 1971; Benevenga et al. 1992; Junghans et al. 1997), its applicability for small mammals was constrained by methodological issues. Speakman and Thomson (1997) overcame these problems by using a single bolus procedure, where a solution of $^{13}C$ labelled bicarbonate was administrated via a single injection. In this procedure, the administration of the solution was easier and the animals were free to perform their routine activities after the injection. Due to the small size of the bicarbonate pool, as compared to the total $CO_2$ flux, the labelled solution reaches equilibrium with the bicarbonate pool quickly, allowing for the measurement of $CO_2$ production right after the injection (Hambly & Voigt 2011). Furthermore, since the $^{13}C$ isotopes has a half-life of 15–20 minutes, this technique allows for the measurement of energy expenditure of very short, but highly intense, activities. For example, Christian Voigt and co-workers has been using this method for measuring the costs of flight in bats and how these costs are affected by a myriad of environmental and biotic factors (e.g. Voigt & Lewanzik 2010; Van Busse et al. 2014).

These methodologies for estimating or measuring FMR each have advantages and disadvantages (Table 5.3; Butler et al. 2008; Tomlinson et al. 2013a,b, 2014). Heart rate and ODBC telemetry/logging require mammals to be large enough to accommodate the mass/volume of the implanted or attached device; the load is generally considered to have to be < 5–10% of body mass (see 5.2.1). These methods also provide essentially instantaneous values for $f_h$/ODBC, hence FMR, over the short term. DLW is generally suitable for medium-sized mammals. It can be problematic for very small mammals (requiring too frequent, hence potentially stressful, capture and bleeding events because of their high mass-specific MR) and very large mammals (because of the high cost for the DLW and slow turnover rate). DLW is also less suitable for aquatic mammals, which potentially have very high rates of water

**Table 5.3** Advantages and limitations of field metabolic rate measurement by doubly labelled water and heart rate monitoring. 1 = particular weakness; 5 = particular strength. Modified from Butler et al. (2008).

| | Doubly labelled water | Heart rate |
|---|---|---|
| Useful in animals > 100 kg | 2 | 5 |
| > 1 kg < 100 kg | 5 | 5 |
| > 50 g > 1 kg | 5 | 5 |
| < 50 g | 2 | 2 |
| Useful in aquatic environments | 1 | 5 |
| Cost per animal > 40 kg | 2 | 5 |
| > 40 kg > 1 kg | 4 | 2 |
| > 1 kg | 5 | 1 |
| Need for calibration/validation | 2 | 1 |
| Short-term stress to animal | 3 | 3 |
| Long-term stress to animal | 5 | 5 |
| Data at temporal scales | 1 | 5 |
| Long-term data | 1 | 4 |
| Accurate individual data | 2 | 2 |
| Accurate group mean data | 4 | 4 |
| Determine standard error of estimate | 1 | 1 |

turnover, increasing measurement error. However, DLW does have the advantage of simultaneously measuring WTR, which is also an interesting and instructive parameter for free-ranging mammals, and also provides estimates of FMR over relatively long time scales (days for small, weeks for large mammals). Neither DLW nor $f_h$ /ODBC provide particularly accurate data for individuals, but both provide good estimates (generally within 5% accuracy) for averages of multiple individuals.

## 5.4 Remote Sensing and Thermal Logging

Remarkable advances are ongoing in the miniaturization of electronics and batteries. There are now many sophisticated methodologies available for remote logging—recording of physiological and ecological data of free-ranging mammals—allowing us to monitor location (using the satellite global positioning system, GPS), temperature (ambient and body), heart rate, $O_2$ partial pressure, solar radiation levels, salinity, pressure, sound, and acceleration. For example, the integration of very fine thermistors attached to temperature loggers allows the measurement of brain and arterial blood temperature in free-ranging antelope (Fuller at al. 2000; Maloney et al. 2002; Hetem et al. 2008; Figure 5.12, left). GPS units and accelerometers can be attached to mammals to monitor their spatial movements and locomotor patterns (Figure 5.12, right).

**Figure 5.12** Left: A greater kudu (*Tragelaphus strepsiceros*) equipped with implanted temperature sensors to measure brain, carotid arterial blood, and abdominal temperature, and an external collar containing GPS and mini-black globe for assessment of microclimate selection, released after surgery. Photograph courtesy of S. K. Maloney. Right: A short-beaked echidna (*Tachyglossus aculeatus*) with a location radio transmitter, GPS, accelerometer, and iButton attached (with an implanted iButton to measure $T_b$). Photograph by C. E. Cooper.

Very high frequency (VHF) radio signals have been used to investigate animal movements and related ecology ever since Cochran and Lord (1963) first described a collar-mounted transmitter that could be affixed to rabbits (*Sylvilagus floridanus*), skunks (*Mephitis mephitis*), and racoons (*Procyon lotor*). The position of a transmitter attached to an animal has often been obtained by triangulation of the signal from several locations. By obtaining fixes at various times of day and across seasons, a picture can be obtained of space use and movement patterns by collared animals; statistical methods are used to estimate the size of the animal's home range (Kochanny et al. 2009). Before these advances, the most common method to calculate an animal's home range was to estimate the area of a polygon (a so-called minimum convex polygon) that incorporated various proportions of the total number of fixes (generally 95 or 50%). Because these polygons often included 'transit' and little-used areas, home ranges were later estimated using kernel methods that provided a probability estimate of the location of fixes, again usually at the 95 or 50% level (Worton 1987).

The advent of satellites to provide GPS coordinates reduced the labour intensity and inconvenience of location and long-term monitoring of home range and movement. Small collars or backpacks can be attached to an animal that log its position at fixed intervals. For example, the spines of echidnas allow various instruments—location radiotransmitters, GPS units, accelerometers, iButtons for temperature logging—to be glued to the animal (Figure 5.12, right). GPS units need to be calibrated for maximum accuracy. For example, Munn et al. (2013b) carefully screened their GPS data for free-ranging kangaroos to exclude fixes with fewer than four satellite fixes and unrealistic altitude measurements, to achieve an accuracy of ±10.4 m (in the open) to ±10.8 m (in the shade). For white-tailed deer, home range estimates during the diurnal hours, as assessed by 95% probability contours, did not differ between VHF manual tracking and GPS collars (Kochanny et al. 2009), even though the GPS method provided more than 20 times as many fixes.

These data suggest that a few fixes a day are sufficient to estimate an animal's range use. But they do not mean that there is no utility in more fixes, because those data can be used for other purposes. For example, fixes obtained several times a day can build a representation of space use and perhaps the distance travelled per day, but are not very useful in estimating the details of animal movement or speed of movement. Fixes obtained at higher temporal resolution (e.g. every 10 minutes) can be used to estimate the total distance travelled by an animal each day (Munn et al. 2013b).

Data from accelerometers (see 5.3.2) can be used to make inferences about behaviour. For example, Lush et al. (2015) fitted accelerometers to free-ranging brown hares (*Lepus europaeus*) to determine their behaviour, calibrated against accelerometry patterns associated with observed behaviours. Particular behaviours could be predicted accurately from accelerometry data (e.g. running, 100% accuracy); feeding, 97.4%; and vigilance, 98.3%). As expected, some behaviours varied in frequency throughout the day (e.g. running, feeding, grooming), whereas others did not (e.g. resting, vigilance).

The ARGOS system, which uses dedicated satellites to locate an Earth-based transmitter via the Doppler effect, has been operating since 1978. While the spatial resolution of ARGOS is of the order of 250–500 m—and so does not have the accuracy of GPS systems—the advantage of ARGOS is that, at the time of writing, the transmitter units are smaller than GPS systems (e.g. Roquet el al. 2014). As GPS systems become ever more miniaturized, that situation may change. ARGOS data are relayed to a user in near real time and so do not require the recapture of a tagged individual to access data. While some versions of GPS-based systems require recapture of the animals and physical downloading of the logged data, there are parallel developments of physiological-monitoring devices, and versions exist that transmit the stored GPS data via either a VHF radio link or via satellite or cellular network to a user (Matthews et al. 2013).

Temperature-sensitive radio transmitters are routinely used in the laboratory and field to remotely record the body temperature of mammals. These transmitters are often coated with inert wax and implanted in the abdominal cavity to record core $T_b$. Körtner and Geiser (1998) used implanted temperature transmitters and remote recording stations to show that eastern pygmy possums (*Cercartetus nanus*) used torpor in summer and winter, and hibernation in winter, with individual bouts lasting up to 5.9 days. Cooper and Withers (2004) used implanted radio-telemeters and recording stations to show that numbats (*Myrmecobius fasciatus*) use torpor in inclement weather to conserve energy; they have a minimum $T_b$ of about 19°C when torpid. For mammals with dense fur, these temperature transmitters can be glued to the skin to approximate the measurement of core $T_b$ (Barclay et al. 1996; Willis & Brigham 2003). For example, Bondarenco et al. (2014) recorded summer torpor by small free-tail bats in a desert environment, showing that they aroused using either passive rewarming or active thermogenesis.

Small programmable temperature loggers are readily available and economical; they are routinely used for measuring $T_a$ and $T_b$. Warnecke et al. (2007) implanted iButtons (small electronic temperature loggers manufactured to monitor the temperature in commercial shipments) encapsulated in inert wax into the abdominal cavity of bandicoots (*Isoodon obesulus*); they recorded a mean $T_b$ of 36.5°C (range: 33.4–39.8°C) with a pronounced nychthemeral pattern having a minimum $T_b$ about 06:00–12:00 and maximum about 16:00–24:00. iButtons can be modified to reduce their mass (Robert & Thompson 2003; Lovegrove 2009b). Passive integrated transponder chips can be made to be temperature-sensitive (PIT tags) and have the advantage of requiring no battery because they are excited by an external electromagnetic field, allowing them to be very small but requiring close contact with a reader to obtain a measurement.

The methods of data recovery from remote-sensing devices are varied (Hooker et al. 2007). The original means was telemetry, sometimes with an amplitude-modulated (AM) signal, but later devices used a frequency-modulated (FM) signal to transmit information. For example, $T_b$ could be remotely sensed from an implanted device that transmitted a pulse at a frequency proportional to the temperature of a temperature-sensitive resistor built into the circuitry, and heart rate could be transmitted via pulses triggered when electrical monitoring of the heart detected the R-wave of the electrocardiogram. By the 1960s, such technology was being used routinely (Bligh & Harthoorn 1965; Bartholomew & Lasiewski 1965). More sophisticated telemetry systems are used today and are common in laboratory studies and in places where transmission distance is not an issue; they can record biopotentials (heart, muscle, brain), blood pressure, temperature, and movement (e.g. Williams et al. 2002b). Some radio-transmitter devices can be manually or automatically monitored in the field by radio receivers for the immediate or subsequent acquisition of data (e.g. Körtner et al. 2010; Cooper & Withers 2004).

During the 1990s, transmission distance became an issue when studies were attempted on animals in their natural habitat, and many researchers moved to technology involving the storage of data on a data logger implanted in, or attached to, an animal. If it is likely that the mammal and its remote-sensing device can be recovered, then data can be stored in memory and downloaded on recapture. For example, the use of implanted iButton temperature loggers relies on storage of data as it is collected, and recapture and recovery of the iButton to download the data (e.g. Warnecke et al. 2007; Lovegrove & Genin 2008). The data capture rate and storage capacity of many data loggers today would have been unheard of even a decade ago. At the time of writing, it is possible, for example, to capture brain, eye, and muscle biopotentials, and to use three-axis accelerometry at 800 Hz for up to 20 days for sleep studies of free-ranging animals (Lesku et al. 2011). The 'daily diary' can record data related to movement, behaviour, energy expenditure, and the physical characteristics of the animal's environment by logging 14 parameters at infra-second frequencies (Wilson et al. 2008). Analysis of remotely obtained data can be complex, however, reflecting the often large rate of data acquisition. Voluminous raw data often need to be filtered to exclude erroneous values (e.g. 'impossible' GPS locations for a moving mammal), reduce inherent signal noise, and consolidate data to a more useful timescale.

A downside of many such early studies was that the device itself had to be retrieved to download the logged data. Later developments meant that data could be stored in digital (rather than transmitted in analogue) form on a memory device attached to or in an animal, and the device could be interrogated via Bluetooth or FM signals to send data at regular intervals to a base station via mobile phone, GPRS, or satellite transmission (Strauss et al. 2015). For example, Polar Bear International provides a web-based tool to allow scientists and the public to watch the movements of polar bears that have a neck collar with a satellite-transmitting GPS (http://www.polarbearsinternational.org/about-polar-bears/tracking/bear-tracker).

The use of such remote-measurement devices requires appropriate ethical experimentation protocols and procedures (Hawkins 2004; Wilson & McMahon 2006; Hooker et al. 2007). Capture stress is the first concern for attachment of external or internal devices for remote sensing. The means of attachment or surgical consequences of implantation is the next consideration. For example, attachment of remote-sensing devices to the fins of marine mammals by single-pin attachment achieves a balance between reducing impacts on the individual to which it is attached, and maximizing transmission success (Balmer & Wells 2013). The mass of the device is also an important consideration. For example, a general requirement for attached loggers is that they do not exceed a fraction of body mass that imposes an unacceptable additional load on the mammal. A transmitter or logger mass of about 5–10% is generally considered acceptable (Hawkins 2004; Wilson & McMahon 2006; Gannon & Sikes 2007), although Rojas et al. (2010) reported no effect of transmitter mass up to 14% of body mass on locomotion in small marsupials. For swimming and flying animals, drag on the device and antenna is also an important consideration. Finally, the potential discomfort to the mammal of carrying the device must be considered; if the normal behaviour is disrupted, the meaningfulness of the obtained data is questionable.

Not only do instrumented mammals provide physiological and ecological data to scientists interested in how they function in their natural environment, but the mammals can provide a broader range of scientists with data. For example, Roquet et al. (2014) used satellite-tagged elephant seals to document their temperature and the depth of dive (via pressure), as well as seawater salinity and temperature (Figure 5.13). Thereby they obtained detailed temporal and spatial hygrographic

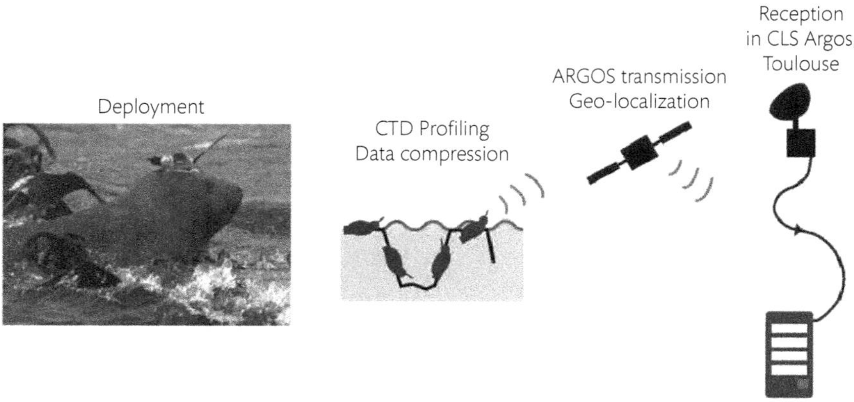


**Figure 5.13** Schematic of satellite-linked data collection procedure for elephant seals instrumented with a CTD tag instrumented for recording of pressure, temperature, and salinity via ARGOS satellite transmission to the laboratory in Toulouse (France). Modified from Roquet et al. (2014) with permission from F. Roquet. Photograph copyright by Christophe Guinet (CEBC/CNRS).

profiles of the Southern Indian Ocean that are of great value to oceanographers as well as the marine mammal investigators themselves.

## 5.5 Circulatory Systems

Advances in electronics have revolutionized the study of the circulatory system; in particular, the remote logging of heart rate and other physiological and environmental variables in the field. For example, heart rate in the field has been used as a proxy for FMR (see 5.3.2) and for the investigation of the physiology of diving mammals (see 4.5).

The first measurements of pressure in an organism's circulatory system were made by Stephen Hales, who used long glass tubes to measure the pressure of sap in trees (Hales 1727). In 1733, he used a 9-foot glass water manometer attached by a brass connection to the trachea of a goose to cannulate and measure the carotid and femoral pressures in a horse (Hales 1733). The process of pressure measurement was made easier when Poiseuille used mercury as the measuring fluid (probably inspired by Torricelli's use of mercury in his barometer) and so reduced the required length by a factor of 13.6 (Comroe 1978). Poiseuille (1828) then made measurements by cannulating sites from the aorta to the smallest arterial branch that he could cannulate; he proved that pressure did not fall linearly from the aorta to the smallest arterioles (as was the accepted wisdom at the time)—rather, most of the decrease occurred at small arterioles.

Non-invasive measurement of blood pressure initially involved compression of an artery (usually the radial) by a mechanical device. A major advance occurred in 1896 when Riva-Rocci developed an easy-to-use sphygmomanometer and combined it with an inflatable cuff (the 'cuff of Riva-Rocci', as it became known) that was placed around the upper arm, and compressed the artery equally from all directions. Inflation of the cuff continued until the radial pulse could no longer be palpated, followed by slow deflation of cuff until the 'breakthrough' pulse was detected. The pressure in the cuff at the moment the breakthrough pulse was detected was a measure of systolic blood pressure (Booth 1977).

The non-invasive measurement of diastolic pressure required a different approach. In 1905, Korotkoff reported (in a 207-word manuscript) on the sounds emitted from an artery compressed by the cuff of Riva-Rocci. He found that a fully compressed artery emitted no sound, and neither did a non-compressed artery, but that in between there appeared short faint tones, indicating that part of the pulse stream had passed under the cuff (Booth 1977). We now know that turbulence created by blood passing through the partially compressed artery causes these sounds. When pressure in the cuff is decreasing, the appearance of the sounds marks the systolic pressure, while their disappearance marks the diastolic pressure.

Strain gauges replaced liquid manometers after Lambert and Wood (1947) were forced to develop an alternative method to the liquid manometer to measure pressures for subjects in their human centrifuge, as they had to make measurements at a distance from the subject (Comroe 1978). This development freed measurements from the necessity of fluid-filled tubes from subject to measurement device, but still required a physical connection (wires). Van Citters and Franklin (1966) made the first measurements of blood pressure in large vessels and chambers of the heart of untethered animals using miniature semiconductor strain gauges as well as a telemetry system (Sarazan & Schweitz 2009). Konigsberg, an engineer at Micro Systems, miniaturized and improved the device; they are still available from Konigsberg Instruments. Later, Brockway and Osgood further refined the design. Brockway went on to found Data Sciences; Osgood started MiniMitter—both companies still market telemetry instruments for electrocardiograms, pressure profiles, and $T_b$ from small conscious animals (Sarazan & Schweitz 2009). At the time of writing, no one has developed a useful logging system for blood pressure.

Franklin also developed the first ultrasonic transit time measurement system for blood flow. The system was used to measure blood flow in the aorta in resting and exercising dogs (Franklin et al. 1959). Through the 1960s, others modified the system, which eventually led to the commercial release of Transonic flow systems that are in use today (Franklin et al. 1959). While relatively easy to use, these measurements require the animal to be anaesthetized or at least heavily restrained. There is also the issue that, if such a probe is used to measure cardiac output via measurement from the aorta, coronary circulation is not captured, because it exits the aorta just above the aortic valve. Thus the measured value is cardiac output minus coronary flow. There are, however, other options for measuring cardiac output (Critchley & Critchley 1999). Grollman (1932) developed an acetylene breathing method that was the gold standard for cardiac output measurement for many years. Eventually it was replaced by the indicator dilution (see later) and direct Fick methods, and it is now known that the acetylene method slightly underestimates cardiac output (Comroe 1978).

Indicator dilution involves injecting an indicator into the right atrium or pulmonary artery and sampling blood from a peripheral artery. Calculations using the time it takes to inject the indicator can be used to estimate the volume of blood that passed the catheter in a given time, hence an estimate of flow through the heart, and hence cardiac output. Probably the most common method of measuring cardiac output now involves Swan-Ganz catheters, which are used routinely in clinical and research settings (Thiele et al. 2015). The Swan-Ganz is inserted via a large vein through the right heart, into the pulmonary artery. This method can involve the injection of cold saline via a port in the right atrium, but more commonly, a heating coil lies in the right atrium, supplying a known amount of heat to the passing blood. The change in temperature of the blood induced by the heating coil is detected by a thermistor further along the catheter in the pulmonary artery.

Non-invasive methods of cardiac output have also been developed. An example is continuous wave Doppler echocardiography of the ascending aorta, although such methods miss the coronary circulation that leaves the aorta just above the aortic valve (Nishimura et al. 1984). Other non-invasive methods include oesophageal Doppler and pulse contour analysis (Berton & Cholley 2002).

The Fick method relies on the Fick principle: oxygen consumption (or carbon dioxide production) of an animal is equal to cardiac output multiplied by the arterio-venous $O_2$ difference (or the veno-arterial $CO_2$ difference; Fagard & Conway 1990). The difference between the arterial and the venous oxygen content of the blood can be estimated from haemoglobin saturation and the haemoglobin concentration, or measured directly in samples of arterial and mixed-venous blood in a Tucker chamber (Tucker 1967). The chamber uses an oxygen electrode to measure the oxygen concentration, to which a small volume of blood is added. The presence of potassium ferricyanide drives oxygen from haemoglobin, resulting in the measurement of the oxygen content in the sample.

Other cardiac output measures include medical imaging techniques that can assess the volume of the left or right ventricle at end-systole and end-diastole, and obtain a measure of stroke volume from the difference; such measures include echocardiography, contrast angiography, electron beam computed tomography, and magnetic resonance imaging (Critchley et al. 2010). For example, Thornton et al. (2005) measured the stroke volume of juvenile elephant seals (*Mirounga angustirostris*) in captivity, using magnetic resonance imaging (MRI) and heart rate using cardiac electrodes. Pre-dive cardiac output, calculated as heart rate times stroke volume, was $6{,}530\,\text{ml}\,\text{min}^{-1}(61.4\,\text{min}^{-1}\times 105\,\text{ml})$; during a simulated dive, output decreased to $4{,}011\,\text{ml}\,\text{min}^{-1}(31.8\,\text{min}^{-1}\times 126\,\text{ml})$.

## 5.6 Molecular Biology

Basic biochemical and cellular functioning, and the adaptations of mammals to varying environments, depend on the genome of the species and the particular pattern of genes that are expressed (or repressed) at any point in time, and which result in protein synthesis. Fundamental advances in molecular techniques (the '-omics' in Figure 5.14) have opened new corridors of understanding of animals and their environment. Such advances have taken place at the level of DNA base sequencing (genomics), which allows a much more detailed understanding of the phylogenetic arrangements of animals (phylogenetics); at the level of protein expression patterns (proteomics); patterns of gene expression; protein synthesis; and at the level of metabolic control between and within species in response to environmental adaptation. The evolution of multicellular animals was accompanied by an immense increase in molecular complexity, reflecting in part the capacity to incorporate other organisms through symbiotic relationships into the basic cellular structure

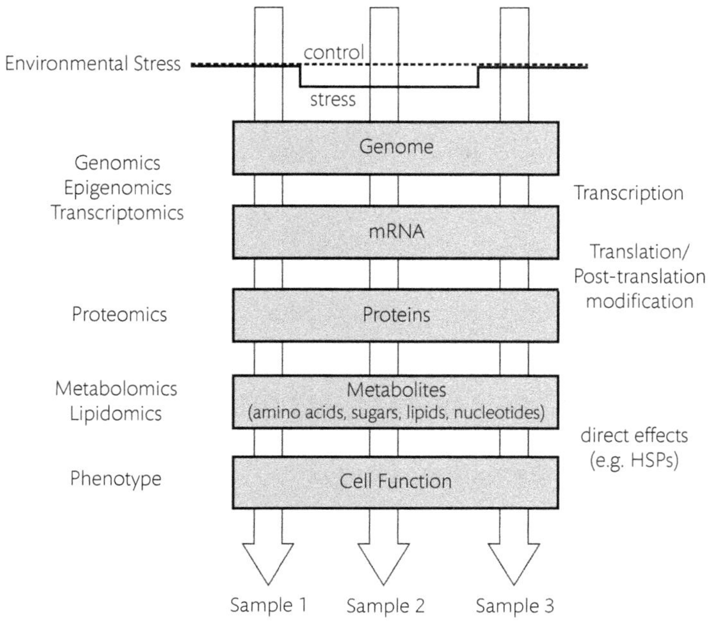


**Figure 5.14** Schematic of the hierarchy of molecular 'omics'—genomics, epigenomics, transcriptomics, proteomics, and metabolomics (including lipidomics)—relative to a change in environmental stress over time (solid upper line) compared to control (dashed line), over multiple sampling points. Modified from Hayward (2014).

(e.g. mitochondria, various symbiotic photo- or chemi-synthetic microorganisms in animals, chloroplasts in plants). This was associated with an increase in the complexity of gene regulatory and transcription mechanisms (Davidson and Erwin 2006): essentially every aspect of messenger ribonucleic acid (mRNA) structure and function is subject to regulation (Moore 2005). The increase in structural complexity and diversity was accompanied by a 'step' increase in metabolic rate (Hemmingsen 1950; Withers 1992).

Many phenotypic differences between and within mammalian species are thought to reflect patterns in gene expression. All species differ in their genetic material, but these differences are surprisingly small. Vertebrate genomes have only about two times the number of genes as invertebrate genomes, and much of the increase is gene duplication rather than new genes (Levine & Tijan 2003). Consequently, the diversity of animals arises largely from differences in patterns of gene expression. Particular patterns of expression, both between and within species under different environmental stress, are due to various levels of control of the parts of the genome that are expressed—the various 'omics'. Thanks to a veritable explosion of advances in molecular biology, our understanding of such 'omics'—our knowledge of cellular and biochemical control underlying phenotypic expression—has increased enormously (Figure 5.14; Hayward 2014).

Genomics is the description of the genetic material (DNA) of animals. The increasing ease and low cost of gene sequencing has resulted in the mapping of the entire human genome. At the time of writing, the entire genomic sequence is available for more than 34 other species of mammal (including 1 monotreme, 4 marsupials, and 29 placentals). Incomplete genomes are available for many other mammalian species. This availability of genomic information allows the reconstruction of molecular phylogenies, which are a valuable tool for comparative methods that analyse ecological and physiological data in a phylogenetic context (see 5.1). Despite the remarkable growth of information regarding millions of gene sequences and complete genomes for many animals, the biological functions of most genes are either unknown or are assumed through homology with the relatively fewer genes whose functions are known.

Epigenomics examines the differential expression of the same genomic material, giving rise to different phenotypes from the same genotype. Epigenetic modifications are permanent effects on translation that can be caused by histones and DNA methylation. The regulation of epigenetic modification will no doubt provide new information on the way that animals adapt to their environment, especially now that it has been shown that epigenetic effects can be inherited (Hayward 2014).

Transcriptomics reflects changes in gene expression by quantification of mRNA, for example, using RNAseq, SAGE (serial analysis of gene expression), DGE (digital gene expression), or microarrays (Hayward 2014). Transcription factors regulate gene expression by either activating or repressing the transcription of genes; they bind to short transcription-binding sites near target genes. These transcription factors are grouped into families based on their common DNA-binding domains (Chen & Rajewsky 2007). Post-transcriptional control of gene expression also exists in the form of microRNAs (miRNAs), a class of approximately 22-nucleotide non-coding RNAs, which are repressors of genome transcription (Bartel 2005). These miRNAs are broad regulators of many aspects of cell physiology, and hence physiological functioning of cells and animals. In addition to the important transcription factors and miRNAs, there are numerous other levels of gene regulation, including cell signalling, mRNA splicing, polyadenylation and localization, chromatin changes and protein modification, localization and degradation (Chen & Rajewsky 2007). Transcriptomics is a powerful tool for examining changes in gene expression resulting from an environmental change. For example, Vermillion et al. (2015) used cDNA (complementary DNA) transcriptomics to identify the expression of distinct cold-tolerance genes in heart and skeletal muscle, as well as increased muscle oxidative capacity, for euthermic and hibernating ground squirrels.

Proteomics examines the detailed pattern of protein expression that reflects, in part, the pattern of gene expression (Hayward 2014). Proteins can be differentiated using simple one- or two-dimensional chromatography; then the various 'spots' can be identified by liquid chromatography and mass spectrophotometry (LC-MS). Typically, a few thousand proteins can be identified, and many can be assigned specific functions (e.g. heat shock proteins; HSPs). Proteomics is limited by the immense range of different proteins in cells, making identification of all proteins impossible, but it can provide crucial insight by quantifying the functional outputs of transcription, since mRNA levels do not necessarily correspond directly to protein levels, and stress adaptation often involves post-translational regulation. For example, during torpor by two distantly related bats, a number of proteins are up - or down- regulated, including proteins involved in amino acid metabolism, nitrogen metabolism, many enzymes related to glucose metabolism, and structural proteins (Pan et al. 2013). Epperson et al. (2010) showed seasonal proteomic changes that preserved and replenished liver proteins in hibernating golden-mantled ground squirrels (*Spermophilus lateralis*). Of a total of 2,485 protein spots, 2,396 were analysed for changes during hibernation; of 51 protein spots that increased with entry into hibernation, 17 proteins of known function were identified; of 84 spots that decreased with hibernation, 51 unique proteins were identified (see Table 5.4).

Metabolomics identifies a large array of low molecular weight metabolites, including amino acids, sugars, lipids, and nucleotides (Hayward 2014). Lipidomics is an important subset of metabolomics because of the important role of cell membranes and steroid hormones in the response to environmental stress. Common analytical approaches include LC-MS and nuclear magnetic resonance (NMR). Metabolomics (like proteomics) is limited by the immense array of different target molecules. Many thousands of biochemicals can be present in cells, and it is currently impossible to identify all of them. Nevertheless, major changes in important metabolites can be identified. For example, some of the many regulated metabolite changes that occur during hibernation include lipids, reflecting the use of lipid fuels during hibernation or as key signalling molecules (Melvin & Andrews 2010; Lang-Ouellette et al. 2014). A variety of non-coding miRNAs are potential regulators of lipid metabolism and are associated with cold adaptation, and there are indications that another family of long non-coding RNAs (lncRNAs) is similarly involved.

Functional genomics examines gene function by the measurement of RNA transcripts extracted from tissue samples, often for animals in control and various experimental states. The RNA is copied and tagged, for subsequent fluorescent detection, and hybridized to a microarray that allows assay of binding to thousands of types of ligand-binding biochemicals (including DNA, RNA, proteins, polysaccharides, lipids, small organic molecules; Holheisel 2006). This microarray technology provides an estimate of the abundance of the target biochemical in the samples. For example, microarrays provide many thousands of measurements of

**Table 5.4** Examples of 7 proteins that increased (of 31 identified to significantly increase) and 6 proteins that decreased (of 84 identified) during re-entry into torpor by a hibernating golden-mantled ground squirrel (*Spermophilus lateralis*). *Gene* is the unique identifier of the human homolog; *ID* is the unique protein identified in the spot; *x* is the fractional increase in spot intensity; *P* is the significance of the change (q-corrected for multiple testing); *Function* is based on the human gene annotation. Data from Epperson et al. (2010).

| Gene | ID | *x* | P | Function |
|---|---|---|---|---|
| **Proteins that increase during entry into hibernation** | | | | |
| A2M | $\alpha_2$-macroglobulin | 1.9 | 0.040 | Protease inhibitor |
| AD11 | Acireductone dioxygenase 1 | 1.5 | 0.024 | Methionine salvage |
| ATP5B | ATP synthetase $\beta$ subunit | 1.5 | 0.011 | ATP synthesis |
| CYB5A | Cytochrome $b_5$ | 1.4 | 0.008 | Electron transport |
| EEF2 | Elongation factor 2 | 1.55 | 0.007 | Protein synthesis |
| PDIA3 | ER-60 protein | 1.3 | 0.026 | Glycoprotein folding |
| SLC27A2 | Solute carrier family 27 | 1.8 | 0.035 | Fatty acid metabolism |
| **Proteins that decrease during entry into hibernation** | | | | |
| AASS | $\alpha$-aminoadipate-semialdehyde synthetase | 0.37 | 0.007 | AA metabolism |
| ACLY | ATP citrate lyase isoform 1 | 0.15 | 0.024 | Cytoplasmic citrate metabolism |
| ATP5H | ATP synthetase subunit d $F_o$ complex | 0.83 | 0.036 | ATP synthesis |
| CAT | Catalase | 0.59 | 0.010 | $H_2O_2$ catalase |
| DHTKD1 | Dehydrogenase E1 transketolase containing | 0.71 | 0.014 | Glycolysis |
| FABP7 | Fatty acid-binding protein 7, brain | 0.48 | 0.018 | Fatty acid transport |

gene expression per sample, which—after normalization relative to control tissue samples or 'housekeeping' genes (genes whose activity is so essential to basic cell functions that their transcription remains constant)—can indicate which genes have increased or decreased expression, or remained stable, under particular experimental conditions (e.g. cold, hypoxia, increased salinity). Identification of the functional significance of these genes (often by inference to similar genes with known functional significance) provides insights into the changes in cell functioning that result from the experimental condition. We can then examine the significance of such changes at physiological, ecological, and behavioural levels.

Limitation of the various 'omics' is that they provide strong evidence of correlation with—but not necessarily causation of—altered phenotypes. Local environmental effects can influence especially metabolomic and lipidomic profiles as well as microbial flora (hence the microbiomics of the animal; Hayward 2014). Nevertheless, the 'omics' have the potential to revolutionize the level of complexity of our understanding of cell functioning and provide the mechanistic link to understanding phenotype and phenotypic plasticity, both in different species and within individuals of a species in response to their environment. As an example, a more complete understanding of the genetic and biochemical adaptations during hibernation might make human hibernation possible (Lee 2008).

## 5.7 Isotopes

Animals are exposed to a range of chemical elements, and to the various isotopes of those elements, that occur in their natural environment (Fraústo da Silva & Williams 2001; Figure 5.15). Of all of the naturally occurring elements, 11 are predominant in biological systems (H, C, N, O, Na, Mg, O, S, Cl, K, Ca; Fraústo da Silva & Williams 2001; Figure 5.15). Many more are essential as trace elements, or are suspected to be essential trace elements. Elements are defined by the number of protons in their nucleus. Isotopes are forms of the same element (e.g. carbon, which has six protons in its nucleus) but with varying numbers of neutrons (e.g. two isotopes of carbon have six and eight neutrons, represented as $^{12}_{6}\mathrm{C}$ and $^{14}_{6}\mathrm{C}$, respectively). Many isotopes are stable over time (e.g. $^{12}_{6}\mathrm{C}$ and $^{13}_{6}\mathrm{C}$), but many are unstable (Cameron & Skofronick 1978) and radioactively decay (e.g. the other 13 isotopes of carbon, including $^{14}_{6}\mathrm{C}$, which has a half-life of 5,700 years, and $^{8}_{6}\mathrm{C}$, which has a half-life of $2\ 10^{-21}$ s). Many radioactive isotopes decay to the same element ($^{21}\mathrm{C}$ decays to $^{20}\mathrm{C}$ by loss of a neutron), whereas other radioactive isotopes decay to a new element ($^{14}_{6}\mathrm{C}$ decays to $^{14}_{7}\mathrm{N}$ by beta emission of an electron). $^{14}\mathrm{C}$ content is a common method for radiometric dating of biological material. $^{13}_{6}\mathrm{C}$ content is used for palaeoclimate dating.

Natural and anthropogenic impacts that change the natural environment have affected ecosystems, and these changes are evident as changes in elemental (e.g. H,

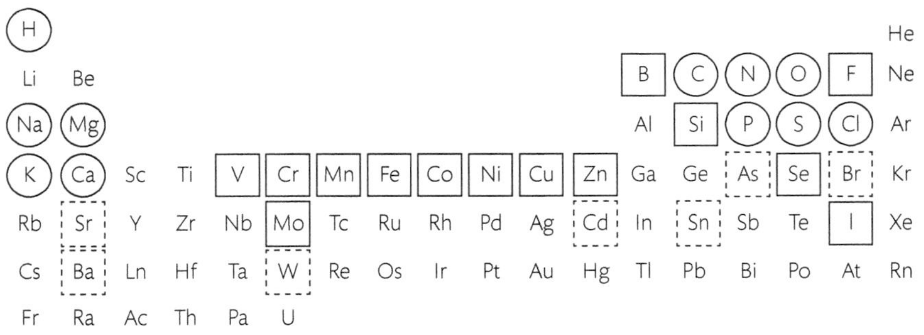


**Figure 5.15** Periodic table of elements showing the 11 common biological ('bulk') elements (circles), various trace elements (squares), and some possible essential trace elements (dashed squares). From Fraústo da Silva and Williams (2001). Reproduced with permission of Oxford University Press.

C, N, O, S, Sr) isotope ratios in atmospheric gases, and in plant and animal tissues (Dawson & Siegwolf 2007). Stable isotope analysis was applied first in chemistry and geochemistry, then biochemistry, physiology, and ecology. Stable isotope levels can be expressed as absolute (but often very small) values, or as relative abundances (which often differ only after the third decimal place), so the widely accepted method for expressing levels is the difference (δ, or 1,000 times the difference) from an internationally accepted standard; the standard has a δ of 0% (see 5.6.2). The levels of stable isotopes in natural systems are described using the δ notation: $^{k}\delta X_{A} = 1000((R_{A} - R_{std}) - 1)$, where k is the rare isotope (e.g. $^{13}C$), X is the element of interest (e.g. carbon, C), A is the sample of interest (e.g. water, tissue, diet), $R_A$ is the ratio of heavy to light isotope (e.g. $^{13}C/^{12}C$), and $R_{std}$ is the standard isotope level (Cerling et al. 2007; Dawson & Siegwolf 2007).

Stable isotopes are typically measured using mass spectrometry (MS), as isotope concentration differences are typically very small (Peterson & Fry 1987). Biological samples are typically collected, washed, dried, weighed, and combusted prior to analysis (e.g. Moritz et al. 2012; Dammhahn et al. 2013). Fractionation and contamination of samples for isotope analysis is a major concern during sample preparation and analysis. For example, about 20% of the H in hair can exchange with environmental water vapour, so appropriate calibration and correction of measurements are required (Cryan et al. 2004). Radioactive isotopes can be measured by mass spectroscopy, but they are often measured by scintillation counting (e.g. $^{3}H$ and $^{14}C$; Knoll 1999). Biological samples are solubilized (if necessary), de-coloured (as much as possible), and added to a scintillant, which absorbs the emission energy of radioactive emissions ($\alpha$, $\beta$, $\gamma$ radiation), then releases this energy as photons of particular frequencies, which is measured by a light detector.

Stable isotopes are commonly used to measure the turnover rate of elements (see 5.7.1), by administering an enriched sample of a low abundance isotope or isotopes and then measuring the decline in the low abundance isotope as it is replaced by the more common isotopes. For example, after an injection of doubly labelled water ($^2H_2{}^{18}O$) into an animal, the $^2H$ and $^{18}O$ equilibrate in the body fluid pool and then decline as the abundant natural isotopes in $^1H_2{}^{16}O$ replace the $^2H$ and $^{18}O$. They are also routinely used for comparing enrichment levels in components of ecological systems; for example, to determine dietary intake patterns or carbon dating of fossils (5.7.2). They are used, too, for biological imaging in clinical and laboratory situations (5.7.3).

## 5.7.1 Isotope Turnover

In biological systems, isotopes of the elements comprising biological tissues and fluids are replaced over time, but different elements can have very different turnover rates. For example, the H in water is turned over relatively rapidly, because water is ingested and lost by evaporation and in urine and faeces. The O in water is turned over even faster, as it is lost as water and also can be transferred to $CO_2$ and lost by respiration; this is the basis for determining animals' field metabolic rate (see 5.3.2, and later). If a stable (or even an unstable) isotope is introduced and equilibrated into a biological system (e.g. $^2H$ and $^{18}O$ by injection of doubly labelled water $^2H_2{}^{18}O$ into an animal), the remaining level of isotope at any point in time decreases exponentially, and is calculated as $N = N_0 e^{-\lambda t}$, where $N_0$ is the initial level, $\lambda$ is the biological rate constant (e.g. $s^{-1}$), and $t$ is the time (Cerling et al. 2007). This exponential relationship can be linearized as $\ln(N / N_0) = -\lambda t$. The biological half-life ($t_{1/2}$) is the time when half of the material has washed out of the system: $t_{1/2} = \ln(2) / \lambda$. If the isotope is radioactively unstable (e.g. $^3H$), its biological half-life needs to be corrected for its radioactive half-life (see 5.7.3), unless the radioactive half-life is substantially longer than the biological half-life. If the washout is normalized by F, where $F = 0$ at the beginning and $F = 1$ at the end, then $\ln(1 - F) = -\lambda t$; $(1 - F)$ describes the reaction progress. The reaction progress model can be more useful than the exponential model for physiological and ecological exchange processes, especially if there are multiple rate constants.

Stable (and radioactive) isotopes have been widely used in animal physiology and ecology. 'Betalights' (a light-emitting glass capsule that contains phosphor, and tritium as an excitant) have been used to track wildlife at night (Thompson 1982). Deuterium ($^2H$) or tritium ($^3H$) incorporated into water have been used to measure water turnover rates. For example, Withers et al. (1980) used tritiated water to measure the water turnover rates of Namib Desert rodents, and showed that they were typically low but increased after the occurrence of an advective fog.

$^{18}O$ in conjunction with $^2H$ or $^3H$ as doubly labelled water (DLW) has been used to determine the field metabolic rate of mammals, birds, and reptiles (e.g. Nagy 1983; Speakman 1997). The $^2H$ (or $^3H$) isotope turned over as water is gained through drinking, in food, and by metabolism, and lost through evaporation and in urine and faeces. The $^{18}O$ isotope is turned over by the same process, but, in addition, the $^{18}O$ label is transferred to $CO_2$ through the action of carbonic anhydrase rapidly catalyzing the reaction of $CO_2$ and $H_2{}^{18}O$ to form labelled carbonic acid, $H_2CO_2{}^{18}O$, which can then transfer the $^{18}O$ when it reverts to $CO^{18}O$ and $H_2O$; the $CO^{18}O$ can be excreted by respiration, enhancing the $^{18}O$ turnover rate. The differences in $^2H$ and $^{18}O$ turnovers can be converted to an estimate of $CO_2$ turnover (ml $CO_2$ $g^{-1}$ $h^{-1}$), which can be converted to an $O_2$ turnover (ml $O_2$ $g^{-1}$ $h^{-1}$) using a RER conversion or joule turnover (e.g. J $g^{-1}$ $h^{-1}$) using a joule coefficient (see Table 2.5). For example, Cooper et al. (2003b) used $^3HO^{18}O$ to determine the FMR (10.8 l CO2 day-1, 269 kJ day-1) and WTR (84 ml H2O day-1) for free-living male numbats (mass 488 g). There is now a considerable database for WTR and FMR of mammals (e.g. Nagy & Peterson 1988; Nagy et al. 1999; Nagy 2005; Riek 2008; Hudson et al. 2013).

## 5.7.2 Isotope Ratios

Stable isotope ratios are routinely used in animal ecology for studying many processes, including food webs, trophic position determination, food sources, and migration (Vander Zanden et al. 2015). Animal tissues that are commonly analysed for stable isotopes (hair, teeth, nails) incorporate the isotopic signature of the foods and water consumed (O'Grady et al. 2012). Thus $\delta^2H$, $\delta^{18}O$, and $\delta^{13}C$ values are commonly used to infer patterns of food consumption, trophic relationships, and the geographic location for animals.

The $\delta^2H$ and $\delta^{18}O$ compositions of animal tissues are expected to reflect those of meteoric water (rainwater) because dietary water and atmospheric H and O sources ($H_2O$ and $O_2$) are generally their main sources of H and O (O'Grady et al. 2012). Folivorous species that obtain much of their water intake from leaves (e.g. giraffe, *Giraffa camelopardalis*) have high $\delta^{18}O$ values, reflecting the relatively high $\delta^{18}O$ of foliage due to the fractionation of O during transpiration from leaves (the lighter $^{16}O$ isotope evaporates more readily). Species that rely more on meteoric water typically have lower $\delta^{18}O$ (since there is less enrichment of $^{18}O$ by fractionation).

Baboons (*Papio* spp.) are obligate drinkers. Two baboon species live sympatrically in Ethiopia: Hamadryas baboons (*P. hamadryas*) and olive baboons (*P. anubis*). Hamadryas baboons are more common in arid thornbush habitats and are inferred to ingest a greater proportion of leaf water and are better able to conserve body water, reducing their requirement to drink meteoric water, hence they are expected to have greater $^{18}O$ enrichment than olive baboons (Moritz et al. 2012). However, based on hair samples, the $\delta^{18}O$ was slightly higher (though not significantly so) in Hamadryas than olive baboons, and hybrids had intermediate $\delta^{18}O$. This lack of difference presumably reflects the longer drinking bouts of Hamadryas baboons, at a common drinking water source with olive baboons.

The $\delta^{13}C$ composition of animal tissues reflects changes in the plant material consumed by herbivores over geological time. For example, there was a major change in the Late Miocene from a predominance of $C_3$ dicotyledonous plants (> 99% of net primary production) to $C_4$ grasses now (> 50%; Cerling et al. 2015). The substantial difference in $\delta^{13}C$ between $C_3$ and $C_4$ plants allowed the determination of the dietary composition from tissue $\delta^{13}C$. For example, analysis of tooth enamel $\delta^{13}C$ in East and Central African mammals indicates that $C_3$ browsers in closed canopy have a $\delta^{13}C$ of less than −14‰, $C_3$ browsers have $-14‰ < \delta^{13}C < -8‰$, mixed $C_3$-$C_4$ diets have $-8‰ < \delta^{13}C < -1‰$, and $C_4$ grazers have $-1‰ < \delta^{13}C$.

The analysis of fossils suggests that many lineages of East and Central African mammals retained essentially the same diet over the last 4 million years (e.g. some antelope, zebra, giraffe, and rhinocerus). Some bovid and suid lineages have switched from a mix-feeding $C_3$-$C_4$ diet to grazing ($C_4$) diet. Some bovid lineages have shifted to more negative $\delta^{13}C$ diets. Elephant and suid lineages were primarily $C_4$ grazers 4–1 MYBP, but are now $C_3$ browsers; their average $\delta^{13}C$ values have decreased from about −1‰ to about −14 to −10‰ now. Similar analyses for hominid fossils (Levin et al. 2015) suggest that hominin diets in the early Pliocene (4 MYBP) included some $C_4$ plant material ($\delta^{13}C$, about −10‰) that has increased overall since, but varies widely from −12 to 0‰ about 1.5 MYBP. For cercopithecid primates, the $\delta^{13}C$ values are generally lower than for hominins, suggesting that they consumed more $C_4$ plant material or had different physiologies with respect to C isotope diet-tissue fractionation.

The $\delta^{15}N$ composition and $\delta^{13}C$ of animal tissues reflects differences in niche differentiation and microhabitat use by mammals. Consumers usually have an enriched $^{15}N$ compared to their diet; $\Delta = \delta^{15}N_{consumer} - \delta^{15}N_{diet}$, and it is higher for animals that excrete urea (e.g. mammals) or uric acid compared to guanine or ammonia. Carnivores, herbivores, and mixed-diet animals have a higher $\delta^{15}N$ than detritivores (Vanderklift & Ponsard 2003). Overall, mammals have a slightly, but not significantly, higher $\Delta$ than birds, fish, and insects, and a significantly higher $\Delta$ than crustaceans and molluscs. Individual tissues of mammals vary from low (kidney) to intermediate (fur, blood cells, plasma, liver) to high $\Delta$ (brain). The trophic $\Delta$ (or $\delta^{15}N$) enrichment is about 3‰ per trophic level for vertebrates.

Examining the combination of $\delta^{15}N$ and $\delta^{13}C$ composition of tissues in animals within communities provides considerable insight into how the animals partition their environment with respect to trophic level and microhabitat. For example, Dammhahn et al. (2013) analysed the trophic diversity and microhabitat

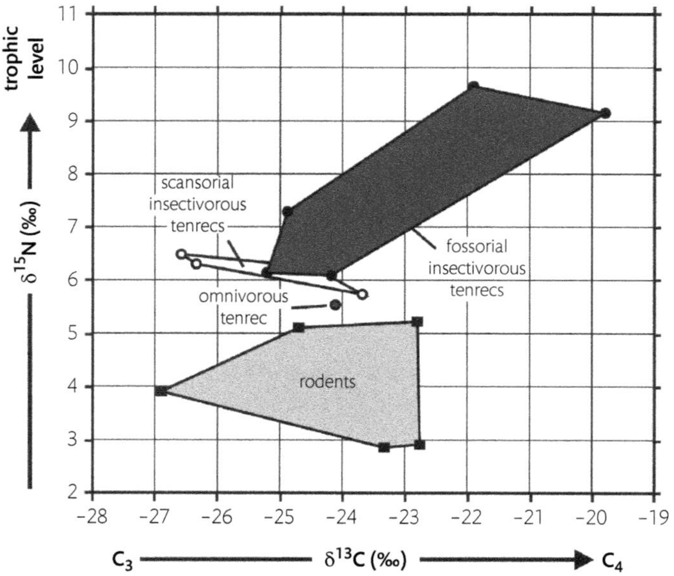


**Figure 5.16** Stable carbon and nitrogen ratios of small mammals in a montane Madagascar community indicate relative trophic levels ($\delta^{15}N$) and $C_3$-$C_4$ content of their plant diet ($\delta^{13}C$). Modified from Dammhahn et al. (2013).

utilization for a small mammal community in montane Eastern Madagascar, using the $\delta^{13}C$ and $\delta^{15}N$ of hair samples (Figure 5.16). Mainly granivorous and frugivorous rodents were the primary consumers, with low $\delta^{15}N$, and fossorial/scansorial/terrestrial insectivorous tenrecs were secondary and tertiary consumers, with about 3‰ enrichment compared to invertebrates. A small (3.6-g) terrestrial tenrec (*Microgale pusilla*) had the highest $\delta^{15}N$, suggesting it was a secondary consumer of predatory insects. A larger (102-g) tenrec (*Hemicentetes nigriceps*) also had a high $\delta^{15}N$, suggesting it was a secondary insect consumer, but also differed from the other small mammals in having a high $\delta^{13}C$, suggesting that it consumed more $C_4$ plants and often foraged in more open habitats. Thus, there was considerable trophic separation of the rodents and tenrecs, as they differed in diet and occurred in different microhabitats. There was also dense spacing (low nearest-neighbour distance) and regular rather than random packing of these species in the community. Thus, the stable isotope analyses explain how these various small mammals coexist in this montane community.

One particular aspect of the $\delta^{13}C$ composition of animal tissues to infer diet is that the isotopic turnover varies as a function of the metabolic activity of the tissue itself. Thus, analysis of the $\delta^{13}C$ composition of tissues with a high turnover rate, such as the liver, is likely to provide information on the diet on a more recent time window than, for example, analysis using muscle or hair (Vander Zanden et al. 2015). Also, a basic assumption of the analysis of $\delta^{13}C$ composition of animal tissues to infer diet is that the carbon content of the foodstuffs will enter a pathway that eventually will result in tissue synthesis. However, some foodstuffs will not follow this nutrient routing and might enter a metabolic pathway; thus, they will not be stored in tissue (Perkins & Speakman 2001). The analysis of $\delta^{13}C$ in the breath potentially overcomes this problem as it provides information on the diet that is being metabolized and, hence, recently eaten, by the animal (Voigt et al. 2008). Although the breath analysis has some limitations, especially for mammals that endure long fasting periods, it allows differentiation between whether endogenous or exogenous energy sources are being used to fuel the metabolism (section 4.7.3). It also allows for a fine-grained assignment of mammals to a particular food web categorization. For example, Voigt et al. (2013) used this method to characterize the diet of the cheetah (*Acinonyx jubatus*) with respect to grazing and browsing herbivores.

### 5.7.3 Radioisotopes in Nuclear Medicine

In contrast to stable isotopes, many elements have unstable, radioactive isotopes (radionuclides). There are more than 1,000 known radionuclides, mostly manufactured. Lighter elements have fewer radionuclides (e.g. hydrogen has one, tritium, ${}^{3}_{1}H$ ), whereas heavier elements have many (e.g. carbon has 13; iodine has 15). Radionuclides vary widely in the energy of emitted particles, light, or rays ($\alpha$, $\beta$, $\gamma$) and in their half-life (e.g. 1.5 $10^{-4}$ s for ${}^{214}_{84}Po$ vs 4.55 $10^{9}$ years for ${}^{238}_{92}U$ ; radionuclide half-life, $t_{1/2}$, is the time it takes for half the material to decay). The SI unit of radioactivity is the Becquerel, defined as the disintegration of one nucleus per second. The older unit is the Curie, originally defined as the radiation emitted from one gram of radium; it is equal to 3.7 $10^{10}$ disintegrations per second.

Radioisotopes are used routinely in nuclear medicine (Cameron & Skofronick 1978). Most radioactive elements emit $\beta$ radiation, which is low energy and has poor penetration of tissues, and is not generally useful, although ${}^{3}_{1}H$, ${}^{14}_{6}C$, and ${}^{32}_{14}P$ are commonly used. Rather, most nuclear medicine radioisotopes emit $\gamma$ rays, which are high energy and highly penetrating. Most have short half-lives, making their synthesis and use problematic. Some manufactured radionuclides emit positrons, which are positively charged beta particles (i.e. positively charged electrons, $\beta^{+}$), whereas beta particles are negatively charged electrons (i.e. $\beta^{-}$). Common uses of radioisotopes in nuclear medicine are medical imaging (e.g. molecular imaging using radioactive markers, CT scans, PET scans) and chemotherapy (MRI and ultrasound are non-radioactive imaging procedures). They are commonly used in biological studies for imaging (e.g. X-rays used in locomotion studies), PET scans (localization of glucose metabolism using fluorodeoxyglucose), and 3D imaging (microCT scanning of biological structures).

## 5.8 Species Geographic Range

Describing the geographic range of a mammalian species is an essential aspect of its biology, but it is often difficult to determine precisely, and it can change (historically, or as a result of climate change). Various methodological tools are available to determine a species' geographic range (Fortin et al. 2005). The question of why a mammalian species occurs where it does and not elsewhere—an important application of its physiology and other features—can be answered, but with difficulty. Explaining why species do not evolve to expand their range limits poses an evolutionary problem that has not been solved (Futuyma 1998). Our understanding of speciation, physiological adaptation, plasticity, and ecological interactions all contribute to our understanding of the distribution and function of biodiversity along geographic gradients and the factors that determine the geographic range limits of species.

Distributional limits must be the product (or by-product) of complex interactions between species-specific physiological, phenological, and ecological traits, as well as dispersal ability and biotic interactions. For example, the distribution of bats in Alaska depends on climate, roost availability, forested habit, photoperiod, prey abundance, and geographic barriers to dispersal, although their relative contributions are not known (Parker et al. 1997). In addition, many of the underlying traits may be phylogenetically conserved amongst related species.

A relatively new approach to investigating physiological function in the context of the Earth's ecosystems and biodiversity has emerged with the integration of physiological ecology and macroecology, termed macrophysiology (Chown et al. 2004). Macrophysiology aims to explain how physiological traits are affected by high levels of environmental variability encountered over large geographic ranges and over long temporal scales. In general, this disciplinary convergence to macrophysiology compares physiological features between individuals with different (in some cases, allopatric) geographical distributions. The macrophysiological approach to biodiversity seeks to understand patterns of geographic physiological variability (physiographic patterns) within the framework of the hierarchical structure of biodiversity, and the mechanisms that underlie these patterns.

A number of climate-based hypotheses for variation in the distribution range of species have emerged (Pither 2003), including the climate variability hypothesis (CVH), the climate extreme hypothesis (CEH), and the optimal climate hypothesis (OCH). Guided by such hypotheses, researchers use data on the physiological traits of the species (focusing on variations in latitude and altitude; Gaston & Spicer 2001) to predict responses to climatic variables, and to understand how these variables may affect the geographic range of assemblages, species, and populations (Bozinovic et al. 2011).

Thermoregulatory and energetic constraints have often been invoked to explain the distribution of species (see Bozinovic et al. 2011), but relatively few studies have examined the actual relationship between physiological capacities and distribution boundaries for endotherms (e.g. McNab 2002). Studies (mostly on birds) suggest that ambient temperature and northern boundaries are related (e.g. Canterbury 2002).

For mammals, Humphries et al. (2002) used a bioenergetic model to predict the range distribution of the little brown bat *Myotis lucifugus* in northern North America. Their results not only suggest that the distribution of *M. lucifugus* is indeed constrained by thermal effects on hibernation energetics, but provide a mechanistic explanation of how energetics, climate, and biogeography are related. Further, the results predict a northward movement over the next 80 years. Lovegrove (2000) reported significant differences in BMR between similar-sized mammals from the six geographical zones (Afrotropical, Australasian, IndoMalayan, Nearctic, Neotropical, and Palearctic). Nearctic and Palearctic mammals had higher BMR than their Afrotropical, Australasian, IndoMalayan, and Neotropical equivalents, and these patterns could be explained with a model describing geographical variation in BMR in terms of the influence of climate variability.

Traditionally, species distribution models (SDMs) use a correlative approach that incorporates the statistical relationships between the actual distribution of species and a particular set of environmental conditions to predict future distributions under different scenarios of climate change (Elith & Leathwick 2009). These correlative models, however, do not incorporate the range of functional traits that are likely to constrain the geographical boundaries of distribution and, thus, might have a poor capacity to forecast the response of the species to novel environments (Elith et al. 2010). Because physiological processes can be paramount to shaping the distribution of animals (Bozinovic et al. 2011), the use of physiological traits has been advocated as a powerful functional tool to predict the responses of species to future scenarios of climate change (Evans et al. 2015). For example, Khaliq et al. (2014) used the thermal tolerance of rodents, bats, and carnivores, as measured by the amplitude of their thermoneutral zone, to predict their vulnerability to different scenarios of future climate change. Mechanistic models (MMs; see Evans et al. 2015) provide a more direct approach, explicitly incorporating physiological traits, than SDMs. The extent by which such mechanistic models perform better, or worse, in predicting shifts in species distribution due to climate changes, when compared to correlative models is still subject to debate. For one mammal, Kearney et al. (2010) found a similar performance of standard correlative models (Bioclim, Maxent) and physiologically based mechanistic models (NicheMapper) to predict the effects of climate change and possible extinction events for the greater glider (*Petauroides volans*).

Rezende et al. (2004) and Bozinovic and Rosenmann (1989) reported a highly significant correlation between the maximum thermoregulatory metabolism (MMR) of rodent species and environmental temperature. Rodriguez-Serrano and Bozinovic (2009) analysed the diversity of physiological responses in non-shivering thermogenesis (NST) amongst species of rodents from different geographical

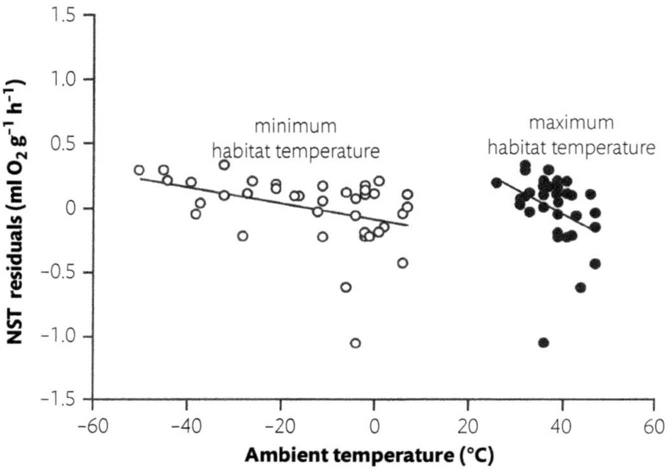


**Figure 5.17** Correlations between mean maximum and minimum geographic temperature and the residuals of non-shivering thermogenesis (NST). Each point represents a rodent species. Modified from Rodriguez-Serrano and Bozinovic (2009). Reproduced with permission of John Wiley & Sons.

areas in the world. They found a negative correlation between NST and geographic temperature, suggesting that selection may act on thermoregulatory performance. The absence of a phylogenetic signal in these traits suggests that interspecific differences in MMR and in NST may be partially explained as adaptations to different thermal environments. The results suggest that temperature imposes a high selective pressure on maximum thermoregulatory capabilities (both MMR and NST) in rodents in particular, and perhaps in small endotherms in general.

The relationship between thermoregulatory capacity and environmental temperature (Figure 5.17) supports the thermal niche expansion model, which proposes that high metabolic rates evolved because they allow animals to increase the range of environmental temperatures in which they can function. Also, many records of species-specific metabolic rates show that these physiological traits are often evolutionarily labile (i.e. there is an absence of phylogenetic signal for mass-independent values) and negatively correlated with environmental temperature. In fact, because higher and lower values of metabolic rates amongst endothermic species are associated with acclimation to cool and short days, or to warm and long days, respectively, a component of phenotypic plasticity may partially account for these results. Indeed, evidence suggests that NST is the most plastic of metabolic variables because its heat-producing machinery (brown fat with sympathetic innervation) increases quickly after cold exposure. Nevertheless—and in spite of the plasticity of the trait—there is a clear geographic pattern: species that evolved in cold climates have higher mass-independent metabolic rate than species from warmer habitats (Rodriguez-Serrano & Bozinovic 2009).

Even though some studies provide good evidence that long-term average temperatures could affect thermogenic capabilities at an evolutionary scale, physiological traits are plastic and dependent on several environmental as well as species-specific factors, which have been only partially taken into account. An assumption of these studies is that metabolic plasticity lies within a limited range that is characteristic of each species or population, though migration could mitigate fundamental differences at this level. Many authors have reported differences in thermogenic capacity after acclimation to cold or warm temperatures. Geiser and Ferguson (2001), for instance, reported geographic and captive-breeding variability in torpor characteristics for feathertail gliders (*Acrobates pygmaeus*). However, developmental acclimation cannot be ruled out as a possible explanation in cases where reported data could reflect either or both genetic adaptation and specialization to a particular thermal environment during ontogeny (i.e. developmental plasticity). For example, Riek and Geiser (2012) showed developmental patterns in response to moderate cold for both morphology and thermal energetics. In short, further work is required to estimate the relative importance of ontogeny on metabolic flexibility.

Another limitation arises from the nature of the study itself; indeed, comparative studies must trade off precision for generality. The use of coarse meteorological variables, for instance, may not accurately reflect the thermal microclimates that species experience *in situ* as well as the whole geographic range of each species. On the other hand, certain species may not be representative of the range of physiological abilities of a given clade. In addition, as pointed out by Spicer and Gaston (1999), an open key question in studies of physiological diversity and range limits is 'How well do between-species patterns in physiological diversity predict patterns across populations?'

# 6
# Conclusions and Future Directions

Ecological and environmental physiology of mammals is an extraordinarily important and dynamic field of biology. Mammals are an iconic group of animals from our human perspective, and they are important components of all ecosystems on Earth. This book has highlighted our current understanding of their physiological function in an environmental and ecological context, and has demonstrated that we currently have an extensive knowledge of how mammals work and how they have exploited the basic mammalian bauplan to inhabit a diverse array of habitats and niches. Research of mammalian ecophysiology has enabled us to address a variety of broad ecological and evolutionary questions, and their importance as companion and agricultural species, and their role as models for human medical research, have informed our broad understanding of their ecophysiology—more so than for any other taxa. Relatively new molecular disciplines are contributing mechanistic and evolutionary background information to environmental and ecological questions.

Despite this extensive body of research, many fundamental questions concerning mammalian ecophysiology remain. Some common examples include our incomplete understanding of the processes associated with mammalian torpor and hibernation, and the necessity of periodic arousals; our limited knowledge of insensible evaporative water loss and its potential regulation; and—despite a huge body of research—the debate concerning the mechanistic explanation for, and scaling exponent of, the allometry of basal metabolic rate.

Clearly, there is still much to learn about the ecological and environmental physiology of extant and past mammals. Advances in knowledge will allow us to improve our fundamental knowledge about the other mammalian species with which we share this planet, to address broad questions of evolutionary and ecological theory and, in more applied contexts, inform medical research, agricultural production, veterinary science, and other human endeavours. For example, studies of mammalian hibernation can inform human medicine by determining how hibernators overcome issues such as muscle atrophy and ischaemia during long periods of inactivity and hypothermia. One day, such studies may even contribute to human space travel (Mitchell 1972; Geiser 2013; Lovegrove et al. 2014).

*Ecological and Environmental Physiology of Mammals*. Philip C. Withers, Christine E. Cooper, Shane K. Maloney, Francisco Bozinovic, & Ariovaldo P. Cruz-Neto.  Published 2016 by Oxford University Press.
DOI 10.1093/acprof:oso/9780199642717.001.0001

One particularly pressing issue is the application of environmental and ecological physiology to conservation. Humans are degrading the environment at an alarming and ever-increasing rate, endangering not just mammals but directly causing extinction of a variety of taxa at an accelerated rate. Many scientists consider that the Earth is now experiencing its sixth mass extinction event, the first five being 'natural' (e.g. volcanism, asteroid collision); the sixth being anthropogenic. Ceballos et al. (2015) estimate that about 338–617 mammalian species have become extinct since 1500 AD (based on conservative estimates), and 198–477 species since 1900 AD (very conservative estimates). This is 8–15 and 22–52 times higher, respectively, than the background, pre-human extinction rate of about 2 E/MSY (extinctions per $10^6$ species per year). The extinction process is no less dire for other vertebrate and invertebrate species, but is probably even less well documented.

## 6.1 Future Directions

Although much future ecological and environmental mammalian research will be devoted to basic aspects of physiology and function, it is likely that this research will become more and more multidisciplinary. Modern molecular and 'omics' approaches will be applied to determine the mechanistic drivers of whole animal responses, as physiology is the interface between genetics, molecular biology, and phenotypic expression (Harrison et al. 2012), and also forms a nexus with more traditional disciplines of anatomy, reproductive biology, and ecology. 'How animals work in their environment' is a fundamental understanding that underpins all these disciplines: physiology gives the cellular studies a broad context, and the environmental disciplines a theoretical and mechanistic basis.

Funding opportunities are likely to have a substantial impact in driving future directions of ecophysiological research. Medical funding dominates other disciplines. For example, in the USA, life sciences receive about 60% of all research and development funding in academic institutions, and of this 69% goes to medical and agricultural research (National Science Board 2014). In Australia, 83% of government research funding is for medical research (Research Australia 2011). Therefore, ecophysiological studies that have a link to, or some direct benefit for, human health or agricultural production may be more likely to proceed. Such human-focused studies can have important implications for improving understanding of how animals work in their natural environment; for example, work on domestic cattle and sheep has contributed greatly to our understanding of ruminant digestion and heat balance under extreme conditions (Beatty et al. 2006, 2008, 2014; Stockman et al. 2011).

A particularly important area for mammalian ecophysiologists to focus on is their contribution to the conservation of extant mammalian species. Effective conservation relies on fundamental knowledge of a species' biology, function, and interaction with its environment; environmental and ecological physiology provides this knowledge. Developing technology might even make it possible to resurrect ('de-extinct') recently extinct mammals; for instance, the quagga (*Equus quagga*) by selective breeding of extant zebra (Heywood 2013), or DNA cloning of the Tasmanian tiger (*Thylacinus cynocephalus*; Weidensaul 2002) and woolly mammoth (*Mammuthus primigenius*; Nicholls 2008). However, the potential for success and the ethics of species resurrection are debatable on various grounds, including species and ecosystem restoration (Oksanen & Siipi 2014; Diehm 2015).

Human-induced environmental change, such as habitat modification and destruction, desertification and climate change, and the resulting impacts on a whole range of mammals will provide opportunities and challenges for environmental physiologists. The responses (positive and negative) of mammalian species to modified environments will provide a plethora of research questions; examining how mammals respond physiologically to changing environmental conditions will provide various natural experiments; the inevitable impacts of environmental change on mammalian diversity and conservation, as well as on human health and food production, will be challenges for which environmental and ecological physiologists are uniquely and ideally placed to address. In this way, mammalian ecophysiology, at its peak in the 1970s to 1990s, is likely to experience a resurgence, as knowledge of how mammals work in their environment becomes essential to identifying and solving perhaps the greatest challenge facing mammals, including humans, in their evolutionary history.

## 6.2 Climate Change

The Earth's climate is changing rapidly. Data from the IPCC (2007, 2014) signal that a trend towards increasing global temperatures is evident, yet not the same, on all six continents (Figure 6.1). Warming is greater over continents than oceans, and is greater for high northern latitudes, where it is predicted to exceed 10°C by 2051–2061 (Delworth et al. 2007). Climate change itself is likely to impact on many mammalian species, and indirect effects consequent on the likely changes in distributions of interacting species (competitors, predators, prey, pathogens) will lead to shifts in the distributions of many species, and potentially extinction.

Poleward and elevational shifts in the distribution of plant and animal species have already been recorded. More than 90% of the tens of thousands of reported trends in biological systems are consistent with climate change and an increase in temperature (Rosenzweig et al. 2008). The causes of these changes are complex and associated with a network of events in which human action appears to have a

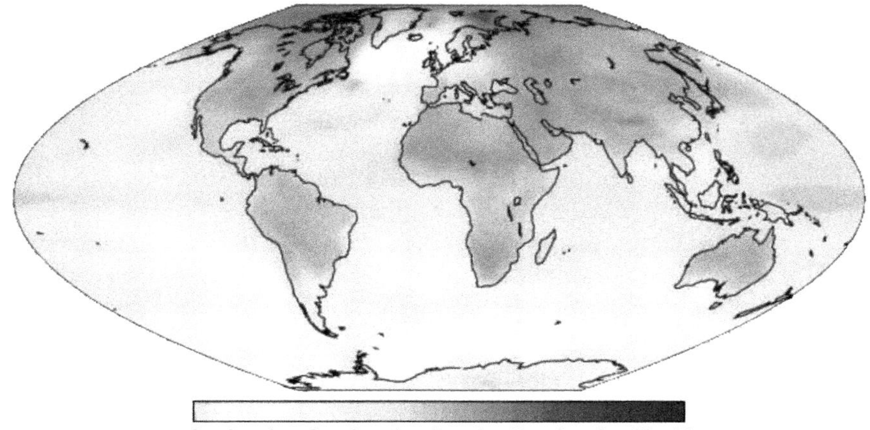


**Figure 6.1** Climate model for mean surface air temperature for 2051–2060 (compared to 1971–2000 average) based on 'middle-of-the-road' estimations (scenario IPCC SRES A1B). Modified from NOAA Geophysical Fluid Dynamics Laboratory (Delworth et al. 2007). (http://www.gfdl.noaa.gov/visualizations-climate-prediction)

determinant role. An unequivocal pattern that involves both natural and human-induced causes is the global, yet heterogeneous, change in the climate of Earth. Observed responses to recent climate change include shifts in phenology, changes in geographic range, disruption of ecological interactions, and changes in primary production (Parmesan & Yohe 2003).

Although these trends are expected to have pervasive effects on the biota, the nature, span, and final consequences of climate change will differ amongst taxa. Diversity in physiological and behavioural traits and ecological associations are evident in the fauna and flora, even amongst individuals. For instance, climate alters the frequency of contact with potentially pathogenic microbial organisms, which can cause new diseases in various species and populations of plants and animals, including humans. The response of mammals to climate change likely will be very different from that of ectotherms, and different between mobile and sedentary species. Thus, understanding the nature of differential effects of climate change on animal and plant species is one of the many urgent interdisciplinary challenges faced by contemporary science.

In terms of species persistence in the face of higher environmental temperatures, we are still not entirely sure what aspect of the environment is the most important in determining survival and persistence—whether it is the mean temperature, or the extremes of high and low. Water is another important aspect, since many mammals show no thermal strain if water is available at even high environmental temperatures, but strain develops rapidly if water intake is limited (Hetem et al. 2016). During heat waves, human morbidity and mortality can be related as much to the lack of a decrease in night-time temperature, and high water vapour pressure, as to high daytime temperatures (Henson 2006).

At least for plants, the combinations of a choice of predictor variables significantly impacts on projections for species persistence and extinction risk in the face of climate change (Pliscoff et al. 2014). Taking the perspective from energetic requirements, Humphries et al. (2002) concluded that the warmer minimum temperatures associated with climate change would place impossible energetic demands on many mammalian species that utilize torpor, because the warmer temperatures will increase the body temperature during torpor and increase their metabolic rate to the extent that a poleward shift in distribution is the only way to maintain energetic equilibrium.

## 6.3 Phenotypic Plasticity and Epigenetics

In responding to environmental pressures, including anthropomorphic challenges such as climate change, habitat modification, and desertification, the options for various species to some extent depend on their life history and especially generation time (Fuller et al. 2010). For species with a short generation time, selection optima will change and genetic adaptation to the altered climate might be possible. For species with a longer generation time (such as large mammals), rapid environmental change may provide insufficient time for genetic selection, so these mammals will be required to tolerate the new conditions, avoid them via migration and shifts in distribution ranges (assuming alternative suitable ranges exist), or face extinction locally or globally.

We are hampered by a lack of information on how much of their potential plasticity mammals use in their present distributions. In the face of new environmental challenges, species may express more of the phenotypic potential that exists, but is not expressed under present environmental conditions. It may be that only those mammalian species inhabiting extreme environments are expressing their full phenotypic potential for that environment. For example, we know that acclimation to heat in humans extends the range of temperature/humidity space in which we can achieve thermal balance by several degrees (Maloney & Forbes 2011). If similar potential exists for other mammals—but is not expressed in a given climate space—there is scope for persistence of species within their current range, even when climate envelope models suggest extinction is likely.

There is some evidence that in wild populations even phenotypic flexibility itself might be heritable and adaptive. The climatic variability hypothesis (CVH) (Gaston 2003) posits that as the range of climatic variability experienced by terrestrial animals increases with latitude, individuals from populations inhabiting higher latitudes will require a broader range of physiological tolerances to persist at that site. In addition, a wider range of physiological tolerance allows species to become more extensively distributed. Given that the range of tolerance of an organism is related to its phenotypic flexibility, the CVH implies that the physiological flexibility of species (or its populations or individuals, or both) should increase with latitude (Chown et al. 2004; Naya et al. 2007, 2008).

Populations may cope with genetic differentiation in fitness-related traits, allowing them, across generations, to adjust their mean phenotypes to prevailing environmental conditions. Further, organisms may show within-generation changes in their phenotypes in response to environmental changes, which is indeed phenotypic plasticity. Plastic responses include modifications in physiological, morphological, developmental, behavioural, and life history traits (Schlichting & Pigliucci 1998). Theoretically, this flexibility allows organisms to adjust to changing biotic and abiotic conditions, through increases in performance and likely Darwinian fitness. Thus it is not surprising that there are documented cases in which most intraspecific and interspecific physiological variation can be attributed to phenotypic plasticity.

A classic macrophysiological example emerging from the CVH is the correlation between metabolic flexibility (absolute metabolic scope = maximum metabolic rate minus basal metabolic rate) and different abiotic variables (e.g. latitude, altitude, rainfall, temperature). Using both conventional and phylogenetically informed analyses, Naya et al. (2012) found a positive correlation between metabolic flexibility and geographic latitude, together with a negative correlation between metabolic flexibility and minimum environmental temperature, for 48 rodent species (Figure 6.2). These findings have implications for the selective forces and mechanisms operating on phenotypic flexibility and the potential for mammalian species to adapt to environmental change.

Although adaptation and variation are key characteristics of life on Earth, it is remarkable that there are few analyses and predictive models concerning the effect of global environmental change on the diversity of potential responses in time and space (within and between individuals, populations, and species, and from molecules to population levels). In this context, few predictions are available to deal with the potentially diverse integrative and cellular physiological responses mammals may employ to cope with long-term local and geographical climate challenges. Denny and Helmuth (2009) argue that a lack of mechanistic insight at the organism level and below is the primary impediment to the successful prediction of the ecological effects of climate change. Through understanding the underlying mechanistic bases of physiological stress and the differences that exist within and amongst species in their abilities to respond to these stressors, a solid foundation can be constructed for developing predictions about the probability of the success or failure of organisms, populations, and species in coping with climate change (Somero 2005, 2011).

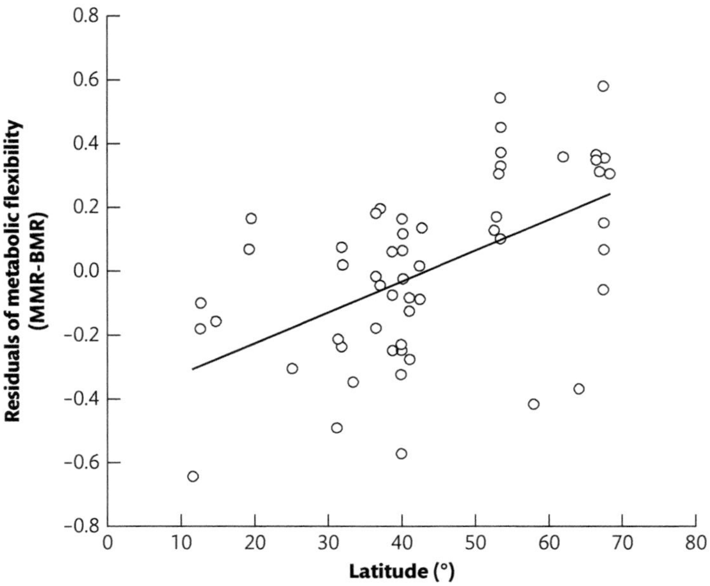


**Figure 6.2** The effect of latitude (corrected for altitude) on metabolic scope (corrected for phylogenetic scope and body mass) for 48 rodent species. Modified from Naya et al. (2012).

For more than half a century, the 'central dogma' of molecular biology, as stated by Francis Crick (1970), has been that DNA makes RNA and RNA makes protein, and that the transfer of residue by residue sequential information in DNA cannot be transferred back from protein to nucleic acid. The central dogma was the mechanistic explanation for the Weismann barrier, which states that inheritance takes place by means only of the germ cells (eggs, sperm), and that somatic cells play no role in evolution other than forming the organism that is the means for propagation of new germ cells over a lifetime. The germ cells give rise to somatic cells but are independent of them. The Weismann barrier sounded the death knell for Lamarckism, the idea that phenotypic characteristics acquired during a life time could be passed on to offspring (Jablonka & Lamb 1999). However, the study of epigenetics has revolutionized our understanding of inheritance. Epigenetics may be defined as 'the structural adaptation of chromosomal regions so as to register, signal or perpetuate altered activity states' (Bird 2007, p. 398); it provides a mechanism for the potential inheritance of acquired traits, via the heritable modification of DNA function. This may occur via the processes of DNA methylation (which 'silences' genes by interfering with transcription-factor binding, or by enabling binding of repressors) or by the variation of the Polycomb/Trithorax system (groups of proteins that maintain repressed or active gene transcription states). In this way, epigenetics may provide a mechanistic approach to examining long-term consequences of environmental conditions on the physiology of mammals and other organisms (Bird 2007).

Although important in a variety of circumstances, environmental change is one particular arena where the study of epigenetics and phenotypic flexibility will make a considerable contribution to mammalian biology. The finding that epigenetic modification of DNA occurs during a lifetime, and that at least some of those changes are heritable (Margueron & Reinberg 2010), have changed the landscape of adaptive processes and evolutionary thought. Epigenetics has broken the Weismann barrier and put physiology squarely at the forefront of evolutionary biology. The realization that a given genotype does not always give rise the same phenotype, but that the genotype/phenotype relationship can be influenced by environmental factors, and can be heritable, is opening entire new vistas of adaptive biology.

## 6.4 Conservation Physiology and Ecology

Ecosystems and their network of biotic and abiotic components face the pressure of human activities. Conservation biologists and managers attempt to identify and mitigate threats, control species extinctions, restore degraded ecosystems, and manage resources sustainably. However, as suggested by Cooke et al. (2013), an explanation of the specific physiological mechanisms underlying conservation problems is becoming increasingly important for decision-making. In part this is because physiological tools and knowledge are especially useful for developing cause–effect relationships and for identifying the optimal range of habitats and stressor thresholds for different species—including, for instance, terrestrial and aquatic mammals. According to Somero (2011), understanding the mechanistic bases of sublethal and lethal stress, and the differences that exist within and amongst species in their abilities to respond to stressors, will help us predict the ability of organisms, populations, and species to cope with environmental change (Root & Hughes 2005; Angilletta 2009; Hofmann & Todgham 2010).

In this view, conservation physiology includes a wide range of applications: for example, eradication of invasive species, refinement of resource management strategies to minimize impacts, and evaluation of restoration plans. This understanding of conservation physiology recognizes the relevance of physiological diversity at different scales. In short, conservation physiology has emerged as an integrative research area that utilizes theoretical and experimental physiological approaches, under both field and laboratory conditions, to understand the mechanisms underlying conservation problems and, as a consequence, to develop conservation strategies.

New directions in the study of mammalian conservation physiology and the impact of environmental change on mammal species and populations may focus on the following concerns: (1) the causal mechanisms underlying environmental adaptation of species encompassing a broad variety of taxa—including alien mammalian species; (2) current environmental issues, as well as a forecasting approach for future issues, in which adaptation and acclimation—or lack thereof—at biochemical, physiological, behavioural, and life history levels would be examined; (3) basic scientific issues in the fields of evolutionary ecology, behaviour, and physiology, as well as in biomedicine and agronomy; (4) research into patterns, processes, and implications of mammalian diversity, including the mechanistic basis of trait expression and the variation of traits at higher levels of organization, over broad geographical and temporal scales.

The main rationale for these particular concerns is that populations exposed to adverse environmental conditions may crash when most individuals deteriorate, and that individuals decline when they reach a state that prevents them from maintaining homeostasis or their ability to defend themselves from biotic (e.g. microbial) or physicochemical (e.g. water availability) stressors—that is, the proper equilibrium through time between internal processes and foreign stimuli. If equilibrium is not maintained, fitness may be reduced, populations may lose genetic variation, and extinction becomes more likely (Deutsch et al. 2008).

Theoretical ecologists are developing different models of how mammalian biodiversity can be impacted by environmental change. Nevertheless, they concur that their models have severe limitations because they lack a mechanistic basis, which may lead to very large under- or overestimations of risks and vulnerabilities for individuals, species, and ecosystems to unprecedented change (Huey & Tewksbury 2009). For example, many studies have used modelling to estimate the likely survival of mammalian species under various climate change mitigation scenarios, with some studies predicting up to 60% species loss this century (Thomas et al. 2004). Many such studies rely on climate envelope models that place a species into a climate niche space—based on ambient temperature, humidity, and rainfall—and project the likely existence of suitable habitat under the various climate change scenarios.

Such models have been criticized on several fronts: (1) they assume that there is no difference between a species' realized niche and its fundamental niche space; (2) they lack mechanistic underpinning, and as such are merely correlations (Kearney & Porter 2009); and (3) they ignore, or at least fail to account for, the potential phenotypic flexibility of species (Fuller et al. 2010). More mechanistic modelling works from the basis that physiological limits are the fundamental constraints on the abundance and distribution of species. Physiological ecology has provided much information on these limits, and has developed mechanistic models of endotherm distributions, such as NicheMapper (Kearney & Porter 2009), contributing to the relatively new field of conservation physiology (Wikelski & Cooke 2006). There is no doubt that physiological knowledge will be crucial to understanding how mammals will be impacted by and respond to environmental change.

Another contribution physiologists can make is to inspire new methodology in experimental, field, and comparative studies. Physiologists can lead the way in rejecting the typical approach—test one variable at a time—in favour of a more integrative approach that takes an interdisciplinary view, incorporating basic and applied perspectives to whole-organism function in the context of environmental biology (Humphries et al. 2002; Helmuth et al. 2005). This is vital, because synergistic and multiple (positive, neutral, or negative) interactions are very likely under climate change scenarios (Reusch & Wood 2007), so we need to understand physiological trade-offs and cost–benefit relationships for mammalian species. Information on the integration of different traits and how to parameterize these traits is fundamental for the modelling of issues such as emerging diseases and determinants of species abundance, survival, range, and limits (Bozinovic & Pörtner 2015). New research programmes should address gaps in the few current approaches concerned with the physiological basis of the response to environmental change, and suggest avenues of restoration, education, and management (Bell & Collins 2008; Cooke & Suski 2008).

## 6.5 Medicine, Veterinary Science, and Agriculture

There has always been a collaborative relationship between the discipline of environmental and ecological physiology and the more applied fields of medicine, veterinary science, and agricultural science. Understanding how wild, free-living mammals function in their natural environment and respond to various selection pressures has informed these human-focused fields by developing evolution-based frameworks to understand the function of physiological processes. In turn, the applied disciplines have contributed greatly to environmental physiology by providing much of the physiological detail and various techniques and technologies that have been exploited by ecophysiologists to better understand how wild mammals 'work' in their natural environments. For example, the use of helox in ecophysiological research—widely used to perturb the thermal and even evaporative environment of mammals to examine their thermoregulatory, metabolic, and hygric capacity— is largely derived from human medical research associated with deep-sea diving and space travel (Behnke 1940; Musacchia 1972; Rosenmann & Morrison 1974a; Lovegrove et al. 2015). In the future, this synergistic, multidisciplinary relationship is likely to strengthen, as the human population faces unprecedented rates of environmental change that will challenge current medical, veterinary, and agricultural knowledge.

Agricultural production is one particular example where the role of ecophysiologists will be paramount. Demand for livestock products is high but relatively stable in developed countries, but is increasing rapidly in developing countries due to population growth, increasing urbanization, and greater wealth, resulting in an ever-growing global demand for animal products, of which mammals comprise the major source (see 1.5.3; Thornton 2010). However, a growing global human population, together with the effects of environmental change (e.g. desertification, climate change), mean that this growing demand will need to be met under conditions of changing resource availability: reduced water availability, altered plant phenology, and competition between food and feed, and in the face of changing socioeconomic factors such as environmental and animal welfare concerns, constraints on carbon emissions, and diminishing fossil fuels (Thornton 2010; Hegarty 2012).

To meet demand under these pressures, improved knowledge and new technologies concerning livestock nutrition and health will be necessary, providing a clear opportunity for mammalian environmental and ecological physiologists to apply their expertise. There are already countless examples of ecophysiological influences on issues associated with livestock production over a broad range of topics (e.g. welfare of livestock involved in live export, methane production by foregut fermenters, livestock resource requirements, competition between native wildlife and introduced livestock, environmental impacts on production; Pethick et al. 2006; Barnes et al. 2008; Munn et al. 2009, 2012, 2013b; Bernabucci et al. 2010; Vendl et al. 2016). We foresee ever-growing opportunities as the challenges increase over time.

An important contribution of physiologists to medical research has been the finding that the thermal environment can influence the expressed phenotype of genetic models of disease. For example, based on the knowledge that uncoupling protein-1 (UCP1) plays a known role in cold defence and also in diet-induced thermogenesis, it was predicted that animals lacking UCP would be cold-sensitive and also obese. When the first UCP1 knockouts were developed, they were indeed susceptible to hypothermia during cold exposure, but, contrary to expectation, were not obese, and even seemed to be protected from developing obesity. It was soon realized that the fact that these experiments were conducted at normal animal house temperature (around 22°C, well below the lower critical temperature of 30°C in most mouse strains) might have played a role. When the same experiments were conducted at 30°C, UCP1 knockout mice did develop obesity, which was exacerbated by high-fat feeding (Maloney et al. 2014). There are now many examples of rodent phenotypes being altered by sub-thermoneutral housing. Thus the interpretation of gene function in humans from experiments on rodent models depends crucially on the use of an appropriate environment.

# Appendix

Classification of the families of Mammalia, including extant and fossil (†) mammals. This taxonomic hierarchy does not directly reflect the phylogenetic arrangement of Atlantogenata (Xenarthra + Afrotheria) and Boreoeutheria (Euarchontoglires and Laurasiatheria); see section 1.1.1 and Table 1.1. Based on McKenna and Bell (1997); Springer et al. (2003); Kielan-Jaworowska et al. (2004); Wilson and Reeder (2005); Nowak and Dickman (2005); Rose (2006); Huchon et al. (2007); and Murphy et al. (2007).

Class MAMMALIA

†*Adelobasileus*, †*Hadroconium*, †*Tricuspes*

†Reigitheriidae

†Kuehneotheriidae

†Sinoconodontidae

Order †MORGANUCODONTA

†*Gondwanadon*, †*Indotherium*

†Morganucodontidae

†Megazostrodontidae

Order †DOCODONTA

†Docodontidae

Order †SHUOTHERIDIA

†Shuotheriidae

Order †EUTRICONODONTA

†Amphilestidae

†Triconodontidae

†Gobiconodontidae

†Austrotriconodontidae

Order †GONDWANATHERIA

†Ferugliotheriidae

†Sudamericidae

Subclass AUSTRALOSPHENIDA

†*Ambondro*, †*Asfalotomylos*

†Kollikodontidae

Order †AUSKTRIBOSPHENIDA

†Ausktribosphenidae

Order MONOTREMATA ('Prototheria' –monotremes)

Ornithorhynchidae: platypus

Tachyglossidae: echidnas

†Steropodontidae

Subclass †ALLOTHERIA

---

Order †HARAMIYIDA

Suborder Theroteinida

†Theroteinidae

Suborder HARAMIYOIDEA

†Haramiyidae

†Haramiyaviidae

†Eleutherodontidae

Order †MULTITUBERCULATA

Suborder PLAGIAULACIDA

†Allodontidae

†Zofiabaataridae

†Paulchoffatidae

†Hahnodontidae

†Pinheirodontidae

†Plagiaulacidae

†Eobaataridae

†Albionbaataridae

†Arginbaataridae

Suborder †CIMOLODONTA

†*Djadochtatherium*

†Sloanbaataridae

†Cimolodontidae

†Ptilodontidae

†Cimolomyidae

†Eucosmodontidae

†Taeniolabididae

†Microcosmodontidae

†Neoplagiaulocidae

†Kogaionida

Subclass TRECHNOTHERIA

Superorder †SYMMETRODONTA

†Amphidontidae

†Tinodontidae

†Spalacotheriidae

Superorder †DRYOLESTOIDEA

†Vincelestidae

Order †DRYOLESTIDA

†Dryolestidae

†Paurodontidae

Order †AMPHITHERIIDA

†Amphitheriidae

Superorder †ZATHERIA

Order †PERAMURA

†Peramuridae

†Arguitheriidae

†Arguimuridae

Subclass BOREOSPHENIDA

Order †AEGIALODONTIA

†Aegialodontidae

Infraclass METATHERIA

†*Holoclemsia*, †*Sinodelphys*

Cohort †DELTATHEROIDA

Order †DELTATHEROIDEA

†Deltatheriidae

Order †ASIADELPHIA

†Asiatheriidae

Cohort MARSUPIALIA (marsupials)

†*Kokopellia*

Magnaorder AMERIDELPHIA

†Alphadontidae

†Pediomyidae

†Stagodontidae

†Protodidelphinae

Order DIDELPHIMORPHIA

†Peradectidae

†Sparassocynidae

Didelphidae: opossums

Order PAUCITUBERCULATA

Superfamily †Caroloameghiniodea

†Caroloameghiniidae

†Glasbiidae

Superfamily Caenolestoidea

†Sternbergiidae

†Palaeothentidae

†Abderitidae

Caenolestidae: rat and shrew opossums

Superfamily †Polydolopoidea

†Sillustaniidae

†Polydolopidae

†Prepidolopidae

†Bonapartheriidae

Superfamily †Argyrolagoidea

†Argyrolagidae

†Patagoniidae

†Groeberiidae

Order †SPARASSODONTA

†Mayulestidae

†Hondadelphidae

†Borhyaenidae

†Proborhyaenidae

†Haliacynidae

Magnaorder AUSTRALIDELPHIA

†*Djarthia*

Superorder MICROBIOTHERIA

Microbiotheriidae: monito del monte

Superorder EOMETATHERIA

Order NOTORYCTEMORPHIA

Notoryctidae: marsupial moles

Grandorder DASYUROMORPHIA

†Thylacinidae: recently extinct Tasmanian tiger

Dasyuridae: Tasmanian devil, quolls, marsupial mice, cats

Myrmecobiidae: numbat

Grandorder SYNDACTYLI

Order PERAMELIA

Peramelidae: bandicoots

Peroryctidae: spiny bandicoots

Order DIPROTODONTIA

†Palorchestidae

†Wynardiidae

†Thylacoleonidae

Tarsipedidae: honey possum

Superfamily Vombatoidea

†Ilariidae

†Diprotodontidae

Vombatidae: wombats

Superfamily Phalangeroidea

Phalangeridae: phalangers

Burramyidae: pygmy possums

Macropodidae: rat kangaroos, kangaroos, wallabies

Petauridae: gliders

†Ektopodontidae

Phascolarctidae: koala

†Pilkipildridae

†Miralinidae

Acrobatidae: feather-tail gliders, pen-tailed phalanger

Infraclass EUTHERIA

†*Eomaia*, †*Montanalestes*

Order †ASIORYCTITHERIA†Asioryctidae

†Kennalestidae

Cohort PLACENTALIA (placental mammals)

Order †BIBYMALAGASIA

Order XENARTHRA (edentates)

Suborder CINGULATA

Superfamily Dasypodoidea

Dasypodidae: armadillos

†Peltephilidae

Superfamily †Glyptodontoidea

†Pampatheriidae

†Palaeopeltidae

†Glyptodontidae

Suborder PILOSA
Infraorder VERMILINGUA
Myrmecophagidae: giant anteaters
Cyclopedidae: pygmy anteater
Infraorder PHYLLOPHAGA
†*Pseudoglyptodon*
†Entelopidae
Bradypodidae: three-toed sloths
Parvorder †MYLODONTA
†Scelidotheriidae
†Mylodontidae
Parvorder MEGATHERIA
†Megatheriidae
†Nothrotheridae
Megalonychidae: two-toed sloths
Superorder INSECTIVORA
Order †LEPTICTIDA
†*Gypsonictops*
†Leptictidae
†Pseudorhyncocyonidae
Order LIPOTYPHLA
†Adapisoriculidae
Suborder ERINACEOMORPHA
Erinaceidae: hedgehogs
†Sespedectidae
†Scenopagidae
†Amphilemuridae
†Adapisoricidae
†Creotarsidae
†Chambilestidae
Suborder SORICOMORPHA
†*Otlestes*, †*Batodon*, †*Paranyctoides*
†Geolabididae
Superfamily Soricoidea
†Nyctitheriidae
Soricidae: shrews

†Plesiosoricidae
†Nesophontidae
Solenodontidae: solenodons
†Micropternodontidae
†Apternodontidae
Superfamily Tenrecoidea
Tenrecidae: tenrecs
Superfamily Talpoidea
†Proscalopidae
Talpidae: moles
†Dimylidae
Suborder CHRYSOCHLOROMORPHA
Chrysochloridae: golden moles
Superorder ANAGALIDA
†Zalambdalestidae
†Anagalidae
†Pseudictopidae
Order MACROSCELIDEA
Macroscelididae: elephant shrews
Grandorder GLIRES
Mirorder DUPLICIDENTATA
†MIMOTONIDA
†Mimotonidae
Order LAGOMORPHA
Ochotonidae: pikas
Leporidae: rabbits and hares
Mirorder SIMPLICIDENTATA
†*Sinomylus*
Order †MIXODONTIA
†Eurymylidae
Order RODENTIA
†Alagomyidae
†Laredomyidae
Superfamily Ctenodactyloidea
†Cocomyidae
†Chapattimyidae
†Yuomyidae

†Tamqauammyidae
†Gobiomyidae
Ctenodactylidae: gundis
Diatomyidae: Laotian rock rat
Suborder SCIUROMORPHA
†Ischyromyidae
†Paramyidae
†Sciuravidae
†Cylindrodontidae
†Theridomyidae
Sciuridae: squirrels
Gliridae: dormice
Superfamily Aplodontoidea
Aplodontidae: mountain beaver
†Allomyidae
†Mylagaulidae
Superfamily Casteroidea
†Eutypomyidae
Castoridae: beavers
Suborder MYOMORPHA
†Protoptychidae
Infraorder MYODONTA
†Armintomyidae
†Simimyidae
Superfamily Dipodoidea
Zapodidae: jumping mice
Dipodidae: jerboas, jumping mice, birch mice
Superfamily Muroidea
Muridae: Old World rats and mice, gerbils, whistling rats
Cricetidae: New World rats and mice, hamsters, voles, lemmings
Infraorder GEOMORPHA
Superfamily †Eomyoidea
†Eomyidae
Superfamily Geomyoidea
†Florentiamyidae
†Heliscomyidae
Geomyidae: pocket gophers
Heteromyidae: pocket mice, kangaroo rats

Suborder ANOMALUROMORPHA

Superfamily Pedetoidea

Pedetidae: springhare

†Parapedetidae

Superfamily Anomaluroidea

†Zegdoumyidae

Anomaluridae: scaly tailed squirrels

Suborder HYSTRICOGNATHA

†Tsaganomyidae

Infraorder HYSTRICHOGNATHI

Hystricidae: Old World porcupines

Erethizontidae: New World porcupines

†Myophiomyidae

†Diamantomyidae

†Phiomyidae

Petromuridae: dassie rats

Thryonomyidae: cane rats

Parvorder BATHYERGOMORPHI

Bathyergidae: mole rats

Parvorder CAVIIDA

Superfamily Cavioidea

Agoutidae: agoutis and pacas

†Eocardiidae

Dinomyidae: pacarana

Caviidae: cavies

Hydrochoeridae: capybara

Superfamily Octodontoidea

Octodontidae: degus, tuco-tucos

Echimyidae: spiny rats, porcupine rats, urares

Capromyidae: hutias, jutias

†Heptaxodontidae

Superfamily Chinchilloidea

Chinchillidae: chinchillas, viscachas

†Neoepiblemidae

Abrocomidae: rat chinchillas

Superorder ARCHONTA

†*Deccanolestes*

Order Chiroptera

Suborder MEGACHIROPTERA

†*Archaeopteropus*

Pteropodidae: flying foxes

Suborder MICROCHIROPTERA

†*Eppsinycteris*, †*Australonycteris*

†Icaronycteridae

†Archaeonycteridae

†Hassianycteridae

†Palaeochiropterygidae

Infraorder YINOCHIROPTERA

Rhinopomatidae: mouse-tailed bats

Craseonycteridae: bumblebee bats

Nycteridae: hispid bats

Megadermatidae: false vampire bats

Rhinolophidae: horseshoe bats

Hipposeridae: Old World leaf-nosed bats

Infraorder YANGOCHIROPTERA

Emballonuridae: sac-winged bats

†Philisidae

Vespertilionidae: vesper bats

Molossidae: free-tailed bats

Mystacinidae: New Zealand short-tailed bats

Noctilionidae: fishing bats

Mormoopidae: spectacled bats

Phyllostomidae: New World leaf-nosed, vampire bats

Myzopodidae: Old World sucker-footed bats

Furipteridae: smoky bats

Thyropteridae: New World sucker-footed bats

Natalidae: funnel-eared bats

Grandorder EUARCHONTA

†Mixodectidae

Order DERMOPTERA

†Plagiomenidae

Galeopithecidae: colugos

Order SCANDENTIA

Tupaiidae: tree shrews

Order PRIMATES

†Purgatoriidae

†Microsyopidae
†Micromomyidae
†Picromomyidae
†Toliapinidae
Superfamily †Paramomyoidea
†Palaechthonidae
†Paromyomyidae
†Picrodontidae
Superfamily †Plesiadapoidea
†Plesiadapidae
†Carpolestidae
†Saxonellidae
Suborder EUPRIMATES
Infraorder STREPSIRRHINI
†Plesiopithecidae
Parvorder LEMURIFORMES
Superfamily †Adapoidea
†Adapidae
†Notharctidae
†Sivaladapidae
Superfamily Lemuroidea
Lemuridae: lemurs
Lepilemuridae: sportive lemur
Daubentoniidae: aye-aye
Indriidae: indris and sifakas
Cheirogaleidae: dwarf lemurs
Superfamily Loroidea
Loridae: lorises
Galagidae: galagos, bushbabies
Infraorder HAPLORHINI
Parvorder TARSIIFORMES
†Omomyidae
Tarsiidae: tarsiers
Parvorder ANTHROPOIDEA
†*Afrotarsius*
†Eosimiidae
†Parapithecidae
†Amphipithecidae

†Proteopithecidae

'PLATYRRHINI'

Superfamily Ceboidea

Callitrichidae: marmosets, tamarins

Cebidae: capuchin and squirrel monkeys

Aotidae: night monkeys

Pitheciidae: titis, saki, uakaris

Atelidae: howler, spider, and woolly monkeys

'CATARRHINI'

†Superfamily Propliopithecoidea

†Oligopithecidae

†Propliopithecidae

†Pliopithecidae

Superfamily Cercopithecoidea

Cercopithecidae: Old World monkeys, colobuses

Superfamily Hominoidea

†Chororapithecidae

†Proconulidae

†Afropithecidae

†Pliobatidae

Hylobatidae: gibbons

Hominidae: humans, great apes (orangutan, gorillas, chimpanzees)

Superorder FERAE

Order †CREODONTA

†Hyaenodontidae

†Oxyaenidae

Order CARNIVORA

†Viverravidae

†Miacidae

Suborder FELIFORMIA

†Nimravidae

Felidae: pantherine and other cats

Viverridae: civets, Asiatic palm civets

Nandiniidae: African palm civets

Herpestidae: mongooses

Hyaenidae: hyaenas, aardwolf

Suborder CANIFORMIA

Infraorder CANOIDEA

Canidae: dogs, wolves, foxes, jackals, dingo

Infraorder ARCTOIDEA

Parvorder URSIDA

Superfamily Ursoidea

Ursidae: bears

†Amphicyonidae

Superfamily Phocoidea (Pinnipedia)

†*Enaliarctos*

Otariidae: eared seals

Phocidae: earless (true) seals

Odobenidae: walrus

Parvorder MUSTELIDA

Mustelidae: weasels, skunks, and relatives

Procyonidae: ringtails, olingos, kinkajou, raccoons, coatis, red panda

Mirorder †CIMOLESTA

†Didymoconidae

†Wyolestidae

Order †DIDELPHODONTA

†Palaeoryctidae

†Cimolestidae

Order †APATOTHERIA

†Apatemyidae

Order †TAENIODONTA

†Stylinodontidae

Order †TILLODONTIA

†Esthonychidae

Order †PANTODONTA

†Harpyodidae

†Bemalambdidae

†Pastoralodontidae

†Pantolambdidae

†Titanoideidae

†Barylambdidae

†Cyriacotheriidae

†Pantolambdodontidae

†Coryphodontidae

Order †PANTOLESTA

†Pantolestidae

†Pentacodontidae

†Paroxyclaenidae

†Ptolemaiidae

Order PHOLIDOTA

†Eomanidae

†Patriomanidae

Manidae: pangolins

Order ?PHOLIDOTA

Suborder †PALAENODONTA

†Escavadodontidae

†Epoicotheriidae

†Metacheiromyidae

Suborder †ERNANODONTA

†Ernanodontidae

Superorder UNGULATOMORPHA

†Zhelestidae

Grandorder UNGULATA (hoofed ungulate mammals)

Order †CONDYLARTHRA

†Arctocyonidae

†Quettacyonidae

†Periptychidae

†Phenacodontidae

†Hyopsodontidae

†Mioclaenidae

†Didolodontidae

Order TUBULIDENTATA

Orycteropodidae: aardvark

Order †DINOCERATA

†Prodinoceratidae

†Uintatheriidae

Order †ARCTOSTYLOPIDA

†Arctostylopidae

Order ARTIODACTYLA (even-toed ungulates)

Suborder SUIFORMES

†Raoellidae

†Choeropotamidae

Superfamily Suoidea

Suidae: pigs

Tayassuidae: peccaries

†Santheriidae

Hippopotamidae: hippos

Superfamily †Dichobunoidea

†Dichobunidae

†Cebochoeridae

†Mixtotheriidae

†Helohyidae

Superfamily †Anthracotherioidea

†Haplobunodontidae

†Anthracotheriidae

Superfamily †Anoplotheroidea

†Dacrytheriidae

†Anoplotheriidae

†Cainotheriidae

Superfamily †Oreodontoidea

†Agriochoeridae

†Oreodontidae

Superfamily †Entelodontoidea

†Entelodontidae

Suborder TYLOPODA

†Xiphodontidae

Superfamily Cameloidea

Camelidae: camels and llamas

†Oromerycidae

Superfamily †Protoceratoidea

†Protoceratidae

Suborder RUMINANTIA

†Amphimerycidae

†Hypertragulidae

Tragulidae: mouse deer

†Leptomerycidae

†Bachitheriidae

†Lophiomerycidae

†Gelocidae

Superfamily Cervoidea
Moschidae: musk deer
Antilocapridae: pronghorn
†Palaeomerycidae
†Hoplitomerycidae
Cervidae: deer
Superfamily Giraffoidea
†Climacoceratidae
Giraffidae: giraffe and okapi
Superfamily Bovoidea
Bovidae: cattle, antelope, and relatives
Mirorder CETE
Order †MESONYCHIA
†Triisodontidae
†Mesonychidae
†Hapalodectidae
Order CETACEA
Suborder †ARCHAEOCETI
†Basilosauridae
†Pakicetidae
†Ambulocetidae
†Protocetidae
†Remingtonocetidae
Suborder ODONTOCETI
Superfamily Physeteroidea
Physeteridae: sperm whales
Superfamily Hyperoodontoidea
Hyperoodontidae: beaked whales
Superfamily Platanistoidea
Platanistidae: river dolphins
Superfamily Delphinoidea
Delphinidae: dolphins
Pontoporiidae: La Plata River dolphin
Lipotidae: baiji
Iniidae: Amazon River dolphin
†Kentridontidae
Monodontidae: beluga and narwhal

†Odobenocetopsidae

†Dalpiazinidae

†Acrodelphinidae

Phocoenidae: porpoises

†Albireonidae

†Hemisyntrachelidae

Suborder MYSTICETI

†Llanocetidae

†Aetiocetidae

†Mammalodontidae

†Kekenodontidae

†Cetotheriidae

Balaenopteridae: rorquals and grey whales

Balaenidae: right and bowhead whales

Mirorder ALTUNGULATA

†*Radinskya*, †*Olbitherium*

Order PERISSODACTYLA

Suborder HIPPOMORPHA

Superfamily Equoidea

Equidae: horses, donkeys, zebra

†Palaeotheriidae

Suborder TAPIROMORPHA

†Isectolophidae

Infraorder CERATOMORPHA

Superfamily Tapiroidea

†Helaletidae

†Deperetellidae

†Lophialetidae

Tapiridae: tapirs

Superfamily Rhinoceratoidea

†Hyrachyidae

†Hyracodontidae

†Amynodontidae

Rhinocerotidae: rhinoceroses

Infraorder †ANCYLOPODA

†Eomoropidae

†Chalicotheriidae

†Lophiodontidae

Suborder †TITANOTHERIOMORPHA

Superfamily †Brontotherioidea

†Brontotheriidae

†Anchilophidae

Order PAENUNGULATA

Suborder HYRACOIDEA

†Pliohyracidae

Procaviidae: hyraxes, dassies

Suborder TETHYTHERIA

Infraorder †EMBRITHOPODA

†Phenacolophidae

†Arsinoitheriidae

Infraorder SIRENIA

†Prorastomidae

†Protosirenidae

Dugongidae: dugongs

Trichechidae: manatees

Infraorder †DESMOSTYLIA

†Desmostylidae

Infraorder PROBOSCIDEA

†Anthracobunidae

†Phosphatheriidae

†Numidotheriidae

†Moeritheriidae

†Barytheriidae

†Deinotheriidae

†Palaeomastodontidae

†Phiomiidae

†Hemimastodontidae

†Mammutidae

†Gomphotheriidae

Elephantidae: elephants

Mirorder †MERIDIUNGULATA

†*Amilnedwardsia*

Order †LITOPTERNA

†Protolipternidae

†Notonychopidae

†Macraucheniidae

†Adianthidae

†Proterotheriidae

Order †NOTOUNGULATA

Suborder †NOTIOPROGONIA

†Henricosborniidae

†Notostylopidae

Suborder †TOXODONTIA

†Isotemnidae

†Notohippidae

†Leontiniidae

†Toxodontidae

†Homalodotheriidae

Suborder †TYPOTHERIA

†Oldfieldthomasiidae

†Interatheriidae

†Archaeopithecidae

†Mesotheriidae

†Campanorchidae

Suborder †HEGETOTHERIA

†Archaeohyracidae

†Hegetotheriidae

Order †ASTRAPOTHERIA

†Eoastrapostylopidae

†Trigonostylopidae

†Astrapotheriidae

Order †XENUNGULATA

†Carodniidae

Order †PYROTHERIA

†Comombitheriidae

†Pyrotheriidae

# References


Aas-Hansen, Ø., Folkow, L. P. & Blix, A. S. (2000). Panting in reindeer (*Rangifer tarandus*). *American Journal of Physiology* **279**, R1190–R1195.

Abaturov, B. D., Kassaye, F., Kuznetsov, G. V., Magomedov, M.-R. D. & Petelin, D. A. (1995). Nutritional estimate of populations of some wild free-ranging African ungulates in grassland (Nechisar National Park, Ethiopia) in dry season. *Ecography* **18**, 164–172.

Abell, R., Thiems, M. L., Revenga, C., Bryer, M., Kottelat, M., Bogutskaya, N., Coad, B., Mandrak, N., Baderas, S. C., Bussing, W., Stiassny, M. L. J., Skelton, P., Allen, G. R., Unmack, P., Neseka, A., Ng, R., Sindorf, N., Robertson, J., Armijo, E., Higgins, J. V., Heibel, T. J., Wiramanayake, E., Olson, D., López, H. L., Reis, R. E., Lundberg, J. G., Pérez, M. H. S. & Petry, P. (2008). Biogeographic units for freshwater biodiversity conservation. *Bioscience* **58**, 403–414.

Abouheif, E. (1999). A method for testing the assumption of phylogenetic independence in comparative data. *Evolutionary and Ecological Research* **1**, 95–909.

Abrams, R., Caton, D., Curet, L. B., Crenshaw, C., Mann, L. & Barron, D. H. (1969). Fetal brain-maternal aorta temperature differences in sheep. *American Journal of Physiology* **217**, 1613–1622.

Adam, P. J. (1999). *Choloepus didactylus*. *Mammalian Species* **621**, 1–8.

Agre, P., King, L., Yasui, M., Guggini, W., Pettersen, O., Fujiyoshi, Y., Engel, A. & Nielsen, S. J. (2002). Aquaporin water channels: from atomic structure to clinical medicine. *Journal of Physiology* **542**, 3–16.

Agosta, S. J., Bernardo, J., Ceballos, G. & Steele, M. A. (2013). A macrophysiological analysis of energetic constraints on geographic range size in mammals. *PLOS ONE* **8**, e72731.

Ahnelt, P. K. & Kolb, H. (2000). The mammalian photoreceptor mosaic—adaptative design. *Progress in Retinal and Eye Research* **19**, 711–777.

Ahnelt, P. K., Hokoç, J. N. & Pal Röhlich, P. (1995). Photoreceptors in a primitive mammal, the South American opossum, *Didelphis marsupialis aurita*: characterization with anti-opsin immunolabeling. *Visual Neuroscience* **12**, 793–804.

Aldridge, H. D. J. N. & Rautenbach, I. L. (1987). Echolocation and resource partitioning in insectivorous bats. *Journal of Animal Ecology* **56**, 763–778.

Alexander, D. E. (2002). *Nature's Flyers: Birds, Insects, and the Biomechanics of Flight*. Baltimore: Johns Hopkins University Press.

Alexander, R. McN. (1981). The gaits of tetrapods: adaptations for stability and economy. *Symposium of the Zoological Society of London* **48**, 269–287.

Alexander, R. McN. (1986). *Locomotion of Animals*. Glasgow and London: Blackie.

Alexander, R. McN. (1989). Optimization and gaits in the locomotion of vertebrates. *Physiological Reviews* **69**, 1199–1227.

Alexander, R. McN. (1999). *Energy for Animal Life*. Oxford: Oxford University Press

Alexander, R. McN. (2002). Tendon elasticity and muscle function. *Comparative Biochemistry and Physiology A* **133**, 1001–1011.

Al-kahtani, M. A., Zuleta, C., Caviedes-Vidal, E. & Garland, T. (2004). Kidney mass and relative medullary thickness of rodents in relation to habitat, body size, and phylogeny. *Physiological and Biochemical Zoology* **77**, 346–365.

Allan, A. P. & Gillooly, G. F. (2009). Towards an integration of ecological stoichiometry and the metabolic theory of ecology to better understand nutrient cycling. *Ecology Letters* **12**, 369–384.

Allen, J. A. (1877). The influence of physical conditions in the genesis of species. *Radical Review* **1**, 108–140.

Alroy, J. (1998). Cope's rule and the dynamics of body mass evolution in North American fossil mammals. *Science* **280**, 731–734.

Amador, G. J., Mao, W., DeMercurio, P., Montero, C., Clewis, J., Alexeev, A. & Hu, D. L. (2015). Eyelashes divert airflow to protect the eye. *Journal of the Royal Society Interface* **12**, 20141294.

Amitai, O., Holtze, S., Barkan, S., Amichai, E., Korine, C., Pinshow, B. & Voigt, C. C. (2010). Fruit bats (Pteropodidae) fuel their metabolism rapidly and directly with exogenous sugars. *Journal of Experimental Biology* **213**, 2693–2699.

Andersson, B. (1963). Aspects of the interrelations between central body temperature and food and water intake. In: *Brain and Behavior* (Ed. Brozier, M. M.). Washington, DC: American Institute of Biological Sciences.

Andrewartha, S. J., Cummings, K. J. & Frappell, P. B. (2014). Acid-base balance in the developing marsupial: from ectotherm to endotherm. *Journal of Applied Physiology* **116**, 1210–1219.

Andrews, R. M. (2004). Patterns of embryonic development. In: *Reptilian Incubation: Environment, Evolution and Behaviour* (Ed. Deeming, D. C.). Nottingham, UK: Nottingham University Press.

Andrews, R. M. & Mathies, T. (2000). Natural history of reptilian development: constraints on the evolution of viviparity. *BioScience* **50**, 227–238.

Angilletta, M. J. (2009). *Thermal Adaptation: A Theoretical and Empirical Synthesis*. Oxford: Oxford University Press.

Apfelbach, R., Blanchard, C. D., Blanchard, R. J., Hayes, R. A. & McGregor, I. S. (2005). The effects of predator odors in mammalian prey species: a review of field and laboratory studies. *Neuroscience and Biobehavioral Reviews* **29**, 1123–1144.

Apitz-Castro, R., Béguin, S., Tablante, A., Bartoli, F., Hold, J. C. & Hemker, H. C. (1995). Purification and partial characterization of draculin, the anticoagulant factor present in the saliva of vampire bats (*Desmodus rotundus*). *Thrombosis and Haemostasis* **73**, 94–100.

Archer, M. (1974). Regurgitation or mercyism in the western native cat, *Dasyurus geoffroii*, and the red-tailed wambenger, *Phascogale calura* (Marsupialia, Dasyuridae). *Journal of Mammalogy* **55**, 448–452.

Archer, M. (1984a). Origins and early radiations of mammals. In: *Vertebrate Zoogeography and Evolution in Australasia* (Eds. Archer, M. & Clayton, G.). Carlisle: Western Australia. Hesperian Press.

Archer, M. (1984b). Origins and early radiations of marsupials. In: *Vertebrate Zoogeography and Evolution in Australasia* (Eds. Archer, M. & Clayton, G.). Carlisle: Western Australia, Hesperian Press.

Archibald, J. D. & Deutschman, D. H. (2001). Quantitative analysis of the timing of the origin and diversification of extant placental orders. *Journal Mammalian Evolution* **8**, 107–124.

Arieli, R. (1979). The atmospheric environment of the fossorial mole rat (*Spalax ehrenbergi*), effects of season, soil texture, rain, temperature and activity. *Comparative Biochemistry and Physiology A* **63**, 569–575.

Arieli, R. (1990). Adaptation of the mammalian gas transport system to subterranean life. In: *Evolution of Subterranean Mammals at the Organismal and Molecular Levels* (Eds. Nevo, E. & Reig, O. A.). New York: Alan R. Liss.

Arieli, R. & Ar, A. (1979). Ventilation of a fossorial mammal (*Spalax ehrenbergi*) in hypoxic and hypercapnic conditions. *Journal of Applied Physiology* 47, 1011–1017.

Arieli, R., Ar, A. & Shkolnik, A. (1977). Metabolic responses of a fossorial rodent (*Spalax ehrenbergi*) to simulated burrow conditions. *Physiological Zoology* **50**, 61–75.

Arita, H. & Fenton, M. B. (1997). Flight and echolocation in the ecology and evolution of bats. *Trends in Ecology and Evolution* **12**: 53–67.

Arman, P. & Field, C. R. (1973). Digestion in the hippopotamus. *East African Wildlife Journal* **11**, 9–17.

Armati, P., Dickman, C. & Hume, I. (2006). *Marsupials*. Cambridge: Cambridge University Press.

Armitage, K. B. (1998). Reproductive strategies of yellow-bellied marmots-energy conservation and differences between the sexes. *Journal of Mammology* 79, 385–393.

Arrese, C., Archer, M., Runham, P., Dunlop, S. A. & Beazley, L. D. (2000). Visual system in a diurnal marsupial, the numbat (*Myrmecobius fasciatus*): retinal organization, visual acuity and visual fields. *Brain, Behaviour and Evolution* **55**, 163–175.

Arrese, C. A., Hart, N. S., Thomas, N., Beazley, L. D. & Shand, J. (2002). Trichromacy in Australian marsupials. *Current Biology* **12**, 657–660.

Aschoff, J. (1981). Thermal conductance in mammals and birds: its dependence on body size and circadian phase. *Comparative Biochemistry and Physiology A* **69**, 611–619.

Aschoff, J. & Pohl, H. (1970). Rhythmic variations in energy metabolism. *Federation Proceedings* **291**, 1541–1552.

Ashton, K. G. & Feldman, C. R. (2003). Bergmann's rule in nonavian reptiles: turtles follow it, lizards and snakes reverse it. *Evolution* 57, 1151–1163.

Atkins, E. (1960). Pathogenesis of fever. *Physiological Reviews* **40**, 580–646.

Atkinson, D. & Sibly, R. M. (1997). Why are organisms usually bigger in colder environments? Making sense of a life history puzzle. *Trends in Ecology and Evolution* **12**, 235–239.

Atkinson, S. N., Nelson, R. A & Ramsay, M. A. (1996). Changes in the body composition of fasting polar bears (*Ursus maritimus*): the effect of relative fatness on protein conservation. *Physiological Zoology* **69**, 304–316.

Au, D. & Weihs, D. (1980). At high speeds dolphins save energy by leaping. *Nature* **284**, 548–550.

Au, J., Marsh, K. J., Wallis, I. R & Foley, W. J. (2013). Whole-body protein turnover reveals the cost of detoxification of secondary metabolites in a vertebrate browser. *Journal of Comparative Physiology B* **183**, 993–1003.

Averianov, A. O. & Lopatin, A. V. (2014). On the phylogenetic position of monotremes (Mammalia, Monotremata). *Paleontological Journal* 48, 426–446.

Axelsson, M., Franklin, C. E., Löfman, C. O., Nilsson, S. & Grigg, G. C. (1996). Dynamic anatomical study of cardiac shunting in crocodiles using high-resolution angioscopy. *Journal of Experimental Biology* **199**, 359–365.

Bacon, C. D., Silvestro, D., Jaramillo, C., Smith, B. T., Chakrabarty, P. & Antonelli, A. (2015). Biological evidence supports an early and complex emergence of the Isthmus of Panama. *Proceedings of the National Academy of Sciences, USA* **112**, 6110–6115.

Bacigalupe, L. D., Nespolo, R. F., Bustamante, D. M. & Bozinovic, F. (2004). The quantitative genetics of sustained energy budget in a wild mouse. *Evolution* **58**, 421–429.

Baillie, J. E. M., Hilton-Taylor, C. & Stuart, S. N. (2004). *IUCN Red List of Threatened Species: A Global Species Assessment*. Glad, Switzerland, & Cambridge, UK: IUCN.

Baird, J. A., Hales, J. R. S. & Lang, W. J. (1974). Thermoregulatory responses to the injection of monoamines, acetylcholine and prostaglandins into a lateral cerebral ventricle of the echidna. *Journal of Physiology* **236**, 539–548.

Baker C. E. & Dunaway P. B. (1975). Elimination of $^{137}Cs$ and $^{59}Fe$ and its relationship to metabolic rates of wild small rodents. *Journal of Experimental Zoology* **192**, 223–236.
Baker, C. V. H., Modrell, M. S. & Gillis, J. A. (2013). The evolution and development of vertebrate lateral line electroreceptors. *Journal of Experimental Biology* **216**, 2515–2522.
Baker, M. A. (1979). A brain-cooling system in mammals. *Scientific American* **240**, 114–122.
Bakken, G. S. (1992). Measurement and application of operative and standard operative temperature in ecology. *American Zoologist* **32**, 194–216.
Bal, N. C., Maurya, S. K., Sopariwala, D. H., Sahoo, S. K., Gupta, S. C., Shaikh, S. A., Pant, M, Rowland, L. A., Bombardier, E., Goonasekera, S. A., Tupling, A. R., Molkentin, J. D. & Periasamy, M. (2012). Sarcolipin is a newly identified regulator of muscle-based thermogenesis in mammals. *Nature Medicine* **18**, 1575–1579.
Baldo, M. B., Antenucci, C. D. & Luna, F. (2015). Effect of ambient temperature on evaporative water loss in the subterranean rodent *Ctenomys talarum*. *Journal of Thermal Biology* **53**, 113–118.
Balinsky, J. B. (1972). Phylogenetic aspects of purine metabolism. *South African Medical Journal* **46**, 993–997.
Ballard, F. J., Hanson, R. W. & Kronfeld, D. S. (1969). Gluconeogenesis and lipogenesis in tissue from ruminant and nonruminant animals. *Federation Proceedings* **28**, 218–231.
Balmer, B. C. & Wells, R. S. (2013). Advances in cetacean telemetry: a review of single-pin transmitter attachment techniques on small cetaceans and development of a new satellite-linked transmitter design. *Marine Mammal Science* **30**, 656–673.
Banavar, J. R., Damuth, J., Maritan, A. & Rinaldo, A. (2002a). Modelling universality and scaling. *Nature* **420**, 626.
Banavar, J. R., Damuth, J., Maritan, A. & Rinaldo, A. (2002b). Supply-demand balance and metabolic scaling. *Proceedings of the National Academy of Sciences, USA* **99**, 10506–10509.
Banavar, J. R., Maritan, A. & Rinaldo, A. (1999). Size and form in efficient transportation networks. *Nature* **399**, 130–132.
Barboza, P. S., Parker, L. L. & Hume, I. D. (2009). *Integrative Wildlife Nutrition*. Heildelberg: Springer-Verlag.
Barclay, R. M. R. & Harder, L. D. (2003). Life history of bats: life in the slow lane. In: *Bat Biology* (Eds. Kunz, T. H. & Fenton, M. B.). Chicago: University of Chicago Press.
Barclay, R. M. R., Kalcounis, M. C., Crampton, L. H., Stefan, C., Vonhof, M. J., Wilkinson, L. & Brigham, R. M. (1996). Can external radiotransmitters be used to assess body temperature and torpor in bats? *Journal of Mammalogy* **77**, 1102–1106.
Barja, G. (2004). Aging in vertebrates, and the effect of caloric restriction: a mitochondrial free radical production-DNA damage mechanism? *Biological Reviews* **79**, 235–251.
Barker, S., Brown, G. D. & Calaby, J. H. (1963). Food regurgitation in Macropodidae. *Australian Journal of Science* **25**, 430–432.
Barnes, B. M. (1989). Freeze avoidance in a mammal: body temperatures below 0°C in an Arctic hibernator. *Science* **244**, 1593–1595.
Barnhardt, M. C. (1983). Gas permeability of the epiphragm of a terrestrial snail, *Otala lactea*. *Physiological Zoology* **56**, 436–444.
Barrett, A., Ndou, T., Hughey, C. A., Straut, C., Howell, A., Dai, Z. & Kaletune, G. (2013). Inhibition of α-amylase and glucoamylase by tannins extracted from cocoa, pomegranites, cranberries, and grapes. *Journal of Agricultural and Food Chemistry* **61**, 1477–1486.
Barriga, O. O. (1999). Evidence and mechanisms of immunosuppression in tick infestations. *Genetic Analysis: Biomolecular Engineering* **15**, 139–142.
Barron, D. H. & Meschia, G. (1954). A comparative study of the exchange of the respiratory gases across the placenta. *Cold Spring Harbor Symposia on Quantitative Biology* **19**, 93–101.

Bartel, D. P. (2005). MicroRNAs: genomics, biogenesis, mechanism, and function. *Cell* **116**, 281–297.

Bartels, H., Bartels, R., Baumann, R., Fons, R., Jurgens, K. D. & Wright, P. (1979). Blood oxygen transport and organ weights of two shrew species (*S. etruscus* and *C. russula*). *American Journal of Physiology* **236**, R221–R224.

Bartels, H., Hilpert, P., Barbey, K., Betke, K., Riegel, K., Lang, E. M. & Metcalfe, J. (1963). Respiratory functions of blood of the yak, llama, camel, Dybowski deer, and African elephant. *American Journal of Physiology* **205**, 331–336.

Bartels, H., Schmelzle, R. & Ulrich, S. (1969). Comparative studies on the respiratory function of mammalian blood. V. Insectivora: shrew, mole and non-hibernating and hibernating hedgehogs. *Respiration Physiology* 7, 278–286.

Bartholomew, G. A. & Hudson, J. W. (1960). Aestivation in the Mohave ground squirrel *Citellus mohavensis*. *Bulletin of the Museum of Comparative Zoology* **124**, 193–208.

Bartholomew, G. A. & Lasiewski, R. C. (1965). Heating and cooling rates, heart rate and simulated diving in the Galapagos marine iguana. *Comparative Biochemistry and Physiology* **16**, 573–582.

Bartholomew, G. A. & Lighton, J. R. B. (1986). Oxygen consumption during hover-feeding in free-ranging Anna hummingbirds. *Journal of Experimental Biology* **123**, 191–199.

Bartholomew, G. A., Vleck, D. & Vleck, C. M. (1981). Instantaneous measurements of oxygen consumption during pre-flight warm-up and post-flight cooling in sphingid and saturniid moths. *Journal of Experimental Biology* **90**, 17–32.

Batzli, G. O. (1985). Nutrition. In: *Biology of New World Microtus* (Ed. Tamarin, R. H.). Special Publication 8. Lawrence Kansas: American Society of Mammalogists.

Batzli, G. O. & Hume, I. D. (1994). Foraging and digestion in herbivores. In: *The Digestive System in Mammals: Food, Form and Function* (Eds. Chivers, D. J. & Langer, P.). Cambridge: Cambridge University Press.

Baudinette, R. V. (1978). Scaling of heart rate during locomotion in mammals. *Journal of Comparative Physiology* **127**, 337–342.

Baudinette, R. V., Gannon, B. J., Runciman, W. B., Wells, S. & Love J. B. (1987). Do cardiorespiratory frequencies show entrainment with hopping in the tammar wallaby? *Journal of Experimental Biology* **129**, 251–263.

Baudinette, R. V., Halpern, E. A. & Hinds, D. S. (1993). Energetic cost of locomotion as a function of ambient temperature and during growth in the marsupial *Potorous tridactylus*. *Journal of Experimental Biology* **174**, 81–95.

Baudinette, R. V. & Schmidt-Nielsen, K. (1974). Energy cost of gliding flight in herring gulls. *Nature* **248**, 83–84.

Baudinette, R. V., Snyder, G. K. & Frappell, P. B. (1992). Energetic cost of locomotion in the tammar wallaby. *American Journal of Physiology* **262**, R771–R778.

Beal, A. M. (1987). Amylase activity, protein and urea in saliva of the red kangaroo (*Macropus rufus*). *Archives of Oral Biology* **32**, 825–832.

Beal, A. M. (1991). Influence of flow rate and aldosterone administration on mandibular salivary composition in the koala (*Phascolarctos cinereus*). *Journal of Zoology* **223**, 265–279.

Bearder, S. K. (1977). Feeding habits of spotted hyaenas in a woodland habitat. *East African Wildlife Journal* **15**, 263–280.

Beasley, D. E., Koltz, A. M., Lambert, J. E., Fierer, N. & Dunn, R. R. (2015). The evolution of stomach acidity and its relevance to the human microbiome. *PLOS ONE* **10**, e0134116.

Beatty, D. T., Barnes, A., Fleming, P. A., Taylor, E. & Maloney, S. K. (2008). The effect of fleece on core and rumen temperature in sheep. *Journal of Thermal Biology* **33**, 437–443.

Beatty, D. T., Barnes, A., Taylor, E., Pethick, D., McCarthy, M. & Maloney, S. K. (2006). Physiological responses of *Bos taurus* and *Bos indicus* cattle to prolonged, continuous heat and humidity. *Journal of Animal Science* **84**, 972–985.
Beatty, D. T., Barnes, A., Taylor, E. & Maloney, S. K. (2014). Do changes in feed intake or ambient temperature cause changes in cattle rumen temperature relative to core temperature? *Journal of Thermal Biology* **33**, 12–19.
Becerra, J. X. (2015). On the factors that promote the diversity of herbivorous insects and plants in tropical forests. *Proceedings of the National Academy of Sciences, USA* **112**, 6098–6103.
Beck, R. M. D. (2008). A dated phylogeny of marsupials using a molecular supermatrix and multiple fossil constraints. *Journal of Mammalogy* **89**, 175–189.
Beck, R. M., Godthelp, D., Weisbecker, H. V., Archer, M. & Hand, S. J. (2008). Australia's oldest marsupial fossils and their biogeographical implications. *PLOS ONE* **3**, 1–8.
Begall, S., Malkemper, E. P., Červeńy, J., Němec, P. & Burda, H. (2013). Magnetic alignment in mammals and other animals. *Mammalian Biology* **78**, 10–20.
Behnke, A. R. (1940). Some physiological considerations of inhalation anaesthesia and helium. *Anesthesia and Analgesia* **19**, 35–41.
Bejder, L. & Hall, B. K. (2002). Limbs in whales and limblessness in other vertebrates: mechanisms of evolutionary and developmental transformation and loss. *Evolution and Development* **4**, 445–458.
Bell, D. P., Tikuisis, P. & Jacobs, I. (1992). Relative intensity of muscular contraction during shivering. *Journal of Applied Physiology* **72**, 2336–2342.
Bell, G. & Collins, S. (2008). Adaptation, extinction and global change. *Evolutionary Applications* **1**, 3–16.
Bell, G. P. (1990). Birds and mammals on an insect diet: a primer on diet composition analysis in relation to ecological energetics. *Studies in Avian Biology* **13**, 416–422.
Belovsky, G. E. & Jordan, P. A. (1978). The time-energy budget of a moose. *Theoretical Population Biology* **14**, 76–104.
Benedict, F. G. (1915). Factors affecting basal metabolism. *Journal of Biological Chemistry* **20**, 263–299.
Benedict, F. G. (1936). *The Physiology of the Elephant*. Washington, DC: Carnegie Institution of Washington.
Benedict, F. G. (1938). Vital energetics. A study in comparative basal metabolism. *Carnegie Institute of Washington, Publication* **503**, 1–215.
Benedict, J. B. & Benedict, A. D. (2001). Subnivean root caching by a montane vole (*Microtus montanus nanus*), Colorado Front Range, USA. *Western North American Naturalist* **61**, 241–244.
Benevenga, N. J., Odle, J. & Asche, G. L. (1992). Comparison of measured carbon dioxide production with that observed by the isotope dilution technique in neonatal pigs: observations on site of infusion. *Journal of Nutrition* **122**, 2174–2182.
Bennett, A. F. & Ruben, J. A. (1979). Endothermy and activity in vertebrates. *Science* **206**, 649–654.
Bennett, S. C. (1996). Aerodynamics and thermoregulatory function of the dorsal sail of *Edaphosaurus*. *Paleobiology* **22**, 496–506.
Bensaude-Vincent, B. (1996). Between history and memory: centennial and bicentennial images of Lavoisier. *Isis* **87**, 481–499.
Bentley, P. J. (1998). *Comparative Vertebrate Endocrinology*. Cambridge: Cambridge University Press.
Bentley, P. J. (2002). *Endocrines and Osmoregulation. A Comparative Account in Vertebrates. Zoophysiology 39*. Berlin: Springer.

Berg, F., Gustafson, U. & Andersson, L. (2006). The uncoupling protein 1 gene UCP1 is disrupted in the pig lineage: a genetic explanation for poor thermoregulation in piglets. *PLoS Genetics* **2**, e129.

Bergmann, C. (1847). Uber die verhaltnisse der warmeokonomie der thiere zu ihrer grosse. *Göttinger Studien* **1**, 595–708.

Bergmann, C. & Leukart, R. (1852). *Anatomischpysiolgische Übersicht des Tierreichs*. Stuttgart: J. B. Mueller.

Bernabucci, U., Lacetera, N., Baumgard, L. H., Rhoads, R. P., Ronchi, B. & Nardone, A. (2010). Metabolic and hormonal acclimation to heat stress in domesticated ruminants. *Animal* **4**, 1167–1183.

Berliner, R. W. (1976). The concentrating mechanism in the renal medulla. *Kidney International* **9**, 214–222.

Berliner, R. W., Levinsky, N. G., Davidson, D. G. & Eden, M. (1958). Dilution and concentration of the urine and the action of antidiuretic hormone. *American Journal of Medicine* **24**, 730–744.

Berta, A., Sumich, J. L. & Kovacs, K. M. (2015). *Marine Mammals. Evolutionary Biology*. Amsterdam: Elsevier.

Berteaux, D. (2000). Energetic cost of heating ingested food in mammalian herbivores. *Journal of Mammalogy* **81**, 683–690.

Berton, C. & Cholley, B. (2002). Equipment review: new techniques for cardiac output measurement—oesophageal Doppler, Fick principle using carbon dioxide, and pulse contour analysis. *Critical Care* **6**, 216–221.

Best, T. L. & Hill Henry, T. (1994). *Lepus arcticus*. *Mammalian Species* **457**, 1–9.

Best, T. L. & Skupski, M. P. (1994). *Perognathus merriami*. *Mammalian Species* **473**, 1–7.

Beuchat, C. A. (1990). Body size, medullary thickness, and urine concentrating ability in mammals. *American Journal of Physiology* **258**, R298–R308.

Beuchat, C. A. (1996). Structure and concentrating ability of the mammalian kidney: correlations with habitat. *American Journal of Physiology* **271**, R157–R179.

Bicego, K. C., Barros, R. C. H. & Branco, L. G. S. (2007). Physiology of temperature regulation: comparative aspects. *Comparative Biochemistry and Physiology A* **147**, 616–639.

Bicudo, J. P. W., Buttemer, W. A., Chappell, M. A., Pearson, J. T. & Bech, C. (2010). *Ecological and Environmental Physiology of Birds*. Oxford: Oxford University Press.

Bieber, C. & Ruf, T. (2009). Summer dormancy in edible dormice (*Glis glis*) without energetic constraints. *Naturwissenchaften* **96**, 165–171.

Bielby, J., Mace, G. M., Bininda-Emonds, O. R. P., Cardillo, M., Gittleman, J. L., Jones, K. E., Orme, C. D. L. & Purvis, A. (2007). The fast-slow continuum in mammalian life history: an empirical reevaluation. *American Naturalist* **169**, 748–757.

Biewener, A., Alexander, R. McN. & Heglund, N. C. (1981). Elastic energy storage in the hopping of kangaroo rats (*Dipodomys spectabilis*). *Journal of Zoology* **195**, 369–383.

Biewener, A. A. (1998). Muscle-tendon stresses and elastic energy storage during locomotion in the horse. *Comparative Biochemistry and Physiology B* **120**, 73–87.

Biewener, A. A. (2003). *Animal Locomotion*. Oxford: Oxford University Press.

Biewener, A. A. &, Baudinette, R. V. (1995). In-vivo muscle force and elastic energy storage during steady speed hopping of tammar wallabies (*Macropus eugenii*). *Journal of Experimental Biology* **198**, 1829–1841.

Bigman, A., Bauchet, M., Pinto, D., Mao, X., Akey, J. M., Mei, R., Scherer, S. W., Julian, C. G., Wilson, M. J., Herráez, D. L., Brutsaert, T., Parra, E. J., Moore, L. G. & Shriver, M. D. (2010). Identifying signature of natural selection in Tibetan and Andean populations use dense genome scan data. *PLoS Genetics* **6**, e1001116.

Bigiani, A., Mucignat-Caretta, C., Montani, G. & Tirindelli, R. (2005). Pheromone reception in mammals. *Reviews of Physiology, Biochemistry and Pharmacology* **155**, 1–35.
Bininda-Emonds, O. R. P., Cardillo, M., Jones, K. E., MacPhee, R. D. E., Beck, R. M. D., Grenyer, R., Price, S. A., Vos, R. A., Gittleman, J. L. & Purvis, A. (2007). The delayed rise of present-day mammals. *Nature* **446**, 507–512.
Bird, A. (2007). Perceptions of epigenetics. *Nature* **447**, 396–398.
Birkhead, T. R. (1993). Sexual selection and the temporal separation of reproductive events: sperm storage data from reptiles, birds and mammals. *Biological Journal of the Linnean Society* **50**, 295–311.
Birnbaumer, M., Seibold, A., Gilbert, S., Ishido, M., Barberis, C., Antaramian, A., Brabet, P. & Rosenthal, W. (1992). Molecular cloning of the receptor for human antidiuretic hormone. *Nature* **357**, 333–335.
Bishop, C. M. (1997). Heart mass and the maximum cardiac output of birds and mammals: implications for estimating the maximum aerobic power input of flying animals. *Philosophical Transactions of the Royal Society of London B* **352**, 447–456.
Bishop, C. M. (1999). The maximal oxygen consumption and factorial aerobic scope of birds and mammals; getting to the heart of the matter. *Proceedings of the Royal Society of London B* **266**, 2275–2281.
Bishop, K. L. (2008). The evolution of flight in bats: narrowing the field of plausible hypotheses. *Quarterly Review of Biology* **83**, 153–169.
Bjorndal, K. A. (1991). Diet mixing: nonadditive interactions of diet items in an omnivorous freshwater turtle. *Ecology* 72, 1234–1241.
Blackburn, D. G. (1991). Evolutionary origins of the mammary gland. *Mammal Review* **21**, 81–96.
Blackburn, D. G. (1993). Lactation: historical patterns and potential for manipulation. *Journal of Dairy Science* 76, 3195–3212.
Blackburn, D. G., Hayssen, V. & Murphy, C. J. (1989). The origins of lactation and the evolution of milk: a review with new hypotheses. *Mammal Review* **19**, 1–26.
Blackburn, T. M. & Gaston, K. J. (1994). The distribution of body sizes of the world's bird species. *Oikos* **70**, 127–130.
Blackburn, T. M., Gaston, K. J. & Loder, N. (1999). Geographic gradients in body size: a clarification of Bergmann's rule. *Diversity and Distribution* **5**, 165–174.
Blake, R. W. (1983). Energetics of leaping in dolphins and other aquatic animals. *Journal of the Marine Biology Association UK* **63**, 61–70.
Blank, I. H., Moloney, A. G., Emslie, A. G. & Simon, I. (1984). The diffusion of water across the stratum corneum as a function of its water content. *Journal of Investigative Dermatology* **82**, 188–192.
Blaxter, K. M. (1989). *Energy Metabolism in Animals and Man*. Cambridge: Cambridge University Press.
Blem, C. R. (1980). The energetics of migration. In: *Migration, Orientation, and Navigation* (Ed. Gauthreaux, S. A.). New York: Academic Press.
Bleuweiss, L., Fox, H., Kudzma, V., Nakashima, D., Peters, R. H. & Sams, R. (1978). Relationships between body size and some life history parameters. *Oecologia* **37**, 257–272.
Bligh, J. (1963). The receptors concerned in the respiratory response to humidity in sheep at high ambient temperature. *Journal of Physiology* **168**, 747–763.
Bligh, J. & Harthoorn, A. M. (1965). Continuous radiotelemetric records of the deep body temperature of some unrestrained African mammals under near-natural conditions. *Journal of Physiology* **176**, 145–162.

Blincoe, C. & Brody, S. (1955). Environmental physiology and shelter engineering. XXXII. The influence of ambient temperature, air velocity, radiation intensity, and starvation on thyroid activity and iodide metabolism in cattle. *Missouri Agricultural Experimental Station Research Bulletin* **576**, 36pp.

Block, B. A. (1994). Thermogenesis in muscle. *Annual Reviews of Physiology* **56**, 535–577

Blomberg, S. P., Garland, T. & Ives, A. R. (2003). Testing for phylogenetic signal in comparative data: behavioral traits are more labile. *Evolution* **57**, 717–745.

Blum, J. J. (1977). On the geometry of four dimensions and the relationship between metabolism and body mass. *Journal of Theoretical Biology* **64**, 599–601.

Bogdanowicz, W., Fenton, M. B. & Daleszczyk, K. (1999). The relationships between echolocation calls, morphology and diet in insectivorous bats. *Journal of Zoology* **247**, 381–393.

Boggs, D. F. (2002). Interactions between locomotion and ventilation in tetrapods. *Comparative and Biochemistry Physiology A* **33**, 269–288.

Boggs, D. F., Kilgore, D. L. & Birchard, G. F. (1984). Respiratory physiology of burrowing mammals and birds. *Comparative Biochemistry and Physiology A* 77, 1–7.

Boggs, D. F., Maginniss, L. A. & Kilgore, D. L. (1999). In vivo blood oxygen binding in hypoxic lesser spear-nosed bats: relationship to control of breathing. *Respiration Physiology* **118**, 193–202.

Bondarenco, A., Körtner, G. & Geiser, F. (2014). Hot bats: extreme thermal tolerance in a desert heat wave. *Naturwissenschaften* **101**, 679–685.

Bonner, J. T. (2006). *Why Size Matters from Bacteria to Blue Whales*. Princeton, New Jersey: Princeton University Press.

Booth, J. (1977). A short history of blood pressure measurement. *Proceedings of the Royal Society of Medicine* **70**, 793–799.

Boulant, J. A. (1997). Thermoregulation. In: *Fever: Basic Mechanisms and Management* (Ed. Mackowiak, P. A.). Philadelphia: Lippincott-Raven.

Bouverot, P. (1985). *Adaptation to Altitude-Hypoxia in Vertebrates*. Berlin: Springer-Verlag.

Boyd, I. L. (2000). Skin temperature during free-ranging swimming and diving in Antarctic fur seals. *Journal of Experimental Biology* **203**, 1907–1914.

Bozinovic, F. (1992). Scaling of basal and maximum metabolic rate in rodents and the aerobic capacity model for the evolution of endothermy. *Physiological Zoology* **65**, 921–932.

Bozinovic, F. (1993). Nutritional ecophysiology of the Andean mouse *Abrothrix andinus*: energy requirements, food quality and turnover time. *Comparative Biochemistry and Physiology* **104**, 601–604.

Bozinovic, F. (1995). Nutritional energetics and digestive responses of an herbivorous rodent (*Octodon degus*) to different levels of dietary fiber. *Journal of Mammalogy* **76**, 627–637.

Bozinovic, F., Calosi, P. & Spicer, J. I. (2011). Physiological correlates of geographic range in animals. *Annual Reviews of Ecology, Evolution and Systematics* **42**, 155–179.

Bozinovic, F., Carter, M. J. & Ebensperger, L. A. (2005). A test of the thermal-stress and the cost-of-burrowing hypotheses among populations of the subterranean rodent *Spalacopus cyanus*. *Comparative Biochemistry and Physiology* **140**, 329–336.

Bozinovic, F. & Gallardo, P. (2006). The water economy of South American desert rodents: from integrative to molecular physiological ecology. *Comparative Biochemistry and Physiology C* **142**, 163–172.

Bozinovic, F., Gallardo, P. A., Visser, R. H. & Cortés, A. (2003). Seasonal acclimatization in water flux rate, urine osmolality and kidney water channels in free-living degus: molecular mechanisms, physiological processes and ecological implications. *Journal of Experimental Biology* **206**, 2959–2966.

Bozinovic, F. & del Rio, C. M. (1996). Animals eat what they should not: why do they reject our foraging models? *Revista Chilena de Historia Natural* **69**, 15–20.

Bozinovic, F., Muñoz, J. L. P., Naya, D. E. & Cruz-Neto, A. P. (2007). Adjusting energy expenditure to energy supply: food availability regulates torpor use and organ size in the Chilean mouse-opossum *Thylamys elegans*. *Journal of Comparative Physiology B* **177**, 393–400.

Bozinovic, F. & Pörtner, H. O. (2015). Physiological ecology meets climate change. *Ecology & Evolution* **5**, 1025–1030.

Bozinovic, F. & Rosenmann, M. (1989). Maximum metabolic rate of rodents: physiological and ecological consequences on distributional limits. *Functional Ecology* **3**, 173–181.

Bozinovic, F., Rosenmann, M., Novoa, F. F. & Medel, R. G., (1995). Mediterranean type of climatic adaptation in the physiological ecology of rodent species. In: *Ecology and Biogeography of Mediterranean Ecosystems in Chile, California, and Australia* (Eds. Arroyo, M. T. K., Zedler, P. H. & Fox, M. D.). Ecological Studies 108. New York: Springer-Verlag.

Bozinovic, F., Ruiz, G. & Rosenmann, M. (2004). Energetics and torpor of a South American living fossil, the microbiotheriid *Dromiciops gliroides*. *Journal of Comparative Physiology B* **174**, 293–297.

Bradley, S. R. & Deavers, D. R. (1980). A re-examination of the relationship between thermal conductance and body weight in mammals. *Comparative Biochemistry and Physiology A* **65**, 465–476.

Bradshaw, S. D. (2003). *Vertebrate Ecophysiology. An Introduction to its Principles and Applications*. Cambridge: Cambridge University Press.

Bramble, D. (1989). Axial-appendicular dynamics and the integration of breathing and gait in mammals. *American Zoologist* **29**, 171–186.

Bramwell, C. D. & Fellgett, P. B. (1973). Thermal regulation in sail lizards. *Nature* **242**, 203–205.

Brand, M. D. & Esteves, T. C. (2005). Physiological functions of the mitochondrial uncoupling proteins UCP2 and UCP3. *Cell Metabolism* **2**, 85–93.

Braun, E. J. & Dantzler, W. H. (1972). Function of mammalian-type and reptilian-type nephrons in the kidney of desert quail. *American Journal of Physiology* **222**, 617–629.

Breidenstein, C. P. (1982). Digestion and assimilation of bovine blood by a vampire bat (*Desmodus rotundus*). *Journal of Mammalogy* **63**, 482–484.

Breukelen, F. & Martin, S. L. (2015). The hibernation continuum: physiological and molecular aspects of metabolic plasticity in mammals. *Physiology* **30**, 273–281.

Brillinger, D. R. & Stewart, B. S. (1998). Elephant-seal movements: modelling migration. *Canadian Journal of Statistics* **26**, 431–443.

Briscoe, N. J. (2015). Tree-hugging behavior beats the heat. *Temperature* **2**, 33–35.

Briscoe, N. J., Handasyde, K. A., Griffiths, S. R., Porter, W. P., Krockenberger, A. & Kearney, M. R. (2014). Tree-hugging koalas demonstrate a novel thermoregulatory mechanism for arboreal mammals. *Biology Letters* **10**, 20140235.

Brockman, R. P. (2005). Glucose and short-chain fatty acid metabolism. In: *Quantitative Aspects of Ruminant Digestion* (Eds. Dijkstra, J., Forbes, J. M. & France, J.). Wellingford, Oxfordshire: CABI Publishing.

Brodal, P. (1992). *The Central Nervous System. Structure and Function*. Oxford: Oxford University Press.

Brody, S. (1945). *Bioenergetics and Growth*. New York: Reinhold.

Brody, S. & Proctor, R. C. (1932). Growth and development, with special reference to domestic animals. XXIII. Relation between basal metabolism and mature body weight in different species of mammals and birds. *Missouri University Agricultural Experimental Station Research Bulletin* **166**, 89–101.

Brook, B. W., Sodhi, N. S. & Bradshaw, C. J. A. (2008). Synergies among extinction drivers under global change. *Trends in Ecology and Evolution* **23**, 453–460.

Broome, L. (2001). Density, home range, seasonal movements and habitat use of the mountain pygmy-possum *Burramys parvus* (Marsupialia: Burramyidae) at Mount Blue Cow. Kosciuszko National Park. *Austral Ecology* **26**, 275–292.

Brown, C. J. (1988). A study of bearded vultures *Gypaetus barbatus* in Southern Africa. PhD Dissertation, University of Pietermaritzburg, South Africa.

Brown, J. H., Gillooly, J. F., Allen, A. P., Savage, V. M. & West, G. B. (2004). Toward a metabolic theory of ecology. *Ecology* **85**, 1771–1789.

Brown, J. H. & West, G. B. (2000). *Scaling in Biology*. Oxford: Oxford University Press.

Brownfield, M. S. & Wunder, B. A. (1976). Relative medullary area: a new structural index for estimating urinary concentrating capacity of mammals. *Comparative Biochemistry and Physiology A* **55**, 69–75.

Brück, K. & Hinckel, P. (1996). Ontogenetic and adaptive adjustments in the thermoregulatory system. In: *Handbook of Physiology, Environmental Physiology* (Eds. Fregly, M. J. & Blatteis, C. M.). New York: Oxford University Press.

Brun, A., Price, E. R., Gontero-Fourcade, M. N., Fernandez-Marinone, G., Cruz-Neto, A. P., Karasov, W. H. & Caviedes-Vidal, E. (2014). Paracellular permeability of perfused intestinal segments is higher in flying than nonflying mammals. *Journal of Experimental Biology* **217**, 3311–3317.

Buchler, E. R. (1976). The use of echolocation by the wandering shrew (*Sorex vagrans*). *Animal Behaviour* **24**, 858–873.

Buckley, L. B., Davies, T. J., Ackerly, B. L., Kraft, N. J. B., Harrison, S. P., Anacker, B. L., Cornell, H. V., Damschen, E. I., Grytnes, J.-A., Hawkins, B. A., McCain, C. M., Stephens, P. R. & Wiens, J. J. (2010). Phylogeny, niche conservatism and the altitudinal diversity gradient in mammals. *Proceedings of the Royal Society of London B* **277**, 2131–2138.

Bueno-Pardo, J. & Lopez-Urrutia, A. (2014). Scaling up the curvature of mammalian metabolism. *Frontiers in Ecology and Evolution* **2**, 61.

Buffenstein, R. (1985). The effect of a high fibre diet on energy and water balance in two Namib desert rodents. *Journal of Comparative Physiology B* **155**, 211–218.

Buffenstein, R. (1985). The effect of starvation, food restriction, and water deprivation on thermoregulation and average daily metabolic rates in *Gerbillurus pusillus*. *Physiological Zoology* **58**, 320–328.

Buffenstein, R. (2000). Ecophysiological responses of subterranean rodents to underground habitats. In: *Life Underground: The Biology of Subterranean Rodents* (Eds. Lacey, E. A., Patton, J. L. & Cameron, G. N.). Chicago: University of Chicago Press.

Buffenstein, R., Campbell, W. E. & Jarvis, J. U. M. (1985). Identification of crystalline allantoin in the urine of African Cricetidae (Rodentia) and its role in their water economy. *Journal of Comparative Physiology B* **155**, 493–499.

Buffenstein, R. & Yahav, S. (1991). Is the naked mole-rat *Heterocephalus glaber* an endothermic yet poikilothermic mammal? *Journal of Thermal Biology* **16**, 227–232.

Bullen, R. B. & McKenzie, N. L. (2004). Bat flight muscle mass: implications for foraging strategies. *Australian Journal of Zoology* **52**, 605–622.

Bullen, R. B. & McKenzie, N. L. (2008). The pelage of bats (Chiroptera) and the presence of aerodynamic riblets: the effect on aerodynamic cleanliness. *Zoology* **111**, 279–286.

Bullen, R. B., McKenzie, N. L. & Cruz-Neto, A. P. (2014). Aerodynamic power and mechanical efficiency of bat airframes using a quasi-steady model. *CEAS Aeronautical Journal* **5**, 253–264.

Bullen, R. D. & McKenzie, N. L. (2001). Bat airframe design: flight performance, stability and control in relation to foraging ecology. *Australian Journal of Zoology* **49**, 235–261.

Bullen, R. D & McKenzie, N. L. (2002). Scaling bat wingbeat frequency and amplitude. *Journal of Experimental Biology* **205**, 2615–2626.

Bundle, M. W & Dial, K. P. (2003). Mechanics of wing-assisted incline running (WAIR). *Journal of Experimental Biology* **206**, 4553–4564.
Bundle, M. W., Hansen, K. S. & Dial, K. P. (2007). Does the metabolic rate-flight speed relationship vary among geometrically similar birds of different mass? *Journal of Experimental Biology* **210**, 1075–1083.
Burda, H., Bruns, V. & Müller, M. (1990b). Sensory adaptations in subterranean mammals. In: *Evolution of Subterranean Mammals at the Organismal and Molecular Levels* (Eds. Nevo, E. & Reig, O.). New York: Alan R. Liss.
Burda, H., Marhold, S., Westenberger, T., Wiltschko, R. & Wiltschko, W. (1990a). Magnetic compass orientation in the subterranean rodent *Cryptomys hottentotus* (Bathyergidae). *Experientia* **46**, 528–530.
Burnett, C. M. L. & Grobe, J. L. (2013). Dietary calorimetry identifies deficiencies in respirometry for the determination of resting metabolic rate in C57B1/6 and FVB mice. *American Journal of Physiology* **305**, E916–E924.
Burnett, C. M. L. & Grobe, J. L. (2014). Dietary effects on resting metabolic rate in C57BL/6 mice are differentially detected by indirect ($O_2/CO_2$ respirometry) and direct calorimetry. *Molecular Metabolism* **3**, 460–464.
Burrows, A. M. & Nash, L. T. (2010). *The Evolution of Exudativory in Primates*. New York: Springer-Verlag.
Burton, A. C. (1934). The application of the theory of heat flow to the study of energy metabolism. *Journal of Nutrition* **7**, 497–533.
Butler, A. B. & Hodos, W. (2005). *Comparative Vertebrate Neuroanatomy. Evolution and Adaptation*. Hoboken, New Jersey: Wiley-Interscience.
Butler, P. J., Green, J. A., Boyd, I. L. & Speakman, J. R. (2008). Measuring metabolic rate in the field: the pros and cons of the doubly labelled water and heart rate methods. *Functional Ecology* **18**, 168–183.
Byrnes, G., Libby, T., Lim, N. T.-L. &. Spence, A. J. (2011). Gliding saves time but not energy in Malayan colugos. *Journal of Experimental Biology* **214**, 2690–2696.
Byrnes, G., Lim, N. T.-L. & Spence, A. J. (2015). Take-off and landing kinetics of a free-ranging gliding mammal, the Malayan colugo (*Galeopterus variegatus*). *Proceedings of the Royal Society of London B* **275**, 1007–1013.
Byrnes, G. & Spence, A. J. (2011). Ecological and biomechanical insights into the evolution of gliding in mammals. *Integrative and Comparative Biology* **51**, 991–1001.
Cain, J. W. III, Krausman, P. R., Rosenstock, S. S. & Turner, J. C. (2006). Mechanisms of thermoregulation and water balance in desert ungulates. *Wildlife Society Bulletin* **34**, 570–581.
Calder, W. A. (1984). *Size, Function and Life History*. Cambridge, Massachusetts: Harvard University Press.
Calder, W. A. & Braun, E. J. (1983). Scaling of osmostic regulation in mammals and birds. *American Journal of Physiology Regulatory* **244**, R601–R606.
Caldwell, W. H. (1887). The embryology of the Monotremata and Marsupialia—Part I. *Philosophical Transactions of the Royal Society of London B* **178**, 463–486.
Calvert, S. J., Holland, R. A. B. & Gemmell, R. T. (1994). Respiratory properties of the neonatal blood of the common brushtail possum (*Trichosurus vulpecula*). *Physiological Zoology* **67**, 407–417.
Cameron, J. R. & Skofronick, J. G. (1978). *Medical Physics*. New York: John Wiley & Sons.
Campbell, G. S. & Norman, J. M. (1998). *An Introduction to Environmental Biophysics*. New York: Springer-Verlag.
Campbell, J. W. (1991). Excretory nitrogen metabolism. In: *Environmental and Metabolic Physiology* (Ed. Prosser, C. L.). New York: Wiley-Liss.

Campbell, K. L., McIntyre, I. W. & MacArthur, R. A. (2000). Postprandial heat increment does not substitute for active thermogenesis in cold challenged star-nosed moles (*Condylura cristata*). *Journal of Experimental Biology* **203**, 301–310.
Campbell, R. M. & Fell, B. (1964). Gastrointestinal hypertrophy in the lactating rat and its relationship to food intake. *Journal of Physiology* **171**, 90–98.
Canals, M., Atala, C., Grossi, B. & Iriarte-Diaz, J. (2005a). Relative size of hearts and lungs of small bats. *Acta Chiropterologica* 7, 65–72.
Canals, M., Atala, C., Olivares, R., Guajardo, F., Figueroa, D., Sabat, P. & Rosenmann, M. (2005b). Functional and structural optimization of the respiratory system of the bat *Tadarida brasiliensis* (Chiroptera; Molossidae): does airway matter? *Journal of Experimental Biology* **208**, 3987–3995.
Canals, M., Iriarte-Diaz, J. & Grossi, B. (2011). Biomechanical, respiratory and cardiovascular adaptations of bats and the case of the small community of bats in Chile. In: *Biomechanics in Applications* (Ed. Klika, V.). Rijeka, Croatia: InTech.
Cannon, B. & Nedergaard, J. (2004). Brown adipose tissue: function and physiological significance. *Physiological Reviews* **84**, 277–359.
Cannon, W. (1932). *Wisdom of the Body*. New York: Norton and Company.
Cannon, W. (1935). Stresses and strains of homeostasis. *American Journal of Medicine* **189**, 1–14.
Canterbury, G. (2002). Metabolic adaptation and climatic constraints on winter bird distribution. *Ecology* **83**, 946–957.
Capellini, I., Venditti, C. & Barton, R. A. (2010). Phylogeny and metabolic scaling in mammals. *Ecology* **91**, 2783–2793.
Capellini, I., Venditti, C. & Barton, R. A. (2011). Placentation and maternal investment in mammals. *American Naturalist* 177, 86–98.
Careau, V., Morand-Ferron, J. & Thomasm, D. (2007). Basal metabolic rate of Canidae from hot deserts to cold arctic climates. *Journal of Mammalogy* **88**, 394–400.
Carey, H. V., Andrews, M. T & Martin, S. L. (2003). Mammalian hibernation: cellular and molecular responses to depressed metabolism and low temperature. *Physiological Reviews* **83**, 1153–1181.
Carey, H. V., Frank, C. L. & Seifert, J. (2000). Hibernation induces oxidative stress and activation of NF-κB in ground squirrel intestine. *Journal of Comparative Physiology B* **170**, 551–559.
Carneiro, A. S., Teixeira, M. N., Rêgo, E. W. & Oliviera, J. B. (2010). Parâmetros hematológicos de *Didelphis albiventris* (Linnaeus, 1847) de vida livre provenientes do Município de Igarassú—Pernambuco—Brasil. *Medecina Veterinária* 4, 9–15.
Carpenter, R. E. (1985). Flight physiology of flying foxes *Pteropus poliocephalus*. *Journal of Experimental Biology* **114**, 619–649.
Carpenter, R. E. (1986). Flight physiology of intermediate sized fruit bats (Pteropodidae). *Journal of Experimental Biology* **120**, 79–103.
Carpenter, R. E. & Graham, J. B. (1967). Physiological responses to temperature in the long-nosed bat, *Leptonycteris sanborni*. *Comparative Biochemistry and Physiology* **22**, 709–722.
Carrick, F. N. & Hughes, R. L. (1978). Reproduction in male monotremes. *Australian Zoologist* **20**, 211–231.
Carroll, R. L. (1988). *Vertebrate Paleontology and Evolution*. New York: Freeman & Co.
Casotti, G., Herrera, M. L. G., Flores, J. J., Mancina, C. A & Braun, E. J. (2006). Relationships between renal morphology and diet in 26 species of new world bats (suborder Microchiroptera). *Zoology* **109**, 196–207

Caspary, W. F. (1992). Physiology and pathophysiology of intestinal absorption. *American Journal of Clinical Nutrition* **55**, 299S–308S.
Castellini, M. (2012). Life under water: physiological adaptations to diving and life at sea. *Comprehensive Physiology* **2**, 1889–1919.
Causey Whittow, G. (2000). *Sturkie's Avian Physiology*. San Diego: Academic Press.
Caviedes-Vidal, E., Karasov, W. H., Chediack, J. G., Fasulo, V., Cruz-Neto, A. P. & Otani, L. (2008). Paracellular absorption: a bat breaks the mammal paradigm. *PLOS ONE* **3**, e1425.
Caviedes-Vidal, E., McWhorter, T. J., Lavin, S. R., Chediack, J. G., Tracy, C. R. & Karasov, W. H. (2007). The digestive adaptation of flying vertebrates: high intestinal paracellular absorption compensates for smaller guts. *Proceedings of the National Academy of Sciences, USA* **104**, 19132–19137.
Ceballos, G., Ehrlich, P. R., Barnosky, A. D., García, A., Pringle, R. M. & Palmer, T. M. (2015). Accelerated modern human-induced species losses: entering the sixth mass extinction. *Science Advances* **1**, e1400253.
Cena, K. & Monteith, J. L. (1975a). Transfer processes in animal coats. II. Conduction and convection. *Proceedings of the Royal Society of London B* **188**, 395–411.
Cena, K. & Monteith, J. L. (1975b). Transfer processes in animal coats. III. Water vapour diffusion. *Proceedings of the Royal Society of London B* **188**, 413–423.
Cerling, T. E., Andanjec, S. A., Blumenthald, S. A., Brown, F. H., Chritz, K. L., Harris, J. M., Hart, J. A., Kirera, F. M., Kaleme, P., Leakey, L. N., Leakey, M. G., Levin, N. E., Manthi, F. K., Passey, B. H., & Uno, K. T. (2015). Dietary changes of large herbivores in the Turkana Basin, Kenya from 4 to 1 Ma. *Proceedings of the National Academy of Sciences, USA* **112**, 11467–11472.
Cerling, T. E., Bowen, GT. J., Ehrlinger, J. R. & Sponheimer, M. (2007). The reaction progress variable and isotope turnover in biological systems. In: *Stable Isotopes as Indicators of Ecological Change* (Eds. Dawson, T. E. & Siegwolf, R. T. W.). Amsterdam: Elsevier.
Cernuda-Cernuda, R., García-Fernández, J. M., Gordijn, M. C. M., Bovee-Geurts, P. H. M. & DeGrip, W. J. (2003). The eye of the African mole-rat *Cryptomys anselli*: to see or not to see? *European Journal of Neuroscience* **17**, 709–720.
Chang, C. L. & Hsu, S. Y. T. (2004). Ancient evolution of stress-regulating peptides in vertebrates. *Peptides* **25**, 1681–1688.
Chappel, R. W. & Hudson, R. J. (1978). Energy cost of feeding in Rocky Mountain bighorn sheep. *Acta Theriologica* **23**, 359–363.
Chappell, M. A. (1980). Thermal energetics and thermoregulatory costs of small Arctic mammals. *Journal of Mammalogy* **61**, 278–291.
Chappell, M. A. (1985). Effects of ambient temperature and altitude on ventilation and gas exchange in deer mice (*Peromyscus maniculatus*). *Journal of Comparative Physiology B* **155**, 751–758.
Chappell, M. A. (1992). Ventilatory accommodation of changing oxygen demand in sciurid rodents. *Journal of Comparative Physiology B* **162**, 722–730.
Chappell, M. A. & Bartholomew, G. A. (1981a). Standard operative temperatures and thermal energetics of the antelope ground squirrel *Ammospermophilus leucurus*. *Physiological Zoology* **54**, 81–93.
Chappell, M. A. & Bartholomew, G. A. (1981b). Activity and thermoregulation of the antelope ground squirrel *Ammospermophilus leucurus* in winter and summer. *Physiological Zoology* **54**, 215–223.
Chappell, M. A. & Roverud, R. C. (1990). Temperature effects on metabolism, ventilation, and oxygen extraction in a Neotropical bat. *Respiration Physiology* **81**, 401–412.
Chapuis-Lardy, L., Contour-Ansel, D. Bernhard-Reversat, F. (2002). High-performance liquid chromatography of water-soluble phenolics in leaf litter of three *Eucalyptus* hybrids (Congo). *Plant Science* **163**, 217–222.

Charles-Dominique, P. (1977). *Ecology and Behaviour of Nocturnal Prosimians*. London: Duckworth.

Charnov, E. L. (1991). Evolution of life history variation among female mammals. *Proceedings of the National Academy of Sciences, USA* **88**, 1134–1137.

Charnov, E. L. (1993). *Life History Invariants: Some Explorations of Symmetry in Evolutionary Ecology*. Oxford: Oxford University Press.

Chassin, P. S., Taylor, C. R., Heglund, N. C. & Seeherman, H. J. (1976). Locomotion in lions: energetic cost and maximum aerobic capacity. *Physiological Zoology* **49**, 1–10.

Chato, J. C. (1980). Heat transfer to blood vessels. *Journal of Biomechanical Engineering* **102**, 110–118.

Chaui-Berlinck, J. G. & Bicudo, J. E. P. W. (2006). *Respirometria—A Técnia*. São Paulo: Santos.

Chen, C. C. W. & Welch, K. C. (2014). Hummingbirds can fuel expensive hovering flight completely with either exogenous glucose or fructose. *Functional Ecology* **28**, 589–600.

Chen, K. & Rajewsky, N. (2007). The evolution of gene regulation by transcription factors and microRNAs. *Nature Reviews Genetics* **8**, 93–103.

Cheney, J. A., Konow, N., Middleton, K. M., Breuer, K. S., Roberts, T. J., Giblin, E. L. & Swartz, S. M. (2014). Membrane muscle function in the compliant wings of bats. *Bioinspiration Biomimetics* **9**, 025007.

Cheng, K., Simpson, S. J. & Raubenheimer, D. (2008). A geometry of regulatory scaling. *American Naturalist* **172**, 681–693.

Cheverud, J. M. & Dow, M. M. (1985). An autocorrelation analysis of genetic variation due to lineal fission in social groups of rhesus macaques. *American Journal of Physical Anthropology* **7**, 113–121.

Chew, R. M. (1955). The skin and respiratory losses of *Peromyscus maniculatus sonoriensis*. *Ecology* **36**, 463–467.

Chew, R. M. & Dammann, A. E. (1961). Evaporative water loss of small vertebrates, as measured with an infrared analyzer. *Science* **133**, 384–385.

Chew, R. M. & White, H. E. (1960). Evaporative water losses of the pallid bat. *Journal of Mammalogy* **41**, 452–458.

Chilcott, M. J. & Hume, I. D. (1985). Coprophagy and selective retention of fluid digesta: their role in the nutrition of the common ringtail possum, *Pseudocheirus peregrinus*. *Australian Journal of Zoology* **33**, 1–15.

Chivers, D. J. (1989). Adaptations of digestive systems in non-ruminant herbivores. *Proceedings of the Nutrition Society* **48**, 59–67.

Chivers, D. J. & Hladik, C. M. (1980). Morphology of the gastrointestinal tract in primates. Comparisons with other mammals in relation to diet. *Journal of Morphology* **166**, 337–386.

Chosniak, I. & Shkolnik, A. (1977). Rapid rehydration in the black Bedouin goats: red blood cell fragility and role of the rumen. *Comparative Biochemistry & Physiology A* **56**, 581–583.

Chown, S. L., Gaston, K. J. & Robinson, D. (2004). Macrophysiology: large-scale patterns in physiological traits and their ecological implications. *Functional Ecology* **18**, 159–167.

Christiansen, P. (2004). Body size in proboscideans. *Zoological Journal of the Linnean Society* **140**, 523–549.

Chudinov, P. K. (1968). Structure of the integument of theriomorphs. *Doklady Akademiia Nauk SSSR, Navaia Seriia* **179**, 207–210.

Cifelli, R. L. & de Muizon, C. (1997). Dentition and jaw of *Kokopellia juddi*, a primitive marsupial or near-marsupial from the medial Cretaceous of Utah. *Journal of Mammalian Evolution* **4**, 241–258.

Cipollini, M. L. (2000). Secondary metabolites of vertebrate-dispersed fruits: evidence for adaptative functions. *Revista Chilena de Historia Natural* **73**, 421–440.

Claridge, A. W. & Cork, S. J. (1994). Nutritional value of hypogeal fungal sporocarps for the long-nosed potoroo (*Potorous tridactylus*), a forest-dwelling mycophagous marsupial. *Australian Journal of Zoology* **42**, 701–710.

Clark, P. (2004). *Haematology of Australian Mammals*. Collingwood, Victoria, Australia: CSIRO Publishing.

Clarke, A. & Pörtner, H. O. (2010). Temperature, metabolic power and the evolution of endothermy. *Biological Reviews* **85**, 703–727.

Clarke, A. & Rothery, P. (2008). Scaling of body temperature in mammals and birds. *Functional Ecology* **2**, 58–67.

Clarke, A., Rothery, P. & Isaac, N. J. B. (2010). Scaling of basal metabolic rate with body mass and temperature in mammals. *Journal of Animal Ecology* **79**, 610–619.

Clauss, M., Frey, R., Kiefer, B., Lechner-Doll, A., Loehlein, W., Polster, C., Rössner, G. E. & Streich, W. J. (2003b). The maximum attainable body size of herbivorous mammals: morphophysiological constraints on foregut, and adaptations of hindgut fermenters. *Oecologia* **136**, 14–27.

Clauss, M., Loehlein, W., Kienzle, E. & Wiesner, H. (2003a). Studies on feed digestibilities in captive Asian elephants (*Elephas maximus*). *Journal of Animal Physiology and Animal Nutrition* **87**, 160–173.

Clauss, M., Schwarm, A., Ortmann, S., Streich, W. J. & Hummel, J. (2007). A case of non-scaling in mammalian physiology? Body size, digestive capacity, food intake, and ingesta passage in mammalian herbivores. *Comparative Biochemistry and Physiology A* **148**, 249–265.

Clauss, M., Steuer, P., Müller, D. W. H., Codron, D. & Hummel, J. (2013). Herbivory and body size: allometries of diet quality and gastrointestinal physiology, and implications for herbivore ecology and dinosaur gigantism. *PLOS ONE* **8**, e68714.

Clavijo-Baquet, S. & Bozinovic, F. (2012). Testing the fitness consequences of the thermoregulatory and parental care models for the origin of endothermy. *PLOS ONE* e37069.

Clemens, W. A. (1968). Origin and early evolution of marsupials. *Evolution* **22**, 1–18.

Clemens, W. A. (1971). Mammalian evolution in the Cretaceous. *Early Mammals, Supplement 1* (Eds. Kermack, D. M. & Kermack, K. A.). *Zoological Journal of the Linnean Society* **50**, 165–180.

Clutton-Brock, T. H. (1989). Mammalian mating systems. *Proceedings of the Royal Society of London B* **236**, 339–372.

Cochran, W. W. & Lord, R. D. (1963). A radio-tracking system for wild animals. *Journal of Wildlife Management* **27**, 9–24.

Codron, J., Lee-Thorp, J. A., Sponheimer, M., Codron, D., Grant, R. C. & De Ruiter, D. J. (2006). Elephant (*Loxodonta africana*) diets in Kruger National Park, South Africa: spatial and landscape differences. *Journal of Mammalogy* **87**, 27–34.

Coleman, J. C. & Downs, C. T. (2010). Characterizing the thermal environment of small mammals: what should we be measuring, and how? *Open Access Animal Physiology* **2**, 47–59.

Collier, R. J., Beede, D. K., Thatcher, W. W., Israel, L. A. & Wilcox, C. J. (1982). Influences of environment and its modification on dairy animal health and production. *Journal of Dairy Science* **65**, 2213–2227.

Collins, J. C., Pilkington, T. C. & Schmidt-Nielsen, K. (1971). A model of respiratory heat transfer in a small mammal. *Biophysical Journal* **11**, 886–914.

Comroe, J. H. (1978). *Retrospectroscope: Insights into Medical Discovery*. Menlo Park, California: Von Gehr Press.

Connolly, M. K. & Cooper, C. E. (2014). How do measurement duration and timing interact to influence estimation of basal physiological variables of a nocturnal rodent? *Comparative Biochemistry and Physiology A* **178**, 24–29.

Conover, D. O., Duffy, T. A. & Hice, L. A. (2009). The covariance between genetic and environmental influences across ecological gradients. Reassessing the evolutionary significance of countergradient and cogradient variation. *Annals of the New York Academy of Sciences* **1168**, 100–129.
Contreras, C., Franco, M., Place, N. J. & Nespolo, R. F. (2014). The effects of poly-unsaturated fatty acids on the physiology of hibernation in a South American marsupial, *Dromiciops gliroides*. *Comparative Biochemistry and Physiology A* 177, 62–69.
Cooke, S. J., Sack, L., Franklin, C. E., Farrell, A. P., Beradall, J., Wihkelski, M. & Chown, S. L. (2013). What is conservation physiology? Perspectives on an increasingly integrated and essential science. *Conservation Physiology* **1**, 1–23.
Cooke, S. J. & Suski, C. D. (2008). Ecological restoration and physiology: an overdue integration. *BioScience* **58**, 957–968.
Cooper, C. E. (2003). Physiological specialisations of the numbat, *Myrmecobius fasciatus* Waterhouse 1836 (Marsupialia: Myrmecobiidae): a unique termitivorous marsupial. PhD Thesis, Department of Zoology, University of Western Australia, Perth.
Cooper, C. E. (2011). *Myrmecobius fasciatus* (Dasyuromorphia: Myrmecobiidae). *Mammalian Species* **43**, 129–140.
Cooper, C. E. (2016). Endocrinology of osmoregulation and thermoregulation of Australian desert tetrapods: a historical perspective. *General and Comparative Endocrinology*. doi: http://dx.doi.org/10.1016/j.ygcen.2015.10.003.
Cooper, C. E. & Cruz-Neto, A. P. (2009). Metabolic, hygric and ventilatory physiology of a hypermetabolic marsupial, the honey possum (*Tarsipes rostratus*). *Journal of Comparative Physiology B* **179**, 773–781.
Cooper, C. E. & Geiser, F. (2008). The 'minimum boundary curve for endothermy' as a predictor of heterothermy in mammals and birds: a review. *Journal of Comparative Physiology B* **178**, 1–8.
Cooper, C. E., McAllan, B. M. & Geiser, F. (2005). Effect of torpor on the water economy of an arid-zone marsupial, the stripe-faced dunnart (*Sminthopsis macroura*). *Journal of Comparative Physiology B* **175**, 323–328.
Cooper, C. E., Walsberg, G. E. & Withers, P. C. (2003a). The biophysical properties of the pelt of a diurnal marsupial, the numbat (*Myrmecobius fasciatus*) and its role in thermoregulation. *Journal of Experimental Biology* **206**, 2771–2777.
Cooper, C. E. & Withers, P. C. (2002). Metabolic physiology of the numbat (*Myrmecobius fasciatus*). *Journal of Comparative Physiology B* **172**, 669–675.
Cooper, C. E. & Withers, P. C. (2004c). Patterns of body temperature variation and torpor in the numbat, *Myrmecobius fasciatus* (Marsupialia: Myrmecobiidae). *Journal of Thermal Biology* **29**, 277–284.
Cooper, C. E. & Withers, P. C. (2004a). Ventilatory physiology of the numbat (*Myrmecobius fasciatus*). *Journal of Comparative Physiology B* **174**, 107–111.
Cooper, C. E. & Withers, P. C. (2004b). Termite digestibility and water and energy contents determine the water economy index of numbats (*Myrmecobius fasciatus*) and other myrmecophages. *Physiological and Biochemical Zoology* 77, 641–650.
Cooper, C. E. & Withers, P. C. (2005). Physiological significance of the microclimate in night refuges of the numbat *Myrmecobius fasciatus*. *Australian Mammalogy* 27, 169–174.
Cooper, C. E. & Withers, P. C. (2008). Allometry of evaporative water loss in marsupials: implications of the effect of ambient relative humidity on the physiology of brushtail possums (*Trichosurus vulpecula*). *Journal of Experimental Biology* **211**, 2759–2766.
Cooper, C. E. & Withers, P. C. (2009). Effects of measurement duration on the determination of basal metabolic rate and evaporative water loss of small marsupials: how long is long enough? *Physiological and Biochemical Zoology* **82**, 438–446.

Cooper, C. E. & Withers, P. C. (2010). Effect of sampling regime on estimation of basal metabolic rate and standard evaporative water loss using flow-through respirometry. *Physiological and Biochemical Zoology* **83**, 385–393.
Cooper, C. E. & Withers, P. C. (2012). Does season or captivity influence the physiology of an endangered marsupial, the numbat (*Myrmecobius fasciatus*)? *Journal of Mammalogy* **93**, 771–777.
Cooper, C. E. & Withers, P. C. (2014a). Physiological responses of a rodent to heliox reveal constancy of evaporative water loss under perturbing environmental conditions. *American Journal of Physiology* **307**, R1042–R1048.
Cooper, C. E. & Withers, P. C. (2014b). Ecological consequences of temperature regulation: why might the mountain pygmy possum *Burramys parvus* need to hibernate near underground streams? *Temperature* **1**, 32–36.
Cooper, C. E., Withers, P. C. & Bradshaw, S. D. (2003b). The field metabolic rate of the numbat (*Myrmecobius fasciatus*). *Journal of Comparative Physiology B* **173**, 687–693.
Cooper, C. E., Withers, P. C. & Cruz-Neto, A. P. (2010). Metabolic, ventilatory and hygric physiology of a South American marsupial, the long-furred woolly mouse opossum. *Journal of Mammalogy* **91**, 1–10.
Cooper, C. E., Withers, P. C., Mawson, P. R., Bradshaw, S. D., Prince, J. & Robertson, H. (2002). Metabolic ecology of cockatoos in the south-west of Western Australia. *Australian Journal of Zoology* **50**, 67–76.
Cooper, H. M., Herbin, M. & Nevo, E. (1993). Visual system of a naturally microphthalmic mammal: the blind mole rat, *Spalax ehrenbergi*. *Journal of Comparative Neurology* **328**, 313–350.
Cope, E. D. (1896). *The Primary Factors of Organic Evolution*. New York: Open Court.
Corbett, J. L., Farrell, D. J., Leng, R. A., McClymont, G. L. & Young, B. A. (1971). Determination of the energy expenditure of penned and grazing sheep from estimates of carbon dioxide entry rates. *British Journal of Nutrition* **26**, 277–291.
Cork, S. J. (1994). Digestive constraints on dietary scope in small and moderately-small mammals: how much do we really understand? In: *The Digestive System in Mammals: Food, Form and Function* (Eds. Chivers, D. J. & Langer, P.). Cambridge: Cambridge University Press.
Cork, S. J., Hume, I. D. & Dawson, T. J. (1983). Digestion and metabolism of a natural foliar diet (*Eucalyptus punctata*) by an arboreal marsupial, the koala (*Phascolarctos cinereus*). *Journal of Comparative Physiology B* **152**, 181–190.
Cork, S. J. & Kenagy, G. J. (1989). Nutritional value of hypogeous fungus for a forest dwelling ground squirrel. *Ecology* **70**, 577–586.
Cornelius, C., Dandrifosse, G. & Jeuniaux, C. (1975). Biosynthesis of chitinases by mammals of the order Carnivora. *Biochemical Systematics and Ecology* **3**, 121–122.
Cortez, M., Wolt, R., Gelwick, F., Osterrieder, S. K. & Davis, R. W. (2016). Development of an altricial mammal at sea: I. Activity budgets of female sea otter and their pups in Simpson Bay, Alaska. *Journal of Experimental Marine Biology and Ecology* **481**, 71–80.
Costa, D. P. & Kooyman, G. L. (1982). Oxygen consumption, thermoregulation, and the effect of fur oiling and washing on the sea otter, *Enhydra lutris*. *Canadian Journal of Zoology* **60**, 2761–2767.
Costa, D. P. & Kooyman, G. L. (1984). Contribution of specific dynamic action to heat balance and thermoregulation in the sea otter *Enhydra lutris*. *Physiological Zoology* **57**, 199–203.
Cotgreave, P. (1993). The relationship between body size and population abundance in animals. *Trends in Ecology and Evolution* **8**, 244–248.

Cowing, J. A., Arrese, C. A., Davies, W. L., Beazley, L. D. & Hunt, D. M. (2008). Cone visual pigments in two marsupial species: the fat-tailed dunnart (*Sminthopsis crassicaudata*) and the honey possum (*Tarsipes rostratus*). *Proceedings of the Royal Society of London B* **275**, 1491–1499.

Cox, C. B. (1970). Migrating marsupials and drifting continents. *Nature* **226**, 767–770.

Cox, C. B. (2000). Plate tectonics, seaways and climate in the historical biogeography of mammals. *Memoirs of the Institute of Oswaldo Cruz, Rio de Janeiro* **95**, 509–516.

Cox, C. B. & Moore, P. D. (2000). *Biogeography: An Evolutionary and Ecological Approach*. Oxford: Blackwell Scientific.

Cracraft, J. (1974). Continental drift and vertebrate distribution. *Annual Review of Ecology and Systematics* **5**, 215–261.

Crane, R. K. (1975). The physiology of the intestinal absorption of sugars. In: *Physiological Effects of Food Carbohydrates* (Eds. Jeans, A. & Hodge. J.). ACS Symposium Series Number 15. Washington, DC: American Chemical Society.

Crête, M. & Larivière, S. (2003). Estimating the costs of locomotion in snow for coyotes. *Canadian Journal of Zoology* **81**, 1808–1814.

Cretekos, C. J., Wang, Y. Green, E. D., Martin, J. F., Rasweiler, J. J. Behringer, R. R. (2008). Regulatory divergence modifies limb length between mammals. *Genes and Development* **22**, 141–151.

Crick, F. (1970). Central dogma of molecular biology. *Nature* **227**, 561–563.

Critchley, L. A., Lee, A. & Ho, A. M.-H. (2010). A critical review of the ability of continuous cardiac output monitors to measure trends in cardiac output. *Anesthesia and Analgesia* **111**, 1180–1192.

Critchley, L. A. H. & Critchley, J. A. J. H. (1999). A meta-analysis of studies using bias and precision statistics to compare cardiac output measurement techniques. *Journal of Clinical Monitoring and Computing* **15**, 85–91.

Croll, D. A., Gaston, A. J., Burger, A. E. & Konnoff, D. (1992). Foraging behavior and physiological adaptation for diving in thick-billed murres. *Ecology* **73**, 344–356.

Crompton, A. W., Taylor, C. R. & Jagger, J. A. (1978). Evolution of homeothermy in mammals. *Nature* **272**, 333–336.

Cross, P. C., Almberg, E. S., Haase, C. G., Hudson, P. J., Maloney, S. K., Metz, M., Munn, A., Nugent, P., Putzeys, O., Stahler, D. R., Stewart, A. C. & Smith, D. W. (2016). Energetic costs of mange in Yellowstone wolves estimated from infrared thermography. *Ecology*. doi:10.1890/15-1346.1.

Crowell, J. W. & Smith, E. E. (1967). Determinant of the optimal haematocrit. *Journal of Applied Physiology* **22**, 501–504.

Cruze, L., Kohno, S., McCoy, M. W. & Guillette, L. J. (2012). Towards an understanding of the evolution of the chorioallantoic placenta: steroid biosynthesis and steroid hormone signalling in the chorioallantoic membrane of an oviparous reptile. *Biology of Reproduction* **87**, 1–11.

Cruz-Neto, A. P. & Bozinovic, F. (2004). The relationship between diet quality and basal metabolic rate in endotherms: insights from intraspecific analysis. *Physiological and Biochemical Zoology* **77**, 877–889.

Cruz-Neto, A. P. & Jones, K. E. (2006). Exploring the evolution of the basal metabolic rate in bats. In: *Functional and Evolutionary Ecology of Bats* (Eds. Zubaid, A., McCracken, G. F. & Kunz, T. H.). New York: Oxford University Press.

Cryan, P. M., Bogan, M. A., Rye, R. O., Landis, G. P. & Kester, C. L. (2004). Stable hydrogen isotope analysis of bat hair as evidence for seasonal molt and long-distance migration. *Journal of Mammalogy* **85**, 995–1001.

Cryan, P. M. & Wolf, B. O. (2003). Sex differences in the thermoregulation and evaporative water loss of a heterothermic bat, *Lasiurus cinereus*, during its spring migration. *Journal of Experimental Biology* **206**, 3381–3390.
Cunningham, E. P. (1991). Breeding programmes for improved dairy production in tropical climates. In: *Animal Husbandry in Warm Climates* (Eds. Ronchi, B., Nardone, A. & Boyazoglu, J. G.). Viterbo, Italy: Pudoc, Wageningen.
Currie, S. E., Körtner, G. & Geiser, F. (2014). Heart rate as a predictor of metabolic rate in heterothermic bats. *Journal of Experimental Biology* **217**, 1519–1524.
Czech-Damal, N. U., Dehnhardt, G., Manger, P. & Hanke, W. (2012a). Passive electroreception in aquatic mammals. *Journal of Comparative Physiology A* **199**, 555–563.
Czech-Damal, N. U., Liebschner, A., Mierisch, L., Klauer, G., Hanke, F. D., Marshall, C., Dehnhardt, G. & Hanke, W. (2012b). Electroreception in the Guiana dolphin (*Sotalia guianensis*). *Proceedings of the Royal Society of London B* **279**, 663–668.
Dammhahn, M., Soarimalala, V. & Goodman, S. M. (2013). Trophic niche differentiation and microhabitat utilization in a species-rich montane forest small mammal community of Eastern Madagascar. *Biotropica* **45**, 111–118.
Damuth, J. (1981). Population density and body size in mammals. *Nature* **290**, 699–700.
Damuth, J. (1987). Interspecific allometry of population density in mammals and other animals: the independence of body mass and population energy use. *Biological Journal of the Linnean Society* **31**, 193–246.
Damuth, J. (2007). The macroevolutionary explanation for energy equivalence in the scaling of body size and population density. *American Naturalist* **169**, 621–631.
Dantzler, W. H. & Braun, E. J. (1980). Comparative nephron function in reptiles, birds, and mammals. *American Journal of Physiology* **239**, R197–R213.
Darling, R. C., Smith, C. A., Asmussen, E. & Cohen, F. M. (1941). Some properties of human fetal and maternal blood. *Journal of Clinical Investigation* **20**, 739–747.
Darlington, P. J. (1957). *Zoogeography: The Geographical Distribution of Animals*. New York: Wiley.
Darlington, P. J. (1965). *Biogeography of the Southern End of the World*. Cambridge: Harvard University Press.
Darden, T. R. (1972). Respiratory adaptations of a fossorial mammal, the pocket gopher (*Thomomys bottae*). *Journal of Comparative Physiology* **78**, 121–137.
Darveau, C. A., Suarez, R. K., Andrews, R. D. Hochachka, P. W. (2002). Allometric cascade as a unifying principle of body mass effects on metabolism. *Nature* **417**, 166–170.
Davidson, E. H. & Erwin, D. H. (2006). Gene regulatory networks and the evolution of animal body plans. *Science* **311**, 796–800.
Dausmann, K. H., Glos, J. & Heldmaier, G. (2009). Energetics of tropical hibernation. *Journal of Comparative Physiology B* **179**, 345–357
Davies, S. J. J. F. (2002). *Ratites and Tinamous*. Oxford: Oxford University Press.
Davis, K. & Argüeso, P. (2014). Composition of terrestrial and marine mammal tears is dependent on species and environment. *Investigative Ophthalmology and Visual Science* **55**, 36.
Davis, K., Doane, M. G., Knop, E., Knop, N., Dubielzig, R. R., Colitz, C. M. H., Argüeso, P. & Sullivan, D. A. (2013). Characterization of ocular gland morphology and tear composition of pinnipeds. *Veterinary Ophthalmology* **16**, 269–275.
Davis, M. J. (1988). Control of bat wing capillary pressure and blood flow during reduced perfusion pressure. *American Journal of Physiology* **255**, H1114–H1128.
Davis, R. W. (2014). A review of the multi-level adaptations for maximising aerobic dive duration in marine mammals: from biochemistry to behavior. *Journal of Comparative Physiology B* **184**, 23–53.

Davis, R. W. & Williams, T. M. (2012). The dive response is exercise modulated to maximise aerobic dive duration. *Journal of Comparative Physiology A* **198**, 583–591.

Dawson, T. E. & Siegwolf, R. T. W. (2007). *Stable Isotopes as Indicators of Ecological Change.* Amsterdam: Elsevier.

Dawson, T. J. (1973a). Primitive mammals. In: *Comparative Physiology of Thermoregulation* (Ed. Whittow, G. C.). New York: Academic Press.

Dawson, T. J. (1973b). Thermoregulatory responses of the arid zone kangaroos, *Megaleia rufa* and *Macropus robustus. Comparative Biochemistry and Physiology A* **46**, 153–169.

Dawson, T. J. (1983). *Monotremes and Marsupials: The Other Mammals.* Studies in Biology 150. London: Edward Arnold Press.

Dawson, T. J., Blaney, C. E., Munn, A. J., Krockenberger, A. & Maloney, S. K. (2001). Thermoregulation by kangaroos from mesic and arid habitats: influence of temperature on routes of heat loss in eastern grey kangaroos (*Macropus giganteus*) and red kangaroos (*Macropus rufus*). *Physiological and Biochemical Zoology* 73, 374–381.

Dawson, T. J. & Dawson, W. R. (1982). Metabolic scope and conductance in response to cold of some dasyurid marsupials and Australian rodents. *Comparative Biochemistry and Physiology A* 71, 59–64.

Dawson, T. J., Finch, E., Freedman, L., Hume, I. D., Renfree, M. B. & Temple-Smith, P. D. (1989). Morphology and physiology of the Metatheria. In: *Fauna of Australia. Mammalia Volume 1B* (Eds. Walton, D. W. & Richardson, B. J.). Canberra: Australian Government Publishing Service.

Dawson, T. J., Grant, T. R. Fanning, D. (1979). Standard metabolism of monotremes and the evolution of endothermy. *Australian Journal of Zoology* 27, 511–515.

Dawson, T. J. & Grant, T. T. (1980). Metabolic capabilities of monotremes and the evolution of homeothermy. In: *Comparative Physiology: Primitive Mammals* (Eds. Schmidt-Nielsen, K., Bolis, L. & Taylor, C. R.). Cambridge: Cambridge University Press.

Dawson, T. J. & Hulbert, A. J. (1970). Standard metabolism, body temperature, and surface areas of Australian marsupials. *American Journal of Physiology* **218**, 1233–1238.

Dawson, T. J., Munn, A. J., Blaney, C. E, Krockenberger, A. & Maloney, S. K. (2000). Ventilatory accommodation of oxygen demand and respiratory water loss in kangaroos from mesic and arid environments, the eastern grey kangaroo (*Macropus giganteus*) and the red kangaroo (*Macropus rufus*). *Physiological Biochemistry and Zoology* 73, 382–388.

Dawson, T. J. & Needham, A. D. (1981). Cardiovascular characteristics of two resting marsupials: an insight into the cardio-respiratory allometry of marsupials. *Journal of Comparative Physiology B* **145**, 95–100.

Dawson, T. J., Robertshaw, D. & Taylor, C. R. (1974). Sweating in the kangaroo: a cooling mechanism during exercise, but not in the heat. *American Journal of Physiology* **227**, 494–498.

Dawson, T. J. & Taylor, C. R. (1973). Energetic cost of locomotion in kangaroos. *Nature* **246**, 313–314.

Dawson, T. J., Webster, K. N. & Maloney, S. K. (2014). The fur of mammals in exposed environments; do crypsis and thermal needs necessarily conflict? The polar bear and marsupial koala compared. *Journal of Comparative Physiology B* **184**, 273–284.

Dawson, T. J., Webster, K. N., Mifsud, B., Raad, E., Lee, E. & Needham, A. D. (2003). Functional capacities of marsupial hearts: size and mitochondrial parameters indicate higher aerobic capabilities than generally seen in placental mammals. *Journal of Comparative Physiology B* **173**, 583–590.

Dearing, D. M. (1997). Effects of *Acomastylis rossii* tannins on a mammalian herbivore, the North American pika, *Ochotona princeps. Oecologia* **109**, 122–131.

DeGabriel, J. L., Moore, B. D., Felton, A. M., Ganzhorn, J. U., Stolter, C., Wallis, I. R., Johnson, C. N. & Foley, W. J. (2014). Translating nutritional ecology from the laboratory to the field: milestones in linking plant chemistry to population regulation in mammalian browsers. *Oikos* **123**, 298–308.

Degen, A. A. (1997). *Ecophysiology of Small Desert Mammals*. Berlin: Springer-Verlag.

Degen, A. A. & Kam, M. (1995). Scaling of field metabolic rate to basal metabolic rate ratio in homeotherms. *Ecoscience* **2**, 48–54.

Dehnel, A. (1949). Studies on the genus *Sorex* L. *Annales Universitatis Mariae Curie-Sklowdowska. Sectio C Biology* **4**, 17–102.

de Longh, H. H., de Jong, C. B., van Goethem, J., Klop, E., Brunsting, A. M. H., Loth, P. E. & Prins, H. H. T. (2011). Resource partitioning among African savanna herbivores in North Cameroon: the importance of diet composition, food quality and body mass. *Journal of Tropical Ecology* **27**, 503–513.

Delworth, T., Stouffer, R. & Winton, M. (2007). Patterns of Greenhouse warming. *GFDL Climate Modelling Research Highlights* **1**. http://www.gfdl.noaa.gov/cms-filesystem-action/user_files/kd/pdf/gfdlhighlight_vol1n6.pdf.

Demment, M. W. & Van Soest, P. J. (1985). A nutritional explanation for body-size patterns of ruminant and nonruminant herbivores. *American Naturalist* **125**, 641–672.

Denny, M. W. (2010). Organismal climatology: analysing environmental variability at scales relevant to physiological stress. *Journal of Experimental Biology* **213**, 995–1003.

Denny, M. & Helmuth, B. (2009). Confronting the physiological bottleneck: a challenge from ecomechanics. *Integrative Comparative Biology* **49**, 197–201.

Depocas, F. & Hart, J. S. (1957). Use of the Pauling oxygen analyser for measurement of oxygen consumption of animals in open-circuit systems and in a short-lag, closed circuit apparatus. *Journal of Applied Physiology* **10**, 388–391.

Deutsch, C. A., Tewksbury, J. J., Huey, R. B., Sheldon, K. S., Ghalambor, C. K., Haak, D. C. & Martin, P. R. (2008). Impacts of climate warming on terrestrial ectotherms across latitude. *Proceedings of the National Academy of Sciences, USA* **105**, 6668–6672.

Dey, S, Roy, U. S. & Chattopadhyay, S. (2015). Effect of heat wave on the Indian flying fox *Pteropus giganteus* (Mammalia: Chiroptera: Pteropodidae) population from Purulia District of West Bengal, India. *Journal of Threatened Taxa* 7, 7029–7033.

Dial, R. (2003a). Energetic savings and the body size distributions of gliding mammals. *Evolutionary Ecology Research* **5**, 1151–1162.

Dial, K. P. (2003b). Wing-assisted incline running and the evolution of flight. *Science* **299**, 402–404.

Diamond, J. (2005). *Guns, Germs and Steel. A Short History of Everybody for the Last 13,000 Years*. London: Vintage Books.

Diehm, C. (2015). Should extinction be forever? Restitution, restoration, and reviving extinct species. *Environmental Ethics* **37**, 131–143.

Dierenfeld, E. S., Hintz, H. F., Robertson, J. B., Van Soest, P. J. & Oftedalt, O. T. (1982). Utilization of bamboo by the giant panda. *Journal of Nutrition* **112**, 636–641.

Dijkstra, J., Forbes, J. M. & France, J. (2005). Introduction. In: *Quantitative Aspects of Ruminant Digestion* (Eds. Dijkstra, J., Forbes, J. M. & France, J.). Wellingford, Oxfordshire: CABI Publishing.

Dill, D. B. (1938). Life, heat and altitude: physiological effects of hot climates and great heights. *American Journal of the Medical Sciences* **196**, 868.

Dill, D., Adolph, E. & Wilber, C. G. (1964). *Adaptation to the Environment*. Washington, DC: American Physiological Society.

Di Marco, M., Cardillo, M., Possingham, H. P., Wilson, K. A., Blomberg, S. P., Boitani, L. & Rondinini, C. (2012). A novel approach for global mammal extinction risk reduction. *Conservation Letters* **5**, 134–141.

Dimiceli, V. E., Piltz, S. F. & Amburn, S. A. (2011). Estimation of black globe temperature for calculation of the wet bulb globe temperature. Proceedings of the World Congress on Engineering and Computer Science 2011 Vol II. World Congress on Engineering and Computer Science: San Francisco.

Dinarello, C. A. (2004). Review: Infection, fever, and exogenous and endogenous pyrogens: some concepts have changed. *Journal of Endotoxin Research* **10**, 201–222.

Ding, X. Z., Liang, C. N., Guo, X., Wu, X. Y., Wang, H. B., Johnson, K. A. & Yana, P. (2014). Physiological insight into the high-altitude adaptations in domesticated yaks (*Bos grunniens*) along the Qinghai-Tibetan Plateau altitudinal gradient. *Livestock Science* **162**, 233–239.

Diniz-Filho, J. A. F., Rangel, T. F., Santos, T. & Bini, L. M. (2012). Exploring patterns of interspecific variation in quantitative traits using sequential phylogenetic eigenvector regressions. *Evolution* **66**, 1079–1090.

Diniz-Filho, J. A. F., Sant'Ana, C. E. R. & Bini, L. M. (1998). An eigenvector method for estimating phylogenetic inertia. *Evolution* **52**, 1247–1262.

Di Prampero, P. E. (2003). Factors limiting maximal performance in humans. *European Journal of Applied Physiology* **90**, 420–429.

Ditrich, H. (2007). The origin of vertebrates. *Zoological Journal of the Linnean Society* **150**, 435–441.

Dixon, J. M. (1989). Thylacinidae. In: *Fauna of Australia. Mammalia Volume 1B* (Eds. Walton, D. W. & Richardson, B. J.). Canberra: Australian Government Publishing Service.

Dlugosz, E. M., Chappell, M. A., Meek, T. H., Szafranska, P. A., Zub, K., Konarzewski, M., Jones, J. H., Bicudo, E. P. W., Nespolo, R. F., Careau, V. & Garland, T. (2013). Phylogenetic analysis of mammalian maximal oxygen consumption during exercise. *Journal of Experimental Biology* **216**, 4712–4721.

Doyle, T. K., Houghton, J. D. R., McDevitt, R., Davenport, J. & Hays, G. C. (2007). The energy density of jellyfish: estimates from bomb-calorimetry and proximate-composition. *Journal of Experimental Marine Biology and Ecology* **343**, 239–252.

Dubin, A. E. & Patapoutian, A. (2010). Nociceptors: the sensors of the pain pathway. *Journal of Clinical Investigation* **120**, 3760–3772.

DuBois, E. F. (1924). *Basal Metabolism in Health and Disease*. Philadelphia: Lea & Febiger.

DuBois, E. F. (1930). Recent advances in the study of basal metabolism. Part I. *Journal of Nutrition* **3**, 217–228.

Dudley, R. (2002). Mechanisms and implications of animal flight manoeuvrability. *Integrative and Comparative Biology* **42**, 135–40.

Dudley, R., Byrnes, G., Yanoviak, S. P., Borrell, B., Brown, R. M. & McGuire, J. A. (2007). Gliding and the functional origins of flight: biomechanical novelty or necessity? *Annual Reviews of Ecology and Systematics* **38**, 179–201.

Dulac, C. & Torello, A. T. (2003). Molecular detection of pheromone signals in mammals: from genes to behaviour. *Nature Reviews Neuroscience* **4**, 551–562.

Duncan, R. P., Forsyth, D. M. & Hone, J. (2007). Testing the metabolic theory of ecology: allometric scaling exponents in mammals. *Ecology* **88**, 324–333.

Dunkin, R. C., Wilson, D., Way, N., Johnson, K. & Williams, T. M. (2013). Climate influences thermal balance and water use in African and Asian elephants: physiology can predict drivers of elephant distribution. *Journal of Experimental Biology* **216**, 2939–2952.

Du Toit, J. T., Jarvis, J. U. M. & Louw, G. N. (1985). Nutrition and burrowing energetics of the Cape mole-rat *Georychus capensis*. *Oecologia* **66**, 81–87.

Dzialowski, E. M. (2005). Use of operative temperature and standard operative temperature models in thermal biology. *Journal of Thermal Biology* **30**, 317–334.

Eastman, J. (2000). *The Eastman Guide to Birds. Natural History Accounts for 150 North American Species*. Mechanicsburg, Pennsylvania: Stackpole Books.

Ebensperger, L. A. & Bozinovic, F. (2000). Energetics and burrowing behaviour in the semifossorial degu, *Octodon degus* (Rodentia: Octodontidae). *Journal of Zoology* **252**, 179–186.

Eberhard, I. H., McNamara, J., Pearse, R. J. & Southwell, I. A. (1975). Ingestion and excretion of *Eucalyptus ounctata* D.c. and its essential oil by the koala, *Phascolarctos Cinereus* (Goldfuss). *Australian Journal of Zoology* **23**, 169–179.

Economos, A. C. (1982). On the origin of biological similarity. *Journal of Theoretical Biology* **94**, 25–60.

Edwards, M. J. & Deakin, J. E. (2013). The marsupial pouch: implications for reproductive success and mammalian evolution. *Australian Journal of Zoology* **61**, 41–47.

Edwards, R. M. & Haines, H. (1978). Effects of ambient water vapour pressure and temperature on evaporative water loss in *Peromyscus maniculatus* and *Mus musculus*. *Journal of Comparative Physiology* **128**, 177–184.

Eisemann, J. H. & Nienaber, J. A. (1990). Tissue and whole-body oxygen uptake in fed and fasted steers. *British Journal of Nutrition* **64**, 399–411.

Eisenberg, J. F. (1981). *The Mammalian Radiations. An Analysis of Trends in Evolution, Adaptation, and Behavior*. Chicago: University of Chicago Press.

Eisenberg, J. F. & Wilson, D. (1978). Relative brain size and feeding strategies in the Chiroptera. *Evolution* **32**, 740–751.

El-Husseini, M. & Haggag, G. (1974). Antidiuretic hormone and water conservation in desert rodents. *Comparative Biochemistry and Physiology A* **47**, 347–350.

Elias, P. M. (2004). The epidermal permeability barrier: from the early days at Harvard to emerging concepts. *Journal of Investigative Dermatology* **122**, xxxvi–xxxix.

Elias, P. M., Cooper, E. R., Korc, A. & Brown, B. E. (1981). Percutaneous transport in relation to stratum corneum structure and lipid composition. *Journal of Investigative Dermatolology* **76**, 297–301.

Elith, J., Kearney, M. & Phillips, S. (2010). The art of modeling range-shifting species. *Methods in Ecology and Evolution* **1**, 330–342.

Elith, J. & Leathwick, J. R. (2009). Species distribution models: ecological explanation and prediction across space and time. *Annual of Review of Ecology, Evolution and Systematics* **40**, 677–697.

Ellington, C. P. (1991). Limitations on animal flight performance. *Journal of Experimental Biology* **160**, 71–91.

Elmhagen, B., Tannerfeldt, M., Verucci, P. & Angerbjörn, A. (2000). The Arctic fox (*Alopex lagopus*): an opportunistic specialist. *Journal of Zoology* **251**, 265–276.

Else, P. (2013). Evolution of 'lactation' in dinosaurs. *Journal of Experimental Biology* **216**, 347–351.

Else, P. L. & Hulbert, A. J. (1985). Mammals: an allometric study of metabolism at tissue and mitochondrial level. *American Journal of Physiology* **248**, R415–R421.

Enders, R. K. (1937). Panniculus carnosus and formation of the pouch in didelphids. *Journal of Morphology* **61**, 1–26.

Endler, J. A. (1986). *Natural Selection in the Wild*. Princeton, New Jersey: Princeton University Press.

Engel, A., Fujiyoshi, Y. & Agre, P. (2000). The importance of aquaporin water channel protein structures. *EMBO Journal* **19**, 800–806.

Epperson, L. E., Rose, J. C., Carey, H. V. Martin, S. L. (2010). Seasonal proteomic changes reveal molecular adaptations to preserve and replenish liver proteins during ground squirrel hibernation. *American Journal of Physiology* **298**, R329–R340.

Erdsack, N., Hanke, F. D., Dehnhardt, G. & Hanke, W. (2012). Control and amount of heat dissipation through thermal windows in harbor seals (*Phoca vitulina*). *Journal of Thermal Biology* **37**, 537–544.

Ernest, S. K. M. (2005). Body size, energy use, and community structure of mammals. *Ecology* **86**, 1407–1413.
Evans, T. G., Diamond, S. E. & Kelly, M. W. (2015). Mechanistic species distribution modelling as a link between physiology and conservation. *Conservation Physiology* **3**, cov056. doi:10.1093/conphys/cov056.
Fagard, R. & Conway, J. (1990). Measurement of cardiac output: Fick principle using catheterization. *European Heart Journal* **11**, 1–5.
Falk, B., Jakobsen, L., Surlykke, A. & Moss, C. F. (2014). Bats coordinate sonar and flight behavior as they forage in open and cluttered environments. *Journal of Experimental Biology* **217**, 4356–4364
Falkenstein, F., Körtner, G., Watson, K. & Geiser, F. (2001). Dietary fats and body lipid composition in relation to hibernation in free-ranging echidnas. *Journal of Comparative Physiology B* **171**, 189–194.
Farmer, C. G. (2000). Parental care: the key to understanding endothermy and other convergent features in birds and mammals. *American Naturalist* **155**, 326–334.
Fasulo, V., Zhang, Z., Chediack, J. G., Cid, F. D., Karasov, W. H. & Caviedes-Vidal, E. (2013). The capacity for paracellular absorption in the insectivorous bat *Tadarida brasiliensis*. *Journal of Comparative Physiology B* **183**, 289–296.
Fedak, M. A., Pullen, M. R. & Kanwisher, J. (1988). Circulatory responses of seals to periodic breathing: heart rate and breathing during exercise and diving in the laboratory and open sea. *Canadian Journal of Zoology* **66**, 53–60.
Fekete, É, M. & Zorilla, E. P. (2007). Physiology, pharmacology, and therapeutic relevance of urocortins in mammals: ancient CRF paralogs. *Frontiers in Neuroendocrinology* **28**, 1–27.
Feldhamer, G. A., Carter, T. C. & Whitaker, J. O. (2009). Prey consumed by eight species of insectivorous bats from southern Illinois. *American Midland Naturalist* **162**, 43–51.
Feldhamer, G. A., Drickamer, L. C., Vessey, S. H., Merritt, J. F. & Krajewski, C. (2015). *Mammalogy: Adaptation, Diversity, Ecology*. Baltimore: Johns Hopkins University Press.
Felsenstein, J. (1985). Phylogenies and the comparative method. *American Naturalist* **125**, 300–311.
Fenton, R. A., Chou, C. L., Stewart, G. S., Smith, C. P. & Knepper, M. A. (2004). Urinary concentrating defect in mice with selective deletion of phloretin-sensitive urea transporters in the renal collecting duct. *Proceedings of the National Academy of Sciences, USA* **101**, 7469–7474.
Fenton, R. A., Flynn. A., Shodeinde, A., Smith, C. P., Schnermann. J. & Knepper, M. A. (2005). Renal phenotype of UT-A urea transporter knockout mice. *Journal of the American Society of Nephrology* **16**, 1583–1592.
Fenton, R. A. & Knepper, M. A. (2007). Urea and renal function in the 21st century: insights from knockout mice. *Journal of the American Society of Nephrology* **18**, 679–688.
FEOW (2013). *Freshwater Ecoregions of the World*. World Wildlife Fund/The Nature Conservancy. http://www.feow.org/mht.php.
Ferguson, J. H. (1985). *Mammalian Physiology*. Columbus, Ohio: Charles E Merrill Publishing Company.
Ferner, K. & Mess, A. (2011). Evolution and development of fetal membranes and placentation in amniote vertebrates. *Respiratory Physiology and Neurobiology* **178**, 39–50.
Ferner, K., Zeller, U. & Renfree, M. B. (2009). Lung development of monotremes: evidence for the mammalian morphotype. *The Anatomical Record: Advances in Integrative Anatomy and Evolutionary Biology* **292**, 190–201.
Fish, F. E. (1998). Comparative kinematics and hydrodynamics of odontocete cetaceans: morphological and ecological correlates with swimming performance. *Journal of Experimental Biology* **201**, 2867–2877.

Fish, F. E. (2000). Biomechanics and energetics in aquatic and semiaquatic mammals: platypus to whale. *Physiological and Biochemical Zoology* **73**, 683–698.
Fish, F. E., Legac, P., Williams, T. M. & Wei, T. (2014). Measurement of hydrodynamic force generation by swimming dolphins using bubble DPIV. *Journal of Experimental Biology* **217**, 252–260.
Fish, F. E., Smelstoys, J., Baudinette, R. V. & Reynolds, P. S. (2002). Fur does not fly, it floats: buoyancy of pelage in semi-aquatic mammals. *Aquatic Mammals* **28**, 103–112.
Fisher, D. O., Owens, I. P. F. & Johnson, C. N. (2001). The ecological basis of life history variation in marsupials. *Ecology* **82**, 3531–3540.
Fleming, T. H. & Eby, P. (2003). Ecology of bat migration. In: *Bat Ecology* (Eds. Kunz, T. H. & Fenton, M. B.). Chicago: Chicago University Press.
Fleming, T. H. & Kress, W. J. (2011). A brief history of fruits and frugivores. *Acta Oecologica* **37**, 521–530.
Fleming, T. H. & Sosa, V. J. (1994). Effects of nectarivorous and frugivorous mammals on reproductive success of plants. *Journal of Mammalogy* **75**, 845–851.
Florant, G. L. (1998). Lipid metabolism in hibernators—the importance of essential fatty acids. *American Zoologist* **38**, 331–340.
Foksinski, M., Rozalski, R., Guz, J., Ruszkowska, B., Sztukowska, P., Piwowarski, M., Klungland, A. & Olinski, R. (2004). Urinary excretion of DNA repair products correlates with metabolic rates as well as with maximum life spans of different mammalian species. *Free Radical Biology and Medicine* **37**, 1449–1454.
Foley, L. E., Gegear, R. J. & Reppert, S. M. (2011). Human cryptochrome exhibits light-dependent magnetosensitivity. *Nature Communications* **2**, 356.
Foley, W. J. & Cork, S. J. (1992). Use of fibrous diets by small herbivores: how far can the rules be 'bent'? *Trends in Ecology and Evolution* **7**, 159–162.
Foley, W. J. & Hume, I. D. (1987). Digestion and metabolism of high-tannin *Eucalyptus* foliage by the brushtail possum (*Trichosurus vulpecula*) (Marsupialia: Phalangeridae). *Journal of Comparative Physiology* **157**, 67–76.
Folk, G. E., Brewer, M. D & Sander, D. (1970). Cardiac physiology of polar bears in winter dens. *Arctic* **23**, 130.
Folk, G. E., Larson, A. & Folk, M. A. (1976). Physiology of hibernating bears. In: *Bears—Their Biology and Management* (Eds. Pelton, M. R., Lentfer, J. W & Folk, G. E.). Morges, Switzerland: International Union for Conservation of Nature and Natural Resources.
Fooden, J. (1972). Breakup of Pangaea and isolation of relict mammals in Australia, South America, and Madagascar. *Science* **175**, 894–898.
Fortelius, M. & Kappelman, J. (1993). The largest land mammal ever imagined. *Zoological Journal of the Linnean Society of London* **107**, 85–101.
Fortin, M.-J., Keitt, T. H., Maurer, B, A., Taper, M. L., Kaufman, D. M. & Blackburn, T. M. (2005). Species' geographic ranges and distributional limits: pattern analysis and statistical issues. *Oikos* **108**, 7–17.
Foster, D. O. & Frydman, M. L. (1978). Nonshivering thermogenesis in the rat. II. Measurements of blood flow with microspheres point to brown adipose tissue as the dominant site of calorigenesis induced by norepinephrine. *Canadian Journal of Physiology and Pharmacology* **56**, 110–122.
Foster, J. B. (1964). Evolution of mammals on islands. *Nature* **202**, 234–235.
Fountaine, T. M. R., Benton, M. J., Dyke, G. J. & Nudds, R. L. (2005). The quality of the fossil record of Mesozoic birds. *Proceedings of the Royal Society of London B* **272**, 289–294.
France, J. & Dijkstra, J., (2005). Volatile fatty acid production. In: *Quantitative Aspects of Ruminant Digestion* (Eds. Dijkstra, J., Forbes, J. M. & France, J.). Wellingford, Oxfordshire: CABI Publishing.

Frank, C. L. (1992). The influence of dietary fatty acids on hibernation by golden-mantled ground squirrels (*Spermophilus lateralis*). *Physiological Zoology* **65**, 906–920.

Frank, C. L. (1994). Polyunsaturate content and diet selection by ground squirrels (*Spermophilus lateralis*). *Ecology* **75**, 458–463.

Frank, C. L., Karpovich, S. & Barnes, B. M. (2008). Dietary fatty acid composition and the hibernation patterns in free-ranging Arctic ground squirrels. *Physiological and Biochemical Zoology* **81**, 486–495.

Frankenfield, D. C. (2010). On heat, respiration, and calorimetry. *Nutrition* **26**, 939–950.

Franklin, D. L., Ellis, R. M. & Rushmer, R. F. (1959). Aortic blood flow in dogs during treadmill exercise. *Journal of Applied Physiology* **14**, 809–812.

Franz, R., Soliva, C. R., Kreuzer, M., Hummel, J. & Clauss, M. (2011). Methane output of rabbits (*Oryctogalus cuniculus*) and guinea pigs (*Cavia porcellus*) fed a hay-only diet: implications for the scaling of methane production with body mass in non-ruminant mammalian herbivores. *Comparative Biochemistry and Physiology A* **158**, 177–181.

Franz, R., Soliva, C. R., Kreuzer, M., Steuer, P., Hummel, J. & Clauss, M. (2010). Methane production in relation to body mass of ruminants and equids. *Evolutionary Ecology Research* **1**, 727–738.

Frappell, P. B. & Baudinette, R. V. (1995). Scaling of respiratory variables and the breathing pattern in adult marsupials. *Respiration Physiology* **100**, 83–90.

Frappell, P. B., Baudinette, R. V., MacFarlane, P. M., Wiggins, P. R. & Shimmin, G. (2002). Ventilation and metabolism in a large semifossorial marsupial: the effect of graded hypoxia and hypercapnia. *Physiological and Biochemical Zoology* **75**, 77–82.

Frappell, P. B., Hinds, D. S. & Boggs, D. F. (2001). Scaling of respiratory variables and the breathing pattern in birds: an allometric and phylogenetic approach. *Physiological and Biochemical Zoology* **74**, 75–89.

Frappell, P. B., Lanthier, C. Baudinette, R. V. & Mortola, J. P. (1992). Metabolism and ventilation in acute hypoxia: a comparative analysis in small mammalian species. *American Journal of Physiology* **262**, R1040–R1046.

Frappell, P. B. & Macfarlane, P. M. (2006). Development of the respiratory system in marsupials. *Respiratory Physiology and Neurobiology* **154**, 252–267.

Frase, B. A. (2002). Hematological parameters of high-elevation bushy-tail woodrats. *Southwestern Naturalist* **47**, 508–510.

Fraser, E. E., McGuire, L. P., Eger, J. L., Longstaffe, F. J. & Fenton, M. B. (2012). Evidence of latitudinal migration in tri-colored bats, *Perimyotis subflavus*. *PLOS ONE* **7**, e31419.

Fraústo da Silva, J. J. R. & Williams, R. J. P. (2001). *The Biological Chemistry of the Elements*. Oxford: Oxford University Press.

Freckleton, R. P, Harvey, P. H. & Pagel, M. (2002). Phylogenetic analysis and comparative data: a test and review of evidence. *American Naturalist* **160**, 712–726.

Frerichs, K. U., Smith, C. B., Brenner, M., Degracia, D. J., Krause, G. S., Marrone, L., Dever, T. E. & Hallenbeck, J. M. (1998). Suppression of protein synthesis in brain during hibernation involves inhibition of protein initiation and elongation. *Proceedings of the National Academy of Sciences, USA* **95**, 14511–14516.

Fullard, J. H., Koehler, C., Surlykke, A. & McKenzie, N. L. (1991). Echolocation ecology and flight morphology of insectivorous bats (Chiroptera) in south-western Australia. *Australian Journal of Zoology* **39**; 427–438.

Fuller, A., Dawson, T., Helmuth, B., Hetem, R. S., Mitchell, D. & Maloney, S. K. (2010). Physiological mechanisms in coping with climate change. *Physiological and Biochemical Zoolology* **83**, 713–720.

Fuller, A., Hetem, R. S., Meyer, L. C. & Maloney, S. K. (2011a). Angularis oculi vein blood flow modulates the magnitude but not the control of selective brain cooling in sheep. *American Journal of Physiology* **300**, R1409–R1417.
Fuller, A., Hetem, R. S., Maloney, S. K. & Mitchell, D. (2014). Adaptation to heat and water shortage in large, arid-zone mammals. *Physiology* **29**, 159–167.
Fuller, A., Kamerman, P. R., Maloney, S. K., Matthee, A., Mitchell, G. & Mitchell, D. (2005). A year in the thermal life of a free-ranging herd of springbok *Antidorcas marsupialis*. *Journal of Experimental Biology* **208**, 2855–2864.
Fuller, A., Maloney, S. K., Kamerman, P. R., Mitchell, G. & Mitchell, D. (2000). Absence of selective brain cooling in free-ranging zebras in their natural habitat. *Experimental Physiology* **85**, 209–217.
Fuller, A., Maloney, S. K. & Mitchell, D. (2011b). No evidence that mammals without a carotid rete can selectively cool their brains. *Journal of Applied Physiology* **110**, 576–577.
Fuller, A., Meyer, L. C. R., Mitchell, D. and Maloney, S. K. (2007). Dehydration increases the magnitude of selective brain cooling independently of core temperature in sheep. *American Journal of Physiology* **293**, R438–R446.
Funakubo, M., Sato, J., Honda, T. & Mizumura, K. (2010). The inner ear is involved in the aggravation of nociceptive behaviour induced by lowering barometric pressure of nerve injured rats. *European Journal of Pain* **14**, 32–39.
Futuyma, D. J. (1998). *Evolutionary Biology*. Sunderland: Sinauer Associates.
Gallardo, P., Cortés, A. & Bozinovic, F. (2005). Phenotypic flexibility at the molecular and organismal level allows desert-dwelling rodents to cope with seasonal water availability. *Physiological and Biochemical Zoology* **78**, 145–152.
Gallivan, G. J. (1992). What are the metabolic rates of cetaceans? *Physiological Zoology* **65**, 1285–1297.
Gallivan, G. J. & Best, R. C. (1986). The influence of feeding and fasting on the metabolic rate and ventilation of the Amazonian manatee (*Trichechus inunguis*). *Physiological Zoology* **59**, 552–557.
Gallivan, G. J. & Ronald, K. (1979). Temperature regulation in freely diving harp seals (*Phoca groenlandica*). *Canadian Journal of Zoology* 57, 2256–2263.
Gallivan, G. J. & Ronald, K. (1981). Apparent specific dynamic action in the harp seal (*Phoca groenlandica*). *Comparative Physiology and Biochemistry* **69A**, 579–581.
Gamble, J. L., McKhann, C. F., Butler, A. M. & Tuthill, E. (1934). An economy of water in renal function referable to urea. *American Journal of Physiology* **109**, 139–154.
Gannon, W. L., Sikes, R. S., & The Animal Care and Use Committee of the American Society of Mammalogists. (2007). Mammalogists guidelines of the American Society of Mammalogists for the use of wild mammals in research. *Journal of Mammalogy* **88**, 809–823.
Gardner, A. L. (1977). Feeding habits. In: *Biology of Bats of the New World Family Phyllostomatidae, Part II*. (Eds. Baker, R. J., Knox Jones, J. & Carter, D. C.). Special Publications of the Museum of Texas Technical University, No. 13. Lubbock, Texas: Texas Technical University.
Garland, T., Dickerman, A. W., Janis, C. M. & Jones, J. A. (1993). Phylogenetic analysis of covariance by computer simulation. *Systematic Biology* **42**, 265–292.
Garland, T. & Ives, A. R. (2000). Using the past to predict the present: confidence intervals for regression equations in phylogenetic comparative methods. *American Naturalist* **155**, 346–364.
Garland, T. & Janis, C. M. (1993). Does metatarsal/femur ratio predict maximal running speed in cursorial mammals? *Journal of Zoology* **229**, 133–151.
Garner, J., Taylor, G. & Thomas, A. (1999). On the origins of birds: the sequence of character acquisition in the evolution of avian flight. *Proceedings of the Royal Society of London B* **266**, 1259–1266.

Gaston, K. J. (2003). *The Structure and Dynamics of Geographic Ranges*. Oxford: Oxford University Press.
Gaston, K. J., Blackburn, T. M. & Spicer, J. I. (1998). Rapoport's rule: time for an epitaph? *Trends in Ecology and Evolution* **13**, 70–74.
Gaston, K. J. & Spicer, J. I. (2001). The relationship between range size and niche breadth: a test using five species of '*Gammarus*' (Amphipoda). *Global Ecology and Biogeography* **10**, 179–188.
Gatenby, R. M. (1977). Conduction of heat from sheep to ground. *Agricultural Meterology* **18**, 387–400.
Gates, G. R. (1978). Vision in the monotreme echidna (*Tachyglossus aculeatus*). *Australian Zoologist* **20**, 147–167.
Gazzola, M., Argentina, M. & Mahadevan, L. (2014). Scaling macroscopic aquatic locomotion. *Nature Physics* **10**, 758–761.
Geiser, F. (1988a). Reduction of metabolism during hibernation and daily torpor in mammals and birds: temperature effect or physiological inhibition? *Journal of Comparative Physiology B* **158**, 25–37.
Geiser, F. (1988b). Daily torpor and thermoregulation in *Antechinus* (Marsupialia): influence of body mass, season, development, reproduction, and sex. *Oecologia* 77, 395–399
Geiser, F. (1990). Influence of polyunsaturated and saturated dietary lipids on adipose tissue, brain and mitochondrial membrane fatty acid composition of a mammalian hibernator. *Biochimica et Biophysical Acta* **1046**, 159–166.
Geiser, F. (1993). Dietary lipids and thermal physiology. In: *Life in the Cold: Ecological, Physiological and Molecular Mechanisms* (Eds. Carey, C., Florant, G. L., Wunder, B. A. & Horwitz, B.). Boulder: Westview Press.
Geiser, F. (1994). Hibernation and daily torpor in marsupials: a review. *Australian Journal of Zoology* **42**, 1–16
Geiser, F. (2004). Metabolic rate and body temperature reduction during hibernation and daily torpor. *Annual Review of Physiology* **66**, 239–274.
Geiser, F. (2007). Yearlong hibernation in a marsupial mammal. *Naturwissenchaften* **94**, 941–944.
Geiser, F. (2010). Aestivation in mammals and birds. In: *Aestivation. Molecular and Physiological Aspects* (Eds. Navas, C. A. & Carvalho, J. E.). Heidelburg: Springer.
Geiser, F. (2013). Hibernation. *Current Biology* **5**, R188–R193.
Geiser, F. & Baudinette, R. V. (1988). Daily torpor and thermoregulation in the small dasyurid marsupials *Planigale gilesi* and *Ningaui yvonnae*. *Australian Journal of Zoology* **36**, 473–481.
Geiser, F. & Brigham, R. M., (2012). The other functions of torpor. In: *Living in a Seasonal World* (Eds. Ruf, T., Bieber, C., Arnold, W. & Millesi, E.). Berlin: Springer-Verlag.
Geiser, F. & Broome, L. S. (1991). Hibernation in the mountain pygmy possum *Burramys parvus* (Marsupialia). *Journal of Zoology* **223**, 593–602.
Geiser, F. & Drury, R. L. (1993). Radiant heat affects thermoregulation and energy expenditure during rewarming from torpor. *Journal of Comparative Physiology B* **173**, 55–60.
Geiser, F. & Ferguson, C. (2001). Intraspecific differences in behaviour and physiological effects of captive breeding on patterns of torpor in feathertail gliders. *Journal of Comparative Physiology B* **171**, 569–576.
Geiser, F. & Kenagy, G. J. (1987). Polyunsaturated lipid diet lengthens torpor and reduces body temperature in a hibernator. *American Journal of Physiology* **252**, R897–R901
Geiser, F. & Masters, P. (1994). Torpor in relation to reproduction in the mulgara, *Dasycercus cristicauda* (Dasyuridae: Marsupialia). *Journal of Thermal Biology* **19**, 33–40.

Geiser, F., Matwiejczyk, L. & Baudinette, R. V. (1986). From ectothermy to heterothermy: the energetics of the kowari, *Dasyuroides byrnei* (Marsupialia: Dasyuridae). *Physiological Zoology* **59**, 220–229.
Geiser, F., Stahl, B. & Learmonth, R. P. (1992). The effect of dietary fatty acids on the pattern of torpor in a marsupial. *Physiological Zoology* **65**, 1236–1245.
Geiser, F. & Drury, R. L. (2003). Radiant heat affects thermoregulation and energy expenditure during rewarming from torpor. *Journal of Comparative Physiology B* **173**, 55–60.
Geiser, F. & Turbill, C. (2009). Hibernation and daily torpor minimize mammalian extinctions. *Naturwissenchaften* **96**, 1235–1240.
Geist, N. R. (2000). Nasal respiratory turbinate function in birds. *Physiological and Biochemical Zoology* **73**, 581–589.
Geist, V. (1987). Bergmann's Rule is invalid. *Canadian Journal of Zoology* **65**, 1035–1038.
Geluso, K. N. (1978). Urine concentrating ability and renal structure of insectivorous bats. *Journal of Mammalogy* **59**, 312–323.
Gerardo Herrera, L. G., Cruz-Neto, A. P., Wojciechowskic, M. S., Larrain, P., Pinshow, B. & Korine, C. (2015). The relationships between food and energy intakes, salt content and sugar types in Egyptian fruit bats. *Mammalian Biology* **80**, 409–413.
Gerlinsky, C. D., Rosen, D. A. & Trites, A. W. (2013). High diving metabolism results in a short aerobic dive limit for Steller sea lions (*Eumetopias jubatus*). *Journal of Comparative Physiology B* **183**, 699–708.
Gerstner, G. & Gerstein, J. B. (2008). Chewing rate allometry among mammals. *Journal of Mammalogy* **89**, 1020–1030.
Giardina, B., Mosca, D. & De Rosa, M. C. (2004). The Bohr effect of haemoglobin in vertebrates: an example of molecular adaptation to different physiological requirements. *Acta Physiologica Scandinavica* **182**, 229–244.
Gibson, L. A. & Hume, I. D. (2002). Nitrogen requirements of the omnivorous greater bilby, *Macrotis lagotis* (Marsupialia: Peramelidae). *Physiological and Biochemical Zoology* **75**, 48–56.
Gikonyo, N. K., Hassanali, A., Njagi, P. G. & Saini, R. K. (2003). Responses of *Glossina morsitans morsitans* to blends of electroantennographically active compounds in the odors of its preferred (Buffalo and ox) and nonpreferred (waterbuck) hosts. *Journal of Chemical Ecology* **29**, 2331–2345
Giladi, I. & Pinshow, B. (1999). Evaporative and excretory water loss during free flight in pigeons. *Journal of Comparative Physiology B* **169**, 311–318.
Gilbert, R. D., Schroder, H., Kawamura, T., Dale, P. S. & Power, G. G. (1985). Heat transfer pathways between fetal lamb and ewe. *Journal of Applied Physiology* **59**, 634–638.
Gill, T. (1872). Arrangement of the families of mammals and synoptical table of the characters of the subdivisions of mammals. *Smithsonian Miscellaneous Collections* **11**, 1–89.
Gillespie, M. J., Stanley, D., Chen, H., Donald, J. A., Nicholas, K. R., Moore, R. J. & Crowley, T. M. (2012). Functional similarities between pigeon 'milk' and mammalian milk: induction of immune gene expression and modification of the microbiota. *PLOS ONE* 7, e48363.
Gillihan, S. W. & Foresman, K. R. (2004). *Sorex vagrans*. *Mammalian Species* **744**, 1–5.
Gillooly, J. F, Brown, J. H, West, G. B, Savage, V. M. & Charnov, E. L. (2001). Effects of size and temperature on metabolic rate. *Science* **293**, 2248–2251.
Ginsberg, J. R. & Huck, U. W. (1989). Sperm competition in mammals. *Trends in Ecology and Evolution* 4, 74–79.
Girardier, L. (1983). Brown fat: an energy dissipating tissue. In: *Mammalian Thermogenesis* (Eds. Girardier, L. & Stock, M. J.). London: Chapman & Hall.

Gittleman, J. L. (1986). Carnivore life history patterns: allometric, phylogenetic, and ecological associations. *American Naturalist* **127**, 744–771.

Gittleman, J. L. (1989). *Carnivore Behavior, Ecology and Evolution*. New York; Comstock Publishing Associates; Ithaca, New York: Cornell University Press.

Gittleman, J. L. & Kot, M. (1990). Adaptation: statistics and a null model for estimating phylogenetic effects. *Systematic Zoology* **39**, 227–241.

Gjedde, A. (2010). Diffusive insights: on the disagreement of Christian Bohr and August Krogh at the Centennial of the Seven Little Devils. *Advances in Physiological Education* **34**, 174–185.

Glazier, D. S. (2005). Beyond the '3/4-power law': variation in the intra- and interspecific scaling of metabolic rate in animals. *Biological Reviews* **80**, 1–52.

Glazier, D. S. (2008). Effects of metabolic level on the body size scaling of metabolic rate in birds and mammals. *Proceedings of the Royal Society of London B* **22**, 1405–1410.

Glazier, D. S. (2010). A unifying explanation for diverse metabolic scaling in animals and plants. *Biological Reviews* **85**, 111–138.

Glazier, D. S. (2015). Body-mass scaling of metabolic rate: what are the relative roles of cellular versus systemic effects? *Biology* **4**, 187–199.

Gleeson, T. T. (1996). Post-exercise lactate metabolism: a comparative review of sites, pathways, and regulation. *Annual Reviews of Physiology* **58**, 565–581.

Gleiss, A. C., Wilson, R. P. & Shepard, E. L. C. (2011). Making overall dynamic body acceleration work: on the theory of acceleration as a proxy for energy expenditure. *Methods in Ecology and Evolution* **2**, 23–33.

Gloger, C. L. (1883). Das abandern der vogel durch einfluss des klimas. Breslau, Germany: A. Schulz.

Godoy-Vitorino, F., Goldfarb, K. C., Karaoz, U., Leal, S., Garcia-Amado, M. A., Hugenholtz, P., Tringe, S. G., Brodie, E. L. & Dominguez-Bello, M. G. (2012). Comparative analyses of foregut and hindgut bacterial communities in hoatzins and cows. *ISME Journal* **6**, 531–541.

Godthelp, H., Archer, M., Cifelli, R., Hand, S. J. & Gilkeson, C. F. (1992). Earliest known Australian Tertiary mammal fauna. *Nature* **356**, 514–516.

Goin, C. J., Goin, O. B. & Zug, G. R. (1978). *Introduction to Herpetology*. San Francisco: Freeman & Co.

Golozoubova, V., Cannon, B. & Nedergaard, J. (2006). UCP1 is essential for adaptive adrenergic nonshivering thermogenesis. *American Journal of Physiology* **291**, E350–E357.

Gomes Rodrigues, H., Marangoni, P., Šumbera, R., Tafforeau, P., Wendelen, W. & Viriot, L. (2011). Continuous dental replacement in a hyper-chisel tooth digging rodent. *Proceedings of the National Academy of Sciences, USA* **108**, 17355–17359.

Gompper, M. E. & Decker, D. M. (1998). *Nasua nasua*. *Mammalian Species* **580**, 1–9.

Gompper, M. E. & Gittleman, J. L. (1991). Home range scaling: intraspecific and comparative trends. *Oecologia* **87**, 343–348.

Gooden, B. A. (1994). Mechanism of the human diving response. *Integrative Physiological and Behavioral Science* **29**, 6–16.

Gopal, P. K. (2013). Morphological adaptations in the kidney and urine concentrating ability in relation to dietary habit in the three species of bats. *World Journal of Zoology* **8**, 198–205.

Gottschalk, C. W. (1987). History of the urinary concentrating mechanism. *Kidney International* **31**, 507–511.

Gould, E. (1965). Evidence for echolocation in the Tenrecidae of Madagascar. *Proceedings of the American Philosophical Society* **109**, 352–360.

Gould, J. K. (2010). Magnetoreception. *Current Biology* **20**, R431–R435.

Grafen, A. (1989). The phylogenetic regression. *Proceedings of the Royal Society of London B* **326**, 119–156.

Grant, T. & Fanning, D. (2007). *Platypus*. Collingwood, Victoria: CSIRO Publishing.
Grant, T. R. (1989). Ornithorhynchidae. In: *Fauna of Australia. Mammalia Volume 1B* (Eds. Walton, D. W. & Richardson, B. J.). Canberra: Australian Government Publishing Service.
Grant, T. R. (1989). Tachyglossidae. In: *Fauna of Australia. Mammalia Volume 1B* (Eds. Walton, D. W. & Richardson, B. J.). Canberra: Australian Government Publishing Service.
Green, B., Anderson, J. & Whateley, T. (1984). Water and sodium turnover and estimated food consumption in free-living lions (*Panthera leo*) and spotted hyaenas (*Crocuta crocuta*). *Journal of Mammalogy* **65**, 593–599.
Green, B. & Eberhard, F. (1983). Water and sodium intake, and estimated food consumption, in free-living eastern quolls *Dasyurus viverrinus*. *Australian Journal of Zoology* **31**, 871–880.
Green, B. (1978). Estimation of food consumption in the dingo, *Canis familiaris dingo*, by means of Na turnover. *Ecology* **59**, 207–210.
Green, H. L. H. H. (1937). The development and morphology of the teeth of *Ornithorhynchus*. *Philosophical Transactions of the Royal Society of London B* **288**, 367–420.
Green, J. A. (2011). The heart rate method for estimating metabolic rate: review and recommendations. *Comparative Biochemistry and Physiology A* **158**, 287–304.
Greenhall, A. M. (1972). The biting and feeding habits of the vampire bat, *Desmodus rotundus*. *Journal of Zoology* **168**, 451–461.
Greenhall, A. M. & Schutt, W. A. (1996). *Diaemus youngi*. *Mammalian Species* **533**, 1–7.
Greenwald, L. Stetson, D. (1988). Urine concentration and the length of the renal papilla. *News in Physiological Science* **3**, 46–49.
Gregory, J. E., Iggo, A., McIntyre, A. K. & Proske, U. (1989). Responses of electroreceptors in the snout of the echidna. *Journal of Physiology* **414**, 521–538.
Gregory, W. K. (1947). The monotremes and the palimpsest theory. *Bulletin of the American Museum of Natural History* **88**, 1–52.
Grenyer, R., Orme, C. D. L., Jackson, S. F., Thomas, G. H., Davies, R. G., Davies, T. J., Jones, K. E., Olson, V. A., Ridgely, R. S., Rasmussen, P. C., Ding, T.-S., Bennett, P. M., Blackburn, T. M., Gaston, K. J., 3, Gittleman, J. L. & Owens, I. P. F. (2006). Global distribution and conservation of rare and threatened vertebrates. *Nature* **444**, 93–96.
Griffith, R. W. (1987). Freshwater or marine origin of vertebrates? *Comparative Biochemistry and Physiology* **87**, 523–531.
Griffiths, M. (1968). *Echidnas*. Oxford: London: Pergamon.
Griffiths, M. (1984). Mammals: Monotremes. In: *Fourth Edition of Marshall's Physiology of Reproduction*, ***Vol. 1*** (Ed. Lamming, G. E.). Edinburgh: Churchill Livingstone.
Griffiths, M. (1989a). Tachyglossidae. In: *Fauna of Australia. Mammalia Volume 1B* (Eds. Walton, D. W. & Richardson, B. J.). Canberra: Australian Government Publishing Service.
Griffiths, M. (1989b). Ornithorhynchidae. In: *Fauna of Australia. Mammalia Volume 1B* (Eds. Walton, D. W. & Richardson, B. J.). Canberra: Australian Government Publishing Service.
Griffiths, M., Green, B., Leckie, R. M. C., Messer, M. & Newgrain, K. W. (1984). Constituents of platypus and echidna milk, with particular reference to the fatty acid complements of the triglycerides. *Australian Journal of Biological Science* **37**, 323–329.
Griffiths, M., McIntosh, D. L. & Coles, R. E. A. (1969). The mammary gland of the echidna, *Tachyglossus aculeatus*, with observations on the incubation of the egg and the newly-hatched young. *Journal of Zoology* **158**, 371–386.
Grigg, G. C., Augee, M. L. & Beard, L. A. (1992). Thermal relations of free-living echidnas during activity and in hibernation in a cold climate. In: *Platypus and Echidnas* (Ed. Augee, M. A.). Sydney: Royal Society of NSW.
Grigg, G. & Beard, L. (1993). Hibernation by Echidnas in mild climates: hints about the evolution of endothermy? In: *Life in the Cold: Ecological, Physiological and Molecular Mechanisms* (Eds. Carey, C., Florant, G. L., Wunder, B. A. & Horwitz, B.). Boulder: Westview Press.

Grigg, G. C., Beard, L. A. & and Augee, M. L. (2004). The evolution of endothermy and its diversity in mammals and birds. *Physiological and Biochemical Zoology* 77, 982–997.
Grodzinski, W. & Wunder, B. A. (1975). Ecological energetics of small mammals. In: *Small Mammals: Their Productivity and Population Dynamics* (Eds. Golley, F. B., Petrusewicz, K. & Ryszkowski, L.). London: Cambridge University Press.
Grollman, A. (1932). *The Cardiac Output of Man in Health and Disease*. Springfield: Charles C. Thomas.
Guglielmo, C. G. (2010). Move that fatty acid: fuel selection and transport in migratory birds and bats. *Integrative and Comparative Biology* **50**, 336–345.
Guillette, L. J. (1993). The evolution of viviparity in lizards. *Bioscience* **43**, 742–751.
Gunn, T. R., Ball, K. T. & Gluckman, P. D. (1993). Withdrawal of placental prostaglandins permits thermogenic responses in fetal sheep brown adipose tissue. *Journal of Applied Physiology* **74**, 998–1004.
Guppy, M. (2007). *Energy Processing in Cells. Recurring Concepts and Themes*. London: Athena Press.
Guppy, M. & Withers, P. C. (1999). Metabolic depression in animals: physiological perspectives and biochemical generalizations. *Biological Reviews* 74, 1–40.
Gutmann, A. K., Lee, D. V. & McGowan, C. P. (2013). Collision-based mechanics of bipedal hopping. *Biology Letters* 9. doi.org/10.1098/rsbl.2013.0418.
Guyton, A. C. & Hall, J. E. (1996). *Textbook of Medical Physiology*. Philadelphia: Saunders College Publishing.
Gwazdauskas, F. C. (1985). Effects of climate on reproduction in cattle. *Journal of Dairy Science* **68**, 1568–1578.
Haeckel, E. (1874). *Anthropogenie, oder, Entwickelungsgeschichte des Menschen: Gemeinverständliche wissenschaftliche Vorträge über die Grundzüge der menschlichen Keimes- und Stammes-Geschichte*. Leipzig: Verlag von Wilhelm Engelmann.
Haig, D. (1993). Genetic conflicts in human pregnancy. *Quarterly Review of Biology* **68**, 495–532.
Hails, C. J. (1979). A comparison of flight energetics in hirundines and other birds. *Comparative Biochemistry and Physiology A* **63**, 581–585.
Hales, J. R. (1973). Effects of exposure to hot environments on the regional distribution of blood flow and on cardiorespiratory function in sheep. *Pflugers Archiv* **344**, 133–148.
Hales, J. R. & Findlay, J. D. (1968). The oxygen cost of thermally-induced and $CO_2$-induced hyperventilation in the ox. *Respiration Physiology* 4, 353–362.
Hales, J. R. S. (1976). Interactions between respiratory and thermoregulatory systems of domestic animals in hot environments. In: *Progress in Biometeorology. Division B Progress in Animal Biometeorology, Volume 1 Effect of temperature on Animals* (Eds. Tromp, S. W. & Bouma, J. J.). Amsterdam: Swets & Zeitlinger.
Hales, S. (1727). *Vegetable Staticks*. London: Innys and Woodward.
Hales, S. (1733). *Statical Essays: Containing Haemastaticks. Vol 2*. London: Innys and Manby.
Haley, M. P., Deutsch, C. J. & Le Boeuf, B. J. (1991). A method for estimating mass of large pinnipeds. *Marine Mammal Science* 7, 157–164.
Haman, F. (2006). Shivering in the cold: from mechanisms of fuel selection to survival. *Journal of Applied Physiology* **100**, 1702–1708.
Hambly, C. & Voigt, C. C. (2011). Measuring energy expenditure in birds using bolus injections of $^{13}$C-labelled Na-bicarbonate. *Comparative Biochemistry and Physiology A* **158**, 323–328.
Hamilton, C. L. (1976). Environmental temperature and feeding behaviour. In: *Progress in Biometeorology. Division B Progress in Animal Biometeorology, Volume 1, Effect of temperature on animals* (Eds. Tromp, S. W. & Bouma, J. J.). Amsterdam: Swets & Zeitlinger.
Hamilton, C. L. & Ciaccia, P. J. (1971). Hypothalamus, temperature regulation, and feeding in the rat. *American Journal of Physiology* **221**, 800–807.

Hammond, K. A. & Diamond, J. (1992). An experimental test for a ceiling on sustained metabolic rate in lactating mice. *Physiological Zoology* **65**, 952–977.
Hanna, J. B. & Schmitt, D. (2011). Locomotor energetics in primates: gait mechanics and their relationship to the energetics of vertical and horizontal locomotion. *American Journal of Physical Anthropology* **145**, 43–54.
Happold, D. C. D. & Happold, M. (1988). Renal form and function in relation to the ecology of bats (Chiroptera) from Malawi, Central Africa. *Journal of Zoology* **215**, 629–655.
Harden, L. M., Kent, S., Pittman, Q. J. & Roth, J. (2015). Fever and sickness behavior: friend or foe? *Brain, Behavior, and Immunity* **50**, 322–333.
Hardy, J. D. & Soderstrom, G. F. (1938). Heat loss from the nude body and peripheral blood flow at temperatures of 22°C to 35°C. *Journal of Nutrition* **16**, 493–510.
Harestad, A. S. Bunnell, F. L. (1979). Home range and body weight—a re-evaluation. *Ecology* **60**, 389–402.
Harned, H. S. & Robinson, R. A. (1940). A note on the temperature variation of the ionisation constants of weak electrolytes. *Transactions of the Faraday Society* **36**, 973–978.
Harris, G., Thirgood, S., Hopcraft, G. C., Cromsigt, J. P. G. M. & Berger, J. (2009). Global decline in aggregated migrations of large terrestrial mammals. *Endangered Species Research* **7**, 55–76.
Harrison, A., Robb, G. N., Bennett, N. C. & Horack, I. G. (2012). Differential feeding success of two paralysis-inducing ticks, *Rhipicephalus warburtoni* and *Ixodes rubicundus* on sympatric small mammal species, *Elephantulus myurus* and *Micaelamys namaquensis*. *Veterinary Parasitology* **188**, 346–354.
Harrison, G. (2015). *Meteorological Measurements and Instrumentation*. Hoboken, New Jersey: Wiley-Blackwell.
Harrison, L. (1921). The migration route of the Australian marsupial fauna. *Australian Zoologist* **3**, 247–263.
Hart, J. S. (1956). Seasonal changes in the insulation of the fur. *Canadian Journal of Zoology* **34**, 53–57.
Harvey, P. H., Clutton-Brock, T. H. & Mace, G. M. (1980). Brain size and ecology in small mammals and primates. *Proceedings of the National Academy of Sciences, USA* 77, 4387.
Harvey, P. H. & Pagel, M. D. (1991). *The Comparative Method in Evolutionary Biology*. Oxford: Oxford University Press.
Harvey, P. H., Read, A. F. & Promislow, D. E. L. (1989). Life history variation in placental mammals: unifying the data with theory. *Oxford Surveys in Evolutionary Biology* **6**, 13–31.
Haskell, J. P., Ritchie, M. E. & Wolff, H. (2002). Fractal geometry predicts varying body size scaling relationships for mammal and bird home ranges. *Nature* **418**, 527–530.
Haswell, W. A. (1914). The animal life of Australia. In: *Federal Handbook, British Association for the Advancement of Science*, 84th *Meeting, Australia* (Ed. Knibb, G. H.). Melbourne: Government Printer.
Have-Opbroek, T. & Antonia, A. W. (1981). The development of the lung in mammals: an analysis of concepts and findings. *American Journal of Anatomy* **162**, 201–219.
Hawkins, B. A., Albuquerque, F. S., Araùjo, M. B., Beck, J., Bini, L. M., Cabreu-Sãnudo, F. J., Castro-Parga, I., Diniz-Filho, J. A. F., Ferrer-Castán, D., Field, R., Gòsnez, J., Hortal, J., Kerr, J. T., Kitching, I. J., León-Corté, J. L., Lobo, J. M., Montoya, D., Moreno, J. C. Olalla-Tárraga, M. A., Rodriguez, M. A., Sanders, N. & Williams, P. (2007). Global evaluation of metabolic theory as an explanation for terrestrial species richness gradients. *Ecology* **88**, 1877–1888.
Hawkins, P. (2004). Bio-logging and animal welfare: practical refinements. *Memoirs of the National Institute of Polar Research* **58**, 58–68.

Hayes, J. P. (1989). Field and maximal metabolic rates of deer mice (*Peromyscus maniculatus*) at low and high altitudes. *Physiological Zoology* **62**, 732–744.
Hayes, J. P. & O'Connor, C. S. (1999). Natural selection on thermogenic capacity of high-altitude deer mice. *Evolution* **53**, 1280–1287.
Hayes, J. P. & Shonkwiler, J. S. (2006). Allometry, antilog transformations, and the perils of prediction on the original scale. *Physiological and Biochemical Zoology* **79**, 665–674.
Hayssen, V. (1993). Empirical and theoretical constraints on the evolution of lactation. *Journal of Dairy Science* **76**, 3213–3233.
Hayssen, V. (2011). *Tamandua tetradactyla (Pilosa: Myrmecophagidae)*. *Mammalian Species* **43** (875), 64–74.
Hayward, J. S. & Lisson, P. A. (1991). Evolution of brown fat: its absence in marsupials and monotremes. *Canadian Journal of Zoology* **70**, 171–179.
Hayward, S. A. L. (2014). Application of functional 'Omics' in environmental stress physiology: insights, limitations, and future challenges. *Current Opinion in Insect Science* **4**, 35–41.
Hedenström, A., Johansson, L. C. & Spedding, G. R. (2009). Bird or bat: comparing airframe design and flight performance. *Bioinspiration and Biomimetics* **4**, 015001.
Hedenstrom, A. & Johansson, L. C. (2015). Bat flight: aerodynamics, kinematics and flight morphology. *Journal of Experimental Biology* **218**, 653–663.
Hedges, S. B., Dudley, J. & Kumar, S. (2006). TimeTree: a public knowledge-base of divergence times among organisms. *Bioinformatics* **22**: 2971–2972.
Hedley, C. (1899). A zoogeographic scheme for the mid-Pacific. *Proceedings of the Linnean Society of New South Wales* **24**, 391–417.
Hedrick, M. S. & Duffield, D. A. (1991). Haematological and rheological characteristics of blood in seven marine mammal species: physiological implications for diving behaviour. *Journal of Zoology* **225**, 273–283.
Hedrick, M. S., Duffield, D. A. & Cornell, L. H. (1986). Blood viscosity and optimal haematocrit in a deep-diving mammal, the northern elephant seal (*Mirounga anguirostris). Canadian Journal of Zoology* **64**, 2081–2085.
Heers, A. M. & Dial, K. P. (2012). From extant to extinct: locomotor ontogeny and the evolution of avian flight. *Trends in Ecology and Evolution* **27**, 296–305.
Hegarty, R. S. (2012). Livestock nutrition—a perspective on future needs in a resource-challenged plant. *Animal Production Science* **52**, 406–415.
Hein, A. M., Hou, C. & Gillooly, J. F. (2012). Energetic and biomechanical constraints on animal migration distance. *Ecology Letters* **15**, 104–110.
Heinrich, B. (1977). Why have some animals evolved to regulate a high body temperature? *American Naturalist* **111**, 623–640.
Heisinger, J. F. & Breitenbach, R. P. (1969). Renal structural characteristics as indexes of renal adaptation for water conservation in the genus *Sylvilagus*. *Physiological Zoology* **42**, 160–172.
Heldmaier, G. (1989). Seasonal acclimatization of energy requirements in mammals: functional significance of body weight control, hypothermia, torpor and hibernation. In: *Energy Transformations in Cells and Organisms* (Eds. Wieser, W. & Gnaiger, E.). Stuttgart: Thieme.
Heller, H. C., Walker, J. M., Florant, G. L., Glotzbach, S. F., Berger, R. J. (1978). Sleep and hibernation: electrophysiological and thermoreguolatory homologies. In: *Strategies in the Cold: Natural Torpidity and Thermogensis* (Eds. Wang, L. C. & Hudson, J. W.). London: Academic Press.
Hellgren, E. C. (1995). Physiology of hibernation in bears. *Ursus* **10**, 467–477.
Helmuth, B., Broitman, B. R., Yamane, L., Gilman, S. E., Mach, K., Mislan, K. A. S. & Denny, M. W. (2010). Organismal climatology: analyzing environmental variability at scales relevant to physiological stress. *Journal of Experimental Biology* **213**, 995–1003.

Helmuth, B., Kingsolver, J. G. & Carrington, E. (2005). Biophysics, physiological ecology, and climate change: does mechanism matter? *Annual Reviews of Physiology* **67**, 177–201.
Hemmingsen, A. M. (1950). The relation of standard (basal) energy metabolism to total fresh weight of living organisms. *Reports of the Steno Memorial Hospital. Nordic Insulin Laboratory* **4**, 7–58.
Henschel, J. R. & Tilson, R. L. (1988). How much does a spotted hyaena eat? Perspective from the Namib Desert. *African Journal of Zoology* **26**, 247–255.
Henshaw, R. E., Underwood, L. S. & Casey, T. M. (1972). Peripheral thermoregulation: foot temperature in two Arctic canines. *Science* **175**, 988–990.
Henson, R. (2006). *The Rough Guide to Climate Change.* London: Penguin.
Hermes-Lima, M. & Zenteno-Savín, T. (2002). Animal response to drastic changes in oxygen availability and physiological oxidative stress. *Comparative Biochemistry and Physiology C* **133**, 537–556.
Herreid, C. F., Bretz, W. L. & Schmidt-Nielsen, K. (1968). Cutaneous gas exchange in bats. *American Journal of Physiology* **215**, 506–508.
Herrera, G. M. L., Cruz-Neto, A. P. & Wojciechowski, M. S. (2015). The relationships between food and energy intakes, salt content and sugar types in Egyptian fruit bats. *Mammalian Biology* **80**, 409–413.
Herrera M, L. G., Ramírez, P. N. & Mirón M, L. (2006). Ammonia excretion increased and urea excretion decreased in urine of a new world nectarivorous bat with decreased nitrogen intake. *Physiological and Biochemical Zoology* **79**, 801–809.
Herring, P. (2002). *The Biology of the Deep Sea.* Oxford: Oxford University Press.
Hesse, R. (1924). *Tiergeographie auf Ökologischer Grundlage.* Jena: Gustav Fischer.
Hetem, R. S., deWitt, B. A., Fick, L. G., Fuller, A., Kerley, G. I. H., Meyer, L. C. R., Mitchell, D. & Maloney, S. K. (2009). Body temperature, thermoregulatory behaviour and pelt characteristics of three colour morphs of springbok (*Antidorcas marsupialis*). *Comparative Biochemistry and Physiology A* **152**; 379–388.
Hetem, R. S., Maloney, S. K., Fuller, A., Meyer, L. C. R. & Mitchell, D. (2007). Validation of a biotelemetric technique, using ambulatory miniature black globe thermometers, to quantify thermoregulatory behaviour in ungulates. *Journal of Experimental Zoology A* **307**, 342–356.
Hetem, R. S., Maloney, S. K., Fuller, A. & Mitchell, D. (2016). Heterothermy in large mammals: inevitable or implemented? *Biological Reviews* **91**, 187–205.
Hetem, R. S., Mitchell, D., Maloney, S. K., Meyer, L. C., Fick, L. G., Kerley, G. I. & Fuller, A. (2008). Fever and sickness behavior during an opportunistic infection in a free-living antelope, the greater kudu (*Tragelaphus strepsiceros*). *American Journal of Physiology* **294**, R246–R254.
Hetem, P. S., Mitchell, D., de Witt, B. A., Fick, L. G., Meyer, L. C. R., Maloney, S. K. & Fuller, R. A. (2013). Cheetah do not abandon hunts because they overheat. *Biology Letters* **9**, 20130472.
Hetem, R. S., Strauss, W. M., Fick, L. G., Maloney, S. K., Meyer, L. C. R., Shobrak, M., Fuller, A. & Mitchell, D. (2012). Activity re-assignment and microclimate selection of free-living Arabian oryx: responses that could minimise the effects of climate change on homeostasis? *Zoology* **115**, 411–416.
Heyning, J. E. & Dahlheim, M. E. (1988). *Orcinus orca. Mammalian Species* **304**, 1–9.
Heywood, P. (2013). The quagga and science: what does the future hold for this extinct species? *Perspectives in Biology and Medicine* **56**, 53–64.
Hicks, J. W. (2002). The physiological and evolutionary significance of cardiovascular shunting patterns in reptiles. *News in Physiological Science* **17**, 241–245.
Higginbotham, A. C. & Koon, W. E. (1955). Temperature regulation in the Virginia opossum. *American Journal of Physiology* **181**, 69–71.

Higgins, D. P., Tobias, G. & Stone, G. M. (2004). Excretion profiles of some reproductive steroids in the faeces of captive female short-beaked echidna (*Tachyglossus aculeatus*) and long-beaked echidna (*Zaglossus* sp.). *Proceedings of the Linnean Society of New South Wales* **125**, 279–286.

Higham, T. E. & Irschick, D. J. (2012). Springs, steroids, and slingshots: the roles of enhancers and constraints in animal movement. *Journal of Comparative Physiology B* **183**, 583–595

Hildebrand, M. (1959). Motions of the running cheetah and horse. *Journal of Mammalogy* **40**, 481–495.

Hildebrand, M. (1961). Further studies of locomotion of the cheetah. *Journal of Mammalogy* **42**, 84–91.

Hildebrand, M. (1995). *Analysis of Vertebrate Structure*. New York: John Wiley & Sons.

Hillenius, W. J. (1992). The evolution of nasal turbinates and mammalian endothermy. *Paleobiology* **18**, 17–29.

Hillenius, W. J. (1994). Turbinates in therapsids: evidence for late Permian origins of mammalian endothermy. *Evolution* **48**, 207–229.

Hillman, S. S., Hancock, T. V. & Hedrick, M. S. (2013). A comparative meta-analysis of maximal aerobic metabolism of vertebrates: implications for respiratory and cardiovascular limits to gas exchange. *Journal of Comparative Physiology B* **183**, 167–179.

Himms-Hagen, J. (1985). Brown adipose tissue metabolism and thermogenesis. *Annual Review of Nutrition* **5**, 69–94.

Hindle, A. G., McIntyre, I. W., Campbell, K. L. & MacArthur, R. A. (2003). The heat increment of feeding and its thermoregulatory implications in the short-tailed shrew (*Blarina brevicauda*). *Canadian Journal of Zoology* **81**, 1445–1453.

Hinds, D. S., Baudinette, R. V., MacMillen, R. E. & Halpern, E. A. (1993). Maximum metabolism and the aerobic factorial scope of endotherms. *Journal of Experimental Biology* **182**, 41–56.

Hinds, D. S. & MacMillen, R. E. (1985). Scaling of energy metabolism and evaporative water loss in heteromyid rodents. *Physiological Zoology* **58**, 282–298.

Hinds, D. S. & MacMillen, R. E. (1986). Scaling of evaporative water loss in marsupials. *Physiological Zoology* **59**, 1–9.

Hittel, D. S. & Storey, K. B. (2002). Differential expression of mitochondria-encoded genes in a hibernating mammal. *Journal of Experimental Biology* **205**, 1625–1631.

Hochachka, P. W. (1985). Exercise limitations at high altitude: the metabolic problem and search for its solution. In: *Circulation, Respiration, and Metabolism* (Ed. Gilles, R.). Berlin: Springer-Verlag.

Hochachka, P. W., Darveau, C. A., Andrews, R. D. & Suarez, R. K. (2003). Allometric cascade: a model for resolving body mass effects on metabolism. *Comparative Biochemistry and Physiology A* **134**, 675–691.

Hoelzl, F., Bieber, C., Cornils, J. S., Gerritsmann, H., Stalder, G. L. Walzer, S. & Ruf, T. (2015). How to spend the summer? Free-living dormice (*Glis glis*) can hibernate for 11 months in non-reproductive years. *Journal of Comparative Physiology B* **185**, 931–939.

Hoffman, E. K., Lambert, I. H. & Pedersen, S. F. (2009). Physiology of cell volume regulation in vertebrates. *Physiological Reviews* **89**, 193–277.

Hofmann, G. E. & Todgham, A. E. (2010). Living in the now: physiological mechanisms to tolerate a rapidly changing environment. *Annual Review of Physiology* **72**, 127–145.

Hofmeyr, M. D. & Louw, G. N. (1987). Thermoregulation, pelage conductance and renal function in the desert-adapted springbok, *Antidorcas marsupialis*. *Journal of Arid Environments* **13**, 137–151.

Hofstetter, R. (1972). Données et hypothèses concernant l' origine et l'histoire biogéographique des marsupiaux. *Comptes Rendu de l'Academie des Sciences Paris D* **274**, 2635–2638.

Hohtola, E. (2004). Shivering thermogenesis in birds and mammals. In: *Life in the Cold: Evolution, Mechanisms, Adaptation, and Application* (Eds. Barnes, B. M. & Carey, H. V.). 12th International Hibernation Symposium. University of Alaska: Fairbanks.

Holheisel, J. (2006). Microarray technology; beyond transcript profiling and genotype analysis. *Nature Reviews in Genetics* 7, 200–210.

Holland, R. A., Borissov, I. & Siemers, B. M. (2010). A nocturnal mammal, the greater mouse-eared bat, calibrates a magnetic compass by the sun. *Proceedings of the National Academy of Sciences, USA* **107**, 6941–6945.

Holland, R. A., Kirschvink, J. L., Doak, T. G. & Wikelski, M. (2008). Bats use magnetite to detect the Earth's magnetic field. *PLOS ONE* **3**, e1676.

Holland, R. A. B., Calvert, S. J., Hope, R. M. & Chesson, C. M. (1994). Blood $O_2$ transport in newborn and adult of a very small marsupial (*Sminthopsis crassicaudata*). *Respiration Physiology* **98**, 69–81.

Holleman, D. F., White, R. G. & Feist, D. D. (1982). Seasonal energy and water metabolism in free-living Alaskan voles. *Journal of Mammalogy* **63**, 293–296.

Holm, E. (1969). Contribution to the knowledge of the biology of the Namib Desert golden mole *Eremitalpa granti namibensis* Bauer & Niethammer 1959. *Scientific Papers of the Namib Desert Research Station* **41**, 37–42.

Holmes, E. B. (1975). A reconsideration of the phylogeny of the tetrapod heart. *Journal of Morphology* **147**, 209–228.

Home Office. (2013). *Annual Statistics of Scientific Procedures on Living Animals. Great Britain. 2012. UK.* Controller of Her Majesty's Stationery Office, UK.

Hooker, S. K., Baird, R. W. & Fahlman, A. (2009). Could beaked whales get the bends? Effect of diving behavior and physiology on modelled gas exchange for three species: *Ziphius cavirostris, Mesoplodon densirostris* and *Hyperoodon ampullatus*. *Respiration Physiology and Neurobiology* **167**, 235–246.

Hooker, S. K., Biuw, M., McConnell, B. J., Miller, J. O. & Sparling, C. E. (2007). Bio-logging science: logging and relaying physical and biological data using animal-attached tags. *Deep Sea Research II* **54**, 177–182.

Hooker, S. K., Fahlman, A., Moore, M. J., Aguilar de Soto, N., Bernaldo de Quirós, Y., Brubakk, A. O., Costa, D. P., Costidis, A. M., Dennison, S., Falke, K. J., Fernandez, A., Ferrigno, M., Fitz-Clarke, J. R., Garner, M. M., Houser, D. S., Jepson, P. D., Ketten, D. R., Kvadsheim, P. H., Madsen, P. T., Pollock, N. W., Rotstein, D. S., Rowles, T. K., Simmons, S. E., Van Bonn, W., Weathersby, P. K., Weise, M. J., Williams, T. M. & Tyack, P. L. (2011). Deadly diving? Physiological and behavioural management of decompression stress in diving mammals. *Proceedings of the Royal Society of London B* **279**, 1041–1050.

Hope, P. J., Pyle, D., Baniels, C. B., Chapman, I., Horowitz, M., Morley, J. E., Trayhurn, P., Kumaratilake, J. & Wittert, G. (1997). Identificaton of brown fat and mechanisms for energy balance in marsupial *Sminthopsis crassicaudata. American Journal of Physiology* **273**, R121–R167.

Hopson, J. A. (1994). Synapsid evolution and the radiation of non-eutherian mammals. In: *Major Features of Vertebrate Evolution. Short Courses in Paleontology No. 7* (Ed. Spencer, R. S.). Knoxville: University of Tennessee.

Hori, T., Nakayama, T., Tokura, H., Hara, F. & Suzuki, M. (1977). Thermoregulation of the Japanese macaque living in a snowy mountain area. *Japanese Journal of Physiology* **27**, 305–319.

Horowitz, M., Argov, D. & Mizrahi, R. (1983). Interrelationships between heat acclimation and salivary cooling mechanism in conscious rats. *Comparative Biochemistry and Physiology A* **74**, 945–949.

Horseman, N. D. & Buntin, J. D. (1995). Regulation of pigeon cropmilk secretion and parental behaviors by prolactin. *Annual Review of Nutrition* **15**, 213–238.

Hosken, D. J. (1997). Sperm competition in bats. *Proceedings of the Royal Society of London B* **264**, 385–392.
Hosken, D. J. & Withers, P. C. (1999). Metabolic physiology of euthermic and torpid lesser long-eared bats, *Nyctophilus geoffroyi* (Chiroptera: Vespertilionidae). *Journal of Mammalogy* **80**, 42–52.
Houston, D. C. & Copsey, J. A. (1994). Bone digestion and intestinal morphology of the bearded vulture. *Journal of Raptor Research* **28**, 73–78.
Howland, H. C., Merola, S. & Basarab, J. R. (2004). The allometry and scaling of the size of vertebrate eyes. *Vision Research* **44**, 2043–2065.
Hoyt, D. F. & Taylor, C. R. (1981). Gait and the energetics of locomotion inhorses. *Nature* **292**, 239–240.
Hu, Y., Wang, Y., Luo, Z. & Li, C. (1997). A new symmetrodont mammal from China and its implications for mammalian evolution. *Nature* **390**, 137–142.
Huchon, D., Chevret, P., Jordan, U., Kilpatrick, C. W., Ranwez, V., Jenkins, P. D., Brosius, J. & Schmitz, J. (2007). Multiple molecular evidences for a living mammalian fossil. *Proceedings of the National Academy of Sciences, USA* **104**, 7495–7499.
Hudson, J. W. & Dawson, T. J. (1975). Role of sweating from the tail in the thermal balance of the rat-kangaroo *Potorous tridactylus*. *Australian Journal of Zoology* **23**, 453–461.
Hudson, L. N., Isaac, N. J. B. & Reuman, D. C. (2013). The relationship between body mass and field metabolic rate among individual birds and mammals. *Journal of Animal Ecology* **82**, 1009–1020.
Huerta-Sánchez, E., Jin, X., Asan, Bianba, Z., Peter, B. M., Vinckenbosch, N., Liang, Y., Yi, X., He, M., Somel, M., Ni, P., BoWang, Ou, X., Huasang, Luosang, J., Cuo, Z. X. P., Li, K., Gao, G., Yin, Y., Wang, W., Zhang, X., Xu, X., Yang, H., Li, Y., Wang, J., Wang, J. & Nielsen, R. (2014). Altitude adaptation in Tibetans caused by introgression of Denisovan-like DNA. *Nature* **512**, 194–197.
Huey, R. B. & Tewksbury, J. J. (2009). Can behavior douse the fire of climate warming? *Proceedings of the National Academy of Sciences, USA* **106**, 3647–3648.
Hughes, R. L. & Carrick, F. N. (1978). Reproduction in female monotremes. *Australian Zoologist* **20**, 233–257.
Hulbert, A. J. (1980). The evolution of energy metabolism in mammals. In: *Primitive Mammals* (Eds. Schmidt-Nielsen, K., Bolis, L. & Taylor, C. R.). Cambridge: Cambridge University Press.
Hulbert, A. J. (1988). Metabolism and the development of endothermy. In: *The Developing Marsupial* (Eds. Tyndale-Biscoe, C. H. & Janssens, P.). Berlin: Springer.
Hulbert, A. J. & Else, P. W. (2000). Mechanisms underlying the cost of living in animals. *Annual Review of Physiology* **62**, 207–235.
Hulbert, A. J. & Else, P. W. (2004). Basal metabolic rate: history, composition, regulation, and usefulness. *Physiological and Biochemical Zoology* 77, 869–876.
Hulbert, A. J. & Grant, T. R. (1983). A seasonal study of body condition and water turnover in a free-living population of platypuses, *Ornithorhynchus anatinus*. *Australian Journal of Zoology* **31**, 109–116.
Hulme, P. E. & Benkman, C. W. (2002). Granivory. In: *Plant-Animal Interactions. An Evolutionary Approach* (Eds. Herrera, C. M. & Pellmyr, O.). Oxford: Blackwell Science.
Hume, I. D. (1982). *Digestive Physiology and Nutrition of Marsupials*. Cambridge: Cambridge University Press.
Hume, I. D. (1989). Optimal digestive strategies in mammalian herbivores. *Physiological Zoology* **62**, 1145–1163.
Hume, I. D. (1995). *Comparative Physiology of the Vertebrate Digestive System*. Cambridge: Cambridge University Press.

Hume, I. D. (2014). Animal and human nutrition. In: *Biology. An Australian Focus* (Eds. Knox, B., Ladiges, P., Evans, B. & Saint, R.). North Ryde, NSW: McGraw-Hill.

Hume, I. D., Morgan, K. R. & Kenagy, G. J. (1993). Digesta retention and digestive performance in sciurid and microtine rodents: effects of hindgut morphology and body size. *Physiological Zoology* **66**, 396–411.

Humphries, M. M., Thomas, D. W. & Kramer, D. L. (2003). The role of energy availability in mammalian hibernation: a cost-benefit approach. *Physiological and Biochemical Zoology* **76**, 165–179.

Humphries, M. M., Thomas, D. W. & Speakman, J. R. (2002). Climate-mediated energetic constraints on the distribution of hibernating mammals. *Nature* **418**, 313–316.

Hunt, D. M., Chan, J., Carvalho, L. S., Hokoc, J. N., Ferguson, M. C., Arrese, C. A. & Beazley, L. D. (2009). Cone visual pigments in two species of South American marsupials. *Gene* **433**, 50–55.

Hutchinson, J. R., Schwerda, D., Famini, D. J., Dale, R. H. I., Fischer, M. S. & Kram, R. (2006). The locomotor kinematics of Asian and African elephants: changes with speed and size. *Journal of Experimental Biology* **209**, 3812–3827.

Huxley, T. H. (1880). On the application of the laws of evolution to the arrangement of the Vertebrata, and more particularly of the Mammalia. *Proceedings of the Royal Society of London* **43**, 649–662.

Iason, G. R., Dicke, M. & Hartley, S. E. (2012). The integrative roles of plant secondary metabolites in natural systems: a synthesis. In: *The Ecology of Plant Secondary Metabolites: From Genes to Global Processes* (Eds. Iason, G. R., Dicke, M. & Hartley, S. E.). Cambridge: Cambridge University Press.

Illius, A. W. & Gordon, I. J. (1992). Modelling the nutritional ecology of ungulate herbivores: evolution of body size and competitive interactions. *Oecologia* **89**, 428–434.

Incropera, F. P. & de Witt, D. P. (1981). *Fundamentals of Heat Transfer*. New York: John Wiley & Sons.

Ingram, D. L. & Weaver, M. E. (1969). A quantitative study of the blood vessels of the pig's skin and the influence of environmental temperature. *Anatomical Record* **163**, 517–524.

IPCC (2007). *Climate Change 2007: Synthesis Report*. Contribution of Working Groups I, II and III to the Fourth Assessment Report of the Intergovernmental Panel on Climate Change (Eds. Core Writing Team, Pachauri, R. K. & Reisinger, A.). Geneva, Switzerland: IPCC.

IPCC (2014). *Climate Change 2014: Synthesis Report*. Contribution of Working Groups I, II and III to the Fifth Assessment Report of the Intergovernmental Panel on Climate Change (Eds. Core Writing Team, Pachauri, R. K. & Reisinger, A.). Geneva, Switzerland: IPCC.

Iriarte-Díaz. J. (2002). Differential scaling of locomotor performance in small and large terrestrial mammals. *Journal of Experimental Biology* **205**, 2897–2908.

Irving, L. & Hart, J. S. (1957). The metabolism and insulation of seals as bare-skinned mammals in cold water. *Canadian Journal of Zoology* **35**, 497–511.

Irving, L., Scholander, P. F. & Grinnell, S. W. (1942). The regulation of arterial blood pressure in the seal during diving. *American Journal of Physiology* **135**, 557–566.

Isler, K. (2011). Energetic trade-offs between brain size and offspring production: marsupials confirm a general mammalian pattern. *BioEssays* **33**, 173–179.

IUPS Thermal Commission. (2003). Glossary of terms for thermal physiology. *Journal of Thermal Biology* **28**, 75–106.

Iwata, S. Lemp, M. A., Holly, F. J. & Dohlman, C. H. (1969). Evaporation rate of water from the precorneal tear film and cornea in the rabbit. *Investigative Ophthalmology and Visual Science* **8**, 613–619.

Jablonka, E. & Lamb, M. J. (1999). *Epigenetic Inheritance and Evolution: The Lamarckian Dimension*. Oxford: Oxford University Press.

Jackson, D. C. & Schmidt-Nielsen, K. (1964). Countercurrent heat exchange in the respiratory passages. *Proceedings of the National Academy of Sciences, USA* **51**, 1192–1197.

Jackson, S. & Schouten, P. (2012). *Gliding Mammals of the World*. Collingwood: CSIRO Publishing.

Jacobs, G. H. (1993). The distribution and nature of colour vision among the mammals. *Biological Reviews* **68**, 413–471.

Jacobs, G. H., Neitz, J. & Deegan, J. F. (1991). Retinal receptors in rodents maximally sensitive to ultraviolet light. *Nature* **353**, 655–656.

Jacobs, P. J. & McKechnie, A. E. (2014). Experimental sources of variation in avian energetics: estimated basal metabolic rate decreases with successive measurements. *Physiological and Biochemical Zoology* **87**, 762–769.

Jaeger, E. C. (1948). Does the poorwill 'hibernate'? *Condor* **50**, 45–46.

Janke, A., Magnell, O., Wieczorek, G., Westerman, M. & Arnason, U. (2002). Phylogenetic analysis of 18S rRNA and the mitochondrial genomes of the wombat, *Vombatus ursinus*, and the spiny anteater, *Tachyglossus aculeatus*: increased support for the Marsupionta hypothesis. *Journal of Molecular Evolution* **54**, 71–80.

Jansky, L. (1973). Non-shivering thermogenesis and its thermoregulatory significance. *Biological Reviews* **48**, 85–132.

Jardine, N. & McKenzie, D. (1972). Continental drift and the dispersal and evolution of organisms. *Nature* **235**, 20–24.

Jastroch, M., Withers, K. W., Taudien, S., Frappell, P. B., Helwig, M., Fromme, T., Hirschber, V., Heldmaier, G., McAllan, B. M., Firth, B. T., Burmester, T., Platzer, M. & Klingenspor, M. (2008). Marsupial uncoupling protein 1 sheds light on the evolution of mammalian nonshivering thermogenesis. *Physiological Genomics* **32**, 161–169.

Jastroch, M., Wuertz, S., Kloas, W. & Klingenspor, M. (2005). Uncoupling protein 1 in fish uncovers an ancient evolutionary history of mammalian nonshivering thermogenesis. *Physiological Genomics* **22**, 150–156

Jelkmann, W., Oberthür, W., Kleinschmidt, T. & Braunitzer, G. (1981). Adaptation of hemoglobin function to subterranean life in the mole, *Talpa europaea*. *Respiration Physiology* **46**, 7–16.

Jenkins, F. A. (1973). Functional anatomy of the humero-ulnar joint. *American Journal of Anatomy* **137**, 281–298.

Jenkinson, D. M. (1973). Comparative physiology of sweating. *British Journal of Dermatology* **88**, 379–406.

Jennings, N. V., Parsons, S., Barllow, K. E. & Gannon, M. R. (2004). Echolocation calls and wing morphology of bats from the West Indies. *Acta Chiropterologica* **6**, 75–90.

Jensen, B., Boukens, B. J. D., Postma, A. V., Gunst, Q. D., van den Hoff, M. J. B, Moorman, A. F. M., Wang, T. & Christoffels, V. M. (2012a). Identifying the evolutionary building blocks of the cardiac conducting system. *PLOS ONE* 7, 1–13.

Jensen, B., Wang, T., Christoffels, V. M. & Moorman, A. F. M. (2012b). Evolution and development of the building plan of the vertebrate heart. *Biochimica et Biophysica Acta* **1883**, 783–794.

Jensen, F. H., Rocco, A., Mansur, R. M., Smith, B. D., Janik, V. M. & Madsen, P. T. (2013). Clicking in shallow rivers: short-range echolocation of Irrawaddy and Ganges river dolphins in a shallow, acoustically complex habitat. *PLOS ONE* **8**, e59284.

Jerison, H. J. (1973). *Evolution of the Brain and Intelligence*. New York: Academic Press.

Jerison, H. J. (2001). The evolution of neural and behavioural complexity. In: *Brain Evolution and Cognition* (Eds. Roth, G. & Wullimann, M. F.). New York: Wiley.

Jessen, C. (1998). Brain cooling: an economy mode of temperature regulation in artiodactyls. *News in Physiological Science* **13**, 281–286.

Jessen, C. (2001a). *Temperature Regulation in Humans and Other Mammals*. Berlin: Springer-Verlag.

Jessen, C. (2001b). Selective brain cooling in mammals and birds. *Japanese Journal of Phyisology* **51**, 291–301.

Jessen, C., Laburn, H. P., Knight, M. H., Kuhnen, G., Goelst, K. and Mitchell, D. (1994). Blood and brain temperatures of free-ranging black wildebeest in their natural environment. *American Journal of Physiology* **36**, R1528–R1536.

Jessen, C. & Pongratz, H. (1979). Air humidity and carotid rete function in thermoregulation of the goat. *Journal of Physiology* **292**, 469–479.

Jetz, W., Carbone, C., Fulford, J. & Brown, J. H. (2004). The scaling of animal space use. *Science* **306**, 266–268.

Ji, Q., Luo, Z.-X., Yuan, C.-L. & Tabrum, A. R. (2006). A swimming mammaliaform from the Middle Jurassic and ecomorphological diversification of early mammals. *Science* **311**, 1123–1127.

Ji, Q., Luo, Z.-X., Yuan, C.-X., Wible, J. R., Zhang, J.-P. & Georg, J. A. (2002). The earliest known eutherian mammal. *Nature* **416**, 816–822.

Ji, Y.-Q., Luo, Z.-X. & Ji, S.-A. (1999). A Chinese triconodont mammal and mosaic evolution of the mammalian skeleton. *Nature* **398**, 326–330.

Johanson, Z. (2006). Origins of mammals: morphology, molecules and a petrosal or two. In: *Evolution and Biogeography of Australian Vertebrates* (Eds. Merrick, J. R., Archer, M., Hickey, G. M. & Lee, M. S. Y.). Oatlands, New South Wales: Auscipub.

Johnson, K. G., Callahan, S. M. & Strack, R. (1988). Temperature and humidity of expired air in sheep. *Australian Journal of Biological Science* **41**, 309–313.

Jones, G. & Holderied, M. W. (2007). Bat echolocation calls: adaptation and convergent evolution. *Proceedings of the Royal Society of London B* **274**, 905–912.

Jones, G. & Teeling, E. C. (2006). The evolution of echolocation in bats. *Trends in Ecology and Evolution* **21**, 149–156.

Jones, J. H. (1998). Optimization of the mammalian respiratory system: symmorphosis versus single species adaptation. *Comparative Biochemistry and Physiology B* **120**, 125–138.

Jones, M. E. & Rose, R. K. (2001). *Dasyurus viverrinus*. *Mammalian Species* **677**, 1–9.

Jordano, P. (2000). Fruits and frugivory. In: *Seeds: The Ecology of Regeneration in Plant Communities* (ed. Fenner, M.). Wallingford, UK: CABI Publications.

Jumars, P. A. (2000). Animal guts as ideal chemical reactors: maximizing absorption rates. *American Naturalist* **155**, 527–543.

Junghans, P., Derno, M., Gehre, M., Hofling, R., Kowski, P., Strauch, G., Jentsch, W. Voigt, J. & Hennig, U. (1997). Calorimetric validation of $^{13}C$ bicarbonate and doubly labeled water method for determining the energy expenditure in goats. *Zeitschrifte für Ernährungswissenschaft* **36**, 268–272.

Jürgens, K. D., Bartels, H. & Bartels, R. (1981). Blood oxygen transport and organ weights of small bats and small non-flying mammals. *Respiration Physiology* **45**, 243–260.


Jürgens, K. D., Bartels, H. & Bartels, R. (1981). Blood oxygen transport and organ weights of small bats and small non-flying mammals. *Respiration Physiology* **45**, 243–260.


Jürgens, K. D. & Prothero, J. (1991). Lifetime energy budgets in mammals and birds. *Comparative Physiology and Biochemistry* **100**, 703–709.

Kabat, A. P., Rose, R. W., Harris, J. & West, A. K. (2003a). Molecular identification of uncoupling proteins (UCP2 and UCP3) and absence of UCP1 in the marsupial Tasmanian bettong, *Bettongia gaimardi*. *Comparative Biochemistry and Physiology B* **134**, 71–77.

Kabat, A. P., Rose, R. W. & West, A. K. (2003b). Non-shivering thermogenesis in a carnivorous marsupial, *Sarcophilus harrisii*, in the absence of UCP1. *Journal of Thermal Biology* **28**, 413–420.
Kaiyala, K. J. & Ramsay, D. S. (2011). Direct animal calorimetry, the underused gold standard for quantifying the fire of life. *Comparative Biochemistry and Physiology A* **158**, 252–264.
Kallen, F. C. (1977). The cardiovascular system of bats: structure and function. In: *Biology of Bats Volume 3* (Ed. Wimsatt, W. A.). New York: Academic Press.
Kamilar, J. M., Muldoon, K. M., Lehman, S. M. & Herrera, J. P. (2012). Testing Bergmann's rule and the resource seasonality hypothesis in Malagasy primates using GIS-based climate data. *American Journal of Physical Anthropology* **147**, 401–408.
Karasov, W. H., Caviedes-Vidal, E., Bakken, B. H., Izhaki, I., Samuni-Blank, M. & Arad, Z. (2012). Capacity for absorption of water-soluble secondary metabolites is greater in birds than in rodents. *PLOS ONE* 7, e32417.
Karasov, W. H. & del Río, C. M. (2007). *Physiological Ecology. How Animals Process Energy, Nutrients, and Toxins*. Princeton, New Jersey: Princeton University Press.
Kardong, K. V. (2009). *Vertebrates. Comparative Anatomy, Function, Evolution*. New York: McGraw Hill.
Kawasaki, K., Lafont, A.-G. & Sire, J.-Y. (2011). The evolution of milk casein genes from tooth genes before the origin of mammals. *Molecular Biology and Evolution* **28**, 2053–2061.
Kazak, L., Chouchani, E. T., Jedrychowski, M. P., Erickson, B. K., Shinoda, K., Cohen, P., Vetrivelan, R., Lu, G. Z., Laznik-Bogoslavski, D., Hasenfuss, S. C., Kajimura, S., Gygi, S. P. & Spiegelma, B. M. (2015). A creatine-driven substrate cycle enhances energy expenditure and thermogenesis in beige fat. *Cell* **163**, 643–655.
Kearney, M. & Porter, W. (2009). Mechanistic niche modelling: combining physiological and spatial data to predict species' ranges. *Ecological Letters* **12**, 334–350.
Kearney, M. R., Wintle, B. A. & Porter, W. P. (2010). Correlative and mechanistic models of species distribution provide congruent forecasts under climate change. *Conservation Letters* **3**, 203–213.
Keast, A. (1971). Continental drift and the evolution of the biota on southern continents. *Quarterly Review of Biology* **46**, 335–378.
Keebaugh, A. C. & Thomas, J. W. (2010). The evolutionary fate of the genes encoding the purine catabolic enzymes in hominoids, birds, and reptiles. *Molecular Biology and Evolution* **27**, 1359–1369.
Kelly, E. M. & Sears, K. E. (2011a). Limb specialization in living marsupial and eutherian mammals: constraints on mammalian limb evolution. *Journal of Mammalogy* **92**, 1038–1049.
Kelly, E. M. & Sears, K. E. (2011b). Reduced phenotypic covariation in marsupial limbs and the implications for mammalian evolution. *Biological Journal of the Linnean Society* **102**, 22–36.
Kemp, R. B. (1999). *Handbook of Thermal Analysis and Calorimetry: From Macromolecules to Man*. Amsterdam: Elsevier.
Kemp, T. S. (2006). The origin of mammalian endothermy: a paradigm for the evolution of complex structure. *Zoological Journal of the Linnean Society* **147**, 437–488.
Kenagy, G. J. & Hoyt, D. F. (1980). Reingestion of feces in rodents and its daily rhythmicity. *Oecologia* **44**, 403–409.
Kenagy, G. J. & Hoyt, D. F. (1989). Source speed and time-energy budget for locomotion in golden-mantled ground squirrels. *Ecology* **70**, 1834–1839.
Kenagy, G. J., Masman, D., Sharbaugh, S. M. & Nagy, K. A. (1990). Energy expenditure during lactation in relation to litter size in free living golden-mantled ground squirrels. *Journal of Animal Ecology* **59**, 73–88.

Kenagy, G. J., Sharbaugh, S. M. & Nagy, K. A. (1989). Annual cycle of energy and time expenditure in a golden mantled ground squirrel population. *Oecologia* **78**, 269–282.

Kenagy, G. J., Veloso, C. & Bozinovic, F. (1999). Daily rhythms of food intake and feces reingestion in the degu, an herbivorous Chilean rodent: optimizing digestion through coprophagy. *Physiological and Biochemical Zoology* **72**, 78–86.

Kennedy, P. M. (2005). Particle dynamics. In: *Quantitative Aspects of Ruminant Digestion* (Eds. Dijkstra, J., Forbes, J. M. & France, J.). Wellingford, Oxfordshire: CABI Publishing.

Kenyon, J. R. (1961). Experimental deep hypothermia. *British Medical Bulletin* **17**, 43–47.

Kenyon, K. W. (1969). The sea otter in the eastern pacific ocean. *North American Fauna* **68**, 1–352.

Khaliq, I., Hof, C., Prinzinger, R., Böhning-Gaese, K. & Pfenninger, M. (2014). Global variation in thermal tolerances and vulnerability of endotherms to climate change. *Proceedings of the Royal Society of London B* **281**, 20141097.

Kibler, H. H., Johnson, H. D., Hahn, G. L. & Shanklin, M. D. (1970). Thermal regulation in cattle at 2° and 35° as influenced by controlled feeding, ad libitum feeding, and fasting. *Missouri Agricultural Experimental Station Research Bulletin* **975**, 23pp.

Kielan-Jaworowska, Z., Cifelli, R. L. & Luo, Z-X. (2004). *Mammals from the Age of Dinosaurs: Origins, Evolution and Structure*. New York: Columbia University Press.

Kiil, F. (1973). Urinary flow and ureteral peristalsis. In: *Urodynamics* (Eds. Lutzeyer, W. & Melchior, H.). Berlin: Springer.

Kikuchia, R. & Vannesteb, M. (2010). A theoretical exercise in the modeling of ground-level ozone resulting from the K–T asteroid impact: its possible link with the extinction selectivity of terrestrial vertebrates. *Palaeogeography, Palaeoclimatology, Palaeoecology* **288**, 14–23.

Kiltie, R. A. (2000). Scaling of visual acuity with body size in mammals and birds. *Functional Ecology* **14**, 226–234.

Kim, G., Ecelbarger, C., Mitchell, C., Packer, R., Wade, J. & Knepper, M. (1999). Vasopressin increases Na-K-2Cl cotransporter expression in thick ascending limb of Henle's loop. *American Journal of Physiology* **276**, F96–F103.

Kimball, S. H., King-Smith, P. E. & Nichols, J. J. (2010). Evidence for the major contribution of evaporation to tear film thinning between blinks. *Investigative Ophthalmology and Visual Science* **51**, 6294–6297.

Kinnear, J. E. & Brown, G. D. (1967). Minimum heart rates of marsupials. *Nature* **215**, 1501.

Kirmiz, J. P. (1962). *Adaptation de la gerboise au milieu desertique*. Alexandria: S.A.E., Societe de Publications Egyptiennes.

Kleiber, M. (1932). Body size and metabolism. *Hilgardia* **6**, 315–353.

Kleiber, M. (1947). Body size and metabolic rate. *Physiological Reviews* **27**, 511–541.

Kleiber, M. (1975). *The Fire of Life. An Introduction to Animal Energetics*. Huntington, New York: Robert E. Krieger Publishing.

Kleinebeckel, D. & Klussmann, F. W. (1990). Shivering. In: *Thermoregulation: Physiology and Pharmacology* (Eds. Schoenbaum, E. & Lomax, P.). New York: Pergamon Press.

Klieve, A. V. (2009). Kangaroo bacteria—increasing productivity and reducing emissions of the greenhouse gas methane. Canberra, ACT, Australia: Meat and Livestock Australia Ltd.

Klingenspor, M., Fromme, T., Huges, D. A., Manzke, L., Polymeropoulos, E., Riemann, T., Trzcionka, M., Hirschber, V. & Jastroch, M. (2008). An ancient look at UCP1. *Biochimica Biophysica Acta* **1777**, 637–641.

Klir, J. J. & Heath, J. E. (1992). An infrared thermographic study of the surface temperature in relation to external thermal stress in three species of foxes: the red fox (*Vulpes vulpes*), Arctic fox (*Alopex lagopus*) and kit fox (*Vulpes macrotis*). *Physiological Zoology* **65**, 1011–1021.

Kluger, M. J. (1978). The evolution and adaptive value of fever: long regarded as a harmful byproduct of infection, fever may instead be an ancient ally against disease, enhancing resistance and increasing chances of survival. *American Scientist* **66**, 38–43.

Kluger, M. J. (1991). Fever: role of pyrogens and cryogens. *Physiological Reviews* 71, 93–127.
Kluger, M. J., Ringler, D. & Anver, M. (1975). Fever and survival. *Science* **188**, 166–168.
Knepper, M. A., Kim, G., Fernández-Llama, P. & Ecerbarger, C. (1999). Regulation of thick ascending limb transport by vasopressin. *Journal of the American Society of Nephrology* **10**, 628–634
Knoll, G. F. (1999). *Radiation Detection and Measurement.* New York: John Wiley & Sons.
Knox, B., Ladiges, P., Evans, B. & Saint, R. (2014). *Biology. An Australian Focus.* North Ryde, NSW: McGraw-Hill.
Kochanny, C. O., Delgiudice, G. D. & Fieberg, J. (2009). Comparing global positioning system and very high frequency telemetry home ranges of white-tailed deer. *Journal of Wildlife Management* 73, 779–787.
Koford, C. B. (1968). Peruvian desert mice: water independence, competition, and breeding cycle near the equator. *Science* **160**, 552–553.
Kohl, K. D., Miller, A. W. & Dearing, M. D. (2014). Evolutionary irony: evidence that 'defensive' plant spines act as a proximate cue to attract a mammalian herbivore. *Oikos* **124**, 835–841.
Kokko, J. P. & Rector, F. C. (1972). Countercurrent multiplication system without active transport in inner medulla. *Kidney International* 2, 214–223.
Kolb, H. & Wang, H. H. (1985). The distribution of photoreceptors, dopaminergic amacrine cells and ganglion cells in the retina of the north american opossum (*Didelphis virginiana*). *Vision Reseach* **25**, 1207–1221.
Kolokotrones, T., Savage, V., Deeds, E. J. & Fontana, W. (2010). Curvature in metabolic scaling. *Nature* **464**, 753–756.
Konarzewski, M., Książek, A. & Łapo, I. B. (2005). Artificial selection on metabolic rates and related traits in rodents. *Integrative and Comparative Biology* **45**, 416–425.
Konow, N., Cheney, J. A., Roberts, T. J., Waldman, R. S. & Swartz, S. M. (2015). Spring or string: does tendon elastic action influence wing muscle mechanics in bat flight? *Proceedings of the Royal Society of London B* **282**, 20151832.
Kooyman, G. (2015). Marine mammals and emperor penguins: a few applications of the Krogh principle. *American Journal of Physiology* **308**, R96–R104.
Kooyman, G. L. & Campbell, W. B. (1972). Heart rate in freely diving Weddell seals (*Leptonychotes weddellii*). *Comparative Biochemistry and Physiology* **43**, 31–36.
Kooyman, G. L., Wahrenbrock, E. A., Castellini, M. A., Davis, R. W. & Sinnett, E. E. (1980). Aerobic and anaerobic metabolism during voluntary diving in Weddell seals: evidence of preferred pathways from blood chemistry and behavior. *Journal of Comparative Physiology* **138**, 335–346.
Korhonen. H. J. (2013). Production and properties of health-promoting proteins and peptides from bovine colostrum and milk. *Cellular and Molecular Biology* **59**, 12–24.
Korine, C., Arad, Z. & Arieli, A. (1996). Nitrogen and energy balance of the fruit bat *Rousettus aegyptlacus* on natural fruit diets. *Physiological Zoology* **69**, 618–634.
Körtner, G. & Geiser, F. (1998). Ecology of natural hibernation in the marsupial mountain pygmy-possum (*Burramys parvus*). *Oecologia* **113**, 170–178.
Körtner, G. & Geiser, F. (2000). The temporal organization of daily torpor and hibernation: circadian and circannual rhythms. *Chronobiology International* 17, 103–128.
Körtner, G. & Geiser, F. (2009). The key to winter survival: daily torpor in a small arid-zone marsupial. *Naturwissenschaften* **96**, 525–530.
Körtner, G., Rojas, A. D. & Geiser, F. (2010). Thermal biology, torpor use and activity patterns of a small diurnal marsupial from a tropical desert: sexual differences. *Journal of Comparative Physiology B* **180**, 869–876.
Koteja, P. (1987). On the relation between basal and maximum metabolic rate in mammals. *Comparative Biochemistry and Physiology A* 87, 205–208.

Koteja, P. (1991). On the relation between basal and field metabolic rates in birds and mammals. *Functional Ecology* **5**, 56–64.
Koteja, P. (2000). Energy assimilation, parental care and the evolution of endothermy. *Proceedings of the Royal Society of London B* **267**, 479–484.
Koteja, P. (2004). The evolution of concepts on the evolution of endother my in birds and mammals. *Physiological and Biochemical Zoology* 77, 1043–1050.
Kottek, M., Grieser, J., Beck, C., Rudolf, B. & Rubel, F. (2006). World Map of the Köppen-Geiger climate classification updated. *Meteorologische Zeitschrift* **15**, 259–263.
Krause, D. W. (2001). Fossil molar from a Madagascan marsupial. *Nature* **412**, 497–498.
Krause, P. F. & Flood, K. L. (1997). *Weather and Climate Extremes*. Alexandria, Virginia: U.S. Army Corps of Engineers, Topgraphic Engineering Center.
Kreienbühl, G., Strittmatter, J. & Ayim, E. (1976). Blood gas analyses of hibernating hamsters and dormice. *Pflügers Archiv* **366**, 167–172.
Kriz, W. (1981). Structural and functional organization of the renal medulla: comparative and functional aspects. *American Journal of Physiology* **241**, R3–R16.
Kriz, W. & Kaissling, B. (2000). Structural organization of the mammalian kidney. In: *The Kidney, Physiology and Pathology Volume I* (Eds. Seldin, D. & Giebisch, G.). Philadelphia, Lippincott: Williams and Wilkins.
Krockenberger, M. B. & Bryden, M. M. (1994). Rate of passage of digesta through the alimentary tract of southern elephant seals (*Mirounga leonina*) (Carnivora: Phocidae). *Journal of Zoology* **234**, 229–237.
Kronfeld-Schor, N., Richardson, C., Silvia, B. A., Kunz, T. H. & Widmaier, E. P. (2000). Dissociation of leptin secretion and adiposity during prehibernatory fattening in little brown bats. *American Journal of Physiology* **279**, R1277–R1281.
Kruuk, H. & Balharry, D. (1990). Effects of sea water on thermal insulation of the otter, *Lutra lutra*. *Journal of Zoology* **222**, 405–415.
Kuenzer, C. & Dech, S. (2013). Theoretical background of thermal infrared remote sensing. In: *Thermal Infrared Remote Sensing: Sensors, Methods, Applications, Remote Sensing and Digital Image Processing* (Eds. Kuenzer, C & Dech, S.). Berlin: Springer.
Kühne, W. G. (1973). The systematic position of monotremes reconsidered (Mammalia). *Zeitschrift für Morphologie der Tiere* 75, 59–64.
Kuhnen, G. (1997). Selective brain cooling reduces respiratory water loss during heat stress. *Comparative Biochemistry and Physiology A* **118**, 891–895.
Kuhnen, G. & Jessen, C. (1994). Thermal signals in control of selective brain cooling. *American Journal of Physiology* **267**, R355–R359.
Kullberg, M., Hallström, B. M., Arnason, U. & Janke, A. (2008). Phylogenetic analysis of 1.5 Mbp and platypus EST data refute the Marsupionta hypothesis and unequivocally support Monotremata as sister group to Marsupialia/Placentalia. *Zoologica Scripta* **37**, 115–127.
Kumar, S. & Hedges, S. B. (1998). A molecular timescale for vertebrate evolution. *Nature* **392**, 917–920.
Kunz, T. H. & Fenton, M. B. (2003). *Bat Ecology*. Chicago: University of Chicago Press.
Kunz, T. H. & Hosken, D. J. (2009). Male lactation: why, why not and is it care? *Trends in Ecology and Evolution* **24**, 80–85.
Kuruppath, S., Bisana, S., Sharp, J. A., Lefevre, C., Kumar, S. & Nicholas, K. R. (2012). Monotremes and marsupials: comparative models to better understand the function of milk. *Journal of Biosciences* **37**, 581–588.
Kushner, I. & Rzewnicki, D. L. (1997). The acute phase response. In: *Fever: Basic Mechanisms and Management* (Ed. Mackowiak, P. A.). Philadelphia: Lippincott-Raven.

Kvadsheim, P. H., Miller, P. J. O., Tyack, P. L., Sivle, L. D., Lam, F. P. A. & Fahlman, A. (2012). Estimated tissue and blood $N_2$ levels and risk of decompression sickness in deep-, intermediate-, and shallow-diving toothed whales during exposure to naval sonar. *Frontiers in Physiology* **3**, 1–14.

Kvist, A. Lindstrom, A., Green, M., Piersma, T. & Visser, G. H. (2002). Carrying large fuel loads during sustained bird flight is cheaper than expected. *Nature* **413**, 730–732.

Labocha, M. K., Sadowska, E. T., Baliga, K., Semer, A. K. & Koteja, P. (2004). Individual variation and repeatability of basal metabolism in the bank vole, *Clethrionomys glareolus*. *Proceedings of the Royal Society of London B* **271**, 367–372.

Laburn, H. P., Mitchell, D. & Goelst, K. (1992). Fetal and maternal body temperatures measured by radiotelemetry in near-term sheep during thermal-stress. *Journal of Applied Physiology* **72**, 894–900.

Lacey, E. A., Patton, J. L. & Cameron, G. N. (2000). *Life Underground: The Biology of Subterranean Rodents*. Chicago: University of Chicago Press.

Lambert, E. H. & Wood, E. H. (1947). The use of a resistance wire, strain gauge manometer to measure intraarterial pressure. *Proceedings of the Society for Experimental Biology and Medicine* **64**, 186–190.

LaMotte, R. H. & Thalhammer, J. G. (1982). Response properties of high-threshold cutaneous cold receptors in the primate. *Brain Research* **244**, 279–287.

Lancaster, W. C., Henson, O. W. & Keating, A. W. (1995). Respiratory muscle activity in relation to vocalization in flying bats. *Journal of Experimental Biology* **198**, 175–191.

Lancaster, W. C., Thomson, S. C. & Speakman, J. R. (1997). Wing temperature in flying bats measured by infrared thermography. *Journal of Thermal Biology* **22**, 109–16.

Landrau-Giovannetti, N., Mignucci-Giannoni, A. A. & Reidenberg, J. S. (2014). Acoustical and anatomical determination of sound production and transmission in West Indian (*Trichechus manatus*) and Amazonian (*T. inunguis*) manatees. *Anatomical Record* **297**, 1896–1907.

Landys, M. M., Piersma, T., Visser, G. H., Jukema, J. & Wijker, A. (2000). Water balance during real and simulated long distance migratory flight in the bar-tailed godwit. *Condor* **102**, 645–652.

Langman, V. A., Maloiy, G. M. O., Schmidt-Nielsen, K. & Schroter, R. C. (1979). Nasal heat exchange in the giraffe and other large mammals. *Respiration Physiology* **37**, 325–333.

Lang-Ouellette, D., Richard, T. G. & Morin, P. (2014). Mammalian hibernation and regulation of lipid metabolism: a focus on non-coding RNAs. *Biochemistry* **79**, 1161–1171.

Langvatn, R. & Albon, S. D. (1986). Geographic clines in body weight of Norwegian red deer: a novel explanation of Bergmann's Rule? *Holarctic Ecology* **9**, 285–293.

Larcombe, A. N. (2002). Effects of temperature on metabolism, ventilation and oxygen extraction in the southern brown bandicoot *Isoodon obesulus* (Marsupialia: Peramelidae). *Physiological and Biochemical Zoology* **75**, 405–411.

Larramendi, A. (2015). Shoulder height, body mass and shape of proboscideans. *Acta Palaeontologica Polonica*. doi:http://dx.doi.org/10.4202/app.00136.2014.

Lasiewski, R. C., Acosta, A. L. & Bernstein, M. H. (1966). Evaporative water loss in birds—I. Characteristics of the open flow method of determination, and their relation to estimates of thermoregulatory ability. *Comparative Biochemistry and Physiology* **19**, 445–457.

Lavin, S. R., Karasov, W. H., Ives, A. R., Middleton, K. M. & Garland, T. (2008). Morphometrics of the avian small intestine compared with that of nonflying mammals: a phylogenetic approach. *Physiological and Biochemical Zoology* **81**, 526–550.

Lawler, J. P. & White, R. G. (2003). Temporal responses in energy expenditure and respiratory quotient following feeding in the muskox: influence of season on energy costs of eating and standing and an endogenous heat increment. *Canadian Journal of Zoology* **81**, 1524–1538.

Laybourne, R. C. (1974). Collision between a vulture and an aircraft at an altitude of 37,000 feet. *Wilson Bulletin* **86**, 461–462.

Lee, A. K. & Cockburn, A. (1985). *Evolutionary Ecology of Marsupials*. Cambridge: Cambridge University Press.

Lee, C. C. (2008). Is human hibernation possible? *Annual Reviews in Medicine* **59**, 177–186.

Lee, R. F., Hirota, J. & Barnett, A. M. (1971). Distribuion and importance of wax esters in marine copepods and other zooplankton. *Deep-Sea Research* **18**, 1147–1165.

Lefévre, C. M., Sharp, J. A. & Nicholas, K. R. (2010). Evolution of lactation: ancient origin and extreme adaptations of the lactation system. *Annual Review of Genomics and Human Genetics* **11**, 219–238.

Leitner, P. & Nelson, J. E. (1967). Body temperature, oxygen consumption and heart rate in the Australian false vampire bat, *Macroderma gigas*. *Comparative Biochemistry and Physiology* **21**, 65–74.

Lekagul, B. & McNeely, J. A. (1977). *Mammals of Thailand*. Bangkok: Association for the Conservation of Wildlife.

Lemons, D. E., Chien, S., Crawshaw, L. I., Weinbaum, S. & Jiji, L. M. (1987). Significance of vessel size and type in vascular heat transfer. *American Journal of Physiology* **253**, R128–R135.

Lesku, J. A., Meyer, L. C. R., Fuller, A., Maloney, S. K., Dell'Omo, G., Vyssotski, A. L. & Rattenborg, N. C. (2011). Ostriches sleep like platypuses. *PLOS ONE* **6**, e23203. doi:10.1371/journal.pone.0023203.

Levick, J. R. (1996). *An Introduction to Cardiovascular Physiology*. Oxford: Butterworth Heinemann.

Levin, N. E., Haile-Selassie, Y., Frost, S. R. & Saylor, B. Z. (2015). Dietary change among hominins and cercopithecids in Ethiopia during the early Pliocene. *Proceedings of the National Academy of Sciences, USA* **112**, 12304–12309.

Levine, M. & Tijan, R. (2003). Transcription regulation and animal diversity. *Nature* **424**, 147–151.

Li, Q., Sun, R., Huang, C., Wang, Z., Liu, X., Hou, J., Liu, J., Cai, L., Li, N., Zhang, S. & Wang, Y. (2001). Cold adaptive thermogenesis in small mammals from different geographical zones of China. *Comparative Biochemistry and Physiology A* **129**, 949–961.

Libardi, G. S. & Percequillo, A. R. (2014). Supernumerary teeth in *Necromys lasiurus* (Rodentia, Cricetidae): the first record in Sigmodontinae. *Mastozoología Neotropical* **21**, 219–229.

Licht, P. & Leitner, P. (1967a). Behavioral responses to high temperatures in three species of Californian bats. *Journal of Mammalogy* **48**, 52–61.

Licht, P. & Leitner, P. (1967b). Physiological responses to high environmental temperatures in three species of micro-chiropteran bats. *Comparative Biochemistry and Physiology A* **22**, 371–387.

Liem, K. F., Bemis, W. E., Walker, W. F. & Grande, L. (2001). *Functional Anatomy of the Vertebrates*. Fort Worth: Harcourt College Publishers.

Lifson, N. & McClintock, R. (1966). Theory of use of the turnover rates of body water for measuring energy and material balance. *Journal of Theoretical Biology* **12**, 46–74.

Lighton, J. R. B. (2008). *Measuring Metabolic Rates. A Manual for Scientists*. Oxford: Oxford University Press.

Lillie, M. A., Piscitelli, M. A., Vogl, A. W., Gosline, J. M. & Shadwick, R. E. (2013). Cardiovascular design in fin whales: high-stiffness arteries protect against adverse pressure gradients at depth. *Journal of Experimental Biology* **216**, 2548–2563.

Lilligraven, J. A. (1987). The origin of eutherian mammals. *Biological Journal of the Linnean Society* **32**, 281–336.

Lillywhite, H. B. (2006). Water relations of tetrapod integument. *Journal of Experimental Biology* **209**, 202–226.

Lindhe Norberg, U. M., Brooke, A. P. & Trewhella, W. J. (2000). Soaring and non-soaring bats of the family Pteropodidae (flying foxes, *Pteropus* spp.): wing morphology and flight performance. *Journal of Experimental Biology* **203**, 651–664.

Lindstedt, S. L. & Calder, W. A. (1981). Body size, physiological time and longevity of homeothermic animals. *Quarterly Review of Biology* **56**, 1–16.

Lindstedt, S. L., Miller, B. J. & Buskirk, S. W. (1986). Home range, time, and body size in mammals. *Ecology* **67**, 413–418.

Lindstedt, S. L. & Swain, S. D. (1988). Body size as a constraint of design and function. In: *Evolution of Life Histories: Pattern and Theory from Mammals* (Ed. Boyce, M. S.). New Haven: Yale University Press.

Lindstedt, S. L., Hokanson, J. F., Wells, D. J., Swain, S. D., Hoppeler, H. & Navarro, V. (1991). Running energetics in the pronghorn antelope. *Nature* **353**, 748–750.

Ling, L. Fuller, D. D., Bach, K. B., Kinkead, R., Olson, E. B. & Mitchell, G. S. (2001). Chronic intermittent hypoxia elicits serotonin-dependent plasticity in the central neural control of breathing. *Journal of Neuroscience* **21**, 5381–5388.

Liow, L. H., Fortelius, M., Lintulaakso, K., Mannila, H. & Stenseth, N. C. (2009). Lower extinction risk in sleep-or-hide mammals. *American Naturalist* **173**, 264–272.

Lissauer, T., Fanaroff, A. A., Rodriguez, R. J. & Weindling, M. (2006). *Neonatalogy at a Glance.* Malden, Massachusetts: Blackwell Publishing.

List, R. J. (1971). *Smithsonian Meteorological Tables.* Washington, DC: Smithsonian Institution Press.

Liu, L., Li, Y., Wang, R., Yin, C., Dong, Q., Hing, H., Kim, C. & Welsh, M. J. (2007). *Drosophila* hygrosensation requires the TRP channels water witch and nanchung. *Nature* **450**, 294–298.

Liwanagi, H. M., Berta, A., Costa, D. P., Budge, S. M. & Williams, T. M. (2012). Morphological and thermal properties of mammalian insulation: the evolutionary transition to blubber in pinnipeds. *Biological Journal of the Linnaean Society* **107**, 774–787.

Lodé, T. (2012). Oviparity or viviparity? That is the question. . . . *Reproductive Biology* **12**, 259–264.

Loeuille, N. & Loreau, M. (2006). Evolution of body size in food webs: does the energetic equivalence rule hold? *Ecological Letters* **9**, 171–178.

Lohmann, K. J. (2010). Magnetic-field perception. *Nature* **464**, 1140–1142.

Long. X., Ye, J., Zhao, D. & Zhang, S.-J. (2015). Magnetogenetics: remote non-invasive magnetic activation of neuronal activity with a magnetoreceptor. *Science Bulletin.* doi: 10.1007/s11434-015-0902-0.

Longman, H. A. (1921). The zoogeography of marsupials. *Memoirs of the Queensland Museum* **8**, 1–15.

Loudon, A., Rothwell, N. & Stock, M. (1985). Brown fat, thermogenesis and physiological birth in a marsupial. *Comparative Biochemistry and Physiology A* **81**, 815–819.

Louw, E., Louw, G. N. & Retief, C. P. (1972). Thermolability, heat tolerance and renal function in the dassie or hyrax, *Procavia capensis*. *Zoologicana Africana* 7, 451–469.

Love, A. H. G. & Shanks, R. G. (1962). The relationship between the onset of sweating and vasodilation in the forearm during body heating. *Journal of Physiology* **162**, 121–128.

Lovegrove, B. G. (1986). The metabolism of social subterranean rodents: adaptation to aridity. *Oecologia* **69**, 551–555.

Lovegrove, B. G. (1989). The cost of burrowing by the social mole rats (Bathyergidae) *Cryptomys damarensis* and *Heterocephalus glaber*: the role of soil moisture. *Physiological Zoology* **62**, 449–469.

Lovegrove, B. G. (2000). The zoogeography of mammalian basal metabolic rate. *American Naturalist* **156**, 201–219.
Lovegrove, B. G. (2001). The evolution of body armor in mammals: plantigrade constraints of large body size. *Evolution* **55**, 1464–1473.
Lovegrove, B. G. (2003). The influence of climate on the basal metabolic rate of small mammals: a slow-fast metabolic continuum. *Journal of Comparative Physiology B* **173**, 87–112.
Lovegrove, B. G. (2004). Locomotor mode, maximum running speed, and basal metabolic rate in placental mammals. *Physiological and Biochemical Zoology* 77, 916–928.
Lovegrove, B. G. (2005). Seasonal thermoregulatory responses in mammals. *Journal of Comparative Physiology B* **175**, 231–247.
Lovegrove, B. G (2009a). Age at first reproduction and growth rate are independent of basal metabolic rate in mammals. *Journal of Comparative Physiology B* **179**, 391–401.
Lovegrove, B. G. (2009b). Modification and miniaturization of Thermochron iButtons for surgical implantation into small animals. *Journal of Comparative Physiology B* **179**, 451–458.
Lovegrove, B. G. (2010). The allometry of rodent intestines. *Journal of Comparative Physiology B* **180**, 741–755.
Lovegrove, B. G. (2012a). The evolution of endothermy in Cenozoic mammals: a plesiomorphic-apomorphic continuum. *Biological Reviews* **87**, 128–162.
Lovegrove, B. G. (2012b). The evolution of mammalian body temperature: the Cenozoic supraendothermic pulse. *Journal of Comparative Physiology* **182**, 579–589.
Lovegrove, B. G. & Génin, F. (2008). Torpor and hibernation in a basal placental mammal, the lesser hedgehog tenrec *Echinops telfairi*. *Journal of Comparative Physiology B* **178**, 691–698.
Lovegrove, B. G., Lawes, M. J. & Roxburgh, L. (1999). Confirmation of pleisiomorphic daily torpor in mammals: the round-eared elephant shrew *Macroscelides proboscideus* (Macroscelidea). *Journal of Comparative Physiology B* **169**, 453–460.
Lovegrove, B. G., Lobban, K. D., Levesque, D. L. (2014). Mammal survival at the Cretaceous—Palaeogene boundary: metabolic homeostasis in prolonged tropical hibernation in tenrecs. *Proceedings of the Royal Society B* **281**, 20141304.
Lovegrove, B. G. & Mowoe, M. O. (2013). The evolution of mammal body sizes: responses to Cenozoic climate change in North American mammals. *Journal of Evolutionary Biology* **26**, 1317–1329.
Lovegrove, B. G. & Mowoe, M. O. (2014). The evolution of micro-cursoriality in mammals. *Journal of Experimental Biology* **217**, 1316–1325.
Lovegrove, B. G. & Wissell, C. (1988). Sociality in mole rats: metabolic scaling and the role of risk sensitivity. *Oecologia* 74, 600–606.
Low, B. S. (1978). Environmental uncertainty and the parental strategies of marsupials and placentals. *American Naturalist* **112**, 197–213.
Lowenstein, J. M. & Ryder, O. A. (1985). Immunological systematics of the extinct quagga (Equidae). *Experientia* **41**, 1192–1193.
Lubbock, J. (1868). On the origin of civilisation and the primitive condition of man. *Transactions of the Ethnological Society of London* **6**, 328–341.
Luckett, W. P. (1993). Ontogenetic staging of the mammalian dentition, and its value for assessment of homology and heterochrony. *Journal of Mammalian Evolution* **1**, 269–282.
Luo, Z-X. (2007). Transformation and diversification in early mammal evolution. *Nature* **450**, 1011–1019.
Luo, Z.-X., Ji, Q., Wible, J. R. & Yuan, C.-X. (2003). An Early Cretaceous tribosphenic mammal and metatherian evolution. *Science* **302**, 1934–1940.
Luo, Z-X., Yuan, C-X., Meng, Q-J. & Ji, Q. (2011). A Jurassic eutherian mammal and divergence of marsupials and placentals. *Nature* **476**, 442–445.

Lush, L., Ellwood, S., Markham, A., Ward, A. I. & Wheeler, P. (2015). Use of tri-axial accelerometers to assess terrestrial mammal behaviour in the wild. *Journal of Zoology*. doi:10.1111/jzo.12308

Ma, T., Yang, B., Gillespie, A., Carlson, E., Epstein, C. & Verkman, A. S. (1998). Severely impaired urinary concentrating ability in transgenic mice lacking aquaporin-1 water channels. *Journal of Biological Chemistry* **273**, 4296–4299.

MacArthur, R. A. (1984). Aquatic thermoregulation in the muskrat (*Ondatra zibethicus*): energy demands of swimming and diving. *Canadian Journal of Zoology* **62**, 241–248.

MacArthur, R. A. & Campbell, K. L. (1994). Heat increment of feeding and its thermoregulatory benefit in the muskrat (*Ondatra zibethicus*). *Journal of Comparative Physiology B* **164**, 141–146.

MacArthur, R. A. & Dyck, A. P. (1990). Aquatic thermoregulation of captive and free-ranging beavers (*Castor canadensis*). *Canadian Journal of Zoology* **68**, 2409–2416.

MacDonald, D. W. (2010). *The Encyclopedia of Mammals*. Oxford: Oxford University Press.

Mackowiak, P. A. & Boulant, J. A. (1997). Fever's upper limit. In: *Fever: Basic Mechanisms and Management* (Ed. Mackowiak, P. A.). Philadelphia: Lippincott-Raven.

MacMillen, R. E. (1965). Aestivation in the cactus mouse *Peromyscus eremicus*. *Comparative Biochemistry and Physiology* **16**, 227–247.

MacMillen, R. E. (1972). Water economy of nocturnal desert rodents. *Symposium of the Zoological Society of London* **31**, 147–174.

MacMillen, R. E. (1983). Water regulation in *Peromyscus*. *Journal of Mammalogy* **64**, 38–47.

MacMillen, R. E. & Hinds, D. S. (1983). Water regulatory efficiency in heteromyid rodents: a model and its application. *Ecology* **64**, 152–164.

MacMillen, R. E. & Lee, A. K. (1967). Australian desert mice: independence of exogenous water. *Science* **158**, 383–385.

Madara, J. L. (1989). Loosening tight junctions; lessons from the intestine. *Journal of Clinical Investigation* **83**, 1089–1094.

Maina, J. N. (2000a). What it takes to fly: the structural and functional respiratory refinements in birds and bats. *Journal of Experimental Biology* **203**, 3045–64.

Maina, J. N. (2000b). Comparative respiratory morphology: themes and principles in the design and construction of the gas exchangers. *Anatomical Record* **261**, 25–44.

Maina, J. N., Thomas, S. P. & Dolls, D. M. (1991). A morphometric study of bats of different size: correlations between structure and function of the chiropteran lung. *Philosophical Transactions of the Royal Society of London B* **333**, 31–50.

Makanya, A. N. & Mortola, J. A. (2007). The structural design of the bat wing web and its possible role in gas exchange. *Journal of Anatomy* **211**, 687–697.

Makanya, A. N., Sparrow, M. P., Warui, C. N., Mwangi, D. K. & Burri, P. H. (2001). Morphological analysis of the postnatally developing marsupial lung: the quokka wallaby. *Anatomical Record* **262**, 253–265.

Malan, A. (1973). Ventilation measured by body plethysmography in hibernating mammals and in poikilotherms. *Respiration Physiology* **17**, 32–44.

Malan, A. (1982). Respiration and acid-base state in hibernation. In: *Hibernation and Torpor in Mammals and Birds* (Eds. Lyman, C. P., Willis, J. S., Malan, A. & Wang, L. C. H.). New York: Academic Press.

Malan, A. (1988). pH and hypometabolism in mammalian hibernation. *Canadian Journal of Zoology* **66**, 95–98.

Malan, A. (2014). The evolution of mammalian hibernation: lessons from comparative acid-base physiology. *Integrative and Comparative Biology*. doi:10.1093/icb/icu002.

Maloiy, G. M. O. (1973). The water metabolism of a small East African antelope: the dik-dik. *Proceedings of the Royal Society of London B* **184**, 167–178.

Maloney, S. & Forbes, C. (2011). What effect will a few degrees of climate change have on human heat balance? Implications for human activity. *International Journal of Biometeorology* 55, 147–160.

Maloney, S. K., Bronner, G. N. & Buffenstein, R. (1999). Thermoregulation in the Angolan free-tailed bat *Mops condylurus*: a small mammal that uses hot roosts. *Physiological and Biochemical Zoology* 72, 385–396.

Maloney, S. K., Fuller, A., Meyer, L. C., Kamerman, P. R., Mitchell, G., & Mitchell, D. (2009). Brain thermal inertia, but no evidence for selective brain cooling, in free-ranging western grey kangaroos (*Macropus fuliginosus*). *Journal of Comparative Physiology B* **179**, 241–251.

Maloney, S. K., Fuller, A., Mitchell, D., Gordon, C. & Overton, J. M. (2014). Translating animal model research: does it matter that our rodents are cold? *Physiology* **29**, 413–420.

Maloney, S. K., Fuller, A., Mitchell, G. & Mitchell, D. (2002). Brain and arterial blood temperatures of free-ranging oryx (*Oryx gazella*). *Pflugers Archiv* **443**, 437–445.

Maloney, S. K. & Mitchell, D. (1996). Regulation of ram scrotal temperature during heat exposure, cold exposure, fever and exercise. *Journal of Physiology* **496**, 421–430.

Maloney, S. K., Moss, G., Cartmell, T. & Mitchell, D. (2005). Alteration in diel activity patterns as a thermoregulatory strategy in black wildebeest (*Connochaetes gnou*). *Journal of Comparative Physiology A* **191**, 1055–1064.

Mançanares, C. A. F., Santos, A. C., Piemonte, M. V., Vasconcelos, B. G., Carvalho, A. F., Miglino, M. A., Ambrósio, C. E. & Neto, A. C. A. (2012). Macroscopic and microscopic analysis of the tongue of the common opossum (*Didelphis marsupialis*). *Microscopy Research and Technique* 75, 1329–1333.

Mancina, C. A., García-Rivera, L. & Miller, B. W. (2012). Wing morphology, echolocation, and resource partitioning in syntopic Cuban mormoopid bats. *Journal of Mammalogy* **93**, 1308–1317.

Manger, P. R. (2006). An examination of cetacean brain structure with a novel hypothesis correlating thermogenesis to the evolution of a big brain. *Biological Reviews* **81**, 293–338.

Mangione, A. M., Dearing, M. D. & Karasov, W. H. (2004). Creosote bush (*Larrea tridentata*) resin increases water demands and reduces energy availability in desert woodrats (*Neotoma lepida*). *Journal of Chemical Ecology* **30**, 1409–1429.

Marchand, P. (2014). *Life in the Cold: An Introduction to Winter Ecology*. Lebanon, New Hampshire: University Press of New England.

Mares, M. A. & Rosenzweig, M. L. (1978). Granivory in North and South American deserts: rodents, birds, and ants. *Ecology* **59**, 235–241.

Margueron, R. & Reinberg, D. (2010). Chromatin structure and the inheritance of epigenetic information. *Nature Reviews Genetics* **11**, 285–296.

Marinello, M. & Bernard, E. (2014). Wing morphology of Neotropical bats: a quantitative and qualitative analysis with implications for habitat use. *Canadian Journal of Zoology* **92**, 141–147.

Markussen, N. H., Ryg, M. & Øritsland, N. A. (1994). The effect of feeding on the metabolic rate in harbour seals (*Phoca vitulina*). *Journal of Comparative Physiology B* **164**, 89–93.

Marquet, P. A., Navarrete, S. A. & Castilla, J. C. (1995). Body size, population density, and the energetic equivalence rule. *Journal of Animal Ecology* **64**, 325–332.

Marsh, K. J., Wallis, I. R., McLean, S., Sorensen, J. S. & Foley, W. J. (2006). Conflicting demands on detoxification pathways influence how common brushtail possums choose their diets. *Ecology* **87**, 2103–2112.

Marsh, O. C. (1880). Odontornithes: a monograph on the extinct toothed birds of North America. *Report of the Geological Exploration of the Fortieth Parallel* 7, 1–201.

Mårtensson, P.-E., Nordøy, E. S. & Blix, A. S. (1994). Digestibility of krill (*Euphausia superba* and *Thysanoessa* sp.) in minke whales (*Balaenoptera acutorostrata*) and crabeater seals (*Lobodon carcinophagus*). *British Journal of Nutrition* 72, 713–716.
Mårtensson, P.-E., NordΦy, E. S., Messelt, E. B. & Blix, A. S. (1998). Gut length, food transit time and diving habit in phocid seals. *Polar Biology* **20**, 213–217.
Martin, J. A. & Hillman, S. S. (2009). The physical movement of urine from the kidneys to the urinary bladder and bladder compliance in two anurans. *Physiological and Biochemical Zoology* **82**, 163–169.
Martin, P. G. (1970). The Darwin Rise hypothesis of the biogeographic dispersal of marsupials. *Nature* **225**, 197–198.
Martin, S. L. & Yoder, A. D. (2014). Theme and variations: heterothermy in mammals. *Integrative and Comparative Biology*. doi:10.1093/icb/icu085.
Martinez, P. A., Marti, D. A., Molina, W. F. & Bidau, C. J. (2013). Bergmann's rule across the equator: a case study in *Cerdocyon thous* (Canidae). *Journal of Animal Ecology* **82**, 997–1008.
Martínez del Río, C., Cork. S. J. & Karasov, W. H. (1994). Modeling gut function: an introduction. In: *The Digestive System in Mammals: Food, Form and Function* (Eds. Chivers, D. J & Langer, P.). Cambridge: Cambridge University Press.
Matheson, A. L., Campbell, K. L. & Willis, C. K. R. (2010). Feasting, fasting and freezing: energetic effects of meal size and temperature on torpor expression by little brown bats *Myotis lucifugus*. *Journal of Experimental Biology* **213**, 2165–2173.
Mathieu-Costello, O. (1993). Comparative aspects of muscle capillary supply. *Annual Review of Physiology* **55**, 503–525.
Mathieu-Costello, O., Szewczak, J. M., Logemann, R. B. & Agey, P. J. (1992). Geometry of blood-tissue exchange in bat flight muscle compared with bat hindlimb and rat soleus muscle. *American Journal of Physiology* **262**, R 955–R 965.
Matthew, W. D. (1915). Climate and evolution. *Annals of the New York Academy of Science* **24**, 171–318.
Matthews, A., Ruykys, L., Ellis, B., FitzGibbon, S., Lunney, D., Crowther, M. S., Glen, A. S., Purcell, B., Moseby, K., Stott, J., Fletcher, D., Wimpenny, C., Allen, B. L., Van Bommel, L., Roberts, M., Davies, N., Green, K., Newsome, T., Ballard, G., Fleming, P., Dickman, C. R., Eberhart, A., Troy, S., McMahon, C. & Wiggins, N. (2013). The success of GPS collar deployments on mammals in Australia. *Australian Mammalogy* **35**, 65–83.
Mauck, B., Bilgmann, K., Jones, D. D., Eysel, U. & Dehnhardt, G. (2003). Thermal windows on the trunk of hauled-out seals: hot spots for thermoregulatory evaporation? *Journal of Experimental Biology* **206**, 1727–1738.
May-Collardo, L. J., Agnarsson, I. & Wartzok, D. (2007). Reexamining the relationship between body size and tonal signals frequency in whales: a comparative approach using a novel phylogeny. *Marine Mammal Science* **23**, 524–552.
Maynard Smith, J. & Savage, R. J. G. (1956). Some locomotory adaptations in mammals. *Journal of the Linnean Society of London, Zoology* **42**, 603–622.
Mayntz, D., Nielsen, V. H., Sørensen, A., Toft, S., Raubenheimer, D., Hejlesen, C. & Simpson, S. J. (2009). Balancing of protein and lipid intake by a mammalian carnivore, the mink, *Mustela vison*. *Animal Behaviour* 77, 349–355.
Mayr, E. (1956). Geographical character gradients and climatic adaptation. *Evolution* **10**, 105–108.
Mayr, E. (1963). *Animal Species and Evolution*. Cambridge: Harvard University Press.
McAllen, B. M. & Geiser, G. (2014). Torpor during reproduction in mammals and birds: dealing with an energetic conundrum. *Integrative and Comparative Biology* **54**, 516–532.

McAllen, R. M., Tanaka, M., Ootsuka, Y. & McKinley, M. J. (2010). Multiple thermoregulatory effectors with independent central controls. *European Journal of Applied Physiology* **109**, 27–33.
McCafferty, D. J., Gilbert, C., Paterson, W., Pomeroy, P. P., Thompson, D., Currie, J. I. & Ancel, A. (2011). Estimating metabolic heat loss in birds and mammals by combining infrared thermography with biophysical modelling. *Comparative Biochemistry and Physiology A* **158**, 337–345.
McCall, P. J., Harding, G., Roberts, J. & Auly, B. (1996). Attraction and trapping of *Aedes aegypti* (Diptera: Culicidae) with host odors in the laboratory. *Journal of Medical Entomology* **33**, 177–179.
McCoy, M. W. & Gillooly, J. F. (2008). Predicting natural mortality rates of plants and animals. *Ecology Letters* **11**, 710–716.
McCracken, G. F. Gillam, E. H., Westbrook, J. K., Lee, Y. U, Jensen, M. L. & Balsley, B. B. (2008). Brazilian free-tailed bats (*Tadarida brasiliensis*: Molossidae, Chiroptera) at high altitude: links to migratory insect populations. *Integrative and Comparative Biology* **48**, 107–118.
McCue, M. D. (2006). Specific dynamic action: a century of investigation. *Comparative Biochemistry and Physiology A* **144**, 381–394.
McCullough, D. R. (1964). Relationship of weather to migratory movements of black-tailed deer. *Ecology* **45**, 249–256.
McDonald, B. I. & Ponganis, P. P. (2012). Lung collapse in the diving sea lion: hold the nitrogen and save the oxygen. *Biology Letters* **8**, 1047–1049.
McFarland, R., Henzi, S. P., Barrett, L., Wanigarantne, A., Coetzee, E., Fuller, A., Hetern, R. S., Mitchell, D. & Maloney, S. K. (2016). Thermal consequences of increased pelt loft infer an additional utilitarian function for grooming. *American Journal of Primatology* **78**, 456–461.
McGhee, R. B. (1968). Some finite state aspects of legged locomotion. *Mathematical Biosciences* **2**, 67–84.
McGilvery, R. W. & Goldstein, G. W. (1983). *Biochemistry. A Functional Approach*. Philadelphia: Saunders College Publishing.
McKechnie, A. E. & Wolf, B. O. (2004). The allometry of avian basal metabolic rate: good predictions need good data. *Physiological and Biochemical Zoology* 77, 502–521.
McKenna, M. C. & Bell, S. K. (1997). *Classification of Mammals above the Species Level*. New York: Columbia University Press.
McLean, D. M. (1981). Size factor in the Late Pleistocene mammalian extinctions. *American Journal of Science* **281**, 1144–1152.
McLean, J. A. & Tobin, G. (1987). *Calorimetry*. Cambridge: Cambridge University Press.
McMahon, T. A. (1973). Size and shape in biology. *Science* **179**, 1201–1204.
McMahon, T. A. & Bonner, J. T. (1983). *On Size and Life*. New York: Scientific American Library.
McNab, B. K. (1963). Bioenergetics and the determination of home range size. *American Naturalist* **97**, 133–140.
McNab, B. K. (1966). The metabolism of fossorial rodents: a study of convergence. *Ecology* **47**, 712–733.
McNab, B. K. (1971). On the ecological significance of Bergmann's Rule. *Ecology* **52**, 845–854.
McNab, B. (1978). The evolution of endothermy in the phylogeny of mammals. *American Naturalist* **112**, 1–21.
McNab, B. K. (1979). The influence of body size on the energetics and distribution of fossorial and burrowing mammals. *Ecology* **60**, 1010–1021.
McNab, B. K. (1980). On estimating thermal conductance in endotherms. *Physiological Zoology* **53**, 145–156.

McNab, B. K. (1983). Energetics, body size, and the limits to endothermy. *Journal of Zoology* **199**, 1–29.
McNab, B. K. (1984). Physiological convergence amongst ant-eating and termite-eating mammals. *Journal of Zoology* **203**, 485–510.
McNab, B. K. (1986). Food habits, energetics, and the reproduction of marsupials. *Journal of Zoology* **208**, 595–614.
McNab, B. K. (1988). Complications inherent in scaling basal rate of metabolism in mammals. *Quarterly Review of Biology* **63**, 25–54.
McNab, B. K. (1992). Energy expenditure: a short history. In: *Mammalian Energetics Interdisciplinary Views of Metabolism and Reproduction* (Eds. Tomasi, T. E. & Horton, T. H.). Ithaca, New York: Cornell University.
McNab, B. K. (1997). On the utility of uniformity in the definition of basal rate of metabolism. *Physiological Zoology* **70**, 718–720.
McNab, B. K. (1999). On the comparative ecological and evolutionary significance of total and mass-specific rates of metabolism. *Physiological and Biochemical Zoology* **72**, 642–644.
McNab, B. K. (2000). Energy constraints on a carnivore diet. *Nature* **407**, 584.
McNab, B. K. (2002). *The Physiological Ecology of Vertebrates: A View from Energetics*. New York: Comstock Publishing Associates, Cornell University Press.
McNab, B. K. (2008). An analysis of the factors that influence the level and scaling of mammalian BMR. *Comparative Biochemistry and Physiology A* **151**, 5–28.
McNab, B. K. (2010). Geographic and temporal correlations of mammalian size reconsidered: a resource rule. *Oecologia* **164**, 13–23.
McNab, B. K. (2012). *Extreme Measures*. Chicago: Chicago University Press.
McNab, B. K. (2015). Behavioral and ecological factors account for variation in the mass-independent energy expenditures of endotherms. *Journal of Comparative Physiology B* **185**, 1–13.
McWhorter, T. J., Bakken, B. H., Karasov, W. H. & Martinez del Rio, C. (2006). Hummingbirds rely on both paracellular and carrier-mediated intestinal glucose absorption to fuel high metabolism. *Biology Letters* **2**, 131–134.
Meagher, E. M., McLellan, W. A., Westgate, A. J., Wells, R. S., Frierson, D. & Pabst, D. A. (2002). The relationship between heat flow and vasculature in the dorsal fin of wild bottlenose dolphins *Tursiops truncatus*. *Journal of Experimental Biology* **205**, 3475–3486.
Meeh, K. (1879). Oberflachenmessungen des menschlichen korpers. *Zeitschrift für Biologie* **15**, 426–458.
Meigal, A. (2002). Gross and fine neuromuscular performance at cold shivering. *International Journal of Circumpolar Health* **61**, 163–172.
Meigal, A., Ivukov, A, Gerasimova. L., Antonen, E. G. & Lupandin, L. V. (2000). The effect of general cooling on the electromyographic characteristics of muscle fatigue evoked by dynamic loading. *Fisiol Cheloveka* **26**, 80–86.
Meiri, S. (2011). Bergmann's Rule—what's in a name? *Global Ecology and Biogeography* **20**, 203–207.
Meissner, H. H., Spreeth, E. B., de Villers, P. A., Pietersen, E. W., Hugo, T. A. & Terblanche, B. F. (1990). Quality of food and voluntary intake by elephants as measured by lignin index. *South African Journal of Wildlife Research* **20**, 104–110.
Mella, V. S. A, Cooper, C. E. & Davies, S. J. J. F. (2016). Effects of historically familiar and novel predator odors on the physiology of an introduced prey. *Current Zoology* **62**, 53–59.
Melvin, R. G. & Andrews, M. T. (2010). Torpor induction in mammals: recent discoveries fueling new ideas. *Trends in Endocrinology and Metabolism* **20**, 490–498.

Meng, J., Hu, Y., Wang, Y., Wang, X. & Li, C. (2006). A Mesozoic gliding mammal from north-eastern China. *Nature* **444**, 889–893.

Menon, G. K., Brown, B. E. & Elias, P. M. (1986). Avian epidermal differentiation: role of lipids in permeability barrier formation. *Tissue and Cell* **18**, 71–82.

Michaeli, G. & Pinshow, B. (2001). Respiratory water loss in free-flying pigeons. *Journal of Experimental Biology* **204**, 3803–3814.

Michilsens, F., D'Août, K., Vereecke, E. E. & Aerts, P. (2012). One step beyond: different step-to-step transitions exist during continuous contact brachiation in siamangs. *Biology Open* **1**, 411–421.

Mildrexler, D. J., Zhao, M. & Running, S. W. (2006). Where are the hottest spots on Earth? *Eos* **87**, 461–476.

Millar, J. S. (1977). Adaptive features of mammalian reproduction. *Evolution* **31**, 370–386.

Millar, J. S. & Hickling, G. J. (1990). Fasting endurance and the evolution of mammalian body size. *Functional Ecology* **4**, 5–12.

Miller, E. M., Christensen, G. C. & Evans, H. E. (1964). *The Anatomy of the Dog*. Philadelphia: W. B. Saunders.

Miller, S. J., Bencini, R. and Hartmann, P. E. (2009). Composition of the milk of the quokka (*Setonix brachyurus*). *Australian Journal of Zoology* **57**, 11–21.

Millesi, E., Prossinger, H., Dittami, J. P. & Fieder, M. (2001). Hibernation effects on memory in European ground squirrels (*Spermophilus citellus*). *Journal of Biological Rhythms* **16**, 264–271.

Minnaar, I. A., Bennett, N. C., Chimimba, C. T & McKechnie, A. E. (2014). Partitioning of evaporative water loss into respiratory and cutaneous pathway in Wahlberg's epauletted fruit bats (*Epomophorus wahlbergi*). *Physiological and Biochemical Zoology* **87**, 475–485.

Mirceta, S., Signore, A. V., Burns, J. M., Cossins, A. R., Campbell, K. L. & Berenbrink, M. (2013). Evolution of mammalian diving capacity traced by myoglobin net surface charge. *Science* **340**, 1303–1310.

Mitchell, G. & Lust, A. (2008). The carotid rete and artiodactyl success. *Biology Letters* **4**, 415–418.

Mitchell, J. H. & Blomqvist, G. (1971). Maximal oxygen uptake. *New England Journal of Medicine* **284**, 1018–1022.

Mitchell, J. W. & Myers, G. E. (1968). An analytical model of the counter-current heat exchange phenomena. *Biophysics Journal* **8**, 897–911.

Mitchell, O. G. (1972). Human hibernation and space travel. *Advances in Space Science and Technology* **11**, 249–265.

Modlin, I. M. & Kidd, M. (2001). Ernest Starling and the discovery of secretin. *Journal of Clinical Gastroenterology* **32**, 187–192.

Monge, C. & León-Velarde, F. (1991). Physiological adaptation to high altitude: oxygen transport in mammals and birds. *Physiological Reviews* **71**, 1135–1172.

Monteith, J. L. & Campbell, G. S. (1980). Diffusion of water vapour through integuments—potential confusion. *Journal of Thermal Biology* **5**, 7–9.

Montell, C. (2008). TRP Channels: it's not the heat, it's the humidity. *Current Biology* **18**, R123–R126.

Moore, B. D. & Foley, W. J. (2005). Tree use by koalas in a chemically complex landscape. *Nature* **435**, 488–490.

Moore, B. D., Wiggins, N. I., Marsh, K. J., Dearing, M. D. & Foley, W. J. (2015). Translating physiological signals to changes in feeding behaviour in mammals and the future effects of global climate change. *Animal Production Science* **55**, 272–283.

Moore, J. C. (1984). The Golgi tendon organ: a review and update. *American Journal of Occupational Therapy* **38**, 227–236.

Moore, M. J. (2005). From birth to death: the complex lives of eukaryotic mRNAs. *Science* **309**, 1514–1518.

Morgan, D. L., Proske, U. & Warren, D. (1978). Measurements of muscle stiffness and the mechanism of elastic storage of energy in hopping kangaroos. *Journal of Physiology* **282**, 253–261.

Moritz, G. I. Fourie, N., Yeake, J. D., Phillips-Conroy, J. E., Jolly, C. J., Koch, P. L. & Dominy, N. J. (2012). Baboons, water, and the ecology of oxygen stable isotopes in an arid hybrid zone. *Physiological and Biochemical Zoology* **85**, 421–430.

Morris, K., Wong, E. S. W. & Belov, K. (2010). Use of genomic information to gain insights into immune function in marsupials: a review of divergent immune genes. In: *Marsupial Genetics and Genomes* (Eds. Deakin, J. E., Waters, P. D. & Marshall Graves, J. A.). Dordrecht: Springer.

Morris, S., Curtin, A. L. & Thompson, M. B. (1994). Heterothermy, torpor, respiratory gas exchange, water balance and the effect of feeding in Gould's long-eared bat *Nyctophilus gouldi*. *Journal of Experimental Biology* **197**, 309–335.

Morrison, P. (1964). Wild animals at high altitudes. *Symposium of the Zoological Society of London* **13**, 49–55.

Morrison, S. F. & Nakamura, K. (2011). Central neural pathways for thermoregulation. *Frontiers in Bioscience* **16**, 74–104.

Morrison, S. F., Nakamura, K. & Madden, C. J. (2008). Central control of thermogenesis in mammals. *Experimental Physiology* **93**, 773–797.

Morrow, G. & Nicol, S. C. (2009). Cool sex? Hibernation and reproduction overlap in the echidna. *PLOS ONE* **4**, e6070.

Mortola, J. P. (2004). Implications of hypoxic hypometabolism during mammalian ontogenesis. *Respiration Physiology and Neurobiology* **141**, 345–56.

Morton, S. R. (1979). Diversity of desert-dwelling mammals: a comparison of Australia and North America. *Journal of Mammalogy* **60**, 253–264.

Motani, R., Jiang, D.-Y., Tintori, A., Rieppel, O. & Chen, G.-B. (2014). Terrestrial origin of viviparity in Mesozoic marine reptiles indicated by early Triassic embryonic fossils. *PLOS ONE* **9**, e88640.

Mount, D. B. (2014). Thick ascending limb of the loop of Henle. *Clinical Journal of the American Society of Nephrology* **9**, 1974–1986.

Moussy, C., Hosken, D. J., Mathews, F., Smith, G. C., Aegerter, J. N. & Bearhop, S. (2013). Migration and dispersal patterns of bats and their influence on genetic structure. *Mammal Review* **43**, 183–195.

Moyes, C. D. & Schulte, P. M. (2008). *Principles of Animal Physiology*. San Francisco: Pearson Education.

Mozo, J., Emre, Y., Bouillaud, F., Ricquier, D. & Criscuolo, F. (2005). Thermoregulation: what role for UCPs in mammals and birds? *Bioscience Reports* **25**, 227–249.

Muijres, F. T., Johansson, L. C., Barfield, R., Wolf, M., Spedding, G. R. & Hedenström, A. (2008). Leading-edge vortex improves lift in slow-flying bats. *Science* **319**, 1250–1253.

Muijres, F. T., Johansson, L. C., Bowlin, M. S., Winter, Y. & Hedenstrom, A. (2012). Comparing aerodynamic efficiency in birds and bats suggests better flight performance in birds. *PLOS ONE* **7**, e37335.

Mueller, P. & Diamond, J. (2001). Metabolic rate and environmental productivity: well-provisioned animals evolved to run and idle fast. *Proceedings of the National Academy of Sciences, USA* **98**, 12550–12554.

Müller, J., Bässler, K., Essbauer, S., Schex, S., Müller, D. W. H., Opgenoorth, L. & Brandl, R. (2014). Relative heart size in two rodent species increases with elevation: reviving Hesse's rule. *Journal of Biogeography* **41**, 2211–2220.

Munch, S. B. & Salinas, S. (2009). Latitudinal variation in lifespan within species is explained by the metabolic theory of ecology. *Proceedings of the National Academy of Sciences, USA* **106**, 13860–13864.

Munn, A. J., Barboza, P. S. & Dehn, J. (2009a). Sensible heat loss from muskoxen (*Ovibos moschatus*) feeding in winter: small calves are not at a thermal disadvantage compared with adult cows. *Physiological and Biochemical Zoology* **82**, 455–467.

Munn, A. J., Cooper, C. E., Russell, B., Dawson, T. J., McLeod, S. R. & Maloney, S. K. (2012). Energy and water use by invasive goats (*Capra hircus*) in an Australian rangeland, and a caution against using broad-scale allometry to predict species-specific requirements. *Comparative Biochemistry and Physiology A* **161**, 216–229.

Munn, A. J., Dawson, T. J. & McLeod, S. R. (2010b). Feeding biology of two functionally different foregut-fermenting mammals, the marsupial red kangaroo and the ruminant sheep: how physiological ecology can inform land management. *Journal of Zoology* **282**, 226–237.

Munn, A. J., Dawson, T. J., McLeod, S. R., Croft, D. B., Thompson, M. B. & Dickman, C. R. (2009b). Field metabolic rate and water turnover of red kangaroos and sheep in an arid rangeland: an empirically derived dry-sheep-equivalent for kangaroos. *Australian Journal of Zoology* **57**, 23–28.

Munn, A. J., Dawson, T. J., McLeod, S. R., Dennis, T. & Maloney, S. K. (2013b). Energy, water and space use by free-living red kangaroos *Macropus rufus* and domestic sheep *Ovis aries* in an Australian rangeland. *Journal of Comparative Physiology B* **183**, 843–858.

Munn, A. J., Dunne, C., Müller, D. W. H. & Clauss, M. (2013a). Energy in-equivalence in Australian marsupials: evidence for disruption of the continent's mammal assemblage, or are rules meant to be broken? *PLOS ONE* **8**, e57449.

Munn, A. J., Kern, P. & McAllan, B. M. (2010a). Coping with chaos: unpredictable food supplies intensify torpor use in an arid-zone marsupial, the fat-tailed dunnart (*Sminthopsis crassicaudata*). *Naturwissenschaften* **97**, 601–605.

Muñoz-Garcia, A., Ben-Hamo, M., Pinshow, B., Williams, J. B. & Korine, C. (2012a). The relationship between cutaneous water loss and thermoregulatory state in Kuhl's pipistrelle *Pipistrellus kuhlii*, a vespertillionid bat. *Physiological and Biochemical Zoology* **85**, 516–525.

Muñoz-Garcia, A., Larraín, P., Ben-Hamo, M. Cruz-Neto, A. Williams, J. B., Pinshow, B. & Korine, C. (2016). Metabolic rate, evaporative water loss and thermoregulatory state in four species of bats in the Negev desert. *Comparative and Biochemistry Physiology A* **191**, 156–165.

Muñoz-Garcia, A., Ro, J., Reichard, J. D., Kunz, T. H. & Williams, J. B. (2012b). Cutaneous water loss and lipids of the stratum corneum in two syntopic species of bats. *Comparative Biochemistry and Physiology A* **61**, 208–215.

Muñoz-Garcia, A. & Williams, J. B. (2005). Basal metabolic rate in carnivores is associated with diet after controlling for phylogeny. *Physiological and Biochemical Zoology* **78**, 1039–1056.

Munro, D. & Thomas, D. W. (2004). The role of polyunsaturated fatty acids in the expression of torpor by mammals: a review. *Zoology* **107**, 29–48.

Murdaugh, H. V., Robin, E. D., Millen, J. E., Drewry, W. F. & Weiss, E. (1966). Adaptations to diving in the harbor seal: cardiac output during diving. *American Journal of Physiology* **210**, 176–180.

Murphy, B. F. & Thompson, M. B. (2011). A review of the evolution of viviparity in squamate reptiles: the past, present and future role of molecular biology and genomics. *Journal of Comparative Physiology B* **181**, 575–594.

Murphy, W. J., Pringle, T. H., Crider, T. A., Springer, M. S. & Miller, W. (2007). Using genomic data to unravel the root of the placental mammal phylogeny. *Genome Research* **17**, 413–421.

Murrish, D. E. (1973). Respiratory heat and water exchange in penguins. *Respiratory Physiology* **19**, 262–270.

Murrish, D. E. & Schmidt-Nielsen, K. (1970). Exhaled air temperature and water conservation in lizards. *Respiratory Physiology* **10**, 151–158.

Musacchia, X. J. (1972). Heat and cold acclimation in helium-cold hypothermia in the hamster. *American Journal of Physiology* **222**, 495–498.

Musser, A. M. (2003). Review of the monotreme fossil record and comparison of palaeontological and molecular data. *Comparative Biochemistry and Physiology A* **136**, 927–942.

Musser, A. M. (2006). Furry egg-layers: monotreme relationships and Radiations. In: *Evolution and Biogeography of Australian Vertebrates* (Eds. Merrick, J. R., Archer, M., Hickey, G. M. & Lee, M. S. Y.). Oatlands, NSW: Auscipub.

Mzilikazi, N. & Lovegrove, B. G. (2004). Daily torpor in free-ranging rock elephant shrews, *Elephantulus myurus*: a year-long study. *Physiological and Biochemical Zoology* 77, 285–296.

Mzilikazi, N., Lovegrove, B. G. & Ribble, D. O. (2002). Exogenous passive heating during torpor arousal in free-ranging elephant shrews, *Elephantulus myurus*. *Oecologia* **133**, 307–314.

Nachtigall, W. (1983). The biophysics of locomotion in water. In: *Biophysics* (Eds. Hoppe, W., Lohmann, W., Markl, H. & Ziegler, H.). Berlin: Springer-Verlag.

Nagy, K. (2005). Field metabolic rate and body size. *Journal of Experimental Biology* **208**, 1621–1625.

Nagy, K. A. (1983). The doubly labeled water ($^3HH^{18}O$) method: a guide to its use. *University of California at Los Angeles Publication* **12**, 1–45.

Nagy, K. A. (2005). Field metabolic rate and body size. *Journal of Experimental Biology* **208**, 1621–1625.

Nagy, K. A. & Bradshaw, S. D. (2000). Scaling of energy and water fluxes in free-living arid-zone Australian marsupials. *Journal of Mammalogy* **81**, 962–970.

Nagy, K. A., Girard, I. A. & Brown, T. K. (1999). Energetics of free-ranging mammals, reptiles, and birds. *Annual Reviews of Nutrition* **19**, 247–277.

Nagy, K. A. & Peterson, C. C. (1988). Scaling of water flux in animals. *University of California Publications in Zoology* **120**, 1–172.

Nagy, K. A., Seymour, R. S., Lee, A. K. & Braithwaite, R. (1978). Energy and water budgets in free-living *Antechinus stuartii* (Marsupialia: Dasyuridae). *Journal of Mammalogy* **59**, 60–68.

National Science Board. (2014). *Science and Engineering Indicators 2014*. http://www.nsf.gov/statistics/seind14/index.cfm/etc/sitemap.htm

Naya, D. E., Bozinovic, F. & Karasov, W. H. (2008). Latitudinal trends in digestive flexibility: testing the climatic variability hypothesis with data on the intestinal length in rodents. *American Naturalist* **172**, 122–134.

Naya, D. E., Karasov, W. H. & Bozinovic, F. (2007). Phenotypic plasticity in laboratory mice and rats: a meta-analysis of current ideas on gut size flexibility. *Evolutionary Ecology Research* **9**, 1363–1374.

Naya, D. E., Spangenberg, l., Naya, H. & Bozinovic, F. (2012). Latitudinal patterns in rodent metabolic flexibility. *American Naturalist* **179**, 172–179.

Naya, D. E., Spangenberg, L., Naya, H. & Bozinovic, F. (2013). How does evolutionary variation in basal metabolic rates arise? A statistical assessment and a mechanistic model. *Evolution* **67**, 1463–1476.

Nedergaard, J., Bengtsson, T. & Cannon, B. (2007). Unexpected evidence for active brown adipose tissue in adult humans. *American Journal of Physiology* **293**, E444–E452.

Nedergaard, J., Golozoubova, V., Matthias, A., Asadi, A., Jacobsson, A. & Cannon, B. (2001). UCP1: the only protein able to mediate adaptive non-shivering thermogenesis and metabolic inefficiency. *Biochimica et Biophysica Acta - Bioenergetics* **1504**, 82–106.

Needham, A. D., Dawson, T. J. & Hales, J. R. S. (1974). Forelimb bloodflow and saliva spreading in the thermoregulation of the red kangaroo, *Megaleia rufa. Comparative Biochemistry and Physiology A* **49**, 555–565.

Nelson, R. A., Folk, G. E., Pfeiffer, E. W., Craighead, J. J., Jonkel, C. J. & Steiger, D. L. (1983). Behaviour, biochemistry and hibernation in black, grizzly and polar bears. *International Conference on Bear Research and Management* **5**, 284–290.

Nelson, R. A., Jones, J. D., Wahner, H. W, McGill, D. B. & Code, C. F. (1975). Nitrogen metabolism in bears: urea metabolism in summer starvation and in winter sleep and role of urinary bladder in water and nitrogen conservation. *Mayo Clinic Proceedings* **50**, 141–146.

Nespolo, R., Bacigalupe, L. D. & Bozinovic, F. (2003). The influence of heat increment of feeding on basal metabolic rate in *Phyllotis darwini* (Muridae). *Comparative Biochemistry and Physiology A* **134**, 139–145.

Nespolo, R. F., Bustamante, D. M., Bacigalupe, L. D. & Bozinovic, F. (2005). Quantitative genetics of bioenergetics and growth-related traits in the wild mammal, *Phyllotis darwini. Evolution* **59**, 1829–1837.

Nespolo, R. F. & Franco, M. (2007). Whole-animal metabolic rate is a repeatable trait: a meta-analysis. *Journal of Experimental Biology* **210**, 2000–2005.

Nevo, E. (1999). *Mosaic Evolution of Subterranean Mammals: Regression, Progression, and Global Convergence*. Oxford: Oxford University Press.

Newsome, A. E. (1971). The ecology of red kangaroo. *Australian Zoologist* **16**, 32–50.

Newsome, A. E. & Coman, B. J. (1989). Canidae. In: *Fauna of Australia. Mammalia Volume 1B* (Eds. Walton, D. W. & Richardson, B. J.). Canberra: Australian Government Publishing Service.

Nice, P., Black, C. P. & Tenney, M. (1980). A comparative study of ventilatory response to hypoxia with reference to hemoglobin $O_2$-affinity in llama, cat, rat, duck, and goose. *Comparative Biochemistry and Physiology A* **66**, 347–350.

Nicholls, H. (2008). Let's make a mammoth. *Nature* **456**, 310–314.

Nicol, S. & Andersen, N. A. (2002). The timing of hibernation in Tasmanian echidnas: why do they do it when they do? *Comparative Biochemistry and Physiology A* **131**, 603–611.

Nicol, S. C., Pavlides, D. & Anderson, N. A. (1997). Nonshivering thermogenesis in marsupials: absence of thermogenic response to β3-adrenergic agonists. *Comparative Biochemistry and Physiology A* **117**, 399–405.

Nielsen, S. Frøkiær, J., Marples, D., Kwon, T., Peter Agre, P., Knepper, M. A. (2002). Aquaporins in the kidney: from molecules to medicine. *Physiological Reviews* **82**, 205–244.

Nielsen, S., di Giovanni, S., Christensen, E., Knepper, M. & Harris, H. (1993). Cellular and subcellular immunolocalization of vasopressin-regulated water channel in the rat kidney. *Proceedings of the National Acadademy of Sciences, USA* **90**, 11663–11667.

Nieuwenhuys, R., Ten Donkelaar, H. J. & Nicholson, C. (1998). *The Central Nervous System of Vertebrates*. Berlin: Springer-Verlag.

Niimura, Y., Matsui, A. & Touhara, K. (2014). Extreme expansion of the olfactory receptor gene repertoire in African elephants and evolutionary dynamics of orthologous gene groups in 13 placental mammals. *Genome Research* **24**, 1485–1496.

Nilssen, K. J., Sundsfijord, J. A. & Blix, A. S. (1984). Regulation of metabolic rate in Svalbard and Norwegian reindeer. *American Journal of Physiology* **247**, R837–R841.

Nishimura, R. A., Callahan, M. J., Schaff, H. V., Ilstrup, D. M., Miller, F. A. & Tajik, A. J. (1984). Noninvasive measurement of cardiac output by continuous-wave Doppler echocardiography: initial experience and review of the literature. *Mayo Clinic Proceedings* **59**, 484–489.

Nobel, P. S. (1999). *Physicochemical and Environmental Plant Physiology*. San Diego: Academic Press.

Nolan, J. V. (1993). Nitrogen kinetics. In: *Quantitative Aspects of Ruminant Digestion and Metabolism* (Eds. Forbes, J. M. & France, J.). Wallington, UK: CAB International.

Nolan, J. V. & Dobos, R. C. (2005). Nitrogen transactions in ruminants. In: *Quantitative Aspects of Ruminant Digestion* (Eds. Dijkstra, J., Forbes, J. M. & France, J.). Wellingford, Oxfordshire: CABI Publishing.

Norberg, U. M. (1985). Evolution of vertebrate flight: an aerodynamic model for the transition from gliding to active flight. *American Naturalist* **126**, 303–327.

Norberg, U. M. (1990). *Vertebrate Flight: Mechanics, Physiology, Morphology, Ecology and Evolution*. Berlin: Springer-Verlag.

Norberg, U. M., Kunz, T., Steffensen, J. F., Winter, Y. & von Helversen, O. (1993). The cost of hovering and forward flight in a nectar-feeding bat, *Glossophaga soricina*, estimated from aerodynamic theory. *Journal of Experimental Biology* **182**, 207–227.

Norberg, U. M. & Norberg, R. A. (2012). Scaling of wingbeat frequency with body mass in bats and limits to maximum bat size. *Journal of Experimental Biology* **215**, 711–722.

Norberg, U. M. & Rayner, J. M. V. (1987). Ecological morphology and flight in bats (Mammalia; Chiroptera): wing adaptations, flight performance, foraging strategy and echolocation. *Philosophical Transactions of the Royal Society of London B* **316**, 335–427.

Nordøy, E. S. (1995). Do minke whales (*Balaenoptera acutorostrata*) digest wax esters? *British Journal of Nutrition* **74**, 717–722.

Noren, S. R., Kendall, T., Cuccurullo, V. & Williams, T. M. (2012). The dive response redefined: underwater behavior influences cardiac variability in freely diving dolphins. *Journal of Experimental Biology* **215**, 2735–2741.

Norris, D. O. (2007). *Vertebrate Endocrinology*. Amsterdam: Elsevier Academic Press.

Norris, K. S. (1964). Some problems of echolocation in cetaceans. In: *Marine Bio-acoustics* (Ed. Tavolga, W. N.). New York: Pergamon Press.

Novacek, M. J. (1992). Mammalian phylogeny: shaking the tree. *Nature* **356**, 121–125.

Novoa, F. F., Ruiz, G. & Rosenmann, M. (2002). El oxígeno y la vida en alta altitud: adaptaciones en vertebrados terrestres. In: *Fisiología Ecológica y Evolutiva* (Ed. Bozinovic, F.). Santiago, Chile: Ediciones Universidad Católica de Chile.

Nowacek, D. P., Johnson, M. P., Tyack, P. L., Shorter, K. A., McLellan, W. A. & Pabst, D. A. (2001). Buoyant balaenids: the ups and downs of buoyancy in right whales. *Proceedings of the Royal Society of London B* **268**, 1811–1816.

Nowack, J., Rojas, A. D., Körtner, G. & Geiser, F. (2015). Snoozing through the storm: torpor use during a natural disaster. *Scientific Reports* **5**, 11243.

Nowak, R. M. & Dickman, C. R. (2005). *Walker's Marsupials of the World*. Baltimore: Johns Hopkins University Press.

Nudds, T. D. (1978). Convergence of group size strategies by mammalian social carnivores. *American Naturalist* **112**, 957–960.

Nummela, S. (2008). Hearing in aquatic mammals. In: *Sensory Evolution on the Threshold. Adaptations in Secondarily Aquatic Vertebrates* (Eds. Thewissenn, J. G. M. & Nummela, S.). Berkeley, California: University of California Press.

Nunn, C. L. & Barton, R. A. (2000). Allometric slopes and independent contrasts: a comparative test of Kleiber's law in primate ranging patterns. *American Naturalist* **156**, 519–533.

Nunn, J. F. (1993). Oxygen. In: *Nunn's Applied Respiratory Physiology* (Ed. Nunn, J. F.). San Diego: Butterworth-Heinmann.

Nyakatura, J. A. & Andrada, E. (2013). A mechanical link model of two-toed sloths: no pendular mechanics during suspensory locomotion. *Acta Theriologica* **58**, 83–93.

Ochoa-Acuna, H. & Kunz, T. (1999). Thermoregulatory behavior in the small island flying fox *Pteropus hypomelanus* (Chiroptera: Phyllostomidae). *Journal of Thermal Biology* **24**, 15–20.

Ochocińska, D. & Taylor, J. R. E. (2003). Bergmann's rule in shrews: geographical variation of body size in Palearctic *Sorex* species. *Biological Journal of the Linnean Society* **78**, 365–381.

O'Connor, M. P., Kemp, S. J., Agosta, S. J., Hansen, F., Sieg, A. E., Wallace, B. P., McNair, J. N. & Dunham, A. E. (2007). Reconsidering the mechanistic basis of the metabolic theory of ecology. *Oikos* **116**, 1058–1072.

O'Connor, S. M., Dawson, T. J., Kram, R. & Donelan, J. M. (2014). The kangaroo's tail propels and powers pentapedal locomotion. *Biology Letters* **10**, 20140381.

Odum, E. P. & Golley, F. B. (1963). Radioactive tracers as an aid to the measurement of energy flow at the population level in nature. In: *Radioecology* (Eds. Schultz, V. & Kement, A. L.). Reinhold: New York.

Oelz, O., Howald, H., Di Prampero, P. E., Hoppeler, H., Claasen, H., Jenni, R., Bühlmann, A., Ferretti, G., Brückner, C., Veicsteinas, A., Gussoni, M. & Cerretelli, P. (1986). Physiological profile of world-class high-altitude climbers. *Journal of Applied Physiology* **60**, 1734–1742.

Oftedal, O. T. (2002). The mammary gland and its origin during synapsid evolution. *Journal of Mammary Gland Biology and Neoplasia* **7**, 225–252.

O'Grady, S. P., Valenzuela, L. O., Remien, C. H., Enright, L. E., Jorgensen, M. J., Kaplan, J. R., Wagner, J. D., Cerling, T. E. & Ehleringer, J. R. (2012). Hydrogen and oxygen isotope ratios in body water and hair: modeling isotope dynamics in nonhuman primates. *American Journal of Primatology* **74**, 651–660.

Ohmura, H., Hiraga, A., Matsui, A., Aida, H., Inoue, Y., Asai, A. & Jones, J. H. (2002). Physiological responses of young thoroughbreds during their first year of race training. *Equine Veterinary Journal Supplement* **34**, 140–146.

Okiea, J. G. & Brown, J. H. (2009). Niches, body sizes, and the disassembly of mammal communities on the Sunda Shelf islands. *Proceedings of the National Academy of Sciences, USA* **106**, 19679–19684.

Oksanen, M. & Siipi, H. (2014). *The Ethics of Animal Re-creation and Modification. Reviving, Rewilding, Restoring*. Basingstoke, Hampshire, UK: Palgrave Macmillan.

Olds, N. & Shoshani, J. (1982). *Procavia capensis*. *Mammalian Species* **171**, 1–7.

O'Leary, M. A., Bloch, J. I., Flynn, J. J., Gaudin, T. J., Giallombardo, A., Giannini, N. P., Goldberg, S. L., Kraatz, B. P., Luo, Z.-X., Meng, J., Ni, X., Novacek, M. J., Perini, F. A., Randall, Z. S., Rougier, G. W., Sargis, E. J., Silcox, M. T., Simmons, N. B., Spaulding, M., Velazco, P. M., Weksler, M., Wible, J. R. & Cirranello, A. L. (2013). The placental mammal ancestor and the post–K-Pg radiation of placentals. *Science* **339**, 662–667.

Oliver, J. A. (1951). 'Gliding' in amphibians and reptiles, with a remark on an arboreal adaptation in the lizard, *Anolis carolinensis* Voight. *American Naturalist* **85**, 171–176.

Olson, J. M., Dawson, W. R. & Camilliere, J. J. (1988). Fat from black-capped chickadees: avian adipose tissue? *Condor* **90**, 529–537.

Olsen, M. A., Blix, A. S., Utsi, T. H. A., Sørmo, W. & Mathiesen, S. D. (1999). Chitinolytic bacteria in the minke whale forestomach. *Canadian Journal of Microbiology* **46**, 85–94.

Olson, D. M., Dinerstein, E., Wikramanyake, E. D., Burgess, N. D., Powell, G. V. N., Underwood, E. C., D'Amico, J. A., Itoua, I., Strand, H. E., Morrison, J. C. Loucks, C. J., Allnutt, T. F., Ricketts, T. H., Kura, Y., Lamoreux, J. F., Wettengel, W. W., Heao, P. & Kassem, K. R. (2001). Terrestrial ecoregions of the world: a new world map of life on earth. *Bioscience* **51**, 933–938.

O'Neill, M. C. O. (2012). Gait-specific metabolic costs and preferred speeds in ring-tailed lemurs (*Lemur catta*), with implications for the scaling of locomotor costs. *American Journal of Physical Anthropology* **149**, 356–364.

O'Neill, P. O. (2013). Magnetoreception and baroreception in birds. *Development, Growth & Differentiation* **55**, 188–197

Opazo, J. C., Nespolo, R. F. & Bozinovic, F. (1999). Arousal from torpor in the Chilean mouse-opossum (*Thylamys elegans*): does non-shivering thermogenesis play a role? *Comparative Biochemistry and Physiology* **123**, 393–397.

Ortiz, R. M. (2001). Osmoregulation in marine mammals. *Journal of Experimental Biology* **204**, 1831–1844.

Ostrom, J. (1974). *Archaeopteryx* and the origin of flight. *Quarterly Review of Biology* **49**, 27–47.

Owen, D. (2003). *The Tragic Tale of the Tasmanian Tiger*. Crows Nest, New South Wales, Australia: Allen & Unwin.

Owerkowicz, T., Musinsky, C., Middleton, K. M. & Crompton, A. W. (2015). Respiratory turbinates and the evolution of endothermy in mammals and birds. In: *Great Transformations in Vertebrate Evolution* (Eds. Dial, K. P., Shubin, N. & Brainerd, E. L.). Chicago: University of Chicago Press.

Pácha, J. (2000). Development of intestinal transport function in mammals. *Physiological Reviews* **80**, 1633–1667.

Packard, G. C. (2012). Is non-loglinear allometry a statistical artefact? *Biological Journal of the Linnean Society* **107**, 764–773.

Packard, G. C. & Birchard, G. F. (2008). Traditional allometric analysis fails to provide a valid predictive model for mammalian metabolic rates. *Journal of Experimental Biology* **211**, 3581–3587.

Packard, G. C. & Boadman, T. J. (1988). The misuse of ratios, indices and percentages in ecological research. *Physiological Zoology* **61**, 1–9.

Packard, G. C. & Boardman, T. J. (2008). Model selection and logarithmic transformation in allometric analysis. *Physiological and Biochemical Zoology* **81**, 496–507.

Page, A. J., Cooper, C. E. & Withers, P. C. (2011). Effects of experiment start time and duration on measurement of standard physiological variables. *Journal of Comparative Physiology B* **181**, 657–665.

Pagel, M. D. (1992). A method for the analysis of comparative data. *Journal of Theoretical Biology* **156**, 431–442.

Paladino, F. V., O'Connor, M. P. & Spotila, J. R. (1990). Metabolism of leatherback turtles, gigantothermy, and thermoregulation of dinosaurs. *Nature* **344**, 858–860.

Pamplona, R., Barja, G. & Portero-Otin, M. (2006). Membrane fatty acid unsaturation, protection against oxidative stress, and maximum life span. *Annals of the New York Academy of Science* **959**, 475–490.

Pan, Y.-H., Zhang, Y., Cui, J., Liu, Y., McAllan, B. M., Liao, C.-C. & Zhang, S. (2013). Adaptation of phenylalanine and tyrosine catabolic pathway to hibernation in bats. *PLOS ONE* **8**, 1–14.

Panneton, W. M. (2013). The mammalian diving response: an enigmatic reflex to preserve life? *Physiology* **28**, 284–297.

Pappas, L. A. (2002). *Taurotragus oryx*. *Mammalian Species* **689**, 1–9.

Parish, O. O. & Putnam, T. W. (1977). Equations for the determination of humidity from dewpoint and psychrometric data. NASA Technical Note D-8401. Washington, DC: National Aeronautics and Space Administration.

Park, H. & Choi, H. (2010). Aerodynamic characteristics of flying fish in gliding flight. *Journal of Experimental Biology* **213**, 3269–3279.

Parker, D. I., Lawhead, B. E. & Cook, J. A. (1997). Distributional limits of bats in Alaska. *Arctic* **50**, 256–265.

Parker, K. L., Robbins, C. T. & Hanley, T. A. (1984). Energy expenditures for locomotion by mule deer and elk. *Journal of Wildlife Management* **48**, 474–488.

Parmesan, C. & Yohe, G. (2003). A globally coherent fingerprint of climate change impacts across natural systems. *Nature* **421**, 37–42.
Parsons, P. E. & Taylor, C. R. (1977). Energetics of brachiation versus walking: a comparison of a suspended and an inverted pendulum mechanism. *Physiological Zoology* **50**, 182–188.
Passmore, R. & Durnin, J. V. G. A. (1955). Human energy expenditure. *Physiological Reviews* **35**, 801–840.
Patterson, B. D. & Pascual, R. (1968). The fossil mammal fauna of South America. *Quarterly Review of Biology* **43**, 409–451.
Peichl, L., Chávez, A. E., Ocampo, A., Mena, W., Bozinovic, F. & Palacios, A. G. (2005). Eye and vision in the subterranean rodent cururo (*Spalacopus cyanus*, Octodontidae). *Journal of Comparative Neurology* **482**, 197–208.
Peichl, L., Künzle, H. & Vogel, P. (2000). Photoreceptor types and distributions in the retinae of insectivores. *Visual Neuroscience* **17**, 937–948.
Peichl, L., Němec, P. & Burda, H. (2004). Unusual cone and rod properties in subterranean African mole-rats (Rodentia, Bathyergidae). *European Journal of Neuroscience* **19**, 1545–1558.
Pengelley, E. T. & Fisher, K. C. (1963). The effect of temperature and photoperiod on the yearly hibernating behavior of captive golden-mantled ground squirrels (*Citellus lateralis tescorum*). *Canadian Journal of Zoology* **41**, 1103–1120.
Pennings, S. C., Nadeau, M. T. & Paul, V. J. (1993). Selectivity and growth of the generalist herbivore *Dalabella auricularia* feeding upon complementary resource. *Ecology* **74**, 879–890.
Penny, D. & Hasegawa, M. (1997). The platypus put in its place. *Nature* **387**, 549–550.
Pennycuick, C. J. (1988). On the reconstruction of pterosaurs and their manner of flight, with notes in vortex wakes. *Biological Review* **63**, 299–331.
Penry, D. L. & Jumars, P. A. (1987). Modeling animal guts as chemical reactors. *American Naturalist* **129**, 69–96.
Perk, K. (1963). The camel's erythrocytes. *Nature* **200**, 272–273.
Perk, K. (1966). Osmotic hemolysis of the camel's erythrocytes. I. A microcinematographic study. *Journal of Experimental Zoology* **163**, 241–246.
Perrin, W. F. (1963). *Stenella longirostris*. *Mammalian Species* **599**, 1–7.
Perry, R. W. (2013). A review of factors affecting cave climates for hibernating bats in temperate North America. *Environmental Reviews* **21**, 28–30.
Perry, S. F. & Duncker, H. R. (1980). Interrelationship of static mechanical factors and anatomical structure in lung evolution. *Journal of Comparative Physiology B* **138**, 321–334.
Perry, S. F. &. Duncker, H.-R. (1978). Lung architecture, volume and static mechanics in five species of lizards. *Respiration Physiology* **34**, 61–81.
Perry, S. F. & Sander, M. (2004). Reconstructing the evolution of the respiratory apparatus in tetrapods. *Respiration Physiology and Neurobiology* **144**, 125–139.
Perry, S. F., Similowski, T., Klein, W. & Codd, J. R. (2010). The evolutionary origin of the mammalian diaphragm. *Respiration Physiology and Neurobiology* **171**, 1–16.
Perttunen, V. & Erkkila, H. (1952). Humidity reaction in *Drosophila melanogaster*. *Nature* **169**, 78.
Peters, R. (1983). *The Ecological Implications of Body Size*. Cambridge: Cambridge University Press.
Peterson, B. J. & Fry, B. (1987). Stable isotopes in ecosystem studies. *Annual Review of Ecology and Systematics* **18**, 293–320.
Petrulis, A. (2011). Pheromones and reproduction in mammals. In: *Hormones and Reproduction of Vertebrates, **Volume 5 Mammals*** (Eds. Norris, D. O. & Lopez, K. H.). London: Elsevier.

Pettigrew, J. D., Jamieson, B. G. M., Robson, S. K., Hall, L. S., McNally, K. I. & Cooper, H. M. (1989). Phylogenetic relations between microbats, megabats and primates (Mammalia: Chiroptera, Primates). *Philosophical Transactions of the Royal Society of London B* **324**, 489–559.

Phillipson, J. (1981). Bioenergetic options and phylogeny. In: *Physiological Ecology: An Evolutionary Approach to Resource Use* (Eds. Townsend, C. R. & Calow, P.). Oxford: Blackwell Scientific.

Pianka, E. R. (1970). Comparative autecology of the lizard *Cnemidophorus tigris* in different parts of its geographic range. *Ecology* **51**, 703–720.

Piersma, T. & Drent, J. (2003). Phenotypic flexibility and the evolution of organismal design. *Trends in Ecology and Evolution* **18**, 228–233.

Pillow, J. J. & Jobe, A. H. (2008). Respiratory disorders of the newborn. In: *Pediatric Respiratory Medicine* (Eds. Taussig, L. M., Landau, L. I., Le Souëf, P. N., Martinez, F. D., Morgan, W. J. & Sly, P. D.). Philadelphia: Mosby Elsevier.

Pineda-Munoz, S. & Alroy, J. (2014). Dietary characterization of terrestrial mammals. *Proceedings of the Royal Society of London B* **281**, 20141173.

Piscitelli, M. A., McLellan. W. A., Rommel, S. A., Blum, J. E., Barco, S. G. & Pabst, D. A. (2010). Lung size and thoracic morphology in shallow and deep-diving cetaceans. *Journal of Morphology* 27, 654–673.

Piscitelli, M. A., Raverty, S. A., Lillie, M. A. & Shadwick, R. E. (2013). A review of cetacean lung morphology and mechanics. *Journal of Morphology* **274**, 1425–1440.

Pither, J. (2003). Climate tolerance and interspecific variation in geographic range size. *Proceedings of the Royal Society of London B* **270**, 475–481.

Pitts, R. F. (1963). *Physiology of the Kidney and Body Fluids*. Chicago: Year Book Medical Publishers.

Place, A. R. (1992). Comparative aspects of lipid digestion and absorption: physiological correlates of wax ester digestion. *American Journal of Physiology* **32**, R464–R471.

Pliscoff, P., Luebert, F., Hilger, H. H. & Guisan, A. (2014). Effects of alternative sets of climatic predictors on species distribution models and associated estimates of extinction risk: a test with plants in an arid environment. *Ecological Modelling* **288**, 166–177.

Pocock, G., Richards, C. D. & Richards, D. A. (2013). *Human Physiology*. Oxford: Oxford University Press.

Pogorzala, L. A., Mishra, S. K. & Hoon, M. A. (2013). The cellular code for mammalian thermosensation. *Journal of Neuroscience* **33**, 5533–5541.

Pohl, H. (1965). Temperature regulation and cold acclimation in the golden hamster. *Journal of Applied Physiology* **20**, 405–410.

Poiseuille, J. L. M. (1828). *Recherches sur la force du coeur aortique*. These No 166. Didot: Paris.

Polymeropoulos, E. T., Jastroch, M. & Frappell, P. B. (2012). Absence of adaptive nonshivering thermogenesis in a marsupial, the fat-tailed dunnart (*Sminthopsis crassicaudata*). *Journal of Comparative Physiology B* **182**, 393–401.

Pond, C. M., Mattacks, C. A., Colby, R. H. & Ramsay, M. A. (1992). The anatomy, chemical composition and metabolism of adipose tissue in wild polar bears (*Ursus maritimus*). *Canadian Journal of Zoology* **70**, 326–341.

Ponganis, P. J., Kooyman, G. L., Castellini, M. A., Ponganis, E. P. & Ponganis, K. V. (1993). Muscle temperature and swim velocity profiles during diving in a Weddell seal, *Leptonychotes weddellii*. *Journal of Experimental Biology* **183**, 341–348.

Ponganis, P. J., Meir, J. U. & Williams, C. L. (2011). In pursuit of Irving and Scholander: a review of oxygen store management in seals and penguins. *Journal of Experimental Biology* **214**, **3325–3339**.

Poole, W. E. (1982). *Macropus giganteus. Mammalian Species* **187**, 1–8.
Postma, M., Tordiffe, A. S. W., Hofmeyr, M. S., Reisinger, R. R., Bester, L. C., Buss, P. E. & De Bruyn, P. J. N. (2015). Terrestrial mammal three-dimensional photogrammetry: multispecies mass estimation. *Ecosphere* **6**, 1–16.
Pough, F. H., Andrews, R. M., Cadle, J. E., Crump, M. L., Savitsky, A. H. & Wells, K. D. (2004). *Herpetology*. Upper Saddle River, New Jersey: Prentice Hall.
Power, M. L. (1996). The other side of callitrichine gummivory. Digestibility and nutritional value. In: *Adaptive Radiations of Neotropical Primates* (Eds. Norconk, M. A., Rosenberger, A. L. & Garber, P. A.). New York: Plenum Press.
Power, M. L. (2010). Nutritional and digestive challenges to being a gum-feeding primate. In: *The Evolution of Exudativory in Primates* (Eds. Burrows, A. M. & Nash, L. T.). New York: Springer-Verlag.
Price, E. R., Brun, A., Caviedes-Vidal, E. & Karasov, W. H. (2015a). Digestive adaptations of aerial lifestyles. *Physiologist* **30**, 69–78.
Price, E. R., Brun, A., Fasulo, V., Karasov, W. H. & Caviedes-Vidal, E. (2013). Intestinal perfusion indicates high reliance on paracellular nutrient absorption in an insectivorous bat *Tadarida brasiliensis. Comparative Biochemistry and Physiology A* **164**, 351–355.
Price, E. R., Brun, A., Gontero-Fourcade, M., Fernández-Marinone, G., Cruz-Neto, A. P., Karasov, W. H. & Caviedes-Vidal, E. (2015b). Intestinal water absorption varies with expected dietary water load among bats but does not drive paracellular nutrient absorption. *Physiological and Biochemical Zoology* **88**, 680–684.
Priestley, C. H. B. (1957). The heat balance of sheep standing in the sun. *Australian Journal of Agricultural Research* **8**, 271–280.
Promislow, D. E. L. (1991). The evolution of mammalian blood parameters: patterns and their interpretation. *Physiological Zoology* **64**, 393–431.
Promislow, D. E. L. (1993). On size and survival: progress and pitfalls in the allometry of life span. *Journal of Gerontology: Biological Sciences* **48**, B115–B123.
Promislow, D. E. L. & Harvey, P. H. (1990). Living fast and dying young: a comparative analysis of life history variation among mammals. *Journal of Zoology* **220**, 417–437.
Proske, U., Gregory, J. E. & Iggo, A. (1998). Sensory receptors in monotremes. *Philosophical Transactions of the Royal Society of London B* **353**, 1187–1198.
Puchalski, W., Bockler, H., Heldmaier, G. & Langefeld, M. (1987). Organ blood flow and brown adipose tissue oxygen consumption during noradrenaline-induced thermogenesis in the Djungarian hamster. *Journal of Experimental Zoology* **242**, 263–271.
Pumo, D. E., Finamore, P. S., Franek, W. R., Phillips, C. J., Tarzami, S. & Balzarano, D. (1998). Complete mitochondrial genome of a neotropical fruit bat, *Artibeus jamaicensis*, and a new hypothesis of relationship of bats to other eutherian mammals. *Journal of Molecular Evolution* **47**, 709–717.
Purvis, A. & Harvey, P. H. (1995). Mammal life-history evolution: a comparative test of Charnov's model. *Journal of Zoology* **237**, 259–283.
Qasem, L., Cardew, A., Wilson, A., Griffiths, I., Halsey, L. G., Shepard, E. L. C., Gleiss, A. C. & Wilson, R. (2012). Tri-axial dynamic acceleration as a proxy for animal energy expenditure; should we be summing values or calculating the vector? *PLOS ONE* 7, e31187.
Qin, S., Yin, H., Yang, C., Dou, Y., Liu, Z., Zhang, P., Yu, H., Huang, Y., Feng, J., Hao, J., Hao, J., Deng, L., Yan, X., Dong, X., Zhao, Z., Jiang, T., Wang, H.-W., Luo, S.-J. & Xie, C. (2015). A magnetic protein biocompass. *Nature Materials*. doi:10.1038/NMAT4484.
Rahn, H. (1974). Body temperature and acid-base regulation. *Pneumonologie* **151**, 87–94.
Rahn, H., Reeves, R. B. & Howell, B. J. (1975). Hydrogen ion regulation, temperature and evolution. *Annual Reviews of Respiratory Disease* **112**, 165–172.

Raichlen, D. A., Gordon, A. D., Muchlinski, M. N. & Snodgrass, J. J. (2010). Causes and significance of variation in mammalian basal metabolism. *Journal of Comparative Physiology B* **180**, 301–311.

Raimbault, S., Dridi, S., Denjean, F., Lachuer, J., Couplan, E., Bouillaud, F., Bordas, A., Duchamp, C., Taouis, M. & Ricquier, D. (2001). An uncoupling protein homologue putatively involved in facultative muscle thermogenesis in birds. *Biochemical Journal* **353**, 441–444.

Rajchard, J. (2013). Kairomones—important substances in interspecific communication in vertebrates: a review. *Veterinarni Medicina* **58**: 561–566.

Rand, R. P., Burton, A. C. & Ing, T. (1965). The tail of the rat, in temperature regulation and acclimatization. *Canadian Journal of Physiology and Pharmacology* **43**, 257–267.

Rapoport, E. H. (1982). *Areography. Geographical Strategies of Species.* (Translated by B. Drausal). Oxford: Pergamon Press.

Rasmussen, K., Palacios, D. M., Calambokidis, J., Saborío, M. T., Rosa, L. D., Secchi, E. R., Steiger, G. H., Allen, J. M. & Stone, G. S. (2007). Southern Hemisphere humpback whales wintering off Central America: insights from water temperature into the longest mammalian migration. *Biology Letters* **3**, 302–305.

Rayner, J. M. V. (1981). Flight adaptations in vertebrates. *Symposium of the Zoological Society of London* **48**, 137–172.

Rayner, J. M. V. (1999). Estimating power curves of flying vertebrates. *Journal of Experimental Biology* **202**, 3449–3461.

R Development Core Team (2013). *R: A Language and Environment for Statistical Computing.* Vienna, Austria: R Foundation for Statistical Computing.

Read, A. F. & Harvey, P. H. (1989). Life history differences among the eutherian radiations. *Journal of Zoology* **219**, 329–353.

Reardon, F. D., Leppik, K. E., Wegmann, R., Webb, P., Ducharme, M. B. & Kenny, G. P. (2006). The Snellen human calorimeter revisited, re-engineered and upgraded: design and performance characteristics. *Medical and Biological Engineering and Computing* **44**, 721–8.

Redford, K. H. (1987). Ants and termites as food: patterns of mammalian myrmecophagy. In: *Current Mammalogy* **Volume 1** (Ed. Genoways, H. H.). New York: Plenum Press.

Redford, K. H. & Dorea, J. G. (1984). The nutritional value of invertebrates with emphasis on ants and termites as food for mammals. *Journal of Zoology* **203**, 385–395.

Reichard, J. D., Fellows, S. R., Frank, A. J. & Kunz, T. H. (2010b). Thermoregulation during flight: body temperature and sensible heat transfer in free-ranging Brazilian free-tailed bats (*Tadarida brasiliensis*). *Physiological and Biochemical Zoology* **83**, 885–897.

Reichard, J. D., Prajapati, S. I., Austad, S. N., Keller, C. & Kunz, T. H. (2010). Thermal windows on Brazilian free-tailed bats facilitate thermoregulation during prolonged flight. *Integrative Comparative Biology* **50**, 358–370.

Renfree, M. B., Suzuki, S. & Kaneko-Ishino, T. (2013). The origin and evolution of genomic imprinting and viviparity in mammals. *Philosophical Transactions of the Royal Society of London B* **368**. doi:10.1098/rstb.2012.0151.

Rensch, B. (1938). Some problems of geographical variation and species formation. *Proceedings of the Linnean Society of London* **150**, 275–285.

Renvoisé, E. & Michon, F. (2014). An evo-devo perspective on ever-growing teeth in mammals and dental stem cell maintenance. *Frontiers in Physiology* 5, 1–12.

Research Australia (2011). *Shaping Up: Trends and Statistics in Funding Health and Medical Research*, Occasional Paper Series 2. Melbourne: Research Australia.

Reusch, T. B. & Wood, T. E. (2007). Molecular ecology of global change. *Molecular Ecology* **16**, 3973–3992.

Revel, F. G., Herwig, A., Garidou, M., Dardente, H., Menet, J. S., Mason-Pevet, M., Simonneaux, V., Saboureau, M. & dn Pevet, P. (2007). The circadian clock stops ticking during deep hibernation in the European hamster. *Proceedings of the National Academy of Sciences, USA* **104**, 13816–13820.

Rewcastle, S. C. (1981). Stance and gait in tetrapods: an evolutionary scenario. *Symposium of the Zoological Society of London* **48**, 239–267.

Reynolds, P. S. (1997). Phylogenetic analysis of surface areas of mammals. *Journal of Mammalogy.* **78**, 859–868.

Rezende, E. L., Bozinovic, F. & Garland, T. (2004). Climatic adaptation and the evolution of basal and maximum rates of metabolism in rodents. *Evolution* **58**, 1361–1374.

Rezende, E. L., Cortés, A., Bacigalupe, L. D., Nespolo, R. F. & Bozinovic, F. (2003). Ambient temperature limits above-ground activity of the subterranean rodent *Spalacopus cyanus. Journal of Arid Environmants* **55**, 63–74.

Rezende, E. L., Gomes, F. R., Ghalambor, C. K., Russell, G. A. & Chappell, M. A. (2005). An evolutionary frame of work to study physiological adaptation to high altitudes. *Revista Chilena de Historia Natural* **78**, 323–336.

Rial, E., Poustie, A. & Nicholls, D. G. (1983). Brown-adipose-tissue mitochondria: the regulation of the 32 000-Mr uncoupling protein by fatty acids and purine nucleotides. *European Journal of Biochemistry* **137**, 197–203.

Richardson, T. Q. & Guyton, A. C. (1959). Effects of polycythemia and anemia on cardiac output and other circulatory factors. *American Journal of Physiology* **197**, 1167–1170.

Richet, C. (1889). *La chaleur animale*. Paris: Felix Alcan.

Richmond, C. R., Landgam, W. H. & Trujillo, T. T. (1962). Comparative metabolism of tritiated water by mammals. *Journal of Cellular and Comparative Physiology* **59**, 45–53.

Ricklefs, R. E. (1996). Morphometry of the digestive tracts of some passerine birds. *Condor* **98**, 279–292.

Riek, A. (2008). Relationship between field metabolic rate and body weight in mammals: effect of the study. *Journal of Zoology* **276**, 187–194.

Riek, A. & Bruggeman, J. (2013). Estimating field metabolic rates for Australian marsupials using phylogeny. *Comparative Biochemistry and Physiology A* **164**, 598–604.

Riek, A. & Geiser, F. (2012). Developmental plasticity in a marsupial. *Journal of Experimental Biology* **215**, 1552–1558.

Riek, A. & Geiser, F. (2013). Allometry of thermal variables in mammals: consequences of body size and phylogeny. *Biological Reviews* **88**, 564–572.

Riek, A. & Geiser, F. (2014). Heterothermy in pouched mammals–a review. *Journal of Zoology* **292**, 74–85.

Rismiller, P. D. & McKelvey, M. W. (2003). Body mass, age and sexual maturity in short-beaked echidnas, *Tachyglossus aculeatus. Comparative Biochemistry and Physiology A* **136**, 851–865.

Robards, M. D. & Reeves, R. R. (2011). The global extent and character of marine mammal consumption by humans. *Biological Conservation* **144**, 2770–2786.

Robbins, C. T. (1993). *Wildlife Feeding and Nutrition*. San Diego: Academic Press.

Robbins, C. T., Lopez-Alfaro, C., Rode, K. D., Tøien, O. & Nelson, L. (2012). Hibernation and seasonal fasting in bears: the energetic costs and consequences for polar bears. *Journal of Mammalogy* **93**, 1493–1503.

Robert, K. A. & Thompson, M. B. (2003). Reconstructing Thermochron ibuttons to reduce size and weight as a new technique in the study of small animal thermal biology. *Herpetological Review* **34**, 130–132.

Roberts, D. L. & Kitchener, A. C. (2006). Inferring extinction from biological records: were we too quick to write off Miss Waldron's Red Colobus monkey (*Piliocolobus badius waldronae*)? *Biological Conservation* **128**, 285–287.

Robertshaw, D. (1975). Catecholamines and control of sweat glands. In: *Handbook of Physiology. Section 7: Endocrinology, Volume VI* (Ed. Geiger, S. R.). Washington, DC: American Physiological Society.

Robertshaw, D. (2006). Mechanisms for the control of respiratory evaporative heat loss in panting animals. *Journal of Applied Physiology* **101**, 664–668.

Rodriguez-Duran, A. & Padilla-Rodriguez, E. (2008). Blood characteristics, heart mass, and wing morphology of Antillean bats. *Caribbean Journal of Science* **44**, 375–379.

Rodriguez-Serrano, E. & Bozinovic, F. (2009). Interplay between global patterns of environmental temperature and variation in non-shivering thermogenesis of rodent species across large spatial scales. *Global Change Biology* **15**, 2116–2122.

Rohlf, F. J. (2001). Comparative methods for the analysis of continuous variables: geometric interpretations. *Evolution* **55**, 2143–2160.

Rojas, A. D., Köertner, G. & Geiser, F. (2010). Do implanted transmitters affect maximum running speed of two small marsupials? *Journal of Mammalogy* **91**, 1360–1364.

Rolfe, D. F. & Brown, G. C. (1997). Cellular energy utilization and molecular origin of standard metabolic rate in mammals. *Physiological Reviews* 77, 731–758.

Romanovsky, A. A. (2007). Thermoregulation: some concepts have changed. Functional architecture of the thermoregulatory system. *American Journal of Physiology* **292**, R37–R46.

Rommel, S. A., Pabst, D. A., McLellan, W. A., Williams, T. M. & Friedl, W. A. (1994). Temperature regulation of the testes of the bottlenose dolphin (*Tursiops truncatus*): evidence from colonic temperatures. *Journal of Comparative Physiology B* **164**, 130–134.

Root, T. L. & Hughes, L. (2005). Present and future phenological changes in wild plants and animals. In: *Climate Change and Biodiversity* (Eds. Lovejoy, T. E. & Hanah, L.). London: Yale Press.

Roquet, F., Williams, R., Hindell, M. A., Harcourt, R., McMahon, C., Guinet, C., Charrassin, J.-B., Reverdin, G., Boehme, L., Lovell, P. & Fedak, M. (2014). A Southern Indian Ocean database of hydrographic profiles obtained with instrumented elephant seals. *Scientific Data* **1**, 140028. doi:10.1038/sdata.2014.28.

Rose, K. D. (2006). *The Beginning of the Age of Mammals*. Baltimore: Johns Hopkins University Press.

Rose, R. W., West, A. K., Ye, J. M., McCormack, G. H. & Colquhoun, E. Q. (1999). Nonshivering thermogenesis in a marsupial (the Tasmanian bettong *Bettongia gaimardi*) is not attributable to brown adipose tissue. *Physiological and Biochemical Zoology* **72**, 699–704.

Rosen, D. A. S. & Trites, A. W. (1997). Heat increment of feeding in Steller sea lions, *Eumetopias jubatus*. *Comparative Biochemistry and Physiology A* **118**, 877–881.

Rosen, D. A. S. & Trites, A. W. (2003). No evidence for bioenergetics interaction between digestion and thermoregulation in Stellar sea lions *Eumetopias jubatus*. *Physiological and Biochemical Zoology* **76**, 899–906.

Rosenmann, M. (1987). La presión crítica de oxígeno. *Archivos de Biología y Medicina Experimental* **20**, 75–78.

Rosenmann, M. & Morrison, P. (1974a). Maximum oxygen consumption and heat loss facilitation in small homeotherms by He-$O_2$. *American Journal of Physiology* **226**, 490–496.

Rosenmann, M. & Morrison, P. R. (1974b). Physiological responses to hypoxia in the tundra vole. *American Journal of Physiology* **227**, 734–739.

Rosenmann, M. & Morrison, P. R. (1975). Metabolic response of highland and lowland rodents to simulated high altitudes and cold. *Comparative Biochemistry and Physiology A* **51**, 523–530.

Rosenmann, M., Morrison, P. & Feist, D. (1975). Seasonal changes in the metabolic capacity of red-backed voles. *Physiological Zoology* **48**, 303–310.

Rosenzweig, C., Karoly, D., Vicarelli, M., Neofotis, P., Wu, Q. G., Casassa, G., Menzel, A., Root, T. L., Estrella, N., Seguin, B., Tryjanowski, P., Liu, C. Z., Rawlins, S. & Imeson, A. (2008). Attributing physical and biological impacts to anthropogenic climate change. *Nature* **453**, 353–358.

Rosenzweig, M. L. (1968). The strategy of body size in mammalian carnivores. *American Midland Naturalist* **80**, 299–315.

Roth, V. L. (1990). Insular dwarf elephants, a case study in body mass estimation and ecological inference. In: *Body Size in Mammalian Paleobiology* (Eds. Damuth, J. & MacFadden, B. J.). Cambridge: Cambridge University Press.

Rothwell, N. J. & Stock, M. J. (1985). Biological distribution and significance of brown adipose tissue. *Comparative Biochemistry and Physiology A* **82**, 745–751.

Rougier, G. W., Ji, Q. & Novacek, M. J. (2003). A new symmetrodont mammal with fur impressions from the Mesozoic of China. *Acta Geologica Sinica* 77, 7–14.

Rowe, T. (1988). Definition, diagnosis, and the origin of Mammalia. *Journal of Paleontology* **8**, 241–264.

Rowe, T. (1993). Phylogenetic systematics and the early history of mammals. In: *Mammal Phylogeny, **Volume 1*** (Eds. Szalay, F. S., Novacek, M. J. &. McKenna, M. C.). New York: Springer-Verlag.

Rowe, T., Rich, T. H., Vickers-Rich, P., Springer, M. &. Woodburne, M. O. (2008). The oldest platypus and its bearing on divergence timing of the platypus and echidna clades. *Proceedings of the National Academy of Sciences, USA* **105**, 1238–1242.

Rowe, T. B., Macrini, T. E. & Luo, Z.-H. (2011). Fossil evidence on origin of the mammalian brain. *Science* **332**, 955–957.

Rubner, M. (1883). Über den einfluss der körpergrösse auf stoff- und kraftwechsel. *Zeitschrift für Biologie* **19**, 536–562.

Rubner, M. (1902). *Die Gesetze des Energieverbrauchs bei der Ernährung*. Lepizig: Franz Deuticke.

Ruf, T., Bieber, C. & Turbill, C. (2012). Survival, aging and life-history tactics in mammalian hibernators. In: *Living in a Seasonal World* (Eds. Ruf, T., Bieber, C., Arnold, W. & Millesi, E.). Berlin: Springer-Verlag.

Ruf, T. & Geiser, F. (2015). Daily torpor and hibernation in birds and mammals. *Biological Reviews* **90**, 891–926.

Ruiz, G., Rosenmann, M. & Cortes, A. (2005). Thermal acclimation and seasonal variations of erythrocyte size in the Andean mouse *Phyllotis xanthopygus rupestris*. *Comparative Biochemistry and Physiology A* **139**, 405–409.

Runciman, S. I. C., Gannon, B. J. & Baudinette, R. V. (1995). Central cardiovascular shunts in the perinatal marsupial. *Anatomical Record* **243**, 71–83.

Rupley, J. A. & Careri, G. (1991). Protein hydration and function. *Advances in Protein Chemistry* **41**, 37–172.

Russell, E. M. (1982). Patterns of parental care and parental investment in marsupials. *Biological Reviews* 57, 423–486.

Russell, E. M. (1985). The metatherians: order Marsupialia. In: *Social Odours in Mammals* (Eds. Brown, R. E. & MacDonald, D. A.). Oxford: Clarendon Press.

Russell, G. A. & Chappell, M. A. (2007). Is BMR repeatable in deer mice? Organ mass correlates and the effects of cold acclimation and natal altitude. *Journal of Comparative Physiology B.* 177, 75–87.

Russell, J., Vidal-Gadea, A. G., Makay, A., Lanam, C. & Pierce-Shimomura, J. T. (2014). Humidity sensation requires both mechanosensory and thermosensory pathways in *Caenorhabditis elegans*. *Proceedings of the National Academy of Sciences, USA* **111**, 8269–8274.

Russell, J. B. & Strobel, H. J. (2005). Microbial energetics. In: *Quantitative Aspects of Ruminant Digestion* (Eds. Dijkstra, J., Forbes, J. M. & France, J.). Wellingford, Oxfordshire: CABI Publishing.

Ruxton, G. (2014). Avian-style respiration allowed gigantism in pterosaurs. *Journal of Experimental Biology* **217**, 2627–2628.

Rydell, J. & Speakman, J. R. (1995). Evolution of nocturnality in bats: potential competitors and predators during their early history. *Biological Journal of the Linnean Society* **54**, 183–191.

Saarinen, J. J., Boyer, A. G., Brown, J. H., Costa, D. P., Morgan Ernest, S. K., Evans, A. R., Fortelius, M., Gittleman, J. K., Hamilton, M. J., Harding, L. E., Lintulaakso, K., Lyons. S. K., Okie, J. G., Sibly, R. M., Stephens, P. R., Theodor, J., Uhen, M. D. & Smith, F. A. (2015). Patterns of maximum body size evolution in Cenozoic land mammals: eco-evolutionary processes and abiotic forcing. *Proceedings of the Royal Society of London B* **281**, 20132049.

Sabogal, C. & Talmaciu, I. (2005). Bronchopulmonary dysplasia (chronic lung disease on infancy). In: *Pediatric Pulmonology* (Ed. Panitch, H. B.). Philadelphia: Elsevier Mosby.

Sacher, G. A. (1959). Relationship of lifespan to brain weight and body weight in mammals. In: *CIBA Foundation Symposium on the Lifespan of Animals* (Eds. Wolstenholme, G. E. W. & O'Connor, M.). Boston: Little, Brown & Co.

Sadowska, E. T., Baliga-Klimczyk, K., Labocha, M. K. & Koteja, P. (2009). Genetic correlations in a wild rodent: grass-eaters and fast-growers evolve high basal metabolic rates. *Evolution* **63**, 1530–1539.

Sadowska, E. T., Stawski, C., Rudolf, A., Dheyongera, G., Chrząścik, K. M., Baliga-Klimczyk, K. & Koteja, P. (2015). Evolution of basal metabolic rate in bank voles from a multidirectional selection experiment. *Proceedings of the Royal Society of London B* **282**. doi:10.1098/rspb.2015.0025.

Saladin, K. S. (2010). *Anatomy and Physiology*. New York: McGraw-Hill.

Sallman, T., Beckman, A. L., Stanton, T. L., Eriksson, K. S., Tarhanen, J., Tuomisto, L. & Panula, P. (1999). Major changes in the brain histamine system of the ground squirrel *Citellus lateralis* during hibernation. *Journal of Neuroscience* **19**, 1824–1835.

Sander, P. M., Christian, A., Clauss, M., Fechner, R., Gee, C. T., Griebeler, E.-M., Gunga, H.-C., Hummel, J., Mallison, Perry, H. S. F., Preuschoft, H., Rauhut, O. W. M., Remes, K., Tütken, T., Wings, O. & Witzel, U. (2011). Biology of the sauropod dinosaurs: the evolution of gigantism. *Biological Reviews* **86**, 117–155.

Sands, J. M. (1999). Regulation of renal urea transporters. *Journal of the American Society of Nephrology* **10**, 635–646.

Sands, J. M. (2007). Critical role of urea in the urine-concentrating mechanism. *Journal of the American Society of Nephrology* **18**, 670–671.

Sands, J. M., Blount, M. A. & Klein, J. D. (2011). Regulation of renal urea transport by vasopressin. *Transactions of the American Clinical and Climatological Association* **122**, 82–92.

Sanyal, S., Jansen, H. G., DeGrip, W. J., Nevo, E. & de Jong, W. W. (1990). The eye of the blind mole rat, *Spalax ehrenbergi*. Rudiment with hidden function? *Investigative Ophthalmology and Vision Science* **31**, 1398–1404.

Sapir, N., Wikelski, M. McCue, M. D., Pinshow, B. & Nathan, R. (2010). Flight modes in migrating European bee-eaters: heart rate may indicate low metabolic rate during soaring and gliding. *PLOS ONE* **5**, e13956.

Sarazan, R. D. & Schweitz, K. T. R. (2009). Standing on the shoulders of giants: Dean Franklin and his remarkable contributions to physiological measurements in animals. *Advances in Physiological Education* **33**, 144–156.

Sarre, S. D., Georges, A. & Quinn, A. (2004). The ends of a continuum: genetic and temperature-dependant sex determination in reptiles. *BioEssays* **26**, 639–645.

Sarrus, F. & Rameaux, J. F. (1839). Application des sciences accessoires et principalement des mathématiques à la physiologie générale. *Bulletin de l'Académie Royale de Médecine* 3, 1094–1100.
Saunders, J. C., Chen, C. S. & Pridmore, P. A. (1971a). Successive habit reversal learning in a monotreme, *Tachyglossus aculeatus*. *Animal Behaviour* **19**, 552–555.
Saunders, J. C., Teague, J., Slonim, D. & Pridmore, P. A. (1971b). A position habit in the monotreme *Tachyglossus aculeatus* (the spiny echidna). *Australian Journal of Psychology* **23**, 47–51.
Saunders, W. B., Harden, L. M., Kent, S., Pittman, Q. J. & Roth, J. (2015). Fever and sickness behavior: friend or foe? *Brain, Behavior, and Immunity* **50**, 322–333.
Savage, V. M., Gillooly, J. F., Brown, A. C., West, G. B. & Charnov, E. L. (2004b). Effects of body size and temperature on population growth. *American Naturalist* **163**, 429–441.
Savage, V. M., Gillooly, J. F., Woodruff, W. H., West, G. B., Allen, A. P., Enquist, B. J. & Brown, A. C. (2004a). The predominance of quarter-power scaling in biology. *Functional Ecology* **18**, 257–282.
Sbarbati, A. & Osculati, F. (2006). Allelochemical communication in vertebrates: kairomones, allomones and synomones. *Cells, Tissues, Organs* **183**, 206–219.
Scantlebury, D. M., Mills, M. G. L., Wilson, R. P., Wilson, J. W., Mills, M. E. J., Durant, S. M., Bennett, N. C., Bradford, P., Marks, N. J. & Speakman, J. R. (2014). Flexible energetics of cheetah hunting strategies provide resistance against kleptoparasitism. *Science* **346**, 79–81.
Schaefer, C. D. & Staples, J. F. (2006). Mitochondrial metabolism in mammalian cold-acclimation: magnitude and mechanisms of fatty-acid uncoupling. *Journal of Thermal Biology* **31**, 355–361.
Schalk, G. & Brigham, R. M. (1995). Prey selection by insectivorous bats: are essential fatty acids important? *Canadian Journal of Zoology* 73, 1855–1895.
Scheibe, J. S., Smith, W. P., Bassham, J. & Magness, D. (2006). Locomotor performance and cost of transport in the northern flying squirrel *Glaucomys sabrinus*. *Acta Theriologica* **51**, 169–178.
Scheich, H., Langner, G., Tidemann, C., Coles, R. B. & Guppy, A. (1986). Electrorecepetion and electrolocation in the platypus. *Nature* **319**, 401–402.
Schinnerl, M., Aydinonat, D., Schwarzenberger, F. & Voight, C. C. (2011). Hematological survey of common neotropical bat species from Costa Rica. *Journal of Zoo and Wildlife Medicine* 42, 382–391.
Schippers, M.-P., Ramirez, O., Arana, M., Pinedo-Bernal, P. & McClelland, G. B. (2012). Increase in carbohydrate utilization in high-altitude Andean mice. *Current Biology* **22**, 2350–2354.
Schitteck, B., Hipfel, R., Sauer, B., Bauer, J., Kalbacher, H., Stevanovic, S., Schirle, M., Schroeder, K., Blin, N., Meier, F., Rassner, G. & Garbe, C. (2001). Dermcidin: a novel human antibiotic peptide secreted by sweat glands. *Nature Immunology* **2**, 1133–1137.
Schlichting, C. D. & Pigliucci, M. (1998). *Phenotypic Evolution. A Reaction Norm Perspective.* Sunderland, Massachusetts: Sinauer Associates.
Schmid, J., Andersen, N. A., Speakman, J. R. & Nicol, S. C. (2003). Field energetics of free-living, lactating and non-lactating echidnas (*Tachyglossus aculeatus*). *Comparative Biochemistry and Physiology A* **136**, 903–909.
Schmid, W. D. (1976). Temperature gradients in the nasal passage of some small mammals. *Comparative Biochemistry and Physiology A* **54**, 305–308.
Schmidt, J., Andersen, N. A., Speakman, J. R. & Nicol. S. C. (2003). Field energetics of free-living, lactating and non-lactating echidnas (*Tachyglossus aculeatus*). *Comparative Biochemistry and Physiology A* **136**, 903–909.
Schmidt-Nielsen, B. & O'Dell, R. (1961). Structure and concentrating mechanism in the mammalian kidney. *American Journal of Physiology* **200**, 1119–1124.

Schmidt-Nielsen, K. (1964). *Desert Animals: Physiological Problems of Heat and Water*. Oxford: Oxford University Press.

Schmidt-Nielsen, K. (1975). Scaling in biology: the consequence of size. *Journal of Experimental Zoology* **194**, 287–308.

Schmidt-Nielsen, K. (1984). *Scaling: Why is Animal Size so Important?* Cambridge: Cambridge University Press.

Schmidt-Nielsen, K. (1997). *Animal Physiology. Adaptation and Environment*. Cambridge: Cambridge University Press.

Schmidt-Nielsen, K., Bretz, W. L. & Taylor, C. R. (1970b). Panting in dogs: unidirectional air flow over evaporative surfaces. *Science* **169**, 1102–1104.

Schmidt-Nielsen, K., Hainsworth, F. R. & Murrish, D. E. (1970a). Countercurrent heat exchange in the respiratory passages: effect on water and heat balance. *Respiration Physiology* **9**, 263–276.

Schmidt-Nielsen, K. & Schmidt-Nielsen, B. (1952). Water metabolism of desert mammals. *Physiological Reviews* **32**, 135–166.

Schmidt-Nielsen, K., Schmidt-Nielsen, B., Jarnum, S. A. & Houpt, T. R. (1957). Body temperature of the camel and its relation to water economy. *American Journal of Physiology* **188**, 103–112.

Schmidt-Nielsen, K., Schroter, R. C. & Shkolnik, A. (1980). Desaturation of the exhaled air in the camel. *Journal of Physiology* **305**, 74P–75P.

Schmidt-Nielsen, K., Schroter, R. C. & Shkolnik, A. (1981). De-saturation of exhaled air in camels. *Proceedings of the Royal Society of London B* **211**, 305–319.

Schmitt, D., Cartmill, M., Griffin, T. M., Hanna, J. B. & Lemelin, P. (2006). Adaptive value of ambling gaits in primates and other mammals. *Journal of Experimental Biology* **209**, 2042–2049.

Scholander, P. F. (1940). Experimental investigations on the respiratory function in diving mammals and birds. *Hvalraadets Skrifter* **22**. Oslo: Det h'orske Videnskaps Akademi I Oslo.

Scholander, P. F. (1957). The wonderful net. *Scientific American* **196**, 96–107.

Scholander, P. F. (1963). The master switch of life. *Scientific American* **209**, 92–106.

Scholander, P. F., Hock, R., Walters, V., Johnson, F. & Irving, L. (1950a). Heat regulation in some Arctic and tropical mammals and birds. *Biological Bulletin* **99**, 237–258.

Scholander, P. F. & Schevill, W. E. (1955). Counter-current vascular heat exchange in the fins of whales. *Journal of Applied Physiology* **8**, 279–282.

Scholander, P. F., Walters, V., Hock, R. & Irving, L. (1950b). Body insulation of some Arctic and tropical mammals and birds. *Biological Bulletin* **99**, 225–236.

Schönbaum, E. & Lomax, P. (1990). *International Encyclopedia of Pharmacology and Therapeutics. Section 131. Thermoregulation, Physiology and Biochemistry*. London: Pergamon Press.

Schondube, J. E., Herrera-M, L. G. & Martinez del Rio, C. (2001). Diet and the evolution of digestion and renal function in phyllostomid bats. *Zoology* **104**, 59–73.

Schorr, G. S., Falcone, E. A., Moretti, D. J. & Russel, D. (2014). First long-term behavioral records from Cuvier's beaked whales (*Ziphius cavirostris*) reveal record-breaking dives. *PLOS ONE* **9**, e92633.

Schröder, H. J. & Power, G. G. (1997). Engine and radiator: fetal and placental interactions for heat dissipation. *Experimental Physiology* **82**, 403–414.

Schroter, R. C., Robertshaw, D. & Zine Filali, R. (1989). Brain cooling and respiratory heat exchange in camels during rest and exercise. *Respiration Physiology* **78**, 95–105.

Schroter, R. C. & Watkins, N. V. (1989). Respiratory heat exchange in mammals. *Respiration Physiology* **78**, 357–368.

Schulte-Hostedde, A. I., Millar, J. S. & Hickling, G. J. (2001). Evaluating body condition in small mammals. *Canadian Journal of Zoology* **79**, 1021–1029.

Sealander, J. A. (1964). The influence of body size, season, sex, age and other factors upon some blood parameters in small mammals. *Journal of Mammalogy* **45**, 598–616.
Sears, K. E., Behringer, R. R., Raswiler, J. J. & Niswander, L. A. (2006). Development of bat flight: morphologic and molecular evolution of bat wing digits. *Proceedings of the National Academy of Sciences, USA* **103**, 6581–6586.
Secor, S. M. (2009). Specific dynamic action: a review of the postprandial metabolic response. *Journal of Comparative Physiology B* **179**, 1–56.
Seebacher, F. (2003). Dinosaur body temperatures: the occurrence of endothermy and ectothermy. *Paleobiology* **28**, 105–122.
Seebeck, J. H. (2001). *Perameles gunni. Mammalian Species* **654**, 1–8.
Selye, H. (1956). *The Stress of Life*. Toronto & New York: McGraw-Hill.
Sernetz, M., Gelleri, B. & Hoffman, J. (1985). The organism as bioreactor: interpretation of the reduction law of metabolism in terms of heterogeneous catalysis and fractal structure. *Journal of Theoretical Biology* **117**, 209–230.
Serrat, M. A. (2013). Allen's rule revisited: temperature influences bone elongation during a critical period of postnatal development. *Anatomical Record* **296**, 1534–1545.
Setchell, B. P. (1977). Reproduction in male marsupials. In: *The Biology of Marsupials* (Eds. Stonehouse, B. & Gilmore, D.). Baltimore: University Park Press.
Seymour, R. S., Bennett-Stamper, C. L., Johnston, S. D., Carrier, D. R. & Grigg, G. C. (2004). Evidence for endothermic ancestors of crocodiles at the stem of archosaur evolution. *Physiological and Biochemical Zoology* 77, 1051–1067.
Seymour, R. S., Withers, P. C. & Weathers, W. W. (1998). Energetics of burrowing, running, and free-living in the Namib Desert golden mole (*Eremitalpa namibensis*). *Journal of Zoology* **244**, 107–117.
Shams, I., Avivi, A. & Nevo, E. (2005). Oxygen and carbon dioxide fluctuations in burrows of subterranean blind mole rats indicate tolerance to hypoxic–hypercapnic stresses. *Comparative Biochemistry and Physiology A* **142**, 376–382.
Shaner, R. F. (1962). Comparative development of the bulbus and ventricles of the vertebrate heart with special reference to Spitzer's theory of heart malformations. *Anatomical Record* **142**, 519–529.
Sharman, G. B. (1970). Reproductive physiology of marsupials. *Science* **167**, 1221–1228.
Sherman, W. M., Plyley, M. J., Sharp, R. L., Van Handel, P. J., McAllister, R. M., Fink, W. J. & Costill, D. L. (1982). Muscle glycogen storage and its relationship with water. *International Journal of Sports Medicine* **3**, 22–24.
Sherwood, L., Klandorf, H. & Yancey, P. H. (2005). *Animal Physiology, From Genes to Organisms*. Belmont, California: Thompson Brooks Cole.
Shield, J. (1966). Oxygen consumption during pouch development of the macropod marsupial *Setonix brachyurus*. *Journal of Physiology* **187**, 257–270.
Shkolnik, A. (1971). Diurnal activity in a small desert rodent. *International Journal of Biometeorology* **15**, 115–120.
Sibly, R. M. & Calow, P. (1986). *Physiological Ecology of Animals: An Evolutionary Approach*. Oxford: Blackwell University Press.
Simon, S. A. & de Araujo, I. E. (2005). The salty and burning taste of capsaicin. *Journal of General Physiology* **125**, 531–534.
Sieg, A. E., O'Connor, M. P., McNair, J. N., Grant, B. W., Agosta, S. J. & Dunham, A. E. (2009). Mammalian metabolic allometry: do intraspecific variation, phylogeny, and regression models matter? *American Naturalist* **174**, 720–733.
Siemers, B. M., Schauermann, G., Turni, H. & von Merten, S. (2009). Why do shrews twitter? Communication or simple echo-based orientation. *Biology Letters* **5**, 593–596.
Siemers, B. M & Schnitlzler, H. U (2004). Echolocation signals reflect niche differentiation in five sympatric congeneric bat species. *Nature* **429**, 657–661.

Silva, I. D. & Kuruwita, V. Y. (1994). The osmotic fragility of erythrocytes of the Asian elephant (*Elephas maximus*). *Gajah* **13**, 25–29.
Silva, M. & Downing, J. A. (1995). The allometric scaling of density and body mass: a nonlinear relationship for terrestrial mammals. *American Naturalist* **145**, 704–727.
Silva, M. A., Prieto, R., Jonsen, I., Baumgartner, M. F. & Santos, R. S. (2013). North Atlantic blue and fin whales suspend their spring migration to forage in middle latitudes: building up energy reserves for the journey? *PLOS ONE* **8**, e76507.
Silva, S. I., Jaksic, F. M. & Bozinovic, F. (2005). Nutritional ecology and digestive response to dietary shift in the large South American fox, *Pseudalopex culpaeus*. *Revista Chilena de Historia Natural* **78**, 239–246.
Simmons, N. B., Seymour, K. L., Habersetzer, J. & Gunnell, G. F. (2008). Primitive early Eocene bat from Wyoming and the evolution of flight and echolocation. *Nature* **451**, 818–821.
Simon, J. C., Delmotte, F., Rispe, C. & Crease, T. (2003). Phylogenetic relationships between parthenogens and their sexual relatives: the possible routes to parthenogenesis in animals. *Biological Journal of the Linnean Society* **79**, 151–163.
Simpson, G. G. (1945). The principles of classification and a classification of mammals. *Bulletin of the American Museum of Natural History* **85**, 1–350.
Simpson, G. G. (1961). Historical zoogeography of Australian mammals. *Evolution* **15**, 431–446.
Sinclair, E. A., Danks, A. & Wayne, A. F. (1996). Rediscovery of Gilbert's potoroo, *Potorous tridactylus* in Western Australia. *Australian Mammalogy* **19**, 69–72.
Singer, M. A. (2001). Of mice and men and elephants: metabolic rate sets glomerular filtration rate. *American Journal of Kidney Diseases*. **37**, 164–178.
Singer, M. A. (2002). Vampire bat, shrew, and bear: comparative physiology and chronic renal failure. *American Journal of Physiology* **282**, R1583–R1592.
Skadhauge, E. (1981). *Osmoregulation in Birds. Zoophysiology 12*. Berlin: Springer-Verlag.
Skelton, H. (1927). The storage of water by various tissues of the body. *Archives of Internal Medicine* **40**, 140–152.
Skrovan, R. C., Williams, T. M., Berry, P. S., Moore, P. W. & Davis, R. W. (1999). The diving physiology of bottlenose dolphins (*Tursiops truncatus*). II. Biomechanics and changes in buoyancy at depth. *Journal of Experimental Biology* **202**, 2749–2761.
Smith, A. P. & Broome, L. (1992). The effects of season, sex and habitat on the diet of the mountain pygmy possum (*Burramys parvus*). *Wildlife Research* **19**, 755–768.
Smith, F. A., Boyer, A. G. Brown, J. H., Costa, D. P., Dayan, T., Morgan Ernest, S. K.,. Evans, A. R., Fortelius, M., Gittleman, J. L., Hamilton, M. J., Harding, L. E., Lintulaakso, K., Lyons, S. K., McCain, C., Okie, J. G., Saarinen, J. J., Sibly, R. M., Stephens, P. R., Theodor, J. & Uhen, M. D. (2010). The evolution of maximum body size of terrestrial mammals. *Science* **330**, 1216–1219.
Smith, F. A., Browning, H. & Shepherd, U. L. (1998). The influence of climate change on the body mass of woodrats *Neotoma* in an arid region of New Mexico, USA. *Ecography* **21**, 140–148.
Smith, F. A., Lyons, S. K., Ernest, S. K. M., Jones, K. E., Kaufman, D. M., Dayan, T., Marquet, P. A., Brown, J. H. & Haskell, J. P. (2003). Body mass of Late Quaternary mammals. *Ecology* **84**, 3403. Ecological Archives E084–E094.
Smith, J. D. & Madkour, G. (1980). Penial morphology and the question of chiropteran phylogeny. In: *Proceedings of the Fifth International Bat Research Conference* (Eds. Wilson, D. E. & Gardner, A. L.). Lubbock, Texas: Texas Tech Press.
Smith, K. (2015). Placental evolution in therian mammals. In: *Great Transformations in Vertebrate Evolution* (Eds. Dial, K. P., Shubin, N. & Brainerd, E. L.). Chicago & London: Chicago University Press.
Snyder, G. K. (1976). Respiratory characteristics of whole blood and selected aspects of circulatory physiology in the common short-nosed fruit bat, *Cynopterus brachyotis*. *Respiration Physiology* **28**, 239–247.

Snyder, G. K. (1983). Respiratory adaptations in diving mammals. *Respiration Physiology* **54**, 269–294.

Snyder, G. K. & Weathers, W. W. (1977). Hematology, viscosity, and respiratory functions of whole blood of the lesser mouse deer, Tragulus javanicus. *Journal of Applied Physiology* **42**, 673–678.

Snyder, L. R. G., Hayes, J. P. & Chappell, M. A. (1988). Alpha-chain hemoglobin polymorphisms are correlated with altitude in the deer mouse, *Peromyscus maniculatus*. *Evolution* **42**, 689–697.

Somero, G. N. (2005). Linking biogeography to physiology: evolutionary and acclimatory adjustments of thermal limits. *Frontiers in Zoology* **2**, 1–9.

Somero, G. N. (2011). Comparative physiology: a 'crystal ball' for predicting consequences of global change. *American Journal of Physiology* **302**, R1–R14.

Song, X. & Geiser, F. (1997). Daily torpor and energy expenditure in *Sminthopsis macroura*: interactions between food and water availability and temperature. *Physiological Zoology* **70**, 331–337.

Souza, C. P., Almeida, B. C., Colwell, R. R. & Rivera, I. N. G. (2011). The importance of chitin in the marine environment. *Marine Biotechnology* **13**, 823–830.

Spaargaren, D. H. (1992). Transport function of branching structures and the 'surface law' for basic metabolic rate. *Journal of Theoretical Biology* **154**, 495–504.

Spalding, M. D., Fox, H. E., Allen, G. R., Davidson, N., Ferdaña, Z. A., Finlayson, M., Halpern, B. S., Jorge, M. A., Lombana, A., Lourie, S. A., Martin, K. D., Mcmanus, E., Molnar, J., Recchia, C. A. & Robertson, J. (2007). Marine ecoregions of the world: a bioregionalization of coastal and shelf areas. *Bioscience* **57**, 573–583.

Spatz, H.-Ch. (1991). Circulation, metabolic rate, and body size in mammals. *Journal of Comparative Physiology B* **161**, 231–236.

Speakman, J. R. (1995). Chiropteran nocturnality. *Symposium of the Zoological Society of London* **67**, 187–201.

Speakman, J. R. (1997). *Doubly Labelled Water. Theory and Practice*. London: Chapman & Hall.

Speakman, J. R. (2005). Correlations between physiology and lifespan—two widely ignored problems with comparative studies. *Aging Cell* **4**, 167–175.

Speakman, J. R. (2014). Should we abandon indirect calorimetry as a tool to diagnose energy expenditure? Not yet. Perhaps not ever. Commentary on Burnett and Grobe (2014). *Molecular Metabolism* **3**, 342–344.

Speakman, J. R., Ergon, T., Cavanagh, R., Reid, K., Scantlebury, D. M. & Lambin, X. (2003). Resting and daily energy expenditures of free-living field voles are positively correlated but reflect extrinsic rather than intrinsic effects. *Proceedings of the National Academy of Sciences, USA* **100**, 14057–14062.

Speakman, J. R. & Król, E. (2010). Maximal heat dissipation capacity and hyperthermia risk: neglected key factors in the ecology of endotherms. *Journal of Animal Ecology* **79**, 726–746.

Speakman, J. R., Król, E. & Johnson, M. S. (2004). The functional significance of individual variation in basal metabolic rate. *Physiological and Biochemical Zoology* **77**, 900–915.

Speakman, J. R., McDevitt, R. M. & Cole, K. R. (1993). Measurement of basal metabolic rates: don't lose sight of reality in the quest for comparability. *Physiological Zoology* **66**, 1045–1049.

Speakman, J. R. & Racey, P. A. (1991). No cost of echolocation for bats in flight. *Nature* **350**, 421–423.

Speakman, J. R. & Thomas, D. W. (2003). Physiological ecology and energetics of bats. In: *Bat Biology* (Eds. Kunz, T. H. & Fenton, M. B.). Chicago: University of Chicago Press.

Speakman, J. R. & Thomson, S. C. (1997). Validation of the labelled bicarbonate technique for measurement of short-term energy expenditure in themouse. *Zeitschrifte für Ernährungswissenschaft* **36**, 273–277.

Speedy, A. (2003). Global production and consumption of animal source foods. *Journal of Nutrition* **133**, 40485–40535.
Spector, N. H., Brobeck, J. R. & Hamilton, C. L. (1968). Feeding and core temperature in albino rats: changes induced by preoptic heating and cooling. *Science* **161**, 286–288.
Sperber, I. (1944). Studies on the mammalian kidney. *Zoologiska Bidrag Från Uppsala* **22**, 249–435.
Spicer, J. I. & Gaston, K. J. (1999). *Physiological Diversity and its Ecological Implications*. Oxford: Blackwell Science.
Springer, M. S., Murphy, W. J., Eizirik, E. & O'Brien, S. J. (2003). Placental mammal diservification and the Cretaceous-Tertiary boundary. *Proceedings of the National Academy of Sciences, USA* **100**, 1056–1061.
Springer, M. S., Stanhope, M. J., Madsen, O. & de Jong, W. W. (2004). Molecules consolidate the placental mammal tree. *Trends in Ecology and Evolution* **18**, 430–438.
Springer, M. S., Teeling, E. C., Madsen, O., Stanhope, M. J. & de Jong, W. W. (2001). Integrated fossil and molecular data reconstruct bat echolocation. *Proceedings of the National Academy of Sciences, USA* **98**, 6241–6246.
Srere, H. K., Wang, C. H. & Martin, S. L. (1992). Central role for differential gene expression in mammalian hibernation. *Proceedings of the National Academy of Sciences, USA* **89**, 7119–7123.
Stahl, R. W. (1967). Scaling of respiratory variables in mammals. *Journal of Applied Physiology* **22**, 453–460.
Stainer, M. W., Mount, L. E. & Bligh, J. (1984). *Energy Balance and Temperature Regulation*. Cambridge: Cambridge University Press.
Stawski, C. & Geiser, F. (2010). Fat and fed: frequent use of summer torpor in a subtropical bat. *Naturwissenschaften* **97**, 29–35.
Stawski, C., Körtner, G., Nowack, J. & Geiser F. (2015). The importance of mammalian torpor for survival in a post-fire landscape. *Biology Letters* **11**, 20150134.
Stearns, S. C. (1992). *The Evolution of Life Histories*. Oxford: Oxford University Press.
Stephenson, J. L. (1972). Concentration of urine in a central core model of the renal counterflow system. *Kidney International* **2**, 85–94.
Stephenson, R., Turner, D. L. & Butler, P. J. (1989). The relationship between diving activity and oxygen storage capacity in the tufted duck (*Aythya fuligula*). *Journal of Experimental Biology* **141**, 265–275.
Stevens, C. E. & Hume, I. D. (1995). *Comparative Physiology of the Vertebrate Digestive System*. Cambridge: Cambridge University Press.
Stevens, G. C. (1989). The latitudinal gradient in geographical range: how so many species coexist in the tropics. *American Naturalist* **133**, 240–256.
Stevens, G. C. (1992). The elevational gradient in altitudinal range: an extension of Rapoport's latitudinal rule to altitude. *American Naturalist.* **140**, 893–911.
Stewart, J. R. & Thompson, M. B. (2000). Evolution of placentation among squamate reptiles: recent research and future directions. *Comparative Biochemistry and Physiology A* **127**, 411–431.
Stibbe, E. P. (1927–1928). A comparative study of the nictitating membrane of birds and mammals. *Journal of Anatomy* **62**, 159–176.
Stockman, C. A., Barnes, A. L., Maloney, S. K., Talyor, E., McCarthy, M. & Pethick, D. (2011). Effect of prolonged exposure to continuous heat and humidity similar to long haul live export voyages in Merino wethers. *Animal Production Science* **51**, 135–143.
Stone, C. & Bacon, P. E. (1994). Relationships among moisture stress, insect herbivory, foliar cineole content and the growth of river red gum *Eucalyptus camaldulensis*. *Journal of Applied Ecology* **31**, 604–612.

Stone, H. O., Thompson, H. K. & Schmidt-Nielsen, K. (1968). Influence of erythrocytes on blood viscosity. *American Journal of Physiology* **214**, 913–918.

Storz, J. F. (2007). Hemoglobin function and physiological adaptation to hypoxia in high-altitude mammals. *Journal of Mammalogy* **88**, 24–31.

Storz, J. F., Balasingh, J., Bhat, H. R., Nathan, P. T., Doss, D. P. S., Prakash, A. A. & Kunz, T. H. (2001). Clinal variation in body size and sexual dimorphism in an Indian fruit bat, *Cynopterus sphinx* (Chiroptera: Pteropodidae). *Biological Journal of the Linnean Society* **72**, 17–31.

Storz, J. F. & Moriyama, H. (2008). Mechanisms of hemoglobin adaptation. *High Altitude Medicine and Biology* **9**, 148–157.

Storz, J. F., Runck, A. M., Sabatino, S. J., Kelly, J. K., Ferrand, N., Moriyama, H., Weber, R. E. & Fago, A. (2009). Evolutionary and functional insights into the mechanism underlying high-altitude adaptation of deer mouse hemoglobin. *Proceedings of the National Academy of Sciences, USA* **106**, 14450–14455.

Strauss, W. M., Hetem, R. S., Mitchell, D., Maloney, S. K., Meyer, L. C. R. & Fuller, A. (2015). Selective brain cooling reduces water turnover in dehydrated sheep. *PLOS ONE* **10**, e0115514.

Streidter, G. F. (2005). *Principles of Brain Evolution*. Sunderland, Massachusetts: Sinauer Associates.

Strobel, S., Roswag, A., Becker, N. I., Trenczek, T. E. & Encarnação. (2013). Insectivorous bats digest chitin in the stomach using acidic mammalian chitinases. *PLOS ONE* **8**, e72770.

Studier, E. H. (1969). Respiratory ammonia filtration, mucous composition and ammonia tolerance in bats. *Journal of Experimental Zoology* **170**, 253–258.

Studier, E. H., Beck, R. L. & Lindeborg, R. G. (1967). Tolerance and initial metabolic response to ammonia intoxication in selected bats and rodents. *Journal of Mammalogy* **48**, 564–572.

Studier, E. H. & Howell, D. J. (1969). Heart rate of female big brown bats in flight. *Journal of Mammalogy* **50**, 842–845.

Studier, E. H. & Wilson. D. E. (1983). Natural urine concentrations and composition in neotropical bats. *Comparative Biochemistry and Physiology A* **75**, 509–515.

Suarez, R. K. & Darveau, C. A. (2005). Multi-level regulation and metabolic scaling. *Journal of Experimental Biology* **208**, 1627–1634.

Suarez, R. K., Lighton, J. R. B., Brown, G. S. & Mathieu-Costello, O. (1991). Mitochondrial respiration in hummingbird flight muscles. *Proceedings of the National Academy of Sciences, USA* **88**, 4870–4873.

Suarez, R. K., Staples, J. F., Lighton, J. R. B. & West, T. G. (1997). Relationships between enzymatic flux capacities and metabolic flux rates in muscles: nonequilibrium reactions in muscle glycolysis. *Proceedings of the National Academy of Sciences, USA* **94**, 7065–7069.

Suarez, R. K., Welch, K. C., Hanna, S. K. & Herrera, M. L. G. (2009). Flight muscle enzymes and metabolic flux rates during hovering flight of the nectar bat, *Glossophaga soricina*: further evidence of convergence with hummingbirds. *Comparative Biochemistry and Physiology A* **153**, 136–140.

Suarez, S. S. (2008a). Control of hyperactivation in sperm. *Human Reproduction Update* **14**, 47–657.

Suarez, S. S. (2008b). Regulation of sperm storage and movement in the mammalian oviduct. *International Journal of Developmental Biology* **52**, 455–462.

Sullivan, R. M. & Best, T. L. (1997). Effects of environment on phenotypic variation and sexual dimorphism in *Dipodomys simulans* (Rodentia: Heteromyidae). *Journal of Mammalogy* **78**, 798–810.

Sumner, P., Arrese, C. A. & Partridge, J. C. (2005). The ecology of visual pigment tuning in an Australian marsupial: the honey possum *Tarsipes rostratus*. *Journal of Experimental Biology* **208**, 1803–1815.

Swaim, Z., Westgate, A. J., Koopman, H. N., Rolland, R. M. & Kraus, S. D. (2009). Metabolism of ingested lipids by North Atlantic right whales. *Endangered Species Research* **6**, 259–271.
Swan, H. (1972). Comparative metabolism: surface versus mass solved by hibernators. In: *International Symposium on Environmental Physiology: Bioenergetics* (Eds. Smith, R. E, Hannon, J. P., Shields, J. L. & Horowitz, B. A.). Bethesda, Maryland: Federation of American Societies for Experimental Biology.
Swartz, S. M., Bennett, M. B. & Carrier, D. R. (1992). Wing bone stresses in free flying bats and the evolution of skeletal design for flight. *Nature* **359**, 726–729.
Swartz, S. M., Bishop, K. L. & Aguirre, I. M. F. (2006). Dynamic complexity of wing form in bats: implications for flight performance. In: *Functional and Evolutionary Ecology of Bats* (Eds. Zubaid, A., McCraken, G. F., & Kunz, T. H.). New York: Oxford University Press.
Sweet, G. (1906). Contributions to our knowledge of the anatomy of *Notoryctes typhlops* Stirling. Part III. The eye. *Quarterly Journal of Microscopical Science* **51**, 333–344.
Szewczak, J. M. & Powell, F. L. (2003). Open-flow plethysmography with pressure-decay compensation. *Respiratory Physiology and Neurobiology* **134**, 57–67.
Tarpley, R. J. & Ridgway, S. H. (1991). Orbital gland structure and secretions in the Atlantic bottlenose dolphin (*Tursiops truncatus*). *Journal of Morphology* **207**, 173–184.
Tattersall, G. J., Leite, C. A. C., Sanders, C. E., Cadena, V., Andrade, D. V., Abe, A. S. & Milsom, W. K. (2016). Seasonal reproductive endothermy in tegu lizards. *Science Advances* **2**, e1500951.
Taylor, C. R. (1968). The minimum water requirements of some East African bovids. *Symposium of the Zoological Society of London* **21**, 195–206.
Taylor, C. R. & Lyman, C. P. (1972). Heat storage in running antelopes: independence of brain and body temperatures. *American Journal of Physiology* **222**, 114–117.
Taylor, C. R., Schmidt-Nielsen, K. & Raab, J. L. (1970). Scaling of energetic cost of running to body size in mammals. *American Journal of Physiology* **219**, 1104–1107.
Taylor, C. R. & Rowntree, V. J. (1973). Temperature regulation and heat balance in running cheetahs: a strategy for sprinters? *American Journal of Physiology* **224**, 848–851.
Taylor, C. R., Shkolnik, A., Dmi'el, R., Baharav, D. & Borut, A. (1974). Running in cheetahs, gazelles, and goats: energy cost and limb configuration. *American Journal of Physiology* **227**, 848–850.
Tazawa, H., Mochizuki, M. & Piiper, J. (1979). Respiratory gas transport by the incompletely separated ventricle in the bullfrog, *Rana catesbeiana*. *Respiration Physiology* **36**, 77–95.
Teeling, E. C., Springer, M. S., Madsen, O., Bates, P. O'Brien, S. J. K. & Murphy, W. J. (2005). A molecular phylogeny for bats illuminates biogeography and the fossil record. *Science* **307**, 580–584.
Telfer, E. S. & Kelsall, J. P. (1984). Adaptation of some large North American mammals for survival in snow. *Ecology* **65**, 1828–1834.
Temple-Smith, P. & Grant, T. (2001). Uncertain breeding: a short history of reproduction in monotremes. *Reproduction, Fertility and Development* **13**, 487–497.
Teplitsky, C. & Millien, V. (2013). Climate warming and Bergmann's rule through time: is there any evidence? *Evolutionary Applications* 7, 156–168.
Thewissenn, J. G. M. & Babcok, S. K. (1992). The origin of flight in bats. *BioScience* **42**, 340–345.
Thiele, R. H., Bartels, K. & Gan, T. J. (2015). Cardiac output monitoring: a contemporary assessment and review. *Critical Care Medicine* **43**, 177–185.
Thomas, C. D., Cameron, A., Green, R. E., Bakkenes, M., Beaumont, L. J., Collingham, Y. C., Erasmus, B. F. N., de Siqueira, M. F., Grainger, A., Hannah, L., Hughes, L., Huntley, B., van Jaarsveld, A. S., Midgley, G. F., Miles, L., Ortega-Huerta, M. A., Peterson, A. T., Phillips, O. L. & Williams, S. E. (2004). Extinction risk from climate change. *Nature* **427**, 145–148.

Thomas, D. W. (1984). Fruit intake and energy budgets of frugivorous bats. *Physiological and Biochemical Zoology* 57, 457–467.
Thomas, S. P. (1975). Metabolism during flight in two species of bats, *Phyllostomus hastatus* and *Pteropus gouldii*. *Journal of Experimental Biology* **63**, 273–293.
Thomas, S. P. (1981). Ventilation and oxygen extraction in the bat *Pteropus gouldii* during rest and steady flight. *Journal of Experimental Biology* **63**, 1–9.
Thomas, S. P. (1987). The physiology of bat flight. In: *Recent Advances in the Study of Bats*, (Ed. Fenton, M. B.). Cambridge: Cambridge University Press.
Thomas, S. P., Follette, D. B. & Farabaugh, A. T. (1991). Influence of air temperature on ventilation rates and thermoregulation of a flying bat. *American Journal of Physiology* **260**, R960–R968.
Thomas, S. P., Lust, M. R. & Van Riper, H. J. (1984). Ventilation and oxygen extraction in the bat *Phyllostomus hastatus* during rest and steady flight. *Physiological Zoology* 57, 237–250.
Thomas, S. P. & Suthers, R. A. (1972). The physiology and energetics of bat flight. *Journal of Experimental Biology* 57, 317–335.
Thompson, M. B. & Speake, B. K. (2006). A review of the evolution of viviparity in lizards: structure, function and physiology of the placenta. *Journal of Comparative Physiology B* **176**, 179–189.
Thompson, S. D. (1982). Optical tracking and telemetry for nocturnal field studies. *Journal of Wildlife Management* **41**, 309–312.
Thompson, S. D., MacMillen, R. E., Burke, E. M. & Taylor, C. R. (1980). The energetic cost of bipedal hopping in small mammals. *Nature* **287**, 223–234.
Thornton, P. K. (2010). Livestock production: recent trends, future prospects. *Philosophical Transactions of the Royal Society of London B* **365**, 2853–2867.
Thornton, S. J., Hochachka, P. W., Crocker, D. E., Costa, D. P., LeBoeuf, B. J., Spielman, D. M. & Pelc, N. J. (2005). Stroke volume and cardiac output in juvenile elephant seals during forced dives. *Journal of Experimental Biology* **208**, 3637–3643.
Thorsteinsson, B., Gliemann, J. & Vinten, J. (1976). The content of water and potassium in fat cells. *Biochimica et Biophysica Acta* **428**, 223–227.
Tibben, E. A., Holland, R. A. B. & Tyndale-Biscoe, H. (1991). Blood oxygen carriage in the marsupial, tammar wallaby (*Macropus eugenii*). *Respiration Physiology* **84**, 93–104.
Tøien, O., Blake, J., Edgar, D. M., Grahn, D. A., Heller, H. C. & Barnes, B. M. (2011). Hibernation in black bears: independence of metabolic suppression from body temperature. *Science* **331**, 906–909.
Tomasco, I. H., del Río, R., Iturriaga, R. & Bozinovic, F. (2010). Comparative respiratory strategies of burrowing and fossorial octodontid rodents to cope with hypoxic and hypercapnic atmospheres. *Journal of Comparative Physiology B* **180**, 877–884.
Tomasi, T. E. (1979). Echolocation by the short-tailed shrew *Blarina brevicauda*. *Journal of Mammalogy* **60**, 751–759.
Tomlinson, S., Arnall, S. G., Munn, A., Bradshaw, S. D., Maloney, S. K., Dixon, K. W. & Didham, R. K. (2014). Applications and implications of ecological energetics. *Trends in Ecology and Evolution* **29**, 280–290.
Tomlinson, S., Maloney, S. K., Withers, P. C., Voigt, C. C. & Cruz-Neto, A. P. (2013a). From doubly labelled water to half-life; validating radio-isotopic rubidium turnover to measure metabolism in small vertebrates. *Methods in Ecology and Evolution* 4, 619–628.
Tomlinson, S., Mathialagan, P. D. & Maloney, S. K. (2013b). Special K: testing the potassium link between radioactive rubidium ($^{86}$Rb) turnover and metabolic rate. *Journal of Experimental Biology* **217**, 1040–1045.

Tomlinson, S. & Withers, P. C. (2008). Biogeographical effects on body mass of native Australian and introduced mice, *Pseudomys hermannsburgensis* and *Mus domesticus*: an inquiry into Bergmann's Rule. *Australian Journal of Zoology* **56**, 423–430.

Torres-Bueno, J. R. (1978). Evaporative cooling and water balance during flight in birds. *Journal of Experimental Biology* 75, 231–236.

Toth, D. M. (1973). Temperature regulation and salivation following preoptic lesions in the rat. *Journal of Comparative and Physiological Psychology* **82**, 480.

Tracy, C. R. (1977). Minimum size of mammalian homeotherms: role of the thermal environment. *Science* **198**, 1034–1035.

Tracy, C. R., McWhorter, T. J., Korine, C., Wojciechowski, M. S., Pinshow, B. & Karasov, W. H. (2007). Absorption of sugars in the Egyptian fruit bat (*Rousettus aegyptiacus*): a paradox explained. *Journal of Experimental Biology* **210**, 1726–1734.

Tracy, R. L. & Walsberg, G. E. (2000). Prevalence of cutaneous evaporation in Merriam's kangaroo rat and its adaptive variation at the subspecific level. *Journal of Experimental Biology* **203**, 773–781.

Trayhurn, P. (1993). Brown adipose tissue: from thermal physiology to bioenergetics. *Journal of Biosciences* **18**, 161–173.

Tseng, Z. J. (2013). Testing adaptive hypotheses of convergence with functional landscapes: a case study of bone-cracking hypercarnivores. *PLOS ONE* **8**, e65305.

Tucker, A. S. & Fraser, G. J. (2014). Evolution and developmental diversity of tooth regeneration. *Seminars in Cell and Developmental Biology* **25–26**, 71–80.

Tucker, V. (1970). Energetic cost of locomotion in animals. *Comparative Biochemistry and Physiology* **34**, 841–846.

Tucker, V. A. (1967). Method for oxygen content and dissociation curves on microliter blood samples. *Journal of Applied Physiology* **23**, 410–4.

Tucker, V. A. (1975). The energetic cost of moving about. *American Scientist* **63**, 413–419.

Tufts, D. M., Revsbech, I. G., Cheviron, Z. A., Weber, R. E., Fago, A. & Storz, J. F. (2013). Phenotypic plasticity in blood–oxygen transport in highland and lowland deer mice. *Journal of Experimental Biology* **216**, 1167–1173.

Turbill, C., Smith, S., Deimel, C. & Ruf, T. (2012). Daily torpor is associated with telomere length change over winter in Djungarian hamsters. *Biology Letters* **8**, 304–307.

Turner, J., Anderson, P., Lachlan-Cope, T., Colwell, S., Phillips, T., Kirchgaessner, A. L., Marshall, G. J., King, J. C., Bracegirdle, T., Vaughan, D. G., Lagun, V. & Orr, A. (2009). Record low surface air temperature at Vostok station, Antarctica. *Journal of Geophysical Research: Atmospheres* **114**, D24102. doi:10.1029/2009JD012104.

Turner, N., Parker, J. & Hudnall, J. (1992). The effect of dry and humid hot air inhalation on expired relative humidity during exercise. *American Industrial Hygiene Association Journal* **53**, 256–260.

Tyndale-Biscoe, H. (2005). *Life of Marsupials*. Collingwood, Victoria: CSIRO Publishing.

Tyndale-Biscoe, H. & Renfree, M. (1987). *Reproductive Physiology of Marsupials*. Cambridge: Cambridge University Press.

Ultsch, G. R. (1973). A theoretical and experimental investigation of the relationships between metabolic rate, body size and oxygen exchange capacity. *Respiration Physiology* **18**, 143–160.

Ultsch, G. R. (1974). The allometric relationship between metabolic rate and body size. The role of the skeleton. *American Midland Naturalist* **92**, 500–504.

Understanding Animal Research (2012). Numbers of animals. http://understandinganimalresearch.org.uk/the-animals/numbers-of-animals.

Vacquié-Garcia, J., Royer, F., Dragon, A.-C., Viviant, M., Bailleul, F. & Guinet, C. (2012). Foraging in the darkness of the southern ocean: influence of bioluminescence on a deep diving predator. *PLOS ONE* 7, e43565.

Van Breukelen, F. & Martin, S. L. (2002). Reversible depression of transcription during hibernation. *Journal of Comparative Physiology B* **172**, 355–361.

Van Citters, R. L. & Franklin, D. L. (1966). Telemetry of blood pressure in free-ranging animals via an intravascular gauge. *Journal of Applied Physiology* **21**, 1633–1636.

Vanderklift, M. A. & Ponsard, S. (2003). Sources of variation in consumer diet $\delta^{15}N$ enrichment: a meta-analysis. *Oecologia* **136**, 169–182.

Vander Zanden, M. J., Clayton, M. K., Moody, E. K., Solomon, C. T. & Weidel, B. C. (2015). Stable isotope turnover and half-life in animal tissues: a literature synthesis. *PLOS ONE* **10**, e0116182.

van Nievelt, A. F. H. & Smith, K. K. (2005). To replace or not to replace: the significance of reduced functional tooth replacement in marsupial and placental mammals. *Paleobiology* **31**, 324–346.

Van Sant, M. J., Oufiero, C. E., Muñoz-Garcia, A., Hammond, K. A. & Williams, J. B. (2012). A phylogenetic approach to total evaporative water loss in mammals. *Physiological and Biochemical Zoology* **85**, 526–532.

Van Soest, P. J. (1982). *Nutritional Ecology of the Ruminant*. Ithaca, New York: Cornell University Press.

Van Valkenburgh, B. (1989). Carnivore dental adaptations and diet: a study of trophic diversity within guilds. In: *Carnivore Behavior, Ecology, and Evolution* (Ed. Gittleman, J. L.). Ithaca, New York: Cornell University Press.

Vaughan, T. A. (1986). *Mammalogy*. Philadelphia: Saunders College Publishing.

Veloso, C. & Bozinovic, F. (1993). Dietary and digestive constraints on basal energy metabolism in a small herbivorous rodent. *Ecology* 74, 2003–2010.

Velten, B. P., Dillaman, R. M., Kinsey, S. T., McLellan, W. A. & Pabst, D. A. (2013). Novel locomotor muscle design in extreme deep-diving whales. *Journal of Experimental Biology* **216**, 1862–1871.

Vendl, C., Clauss, C., Stewart, M., Leggett, K., Hummel, J., Kreuzer, M. & Munn, A. (2015a). Decreasing methane yield with increasing food intake keeps daily methane emissions constant in two foregut fermenting marsupials, the western grey kangaroo and red kangaroo. *Journal of Experimental Biology* **218**, 3425–3434.

Vendl, C., Frei, S., Dittmann, M. T., Furrer, S., Ortmann, S., Lawrenz, A., Lange, B., Munn, A., Kreuzer, M. & Clauss, M. (2016). Methane production by two non-ruminant foregut-fermenting herbivores: the collared peccary (*Pecari tajacu*) and the pygmy hippopotamus (*Hexaprotodon liberiensis*). *Comparative Biochemistry and Physiology A* **191**, 107–114.

Vendl, C., Frei, S., Dittmann, M. T., Furrer, S., Osmann, C., Ortmann, S., Munn, A., Kreuzer, M. & Claus, M. (2015b). Digestive physiology, metabolism and methane production of captive Linne's two-toed sloths (*Choloepus didactylus*). *Journal of Animal Physiology and Animal Nutrition*. doi:10.1111/jpn.12356.

Venkatachalam, K. & Montell, C. (2007). TRP channels. *Annual Reviews in Biochemistry* **76**, 387–417.

Vereecke, E. E. & Channon, A. J. (2013). The role of hind limb tendons in gibbon locomotion: springs or strings? *Journal of Experimental Biology* **216**, 3971–3980.

Verkman. A., van Hoek, A., Ma, T., Frigeri, A., Skach, W., Mitra, A. & Farinas, J. (1996). Water transport across mammalian cell membranes. *American Journal of Physiology* **270**, C12–C30.

Vermillion, K. L., Anderson, K. J., Hampton, M. & Andrews, M. T. (2015). Gene expression changes controlling distinct adaptations in the heart and skeletal muscle of a hibernating mammal. *Physiological Genomics* 47, 58–74.

Vesterdorf, K., Blache, D. & Maloney, S. K. (2011). The cranial arterio-venous temperature difference is related to respiratory evaporative heat loss in a panting species, the sheep (*Ovis aries*). *Journal of Comparative Physiology B* **181**, 277–288.

Villarreal, J. A., Schlegel, W. M. & Prange, H. D. (2007). Thermal environment affects morphological and behavioral development of *Rattus norvegicus*. *Physiology and Behaviour* **91**, 26–35.

Vitt, L. J. & Caldwell, J. P. (2009). *Herpetology*. New York: Academic Press.

Vleck, D. (1979). The energy cost of burrowing by the pocket gopher *Thomomys bottae*. *Physiological Zoology* **52**, 122–136.

Vleck, D. (1981). Burrow structure and foraging costs in the fossorial rodent, *Thomomys bottae*. *Oecologia* **49**, 391–396.

Vogel, S. (1994). *Life in Moving Fluids*. Princeton, New Jersey: Princeton University Press.

Vogel, S., Ellington, C. P. & Kilgore, D. L. (1973). Wind-induced ventilation of the burrow of the prairie-dog, *Cynomys ludovicianus*. *Journal of Comparative Physiology* **85**, 1–14.

Voigt, C. C. & Holderied, M. W. (2012). High manoeuvring costs force narrow-winged molossid bats to forage in open space. *Journal of Comparative Physiology B* **82**, 415–424.

Voigt, C. C., Sörgel, K., Śuba, J., Keišs, O. & Pētersons, G. (2012). The insectivorous bat *Pipistrellus nathusii* uses a mixed-fuel strategy to power autumn migration. *Proceedings of the Royal Society of London B* **279**, 3772–3778.

Voigt, C. C. & Lewanzik, D. (2010). Trapped in the darkness of the night: thermal and energetic constraints of daylight flight in bats. *Proceeding of the Royal Society B* **1716**, 2311–2317.

Voigt, C. C. & Lewanzik, D. (2012). 'No cost of echolocation for flying bats' revisited. *Journal of Comparative Physiology B* **182**, 831–840.

Voigt, C. C., Rex, K., Michener, R. H. & Speakman, J. R. (2008). Nutrient routing in omnivorous animals tracked by stable carbon isotopes in tissue and exhaled breath. *Oecologia* **157**, 31–40.

Voigt, C. C., Sörgel, K. & Dechmann, D. K. N. (2010). Refuelling while flying: foraging bats combust food rapidly and directly to fuel flight. *Ecology* **91**, 2908–2917.

Voigt, C. C. & Speakman, J. (2007). Nectar-feeding bats fuel their high metabolism directly with exogenous carbohydrates. *Functional Ecology* **21**, 913–921.

Voigt, C. C. & Winter, Y. (1999). Energetic cost of hovering flight in nectar-feeding bats (Phyllostomidae: Glossophaginae) and its scaling in moths, birds and bats. *Journal of Comparative Physiology B* **169**, 38–48.

Voigt-Heucke, S. L., Taborsky, M. & Dechmann, D. K. N. (2010). A dual function of echolocation: bats use echolocation calls to identify familiar and unfamiliar individuals. *Animal Behaviour* **80**, 59–67.

von Busse, R., Swartz, S. M. & Voigt, C. C. (2013). Flight metabolism in relation to speed in Chiroptera: testing the U-shape paradigm in the short-tailed fruit bat *Carollia perspicillata*. *Journal of Experimental Biology* **216**, 2073–2080.

von Busse, R., Waldman, R. M., Swartz, S. M., Voigt, C. C. & Breuer, K. S. (2014). The aerodynamic cost of flight in the short-tailed fruit bat (*Carollia perspicillata*): comparing theory with measurement. *Journal of Royal Society Interface* **11**, 2014.0147.

Waites, G. M. (1976). Temperature regulation and fertility in male and female mammals. *Israel Journal of Medical Sciences* **12**, 982–93.

Waites, G. M. G. (1970). Temperature regulation and the testis. In: *The Testis: Development, Anatomy and Physiology* (Eds. Johnson, A. D., Gomes, W. R. & Vandemark, N. L.). New York: Academic Press.

Waites, G. M. H. & Moule, G. R. (1961). Relation of vascular heat exchange to temperature regulation in the testis of the ram. *Journal of Reproduction and Fertility* **2**, 213–224.

Walker, J. E. C., Wells, R. E. & Merrill, E. W. (1961). Heat and water exchange in the respiratory tract. *American Journal of Physiology* **30**, 259–267.

Wallace, A. R. (1876). *The Geographical Distribution of Animals*. New York: Harper.
Walls, G. L. (1939). Origin of the vertebrate eye. *Archives of Ophthalmology* **22**, 452–486.
Walsberg, G. E. (1983). Coat color and solar heat gain in animals. *BioScience* **33**; 88–91.
Walsberg, G. E. (1988a). Consequences of skin colour and fur properties for solar heat gain and ultraviolet irradiance in two mammals. *Journal of Comparative Physiology B* **158**, 213–221.
Walsberg, G. E. (1988b). The significance of fur structure for solar heat gain in the rock squirrel (*Spermophilus variegates*). *Journal of Experimental Biology* **138**, 243–257.
Walsberg, G. E. (1990). Convergence of solar heat gain in two squirrel species with contrasting coat colours. *Physiological Zoology* **63**, 1025–1042.
Walsberg, G. E. (2000). Small mammals in hot deserts: some generalizations revisited. *Bioscience* **50**, 109–120.
Walsberg, G. E. & Hoffman, T. C. M. (2005). Direct calorimetry reveals large errors in respirometric estimates of energy expenditure. *Journal of Experimental Biology* **208**, 1035–1043.
Walsberg, G. E. & Hoffman, T. C. M. (2006). Using direct calorimetry to test the accuracy of indirect calorimetry in an ectotherm. *Physiological and Biochemical Zoology* **79**, 830–835.
Walsberg, G. E. & King, J. R. (1978). The relationship of the external surface area of birds to skin surface area and body mass. *Journal of Experimental Biology* **76**, 185–189.
Walsberg, G. E. & Schmidt, C. A. (1989). Seasonal adjustment of solar heat gain in a desert mammal by altering coat properties independently of surface coloration. *Journal of Experimental Biology* **142**, 387–400.
Walsberg, G. E., Tracy, R. L. & Hoffman, T. C. (1997). Do metabolic responses to solar radiation scale directly with intensity of irradiance? *Journal of Experimental Biology* **200**, 2115–2121.
Walsberg, G. E. & Wolf, B. O. (1995a). Solar heat gain in a desert rodent: unexpected increases with wind speed and implications for estimating the heat balance of free-living animals. *Journal of Comparative Physiology* **165**, 306–314.
Walsberg, G. E. & Wolf, B. O. (1995b). Effects of solar radiation and wind speed on metabolic heat production by two mammals with contrasting coat colours. *Journal of Experimental Biology* **198**, 1499–1507.
Walsh, J. P., Boggs, D. F. & Kilgore, D. L. (1996). Ventilatory and metabolic responses of a bat, *Phyllostomus discolor*, to hypoxia and $CO_2$: implications for the allometry of respiratory control. *Journal of Comparative Physiology B* **166**, 351–358.
Warburton, N., Wood, C., Lloyd, C., Song, S. & Withers, P. (2003). The 3-dimensional anatomy of the north-western marsupial mole (*Notoryctes caurinus* Thomas 1920) using computed tomography, X-ray and magnetic resonance imaging. *Records of the Western Australian Museum* **22**, 1–7.
Warburton, N. M. (2006). Functional morphology of marsupial moles (Marsupialia, Notoryctidae). *Verhandlungen des Naturwissenschaftlichen Vereins Hamburg* **42**, 39–149.
Ward, S., Möller, U., Rayner, J. M. V., Jackson, D. M., Bilo, D., Nachtigall, W. & Speakman, J. R. (2001). Metabolic power, mechanical power and efficiency during wind tunnel flight by the European starling *Sturnus vulgaris*. *Journal of Experimental Biology* **204**, 3311–3322.
Warnecke, L. & Geiser, F. (2009). Basking behaviour and torpor use in free-ranging *Planigale gilesi*. *Australian Journal of Zoology* **57**, 373–375.
Warnecke, L. & Geiser, F. (2010). The energetics of basking behaviour and torpor in a small marsupial exposed to simulated natural conditions. *Journal of Comparative Physiology B* **180**, 437–445.
Warnecke, L., Withers, P. C., Schleucher, E. & Maloney, S. K. (2007). Body temperature variation of free-ranging and captive southern brown bandicoots *Isoodon obesulus* (Marsupialia: Peramelidae). *Journal of Thermal Biology* **32**, 72–77.
Waymouth, C. (1970). Osmolality of mammalian blood and of media for culture of mammalian cells. *In Vitro* **6**, 109–127.

Webb, P. I. (1995). The comparative ecophysiology of water balance in microchiropteran bats. *Symposium Zoological Society London* **67**, 203–218.

Webb, P. I. Hays, G. C., Speakman, J. R. & Racey, P. A. (1992). The functional significance of ventilation frequency, and its relationship to oxygen demand in the resting brown long-eared bat, *Plecotus auritus*. *Journal of Comparative Physiology B* **162**, 144–147.

Weber, R. E. (2007). High-altitude adaptations in vertebrate hemoglobins. *Respiratory Physiology and Neurobiology* **158**, 132–142.

Webster, K. N. & Dawson, T. J. (2003). Locomotion energetics and gait characteristics of a rat-kangaroo, *Bettongia penicillata*, have some kangaroo-like features. *Journal of Comparative Physiology B* **173**, 549–557.

Webster, K. N. & Dawson, T. J. (2012). The high aerobic capacity of a small, marsupial rat-kangaroo (*Bettongia penicillata*) is matched by the mitochondrial and capillary morphology of its skeletal muscles. *Journal of Experimental Biology* **215**, 3223–3230.

Webster, M. D., Campbell, G. S. & King. J. R. (1985). Cutaneous resistance to water-vapor diffusion in pigeons and the role of the plumage. *Physiological Zoology* **58**, 58–70.

Wegener, A. (1924). *The Origin of Continents and Oceans*. London: Methuen.

Weibel, E. R. (1964). *The Pathway for Oxygen, Structure and Function in the Mammalian Respiratory System*. Cambridge, Massachusetts: Harvard University Press.

Weibel, E. R., Bacigalupe, L. D., Schmitt, B. & Hoppeler, H. (2004). Allometric scaling of maximal metabolic rate in mammals: muscle aerobic capacity as determinate factor. *Respiratory Physiology and Neurobiology* **140**, 115–132.

Weibel, E. R. & Hoppeler, H. (2004). Modelling design and functional integration in the oxygen and fuel pathways to working muscle. *Cardiovascular Engineering* **4**, 5–18.

Weibel, E. R. & Hoppeler, H. (2005). Exercise-induced maximal metabolic rate scales with muscle aerobic capacity. *Journal of Experimental Biology* **208**, 1635–1644.

Weibel, E. R., Taylor, C. R. & Hoppeler, H. (1992). Variations in function and design: testing symmorphosis in the respiratory system. *Respiration Physiology* **87**, 325–348.

Weibel, E. R., Taylor, C. R. & Hoppler, H. (1991). The concept of symmorphosis: a testable hypothesis of structure-function relationship. *Proceedings of the National Academy of Sciences, USA* **88**, 10357–10361.

Weidensaul, S. (2002). Raising the dead. *Audubon* **May-June**, 58–67.

Weilgart, L. S. (2007). The impacts of anthropogenic ocean noise on cetaceans and implications for management. *Canadian Journal of Zoology* **85**, 1091–1116.

Weir, J. T., Bermingham, E. & Schluter, D. (2009). The Great American Biotic Interchange in birds. *Proceedings of the National Academy of Sciences, USA* **106**, 21737–21742.

Weissenböck, N. M., Weiss, C. M., Schwammer, H. M. & Kratochvil, H. (2010). Thermal windows on the body surface of African elephants (*Loxodonta africana*) studied by infrared thermography. *Journal of Thermal Biology* **35**, 182–188.

Weissenborn, J. (1906). Animal-worship in Africa. *Journal of the Royal African Society* **5**, 167–181.

Welbergen, J. A., Klose, S. M., Markus, N. & Ebby, P. (2008). Climate change and the effects of temperature extremes on Australian flying-foxes. *Proceeding of the Royal Society of London B* **275**, 419–425.

Welch, K. C. & Chen, C. C. C. (2014). Sugar flux through the flight muscles of hovering vertebrate nectarivores: a review. *Journal of Comparative Physiology of London B* **184**, 945–959.

Welch, K. C., Herrera, M. L. G. & Suarez, R. K. (2008). Dietary sugar as a direct fuel for flight in the nectarivorous bat *Glossophaga soricina*. *Journal of Experimental Biology* **211**, 310–316.

Welker, W. I. & Campos, G. B. (1963). Physiological significance of sulci in somatic sensory cerebral cortex in mammals of the family Procyonidae. *Journal of Comparative Neurology* **120**, 19–36.

Wells, R. T. (1989). Vombatidae. (1989). In: *Fauna of Australia. Mammalia Volume 1B* (Eds. Walton, D. W. & Richardson, B. J.). Canberra: Australian Government Publishing Service.
Welsh, K. C., Otarola-Ardilla, A., Herrera, M. L. G. & Flores-Martinez, J. J. (2015). The cost of digestion in the fish-eating myotis (*Myotis vivesi*). *Journal of Experimental Biology* **218**, 1180–1187.
Wenninger, P. S. & Shipley, L. A. (2000). Harvesting, rumination, digestion, and passage of fruit and leaf diets by a small ruminant, the blue duiker. *Oecologia* **123**, 466–474.
Werdelin, L. & Nilsonne, Å. (1999). The evolution of the scrotum and testicular descent in mammals: a phylogenetic view. *Journal of Theoretical Biology* **196**, 61–72.
Werth, A. J. (2006). Mandibular and dental variation and the evolution of suction feeding in Odontoceti. *Journal of Mammalogy* **87**, 579–588.
Wesolowski, T., Schaarschrnidt, B. & Lamprecht, L. (1985). A poor man's calorimeter (PMC) for small animals. *Journal of Thermal Analysis* **30**, 1403–1413.
West, G. C. (1965). Shivering and heat production in wild birds. *Physiological Zoology* **38**, 111–120.
West, G. B. (1999). The origin of universal scaling laws in biology. *Physica A* **263**, 104–113.
West, G. B. & Brown J. H. (2004). Life's universal scaling laws. *Physics Today* **57**, 36.
West, G. B., Brown, J. H. & Enquist, B. J. (1997). A general model for the origin of allometric scaling laws in biology. *Science* **276**, 122–126.
West, G. B., Brown, J. H. & Enquist, B. J. (1999). The fourth dimension of life: fractal geometry and allometric scaling of organisms. *Science* **284**, 1677–1679.
West, G. B., Enquist, B. J. & Brown, J. H. (2002a). Modelling universality and scaling—reply. *Nature* **420**, 626–627.
West, G. B., Savage, V. M., Gillooly, J. F., Enquist, B. J., Woodruff, W. & Brown, J. H. (2003). Why does metabolic rate scale with body size? *Nature* **421**, 713.
West, G. B., Woodruff, W. H. & Brown, J. H. (2002b). Allometric scaling of metabolic rate from molecules and mitochondria to cells and mammals. *Proceedings of the National Academy of Sciences, USA* **99**, 2473–2478.
West, J. W. (2003). Effects of heat-stress on production in dairy cattle. *Journal of Dairy Science* **86**, 2131–2144.
Western, D. (1975). Water availability and its influence on the structure and dynamics of a savannah large mammal community. *East African Wildlife Journal* **13**, 265–286.
Western, D. (1979). Size, life history and ecology in mammals. *African Journal of Ecology* **17**, 185–204.
Western, D. & Ssemakula, J. (1982). Life history patterns in birds and mammals and their evolutionary interpretation. *Oecologia* **54**, 281–290.
Westoby, M. (1978). What are the biological bases of varied diets? *American Naturalist* **112**, 627–631.
Wexler, A. & Hasegawa, S. (1954). Relative humidity-temperature relationships of some saturated salt solutions in the temperature range 0° to 50°C. *Journal of Research of the National Bureau of Standards* **53**, 19–26.
White, C. R. (2003). The influence of foraging mode and arid adaptation on the basal metabolic rates of burrowing mammals. *Physiological and Biochemical Zoology* **76**, 122–134.
White, C. R. (2011). Allometric estimation of metabolic rates in animals. *Comparative Biochemistry and Physiology A* **158**, 346–357.
White, C. R., Blackburn, T. M. & Seymour, R. S. (2009). Phylogenetically informed analysis of the allometry of mammalian basal metabolic rate supports neither geometric nor quarter-power scaling. *Evolution* **63**, 2658–2667.
White, C. R., Cassey, P. & Blackburn, T. M. (2007). Allometric exponents do not support a universal metabolic allometry. *Ecology* **88**, 315–323.

White, C. R. & Kearney, M. R. (2014). Metabolic scaling in animals: methods, empirical results, and theoretical explanations. *Comprehensive Physiology* 4, 231–256.
White, C. R., Schimpf, N. G. & Cassey, P. (2013). The repeatability of metabolic rate declines with time. *Journal of Experimental Biology* **216**, 1763–1765.
White, C. R. & Seymour, R. S. (2003). Mammalian basal metabolic rate is proportional to body mass$^{2/3}$. *Proceedings of the National Academy of Sciences, USA* **100**, 4045–4049.
White, C. R. & Seymour, R. S. (2004). Does basal metabolic rate contain a useful signal? Mammalian BMR allometry and correlations with a selection of physiological, ecological, and life-history variables. *Physiological and Biochemical Zoology* 77, 929–941.
White, C. R. & Seymour, R. S. (2005). Allometric scaling of mammalian metabolism. *Journal of Experimental Biology* **208**, 1611–1619.
White, F. N. (1968). Functional anatomy of the heart of reptiles. *American Zoologist* **8**, 211–219.
Whiteman, J. P., Harlow, H. J., Durner, G. M., Anderson-Sprecher, R., Albeke, S. E., Regehr, E. V., Amstrup, C. & Ben-David, M. (2015). Summer declines in activity and body temperature offer polar bears limited energy savings. *Science* **349**, 295–298.
Whitney, P. (1977). Seasonal maintenance and net production of two sympatric species of subarctic microtine rodents. *Ecology* **58**, 314–325.
Whittingham, D. G. (1980). Parthenogenesis in mammals. In: *Oxford Reviews of Reproductive Biology. Volume 2* (Ed. Finn, C. A.). Oxford: Clarendon Press.
Wible, J. R., Rougier, G. W., Novacek, M. J. & Asher, R. J. (2007). Cretaceous eutherians and Laurasian origin for placental mammals near the K/T boundary. *Nature* **447**, 1003–1006.
Wickstrom, M. L., Robbins, C. T., Hanley, T. A., Spalinger, D. E. & Parish, S. M. (1984). Food intake and foraging energetics of elk and mule deer. *Journal of Wildlife Management* **48**, 1285–1301.
Wiener, G., Jianlin, H. & Ruijun, L. (2006). *The Yak in Relation to Its Environment*. Bangkok: Regional Office for Asia and the Pacific Food and Agriculture Organization of the United Nations.
Wikelski, M. & Cooke, S. J. (2006). Conservation physiology. *Trends in Ecology and Evolution* **21**, 38–46.
Wildman, D. E., Chen, C., Erez, O., Grossman, L. I., Goodman, M. & Romero, R. (2006). Evolution of the mammalian placenta revealed by phylogenetic analysis. *Proceedings of the National Academy of Sciences, USA* **103**, 3203–3208.
Williams, C. T., Barnes, B. M. & Buck, C. L. (2012). Daily body temperature rhythms persist under the midnight sun but are absent during hibernation in free-living Arctic ground squirrels. *Biology Letters* **8**, 31–34.
Williams, C. T., Barnes, B. M., Kenagy, G. J. & Buck, C. L. (2014a). Phenology of hibernation and reproduction in ground squirrels: integration of environmental cues with endogenous programming. *Journal of Zoology* **292**, 112–124.
Williams, J. B. (1996). A phylogenetic perspective of evaporative water loss in birds. *Auk* **113**, 457–472.
Williams, J. B., Lenain, D., Ostrowski, S. & Tieleman, B. I. (2002a). Energy expenditure and water flux of Rüppell's foxes in Saudi Arabia. *Physiological and Biochemical Zoology* **75**, 33–42.
Williams, J. B., Muñoz-Garcia, A., Ostrowski, S. & Tieleman, B. I. (2004). A phylogenetic analysis of basal metabolism, total evaporative water loss, and life-history among foxes from desert and mesic regions. *Journal of Comparative Physiology B* **174**, 29–39.
Williams, M. P. L. & Hutson, J. M. (1991). The phylogeny of testicular descent. *Pediatric Surgery International* **6**, 162–166.
Williams, T. D., Allen, D. D., Groff, J. M. & Glass, R. L. (1992). An analysis of California sea otter (*Enhydra lutris*) pelage and integument. *Marine Mammal Science* **8**, 1–18.

Williams, T. D., Chambers, J. B., Henderson, R. P., Rashotte, M. E. & Overton, J. M. (2002b). Cardiovascular responses to caloric restriction and thermoneutrality in C57BL/6J mice. *American Journal of Physiology* **282**, R1459–R1467.

Williams, T. M. (1986). Thermoregulation of the North American mink during rest and activity in the aquatic environment. *Physiological Zoology* **59**, 293–305.

Williams, T. M. (1989). Swimming by sea otters: adaptations for low energetic cost locomotion. *Journal of Comparative Physiology A* **164**, 815–824.

Williams, T. M. (1990). Heat transfer in elephants: thermal partitioning based on skin temperature profiles. *Journal of Zoology* **222**, 235–245.

Williams, T. M. (2001). Intermittent swimming by mammals: a strategy for increasing energetic efficiency during diving. *American Zoologist* **41**, 166–176.

Williams, T. M., Davis, R. W., Fuiman, L. A., Francis, J., Le Boef, B. J., Horning, M., Calambokidis, J. & Croll, D. A. (2000). Sink or swim: strategies for cost-efficient diving by marine mammals. *Science* **288**, 133–136.

Williams, T. M., Wolfe, L., Davis, T., Kendall, T., Richter, B., Wang, Y., Bryce, C., Elkaim, G. H. & Wilmers, C. C. (2014b). Instantaneous energetics of puma kills reveal advantage of felid sneak attacks. *Science* **346**, 81–85.

Williams, T. M. & Worthy, G. A. J. (2009). Anatomy and physiology: the challenge of aquatic living. In: *Marine Mammal Biology: An Evolutionary Approach* (Ed. Hoelzel, A. R.). Hoboken: Wiley.

Willis, C. R. & Brigham, R. M. (2003). Defining torpor in free-ranging bats: experimental evaluation of external temperature-sensitive radiotransmitters and the concept of active temperature. *Journal of Comparative Physiology B* **173**, 379–389.

Willis, C. R. & Cooper, C. E. (2009). Techniques for studying thermoregulation and thermal biology in bats. In: *Ecological and Behavioural Methods for the Study of Bats* (Eds. Kunz, T. H. & Parsons, S.). Baltimore: Johns Hopkins University Press.

Williston, S. W. (1879). Are birds derived from dinosaurs? *Kansas City Review of Science* **3**, 457–460.

Willmer, P., Stone, G. & Johnston, I. (2009). *Environmental Physiology of Animals*. Oxford: Blackwell Science.

Wilson, A. M., Lowe, J. C., Roskilly, K., Hudson, P. E., Golabek, K. A. & McNutt, J. W. (2013a). Locomotion dynamics of hunting in wild cheetahs. *Nature* **498**, 185–189.

Wilson, D. E. & Reeder, D. M. (2005). *Mammal Species of the World*. Baltimore: Johns Hopkins University Press.

Wilson, J. W., Mills, M. G. L., Wilson, R. P., Peters, G., Mills, M. E. J., Speakman, J. R., Durant, S. M., Bennett, N. C, Marks, N. J. & Scantlebury, M. (2013b). Cheetahs, *Acinonyx jubatus*, balance turn capacity with pace when chasing prey. *Biology Letters* **9**, 20130620.

Wilson, K. J. & Kilgore, D. L. (1978). The effects of location and design on the diffusion of respiratory gases in mammal burrows. *Journal of Theoretical Biology* **71**, 73–101.

Wilson, R. P. & McMahon, C. R. (2006). Measuring devices on wild animals: what constitutes acceptable practice? *Frontiers in Ecology and Evolution* **4**, 147–154.

Wilson, R. P., Shepard, E. L. C. & Liebsch, N. (2008). Prying into the intimate details of animal lives: use of a daily diary on animals. *Endangered Species* **4**, 123–137.

Wilson, R. P., White, C. R., Quintana, F., Halsey, L. G., Liebsch, N., Martin, G. R. & Butler, P. J. (2006). Moving towards acceleration for estimates of activity-specific metabolic rate in free-living animals: the case of the cormorant. *Journal of Animal Ecology* **75**, 1081–1090.

Wilz, M. & Heldmaier, G. (2000). Comparison of hibernation, estivation and daily torpor in the edible dormouse, *Glis glis*. *Journal of Comparative Physiology B* **170**, 511–521.

Wimsatt, W. A. (1969). Transient behaviour, nocturnal activity patterns and feeding efficiency of vampire bats (*Desmodus rotundus*) under natural conditions. *Journal of Mammalogy* **50**, 233–244.

Winter, Y. (1998). Energetic cost of hovering flight in a nectar-feeding bat measured with fast-response respirometry. *Journal of Comparative Physiology B* **168**, 434–444.

Winter, Y. & von Helversen, O. (1998). The energy cost of flight: do small bats fly more cheaply than birds? *Journal of Comparative Physiology B* **168**, 105–111.

Wislocki, G. B. (1933). Location of the testes and body temperature in mammals. *Quarterly Review of Biology* **8**, 385–396.

Withers, P. C. (1977a). Metabolic, respiratory and haematological adjustments of the little pocket mouse to circadian torpor cycles. *Respiration Physiology* **31**, 295–307.

Withers, P. C. (1977b). Measurement of $VO_2$, $VCO_2$, and evaporative water loss with a flow-through mask. *Journal of Applied Physiology* **42**, 120–123.

Withers, P. C. (1978). Models of diffusion-mediated gas exchange in animal burrows. *American Naturalist* **112**, 1101–1112.

Withers, P. C. (1982). Effect of diet and assimilation efficiency on water balance for two desert rodents. *Journal of Arid Environments* **5**, 375–384.

Withers, P. C. (1992). *Comparative Animal Physiology*. Philadelphia: Saunders College Publishing.

Withers, P. C. (1998a). Urea: diverse functions of a 'waste' product. *Clinical and Experimental Pharmacology and Physiology* **25**, 722–727.

Withers, P. C. (1998b). Evaporative water loss and the role of cocoon formation in Australian frogs. *Australian Journal of Zoology* **46**, 405–418.

Withers, P. C. (2001). Design, calibration and calculation for flow-through respirometry systems. *Australian Journal of Zoology* **49**, 445–461.

Withers, P. C., Casey, T. M. & Casey, K. K. (1979a). Allometry of respiratory and haematological parameters of Arctic mammals. *Comparative Biochemistry and Physiology A* **64**, 343–350.

Withers, P. C. & Cooper, C. E. (2009a). Thermal, metabolic, and hygric physiology of the little red kaluta, *Dasykaluta rosamondae* (Dasyuromorphia: Dasyuridae). *Journal of Mammalogy* **90**, 752–760.

Withers, P. C. & Cooper, C. E. (2009b). Thermal, metabolic, hygric and ventilatory physiology of the sandhill dunnart (*Sminthopsis psammophila*; Marsupialia, Dasyuridae). *Comparative Biochemistry and Physiology A* **153**, 317–323.

Withers, P. C. & Cooper, C. E. (2010). Metabolic depression: a historical perspective. In: *Aestivation. Molecular and Physiological Aspects* (Eds. Navas, C. A. & Carvalho, J. E.). Heidelburg: Springer.

Withers, P. C. & Cooper, C. E. (2014). Physiological regulation of evaporative water loss in endotherms: is the little red kaluta (*Dasykaluta rosamondae*) an exception or the rule? *Proceedings of the Royal Society of London B* **281**, 20140149.

Withers, P. C., Cooper, C. E. & Larcombe, A. N. (2006). Environmental correlates of physiological variables in marsupials. *Physiological and Biochemical Zoology* **79**, 437–453.

Withers, P. C., Cooper, C. E. & Nespolo, R. F. (2012). Evaporative water loss, relative water economy and evaporative partitioning of a heterothermic marsupial, the monito del monte (*Dromiciops gliroides*). *Journal of Experimental Biology* **215**, 2806–2813.

Withers, P. C. & Dickman, C. R. (1995). The role of diet in determining water, energy and salt intake in the thorny devil *Moloch horridus* (Lacertilia: Agamidae). *Journal of the Royal Society of Western Australia* **78**, 3–11.

Withers, P. C. & Hillman, S. S. (1988). A steady-state model of maximal oxygen and carbon dioxide transport in anuran amphibians. *Journal of Applied Physiology* **64**, 860–868.
Withers, P. C. & Jarvis, J. U. M. (1980). The effect of huddling on thermoregulation and oxygen consumption for the naked mole-rat. *Comparative Biochemistry and Physiology A* **66**, 215–219.
Withers, P. C., Lea, M., Solberg, T. C., Baustian, M. & Hedrick, M. (1988). Metabolic fates of lactate during recovery from activity in an anuran amphibian, *Bufo americanus*. *Journal of Experimental Zoology* **246**, 236–243.
Withers, P. C., Lee, A. K. & Martin, R. W. (1979b). Metabolism, respiration and evaporative water loss in the Australian hopping mouse *Notomys alexis* (Rodentia: Muridae). *Australian Journal of Zoology* **27**, 195–204.
Withers, P. C., Louw, G. N. & Henschel, J. (1980). Energetics and water relations in Namib Desert rodents. *South African Journal of Zoology* **15**, 131–137.
Withers, P. C., Richardson, K. C. & Wooller, R. D. (1990). Metabolic physiology of the euthermic and torpid honey possums, *Tarsipes rostratus*. *Australian Journal of Zoology* **37**, 685–693.
Withers, P. C., Siegfried, W. R. & Louw, G. N. (1981). Desert ostrich exhales unsaturated air. *South African Journal of Science* **77**, 569–570.
Withers, P. C., Thompson, G. G. & Seymour, R. S. (2000). Metabolic physiology of the north-western marsupial mole, *Notoryctes caurinus* (Marsupialia: Notoryctidae). *Australian Journal of Zoology* **48**, 241–258.
Witmer, L. M. (1995). Homology of facial structures in extant archosaurs (birds and crocodilians), with special reference to paranasal pneumaticity and nasal conchae. *Journal of Morphology* **225**, 269–327.
Wolf, B. & Walsberg, G. E. (2000). The role of the plumage in heat transfer processes of birds. *American Zoologist* **40**, 575–584.
Wolff, J. O. & Lidicker, W. Z. (1981). Communal winter nesting and food sharing in taiga voles. *Behavioural Ecology and Sociobiology* **9**, 237–240.
Wong, R. (2000). Breath-holding diving can cause decompression sickness. *South Pacific Underwater Medicine Society Journal* **30**, 2–6.
Wood, G. (1983). *The Guinness Book of Animal Facts and Feats*. New York: Sterling Publishing Company.
Wooller, R. & Wooller, S. (2013). *Sugar and Sand: The World of the Honey Possum*. Cottesloe, Western Australia: Swanbrae Press.
Worton, B. J. (1987). A review of models of home range for animal movement. *Ecological Modelling* **38**, 277–298.
Wright, A., Chester-Jones, I. & Phillips, J. G. (1957). The histology of the adrenal glands of Prototheria. *Journal of Endocrinology* **15**, 100–107.
Wroe, S. & Archer, M. (2006). Origins and early radiations of marsupials. In: *Evolution and Biogeography of Australian Vertebrates* (Eds. Merrick, J. R., Archer, M., Hickey, G. M. & Lee, M. S. Y.). Oatlands, New South Wales, Australia: Auscipub.
Wroe, S., Crowther, M., Dortch, J. & Chong, J. (2004). The size of the largest marsupial and why it matters. *Proceedings of the Royal Society of London B (Supplement)* **271**, S34–S36.
Wroe, S., Field, J. H., Archer, M., Grayson, D. K., Price, G. J., Louys, J., Faith, J. T., Webb, G. E., Davidson, I. & Mooney, S. D. (2013). Climate change frames debate over the extinction of megafauna in Sahul (Pleistocene Australia-New Guinea). *Proceedings of the National Academy of Sciences, USA* **110**, 8777–8781.
Wroe, S., McHenry, C. & Thomason, J. (2005). Bite club: comparative bite force in big biting mammals and the prediction of predatory behaviour in fossil taxa. *Proceedings of the Royal Society of London B* **272**, 619–625.

Xu, L., Snelling, E. P. & Seymour, R. S. (2014). Burrowing energetics of the giant burrowing cockroach *Macropanesthia rhinoceros*: an allometric study. *Journal of Insect Physiology* **70**, 81–87.

Xu, Y., Shao, C., Fedorov, V. B., Gorophashnaya, A. V., Barnes, B. M. and Yan, J. (2013). Molecular signatures of mammalian hibernation: comparisons with alternative phenotypes. BMC Genomics **14**: 567.

Yancey, P. H. (1988). Osmotic effectors in kidneys of xeric and mesic rodents: corticomedullary distributions and changes with water availability. *Journal of Comparative Physiology B* ***158***, 369–380.

Yarmolinsky, D. A., Zuker, C. S. & Ryba, N. J. P. (2009). Common sense about taste: from mammals to insects. *Cell* **139**, 234–244.

Ye, J. M., Edwards, S. J., Rose, R. W., Steen, J. T., Clark, M. G. & Colquhoun, E. Q. (1996). Alpha-adrenergic stimulation of thermogenesis in a rat kangaroo (Marsupialia, *Bettongia gaimardi*). *American Journal of Physiology* **271**, R586–R592.

Yeates, L. C., Williams, T. M. & Fink, T. L. (2007). Diving and foraging energetics of the smallest marine mammal, the sea otter (*Enhydra lutris*). *Journal of Experimental Biology* **210**, 1960–1979.

Yi, X., Liang, Y., Huerta-Sanchez, E., Jin, X., Cuo, Z. H. P., Pool, J. E., Xu, X., Jiang, H., Vinckenbosch, N., Korneliussen, T. S., Zheng, H., Liu, T., He, W., Li, K., Luo, R., Nie, X., Wu, H., Zhao, M., Cao, H., Zou, J., Shan, Y., Li, S., Yang, Q., Asan, Ni, P., Tian, G., Xu, J., Liu, X., Jiang, T., Wu, R., Zhou, G., Tang, M., Qin, J., Wang, T., Feng, S., Li, G., Huasang, Luosang, J., Wang, W., Chen, F., Wang, Y., Zheng, X., Li, Z., Bianba, Z., Yang, G., Wang, X., Tang, S., Gao, G., Chen, Y., Luo, Z., Gusang, L., Cao, Z., Zhang, Q., Ouyang, W., Ren, X., Liang, H., Zheng, H., Huang, Y., Li, J., Bolund, L., Kristiansen, K., Li, Y., Zhang, Y., Zhang, X., Li, R., Li, S., Yang, H., Nielsen, R., Wang, J. & Wang, J. (2010). Sequencing of 50 human exomes reveals adaptation to high altitude. *Science* **329**, 75–78.

Yokota, S. D., Benyajati, S. & Dantzler, W. H. (1985). Comparative aspects of glomerular filtration in vertebrates. *Renal Physiology* **8**, 193–221.

Yousef, M. K., Hahn, L. & Johnson, H. D. (1968). Adaptation of cattle. In: *Adaptation of Domestic Animals* (Ed. Hafez, E. S. E.). Philadelphia: Lea and Febiger.

Yousef, M. K. & Johnson, H. D. (1966). Blood thyroxine degradation rate of cattle as influenced by temperature and feed intake. *Life Sciences* **5**, 1349–1363.

Yousef, M. K. & Johnson, H. D. (1968). Effects of heat and feed restriction during growth on thyroxine secretion rate of male rats. *Endocrinology* **82**, 353–358.

Yousef, M. K., Kibler, H. H. & Johnson, H. D. (1967). Thyroid activity and heat production in cattle following sudden ambient temperature changes. *Journal of Animal Science* **26**, 142–148.

Yousef, M. K., Robertson, W. D., Johnson, H. D. & Hahn, L. (1968). Effect of ruminal heating on thyroid function and heat production of cattle. *Journal of Animal Science* **27**, 677.

Zakynthinos, S. and Roussos, C. (1991). Oxygen cost of breathing. In: *Tissue Oxygen Utilization* (Eds. Gutierrez, G. & Vincent, J.). Berlin: Spring-Verlag.

Zapol, W. M., Liggins, G. C., Schneider, R. C., Qvist, J., Snider, M. T., Creasy, R. K. & Hochachka, P. W. (1979). Regional blood flow during simulated diving in the conscious Weddell seal. *Journal of Applied Physiology* **47**, 968–973, 1979.

Zardoya, R. & Meyer, A. (1998). Complete mitochondrial genome suggests diapsid affinities of turtles. *Proceedings of the National Association of Sciences, USA* **95**, 14226–14231.

Zenteno-Savín, T., Vázquez-Medina, J. P., Cantú-Medellín, N., Ponganis, P. J. & Elsner, R. (2012). Ischemia/reperfusion in diving birds and mammals: how they avoid oxidative damage. In: *Oxidative Stress in Aquatic Ecosystems* (Eds. Abele, D., Vázquez-Medina, J. P. & Zenteno-Savín, T.). Chichester: John Wiley & Sons.

Zhang, P., Watanabe, K., & Eishi, T. (2007). Habitual hot-spring bathing by a group of Japanese macaques (*Macaca fuscata*) in their natural habitat. *American Journal of Primatology* **69**, 1425–1430.

Zhang, Z., Brun, A., Price, E., Cruz-Neto, A. P., Karasov, W. H. & Caviedes-Vidal, E. (2014). A comparison of mucosal surface area and villous histology in small intestines of the Brazilian free-tailed bat (*Tadarida brasiliensis*) and the mouse (*Mus musculus*). *Journal of Morphology* **276**: 102–108.

Zhong, W., Wang, G., Zhou, Q. & Wang, G. (2007). Communal food caches and social groups of Brandt's voles in the typical steppes of Inner Mongolia, China. *Journal of Arid Environments* **68**: 398–407.

Zhou, L.-M., Xia, S.-S., Chen, Q., Wang, R.-M., Zheng, W.-H. & Liu, J.-S. (2015). Phenotypic flexibility of thermogenesis in the Hwamei (*Garrulax canorus*): responses to cold acclimation. *American Journal of Physiology*. doi:10.1152/ajpregu.00259.2015.

Ziegler, A. C. (1982). The Australo-Papuan genus *Syconycteris* (Chiroptera: Pteropodidae) with the description of a new Papua New Guinea species. *Occasional Papers of Bernice P. Bishop Museum* **25**, 1–22.

# Index

## B

## C

## D

## E

## F

## G

## N

## T

## U



## V